D0164265

Linear Algebra in Action

Harry Dym

Graduate Studies in Mathematics

Volume 78

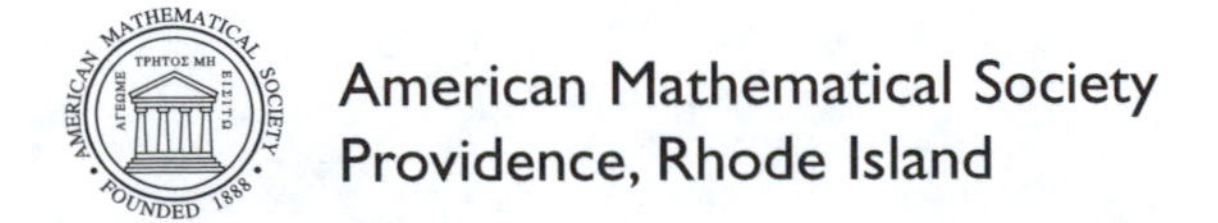

American Mathematical Society
Providence, Rhode Island

2000 *Mathematics Subject Classification.* Primary 15-01, 30-01, 34-01, 39-01, 52-01, 93-01.

For additional information and updates on this book, visit
www.ams.org/bookpages/gsm-78

Library of Congress Cataloging-in-Publication Data

Dym, H. (Harry), 1938–.
Linear algebra in action / Harry Dym.
p. cm. — (Graduate studies in mathematics, ISSN 1065-7339 ; v. 78)
Includes bibliographical references and index.
ISBN-13: 978-0-8218-3813-6 (alk. paper)
ISBN-10: 0-8218-3813-X (alk. paper)
1. Algebras, Linear. I. Title.

QA184.2.D96 2006
512′.5—dc22 2006049906

∞ The paper used in this book is acid-free and falls within the guidelines
established to ensure permanence and durability.
Visit the AMS home page at http://www.ams.org/

10 9 8 7 6 5 4 3 2 1 12 11 10 09 08 07

Dedicated to the memory of our oldest son Jonathan Carroll Dym and our first granddaughter Avital Chana Dym, who were recalled prematurely for no apparent reason, he but 44 and she but 12.

Yhi zichram baruch

Contents

Preface

A foolish consistency is the hobgoblin of little minds,...

Ralph Waldo Emerson, *Self Reliance*

This book is based largely on courses that I have taught at the Feinberg Graduate School of the Weizmann Institute of Science over the past 35 years to graduate students with widely varying levels of mathematical sophistication and interests. The objective of a number of these courses was to present a user-friendly introduction to linear algebra and its many applications. Over the years I wrote and rewrote (and then, more often than not, rewrote some more) assorted sets of notes and learned many interesting things en route. This book is the current end product of that process. The emphasis is on developing a comfortable familiarity with the material. Many lemmas and theorems are made plausible by discussing an example that is chosen to make the underlying ideas transparent in lieu of a formal proof; i.e., I have tried to present the material in the way that most of the mathematicians that I know work rather than in the way they write. The coverage is not intended to be exhaustive (or exhausting), but rather to indicate the rich terrain that is part of the domain of linear algebra and to present a decent sample of some of the tools of the trade of a working analyst that I have absorbed and have found useful and interesting in more than 40 years in the business. To put it another way, I wish someone had taught me this material when I was a graduate student. In those days, in the arrogance of youth, I thought that linear algebra was for boys and girls and that real men and women worked in functional analysis. However, this is but one of many opinions that did not stand the test of time.

In my opinion, the material in this book can (and has been) used on many levels. A core course in classical linear algebra topics can be based on the first six chapters, plus selected topics from Chapters 7–9 and 13. The latter treats difference equations, differential equations and systems thereof. Chapters 14–16 cover applications to vector calculus, including a proof of the implicit function based on the contractive fixed point theorem, and extremal problems with constraints. Subsequent chapters deal with matrix valued holomorphic functions, matrix equations, realization theory, eigenvalue location problems, zero location problems, convexity, and matrices with nonnegative entries. I have taken the liberty of straying into areas that I consider significant, even though they are not usually viewed as part of the package associated with linear algebra. Thus, for example, I have added short sections on complex function theory, Fourier analysis, Lyapunov functions for dynamical systems, boundary value problems and more. A number of the applications are taken from control theory.

I have adapted material from many sources. But the one which was most significant for at least the starting point of a number of topics covered in this work is the wonderful book [**45**] by Lancaster and Tismenetsky.

A number of students read and commented on substantial sections of assorted drafts: Boris Ettinger, Ariel Ginis, Royi Lachmi, Mark Kozdoba, Evgeny Muzikantov, Simcha Rimler, Jonathan Ronen, Idith Segev and Amit Weinberg. I thank them all, and extend my appreciation to two senior readers: Aad Dijksma and Andrei Iacob for their helpful insightful remarks. A special note of thanks goes to Deborah Smith, my copy editor at AMS, for her sharp eye and expertise in the world of commas and semicolons.

On the production side, I thank Jason Friedman for typing an early version, and our secretaries Diana Mandelik, Ruby Musrie, Linda Alman, Terry Debesh, all of whom typed selections and to Diana again for preparing all the figures and clarifying numerous mysterious intricacies of Latex. I also thank Barbara Beeton of AMS for helpful advice on AMS Latex.

One of the difficulties in preparing a manuscript for a book is knowing when to let go. It is always possible to write it better.[1] Fortunately AMS maintains a web page: http://www.ams.org/bookpages/gsm-78, for sins of omission and commission (or just plain afterthoughts).

TAM, ACH TEREM NISHLAM,...
October 18, 2006
Rehovot, Israel

[1]Israel Gohberg tells of a conversation with Lev Sakhnovich that took place in Odessa many years ago: Lev: Israel, how is your book with Mark Gregorovic (Krein) progressing? Israel: It's about 85% done. Lev: That's great! Why so sad? Israel: If you would have asked me yesterday, I would have said 95%.

Chapter 1

Vector spaces

The road to wisdom? Well it's plain and simple to express.
Err and err and err again, but less and less and less.

Cited in [**43**]

1.1. Preview

One of the fundamental issues that we shall be concerned with is the solution of linear equations of the form

$$\begin{aligned} a_{11}x_1 + a_{12}x_2 + \cdots + a_{1q}x_q &= b_1 \\ a_{21}x_1 + a_{22}x_2 + \cdots + a_{2q}x_q &= b_2 \\ \vdots \qquad \vdots \qquad\quad \vdots \quad &\quad \vdots \\ a_{p1}x_1 + a_{p2}x_2 + \cdots + a_{pq}x_q &= b_p \,, \end{aligned}$$

where the a_{ij} and the b_i are given numbers (either real or complex) for $i = 1, \ldots, p$ and $j = 1, \ldots, q$, and we are looking for the x_j for $j = 1, \ldots, q$. Such a system of equations is equivalent to the matrix equation

$$A\mathbf{x} = \mathbf{b}\,,$$

where

$$A = \begin{bmatrix} a_{11} & \cdots & a_{1q} \\ \vdots & & \vdots \\ a_{p1} & \cdots & a_{pq} \end{bmatrix}, \quad \mathbf{x} = \begin{bmatrix} x_1 \\ \vdots \\ x_q \end{bmatrix} \quad \text{and } \mathbf{b} = \begin{bmatrix} b_1 \\ \vdots \\ b_p \end{bmatrix}.$$

- **RC Cola**: The term a_{ij} in the matrix A sits in the i'th row and the j'th column of the matrix; i.e., the first index stands for the number of the row and the second for the number of the column. The order is **rc** as in the popular drink by that name.

Given A and $\mathbf{b}$, the basic questions are:

1. When does there exist at least one solution $\mathbf{x}$?
2. When does there exist at most one solution $\mathbf{x}$?
3. How to calculate the solutions, when they exist?
4. How to find approximate solutions?

The answers to these questions are part and parcel of the theory of vector spaces.

1.2. The abstract definition of a vector space

This subsection is devoted to the abstract definition of a vector space. Even though the emphasis in this course is definitely computational, it seems advisable to start with a few abstract definitions which will be useful in future situations as well as in the present.

A **vector space** $\mathcal{V}$ over the real numbers is a nonempty collection of objects called **vectors**, together with an operation called **vector addition**, which assigns a new vector $\mathbf{u} + \mathbf{v}$ in $\mathcal{V}$ to every pair of vectors $\mathbf{u}$ in $\mathcal{V}$ and $\mathbf{v}$ in $\mathcal{V}$, and an operation called **scalar multiplication**, which assigns a vector $\alpha\mathbf{v}$ in $\mathcal{V}$ to every real number α and every vector $\mathbf{v}$ in $\mathcal{V}$ such that the following hold:

1. For every pair of vectors $\mathbf{u}$ and $\mathbf{v}$, $\mathbf{u}+\mathbf{v} = \mathbf{v}+\mathbf{u}$; i.e., **vector addition is commutative**.
2. For any three vectors $\mathbf{u}$, $\mathbf{v}$ and $\mathbf{w}$, $\mathbf{u} + (\mathbf{v} + \mathbf{w}) = (\mathbf{u} + \mathbf{v}) + \mathbf{w}$; i.e., **vector addition is associative**.
3. There is a **zero vector** (or, in other terminology, **additive identity**) $\mathbf{0} \in \mathcal{V}$ such that $\mathbf{0} + \mathbf{v} = \mathbf{v} + \mathbf{0} = \mathbf{v}$ for every vector $\mathbf{v}$ in $\mathcal{V}$.
4. For every vector $\mathbf{v}$ there is a vector $\mathbf{w}$ (an **additive inverse** of $\mathbf{v}$) such that $\mathbf{v} + \mathbf{w} = \mathbf{0}$.
5. For every vector $\mathbf{v}$, $1\mathbf{v} = \mathbf{v}$.
6. For every pair of real numbers α and β and every vector $\mathbf{v}$, $\alpha(\beta\mathbf{v}) = (\alpha\beta)\mathbf{v}$.
7. For every pair of real numbers α and β and every vector $\mathbf{v}$, $(\alpha+\beta)\mathbf{v} = \alpha\mathbf{v} + \beta\mathbf{v}$.
8. For every real number α and every pair of vectors $\mathbf{u}$ and $\mathbf{v}$, $\alpha(\mathbf{u}+\mathbf{v}) = \alpha\mathbf{u} + \alpha\mathbf{v}$.

Because of Item 2, we can write $\mathbf{u}+\mathbf{v}+\mathbf{w}$ without brackets; similarly, because of Item 6 we can write $\alpha\beta\mathbf{v}$ without brackets. It is also easily checked that there is **exactly one zero vector** $\mathbf{0} \in \mathcal{V}$: If $\widetilde{\mathbf{0}}$ is a second zero vector, then $\mathbf{0} = \mathbf{0} + \widetilde{\mathbf{0}} = \widetilde{\mathbf{0}}$. A similar argument shows that each vector $\mathbf{v} \in \mathcal{V}$ has **exactly one additive inverse**, $-\mathbf{v} = (-1)\mathbf{v}$ in $\mathcal{V}$. Correspondingly, we write $\mathbf{u} + (-\mathbf{v}) = \mathbf{u} - \mathbf{v}$.

From now on we shall use the symbol $\mathbb{R}$ to designate the real numbers, the symbol $\mathbb{C}$ to designate the complex numbers and the symbol $\mathbb{F}$ when the statement in question is valid for both $\mathbb{R}$ and $\mathbb{C}$ and there is no need to specify. Numbers in $\mathbb{F}$ are often referred to as **scalars**.

A vector space $\mathcal{V}$ over $\mathbb{C}$ is defined in exactly the same way as a vector space $\mathcal{V}$ over $\mathbb{R}$ except that the numbers α and β which appear in the definition above are allowed to be complex.

Exercise 1.1. Show that if $\mathcal{V}$ is a vector space over $\mathbb{C}$, then $0\mathbf{v} = \mathbf{0}$ for every vector $\mathbf{v} \in \mathcal{V}$.

Exercise 1.2. Let $\mathcal{V}$ be a vector space over $\mathbb{F}$. Show that if $\alpha, \beta \in \mathbb{F}$ and if $\mathbf{v}$ is a nonzero vector in $\mathcal{V}$, then $\alpha\mathbf{v} = \beta\mathbf{v} \Longleftrightarrow \alpha = \beta$. [HINT: $\alpha - \beta \neq 0 \Longrightarrow \mathbf{v} = (\alpha - \beta)^{-1}(\alpha - \beta)\mathbf{v})$.]

Example 1.1. The set of column vectors

$$\mathbb{F}^p = \left\{ \begin{bmatrix} x_1 \\ \vdots \\ x_p \end{bmatrix} : x_i \in \mathbb{F},\ i = 1, \ldots, p \right\}$$

of height p with entries $x_i \in \mathbb{F}$ that are subject to the natural rules of vector addition

$$\begin{bmatrix} x_1 \\ \vdots \\ x_p \end{bmatrix} + \begin{bmatrix} y_1 \\ \vdots \\ y_p \end{bmatrix} = \begin{bmatrix} x_1 + y_1 \\ \vdots \\ x_p + y_p \end{bmatrix}$$

and multiplication

$$\alpha \begin{bmatrix} x_1 \\ \vdots \\ x_p \end{bmatrix} = \begin{bmatrix} \alpha x_1 \\ \vdots \\ \alpha x_p \end{bmatrix}$$

of the vector $\mathbf{x}$ by a number $\alpha \in \mathbb{F}$ is the most basic example of a vector space. Note the difference between the number 0 and the vector $\mathbf{0} \in \mathbb{F}^p$. The latter is a column vector of height p with all p entries equal to the number zero.

The set $\mathbb{F}^{p\times q}$ of $p \times q$ matrices with entries in $\mathbb{F}$ is a vector space with respect to the rules of vector addition:

$$\begin{bmatrix} x_{11} & \cdots & x_{1q} \\ \vdots & & \vdots \\ x_{p1} & \cdots & x_{pq} \end{bmatrix} + \begin{bmatrix} y_{11} & \cdots & y_{1q} \\ \vdots & & \vdots \\ y_{p1} & \cdots & y_{pq} \end{bmatrix} = \begin{bmatrix} x_{11}+y_{11} & \cdots & x_{1q}+y_{1q} \\ \vdots & & \vdots \\ x_{p1}+y_{p1} & \cdots & x_{pq}+y_{pq} \end{bmatrix},$$

and multiplication by a scalar $\alpha \in \mathbb{F}$:

$$\alpha \begin{bmatrix} x_{11} & \cdots & x_{1q} \\ \vdots & & \vdots \\ x_{p1} & \cdots & x_{pq} \end{bmatrix} = \begin{bmatrix} \alpha x_{11} & \cdots & \alpha x_{1q} \\ \vdots & & \vdots \\ \alpha x_{p1} & \cdots & \alpha x_{pq} \end{bmatrix}.$$

Notice that the vector space $\mathbb{F}^p$ dealt with a little earlier coincides with the vector space that is designated $\mathbb{F}^{p\times 1}$ in the current example.

Exercise 1.3. Show that the space $\mathbb{R}^3$ endowed with the rule

$$\mathbf{x} \,\square\, \mathbf{y} = \begin{bmatrix} \max(x_1, y_1) \\ \max(x_2, y_2) \\ \max(x_3, y_3) \end{bmatrix}$$

for vector addition and the usual rule for scalar multiplication is not a vector space over $\mathbb{R}$. [HINT: Show that this "addition" rule does not admit a zero element; i.e., there is no vector $\mathbf{a} \in \mathbb{R}^3$ such that $\mathbf{a} \,\square\, \mathbf{x} = \mathbf{x} \,\square\, \mathbf{a} = \mathbf{x}$ for every $\mathbf{x} \in \mathbb{R}^3$.]

Exercise 1.4. Let $\mathcal{C} \subset \mathbb{R}^3$ denote the set of vectors $\mathbf{a} = \begin{bmatrix} a_1 \\ a_2 \\ a_3 \end{bmatrix}$ such that the polynomial $a_1 + a_2 t + a_3 t^2 \geq 0$ for every $t \in \mathbb{R}$. Show that it is closed under vector addition (i.e., $\mathbf{a}, \mathbf{b} \in \mathcal{C} \Longrightarrow \mathbf{a}+\mathbf{b} \in \mathcal{C}$) and under multiplication by positive numbers (i.e., $\mathbf{a} \in \mathcal{C}$ and $\alpha > 0 \Longrightarrow \alpha\mathbf{a} \in \mathcal{C}$), but that $\mathcal{C}$ is not a vector space over $\mathbb{R}$. [REMARK: A set $\mathcal{C}$ with the indicated two properties is called a **cone**.]

Exercise 1.5. Show that for each positive integer n, the space of polynomials

$$p(\lambda) = \sum_{j=0}^{n} a_j \lambda^j \quad \text{of degree} \quad n$$

with coefficients $a_j \in \mathbb{C}$ is a vector space over $\mathbb{C}$ under the natural rules of addition and scalar multiplication. [REMARK: You may assume that $\sum_{j=0}^{n} a_j \lambda^j = 0$ for every $\lambda \in \mathbb{C}$ if and only if $a_0 = a_1 = \cdots = a_n = 0$.]

Exercise 1.6. Let $\mathcal{F}$ denote the set of continuous real-valued functions $f(x)$ on the interval $0 \leq x \leq 1$. Show that $\mathcal{F}$ is a vector space over $\mathbb{R}$ with respect to the natural rules of vector addition ($(f_1 + f_2)(x) = f_1(x) + f_2(x)$) and scalar multiplication ($(\alpha f)(x) = \alpha f(x)$).

1.3. Some definitions

- **Subspaces**: A subspace $\mathcal{M}$ of a vector space $\mathcal{V}$ over $\mathbb{F}$ is a nonempty subset of $\mathcal{V}$ that is **closed under vector addition and scalar multiplication**. In other words if $\mathbf{x}$ and $\mathbf{y}$ belong to $\mathcal{M}$, then $\mathbf{x}+\mathbf{y} \in \mathcal{M}$ and $\alpha\mathbf{x} \in \mathcal{M}$ for every scalar $\alpha \in \mathbb{F}$. A subspace of a vector space is automatically a vector space in its own right.

Exercise 1.7. Let $\mathcal{F}_0$ denote the set of continuous real-valued functions $f(x)$ on the interval $0 \leq x \leq 1$ that meet the auxiliary constraints $f(0) = 0$ and $f(1) = 0$. Show that $\mathcal{F}_0$ is a vector space over $\mathbb{R}$ with respect to the natural rules of vector addition and scalar multiplication that were introduced in Exercise 1.6 and that $\mathcal{F}_0$ is a subspace of the vector space $\mathcal{F}$ that was considered there.

Exercise 1.8. Let $\mathcal{F}_1$ denote the set of continuous real-valued functions $f(x)$ on the interval $0 \leq x \leq 1$ that meet the auxiliary constraints $f(0) = 0$ and $f(1) = 1$. Show that $\mathcal{F}_1$ is not a vector space over $\mathbb{R}$ with respect to the natural rules of vector addition and scalar multiplication that were introduced in Exercise 1.6.

- **Span**: If $\mathbf{v_1}, \ldots, \mathbf{v_k}$ is a given set of vectors in a vector space $\mathcal{V}$ over $\mathbb{F}$, then

$$\text{span}\,\{\mathbf{v_1}, \ldots, \mathbf{v_k}\} = \left\{ \sum_{j=1}^{k} \alpha_j \mathbf{v}_j : \alpha_1, \ldots, \alpha_k \in \mathbb{F} \right\}.$$

In words, the span is the set of all **linear combinations** $\alpha_1\mathbf{v}_1 + \cdots + \alpha_k\mathbf{v}_k$ of the indicated set of vectors, with coefficients $\alpha_1, \ldots, \alpha_k$ in $\mathbb{F}$. It is important to keep in mind that $\text{span}\{\mathbf{v_1}, \ldots, \mathbf{v_k}\}$ may be small in some sense. In fact, $\text{span}\,\{\mathbf{v_1}, \ldots, \mathbf{v_k}\}$ is the smallest vector space that contains the vectors $\mathbf{v_1}, \ldots, \mathbf{v_k}$. The number of vectors k that were used to define the span is not a good indicator of the size of this space. Thus, for example, if

$$\mathbf{v}_1 = \begin{bmatrix} 1 \\ 2 \\ 1 \end{bmatrix}, \mathbf{v}_2 = \begin{bmatrix} 2 \\ 4 \\ 2 \end{bmatrix} \text{ and } \mathbf{v}_3 = \begin{bmatrix} 3 \\ 6 \\ 3 \end{bmatrix},$$

then

$$\text{span}\{\mathbf{v}_1, \mathbf{v}_2, \mathbf{v}_3\} = \text{span}\{\mathbf{v}_1\}.$$

To clarify the notion of the *size of the span* we need the concept of linear dependence.

- **Linear dependence**: A set of vectors $\mathbf{v}_1, \ldots, \mathbf{v}_k$ in a vector space $\mathcal{V}$ over $\mathbb{F}$ is said to be **linearly dependent over** $\mathbb{F}$ if there exists a set of scalars $\alpha_1, \ldots, \alpha_k \in \mathbb{F}$, not all of which are zero, such that

$$\alpha_1\mathbf{v}_1 + \cdots + \alpha_k\mathbf{v}_k = \mathbf{0}\,.$$

Notice that this permits you to express one or more of the given vectors in terms of the others. Thus, if $\alpha_1 \neq 0$, then

$$\mathbf{v}_1 = -\frac{\alpha_2}{\alpha_1}\mathbf{v}_2 - \cdots - \frac{\alpha_k}{\alpha_1}\mathbf{v}_k$$

and hence

$$\text{span}\{\mathbf{v}_1, \ldots, \mathbf{v}_k\} = \text{span}\{\mathbf{v}_2, \ldots, \mathbf{v}_k\}\,.$$

Further reductions are possible if the vectors $\mathbf{v}_2, \ldots, \mathbf{v}_k$ are still linearly dependent.

- **Linear independence**: A set of vectors $\mathbf{v_1}, \ldots, \mathbf{v_k}$ in a vector space $\mathcal{V}$ over $\mathbb{F}$ is said to be **linearly independent over** $\mathbb{F}$ if the only scalars $\alpha_1, \ldots, \alpha_k \in \mathbb{F}$ for which

$$\alpha_1\mathbf{v}_1 + \cdots + \alpha_k\mathbf{v}_k = \mathbf{0}$$

are $\alpha_1 = \ldots = \alpha_k = 0$. This is just another way of saying that you cannot express one of these vectors in terms of the others. Moreover, if $\{\mathbf{v}_1, \ldots, \mathbf{v}_k\}$ is a set of linearly independent vectors in a vector space $\mathcal{V}$ over $\mathbb{F}$ and if

$$\mathbf{v} = \alpha_1\mathbf{v}_1 + \cdots + \alpha_k\mathbf{v}_k \quad \text{and} \quad \mathbf{v} = \beta_1\mathbf{v}_1 + \cdots + \beta_k\mathbf{v}_k \tag{1.1}$$

for some choice of constants $\alpha_1, \ldots, \alpha_k, \beta_1, \ldots, \beta_k \in \mathbb{F}$, then $\alpha_j = \beta_j$ for $j = 1, \ldots, k$.

Exercise 1.9. Verify the last assertion; i.e., if (1.1) holds for a linearly independent set of vectors, $\{\mathbf{v}_1, \ldots, \mathbf{v}_k\}$, then $\alpha_j = \beta_j$ for $j = 1, \ldots, k$. Show by example that this conclusion is false if the given set of k vectors is not linearly independent.

- **Basis**: A set of vectors $\mathbf{v}_1, \ldots, \mathbf{v}_k$ is said to form a **basis** for a vector space $\mathcal{V}$ over $\mathbb{F}$ if
  (1) $\text{span}\{\mathbf{v}_1, \ldots, \mathbf{v}_k\} = \mathcal{V}$.
  (2) The vectors $\mathbf{v}_1, \ldots, \mathbf{v}_k$ are linearly independent.

  Both of these conditions are essential. The first guarantees that the given set of k vectors is big enough to express every vector $\mathbf{v} \in \mathcal{V}$ as a linear combination of $\mathbf{v}_1, \ldots, \mathbf{v}_k$; the second that you cannot achieve this with less than k vectors.

  A nontrivial vector space $\mathcal{V}$ has many bases. However, the number of elements in each basis for $\mathcal{V}$ is exactly the same and is referred to as the **dimension** of $\mathcal{V}$ and will be denoted $\dim V$. A proof of this statement will be furnished later. The next example should make it plausible.

Example 1.2. It is readily checked that the vectors

$$\begin{bmatrix}1\\0\\0\end{bmatrix}, \begin{bmatrix}0\\1\\0\end{bmatrix} \text{ and } \begin{bmatrix}0\\0\\1\end{bmatrix}$$

form a basis for the vector space $\mathbb{F}^3$ over the field $\mathbb{F}$. It is also not hard to show that no smaller set of vectors will do. (Thus, $\dim \mathbb{F}^3 = 3$, and, of course, $\dim \mathbb{F}^k = k$ for every positive integer k.)

In a similar vein, the $p \times q$ matrices $E_{ij}, i = 1, \ldots, p, j = 1, \ldots, q$, that are defined by setting every entry in E_{ij} equal to zero except for the ij entry, which is set equal to one, form a basis for the vector space $\mathbb{F}^{p\times q}$.

- **Matrix multiplication**: Let $A = [a_{ij}]$ be a $p\times q$ matrix and $B = [b_{st}]$ be a $q \times r$ matrix. Then the product AB is the $p \times r$ matrix $C = [c_{kl}]$ with entries

$$c_{k\ell} = \sum_{j=1}^{q} a_{kj} b_{j\ell}, \quad k = 1, \ldots, p; \ell = 1 \ldots, r.$$

Notice that $c_{k\ell}$ is the matrix product of the the k'th row $\vec{\mathbf{a}}_k$ of A with the ℓ'th column $\mathbf{b}_\ell$ of B:

$$c_{k\ell} = \vec{\mathbf{a}}_k \mathbf{b}_\ell = [a_{k1} \ \cdots \ a_{kq}] \begin{bmatrix} b_{1\ell} \\ \vdots \\ b_{q\ell} \end{bmatrix}.$$

Thus, for example, if

$$A = \begin{bmatrix} 1 & 3 & 5 \\ 2 & 1 & 0 \end{bmatrix} \quad \text{and} \quad B = \begin{bmatrix} 1 & 2 & 3 & 4 \\ 1 & 0 & -1 & 1 \\ 0 & 1 & 2 & -1 \end{bmatrix},$$

then

$$AB = \begin{bmatrix} 4 & 7 & 10 & 2 \\ 3 & 4 & 5 & 9 \end{bmatrix}.$$

Moreover, if $A \in \mathbb{F}^{p\times q}$ and $\mathbf{x} \in \mathbb{F}^q$, then $\mathbf{y} = A\mathbf{x}$ is the vector in $\mathbb{F}^p$ with components $y_i = \sum_{j=1}^{q} a_{ij} x_j$ for $i = 1, \ldots, p$.

- **Identity matrix**: We shall use the symbol I_n to denote the $n \times n$ matrix $A = [a_{ij}]$, $i, j = 1, \ldots, n$, with $a_{ii} = 1$ for $i = 1, \ldots, n$ and $a_{ij} = 0$ for $i \neq j$. Thus,

$$I_3 = \begin{bmatrix} 1 & 0 & 0 \\ 0 & 1 & 0 \\ 0 & 0 & 1 \end{bmatrix}.$$

The matrix I_n is referred to as the $n \times n$ identity matrix, or just the identity matrix if the size is clear from the context. The name stems from the fact that $I_n\mathbf{x} = \mathbf{x}$ for every vector $\mathbf{x} \in \mathbb{F}^n$.

- **Zero matrix**: We shall use the symbol $O_{p\times q}$ for the matrix in $\mathbb{F}^{p\times q}$ all of whose entries are equal to zero. The subscript $p \times q$ will be dropped if the size is clear from the context.

The definition of matrix multiplication is such that:

- Matrix multiplication is **not commutative**, i.e., even if A and B are both $p \times p$ matrices, in general $AB \neq BA$. In fact, if $p > 1$, then one can find A and B such that $AB = O_{p\times p}$, but $BA \neq O_{p\times p}$.

Exercise 1.10. Find a pair of 2×2 matrices A and B such that $AB = O_{2\times 2}$ but $BA \neq O_{2\times 2}$.

- Matrix multiplication is **associative**: If $A \in \mathbb{F}^{p\times q}$, $B \in \mathbb{F}^{q\times r}$ and $C \in \mathbb{F}^{r\times s}$, then

$$(AB)C = A(BC).$$

- Matrix multiplication is **distributive**: If $A, A_1, A_2 \in \mathbb{F}^{p\times q}$ and $B, B_1, B_2 \in \mathbb{F}^{q\times r}$, then

$$(A_1 + A_2)B = A_1B + A_2B \quad \text{and} \quad A(B_1 + B_2) = AB_1 + AB_2.$$

- If $A \in \mathbb{F}^{p\times q}$ is expressed both as an array of p **row vectors** of length q and as an array of q **column vectors** of height p:

$$A = \begin{bmatrix} \vec{\mathbf{a}}_1 \\ \vdots \\ \vec{\mathbf{a}}_p \end{bmatrix} = \begin{bmatrix} \mathbf{a}_1 & \cdots & \mathbf{a}_q \end{bmatrix},$$

and if $B \in \mathbb{F}^{q\times r}$ is expressed both as an array of q row vectors of length r and as an array of r column vectors of height q:

$$B = \begin{bmatrix} \vec{\mathbf{b}}_1 \\ \vdots \\ \vec{\mathbf{b}}_q \end{bmatrix} = \begin{bmatrix} \mathbf{b}_1 & \cdots & \mathbf{b}_r \end{bmatrix},$$

then the product AB can be expressed in the following three ways:

$$AB = \begin{bmatrix} \vec{\mathbf{a}}_1 B \\ \vdots \\ \vec{\mathbf{a}}_p B \end{bmatrix} = \begin{bmatrix} A\mathbf{b}_1 & \cdots & A\mathbf{b}_r \end{bmatrix} = \sum_{i=1}^{q} \mathbf{a}_i \vec{\mathbf{b}}_i. \tag{1.2}$$

Exercise 1.11. Show that if

$$A = \begin{bmatrix} a_{11} & a_{12} & a_{13} \\ a_{21} & a_{22} & a_{23} \end{bmatrix} \quad \text{and} \quad B = \begin{bmatrix} b_{11} & b_{12} & b_{13} & b_{14} \\ b_{21} & b_{22} & b_{23} & b_{24} \\ b_{31} & b_{32} & b_{33} & b_{34} \end{bmatrix},$$

then

$$AB = \begin{bmatrix} a_{11} & 0 & 0 \\ a_{21} & 0 & 0 \end{bmatrix} B + \begin{bmatrix} 0 & a_{12} & 0 \\ 0 & a_{22} & 0 \end{bmatrix} B + \begin{bmatrix} 0 & 0 & a_{13} \\ 0 & 0 & a_{23} \end{bmatrix} B$$

and hence that

$$AB = \begin{bmatrix} a_{11} \\ a_{21} \end{bmatrix} [b_{11} \ \cdots \ b_{14}] + \begin{bmatrix} a_{12} \\ a_{22} \end{bmatrix} [b_{21} \ b_{22} \ b_{23} \ b_{24}] + \begin{bmatrix} a_{13} \\ a_{23} \end{bmatrix} [b_{31} \ \cdots \ b_{34}] .$$

Exercise 1.12. Verify the three ways of writing a matrix product in formula (1.2). [HINT: Let Exercise 1.11 serve as a guide.]

- **Block multiplication**: It is often convenient to express a large matrix as an array of sub-matrices (i.e., blocks of numbers) rather than as an array of numbers. Then the rules of matrix multiplication still apply (block by block) provided that the block decompositions are compatible. Thus, for example, if

$$A = \begin{bmatrix} A_{11} & A_{12} \\ A_{21} & A_{22} \\ A_{31} & A_{32} \end{bmatrix} \quad \text{and} \quad B = \begin{bmatrix} B_{11} & B_{12} & B_{13} & B_{14} \\ B_{21} & B_{22} & B_{23} & B_{24} \end{bmatrix}$$

with entries $A_{ij} \in \mathbb{F}^{p_i \times q_j}$ and $B_{jk} \in \mathbb{F}^{q_j \times r_k}$, then

$$C = AB = \left[C_{ij}\right], i = 1, \ldots, 3, \ j = 1, \ldots, 4,$$

where

$$C_{ij} = A_{i1}B_{1j} + A_{i2}B_{2j},$$

is a $p_i \times r_j$ matrix.

- **Transposes**: The transpose of a $p \times q$ matrix A is the $q \times p$ matrix whose k'th row is equal to the k'th column of A laid sideways, $k = 1, \ldots, q$. In other words, the ij entry of A is equal to the ji entry of its transpose. The symbol A^T is used to designate the transpose of A. Thus, for example, if

$$A = \begin{bmatrix} 1 & 3 & 5 \\ 4 & 2 & 6 \end{bmatrix}, \text{ then } A^T = \begin{bmatrix} 1 & 4 \\ 3 & 2 \\ 5 & 6 \end{bmatrix} .$$

It is readily checked that

$$(A^T)^T = A \quad \text{and} \quad (AB)^T = B^T A^T . \tag{1.3}$$

- **Hermitian transposes**: The Hermitian transpose A^H of a $p \times q$ matrix A is the same as the transpose A^T of A, except that all the entries in the transposed matrix are replaced by their complex conjugates. Thus, for example, if

$$A = \begin{bmatrix} 1 & 3i & 5+i \\ 4 & 2-i & 6i \end{bmatrix}, \text{ then } A^H = \begin{bmatrix} 1 & 4 \\ -3i & 2+i \\ 5-i & -6i \end{bmatrix}.$$

It is readily checked that

$$(A^H)^H = A \quad \text{and} \quad (AB)^H = B^H A^H . \tag{1.4}$$

- **Inverses**: Let $A \in \mathbb{F}^{p\times q}$. Then:
  (1) A matrix $C \in \mathbb{F}^{q\times p}$ is said to be a **left inverse** of A if $CA = I_q$.
  (2) A matrix $B \in \mathbb{F}^{p\times q}$ is said to be a **right inverse** of A if $AB = I_p$.

  In the first case A is said to be **left invertible**. In the second case A is said to be **right invertible**. It is readily checked that if a matrix $A \in \mathbb{F}^{p\times q}$ has both a left inverse C and a right inverse B, then $B = C$:

$$C = CI_p = C(AB) = (CA)B = I_qB = B.$$

  Notice that this implies that if A has both a left and a right inverse, then it has exactly one left inverse and exactly one right inverse and (as shown just above) the two are equal. In this instance, we shall say that A is **invertible** and refer to $B = C$ as the **inverse** of A and denote it by A^{-1}. In other words, a matrix $A \in \mathbb{F}^{p\times q}$ is invertible if and only if there exists a matrix $B \in \mathbb{F}^{q\times p}$ such that $AB = I_p$ and $BA = I_q$. In fact, as we shall see later, we must also have $q = p$ in this case.

Exercise 1.13. Show that if A and B are invertible matrices of the same size, then AB is invertible and $(AB)^{-1} = B^{-1}A^{-1}$.

Exercise 1.14. Show that the matrix $A = \begin{bmatrix} 1 & 0 & 1 \\ 1 & 1 & 0 \\ 1 & 1 & 0 \end{bmatrix}$ has no left inverses and no right inverses.

Exercise 1.15. Show that the matrix $A = \begin{bmatrix} 1 & 0 & 1 \\ 0 & 1 & 1 \end{bmatrix}$ has at least two right inverses, but no left inverses.

Exercise 1.16. Show that if a matrix $A \in \mathbb{C}^{p\times q}$ has two right inverses B_1 and B_2, then $\lambda B_1 + (1-\lambda)B_2$ is also a right inverse for every choice of $\lambda \in \mathbb{C}$.

Exercise 1.17. Show that a given matrix $A \in \mathbb{F}^{p\times q}$ has either 0, 1 or infinitely many right inverses and that the same conclusion prevails for left inverses.

Exercise 1.18. Let $A_{11} \in \mathbb{F}^{p\times p}$, $A_{12} \in \mathbb{F}^{p\times q}$ and $A_{21} \in \mathbb{F}^{q\times p}$. Show that if A_{11} is invertible, then

$$[A_{11}\ A_{12}] \quad \text{is right invertible and} \quad \begin{bmatrix} A_{11} \\ A_{21} \end{bmatrix} \quad \text{is left invertible.}$$

1.4. Mappings

- **Mappings**: A mapping (or transformation) T from a subset $\mathcal{D}_T$ of a vector space $\mathcal{U}$ into a vector space $\mathcal{V}$ is a rule that assigns exactly one vector $\mathbf{v} \in \mathcal{V}$ to each $\mathbf{u} \in \mathcal{D}_T$. The set $\mathcal{D}_T$ is called the **domain** of T.

  The following three examples give some idea of the possibilities:

  (a) $T: \begin{bmatrix} x_1 \\ x_2 \end{bmatrix} \in \mathbb{R}^2 \mapsto \begin{bmatrix} 3x_1^2 + 4x_2 \\ x_2 - x_1 \\ x_1 + 2x_2 + 6 \end{bmatrix} \in \mathbb{R}^3.$

  (b) $T: \left\{ \begin{bmatrix} x_1 \\ x_2 \end{bmatrix} \in \mathbb{R}^2 : x_1 - x_2 \neq 0 \right\} \mapsto [1/(x_1 - x_2)] \in \mathbb{R}^1.$

  (c) $T: \begin{bmatrix} x_1 \\ x_2 \end{bmatrix} \in \mathbb{R}^2 \mapsto \begin{bmatrix} 3x_1 + x_2 \\ x_1 - x_2 \\ 3x_1 + x_2 \end{bmatrix} \in \mathbb{R}^3.$

  The restriction on the domain in the second example is imposed in order to insure that the definition is meaningful. In the other two examples the domain is taken equal to the full vector space.

  In this framework we shall refer to the set

  $$\mathcal{N}_T = \{\mathbf{u} \in \mathcal{D}_T : T\mathbf{u} = \mathbf{0}_\mathcal{V}\}$$

  as the **nullspace** (or **kernel**) of T and the set

  $$\mathcal{R}_T = \{T\mathbf{u} : \mathbf{u} \in \mathcal{D}_T\}$$

  as the **range** (or **image**) of T. The subscript $\mathcal{V}$ is added to the symbol $\mathbf{0}$ in the first definition to emphasize that it is the zero vector in $\mathcal{V}$, not in $\mathcal{U}$.

- **Linear mapping**: A mapping T from a vector space $\mathcal{U}$ over $\mathbb{F}$ into a vector space $\mathcal{V}$ over the same number field $\mathbb{F}$ is said to be **a linear mapping (or a linear transformation)** if for every choice of $\mathbf{u}, \mathbf{v} \in \mathcal{U}$ and $\alpha \in \mathbb{F}$ the following two conditions are met:

  (1) $T(\mathbf{u} + \mathbf{v}) = T\mathbf{u} + T\mathbf{v}$.

  (2) $T(\alpha\mathbf{u}) = \alpha T\mathbf{u}$.

  It is readily checked that if T is a linear mapping from a vector space $\mathcal{U}$ over $\mathbb{F}$ into a vector space $\mathcal{V}$ over $\mathbb{F}$, then $\mathcal{N}_T$ is a subspace of $\mathcal{U}$ and $\mathcal{R}_T$ is a subspace of $\mathcal{V}$. Moreover, in the preceding set of three examples, T is linear only in case (c).

- **The identity**: Let $\mathcal{U}$ be a vector space over $\mathbb{F}$. The special linear transformation from $\mathcal{U}$ into $\mathcal{U}$ that maps each vector $\mathbf{u} \in \mathcal{U}$ into itself is called the **identity** mapping. It is denoted by the symbol I_n if $\mathcal{U} = \mathbb{F}^n$ and by $I_{\mathcal{U}}$ otherwise, though, more often than not, when the underlying space $\mathcal{U}$ is clear from the context, the subscript $\mathcal{U}$ will be dropped and I will be written in place of $I_{\mathcal{U}}$. Thus, $I_{\mathcal{U}}\mathbf{u} = I\mathbf{u} = \mathbf{u}$ for every vector $\mathbf{u} \in \mathcal{U}$.

Exercise 1.19. Compute $\mathcal{N}_T$ and $\mathcal{R}_T$ for each of the three cases (a), (b) and (c) considered above and say which are subspaces and which are not.

Linear transformations are intimately connected with matrix multiplication:

Exercise 1.20. Show that if T is a linear transformation from a vector space $\mathcal{U}$ over $\mathbb{F}$ with basis $\{\mathbf{u}_1, \ldots, \mathbf{u}_q\}$ into a vector space $\mathcal{V}$ over $\mathbb{F}$ with basis $\{\mathbf{v}_1, \ldots, \mathbf{v}_p\}$, then there exists a unique set of scalars $a_{ij} \in \mathbb{F}$, $i = 1, \ldots, p$ and $j = 1, \ldots, q$ such that

$$T\mathbf{u}_j = \sum_{i=1}^{p} a_{ij}\mathbf{v}_i \quad \text{for} \quad j = 1, \ldots, q \tag{1.5}$$

and hence that

$$T(\sum_{j=1}^{q} x_j\mathbf{u}_j) = \sum_{i=1}^{p} y_i\mathbf{v}_i \Longleftrightarrow A\mathbf{x} = \mathbf{y}\,, \tag{1.6}$$

where $\mathbf{x} \in \mathbb{F}^q$ has components $x_1, \ldots, x_q$, $\mathbf{y} \in \mathbb{F}^p$ has components $y_1, \ldots, y_p$ and the entries a_{ij} of $A \in \mathbb{F}^{p\times q}$ are determined by formula (1.5).

- **WARNING:** If $A \in \mathbb{C}^{p\times q}$, then matrix multiplication defines a linear map from $\mathbf{x} \in \mathbb{C}^q$ to $A\mathbf{x} \in \mathbb{C}^p$. Correspondingly, the nullspace of this map,

  $$\mathcal{N}_A = \{\mathbf{x} \in \mathbb{C}^q : A\mathbf{x} = \mathbf{0}\}\,, \quad \text{is a subspace of} \quad \mathbb{C}^q\,,$$

  and the range of this map,

  $$\mathcal{R}_A = \{A\mathbf{x} : \mathbf{x} \in \mathbb{C}^q\}\,, \quad \text{is a subspace of} \quad \mathbb{C}^p\,.$$

  However, if $A \in \mathbb{R}^{p\times q}$, then matrix multiplication also defines a linear map from $\mathbf{x} \in \mathbb{R}^q$ to $A\mathbf{x} \in \mathbb{R}^p$; and in this setting

  $$\mathcal{N}_A = \{\mathbf{x} \in \mathbb{R}^q : A\mathbf{x} = \mathbf{0}\} \quad \text{is a subspace of} \quad \mathbb{R}^q\,,$$

  and the range of this map,

  $$\mathcal{R}_A = \{A\mathbf{x} : \mathbf{x} \in \mathbb{R}^q\}\,, \quad \text{is a subspace of} \quad \mathbb{R}^p\,.$$

  In short, it is important to clarify the space on which A is acting, i.e., the **domain** of A. This will usually be clear from the context.

1.5. Triangular matrices

An $n \times n$ matrix $A = [a_{ij}]$ is said to be

- **upper triangular** if all its nonzero entries sit either on or above the diagonal, i.e., if $a_{ij} = 0$ when $i > j$.
- **lower triangular** if all its nonzero entries sit either on or below the diagonal, i.e., if A^T is upper triangular.
- **triangular** if it is either upper triangular or lower triangular.
- **diagonal** if $a_{ij} = 0 \quad \text{when} \quad i \neq j$.

Systems of equations based on a triangular matrix are particularly convenient to work with, even if the matrix is not invertible.

Example 1.3. Let $A \in \mathbb{F}^{4\times 4}$ be a 4×4 upper triangular matrix with nonzero diagonal entries and let $\mathbf{b}$ be any vector in $\mathbb{F}^4$. Then the vector $\mathbf{x}$ is a solution of the equation

$$A\mathbf{x} = \mathbf{b} \tag{1.7}$$

if and only if

$$\begin{aligned} a_{11}x_1 + a_{12}x_2 + a_{13}x_3 + a_{14}x_4 &= b_1 \\ a_{22}x_2 + a_{23}x_3 + a_{24}x_4 &= b_2 \\ a_{33}x_3 + a_{34}x_4 &= b_3 \\ a_{44}x_4 &= b_4 \,. \end{aligned}$$

Therefore, since the diagonal entries of A are nonzero, it is readily seen that these equations admit a (unique) solution, by working from the bottom up:

$$\begin{aligned} x_4 &= a_{44}^{-1}b_4 \\ x_3 &= a_{33}^{-1}(b_3 - a_{34}x_4) \\ x_2 &= a_{22}^{-1}(b_2 - a_{23}x_3 - a_{24}x_4) \\ x_1 &= a_{11}^{-1}(b_1 - a_{12}x_2 - a_{13}x_3 - a_{14}x_4) \,. \end{aligned}$$

Thus, we have shown that for any right-hand side $\mathbf{b}$, the equation (1.7) admits a (unique) solution $\mathbf{x}$.

Exploiting the freedom in the choice of $\mathbf{b}$, let $\mathbf{e}_j$, $j = 1, \ldots, 4$, denote the j'th column of the identity matrix I_4 and let $\mathbf{x}_j$ denote the solution of the equation $A\mathbf{x}_j = \mathbf{e}_j$ for $j = 1, \ldots, 4$. Then the 4×4 matrix

$$X = [\mathbf{x}_1 \quad \mathbf{x}_2 \quad \mathbf{x}_3 \quad \mathbf{x}_4]$$

with columns $\mathbf{x}_1, \ldots, \mathbf{x}_4$ is a right inverse of A:

$$\begin{aligned} AX = A[\mathbf{x}_1 \cdots \mathbf{x}_4] &= [A\mathbf{x}_1 \cdots A\mathbf{x}_4] \\ &= [\mathbf{e}_1 \cdots \mathbf{e}_4] = I_4 \,. \end{aligned}$$

Analogous examples can be built for $p \times p$ lower triangular matrices. The only difference is that now it is advantageous to work from the top down. The existence of a left inverse can also be obtained by writing down the requisite equations that must be solved. It is easier, however, to play with transposes. This works because A is a triangular matrix with nonzero diagonal entries if and only if A^T is a triangular matrix with nonzero diagonal entries and

$$YA = I_p \iff A^T Y^T = I_p \, .$$

Exercise 1.21. Show that the right inverse X of the upper triangular matrix A that is constructed in the preceding example is also a left inverse and that it is upper triangular.

Lemma 1.4. *Let A be a $p \times p$ triangular matrix. Then*

(1) *A is invertible if and only if all its diagonal entries are different from zero.*

Moreover, if A is an invertible triangular matrix, then

(2) *A is upper triangular $\iff$ A^{-1} is upper triangular.*

(3) *A is lower triangular $\iff$ A^{-1} is lower triangular.*

Proof. Suppose first that

$$A = \begin{bmatrix} a_{11} & a_{12} \\ 0 & a_{22} \end{bmatrix}$$

is a 2 x 2 upper triangular matrix with nonzero diagonal entries a_{11} and a_{22}. Then it is readily checked that the matrix equation

$$A \begin{bmatrix} x_{11} & x_{12} \\ x_{21} & x_{22} \end{bmatrix} = \begin{bmatrix} 1 & 0 \\ 0 & 1 \end{bmatrix},$$

which is equivalent to the pair of equations

$$A \begin{bmatrix} x_{11} \\ x_{21} \end{bmatrix} = \begin{bmatrix} 1 \\ 0 \end{bmatrix} \text{ and } A \begin{bmatrix} x_{12} \\ x_{22} \end{bmatrix} = \begin{bmatrix} 0 \\ 1 \end{bmatrix},$$

has exactly one solution

$$X = \begin{bmatrix} x_{11} & x_{12} \\ x_{21} & x_{22} \end{bmatrix} = \begin{bmatrix} a_{11}^{-1} & -a_{11}^{-1}a_{12}a_{22}^{-1} \\ 0 & a_{22}^{-1} \end{bmatrix}$$

and that this solution is also a left inverse of A:

$$\begin{bmatrix} a_{11}^{-1} & -a_{11}^{-1}a_{12}a_{22}^{-1} \\ 0 & a_{22}^{-1} \end{bmatrix} A = \begin{bmatrix} 1 & 0 \\ 0 & 1 \end{bmatrix}.$$

Thus, every 2×2 upper triangular matrix A with nonzero diagonal entries is invertible and

$$A^{-1} = \begin{bmatrix} a_{11}^{-1} & -a_{11}^{-1}a_{12}a_{22}^{-1} \\ 0 & a_{22}^{-1} \end{bmatrix} \tag{1.8}$$

is also upper triangular.

Now let A and B be upper triangular $k \times k$ matrices such that $AB = BA = I_k$. Then for every choice of $\mathbf{a}, \mathbf{b}, \mathbf{c} \in \mathbb{C}^k$ and $\alpha, \beta \in \mathbb{C}$ with $\alpha \neq 0$,

$$\begin{aligned} \begin{bmatrix} A & \mathbf{a} \\ O & \alpha \end{bmatrix} \begin{bmatrix} B & \mathbf{b} \\ \mathbf{c}^T & \beta \end{bmatrix} &= \begin{bmatrix} AB + \mathbf{a}\mathbf{c}^T & A\mathbf{b} + \mathbf{a}\beta \\ \alpha\mathbf{c}^T & \alpha\beta \end{bmatrix} \\ &= \begin{bmatrix} I_k + \mathbf{a}\mathbf{c}^T & A\mathbf{b} + \beta\mathbf{a} \\ \alpha\mathbf{c}^T & \alpha\beta \end{bmatrix}. \end{aligned}$$

Consequently, the product of these two matrices will be equal to I_{k+1} if and only if $\mathbf{c} = \mathbf{0}$, $A\mathbf{b} + \beta\mathbf{a} = \mathbf{0}$ and $\alpha\beta = 1$, that is, if and only if $\mathbf{c} = \mathbf{0}$, $\mathbf{b} = -\beta B\mathbf{a}$ and $\beta = 1/\alpha$. Moreover, if $\mathbf{c}$, $\mathbf{b}$ and β are chosen to meet these conditions, then

$$\begin{bmatrix} B & \mathbf{b} \\ O & \beta \end{bmatrix} \begin{bmatrix} A & \mathbf{a} \\ O & \alpha \end{bmatrix} = \begin{bmatrix} BA & B\mathbf{a} + \alpha\mathbf{b} \\ O & \beta\alpha \end{bmatrix} = I_{k+1},$$

since

$$B\mathbf{a} + \alpha\mathbf{b} = B\mathbf{a} + \alpha(-\beta B\mathbf{a}) = \mathbf{0}.$$

Thus, we have shown if $k \times k$ upper triangular matrices with nonzero entries on the diagonal are invertible, then the same holds true for $(k+1) \times (k+1)$ upper triangular matrices with nonzero entries on the diagonal. Therefore, since we already know that 2×2 upper triangular matrices with nonzero entries on the diagonal are invertible, it follows by induction that every upper triangular matrix with nonzero entries on the diagonal is invertible and that the inverse is upper triangular.

Suppose next that $A \in \mathbb{C}^{p \times p}$ is an invertible upper triangular matrix with inverse $B \in \mathbb{C}^{p \times p}$. Then, upon expressing the identity $AB = I_p$ in block form as

$$\begin{bmatrix} A_1 & \mathbf{a}_1 \\ O & \alpha_1 \end{bmatrix} \begin{bmatrix} B_1 & \mathbf{b}_1 \\ \mathbf{c}_1^T & \beta_1 \end{bmatrix} = \begin{bmatrix} I_{p-1} & O \\ O & 1 \end{bmatrix}$$

with diagonal blocks of size $(p-1) \times (p-1)$ and 1×1, respectively, it is readily seen that $\alpha_1\beta_1 = 1$. Therefore, $\alpha_1 \neq 0$. The next step is to play the same game with A_1 to show that its bottom diagonal entry is nonzero and, continuing this way down the line, to conclude that the diagonal entries of A are nonzero and that the inverse matrix B is also automatically upper triangular. The details are left to the reader.

This completes the proof of the asserted statements for upper triangular matrices. The proof for lower triangular matrices may be carried out in much the same way or, what is simpler, by taking transposes. $\square$

Exercise 1.22. Show that if $A \in \mathbb{C}^{n\times n}$ and $A^k = O_{n\times n}$ for some positive integer k, then $I_n - A$ is invertible. [HINT: It's enough to show that

$$(I_n - A)(I_n + A + A^2 + \cdots + A^{k-1}) = (I_n + A + A^2 + \cdots + A^{k-1})(I_n - A) = I_n\,.]$$

Exercise 1.23. Show that even though all the diagonal entries of the matrix

$$A = \begin{bmatrix} 0 & 1 & 1 \\ 1 & 0 & 1 \\ 1 & 0 & 0 \end{bmatrix}$$

are equal to zero, A is invertible, and find A^{-1}.

Exercise 1.24. Use Exercise 1.22 to show that a triangular $n \times n$ matrix A with nonzero diagonal entries is invertible by writing

$$A = D + (A - D) = D(I_n + D^{-1}(A - D))\,,$$

where D is the diagonal matrix with $d_{jj} = a_{jj}$ for $j = 1, \ldots, n$. [HINT: The key observation is that $(D^{-1}(A - D))^n = O$.]

1.6. Block triangular matrices

A matrix $A \in \mathbb{F}^{n\times n}$ with block decomposition

$$A = \begin{bmatrix} A_{11} & \cdots & A_{1k} \\ \vdots & & \vdots \\ A_{k1} & \cdots & A_{kk} \end{bmatrix},$$

where $A_{ij} \in \mathbb{F}^{p_i\times q_j}$ for $i, j = 1, \ldots, k$ and $p_1 + \cdots + p_k = q_1 + \cdots + q_k = n$ is said to be

- **upper block triangular** if $p_i = q_i$ for $i = 1, \ldots, k$ and $A_{ij} = O$ for $i > j$.
- **lower block triangular** if $p_i = q_i$ for $i = 1, \ldots, k$ and $A_{ij} = O$ for $i < j$.
- **block triangular** if it is either upper block triangular or lower block triangular.
- **block diagonal** if $p_i = q_i$ for $i = 1, \ldots, k$ and $A_{ij} = O$ for $i \neq j$.

Note that the blocks A_{ii} in a block triangular decomposition need not be triangular.

Exercise 1.25. Let $A = \begin{bmatrix} A_{11} & A_{12} \\ O_{q\times p} & A_{22} \end{bmatrix}$ be an upper block triangular matrix with invertible diagonal blocks A_{11} of size $p \times p$ and A_{22} of size $q \times q$. Show that A is invertible and that

$$A^{-1} = \begin{bmatrix} A_{11}^{-1} & -A_{11}^{-1}A_{12}A_{22}^{-1} \\ O_{q\times p} & A_{22}^{-1} \end{bmatrix}, \tag{1.9}$$

which generalizes formula (1.8).

Exercise 1.26. Use formula (1.9) to calculate the inverse of the matrix $A = \begin{bmatrix} 2 & 1 & 0 \\ 1 & 2 & 1 \\ 0 & 0 & 5 \end{bmatrix}$.

Exercise 1.27. Let $A = \begin{bmatrix} A_{11} & O_{p\times q} \\ A_{21} & A_{22} \end{bmatrix}$ be a lower block triangular matrix with invertible diagonal blocks A_{11} of size $p \times p$ and A_{22} of size $q \times q$. Find a matrix B of the same form as A such that $AB = BA = I_{p+q}$.

1.7. Schur complements

Let

$$E = \begin{bmatrix} A & B \\ C & D \end{bmatrix}, \tag{1.10}$$

where $A \in \mathbb{C}^{p\times p}$, $B \in \mathbb{C}^{p\times q}$, $C \in \mathbb{C}^{q\times p}$ and $D \in \mathbb{C}^{q\times q}$. Then the following two factorization formulas are extremely useful:

(1) If A is an invertible matrix, then

$$E = \begin{bmatrix} I_p & O \\ CA^{-1} & I_q \end{bmatrix} \begin{bmatrix} A & O \\ O & D - CA^{-1}B \end{bmatrix} \begin{bmatrix} I_p & A^{-1}B \\ O & I_q \end{bmatrix} \tag{1.11}$$

and $D - CA^{-1}B$ is referred to as the **Schur complement of A with respect to E.**

(2) If D is an invertible matrix, then

$$E = \begin{bmatrix} I_p & BD^{-1} \\ O & I_q \end{bmatrix} \begin{bmatrix} A - BD^{-1}C & O \\ O & D \end{bmatrix} \begin{bmatrix} I_p & O \\ D^{-1}C & I_q \end{bmatrix} \tag{1.12}$$

and $A - BD^{-1}C$ is referred to as the **Schur complement of D with respect to E.**

At this point, these two formulas may appear to be simply tedious exercises in block matrix multiplication. However, they are extremely useful. Another proof based on block Gaussian elimination, which leads to even more general factorization formulas, will be presented in Chapter 3. Notice that the first formula exhibits E as the product of an invertible lower triangular matrix times a block diagonal matrix times an invertible upper triangular matrix, whereas the second formula exhibits E as the product of an invertible upper triangular matrix times a block diagonal matrix times an invertible lower triangular matrix.

Exercise 1.28. Verify formulas (1.11) and (1.12) under the stated conditions.

Exercise 1.29. Show that if $B \in \mathbb{C}^{p\times q}$ and $C \in \mathbb{C}^{q\times p}$, then

$$I_p - BC \quad \text{is invertible} \quad \Longleftrightarrow \quad I_q - CB \quad \text{is invertible} \tag{1.13}$$

and that if these two matrices are invertible, then

$$C(I_p - BC)^{-1} = (I_q - CB)^{-1}C\,. \tag{1.14}$$

[HINT: Exploit formulas (1.11) and (1.12).]

Exercise 1.30. Let the matrix E be defined by formula (1.10). Show that:

$$A \quad \text{and} \quad D - CA^{-1}B \quad \text{invertible} \Longrightarrow E \quad \text{is invertible}\,,$$

and construct an example to show that the opposite implication is false.

Exercise 1.31. Show that if the matrix E is defined by formula (1.10), then

$$D \quad \text{and} \quad A - BD^{-1}C \quad \text{invertible} \Longrightarrow E \quad \text{is invertible}\,,$$

and show by example that the opposite implication is false.

Exercise 1.32. Show that if the blocks A and D in the matrix E defined by formula (1.10) are invertible, then

$$\begin{aligned} E \text{ is invertible} \quad &\Longleftrightarrow \quad D - CA^{-1}B \text{ is invertible} \\ &\Longleftrightarrow \quad A - BD^{-1}C \text{ is invertible}\,. \end{aligned}$$

Exercise 1.33. Show that if blocks A and D in the matrix E defined by formula (1.10) are invertible and $A - BD^{-1}C$ is invertible, then

$$(A - BD^{-1}C)^{-1} = A^{-1} + A^{-1}B(D - CA^{-1}B)^{-1}CA^{-1}\,. \tag{1.15}$$

[HINT: Multiply both sides of the asserted identity by $A - BD^{-1}C$.]

Exercise 1.34. Show that if if blocks A and D in the matrix E defined by formula (1.10) are invertible and $D - CA^{-1}B$ is invertible, then

$$(D - CA^{-1}B)^{-1} = D^{-1} + D^{-1}C(A - BD^{-1}C)^{-1}BD^{-1}\,. \tag{1.16}$$

[HINT: Multiply both sides of the asserted identity by $D - CA^{-1}B$.]

Exercise 1.35. Show that if $A \in \mathbb{C}^{p\times p}$, $B \in \mathbb{C}^{p\times q}$, $C \in \mathbb{C}^{q\times p}$ and the matrices A and $A + BC$ are both invertible, then the matrix $I_q + CA^{-1}B$ is invertible and $(I_q + CA^{-1}B)^{-1} = I_q - C(A + BC)^{-1}B$.

Exercise 1.36. Show that if $A \in \mathbb{C}^{p\times p}$, $B \in \mathbb{C}^{p\times q}$, $C \in \mathbb{C}^{q\times p}$ and the matrix $A+BC$ is invertible, then the matrix $\begin{bmatrix} A & B \\ C & -I_q \end{bmatrix}$ is invertible, and find its inverse.

Exercise 1.37. Let $A \in \mathbb{C}^{p\times p}$, $\mathbf{u} \in \mathbb{C}^p$, $\mathbf{v} \in \mathbb{C}^p$ and assume that A is invertible. Show that

$$\begin{bmatrix} A & -\mathbf{u} \\ \mathbf{v}^H & 1 \end{bmatrix} \text{ is invertible} \iff A+\mathbf{u}\mathbf{v}^H \text{ is invertible} \iff 1+\mathbf{v}^H A^{-1}\mathbf{u} \neq 0$$

and that if these conditions are met, then

$$(I_p + \mathbf{u}\mathbf{v}^H A^{-1})^{-1}\mathbf{u} = \mathbf{u}(1+\mathbf{v}^H A^{-1}\mathbf{u})^{-1}.$$

Exercise 1.38. Show that if in the setting of Exercise 1.37 the condition $1+\mathbf{v}^H A^{-1}\mathbf{u} \neq 0$ is met, then the Sherman-Morrison formula

$$(A+\mathbf{u}\mathbf{v}^H)^{-1} = A^{-1} - \frac{A^{-1}\mathbf{u}\mathbf{v}^H A^{-1}}{1+\mathbf{v}^H A^{-1}\mathbf{u}} \tag{1.17}$$

holds.

Exercise 1.39. Show that if A is a $p \times q$ matrix and C is a $q \times q$ invertible matrix, then $\mathcal{R}_{AC} = \mathcal{R}_A$.

Exercise 1.40. Show that the upper block triangular matrix

$$A = \begin{bmatrix} A_{11} & A_{12} & A_{13} \\ O & A_{22} & A_{23} \\ O & O & A_{33} \end{bmatrix}$$

with entries A_{ij} of size $p_i \times p_j$ is invertible if the diagonal blocks A_{11}, A_{22} and A_{33} are invertible, and find a formula for A^{-1}. [HINT: Look for a matrix B of the same form as A such that $AB = I_{p_1+p_2+p_3}$.]

1.8. Other matrix products

Two other product rules for matrices that arise in assorted applications are:

- The **Schur product** $C = A \circ B$ of $A = [a_{ij}] \in \mathbb{C}^{n\times n}$ with $B = [b_{ij}] \in \mathbb{C}^{n\times n}$ is defined as the $n \times n$ matrix $C = [c_{ij}]$ with entries $c_{ij} = a_{ij}b_{ij}$ for $i, j = 1, \ldots, n$.
- The **Kronecker product** $A \otimes B$ of $A = [a_{ij}] \in \mathbb{C}^{p\times q}$ with $B = [b_{ij}] \in \mathbb{C}^{n\times m}$ is defined by the formula

$$A \otimes B = \begin{bmatrix} a_{11}B & \cdots & a_{1q}B \\ \vdots & & \vdots \\ a_{p1}B & \cdots & a_{pq}B \end{bmatrix}.$$

The Schur product of two square matrices of the same size is clearly commutative. It is also readily checked that the Kronecker product of real (or complex) matrices is associative:

$$(A \otimes B) \otimes C = A \otimes (B \otimes C)$$

and satisfies the rules

$$(A \otimes B)^T = A^T \otimes B^T \,,$$
$$(A \otimes B)(C \otimes D) = AC \otimes BD \,,$$

when the indicated matrix multiplications are meaningful. If $\mathbf{x} \in \mathbb{F}^k$, $\mathbf{u} \in \mathbb{F}^k$, $\mathbf{y} \in \mathbb{F}^\ell$ and $\mathbf{v} \in \mathbb{F}^\ell$, then the last rule implies that

$$(\mathbf{x}^T\mathbf{u})(\mathbf{y}^T\mathbf{v}) = (\mathbf{x}^T \otimes \mathbf{y}^T)(\mathbf{u} \otimes \mathbf{v}) \,.$$

Chapter 2

Gaussian elimination

... People can tell you... do it like this. But that ain't the way to learn. You got to do it for yourself.

Willie Mays, cited in Kahn [**40**], p. 163

Gaussian elimination is a way of passing from a given system of equations to a new system of equations that is easier to analyze. The passage from the given system to the new system is effected by multiplying both sides of the given equation, say

$$A\mathbf{x} = \mathbf{b},$$

successively on the left by appropriately chosen invertible matrices. The restriction to invertible multipliers is essential. Otherwise, the new system will not have the same set of solutions as the given one. In particular, the left multipliers will be either permutation matrices (which are defined below) or lower triangular matrices with ones on the diagonal. Both types are invertible. The first operation serves to interchange (i.e., permute) rows, whereas the second serves to add a multiple of one row to other rows. Thus, for example,

$$\begin{bmatrix} 0 & 1 & 0 \\ 1 & 0 & 0 \\ 0 & 0 & 1 \end{bmatrix} \begin{bmatrix} a_{11} & a_{12} & \cdots & a_{1n} \\ a_{21} & a_{22} & \cdots & a_{2n} \\ a_{31} & a_{32} & \cdots & a_{3n} \end{bmatrix} = \begin{bmatrix} a_{21} & a_{22} & \cdots & a_{2n} \\ a_{11} & a_{12} & \cdots & a_{1n} \\ a_{31} & a_{32} & \cdots & a_{3n} \end{bmatrix},$$

whereas

$$\begin{bmatrix} 1 & 0 & 0 \\ \alpha & 1 & 0 \\ \beta & 0 & 1 \end{bmatrix} \begin{bmatrix} a_{11} & \cdots & a_{1n} \\ a_{21} & \cdots & a_{2n} \\ a_{31} & \cdots & a_{3n} \end{bmatrix} = \begin{bmatrix} a_{11} & a_{12} & \cdots & a_{1n} \\ \alpha a_{11} + a_{21} & \alpha a_{12} + a_{22} & \cdots & \alpha a_{1n} + a_{2n} \\ \beta a_{11} + a_{31} & \beta a_{12} + a_{32} & \cdots & \beta a_{1n} + a_{3n} \end{bmatrix}.$$

2.1. Some preliminary observations

The operation of adding (or subtracting) a constant multiple of one row of a $p \times q$ matrix from another row of that matrix can always be achieved by multiplying on the left by a $p \times p$ matrix with ones on the diagonal and one other nonzero entry. Every such matrix can be expressed in the form

$$E_\alpha = I_p + \alpha \mathbf{e}_i \mathbf{e}_j^T \quad \text{with } i \text{ and } j \text{ fixed and} \quad i \neq j\,, \tag{2.1}$$

where the vectors $\mathbf{e}_1 \ldots, \mathbf{e}_p$ denote the standard basis for $\mathbb{F}^p$ (i.e., the columns in the identity matrix I_p) and $\alpha \in \mathbb{F}$.

It is readily seen that the following conclusions hold for the class of matrices $\mathcal{E}_{ij}$ of the form (2.1):

(1) $\mathcal{E}_{ij}$ is closed under multiplication: $E_\alpha E_\beta = E_{\alpha+\beta}$.

(2) The identity belongs to $\mathcal{E}_{ij}$: $E_0 = I_p$.

(3) Every matrix in $\mathcal{E}_{ij}$ is invertible: E_α is invertible and $E_\alpha^{-1} = E_{-\alpha}$.

(4) Multiplication is commutative in $\mathcal{E}_{ij}$: $E_\alpha E_\beta = E_\beta E_\alpha$.

Thus, the class of matrices of the form (2.1) is a **commutative group** with respect to matrix multiplication. The same conclusion holds for the more general class of $p \times p$ matrices of the form

$$E_{\mathbf{u}} = I_p + \mathbf{u}\mathbf{e}_i^T\,, \quad \text{with} \quad \mathbf{u} \in \mathbb{F}^p \quad \text{and} \quad \mathbf{e}_i^T \mathbf{u} = 0\,. \tag{2.2}$$

The trade secret is the identity, which is considered in the next exercise, or, in less abstract terms, the observation that

$$\begin{bmatrix} 1 & 0 & 0 & 0 \\ 0 & 1 & 0 & 0 \\ 0 & a & 1 & 0 \\ 0 & b & 0 & 1 \end{bmatrix} \begin{bmatrix} 1 & 0 & 0 & 0 \\ 0 & 1 & 0 & 0 \\ 0 & c & 1 & 0 \\ 0 & d & 0 & 1 \end{bmatrix} = \begin{bmatrix} 1 & 0 & 0 & 0 \\ 0 & 1 & 0 & 0 \\ 0 & a+c & 1 & 0 \\ 0 & b+d & 0 & 1 \end{bmatrix}$$

and the realization that there is nothing special about the size of this matrix or the second column.

Exercise 2.1. Let $\mathbf{u}, \mathbf{v} \in \mathbb{F}^p$ be such that $\mathbf{e}_i^T \mathbf{u} = 0$ and $\mathbf{e}_i^T \mathbf{v} = 0$. Show that

$$(I_p + \mathbf{u}\mathbf{e}_i^T)(I_p + \mathbf{v}\mathbf{e}_i^T) = (I_p + \mathbf{v}\mathbf{e}_i^T)(I_p + \mathbf{u}\mathbf{e}_i^T) = I_p + (\mathbf{v} + \mathbf{u})\mathbf{e}_i^T\,.$$

- **Permutation matrices:** Every $n \times n$ permutation matrix P is obtained by taking the identity matrix I_n and interchanging some of the rows. Consequently, P can be expressed in terms of the columns $\mathbf{e}_j$, $j = 1, \ldots, n$ of I_n and a one to one mapping σ of the set of integers $\{1, \ldots, n\}$ onto itself by the formula

$$P = P_\sigma = \sum_{j=1}^{n} \mathbf{e}_j \mathbf{e}_{\sigma(j)}^T\,. \tag{2.3}$$

Thus, for example, if $n = 4$ and $\sigma(1) = 3$, $\sigma(2) = 2$, $\sigma(3) = 4$ and $\sigma(4) = 1$, then

$$P_\sigma = \mathbf{e}_1\mathbf{e}_3^T + \mathbf{e}_2\mathbf{e}_2^T + \mathbf{e}_3\mathbf{e}_4^T + \mathbf{e}_4\mathbf{e}_1^T = \begin{bmatrix} 0 & 0 & 1 & 0 \\ 0 & 1 & 0 & 0 \\ 0 & 0 & 0 & 1 \\ 1 & 0 & 0 & 0 \end{bmatrix}.$$

The set of $n \times n$ permutation matrices also forms a group under multiplication, but this group is not commutative (i.e., conditions (1)–(3) in the list given above are satisfied, but not (4)).

- **Orthogonal matrices**: An $n \times n$ matrix V with real entries is said to be an orthogonal matrix if $V^TV = I_n$.

Exercise 2.2. Show that every permutation matrix is an orthogonal matrix. [HINT: Use formula (2.3).]

The following notions will prove useful:

- **Upper echelon:** A $p \times q$ matrix U is said to be an upper echelon matrix if the first nonzero entry in row i lies to the left of the first nonzero entry in row $i+1$. Thus, for example, the first of the following two matrices is an upper echelon matrix, while the second is not.

$$\begin{bmatrix} 3 & 6 & 2 & 4 & 1 & 0 \\ 0 & 0 & 1 & 0 & 5 & 0 \\ 0 & 0 & 0 & 0 & 2 & 0 \\ 0 & 0 & 0 & 0 & 0 & 0 \end{bmatrix}, \quad \begin{bmatrix} 4 & 2 & 3 & 1 \\ 0 & 0 & 6 & 0 \\ 0 & 5 & 0 & 5 \\ 0 & 0 & 0 & 0 \end{bmatrix}.$$

- **Pivots:** The first nonzero entry in each row of an upper echelon matrix is termed a pivot. The pivots in the matrix on the left just above are 3, 1 and 2.

- **Pivot columns:** A column in an upper echelon matrix U will be referred to as a **pivot column** if it contains a pivot. Thus, the first, third and fifth columns of the matrix considered in the preceding paragraph are pivot columns. If $GA = U$, where G is invertible and $U \in \mathbb{F}^{p\times q}$ is in upper echelon form with k pivots, then the columns $\mathbf{a}_{i_1}, \ldots, \mathbf{a}_{i_k}$ of A that correspond in position to the pivot columns $\mathbf{u}_{i_1}, \ldots, \mathbf{u}_{i_k}$ of U will also be called pivot columns (even though the pivots are in U not in A) and the entries $x_{i_1}, \ldots, x_{i_k}$ in $\mathbf{x} \in \mathbb{F}^q$ will be referred to as **pivot variables**.

2.2. Examples

Example 2.1. Consider the equation $A\mathbf{x} = \mathbf{b}$, where

$$A = \begin{bmatrix} 0 & 2 & 3 & 1 \\ 1 & 5 & 3 & 4 \\ 2 & 6 & 3 & 2 \end{bmatrix} \quad \text{and } \mathbf{b} = \begin{bmatrix} 1 \\ 2 \\ 1 \end{bmatrix} . \tag{2.4}$$

1. Construct the augmented matrix

$$\widetilde{A} = \begin{bmatrix} 0 & 2 & 3 & 1 & 1 \\ 1 & 5 & 3 & 4 & 2 \\ 2 & 6 & 3 & 2 & 1 \end{bmatrix} \tag{2.5}$$

that is formed by adding $\mathbf{b}$ as an extra column to the matrix A on the far right. The augmented matrix is introduced to insure that the row operations that are applied to the matrix A are also applied to the vector $\mathbf{b}$.

2. Interchange the first two rows of $\widetilde{A}$ to get

$$\begin{bmatrix} 1 & 5 & 3 & 4 & 2 \\ 0 & 2 & 3 & 1 & 1 \\ 2 & 6 & 3 & 2 & 1 \end{bmatrix} = P_1\widetilde{A}, \tag{2.6}$$

where

$$P_1 = \begin{bmatrix} 0 & 1 & 0 \\ 1 & 0 & 0 \\ 0 & 0 & 1 \end{bmatrix}$$

has been chosen to obtain a nonzero entry in the upper left-hand corner of the new matrix.

3. Subtract two times the top row of the matrix $P_1\widetilde{A}$ from its bottom row to get

(2.7)

$$\begin{bmatrix} 1 & 5 & 3 & 4 & 2 \\ 0 & 2 & 3 & 1 & 1 \\ 0 & -4 & -3 & -6 & -3 \end{bmatrix} = E_1P_1\widetilde{A}, \quad \text{where} \quad E_1 = \begin{bmatrix} 1 & 0 & 0 \\ 0 & 1 & 0 \\ -2 & 0 & 1 \end{bmatrix}$$

is chosen to obtain all zeros below the pivot in the first column.

4. Add two times the second row of $E_1P_1\widetilde{A}$ to its third row to get

$$\begin{bmatrix} 1 & 5 & 3 & 4 & 2 \\ 0 & 2 & 3 & 1 & 1 \\ 0 & 0 & 3 & -4 & -1 \end{bmatrix} = E_2E_1P_1\widetilde{A} = \begin{bmatrix} U & \mathbf{c} \end{bmatrix} , \tag{2.8}$$

where

$$E_2 = \begin{bmatrix} 1 & 0 & 0 \\ 0 & 1 & 0 \\ 0 & 2 & 1 \end{bmatrix}$$

is chosen to obtain all zeros below the pivot in the second column, $U = E_2E_1P_1A$ is in upper echelon form and $\mathbf{c} = E_2E_1P_1\mathbf{b}$. It was not necessary to permute the rows, since the upper left-hand corner of the block $\begin{bmatrix} 2 & 3 & 1 & 1 \\ 0 & 3 & -4 & -1 \end{bmatrix}$ was already nonzero.

5. Try to solve the new system of equations

$$U\mathbf{x} = \begin{bmatrix} 1 & 5 & 3 & 4 \\ 0 & 2 & 3 & 1 \\ 0 & 0 & 3 & -4 \end{bmatrix} \begin{bmatrix} x_1 \\ x_2 \\ x_3 \\ x_4 \end{bmatrix} = \begin{bmatrix} 2 \\ 1 \\ -1 \end{bmatrix} \tag{2.9}$$

by solving for the pivot variables from the bottom row up: The bottom row equation is

$$3x_3 - 4x_4 = -1,$$

and hence for the third pivot variable x_3 we obtain the formula

$$3x_3 = 4x_4 - 1\,.$$

The second row equation is

$$2x_2 + 3x_3 + x_4 = 1\,,$$

and hence for the second pivot variable x_2 we obtain the formula

$$2x_2 = -3x_3 - x_4 + 1 = -5x_4 + 2\,.$$

Finally, the top row equation is

$$x_1 + 5x_2 + 3x_3 + 4x_4 = 2,$$

and hence for the first pivot variable x_1 we get

$$\begin{aligned} x_1 &= -5x_2 - 3x_3 - 4x_4 + 2 \\ &= \frac{-5(-5x_4 + 2)}{2} - (4x_4 - 1) - 4x_4 + 2 \\ &= \frac{9}{2}x_4 - 2. \end{aligned}$$

Thus, we have expressed each of the pivot variables x_1, x_2, x_3 in terms of the variable x_4. In vector notation,

$$\mathbf{x} = \begin{bmatrix} x_1 \\ x_2 \\ x_3 \\ x_4 \end{bmatrix} = \begin{bmatrix} -2 \\ 1 \\ -1/3 \\ 0 \end{bmatrix} + x_4 \begin{bmatrix} 9/2 \\ -5/2 \\ 4/3 \\ 1 \end{bmatrix}$$

is a solution of the system of equations (2.9), or equivalently,

$$E_2E_1P_1A\mathbf{x} = E_2E_1P_1\mathbf{b} \tag{2.10}$$

(with A and $\mathbf{b}$ as in (2.4)) for every choice of x_4. However, since the matrices E_2, E_1 and P_1 are invertible, $\mathbf{x}$ is a solution of (2.10) if and only if $A\mathbf{x} = \mathbf{b}$, i.e., if and only if $\mathbf{x}$ is a solution of the original equation.

6. Check that the computed solution solves the original system of equations. Strictly speaking, this step is superfluous, because the construction guarantees that every solution of the new system is a solution of the old system, and vice versa. Nevertheless, this is an **extremely important step**, because it gives you a way of verifying that your calculations are correct.

Conclusions: Since U is a 3×4 matrix with 3 pivots, much the same sorts of calculations as those carried out above imply that for each choice of $\mathbf{b} \in \mathbb{F}^3$, the equation $A\mathbf{x} = \mathbf{b}$ considered in this example has at least one solution $\mathbf{x} \in \mathbb{F}^4$. Therefore, $\mathcal{R}_A = \mathbb{F}^3$. Moreover, for any given $\mathbf{b}$, there is a family of solutions of the form $\mathbf{x} = \mathbf{u} + x_4\mathbf{v}$ for every choice of $x_4 \in \mathbb{F}$. But this implies that $A\mathbf{x} = A\mathbf{u} + x_4 A\mathbf{v} = A\mathbf{u}$ for every choice of $x_4 \in \mathbb{F}$, and hence that $\mathbf{v} \in \mathcal{N}_A$. In fact,

$$\mathcal{N}_A = \text{span}\left\{\begin{bmatrix} 9/2 \\ -5/2 \\ 4/3 \\ 1 \end{bmatrix}\right\} \quad \text{and } \mathcal{R}_A = \mathbb{F}^3.$$

This, as we shall see shortly, is a consequence of the number of pivots and their positions. (In particular, anticipating a little, it is not an accident that the dimensions of these two spaces sum to the number of columns of A.)

Example 2.2. Consider the equation $A\mathbf{x} = \mathbf{b}$ with

$$A = \begin{bmatrix} 0 & 0 & 4 & 3 \\ 1 & 2 & 4 & 1 \\ 1 & 2 & 8 & 4 \end{bmatrix} \quad \text{and } \mathbf{b} = \begin{bmatrix} b_1 \\ b_2 \\ b_3 \end{bmatrix}.$$

1. Form the augmented matrix

$$\widetilde{A} = \begin{bmatrix} 0 & 0 & 4 & 3 & b_1 \\ 1 & 2 & 4 & 1 & b_2 \\ 1 & 2 & 8 & 4 & b_3 \end{bmatrix}.$$

2. Interchange the first two rows to get

$$\begin{bmatrix} 1 & 2 & 4 & 1 & b_2 \\ 0 & 0 & 4 & 3 & b_1 \\ 1 & 2 & 8 & 4 & b_3 \end{bmatrix} = P_1\widetilde{A}$$

with P_1 as in Step 2 of the preceding example.

3. Subtract the top row of $P_1\widetilde{A}$ from its bottom row to get

$$\begin{bmatrix} 1 & 2 & 4 & 1 & b_2 \\ 0 & 0 & 4 & 3 & b_1 \\ 0 & 0 & 4 & 3 & b_3 - b_2 \end{bmatrix} = E_1P_1\widetilde{A},$$

where

$$E_1 = \begin{bmatrix} 1 & 0 & 0 \\ 0 & 1 & 0 \\ -1 & 0 & 1 \end{bmatrix}.$$

4. Subtract the second row of $E_1P_1\widetilde{A}$ from its third row to get

$$\begin{bmatrix} 1 & 2 & 4 & 1 & b_2 \\ 0 & 0 & 4 & 3 & b_1 \\ 0 & 0 & 0 & 0 & b_3 - b_2 - b_1 \end{bmatrix} = E_2E_1P_1\widetilde{A} = \begin{bmatrix} U & \mathbf{c} \end{bmatrix},$$

where

$$E_2 = \begin{bmatrix} 1 & 0 & 0 \\ 0 & 1 & 0 \\ 0 & -1 & 1 \end{bmatrix}, \quad U = \begin{bmatrix} 1 & 2 & 4 & 1 \\ 0 & 0 & 4 & 3 \\ 0 & 0 & 0 & 0 \end{bmatrix} \quad \text{and} \quad \mathbf{c} = \begin{bmatrix} b_2 \\ b_1 \\ b_3 - b_2 - b_1 \end{bmatrix}.$$

5. Try to solve the new system of equations

$$\begin{bmatrix} 1 & 2 & 4 & 1 \\ 0 & 0 & 4 & 3 \\ 0 & 0 & 0 & 0 \end{bmatrix} \begin{bmatrix} x_1 \\ x_2 \\ x_3 \\ x_4 \end{bmatrix} = \begin{bmatrix} b_2 \\ b_1 \\ b_3 - b_2 - b_1 \end{bmatrix},$$

working from the bottom up.

To begin with, the bottom row yields the equation $0 = b_3 - b_2 - b_1$. Thus, it is clear that there are no solutions unless $b_3 = b_1 + b_2$. If this restriction is in force, then the second row gives us the equation

$$4x_3 + 3x_4 = b_1$$

and hence, the pivot variable,

$$x_3 = \frac{b_1 - 3x_4}{4}.$$

Next, the first row gives us the equation

$$x_1 + 2x_2 + 4x_3 + x_4 = b_2$$

and hence, the other pivot variable,

$$\begin{aligned} x_1 &= b_2 - 2x_2 - 4x_3 - x_4 \\ &= b_2 - 2x_2 - (b_1 - 3x_4) - x_4 \\ &= b_2 - b_1 - 2x_2 + 2x_4\,. \end{aligned}$$

Consequently, if $b_3 = b_1 + b_2$, then

$$\mathbf{x} = \begin{bmatrix} x_1 \\ x_2 \\ x_3 \\ x_4 \end{bmatrix} = \begin{bmatrix} b_2 - b_1 \\ 0 \\ b_1/4 \\ 0 \end{bmatrix} + x_2 \begin{bmatrix} -2 \\ 1 \\ 0 \\ 0 \end{bmatrix} + x_4 \begin{bmatrix} 2 \\ 0 \\ -3/4 \\ 1 \end{bmatrix}$$

is a solution of the given system of equations for every choice of x_2 and x_4 in $\mathbb{F}$.

6. Check that the computed solution solves the original system of equations.

Conclusions: The preceding calculations imply that the equation $A\mathbf{x} = \mathbf{b}$ is solvable if and only if

$$\mathbf{b} = \begin{bmatrix} b_1 \\ b_2 \\ b_1 + b_2 \end{bmatrix} ; \quad \text{i.e., if and only if} \quad \mathbf{b} \in \text{span} \left\{ \begin{bmatrix} 1 \\ 0 \\ 1 \end{bmatrix}, \begin{bmatrix} 0 \\ 1 \\ 1 \end{bmatrix} \right\} .$$

Moreover, for each such $\mathbf{b} \in \mathbb{F}^3$ there exists a solution of the form $\mathbf{x} = \mathbf{u} + x_2\mathbf{v}_1 + x_4\mathbf{v}_2$ for every $x_2, x_4 \in \mathbb{F}$. In particular, $x_2A\mathbf{v}_1 + x_4A\mathbf{v}_2 = \mathbf{0}$ for every choice of x_2 and x_4. But this is possible only if $A\mathbf{v}_1 = \mathbf{0}$ and $A\mathbf{v}_2 = \mathbf{0}$.

Exercise 2.3. Check that for the matrix A in Example 2.2, $\mathcal{R}_A$ is the span of the pivot columns of A:

$$\mathcal{R}_A = \text{span} \left\{ \begin{bmatrix} 0 \\ 1 \\ 1 \end{bmatrix}, \begin{bmatrix} 4 \\ 4 \\ 8 \end{bmatrix} \right\}, \quad \text{and} \quad \mathcal{N}_A = \text{span} \left\{ \begin{bmatrix} -2 \\ 1 \\ 0 \\ 0 \end{bmatrix}, \begin{bmatrix} 2 \\ 0 \\ -3/4 \\ 1 \end{bmatrix} \right\} .$$

The next example is carried out more quickly.

Example 2.3. Let

$$A = \begin{bmatrix} 0 & 0 & 3 & 4 & 7 \\ 0 & 1 & 0 & 0 & 0 \\ 0 & 2 & 3 & 6 & 8 \\ 0 & 0 & 6 & 8 & 14 \end{bmatrix} \quad \text{and } \mathbf{b} = \begin{bmatrix} b_1 \\ b_2 \\ b_3 \\ b_4 \end{bmatrix} .$$

Then a vector $\mathbf{x} \in \mathbb{F}^5$ is a solution of the equation $A\mathbf{x} = \mathbf{b}$ if and only if

$$\begin{bmatrix} 0 & 1 & 0 & 0 & 0 \\ 0 & 0 & 3 & 4 & 7 \\ 0 & 0 & 0 & 2 & 1 \\ 0 & 0 & 0 & 0 & 0 \end{bmatrix} \begin{bmatrix} x_1 \\ x_2 \\ x_3 \\ x_4 \\ x_5 \end{bmatrix} = \begin{bmatrix} b_2 \\ b_1 \\ b_3 - 2b_2 - b_1 \\ b_4 - 2b_1 \end{bmatrix} .$$

The pivots of the upper echelon matrix on the left are in columns 2, 3 and 4. Therefore, upon solving for the pivot variables x_2, x_3 and x_4 in terms of x_1, x_5 and $b_1, \ldots, b_4$ from the bottom row up, we obtain the formulas

$$\begin{aligned} 0 &= b_4 - 2b_1 \\ 2x_4 &= b_3 - 2b_2 - b_1 - x_5 \\ 3x_3 &= b_1 - 4x_4 - 7x_5 \\ &= 3b_1 + 4b_2 - 2b_3 - 5x_5 \\ x_2 &= b_2 \, . \end{aligned}$$

But this is the same as

$$\begin{bmatrix} x_1 \\ x_2 \\ x_3 \\ x_4 \\ x_5 \end{bmatrix} = \begin{bmatrix} x_1 \\ b_2 \\ (-5x_5 + 3b_1 + 4b_2 - 2b_3)/3 \\ (-x_5 + b_3 - 2b_2 - b_1)/2 \\ x_5 \end{bmatrix} = x_1 \begin{bmatrix} 1 \\ 0 \\ 0 \\ 0 \\ 0 \end{bmatrix} + x_5 \begin{bmatrix} 0 \\ 0 \\ -5/3 \\ -1/2 \\ 1 \end{bmatrix}$$

$$+ b_1 \begin{bmatrix} 0 \\ 0 \\ 1 \\ -1/2 \\ 0 \end{bmatrix} + b_2 \begin{bmatrix} 0 \\ 1 \\ 4/3 \\ -1 \\ 0 \end{bmatrix} + b_3 \begin{bmatrix} 0 \\ 0 \\ -2/3 \\ 1/2 \\ 0 \end{bmatrix}$$

$$= \quad x_1\mathbf{u}_1 + x_5\mathbf{u}_2 + b_1\mathbf{u}_3 + b_2\mathbf{u}_4 + b_3\mathbf{u}_5 \, ,$$

where $\mathbf{u}_1, \ldots, \mathbf{u}_5$ denote the five vectors in $\mathbb{F}^5$ of the preceding line. Thus, we have shown that for each vector $\mathbf{b} \in \mathbb{F}^4$ with $b_4 = 2b_1$, the vector

$$\mathbf{x} = x_1\mathbf{u}_1 + x_5\mathbf{u}_2 + b_1\mathbf{u}_3 + b_2\mathbf{u}_4 + b_3\mathbf{u}_5$$

is a solution of the equation $A\mathbf{x} = \mathbf{b}$ for every choice of x_1 and x_5. Therefore, $x_1\mathbf{u}_1 + x_5\mathbf{u}_2$ is a solution of the equation $A\mathbf{x} = \mathbf{0}$ for every choice of $x_1, x_5 \in \mathbb{F}$. Thus, $\mathbf{u}_1, \mathbf{u}_2 \in \mathcal{N}_A$ and, as

$$A\mathbf{x} = x_1A\mathbf{u}_1 + x_5A\mathbf{u}_2 + b_1A\mathbf{u}_3 + b_2A\mathbf{u}_4 + b_3A\mathbf{u}_5 = b_1A\mathbf{u}_3 + b_2A\mathbf{u}_4 + b_3A\mathbf{u}_5 \, ,$$

the vectors

$$\mathbf{v}_1 = A\mathbf{u}_3 = \begin{bmatrix} 1 \\ 0 \\ 0 \\ 2 \end{bmatrix}, \quad \mathbf{v}_2 = A\mathbf{u}_4 = \begin{bmatrix} 0 \\ 1 \\ 0 \\ 0 \end{bmatrix} \text{ and } \mathbf{v}_3 = A\mathbf{u}_5 = \begin{bmatrix} 0 \\ 0 \\ 1 \\ 0 \end{bmatrix}$$

belong to $\mathcal{R}_A$.

Exercise 2.4. Let $\mathbf{a}_j$, $j = 1, \ldots, 5$, denote the j'th column vector of the matrix A considered in the preceding example. Show that

(1) $\operatorname{span}\{\mathbf{v}_1, \mathbf{v}_2, \mathbf{v}_3\} = \operatorname{span}\{\mathbf{a}_2, \mathbf{a}_3, \mathbf{a}_4\}$ i.e., the span of the pivot columns of A.

(2) $\mathrm{span}\{\mathbf{u}_1, \mathbf{u}_2\} = \mathcal{N}_A$ and $\mathrm{span}\{\mathbf{v}_1, \mathbf{v}_2, \mathbf{v}_3\} = \mathcal{R}_A$.

2.3. Upper echelon matrices

The examples in the preceding section serve to illustrate the central role played by the number of pivots in an upper echelon matrix U and their positions when trying to solve systems of equations by Gaussian elimination.

Our next main objective is to exploit the special structure of upper echelon matrices in order to draw some general conclusions for matrices in this class. Extensions to general matrices will then be made on the basis of the following lemma:

Lemma 2.4. *Let $A \in \mathbb{F}^{p\times q}$ and assume that $A \neq O_{p\times q}$. Then there exists an invertible matrix $G \in \mathbb{F}^{p\times p}$ such that*

$$GA = U \tag{2.11}$$

is in upper echelon form.

Proof. By Gaussian elimination there exists a sequence $P_1, P_2, \ldots, P_k$ of $p\times p$ permutation matrices and a sequence $E_1, E_2, \ldots, E_k$ of lower triangular matrices with ones on the diagonal such that

$$E_k P_k \cdots E_2 P_2 E_1 P_1 A = U$$

is in upper echelon form. Consequently the matrix $G = E_k P_k \cdots E_2 P_2 E_1 P_1$ fulfills the asserted conditions, since it is the product of invertible matrices. □

Lemma 2.5. *Let $U \in \mathbb{F}^{p\times q}$ be an upper echelon matrix with k pivots and let $\mathbf{e}_j$ denote the j'th column of I_p for $j = 1, \ldots, p$. Then:*

(1) $k \leq \min\{p, q\}$.

(2) *The pivot columns of U are linearly independent.*

(3) *The span of the pivot columns $=$ span$\{\mathbf{e}_1, \ldots, \mathbf{e}_k\} = \mathcal{R}_U$; i.e.,*

 (a) *If $k < p$, then*

$$\mathcal{R}_U = \left\{ \begin{bmatrix} \mathbf{b} \\ \mathbf{0} \end{bmatrix} : \mathbf{b} \in \mathbb{F}^k \quad \text{and} \quad \mathbf{0} \in \mathbb{F}^{p-k} \right\}.$$

 (b) *If $k = p$, then $\mathcal{R}_U = \mathbb{F}^p$.*

(4) *The first k columns of U^T form a basis for $\mathcal{R}_{U^T}$.*

Proof. The first assertion follows from the fact there is at most one pivot in each column and at most one pivot in each row. Next, let $\mathbf{u}_1, \ldots, \mathbf{u}_q$

denote the columns of U and let $\mathbf{u}_{i_1},\ldots,\mathbf{u}_{i_k}$ (with $i_1 < \cdots < i_k$) denote the pivot columns of U. Then clearly

(2.12)

$$\operatorname{span}\{\mathbf{u}_{i_1},\ldots,\mathbf{u}_{i_k}\} \subseteq \operatorname{span}\{\mathbf{u}_1,\ldots,\mathbf{u}_q\} \subseteq \left\{\begin{bmatrix}\mathbf{b}\\ \mathbf{0}\end{bmatrix} : \mathbf{b}\in\mathbb{F}^k \text{ and } \mathbf{0}\in\mathbb{F}^{p-k}\right\},$$

if $k < p$. On the other hand, the matrix formed by arraying the pivot columns one after the other is of special form:

$$\begin{bmatrix}\mathbf{u}_{i_1} & \cdots & \mathbf{u}_{i_k}\end{bmatrix} = \begin{bmatrix}U_{11}\\ U_{21}\end{bmatrix},$$

where U_{11} is a $k\times k$ upper triangular matrix with the pivots as diagonal entries and $U_{21} = O_{(p-k)\times k}$. Therefore, U_{11} is invertible, and, for any choice of $\mathbf{b}\in\mathbb{F}^k$, the formulas

$$\begin{bmatrix}\mathbf{u}_{i_1} & \cdots & \mathbf{u}_{i_k}\end{bmatrix} U_{11}^{-1}\mathbf{b} = \begin{bmatrix}U_{11}\\ U_{21}\end{bmatrix} U_{11}^{-1}\mathbf{b} = \begin{bmatrix}\mathbf{b}\\ \mathbf{0}\end{bmatrix}$$

imply (2) and that

(2.13)

$$\left\{\begin{bmatrix}\mathbf{b}\\ \mathbf{0}\end{bmatrix} : \mathbf{b}\in\mathbb{F}^k \text{ and } \mathbf{0}\in\mathbb{F}^{p-k}\right\} \subseteq \{U\mathbf{x} : \mathbf{x}\in\mathbb{F}^q\} \subseteq \operatorname{span}\{\mathbf{u}_{i_1},\ldots,\mathbf{u}_{i_k}\}.$$

The two inclusions (2.12) and (2.13) yield the equality advertised in (a) of (3). The same argument (but with $U = U_{11}$) serves to justify (b) of (3).

Item (4) is easy and is left to the reader. □

Exercise 2.5. Verify (4) of Lemma 2.5.

Exercise 2.6. Let $U \in \mathbb{F}^{p\times q}$ be an upper echelon matrix with k pivots. Show that there exists an invertible matrix $K\in\mathbb{F}^{q\times q}$ such that:

(1) If $k < q$, then

$$\mathcal{R}_{U^T} = \left\{K\begin{bmatrix}\mathbf{b}\\ \mathbf{0}\end{bmatrix} : \mathbf{b}\in\mathbb{F}^k \quad \text{and} \quad \mathbf{0}\in\mathbb{F}^{q-k}\right\}.$$

(2) If $k = q$, then $\mathcal{R}_{U^T} = \mathbb{F}^q$.

[HINT: In case of difficulty, try some numerical examples for orientation.]

Exercise 2.7. Let U be a 4×5 matrix of the form

$$U = \begin{bmatrix}\mathbf{u}_1 & \mathbf{u}_2 & \mathbf{u}_3 & \mathbf{u}_4 & \mathbf{u}_5\end{bmatrix} = \begin{bmatrix} u_{11} & u_{12} & u_{13} & u_{14} & u_{15}\\ 0 & 0 & u_{23} & u_{24} & u_{25}\\ 0 & 0 & 0 & u_{34} & u_{35}\\ 0 & 0 & 0 & 0 & 0\end{bmatrix}$$

with u_{11}, u_{23} and u_{34} all nonzero. Show that $\operatorname{span}\{\mathbf{u}_1, \mathbf{u}_3, \mathbf{u}_4\} = \mathcal{R}_U$.

Exercise 2.8. Find a basis for the null space $\mathcal{N}_U$ of the 4×5 matrix U considered in Exercise 2.7 in terms of its entries u_{ij}, when the pivots of U are all set equal to one.

Lemma 2.6. *Let $U \in \mathbb{F}^{p\times q}$ be in upper echelon form with k pivots. Then:*

(1) $k \le \min\{p, q\}$.

(2) $k = q \Longleftrightarrow U$ *is left invertible* $\Longleftrightarrow \mathcal{N}_U = \{0\}$.

(3) $k = p \Longleftrightarrow U$ *is right invertible* $\Longleftrightarrow \mathcal{R}_U = \mathbb{F}^p$.

Proof. The first assertion is established in Lemma 2.5 (and is repeated here for perspective).

Suppose next that U has q pivots. Then

$$U = \begin{bmatrix} U_{11} \\ O_{(p-q)\times q} \end{bmatrix} \quad \text{if} \quad q < p \quad \text{and} \quad U = U_{11} \quad \text{if} \quad q = p,$$

where U_{11} is a $q \times q$ upper triangular matrix with nonzero diagonal entries. Thus, if $q < p$ and $V \in \mathbb{F}^{q\times p}$ is written in block form as

$$V = [V_{11} \quad V_{12}]$$

with $V_{11} = U_{11}^{-1}$ and $V_{12} \in \mathbb{F}^{q\times(p-q)}$, then V is a left inverse of U for every choice of $V_{12} \in \mathbb{F}^{q\times(p-q)}$; i.e., $k = q \Longrightarrow U$ is left invertible.

Suppose next that U is left invertible with a left inverse V. Then

$$\mathbf{x} \in \mathcal{N}_U \Longrightarrow U\mathbf{x} = \mathbf{0} \Longrightarrow \mathbf{0} = V(U\mathbf{x}) = (VU)\mathbf{x} = \mathbf{x},$$

i.e., U left invertible $\Longrightarrow \mathcal{N}_U = \{\mathbf{0}\}$.

To complete the proof of (2), observe that: The span of the pivot columns of U is equal to the span of all the columns of U, alias $\mathcal{R}_U$. Therefore, every column of U can be expressed as a linear combination of the pivot columns. Thus, as

$$\mathcal{N}_U = \{\mathbf{0}\} \Longrightarrow \quad \text{the } q \text{ columns of } U \text{ are linearly independent},$$

it follows that

$$\mathcal{N}_U = \{\mathbf{0}\} \Longrightarrow U \quad \text{has } q \text{ pivots}.$$

Finally, even though the equivalence $k = p \Longleftrightarrow \mathcal{R}_U = \mathbb{F}^p$ is known from Lemma 2.5, we shall present an independent proof of all of (3), because it is instructive and indicates how to construct right inverses, when they exist. We proceed in three steps:

(a) $k = p \Longrightarrow U$ is right invertible: If $k = p = q$, then U is right (and left) invertible by Lemma 1.4. If $k = p$ and $q > p$, then there exists a

$q \times q$ permutation matrix P that (multiplying U on the right) serves to interchange the columns of U so that the pivots are concentrated on the left, i.e.,

$$UP = [U_{11} \quad U_{12}] \,,$$

where U_{11} is a $p \times p$ upper triangular matrix with nonzero diagonal entries. Thus, if $q > p$ and $V \in \mathbb{F}^{q\times p}$ is written in block form as

$$V = \begin{bmatrix} V_{11} \\ V_{21} \end{bmatrix}$$

with $V_{11} \in \mathbb{F}^{p\times p}$ and $V_{21} \in \mathbb{F}^{(q-p)\times p}$, then

$$UPV = I_p \iff U_{11}V_{11} + U_{12}V_{21} = I_p \iff V_{11} = U_{11}^{-1}(I_p - U_{12}V_{21}) \,.$$

Consequently, for any choice of $V_{21} \in \mathbb{F}^{(q-p)\times p}$, the matrix PV will be a right inverse of U if V_{11} is chosen as indicated just above; i.e., (a) holds.

(b) U is right invertible $\Longrightarrow \mathcal{R}_U = \mathbb{F}^p$: If U is right invertible and V is a right inverse of U, then for each choice of $\mathbf{b} \in \mathbb{F}^p$, $\mathbf{x} = V\mathbf{b}$ is a solution of the equation $U\mathbf{x} = \mathbf{b}$:

$$UV = I_p \Longrightarrow U(V\mathbf{b}) = (UV)\mathbf{b} = \mathbf{b}\,;$$

i.e., (b) holds.

(c) $\mathcal{R}_U = \mathbb{F}^p \Longrightarrow k = p$: If $\mathcal{R}_U = \mathbb{F}^p$, then there exists a vector $\mathbf{v} \in \mathbb{F}^q$ such that $U\mathbf{v} = \mathbf{e}_p$, where $\mathbf{e}_p$ denotes the p'th column of I_p. If U has less than p pivots, then the last row of U, $\mathbf{e}_p^T U = \mathbf{0}^T$, i.e.,

$$1 = \mathbf{e}_p^T\mathbf{e}_p = \mathbf{e}_p^T(U\mathbf{v}) = (\mathbf{e}_p^T U)\mathbf{v} = \mathbf{0}^T\mathbf{v} = 0\,,$$

which is impossible. Therefore, $\mathcal{R}_U = \mathbb{F}^p \Longrightarrow U$ has p pivots and (c) holds. $\square$

Exercise 2.9. Let $A = \begin{bmatrix} 1 & 1 & 0 \\ 2 & 0 & 1 \\ 0 & 0 & 0 \end{bmatrix}$ and $B = \begin{bmatrix} 0 & 0 & 1 \\ 0 & 1 & 0 \\ 1 & 0 & 0 \end{bmatrix}$. Find a basis for each of the spaces $\mathcal{R}_{BA}$, $\mathcal{R}_A$ and $\mathcal{R}_{AB}$.

Exercise 2.10. Find a basis for each of the spaces $\mathcal{N}_{BA}$, $\mathcal{N}_A$ and $\mathcal{N}_{AB}$ for the matrices A and B that are given in the preceding exercise.

Exercise 2.11. Show that if $A \in \mathbb{F}^{p\times q}$, $B \in \mathbb{F}^{p\times p}$ and $\mathbf{u}_1, \ldots, \mathbf{u}_k$ is a basis for $\mathcal{R}_A$, then $\operatorname{span}\{B\mathbf{u}_1, \ldots, B\mathbf{u}_k\} = \mathcal{R}_{BA}$ and that this second set of vectors will be a basis for $\mathcal{R}_{BA}$ if B is left invertible.

Exercise 2.12. Show that if A is a $p \times q$ matrix and C is a $q \times q$ invertible matrix, then $\mathcal{R}_{AC} = \mathcal{R}_A$.

Exercise 2.13. Show that if $U \in \mathbb{F}^{p\times q}$ is a $p \times q$ matrix in upper echelon form with p pivots, then U has exactly one right inverse if and only if $p = q$.

If $A \in \mathbb{F}^{p\times q}$ and $\mathcal{U}$ is a subspace of $\mathbb{F}^q$, then

$$A\mathcal{U} = \{A\mathbf{u} : \mathbf{u} \in \mathcal{U}\}. \tag{2.14}$$

Exercise 2.14. Show that if $GA = B$ and G is invertible (as is the case in formula (2.11) with $U = B$), then

$$\mathcal{R}_B = G\mathcal{R}_A,\ \mathcal{N}_B = \mathcal{N}_A,\ \mathcal{R}_{B^T} = \mathcal{R}_{A^T} \text{ and } G^T\mathcal{N}_{B^T} = \mathcal{N}_{A^T}.$$

Exercise 2.15. Let $U \in \mathbb{F}^{p\times q}$ be an upper echelon matrix with k pivots, where $1 \leq k \leq p < q$. Show that $\mathcal{N}_U \neq \{\mathbf{0}\}$. [HINT: There exists a $q \times q$ permutation matrix P (that is introduced to permute the columns of U, if need be) such that

$$UP = \begin{bmatrix} U_{11} & U_{12} \\ U_{21} & U_{22} \end{bmatrix},$$

where U_{11} is a $k \times k$ upper triangular matrix with nonzero diagonal entries, $U_{12} \in \mathbb{F}^{k\times(q-k)}$, $U_{21} = O_{(p-k)\times k}$ and $U_{22} = O_{(p-k)\times(q-k)}$ and hence that

$$\mathbf{x} = P\begin{bmatrix} U_{11}^{-1}U_{12} \\ -I_{q-k} \end{bmatrix}\mathbf{y}$$

is a nonzero solution of the equation $U\mathbf{x} = \mathbf{0}$ for every nonzero vector $\mathbf{y} \in \mathbb{F}^{q-k}$.]

Exercise 2.16. Let $n_L = n_L(U)$ and $n_R = n_R(U)$ denote the number of left and right inverses, respectively, of an upper echelon matrix $U \in \mathbb{F}^{p\times q}$. Show that the combinations ($n_L = 0$, $n_R = 0$), ($n_L = 0$, $n_R = \infty$), ($n_L = 1$, $n_R = 1$) and ($n_L = \infty$, $n_R = 0$) are possible.

Exercise 2.17. In the notation of the previous exercise, show that the combinations ($n_L = 0$, $n_R = 1$), ($n_L = 1$, $n_R = 0$), ($n_L = \infty$, $n_R = 1$), ($n_L = 1$, $n_R = \infty$) and ($n_L = \infty$, $n_R = \infty$) are impossible.

Lemma 2.7. *Let $A \in \mathbb{F}^{p\times q}$ and assume that $\mathcal{N}_A = \{\mathbf{0}\}$. Then $p \geq q$.*

Proof. Lemma 2.4 guarantees the existence of an invertible matrix $G \in \mathbb{F}^{p\times p}$ such that formula (2.11) is in force and hence that

$$\mathcal{N}_A = \{\mathbf{0}\} \iff \mathcal{N}_U = \{\mathbf{0}\}.$$

Moreover, in view of Lemma 2.6,

$$\mathcal{N}_U = \{\mathbf{0}\} \iff U \quad \text{has } q \text{ pivots}.$$

Therefore, by another application of Lemma 2.6, $q \leq p$. □

Theorem 2.8. *Let $\mathbf{v}_1, \ldots, \mathbf{v}_\ell$ be a basis for a vector space $\mathcal{V}$ over $\mathbb{F}$ and let $\mathbf{u}_1, \ldots, \mathbf{u}_k$ be a basis for a subspace $\mathcal{U}$ of $\mathcal{V}$. Then:*

(1) $k \le \ell$.

(2) $k = \ell \Longleftrightarrow \mathcal{U} = \mathcal{V}$.

(3) *If* $k < \ell$, *then there exist a set of vectors* $\{\mathbf{w}_1, \ldots, \mathbf{w}_{\ell-k}\}$ *in* $\mathcal{V}$ *such that* $\{\mathbf{u}_1, \ldots, \mathbf{u}_k, \mathbf{w}_1, \ldots, \mathbf{w}_{\ell-k}\}$ *is a basis for* $\mathcal{V}$.

Proof. Since $\mathcal{U}$ is a subspace of $\mathcal{V}$, each of the vectors $\mathbf{u}_j$ can be expressed as a unique linear combination of the vectors in the basis for $\mathcal{V}$, i.e.,

$$\mathbf{u}_j = \sum_{i=1}^{\ell} a_{ij}\mathbf{v}_i \,, \quad \text{for} \quad j = 1, \ldots, k \,.$$

Thus, the $\ell \times k$ matrix

$$A = \begin{bmatrix} a_{11} & \cdots & a_{1k} \\ \vdots & & \vdots \\ a_{\ell 1} & \cdots & a_{\ell k} \end{bmatrix}$$

that is based on these coefficients is uniquely defined by these two sets of vectors. The next objective is to show that $\mathcal{N}_A = \{\mathbf{0}\}$. To this end, suppose that

$$\sum_{j=1}^{k} a_{ij}c_j = 0 \quad \text{for} \quad i = 1, \ldots, \ell \,.$$

Then

$$\sum_{j=1}^{k} c_j\mathbf{u}_j = \sum_{j=1}^{k} c_j \left(\sum_{i=1}^{\ell} a_{ij}\mathbf{v}_i \right) = \sum_{i=1}^{\ell} \left(\sum_{j=1}^{k} a_{ij}c_j \right) \mathbf{v}_i = \mathbf{0} \,.$$

Thus, $c_1 = \cdots = c_k = 0$, since the vectors $\mathbf{u}_1, \ldots, \mathbf{u}_k$ are linearly independent. Therefore, $\mathcal{N}_A = \{\mathbf{0}\}$ and, by Lemma 2.7, $k \le \ell$.

Suppose next that $k = \ell$ and $\operatorname{span}\{\mathbf{u}_1, \ldots, \mathbf{u}_k\} \neq \mathcal{V}$. Then there exists a vector $\mathbf{w} \in \mathcal{V}$ that is linearly independent of $\{\mathbf{u}_1, \ldots, \mathbf{u}_k\}$. But this in turn implies that $\operatorname{span}\{\mathbf{u}_1, \ldots, \mathbf{u}_k, \mathbf{w}\}$ is a subspace of $\mathcal{V}$ and hence, by the argument furnished to justify assertion (1), that $\ell \ge k + 1$. But this is impossible if $k = \ell$. Therefore,

$$k = \ell \Longrightarrow \mathcal{U} = \mathcal{V} \,.$$

Conversely, if $\mathcal{U} = \mathcal{V}$, then every vector $\mathbf{v}_j$, $j = 1, \ldots, \ell$, can be written as a linear combination of $\mathbf{u}_1, \ldots, \mathbf{u}_k$, and hence the argument used to verify (1) implies that the inequality $\ell \le k$ must also be in force. Therefore, $k = \ell$.

Finally, if $k < \ell$, then (2) implies that $\mathcal{U} \neq \mathcal{V}$ and hence that there exists a vector $\mathbf{w}_1 \notin \operatorname{span}\{\mathbf{u}_1, \ldots, \mathbf{u}_k\}$. But this implies that $\{\mathbf{u}_1, \ldots, \mathbf{u}_k, \mathbf{w}_1\}$ is a linearly independent set of vectors. Moreover, if $k + 1 < \ell$, then there exists a vector $\mathbf{w}_2 \notin \operatorname{span}\{\mathbf{u}_1, \ldots, \mathbf{u}_k, \mathbf{w}_1\}$. Therefore, $\{\mathbf{u}_1, \ldots, \mathbf{u}_k, \mathbf{w}_1, \mathbf{w}_2\}$ is a linearly independent set of vectors inside the vector space $\mathcal{V}$. If $k + 2 < \ell$, then the procedure continues until $\ell - k$ new vectors $\{\mathbf{w}_1, \dots, \mathbf{w}_{\ell-k}\}$ have been added to the original set $\{\mathbf{u}_1, \dots, \mathbf{u}_k\}$ to form a basis for $\mathcal{V}$. $\square$

- **dimension:** The preceding theorem guarantees that if $\mathcal{V}$ is a vector space over $\mathbb{F}$ with a finite basis, then **every basis of $\mathcal{V}$ has exactly the same number of vectors.** This number is called the **dimension** of the vector space.
- **zero dimension:** The dimension of the vector space $\{\mathbf{0}\}$ will be assigned the number zero.

Exercise 2.18. Show that $\mathbb{F}^k$ is a k dimensional vector space over $\mathbb{F}$.

Exercise 2.19. Show that if $\mathcal{U}$ and $\mathcal{V}$ are finite dimensional subspaces of a vector space $\mathcal{W}$ over $\mathbb{F}$, then the set $\mathcal{U} + \mathcal{V}$ that is defined by the formula

$$\mathcal{U} + \mathcal{V} = \{\mathbf{u} + \mathbf{v} : \mathbf{u} \in \mathcal{U} \text{ and } \mathbf{v} \in \mathcal{V}\} \tag{2.15}$$

is a vector space over $\mathbb{F}$ and

$$\dim(\mathcal{U} + \mathcal{V}) = \dim \mathcal{U} + \dim \mathcal{V} - \dim \mathcal{U} \cap \mathcal{V}\,. \tag{2.16}$$

Exercise 2.20. Let T be a linear mapping from a finite dimensional vector space $\mathcal{U}$ over $\mathbb{F}$ into a vector space $\mathcal{V}$ over $\mathbb{F}$. Show that $\dim \mathcal{R}_T \le \dim \mathcal{U}$. [HINT: If $\mathbf{u}_1, \dots, \mathbf{u}_n$ is a basis for $\mathcal{U}$, then the vectors $T\mathbf{u}_j$, $j = 1, \dots, n$ span $\mathcal{R}_T$.]

Exercise 2.21. Construct a linear mapping T from a vector space $\mathcal{U}$ over $\mathbb{F}$ into a vector space $\mathcal{V}$ over $\mathbb{F}$ such that $\dim \mathcal{R}_T < \dim \mathcal{U}$.

2.4. The conservation of dimension

Theorem 2.9. *Let T be a linear mapping from a finite dimensional vector space $\mathcal{U}$ over $\mathbb{F}$ into a vector space $\mathcal{V}$ over $\mathbb{F}$ (finite dimensional or not). Then*

$$\dim \mathcal{N}_T + \dim \mathcal{R}_T = \dim \mathcal{U}. \tag{2.17}$$

Proof. In view of Exercise 2.20, $\mathcal{R}_T$ is automatically a finite dimensional space regardless of the dimension of $\mathcal{V}$. Suppose first that $\mathcal{N}_T \ne \{\mathbf{0}\}$, $\mathcal{R}_T \ne \{\mathbf{0}\}$ and let $\mathbf{u}_1, \dots, \mathbf{u}_k$ be a basis for $\mathcal{N}_T$, $\mathbf{v}_1, \dots, \mathbf{v}_l$ be a basis for $\mathcal{R}_T$ and choose vectors $\mathbf{y}_j \in \mathcal{U}$ such that

$$T\mathbf{y}_j = \mathbf{v}_j, \quad j = 1, \dots, l.$$

The first item of business is to show that the vectors $\mathbf{u}_1, \dots, \mathbf{u}_k$ and $\mathbf{y}_1, \dots, \mathbf{y}_l$ are linearly independent over $\mathbb{F}$. Suppose, to the contrary, that there exists scalars $\alpha_1, \dots, \alpha_k$ and $\beta_1, \dots, \beta_l$ such that

$$\sum_{i=1}^{k} \alpha_i \mathbf{u}_i + \sum_{j=1}^{l} \beta_j \mathbf{y}_j = \mathbf{0}\,. \tag{2.18}$$

Then

$$T\left(\sum_{i=1}^{k}\alpha_i\mathbf{u}_i+\sum_{j=1}^{l}\beta_j\mathbf{y}_j\right)=T(\mathbf{0})=\mathbf{0}\,.$$

But the left-hand side of the last equality can be reexpressed as

$$\sum_{i=1}^{k}\alpha_i T\mathbf{u}_i+\sum_{j=1}^{l}\beta_j T\mathbf{y}_j=\mathbf{0}+\sum_{j=1}^{l}\beta_j\mathbf{v}_j\,.$$

Therefore, $\beta_1=\cdots=\beta_l=0$ and so too, by (2.18), $\alpha_1=\cdots=\alpha_k=0$. This completes the proof of the asserted linear independence.

The next step is to verify that the vectors $\mathbf{u}_1,\ldots,\mathbf{u}_k$, $\mathbf{y}_1,\ldots,\mathbf{y}_l$ span $\mathcal{U}$ and thus that this set of vectors is a basis for $\mathcal{U}$. To this end, let $\mathbf{w}\in\mathcal{U}$. Then, since

$$T\mathbf{w}=\sum_{j=1}^{l}\beta_j\mathbf{v}_j=\sum_{j=1}^{l}\beta_j T\mathbf{y}_j\,,$$

for some choice of $\beta_1,\ldots,\beta_\ell\in\mathbb{F}$, it follows that

$$T\left(\mathbf{w}-\sum_{j=1}^{l}\beta_j\mathbf{y}_j\right)=\mathbf{0}\,.$$

This means that

$$\mathbf{w}-\sum_{j=1}^{l}\beta_j\mathbf{y}_j\in\mathcal{N}_T$$

and, consequently, this vector can be expressed as a linear combination of $\mathbf{u}_1,\ldots,\mathbf{u}_k$. In other words,

$$\mathbf{w}=\sum_{i=1}^{k}\alpha_i\mathbf{u}_i+\sum_{j=1}^{l}\beta_j\mathbf{y}_j$$

for some choice of scalars $\alpha_1,\ldots,\alpha_k$ and $\beta_1,\ldots,\beta_l$ in $\mathbb{F}$. But this means that

$$\operatorname{span}\{\mathbf{u}_1,\ldots,\mathbf{u}_k,\mathbf{y}_1,\ldots,\mathbf{y}_l\}=\mathcal{U}$$

and hence, in view of the already exhibited linear independence, that

$$\begin{aligned}\dim\mathcal{U}&=k+l\\&=\dim\mathcal{N}_T+\dim\mathcal{R}_T,\end{aligned}$$

as claimed.

Suppose next that $\mathcal{N}_T=\{\mathbf{0}\}$ and $\mathcal{R}_T\neq\{\mathbf{0}\}$. Then much the same sort of argument serves to prove that if $\mathbf{v}_1,\ldots,\mathbf{v}_l$ is a basis for $\mathcal{R}_T$ and if $\mathbf{y}_j\in\mathcal{U}$ is such that $T\mathbf{y}_j=\mathbf{v}_j$ for $j=1,\ldots,l$, then the vectors $\mathbf{y}_1,\ldots,\mathbf{y}_l$ are linearly independent and span $\mathcal{U}$. Thus, $\dim \mathcal{U} = \dim \mathcal{R}_T = \ell$, and hence formula (2.17) is still in force, since $\dim \mathcal{N}_T = 0$.

It remains only to consider the case $\mathcal{R}_T = \{\mathbf{0}\}$. But then $\mathcal{N}_T = \mathcal{U}$, and formula (2.17) is still valid. $\square$

Remark 2.10. We shall refer to formula (2.17) as the **principle of conservation of dimension**. Notice that it is correct as stated if $\mathcal{U}$ is a finite dimensional subspace of some other vector space $\mathcal{W}$.

2.5. Quotient spaces

This section is devoted to a brief sketch of another approach to establishing Theorem 2.9 that is based on quotient spaces. It can be skipped without loss of continuity.

- **Quotient spaces:** Let $\mathcal{V}$ be a vector space over $\mathbb{F}$ and let $\mathcal{M}$ be a subspace of $\mathcal{V}$ and, for $\mathbf{u} \in \mathcal{V}$, let $\mathbf{u}_{\mathcal{M}} = \{\mathbf{u} + \mathbf{m} : \mathbf{m} \in \mathcal{M}\}$. Then $\mathcal{V}/\mathcal{M} = \{\mathbf{u}_{\mathcal{M}} : \mathbf{u} \in \mathcal{V}\}$ is a vector space over $\mathbb{F}$ with respect to the rules $\mathbf{u}_{\mathcal{M}} + \mathbf{v}_{\mathcal{M}} = (\mathbf{u} + \mathbf{v})_{\mathcal{M}}$ and $\alpha(\mathbf{u}_{\mathcal{M}}) = (\alpha\mathbf{u})_{\mathcal{M}}$ of vector addition and scalar multiplication, respectively. The details are easily filled in with the aid of Exercises 2.22–2.24.

Exercise 2.22. Let $\mathcal{M}$ be a proper nonzero subspace of a vector space $\mathcal{V}$ over $\mathbb{F}$ and, for $\mathbf{u} \in \mathcal{V}$, let $\mathbf{u}_{\mathcal{M}} = \{\mathbf{u} + \mathbf{m} : \mathbf{m} \in \mathcal{M}\}$. Show that if $\mathbf{x}, \mathbf{y} \in \mathcal{V}$, then

$$\mathbf{x}_{\mathcal{M}} = \mathbf{y}_{\mathcal{M}} \Longleftrightarrow \mathbf{x} - \mathbf{y} \in \mathcal{M}$$

and use this result to describe the set of vectors $\mathbf{u} \in \mathcal{V}$ such that $\mathbf{u}_{\mathcal{M}} = \mathbf{0}_{\mathcal{M}}$.

Exercise 2.23. Show that if, in the setting of Exercise 2.22, $\mathbf{u}, \mathbf{v}, \mathbf{x}, \mathbf{y} \in \mathcal{V}$ and if also $\mathbf{u}_{\mathcal{M}} = \mathbf{x}_{\mathcal{M}}$ and $\mathbf{v}_{\mathcal{M}} = \mathbf{y}_{\mathcal{M}}$, then $(\mathbf{u} + \mathbf{v})_{\mathcal{M}} = (\mathbf{x} + \mathbf{y})_{\mathcal{M}}$.

Exercise 2.24. Show that if, in the setting of Exercise 2.22, $\alpha, \beta \in \mathbb{F}$ and $\mathbf{u} \in \mathcal{V}$, but $\mathbf{u} \notin \mathcal{M}$, then $(\alpha\mathbf{u})_{\mathcal{M}} = (\beta\mathbf{u})_{\mathcal{M}}$ if and only if $\alpha = \beta$.

Exercise 2.25. Let $\mathcal{U}$ be a finite dimensional vector space over $\mathbb{F}$ and let $\mathcal{V}$ be a subspace of $\mathcal{U}$. Show that $\dim\mathcal{U} = \dim\,(\mathcal{U}/\mathcal{V}) + \dim \mathcal{V}$.

Exercise 2.26. Establish the principle of conservation of dimension with the aid of Exercise 2.25.

2.6. Conservation of dimension for matrices

One of the prime applications of the principle of conservation of dimension is to the particular linear transformation T from $\mathbb{F}^q$ into $\mathbb{F}^p$ that is defined by multiplying each vector $\mathbf{x} \in \mathbb{F}^p$ by a given matrix $A \in \mathbb{F}^{p\times q}$. Because of its importance, the main conclusions are stated as a theorem, even though they are easily deduced from the definitions of the requisite spaces and Theorem 2.9.

Theorem 2.11. *If $A \in \mathbb{F}^{p\times q}$, then*

$$\begin{aligned} (2.19)\qquad \mathcal{N}_A &= \{\mathbf{x}\in\mathbb{F}^q : A\mathbf{x}=\mathbf{0}\} \quad \textit{is a subspace of}\quad \mathbb{F}^q\,, \\ (2.20)\qquad \mathcal{R}_A &= \{A\mathbf{x} : \mathbf{x}\in\mathbb{F}^q\} \quad \textit{is a subspace of}\quad \mathbb{F}^p \quad \textit{and} \\ (2.21)\qquad q &= \dim\mathcal{N}_A + \dim\mathcal{R}_A. \end{aligned}$$

- **rank:** If $A \in \mathbb{F}^{p\times q}$, then the dimension of $\mathcal{R}_A$ is termed the **rank** of A:

$$\operatorname{rank} A = \dim \mathcal{R}_A\,.$$

Exercise 2.27. Let $A \in \mathbb{F}^{p\times q}$, $B \in \mathbb{F}^{p\times p}$ and $C \in \mathbb{F}^{q\times q}$. Show that:

(1) $\operatorname{rank} BA \le \operatorname{rank} A$, with equality if B is invertible.

(2) $\operatorname{rank} AC \le \operatorname{rank} A$, with equality if C is invertible.

Theorem 2.12. *If $A \in \mathbb{F}^{p\times q}$, then*

$$(2.22)\qquad \operatorname{rank} A = \operatorname{rank} A^T = \operatorname{rank} A^H \le \min\{p,\,q\}\,.$$

Proof. The statement is obvious if $A = O_{p\times q}$. If $A \neq O_{p\times q}$, then there exists an invertible matrix $G \in \mathbb{F}^{p\times p}$ such that $GA = U$ is in upper echelon form. Thus,

$$\operatorname{rank} A = \operatorname{rank} GA = \operatorname{rank} U = \text{ the number of pivots of } U\,,$$

whereas,

$$\operatorname{rank} A^T = \operatorname{rank} A^T G^T = \operatorname{rank} U^T = \text{the number of pivots of } U\,.$$

The proof that $\operatorname{rank} A = \operatorname{rank} A^H$ is left to the reader as an exercise. □

Exercise 2.28. Show that if $A \in \mathbb{C}^{p\times q}$, then $\operatorname{rank} A = \operatorname{rank} A^H$.

Exercise 2.29. Show that if $A \in \mathbb{C}^{p\times q}$ and $C \in \mathbb{C}^{k\times q}$, then

$$(2.23)\qquad \operatorname{rank}\begin{bmatrix} A \\ C \end{bmatrix} = q \Longleftrightarrow \mathcal{N}_A \cap \mathcal{N}_C = \{\mathbf{0}\}\,.$$

Exercise 2.30. Show that if $A \in \mathbb{C}^{p\times q}$ and $B \in \mathbb{C}^{p\times r}$, then

$$(2.24)\qquad \operatorname{rank}\begin{bmatrix} A & B \end{bmatrix} = p \Longleftrightarrow \mathcal{N}_{A^H} \cap \mathcal{N}_{B^H} = \{\mathbf{0}\}\,.$$

Exercise 2.31. Show that if A is a triangular matrix (either upper or lower), then $\operatorname{rank} A$ is bigger than or equal to the number of nonzero diagonal entries in A. Give an example of an upper triangular matrix A for which the inequality is strict.

Exercise 2.32. Calculate $\dim \mathcal{N}_A$ and $\dim \mathcal{R}_A$ in the setting of Exercise 2.4 and confirm that these numbers are consistent with the principle of conservation of dimension.

2.7. From U to A

The next theorem is an analogue of Lemma 2.6 that is stated for general matrices $A \in \mathbb{F}^{p\times q}$; i.e., the conclusions are not restricted to upper echelon matrices. It may be obtained from Lemma 2.6 by exploiting formula (2.11). However, it is more instructive to give a direct proof.

Theorem 2.13. *Let $A \in \mathbb{F}^{p\times q}$. Then*

(1) $\operatorname{rank} A = p \Longleftrightarrow A$ *is right invertible* $\Longleftrightarrow \mathcal{R}_A = \mathbb{F}^p$.

(2) $\operatorname{rank} A = q \Longleftrightarrow A$ *is left invertible* $\Longleftrightarrow \mathcal{N}_A = \{\mathbf{0}\}$.

(3) *If A has both a left inverse $B \in \mathbb{F}^{q\times p}$ and a right inverse $C \in \mathbb{F}^{q\times p}$, then $B = C$ and $p = q$.*

Proof. Since $\mathcal{R}_A \subseteq \mathbb{F}^p$, it is clear that

$$\mathcal{R}_A = \mathbb{F}^p \Longleftrightarrow \operatorname{rank} A = p\,.$$

Suppose next that $\mathcal{R}_A = \mathbb{F}^p$. Then the equations

$$A\mathbf{x}_j = \mathbf{b}_j, \quad j = 1, \dots, p,$$

are solvable for every choice of the vectors $\mathbf{b}_j$. If, in particular, $\mathbf{b}_j$ is set equal to the j'th column of the identity matrix I_p, then

$$A\begin{bmatrix}\mathbf{x}_1 & \cdots & \mathbf{x}_p\end{bmatrix} = \begin{bmatrix}\mathbf{b}_1 & \cdots & \mathbf{b}_p\end{bmatrix} = I_p.$$

This exhibits the $q \times p$ matrix

$$X = \begin{bmatrix}\mathbf{x}_1 & \cdots & \mathbf{x}_p\end{bmatrix}$$

with columns $\mathbf{x}_1, \dots, \mathbf{x}_p$ as a right inverse of A.

Conversely, if $AC = I_p$ for some matrix $C \in \mathbb{F}^{q\times p}$, then $\mathbf{x} = C\mathbf{b}$ is a solution of the equation $A\mathbf{x} = \mathbf{b}$ for every choice of $\mathbf{b} \in \mathbb{F}^p$, i.e., $\mathcal{R}_A = \mathbb{F}^p$. This completes the proof of (1).

Next, (2) follows from (1) and the observation that

$$\begin{aligned}\mathcal{N}_A = \{\mathbf{0}\} &\Longleftrightarrow \operatorname{rank} A = q \quad \text{(by Theorem 2.11)}\\ &\Longleftrightarrow \operatorname{rank} A^T = q \quad \text{(by Theorem 2.12)}\\ &\Longleftrightarrow A^T \text{ is right invertible} \quad \text{(by part (1))}\\ &\Longleftrightarrow A \text{ is left invertible}\,.\end{aligned}$$

Moreover, (1) and (2) imply that if A is both left invertible and right invertible, then $p = q$ and, as has already been shown, the two one-sided inverses coincide:

$$B = BI_p = B(AC) = (BA)C = I_qC = C.$$

□

Exercise 2.33. Find the null space $\mathcal{N}_A$ and the range $\mathcal{R}_A$ of the matrix

$$A = \begin{bmatrix} 3 & 1 & 0 & 2 \\ 4 & 1 & 0 & 2 \\ 5 & 2 & 0 & 4 \end{bmatrix} \quad \text{acting on} \quad \mathbb{R}^4$$

and check that the principle of conservation of dimension holds.

2.8. Square matrices

Theorem 2.14. *Let $A \in \mathbb{F}^{p\times p}$. Then the following statements are equivalent:*

(1) *A is left invertible.*

(2) *A is right invertible.*

(3) $\mathcal{N}_A = \{\mathbf{0}\}$.

(4) $\mathcal{R}_A = \mathbb{F}^p$.

Proof. This is an immediate corollary of Theorem 2.13. □

Remark 2.15. The equivalence of (3) and (4) in Theorem 2.14 is a special case of the **Fredholm alternative**, which, in its most provocative form, states that *if the solution to the equation $A\mathbf{x} = \mathbf{b}$ is unique, then it exists*, or to put it better:

If $A \in \mathbb{F}^{p\times p}$ and the equation $A\mathbf{x} = \mathbf{b}$ has at most one solution, then it has exactly one.

Lemma 2.16. *If $A \in \mathbb{F}^{p\times p}$, $B \in \mathbb{F}^{p\times p}$ and AB is invertible, then both A and B are invertible.*

Proof. Clearly, $\mathcal{R}_{AB} \subseteq \mathcal{R}_A$ and $\mathcal{N}_{AB} \supseteq \mathcal{N}_B$. Therefore,

$$p = \operatorname{rank} AB \le \operatorname{rank} A \le p \quad \text{and} \quad 0 = \dim \mathcal{N}_{AB} \ge \dim \mathcal{N}_B \ge 0\,.$$

The rest is immediate from Theorem 2.14. □

Lemma 2.17. *If $V \in \mathbb{F}^{p\times q}$, then*

$$\mathcal{N}_{V^HV} = \mathcal{N}_V \quad \textit{and} \quad \operatorname{rank} V^H V = \operatorname{rank} V\,. \tag{2.25}$$

Proof. It is easily seen that $\mathcal{N}_V \subseteq \mathcal{N}_{V^HV}$, since $V\mathbf{x} = \mathbf{0}$ clearly implies that $V^HV\mathbf{x} = \mathbf{0}$. On the other hand, if $V^HV\mathbf{x} = \mathbf{0}$, then $\mathbf{x}^HV^HV\mathbf{x} = 0$ and hence the vector $\mathbf{y} = V\mathbf{x}$, with entries $y_, \ldots, y_p$ is subject to the constraints

$$0 = \mathbf{y}^H\mathbf{y} = \sum_{j=1}^{p} |y_j|^2\,.$$

Therefore, $V^H V\mathbf{x} = \mathbf{0} \Longrightarrow \mathbf{y} = V\mathbf{x} = \mathbf{0}$. This completes the proof of the first assertion in (2.25). The second then follows easily from the principle of conservation of dimension, since $V^H V$ and V both have q columns. □

Exercise 2.34. Show that if $A \in \mathbb{F}^{p\times q}$ and $B \in \mathbb{F}^{q\times p}$, then $\mathcal{N}_{AB} = \{\mathbf{0}\}$ if and only if $\mathcal{N}_A \cap \mathcal{R}_B = \{\mathbf{0}\}$ and $\mathcal{N}_B = \{\mathbf{0}\}$.

Exercise 2.35. Find a $p \times q$ matrix A and a vector $\mathbf{b} \in \mathbb{R}^p$ such that $\mathcal{N}_A = \{\mathbf{0}\}$ and yet the equation $A\mathbf{x} = \mathbf{b}$ has no solutions.

Exercise 2.36. Let $B \in \mathbb{F}^{n\times p}$, $A \in \mathbb{F}^{p\times q}$ and let $\{\mathbf{u}_1, \ldots, \mathbf{u}_k\}$ be a basis for $\mathcal{R}_A$. Show that $\{B\mathbf{u}_1, \ldots, B\mathbf{u}_k\}$ is a basis for $\mathcal{R}_{BA}$ if and only if $\mathcal{R}_A \cap \mathcal{N}_B = \{\mathbf{0}\}$.

Exercise 2.37. Find a basis for $\mathcal{R}_A$ and $\mathcal{N}_A$ if $A = \begin{bmatrix} 1 & 3 & 1 & 8 & 2 \\ 0 & 1 & 2 & 1 & 3 \\ 1 & -2 & 3 & 3 & 1 \\ 1 & 6 & 11 & 5 & 9 \end{bmatrix}$.

Exercise 2.38. Let $B \in \mathbb{F}^{n\times p}$, $A \in \mathbb{F}^{p\times q}$ and let $\{\mathbf{u}_1, \ldots, \mathbf{u}_k\}$ be a basis for $\mathcal{R}_A$. Show that:

(a) $\text{span}\{B\mathbf{u}_1, \ldots, B\mathbf{u}_k\} = \mathcal{R}_{BA}$.

(b) If B is left invertible, then $\{B\mathbf{u}_1, \ldots, B\mathbf{u}_k\}$ is a basis for $\mathcal{R}_{BA}$.

Exercise 2.39. Let $A \in \mathbb{F}^{4\times 5}$, let $\mathbf{v}_1, \mathbf{v}_2, \mathbf{v}_3$ be a basis for $\mathcal{R}_A$ and let $V = [\mathbf{v}_1\, \mathbf{v}_2\, \mathbf{v}_3]$. Show that $V^H V$ is invertible, that $B = V(V^H V)^{-1} V^H$ is not left invertible and yet $\mathcal{R}_B = \mathcal{R}_{BA}$.

Exercise 2.40. Let $\mathbf{u}_1, \mathbf{u}_2, \mathbf{u}_3$ be linearly independent vectors in a vector space $\mathcal{U}$ over $\mathbb{F}$ and let $\mathbf{u}_4 = \mathbf{u}_1 + 2\mathbf{u}_2 + \mathbf{u}_3$.

(a) Show that the vectors $\mathbf{u}_1, \mathbf{u}_2, \mathbf{u}_4$ are linearly independent and that span $\{\mathbf{u}_1, \mathbf{u}_2, \mathbf{u}_3\} =$ span $\{\mathbf{u}_1, \mathbf{u}_2, \mathbf{u}_4\}$.

(b) Express the vector $7\mathbf{u}_1 + 13\mathbf{u}_2 + 5\mathbf{u}_3$ as a linear combination of the vectors $\mathbf{u}_1, \mathbf{u}_2, \mathbf{u}_4$. [Note that the coefficients of all three vectors change.]

Exercise 2.41. For which values of x is matrix $\begin{bmatrix} 1 & 3 & 2 \\ 2 & 4 & 1 \\ 0 & 4 & x \end{bmatrix}$ invertible?

Exercise 2.42. Show that the matrix $A = \begin{bmatrix} 1 & 3 & 2 \\ 2 & 4 & 1 \\ 0 & 4 & 2 \end{bmatrix}$ is invertible and find its inverse by solving the system of equations $A[\mathbf{x}_1 \quad \mathbf{x}_2 \quad \mathbf{x}_3] = I_3$ column by column.

Exercise 2.43. Show that if $A \in \mathbb{C}^{p\times p}$ is invertible, $B \in \mathbb{C}^{p\times q}$, $C \in \mathbb{C}^{q\times p}$, $D \in \mathbb{C}^{q\times q}$ and

$$E = \begin{bmatrix} A & B \\ C & D \end{bmatrix},$$

then $\dim \mathcal{R}_E = \dim \mathcal{R}_A + \dim \mathcal{R}_{(D-CA^{-1}B)}$.

Exercise 2.44. Show that if, in the setting of the previous exercise, D is invertible, then $\operatorname{rank} E = \operatorname{rank} D + \operatorname{rank}(A - BD^{-1}C)$.

Exercise 2.45. Use the method of Gaussian elimination to solve the equation $A\mathbf{x} = \mathbf{b}$ when $A = \begin{bmatrix} 1 & 4 & 2 & 3 \\ 2 & 0 & 1 & 0 \\ 3 & 0 & 1 & 2 \end{bmatrix}$, $\mathbf{b} = \begin{bmatrix} 2 \\ 1 \\ 1 \end{bmatrix}$, if possible, and find a basis for $\mathcal{R}_A$ and a basis for $\mathcal{N}_A$.

Exercise 2.46. Use the method of Gaussian elimination to solve the equation $A\mathbf{x} = \mathbf{b}$ when $A = \begin{bmatrix} 5 & 2 & 4 \\ 1 & 3 & 0 \\ 0 & 1 & 0 \\ 0 & 2 & 4 \end{bmatrix}$, $\mathbf{b} = \begin{bmatrix} 1 \\ 1 \\ 0 \\ 1 \end{bmatrix}$, if possible, and find a basis for $\mathcal{R}_A$ and a basis for $\mathcal{N}_A$.

Exercise 2.47. Use the method of Gaussian elimination to solve the equation $A\mathbf{x} = \mathbf{b}$ when $A = \begin{bmatrix} 1 & 0 & 2 \\ 3 & 1 & 0 \\ 2 & 5 & 0 \end{bmatrix}$, $\mathbf{b} = \begin{bmatrix} 1 \\ 3 \\ 1 \end{bmatrix}$, if possible, and find a basis for $\mathcal{R}_A$ and a basis for $\mathcal{N}_A$.

Exercise 2.48. Find lower triangular matrices with ones on the diagonal $E_1, E_2, \ldots$ and permutation matrices $P_1, P_2, \ldots$ such that

$$E_k P_k E_{k-1} P_{k-1} \cdots E_1 P_1 A$$

is in upper echelon form for any two of the three preceding exercises.

Exercise 2.49. Use Gaussian elimination to find at least two right inverses to the matrix A given in Exercise 2.45. [HINT: Try to solve the equation

$$A \begin{bmatrix} x_{11} & x_{12} & x_{13} \\ x_{21} & x_{22} & x_{23} \\ x_{31} & x_{32} & x_{33} \\ x_{41} & x_{42} & x_{43} \end{bmatrix} = I_3,$$

column by column.]

Exercise 2.50. Use Gaussian elimination to find at least two left inverses to the matrix A given in Exercise 2.46. [HINT: Find right inverses to A^T.]

Chapter 3

Additional applications of Gaussian elimination

I was working on the proof of one of my poems all morning, and took out a comma. In the afternoon I put it back again.

Oscar Wilde

This chapter is devoted to a number of applications of Gaussian elimination, both theoretical and computational. There is some overlap with conclusions reached in the preceding chapter, but the methods of obtaining them are usually different.

3.1. Gaussian elimination redux

Recall that the method of Gaussian elimination leads to the following conclusion:

Theorem 3.1. *Let $A \in \mathbb{F}^{p\times q}$ be a nonzero matrix. Then there exists a set of lower triangular $p \times p$ matrices $E_1, \ldots, E_k$ with ones on the diagonal and a set of $p \times p$ permutation matrices $P_1, \ldots, P_k$ such that*

$$E_k P_k \cdots E_1 P_1 A = U \tag{3.1}$$

is in upper echelon form. Moreover, in this formula, P_j acts only (if at all) on rows $j, \ldots, p$ and $E_j - I_p$ has nonzero entries in at most the j'th column.

The extra information on the structure of the permutation matrices may seem tedious, but, as we shall see shortly, it has significant implications: it enables us to slide all the permutations to the right in formula (3.1) without changing the form of the matrices $E_1, \ldots, E_k$.

Theorem 3.2. *Let $A \in \mathbb{F}^{p\times q}$ be any nonzero matrix. Then there exists a lower triangular $p \times p$ matrix E with ones on the diagonal and a $p \times p$ permutation matrix P such that*

$$EPA = U$$

is in upper echelon form.

Discussion. To understand where this theorem comes from, suppose first that A is a nonzero 4×5 matrix and let $\mathbf{e}_1, \ldots, \mathbf{e}_4$ denote the columns of I_4. Then there exists a choice of permutation matrices P_1, P_2, P_3 and lower triangular matrices

$$E_1 = \begin{bmatrix} 1 & 0 & 0 & 0 \\ a & 1 & 0 & 0 \\ b & 0 & 1 & 0 \\ c & 0 & 0 & 1 \end{bmatrix} = I_4 + \mathbf{u}_1\mathbf{e}_1^T \quad \text{with} \quad \mathbf{u}_1 = \begin{bmatrix} 0 \\ a \\ b \\ c \end{bmatrix},$$

$$E_2 = \begin{bmatrix} 1 & 0 & 0 & 0 \\ 0 & 1 & 0 & 0 \\ 0 & d & 1 & 0 \\ 0 & e & 0 & 1 \end{bmatrix} = I_4 + \mathbf{u}_2\mathbf{e}_2^T \quad \text{with} \quad \mathbf{u}_2 = \begin{bmatrix} 0 \\ 0 \\ d \\ e \end{bmatrix},$$

$$E_3 = \begin{bmatrix} 1 & 0 & 0 & 0 \\ 0 & 1 & 0 & 0 \\ 0 & 0 & 1 & 0 \\ 0 & 0 & f & 1 \end{bmatrix} = I_4 + \mathbf{u}_3\mathbf{e}_3^T \quad \text{with} \quad \mathbf{u}_3 = \begin{bmatrix} 0 \\ 0 \\ 0 \\ f \end{bmatrix},$$

such that

$$E_3P_3E_2P_2E_1P_1A = U \tag{3.2}$$

is in upper echelon form. In fact, since P_2 is chosen so that it interchanges the second row of E_1P_1A with its third or fourth row, if necessary, and P_3 is chosen so that it interchanges the third row of $E_2P_2E_1P_1A$ with its fourth row, if necessary, these two permutation matrices have a special form:

$$P_2 = \begin{bmatrix} 1 & \mathbf{0}^T \\ \mathbf{0} & \Pi_1 \end{bmatrix}, \quad \text{where} \quad \Pi_1 \quad \text{is a } 3\times 3 \text{ permutation matrix}$$

$$P_3 = \begin{bmatrix} I_2 & O \\ O & \Pi_2 \end{bmatrix}, \quad \text{where} \quad \Pi_2 \quad \text{is a } 2\times 2 \text{ permutation matrix}.$$

This exhibits the pattern, which underlies the fact that

$$P_iE_j = E_j'P_i \quad \text{if} \quad i > j,$$

where E'_j denotes a matrix of the same form as E_j. Thus, for example, since $\mathbf{e}_2^T P_3 = \mathbf{e}_2^T$ and $\mathbf{v}_2 = P_3\mathbf{u}_2$ is a vector of the same form as $\mathbf{u}_2$, it follows that

$$\begin{aligned} P_3E_2 &= P_3(I_4 + \mathbf{u}_2\mathbf{e}_2^T) \\ &= P_3 + \mathbf{v}_2\mathbf{e}_2^T \\ &= P_3 + \mathbf{v}_2\mathbf{e}_2^T P_3 = E'_2 P_3 \,, \end{aligned}$$

where

$$E'_2 = I_4 + \mathbf{v}_2\mathbf{e}_2^T$$

is a matrix of the same form as E_2. In a similar vein

$$P_3P_2E_1 = P_3E'_1P_2 = E''_1P_3P_2$$

and consequently,

$$E_3P_3E_2P_2E_1P_1 = E_3E'_2E''_1P_3P_2P_1 = EP\,,$$

with

$$E = E_3E'_2E''_1 \quad \text{and} \quad P = P_3P_2P_1\,.$$

Much the same argument works in the general setting of Theorem 3.2. You have only to exploit the fact that Gaussian elimination corresponds to multiplication on the left by $E_kP_k\cdots E_1P_1$, where

$$E_j = I_p + \begin{bmatrix} \mathbf{0} \\ \mathbf{b}_j \end{bmatrix} \mathbf{e}_j^T \quad \text{with} \quad \mathbf{0} \in \mathbb{F}^j \quad \text{and} \quad \mathbf{b}_j \in \mathbb{F}^{p-j}\,,$$

and that the $p \times p$ permutation matrix P_i may be written in block form as

$$P_i = \begin{bmatrix} I_{i-1} & O \\ O & \Pi_{i-1} \end{bmatrix},$$

where Π_{i-1} is a $(p-i+1) \times (p-i+1)$ permutation matrix. Then, letting $\mathbf{c}_j = \Pi_{i-1}\mathbf{b}_j$,

$$\begin{aligned} P_iE_j &= P_i\left(I_p + \begin{bmatrix} \mathbf{0} \\ \mathbf{b}_j \end{bmatrix} \mathbf{e}_j^T\right) \\ &= P_i + \begin{bmatrix} \mathbf{0} \\ \mathbf{c}_j \end{bmatrix} \mathbf{e}_j^T \\ &= \left(I_p + \begin{bmatrix} \mathbf{0} \\ \mathbf{c}_j \end{bmatrix} \mathbf{e}_j^T\right) P_i = E'_jP_i \quad \text{for} \quad i > j\,, \end{aligned}$$

since

$$\mathbf{e}_j^T P_i = \mathbf{e}_j^T \quad \text{for} \quad i > j\,.$$

Remark 3.3. Theorem 3.2 has interesting implications. However, we wish to emphasize that when Gaussian elimination is used in practice to study the equation $A\mathbf{x} = \mathbf{b}$, it is not necessary to go through all this theoretical analysis. It suffices to carry out all the row operations on the augmented matrix

$$\begin{bmatrix} a_{11} & \cdots & a_{1q} & b_1 \\ \vdots & & \vdots & \vdots \\ a_{p1} & \cdots & a_{1q} & b_q \end{bmatrix}$$

and then to solve for the "pivot variables", just as in the examples. **But, do check that your answer works.**

In what follows, we shall reap some extra dividends from the representation formula

$$EPA = U \tag{3.3}$$

(that is valid for both real and complex matrices) by exploiting the special structure of the upper echelon matrix U.

Exercise 3.1. Show that if $A \in \mathbb{F}^{n\times n}$ is an invertible matrix, then there exists a permutation matrix P such that

$$PA = LDU\,, \tag{3.4}$$

where L is lower triangular with ones on the diagonal, U is upper triangular with ones on the diagonal and D is a diagonal matrix.

Exercise 3.2. Show that if $L_1D_1U_1 = L_2D_2U_2$, where L_j, D_j and U_j are $n\times n$ matrices of the form exhibited in Exercise 3.1, then $L_1 = L_2$, $D_1 = D_2$ and $U_1 = U_2$. [HINT: Consider $L_2^{-1}L_1D_1 = D_2U_2U_1^{-1}$.]

Exercise 3.3. Show that there exists a 3×3 permutation matrix P and a lower triangular matrix

$$B = \begin{bmatrix} 1 & 0 & 0 \\ b_{21} & 1 & 0 \\ b_{31} & b_{32} & 1 \end{bmatrix} \quad \text{such that} \quad \begin{bmatrix} 0 & 1 & 0 \\ 1 & 0 & 0 \\ 0 & 0 & 1 \end{bmatrix}\begin{bmatrix} 1 & 0 & 0 \\ \alpha & 1 & 0 \\ \beta & 0 & 1 \end{bmatrix} = BP$$

if and only if $\alpha = 0$.

Exercise 3.4. Find a permutation matrix P such that $PA = LU$, where L is a lower triangular invertible 3×3 matrix and U is an upper triangular invertible 3×3 matrix for the matrix $A = \begin{bmatrix} 0 & 1 & 1 \\ 1 & 1 & 1 \\ 1 & 1 & 0 \end{bmatrix}$.

3.2. Properties of BA and AC

In this section a number of basic properties of the product of two matrices in terms of the properties of their factors are reviewed for future use.

Lemma 3.4. *Let $A \in \mathbb{F}^{p\times q}$ and let $B \in \mathbb{F}^{p\times p}$ be invertible. Then:*

(1) *A is left invertible if and only if BA is left invertible.*

(2) *A is right invertible if and only if BA is right invertible.*

(3) $\mathcal{N}_A = \mathcal{N}_{BA}$.

(4) $B\mathcal{R}_A = \mathcal{R}_{BA}$.

(5) $\operatorname{rank} BA = \operatorname{rank} A$.

(6) $\mathcal{N}_A = \{\mathbf{0}\} \iff \mathcal{N}_{BA} = \{\mathbf{0}\}$.

(7) $\mathcal{R}_A = \mathbb{F}^p \iff \mathcal{R}_{BA} = \mathbb{F}^p$.

Proof. The first assertion follows easily from the observation that if C is a left inverse of A, then

$$CA = I_q \Longrightarrow (CB^{-1})(BA) = I_q\,.$$

Conversely, if C is a left inverse of BA, then

$$C(BA) = I_q \Longrightarrow (CB)A = I_q\,.$$

This completes the proof of (1).

Next, to verify (2), notice that if C is a right inverse of A, then

$$AC = I_p \Longrightarrow (BA)C = B(AC) = B \Longrightarrow BA(CB^{-1}) = I_p\,;$$

i.e., (CB^{-1}) is a right inverse of BA. Conversely, if C is a right inverse of BA, then

$$(BA)C = I_p \Longrightarrow B(AC) = I_p \Longrightarrow AC = B^{-1} \Longrightarrow A(CB) = I_p\,;$$

i.e., CB is a right inverse of A. Items (3) and (4) are easy and are left to the reader.

To verify (5), let $\{\mathbf{u}_1, \ldots, \mathbf{u}_k\}$ be a basis for $\mathcal{R}_A$. Then clearly

$$\operatorname{span}\{B\mathbf{u}_1, \ldots, B\mathbf{u}_k\} = \mathcal{R}_{BA}\,.$$

Moreover, the vectors $B\mathbf{u}_1, \ldots, B\mathbf{u}_k$ are linearly independent, since

$$\sum_{j=1}^{k} \alpha_j\,(B\mathbf{u}_j) = \mathbf{0} \Longrightarrow B\left(\sum_{j=1}^{k} \alpha_j \mathbf{u}_j\right) = \mathbf{0} \Longrightarrow \sum_{j=1}^{k} \alpha_j \mathbf{u}_j \in \mathcal{N}_B\,,$$

and $\mathcal{N}_B = \{\mathbf{0}\}$, which forces all the coefficients $\alpha_1, \ldots, \alpha_k$ to be zero, because the vectors $\mathbf{u}_1, \ldots, \mathbf{u}_k$ are linearly independent. Thus, the vectors $B\mathbf{u}_1, \ldots, B\mathbf{u}_k$ are also linearly independent and hence

$$\dim\,\mathcal{R}_{BA} = k = \dim\,\mathcal{R}_A\,,$$

which proves (5).

Finally, (6) is immediate from (3), and (7) is immediate from (5). □

Exercise 3.5. Verify items (3), (4), and (7) of Lemma 3.4.

Exercise 3.6. Verify item (5) of Lemma 3.4 on the basis of (3) and the law of conservation of dimension. [HINT: The matrices A and BA have the same number of columns.]

Lemma 3.5. *Let $A \in \mathbb{F}^{p\times q}$ and let $C \in \mathbb{F}^{q\times q}$ be invertible. Then:*

(1) *A is left invertible if and only if AC is left invertible.*

(2) *A is right invertible if and only if AC is right invertible.*

(3) $\mathcal{N}_A = C\mathcal{N}_{AC}$.

(4) $\mathcal{R}_A = \mathcal{R}_{AC}$.

(5) $\dim \mathcal{N}_A = \dim \mathcal{N}_{AC}$.

(6) $\mathcal{N}_A = \{\mathbf{0}\} \iff \mathcal{N}_{AC} = \{\mathbf{0}\}$.

(7) $\mathcal{R}_A = \mathbb{F}^p \iff \mathcal{R}_{AC} = \mathbb{F}^p$.

Exercise 3.7. Verify Lemma 3.5. [HINT: The fact that $\{\mathbf{u}_1, \ldots, \mathbf{u}_k\}$ is a basis for $\mathcal{N}_A \iff \{C^{-1}\mathbf{u}_1, \ldots, C^{-1}\mathbf{u}_k\}$ is a basis for $\mathcal{N}_{AC}$ is helpful.]

Exercise 3.8. Give an example of a pair of matrices $A \in \mathbb{F}^{p\times q}$ and $C \in \mathbb{F}^{q\times q}$ such that C is invertible, but $\mathcal{N}_A \neq \mathcal{N}_{AC}$.

Exercise 3.9. Let $A \in \mathbb{F}^{p\times q}$, $B \in \mathbb{F}^{p\times p}$ and let $\{\mathbf{u}_1, \ldots, \mathbf{u}_k\}$ be a basis for $\mathcal{R}_A$. Show that if B is left invertible, then $\{B\mathbf{u}_1, \ldots, B\mathbf{u}_k\}$ is a basis for $\mathcal{R}_{BA}$.

Exercise 3.10. Find a pair of matrices $A \in \mathbb{F}^{p\times q}$ and $B \in \mathbb{F}^{p\times p}$ such that B is not left invertible and yet $\{B\mathbf{u}_1, \ldots, B\mathbf{u}_k\}$ is a basis for $\mathcal{R}_{BA}$ for every basis $\{\mathbf{u}_1, \ldots, \mathbf{u}_k\}$ of $\mathcal{R}_A$.

3.3. Extracting a basis

Let $\{\mathbf{v}_1, \ldots, \mathbf{v}_k\}$ be a set of vectors in $\mathbb{F}^m$. A problem of interest is to find a basis for the subspace

$$\mathcal{V} = \text{span}\{\mathbf{v}_1, \ldots, \mathbf{v}_k\}.$$

This problem may be solved by the following sequence of steps:

(1) Let $A = \begin{bmatrix}\mathbf{v}_1 & \cdots & \mathbf{v}_k\end{bmatrix}$.

(2) Use Gaussian elimination to reduce A to an upper echelon matrix U.

(3) The pivot columns of A form a basis for $\mathcal{V}$.

Example 3.6. Let

$$\mathbf{v}_1 = \begin{bmatrix}1\\3\\1\end{bmatrix}, \mathbf{v}_2 = \begin{bmatrix}2\\6\\2\end{bmatrix}, \mathbf{v}_3 = \begin{bmatrix}2\\10\\4\end{bmatrix} \text{ and } \mathbf{v}_4 = \begin{bmatrix}0\\2\\1\end{bmatrix}.$$

Then, following the indicated strategy, we first set

$$A = \begin{bmatrix} 1 & 2 & 2 & 0 \\ 3 & 6 & 10 & 2 \\ 1 & 2 & 4 & 1 \end{bmatrix}.$$

Then, by Gaussian elimination, a corresponding upper echelon matrix is

$$U = \begin{bmatrix} 1 & 2 & 2 & 0 \\ 0 & 0 & 4 & 2 \\ 0 & 0 & 0 & 0 \end{bmatrix}.$$

The pivot columns of U are the first and the third. Therefore, by the recipe furnished above,

$$\mathcal{R}_A = \text{span}\{\mathbf{v}_1, \mathbf{v}_2, \mathbf{v}_3, \mathbf{v}_4\} = \text{span}\{\mathbf{v}_1, \mathbf{v}_3\},$$

and $\dim \mathcal{R}_A = 2$.

Exercise 3.11. Use Gaussian elimination to find $\mathcal{N}_A$ and $\mathcal{R}_A$ for each of the following choices of the matrix A:

$$\begin{bmatrix} 3 & 1 & 2 & 4 \\ 2 & 1 & 8 & 7 \\ 3 & 2 & 6 & 1 \end{bmatrix}, \quad \begin{bmatrix} 1 & 2 & 0 & 2 & 1 \\ -1 & -2 & 1 & 1 & 0 \\ 1 & 2 & -3 & -7 & -2 \end{bmatrix}, \quad \begin{bmatrix} 0 & 0 & 8 & 1 \\ 1 & 2 & 4 & 1 \\ 2 & 3 & 0 & 0 \end{bmatrix}.$$

Exercise 3.12. Find a basis for the span of the vectors

$$\begin{bmatrix} 2 \\ 3 \\ 1 \\ 4 \end{bmatrix}, \quad \begin{bmatrix} 1 \\ 0 \\ 2 \\ 1 \end{bmatrix}, \quad \begin{bmatrix} 0 \\ 3 \\ -3 \\ 2 \end{bmatrix}, \quad \begin{bmatrix} 3 \\ -3 \\ 9 \\ 1 \end{bmatrix}.$$

3.4. Computing the coefficients in a basis

Let $\{\mathbf{u}_1, \ldots, \mathbf{u}_k\}$ be a basis for a k dimensional subspace $\mathcal{U}$ of $\mathbb{F}^n$. Then every vector $\mathbf{b} \in \mathcal{U}$ can be expressed as a unique linear combination of the vectors $\{\mathbf{u}_1, \ldots, \mathbf{u}_k\}$; i.e., there exists a unique set of coefficients $c_1, \ldots, c_k$ such that

$$\mathbf{b} = \sum_{j=1}^{k} c_j \mathbf{u}_j .$$

The problem of computing these coefficients is equivalent to the problem of solving the equation

$$A\mathbf{c} = \mathbf{b},$$

where $A = \begin{bmatrix} \mathbf{u}_1 & \cdots & \mathbf{u}_k \end{bmatrix}$ is the $n \times k$ matrix with columns $\mathbf{u}_1, \ldots, \mathbf{u}_k$ and $\mathbf{c} = \begin{bmatrix} c_1 & \cdots & c_k \end{bmatrix}^T$. This problem, too, can be solved efficiently by Gaussian elimination.

Exercise 3.13. Let

$$U = \begin{bmatrix} \mathbf{u}_1 & \mathbf{u}_2 & \mathbf{u}_3 & \mathbf{u}_4 & \mathbf{u}_5 \end{bmatrix} = \begin{bmatrix} 1 & 3 & 5 & 7 & 4 \\ 0 & 0 & 2 & 1 & 6 \\ 0 & 0 & 0 & 4 & 2 \\ 0 & 0 & 0 & 0 & 0 \end{bmatrix}.$$

Show that the pivot columns $\mathbf{u}_1$, $\mathbf{u}_3$ and $\mathbf{u}_4$ form a basis for $\mathcal{R}_U$.

Exercise 3.14. Show that in the setting of the previous exercise, the columns $\mathbf{u}_1$, $\mathbf{u}_3$ and $\mathbf{u}_5$ also form a basis for $\mathcal{R}_U$ and calculate the coefficients a, b, c, d, e, f in the two representations

$$\begin{bmatrix} 3 \\ 1 \\ 1 \\ 0 \end{bmatrix} = a\mathbf{u}_1 + b\mathbf{u}_3 + c\mathbf{u}_4 = d\mathbf{u}_1 + e\mathbf{u}_3 + f\mathbf{u}_4$$

of the vector given on the left.

3.5. The Gauss-Seidel method

Gaussian elimination can be used to find the inverse of a $p \times p$ invertible matrix A by solving each of the equations

$$A\mathbf{x}_j = \mathbf{e}_j, \; j = 1, \dots, p,$$

where the right-hand side $\mathbf{e}_j$ is the j'th column of the identity matrix I_p. Then the formula

$$A \begin{bmatrix} \mathbf{x}_1 & \cdots & \mathbf{x}_p \end{bmatrix} = \begin{bmatrix} \mathbf{e}_1 & \cdots & \mathbf{e}_p \end{bmatrix} = I_p$$

identifies $X = \begin{bmatrix} \mathbf{x}_1 & \cdots & \mathbf{x}_p \end{bmatrix}$ as the inverse A^{-1} of A.

The Gauss-Seidel method is a systematic way of organizing all p of these separate calculations into one more efficient calculation by proceeding as follows, given $A \in \mathbb{F}^{p \times p}$, invertible or not:

1. Construct the $p \times 2p$ augmented matrix

$$\widetilde{A} = \begin{bmatrix} A & I_p \end{bmatrix}.$$

2. Carry out elementary row operations on $\widetilde{A}$ that are designed to bring A into upper echelon form. This is equivalent to choosing E and P so that

$$\begin{aligned} EP\widetilde{A} &= \begin{bmatrix} EPA & EPI_p \end{bmatrix} \\ &= \begin{bmatrix} U & EP \end{bmatrix}. \end{aligned}$$

3. Observe that U is a $p \times p$ upper triangular matrix with k pivots. If $k < p$, then A is not invertible and the procedure grinds to a halt. If $k = p$, then $u_{ii} \neq 0$ for $i = 1, \dots, p$. Therefore, by Lemma 1.4, there exists an upper triangular matrix F such that $FU = I_p$ and hence $A^{-1} = FEP$. To obtain

A^{-1} numerically, go on to the next steps.

4. Multiply $\begin{bmatrix} U & EP \end{bmatrix}$ on the left by a diagonal matrix $D = \begin{bmatrix} d_{ij} \end{bmatrix}$ with $d_{ii} = (u_{ii})^{-1}$ for $i = 1, \ldots, p$ to obtain

$$\begin{bmatrix} \widetilde{U} & DEP \end{bmatrix},$$

where now $\widetilde{U} = DU$ is an upper triangular matrix with ones on the diagonal.

5. Carry out elementary row manipulations on $\begin{bmatrix} \widetilde{U} & DEP \end{bmatrix}$ that are designed to bring $\widetilde{U}$ to the identity. This is equivalent to choosing an upper triangular matrix $\widetilde{F}$ such that $\widetilde{F}\widetilde{U} = I_p$. Then

$$\begin{bmatrix} \widetilde{F}\widetilde{U} & \widetilde{F}DEP \end{bmatrix} = \begin{bmatrix} I_p & \widetilde{F}DEP \end{bmatrix},$$

and hence, as $\widetilde{F}DEPA = I_p$, the second block on the right

$$\widetilde{F}DEP = FEP = A^{-1}\,.$$

6. Check! Multiply your candidate for A^{-1} by A to see if you really get the identity matrix I_p as an answer.

Thus, for example, if

$$A = \begin{bmatrix} 1 & 3 & 1 \\ 2 & 8 & 4 \\ 0 & 4 & 7 \end{bmatrix},$$

then

$$\widetilde{A} = \begin{bmatrix} 1 & 3 & 1 & 1 & 0 & 0 \\ 2 & 8 & 4 & 0 & 1 & 0 \\ 0 & 4 & 7 & 0 & 0 & 1 \end{bmatrix},$$

and two steps of Gaussian elimination lead in turn to the forms

$$\widetilde{A}_1 = \begin{bmatrix} 1 & 3 & 1 & 1 & 0 & 0 \\ 0 & 2 & 2 & -2 & 1 & 0 \\ 0 & 4 & 7 & 0 & 0 & 1 \end{bmatrix}$$

and

$$\widetilde{A}_2 = \begin{bmatrix} 1 & 3 & 1 & 1 & 0 & 0 \\ 0 & 2 & 2 & -2 & 1 & 0 \\ 0 & 0 & 3 & 4 & -2 & 1 \end{bmatrix}.$$

Next, let

$$\widetilde{A}_3 = \begin{bmatrix} 1 & 0 & 0 \\ 0 & \frac{1}{2} & 0 \\ 0 & 0 & \frac{1}{3} \end{bmatrix} \widetilde{A}_2 = \begin{bmatrix} 1 & 3 & 1 & 1 & 0 & 0 \\ 0 & 1 & 1 & -1 & \frac{1}{2} & 0 \\ 0 & 0 & 1 & \frac{4}{3} & -\frac{2}{3} & \frac{1}{3} \end{bmatrix},$$

and then subtract the bottom row of $\widetilde{A}_3$ from the second and first rows to obtain

$$\widetilde{A}_4 = \begin{bmatrix} 1 & 3 & 0 & -\frac{1}{3} & \frac{2}{3} & -\frac{1}{3} \\ 0 & 1 & 0 & -\frac{7}{3} & \frac{7}{6} & -\frac{1}{3} \\ 0 & 0 & 1 & \frac{4}{3} & -\frac{2}{3} & \frac{1}{3} \end{bmatrix}.$$

The next to last step is to subtract three times the second row of $\widetilde{A}_4$ from the first to obtain

$$\widetilde{A}_5 = \begin{bmatrix} 1 & 0 & 0 & \frac{20}{3} & -\frac{17}{6} & \frac{2}{3} \\ 0 & 1 & 0 & -\frac{7}{3} & \frac{7}{6} & -\frac{1}{3} \\ 0 & 0 & 1 & \frac{4}{3} & -\frac{2}{3} & \frac{1}{3} \end{bmatrix}.$$

The matrix built from the last 3 columns of $\widetilde{A}_5$ A^{-1}. The final step is to check that this matrix, which is conveniently written as

$$B = \frac{1}{6} \begin{bmatrix} 40 & -17 & 4 \\ -14 & 7 & -2 \\ 8 & -4 & 2 \end{bmatrix},$$

is indeed the inverse of A, i.e., that

$$AB = I_3 .$$

Exercise 3.15. Find the inverse of the matrix $\begin{bmatrix} 1 & 3 & 2 \\ 2 & 4 & 1 \\ 0 & 4 & 2 \end{bmatrix}$ by the Gauss-Seidel method.

Exercise 3.16. Use the Gauss-Seidel method to find the inverse of the matrix

$$\begin{bmatrix} 1 & 2 & 1 & 2 \\ 2 & 1 & 2 & 1 \\ 1 & 3 & 3 & 1 \\ 3 & 1 & 1 & 4 \end{bmatrix}.$$

3.6. Block Gaussian elimination

Gaussian elimination can also be carried out in block matrices providing that appropriate range conditions are fulfilled. Thus, for example, if

$$A = \begin{bmatrix} A_{11} & A_{12} & A_{13} \\ A_{21} & A_{22} & A_{23} \\ A_{31} & A_{32} & A_{33} \end{bmatrix}$$

is a block matrix and if there exists a pair of matrices K_1 and K_2 such that

$$A_{21} = K_1 A_{11} \quad \text{and} \quad A_{31} = K_2 A_{11}\,, \tag{3.5}$$

then

$$\begin{bmatrix} I & O & O \\ -K_1 & I & O \\ -K_2 & O & I \end{bmatrix} A = \begin{bmatrix} A_{11} & A_{12} & A_{13} \\ O & -K_1A_{11} + A_{22} & -K_1A_{13} + A_{23} \\ O & -K_2A_{11} + A_{32} & -K_2A_{13} + A_{33} \end{bmatrix}.$$

This operation is the block matrix analogue of clearing the first column in conventional Gaussian elimination. The implementation of such a step depends critically on the existence of matrices K_1 and K_2 that fulfill the conditions in (3.5). If A_{11} is invertible, then clearly $K_1 = A_{21}A_{11}^{-1}$ and $K_2 = A_{31}A_{11}^{-1}$ meet the requisite conditions. However, matrices K_1 and K_2 that satisfy the conditions in (3.5) may exist even if A_{11} is not invertible.

Lemma 3.7. *Let $A \in \mathbb{F}^{p\times q}$, $B \in \mathbb{F}^{p\times r}$ and $C \in \mathbb{F}^{p\times r}$. Then:*

(1) *There exists a matrix $K \in \mathbb{F}^{r\times q}$ such that $A = BK$ if and only if $\mathcal{R}_A \subseteq \mathcal{R}_B$.*

(2) *There exists a matrix $L \in \mathbb{F}^{r\times q}$ such that $A = LC$ if and only if $\mathcal{R}_{A^T} \subseteq \mathcal{R}_{C^T}$.*

Proof. Suppose first that $\mathcal{R}_A \subseteq \mathcal{R}_B$ and let $\mathbf{e}_j$ denote the j'th column of I_q. Then the presumed range inclusion implies that $A\mathbf{e}_j = B\mathbf{u}_j$ for some vector $\mathbf{u}_j \in \mathbb{F}^r$ for $j = 1, \dots, q$ and hence that $A = A\begin{bmatrix}\mathbf{e}_1 & \cdots & \mathbf{e}_q\end{bmatrix} = B\begin{bmatrix}\mathbf{u}_1 & \cdots & \mathbf{u}_q\end{bmatrix} = BK$, with $K = \begin{bmatrix}\mathbf{u}_1 & \cdots & \mathbf{u}_q\end{bmatrix}$. This proves half of (1). The other half is easy and is left to the reader together with (2). □

Exercise 3.17. Complete the proof of Lemma 3.7.

The Schur complement formulas (1.11) and (1.12) that were furnished in Chapter 1 and generalizations that are valid under less restrictive assumptions can be obtained by a double application of block Gaussian elimination.

Theorem 3.8. *If $A_{11} \in \mathbb{F}^{p\times p}$, $A_{12} \in \mathbb{F}^{p\times q}$, $A_{21} \in \mathbb{F}^{q\times p}$, $A_{22} \in \mathbb{F}^{q\times q}$ and the* **range conditions**

$$\mathcal{R}_{A_{12}} \subseteq \mathcal{R}_{A_{11}} \quad \textit{and} \quad \mathcal{R}_{A_{21}^T} \subseteq \mathcal{R}_{A_{11}^T} \tag{3.6}$$

are in force, then there exists a pair of matrices $K \in \mathbb{F}^{p\times q}$ and $L \in \mathbb{F}^{q\times p}$ such that

$$A_{12} = A_{11}K \quad \textit{and} \quad A_{21} = LA_{11} \tag{3.7}$$

and

$$\begin{bmatrix} A_{11} & A_{12} \\ A_{21} & A_{22} \end{bmatrix} = \begin{bmatrix} I_p & O \\ L & I_q \end{bmatrix} \begin{bmatrix} A_{11} & O \\ O & A_{22} - LA_{11}K \end{bmatrix} \begin{bmatrix} I_p & K \\ O & I_q \end{bmatrix}. \tag{3.8}$$

Proof. Lemma 3.7 guarantees the existence of a pair of matrices $L \in \mathbb{F}^{q\times p}$ and $K \in \mathbb{F}^{p\times q}$ that meet the conditions in (3.7). Thus,

$$\begin{bmatrix} I_p & O \\ -L & I_q \end{bmatrix} \begin{bmatrix} A_{11} & A_{12} \\ A_{21} & A_{22} \end{bmatrix} = \begin{bmatrix} A_{11} & A_{12} \\ O & A_{22} - LA_{12} \end{bmatrix}$$

and

$$\begin{bmatrix} A_{11} & A_{12} \\ O & A_{22} - LA_{12} \end{bmatrix} \begin{bmatrix} I_p & -K \\ O & I_q \end{bmatrix} = \begin{bmatrix} A_{11} & O \\ O & A_{22} - LA_{12} \end{bmatrix},$$

which in turn leads easily to formula (3.8). □

Exercise 3.18. Show that formula (3.8) coincides with the first Schur complement formula (1.11) in the special case that A_{11} is invertible.

Similar considerations lead to a generalization of the second Schur complement formula (1.12).

Exercise 3.19. Let $A \in \mathbb{F}^{n\times n}$ be a four block matrix with entries $A_{11} \in \mathbb{F}^{p\times p}$, $A_{12} \in \mathbb{F}^{p\times q}$, $A_{21} \in \mathbb{F}^{q\times p}$, $A_{22} \in \mathbb{F}^{q\times q}$, where $n = p + q$. Show that if the range conditions

$$\mathcal{R}_{A_{12}^T} \subseteq \mathcal{R}_{A_{22}^T} \quad \text{and} \quad \mathcal{R}_{A_{21}} \subseteq \mathcal{R}_{A_{22}} \tag{3.9}$$

are in force, then A admits a factorization of the form

$$\begin{bmatrix} A_{11} & A_{12} \\ A_{21} & A_{22} \end{bmatrix} = \begin{bmatrix} I_p & M \\ O & I_q \end{bmatrix} \begin{bmatrix} A_{11} - MA_{22}N & O \\ O & A_{22} \end{bmatrix} \begin{bmatrix} I_p & O \\ N & I_q \end{bmatrix}. \tag{3.10}$$

3.7. $\{0,\ 1,\ \infty\}$

Theorem 3.9. *The equation $A\mathbf{x} = \mathbf{b}$ has either 0, 1, or infinitely many solutions.*

Proof. There are three possibilitities to consider:

(1) $\mathbf{b} \notin \mathcal{R}_A$.

(2) $\mathbf{b} \in \mathcal{R}_A$ and $\mathcal{N}_A = \{\mathbf{0}\}$.

(3) $\mathbf{b} \in \mathcal{R}_A$ and $\mathcal{N}_A \neq \{\mathbf{0}\}$.

In case (1) the equation $A\mathbf{x} = \mathbf{b}$ has no solutions. Suppose next that $\mathbf{b} \in \mathcal{R}_A$ and that $\mathbf{x}_1$ and $\mathbf{x}_2$ are both solutions to the given equation. Then the identities

$$\mathbf{0} = \mathbf{b} - \mathbf{b} = A\mathbf{x}_1 - A\mathbf{x}_2 = A(\mathbf{x}_1 - \mathbf{x}_2)$$

imply that $(\mathbf{x}_1 - \mathbf{x}_2) \in \mathcal{N}_A$. Thus, in case (2) $(\mathbf{x}_1 - \mathbf{x}_2) = \mathbf{0}$; i.e., the equation has exactly one solution, whereas in case (3) it has infinitely many solutions: If $A\mathbf{x} = \mathbf{b}$ and $\mathbf{u} \in \mathcal{N}_A$, then $A(\mathbf{x} + \alpha\mathbf{u}) = \mathbf{b}$ for every $\alpha \in \mathbb{F}$. $\square$

Exercise 3.20. Find a system of 5 equations and 3 unknowns that has exactly one solution and a system of 3 equations and 5 unknowns that has no solutions.

Exercise 3.21. Let $n_L = n_L(A)$ and $n_R = n_R(A)$ denote the number of left and right inverses, respectively, of a matrix $A \in \mathbb{F}^{p\times q}$. Show that the combinations $(n_L = 0,\, n_R = 0)$, $(n_L = 0,\, n_R = \infty)$, $(n_L = 1,\, n_R = 1)$ and $(n_L = \infty,\, n_R = 0)$ are possible. [HINT: To warm up, consider upper echelon matrices first.]

Exercise 3.22. In the notation of the previous exercise, show that the combinations $(n_L = 0,\, n_R = 1)$, $(n_L = 1,\, n_R = 0)$, $(n_L = \infty,\, n_R = 1)$, $(n_L = 1,\, n_R = \infty)$ and $(n_L = \infty,\, n_R = \infty)$ are impossible.

3.8. Review

Before we go on, let us review a few of the main facts that have been established in the first three chapters. To this end, let $A \in \mathbb{F}^{p\times q}$ be a nonzero matrix, let E be a $p \times p$ lower triangular matrix with ones on the diagonal and let P be a $p \times p$ permutation matrix such that

$$EPA = U \tag{3.11}$$

is in upper echelon form with k pivots. Then:

(1) $\mathcal{N}_A = \mathcal{N}_U$.

(2) $\dim \mathcal{N}_A = \dim \mathcal{N}_U = q - k$.

(3) $\dim \mathcal{R}_A = \dim \mathcal{R}_U = k$.

(4) $\mathcal{R}_A =$ the span of the pivot columns of A.

Moreover,

- The following are equivalent:
  1. $\mathcal{N}_A = \{\mathbf{0}\}$.
  2. A is left invertible.
  3. The equation $A\mathbf{x} = \mathbf{b}$ has at most one solution $\mathbf{x}$ for each right-hand side $\mathbf{b}$.
  4. $k = q$.
  5. $\operatorname{rank} A = q$.

- The following are equivalent:
  1. $\mathcal{R}_A = \mathbb{F}^p$.
  2. A is right invertible.
  3. The equation $A\mathbf{x} = \mathbf{b}$ has at least one solution $\mathbf{x}$ for each right-hand side $\mathbf{b}$.
  4. $k = p$.
  5. $\operatorname{rank} A = p$.
- Law of conservation of dimension:

$$q = \dim \mathcal{N}_A + \dim \mathcal{R}_A.$$

- $\operatorname{rank} A = \operatorname{rank} A^T$.
- If $p = q$, then $\mathcal{N}_A = \{\mathbf{0}\}$ if and only if $\mathcal{R}_A = \mathbb{F}^p$. Therefore, if $p = q$, then A is left invertible if and only if it is right invertible. (Moreover, in this case, as we have already noted in Theorem 2.13, there is exactly one left inverse and exactly one right inverse and they coincide.)

Exercise 3.23. Use the method of Gaussian elimination to solve each of the following systems of linear equations when possible:

$$\begin{array}{rcrcrcr} 2x_1 & + & x_2 & + & x_3 & = & 0 \\ x_1 & - & x_2 & + & x_3 & = & -1 \\ 3x_1 & + & 2x_2 & + & 2x_3 & = & 1 \end{array}, \qquad \begin{array}{rcrcr} 3x_1 & + & 5x_2 & = & 9 \\ 2x_1 & + & 4x_2 & = & 7 \\ 3x_1 & + & 6x_2 & = & 5 \end{array},$$

$$\begin{array}{rcrcrcrcr} 5x_1 & + & 4x_2 & + & 6x_3 & + & 2x_4 & = & 8 \\ 3x_1 & + & 2x_2 & + & 5x_3 & + & 7x_4 & = & 6 \\ 2x_1 & + & x_2 & + & 3x_3 & + & 5x_4 & = & 1 \end{array}.$$

Exercise 3.24. Discuss the answers obtained in the preceding exercise in terms of $\mathcal{N}_A$ and $\mathcal{R}_A$.

Exercise 3.25. Let $\mathcal{U}$ and $\mathcal{V}$ be subspaces of $\mathbb{F}^k$ and $\mathbb{F}^\ell$, respectively. Show that

$$\mathcal{W} = \left\{ \begin{bmatrix} \mathbf{u} \\ \mathbf{v} \end{bmatrix} : \mathbf{u} \in \mathcal{U} \quad \text{and} \quad \mathbf{v} \in \mathcal{V} \right\}$$

is a subspace of $\mathbb{F}^{k+\ell}$ and that $\dim \mathcal{W} = \dim \mathcal{U} + \dim \mathcal{V}$.

Exercise 3.26. Let $\mathcal{U}_j$ be subspaces of $\mathbb{F}^{k_j}$ for $j = 1, \ldots, \ell$. Show that

$$\mathcal{W} = \left\{ \begin{bmatrix} \mathbf{u}_1 \\ \vdots \\ \mathbf{u}_\ell \end{bmatrix} : \mathbf{u}_j \in \mathcal{U}_j \right\}$$

is a subspace of $\mathbb{F}^{k_1+\cdots+k_\ell}$ and that $\dim \mathcal{W} = \dim \mathcal{U}_1 + \cdots + \dim \mathcal{U}_\ell$, and exhibit a basis for $\mathcal{W}$.

Exercise 3.27. Let $A \in \mathbb{C}^{p\times n}$ and $C \in \mathbb{C}^{p\times q}$. Show that there exists at most one matrix $B \in \mathbb{C}^{n\times q}$ such that $AB = C$ and $B^H\mathbf{v} = \mathbf{0}$ for every vector $\mathbf{v} \in \mathcal{N}_A$. [HINT: The second fact in (2.25) may be helpful.]

Exercise 3.28. Let $A \in \mathbb{C}^{p\times n}$ and $C \in \mathbb{C}^{p\times q}$ and let $V \in \mathbb{C}^{n\times k}$ be a matrix whose columns form a basis for $\mathcal{N}_A$. Show that if $AB = C$, then $\widetilde{B} = (I_n - V(V^HV)^{-1}V^H)B$ meets the conditions $A = \widetilde{B}C$ and $\widetilde{B}^H\mathbf{v} = \mathbf{0}$ for every vector $\mathbf{v} \in \mathcal{N}_A$ and is the only matrix in $\mathbb{C}^{n\times q}$ to do so.

Exercise 3.29. Let A, X, $B \in \mathbb{C}^{n\times n}$ be such that $AX = XB$. Show that if A is invertible, then there exists a matrix $C \in \mathbb{C}^{n\times n}$ such that $AX = XC$ and C is invertible. [HINT: If X is not invertible, then, without loss of generality, you may assume that $X = \begin{bmatrix} X_1 & X_2 \end{bmatrix}$, where the columns of X_1 form a basis for $\mathcal{R}_X$ and hence $AX_1 = X_1K$ and $X_2 = X_1L$ for suitably chosen matrices K and L.]

Exercise 3.30. Show that if $A \in \mathbb{C}^{n\times n}$ meets the condition $A\mathbf{x} = \varphi(\mathbf{x})\mathbf{x}$ for every vector $\mathbf{x} \in \mathbb{C}^n$, where $\varphi(\mathbf{x})$ is a scalar valued function of $\mathbf{x}$, then $A = \lambda I_n$ for some constant $\lambda \in \mathbb{C}$.

Exercise 3.31. Let $A \in \mathbb{C}^{n\times n}$ and suppose that $A^{k-1} \neq O$, but $A^k = O$. Show that

$$\operatorname{rank} \begin{bmatrix} A^{k-1} & A^{k-2} & \cdots & I_n \\ O & A^{k-1} & \cdots & A \\ \vdots & & \ddots & \vdots \\ O & O & \cdots & A^{k-1} \end{bmatrix} = n\,.$$

[HINT: The given matrix can be expressed as the product of its last block column with its first block row.]

Exercise 3.32. If $|\alpha| \neq 1$, then the matrix equation

$$\begin{bmatrix} 1 & \overline{\alpha} & \cdots & \overline{\alpha}^k \\ \alpha & 1 & \cdots & \overline{\alpha}^{k-1} \\ \vdots & & & \vdots \\ \alpha^k & \alpha^{k-1} & \cdots & 1 \end{bmatrix} \begin{bmatrix} x_0 \\ x_1 \\ \vdots \\ x_k \end{bmatrix} = \begin{bmatrix} 0 \\ \vdots \\ 0 \\ 1 \end{bmatrix} \tag{3.12}$$

admits a unique solution $\mathbf{x} \in \mathbb{C}^{k+1}$ with bottom entry $x_k = (1 - |\alpha|^2)^{-1}$. Use Gaussian elimination to verify this statement when $k = 2$ and $k = 3$.

Exercise 3.33. Show that if $A \in \mathbb{F}^{p\times q}$, $B \in \mathbb{F}^{q\times p}$ and AB is invertible, then A is right invertible and B is left invertible, however, the converse is false.

Chapter 4

Eigenvalues and eigenvectors

Can you imagine a mathematician writing Moby Dick*? Let my name be Ishmael, let the captain's name be Ahab, let the boats name be Pequod, and let the whale's name be as in the title.*

B. A. Cipra [**17**]

This chapter takes the first steps towards establishing the Jordan decomposition theorem for matrices $A \in \mathbb{C}^{n\times n}$. Computational techniques and a variant for $A \in \mathbb{R}^{n\times n}$ in which all the factors in the Jordan decomposition are also constrained to belong to $\mathbb{R}^{n\times n}$ will be considered in Chapter 6. It is convenient to begin the story in the setting of arbitrary linear transformations acting from one finite dimensional vector space into another. At first glance this might appear to be a much too abstract setting for the stated goal of decomposing a matrix. However, it isn't really, because, as was already indicted in Exercise 1.20, every linear transformation T from a q dimensional vector space $\mathcal{U}$ over $\mathbb{F}$ into a p dimensional vector space $\mathcal{V}$ over $\mathbb{F}$ can be fully described in terms of a $p \times q$ matrix whose entries are determined by specifying a basis for each space. Thus, to recap:

If $\mathbf{u}_1, \ldots, \mathbf{u}_q$ is a basis for $\mathcal{U}$ and $\mathbf{v}_1, \ldots, \mathbf{v}_p$ is a basis for $\mathcal{V}$, then

$$T\mathbf{u}_i = \sum_{j=1}^{p} a_{ji}\mathbf{v}_j$$

for some unique choice of $a_{ji} \in \mathbb{F}$, $i = 1, \ldots, q$, $j = 1, \ldots, p$. Moreover, since

$$\mathbf{u} \in \mathcal{U} \Longrightarrow \mathbf{u} = \sum_{i=1}^{q} \alpha_i \mathbf{u}_i \quad \text{for some choice of coefficients} \quad \alpha_i \in \mathbb{F}$$

and T is linear,

$$\mathbf{v} = T\mathbf{u} = T\sum_{i=1}^{q} \alpha_i \mathbf{u}_i = \sum_{i=1}^{q} \alpha_i (T\mathbf{u}_i) = \sum_{j=1}^{p} \left(\sum_{i=1}^{q} a_{ji}\alpha_i \right) \mathbf{v}_j = \sum_{j=1}^{p} \beta_j \mathbf{v}_j ,$$

where

$$\begin{bmatrix} \beta_1 \\ \vdots \\ \beta_p \end{bmatrix} = \begin{bmatrix} a_{11} & \cdots & a_{1q} \\ \vdots & & \vdots \\ a_{p1} & \cdots & a_{pq} \end{bmatrix} \begin{bmatrix} \alpha_1 \\ \vdots \\ \alpha_q \end{bmatrix} .$$

Thus, the coefficients $\beta_1, \ldots, \beta_p$ of $\mathbf{v} = T\mathbf{u}$, stacked as a column vector, are obtained by multiplying the stacked coefficients $\alpha_1, \ldots, \alpha_q$ of $\mathbf{u}$ by the $p \times q$ matrix A with entries a_{ij}. In other words, every linear transformation T from $\mathbb{F}^q$ to $\mathbb{F}^p$ corresponds to multiplication by a $p \times q$ matrix A. The converse is also true; i.e., multiplication of vectors in $\mathbb{F}^q$ by a matrix $A \in \mathbb{F}^{p\times q}$ defines a linear transformation from $\mathbb{F}^q$ to $\mathbb{F}^p$. There is in fact a one to one correspondence between linear transformations and matrix multiplication. Moreover, we can, as in the preceeding discussion, either think of the given column vectors as an encoding of some specified basis or more simply as the coefficients with respect to the standard basis for the spaces $\mathbb{F}^q$ and $\mathbb{F}^p$, respectively.

In this chapter we shall focus on linear mappings from a finite dimensional vector space $\mathcal{U}$ over $\mathbb{F}$ into itself and shall take the same basis, say $\mathbf{u}_1, \ldots, \mathbf{u}_n$, for both $\mathcal{D}_T$, the domain of T, and $\mathcal{R}_T$, the range of T. Correspondingly the matrix $A \in \mathbb{F}^{n\times n}$, with entries $a_{ij}, i, j = 1, \ldots, n$, is defined by the rule

$$T\mathbf{u}_i = \sum_{j=1}^{n} a_{ji}\mathbf{u}_j , \quad i = 1, \ldots, n. \tag{4.1}$$

There are many different choices that one can make for the basis. Some will turn out to be more convenient than others.

4.1. Change of basis and similarity

- **similarity:** A pair of matrices $A, B \in \mathbb{C}^{n\times n}$ are said to be **similar** if there exists an invertible matrix $C \in \mathbb{C}^{n\times n}$ such that $A = CBC^{-1}$.
- **change of basis:** The matrix representation (4.1) of a given linear transformation T from a vector space $\mathcal{U}$ into itself changes as the basis changes. However, any two such matrices are similar.

Theorem 4.1. *Let T be a linear mapping from an n dimensional vector space $\mathcal{U}$ over $\mathbb{F}$ into itself. Let $\mathbf{u}_1, \ldots, \mathbf{u}_n$ and $\mathbf{w}_1, \ldots, \mathbf{w}_n$ be two bases for $\mathcal{U}$ and suppose that*

$$T\mathbf{u}_i = \sum_{j=1}^{n} a_{ji}\mathbf{u}_j, \quad i = 1, \ldots, n \text{ and } T\mathbf{w}_i = \sum_{j=1}^{n} b_{ji}\mathbf{w}_j, \quad i = 1, \ldots, n.$$

Then the matrix $A = [a_{ji}]$ is similar to the matrix $B = [b_{ji}]$; i.e., there exists an invertible $n \times n$ matrix C such that

$$A = CBC^{-1}.$$

Proof. Let C be the $n \times n$ matrix with entries $c_{ij}, i, j = 1, \ldots, n$, that are defined by the rule

$$\mathbf{w}_i = \sum_{s=1}^{n} c_{si}\mathbf{u}_s, \quad i = 1, \ldots, n.$$

Then, on the one hand,

$$T\mathbf{w}_i = \sum_{s=1}^{n} c_{si}T\mathbf{u}_s = \sum_{s=1}^{n} c_{si}\sum_{j=1}^{n} a_{js}\mathbf{u}_j = \sum_{j=1}^{n}(\sum_{s=1}^{n} a_{js}c_{si})\mathbf{u}_j,$$

whereas, on the other hand,

$$T\mathbf{w}_i = \sum_{t=1}^{n} b_{ti}\mathbf{w}_t = \sum_{t=1}^{n} b_{ti}\sum_{j=1}^{n} c_{jt}\mathbf{u}_j = \sum_{j=1}^{n}(\sum_{t=1}^{n} c_{jt}b_{ti})\mathbf{u}_j.$$

Therefore, upon comparing the coefficients of $\mathbf{u}_j$, we see that

$$\sum_{s=1}^{n} a_{js}c_{si} = \sum_{t=1}^{n} c_{jt}b_{ti} \quad \text{i.e.,} \quad AC = CB.$$

Therefore, in order to complete the proof it remains only to show that C is invertible. But if C were not invertible, then there would exist a nonzero vector $\mathbf{x}$ such that $C\mathbf{x} = 0$. But this in turn implies that

$$\sum_{i=1}^{n} x_i\mathbf{w}_i = \sum_{i=1}^{n}(\sum_{j=1}^{n} c_{ji}x_i)\mathbf{u}_j = \mathbf{0},$$

which contradicts the presumed linear independence of $\mathbf{w}_1, \ldots, \mathbf{w}_n$. □

Exercise 4.1. Show that similarity is an equivalence relation, i.e., denoting similarity by $\sim$: (1) $A \sim A$; (2) $A \sim B \Longrightarrow B \sim A$; (3) $A \sim B$ and $B \sim C \Longrightarrow A \sim C$.

4.2. Invariant subspaces

Let T be a linear mapping from a vector space $\mathcal{U}$ over $\mathbb{F}$ into itself. Then a subspace $\mathcal{M}$ of $\mathcal{U}$ is said to be **invariant** under T if $T\mathbf{u} \in \mathcal{M}$ whenever $\mathbf{u} \in \mathcal{M}$.

The simplest invariant subspaces are the one dimensional ones, **if they exist**. Clearly, a one dimensional invariant subspace $\mathcal{M} = \{\alpha\mathbf{u} : \alpha \in \mathbb{F}\}$ based on a nonzero vector $\mathbf{u} \in \mathcal{U}$ is invariant under T if and only if there exists a constant $\lambda \in \mathbb{F}$ such that

$$T\mathbf{u} = \lambda\mathbf{u}\,, \quad \mathbf{u} \neq \mathbf{0}\,, \tag{4.2}$$

or, equivalently, if and only if

$$\mathcal{N}_{(T-\lambda I)} \neq \{\mathbf{0}\}\,;$$

i.e., the nullspace of $T - \lambda I$ is not just the zero vector. In this instance, the number λ is said to be an **eigenvalue** of T and the vector $\mathbf{u}$ is said to be an **eigenvector** of T. In fact, every nonzero vector in $\mathcal{N}_{(T-\lambda I)}$ is said to be an eigenvector of T.

It turns out that if $\mathbb{F} = \mathbb{C}$ and $\mathcal{U}$ is finite dimensional, then a one dimensional invariant subspace always exists. However, if $\mathbb{F} = \mathbb{R}$, then T may not have any one dimensional invariant subspaces. The best that you can guarantee for general T in this case is that there exists a two dimensional invariant subspace. As we shall see shortly, this is connected with the fact that a polynomial with real coefficients (of even degree) may not have any real roots.

Exercise 4.2. Show that if T is a linear transformation from a vector space $\mathcal{V}$ over $\mathbb{F}$ into itself, then the vector spaces $\mathcal{N}_{(T-\lambda I)}$ and $\mathcal{R}_{(T-\lambda I)}$ are both invariant under T for each choice of $\lambda \in \mathbb{F}$.

Exercise 4.3. The set $\mathcal{V}$ of polynomials $p(t)$ with complex coefficients is a vector space over $\mathbb{C}$ with respect to the natural rules of vector addition and scalar multiplication. Let $Tp = p''(t) + tp'(t)$ and $Sp = p''(t) + t^2p'(t)$. Show that the subspace $\mathcal{U}_k$ of $\mathcal{V}$ of polynomials $p(t) = c_0 + c_1t + \cdots + c_kt^k$ of degree less than or equal to k is invariant under T but not under S. Find a nonzero polynomial $p \in \mathcal{U}_3$ and a number $\lambda \in \mathbb{C}$ such that $Tp = \lambda p$.

Exercise 4.4. Show that if T is a linear transformation from a vector space $\mathcal{V}$ over $\mathbb{F}$ into itself, then $T^2 + 5T + 6I = (T + 3I)(T + 2I)$.

4.3. Existence of eigenvalues

The first theorem in this section serves to establish the existence of at least one eigenvalue $\lambda \in \mathbb{C}$ for a linear transformation that maps a finite dimensional vector space over $\mathbb{C}$ into itself. The second theorem serves to bound the number of distinct eigenvalues of such a transformation by the dimension of the space.

Theorem 4.2. *Let T be a linear transformation from a vector space $\mathcal{V}$ over $\mathbb{C}$ into itself and let $\mathcal{U} \neq \{\mathbf{0}\}$ be a finite dimensional subspace of $\mathcal{V}$ that is invariant under T. Then there exists a nonzero vector $\mathbf{w} \in \mathcal{U}$ and a number $\lambda \in \mathbb{C}$ such that*

$$T\mathbf{w} = \lambda\mathbf{w}\,.$$

Proof. By assumption, $\dim \mathcal{U} = \ell$ for some positive integer ℓ. Consequently, for any nonzero vector $\mathbf{u} \in \mathcal{U}$ the set of $\ell + 1$ vectors

$$\mathbf{u}, T\mathbf{u}, \ldots, T^{\ell}\mathbf{u}$$

is linearly dependent over $\mathbb{C}$; i.e., there exists a set of complex numbers $c_0, \ldots, c_\ell$, not all of which are zero, such that

$$c_0\mathbf{u} + \cdots + c_\ell T^\ell \mathbf{u} = \mathbf{0}\,.$$

Let $k = \max\{j : c_j \neq 0\}$. Then, by the fundamental theorem of algebra, the polynomial

$$p(x) = c_0 + c_1 x + \cdots + c_\ell x^\ell = c_0 + c_1 x + \cdots + c_k x^k$$

can be factored as a product of k polynomial factors of degree one with roots $\mu_1, \ldots, \mu_k \in \mathbb{C}$:

$$p(x) = c_k(x - \mu_k)\cdots(x - \mu_1)\,.$$

Correspondingly,

$$\begin{aligned} c_0\mathbf{u} + \cdots + c_\ell T^\ell \mathbf{u} &= c_0\mathbf{u} + \cdots + c_k T^k \mathbf{u} \\ &= c_k(T - \mu_k I)\cdots(T - \mu_2 I)(T - \mu_1 I)\mathbf{u} = \mathbf{0}\,. \end{aligned}$$

This in turn implies that there are k possibilities:

(1) $(T - \mu_1 I)\mathbf{u} = \mathbf{0}$.

(2) $(T - \mu_1 I)\mathbf{u} \neq \mathbf{0}$ and $(T - \mu_2 I)(T - \mu_1 I)\mathbf{u} = \mathbf{0}$.

$\vdots$

(k) $(T - \mu_{k-1} I)\cdots(T - \mu_1 I)\mathbf{u} \neq \mathbf{0}$ and $(T - \mu_k I)\cdots(T - \mu_1 I)\mathbf{u} = \mathbf{0}$.

In the first case, μ_1 is an eigenvalue and $\mathbf{u}$ is an eigenvector.
In the second case, the vector $\mathbf{w}_1 = (T - \mu_1 I)\mathbf{u}$ is a nonzero vector in $\mathcal{U}$ and $T\mathbf{w}_1 = \mu_2\mathbf{w}_1$. Therefore, $(T - \mu_1 I)\mathbf{u}$ is an eigenvector of T corresponding to the eigenvalue μ_2.

$\vdots$

In the k'th case, the vector $\mathbf{w}_{k-1} = (T - \mu_{k-1} I)\cdots(T - \mu_1 I)\mathbf{u}$ is a nonzero vector in $\mathcal{U}$ and $T\mathbf{w}_{k-1} = \mu_k\mathbf{w}_{k-1}$. Therefore, $(T - \mu_{k-1} I)\cdots(T - \mu_1 I)\mathbf{u}$ is an eigenvector of T corresponding to the eigenvalue μ_k. □

Notice that the proof **does not guarantee the existence of real eigenvalues** for linear transformations T from a vector space $\mathcal{V}$ over $\mathbb{R}$ into itself because the polynomial $p(x) = c_0 + c_1x + \cdots + c_kx^k$ may have only complex roots $\mu_1, \ldots, \mu_k$ even if the coefficients $c_1, \ldots, c_k$ are real; see e.g., Exercise 4.5.

Exercise 4.5. Let T be a linear transformation from a vector space $\mathcal{V}$ over $\mathbb{R}$ into itself and let $\mathcal{U}$ be a two dimensional subspace of $\mathcal{V}$ with basis $\{\mathbf{u}_1, \mathbf{u}_2\}$. Show that if $T\mathbf{u}_1 = \mathbf{u}_2$ and $T\mathbf{u}_2 = -\mathbf{u}_1$, then $T^2\mathbf{u} + \mathbf{u} = \mathbf{0}$ for every vector $\mathbf{u} \in \mathcal{U}$ but that there are no one dimensional subspaces of $\mathcal{U}$ that are invariant under T. Why? [HINT: A one dimensional subspace of $\mathcal{U}$ is equal to $\{\alpha(c_1\mathbf{u}_1 + c_2\mathbf{u}_2) : \alpha \in \mathbb{R}\}$ for some choice of $c_1, c_2 \in \mathbb{R}$ with $|c_1| + |c_2| > 0$.]

Theorem 4.3. *Let T be a linear transformation from an n dimensional vector space $\mathcal{U}$ over $\mathbb{C}$ into itself and let $\mathbf{u}_1, \ldots, \mathbf{u}_k \in \mathcal{U}$ be eigenvectors of T corresponding to a distinct set of eigenvalues $\lambda_1, \ldots, \lambda_k$. Then the vectors $\mathbf{u}_1, \ldots, \mathbf{u}_k$ are linearly independent and hence $k \leq n$.*

Proof. Let $\alpha_1, \ldots, \alpha_k$ be a set of numbers such that

$$\alpha_1\mathbf{u}_1 + \cdots + \alpha_k\mathbf{u}_k = \mathbf{0}\,. \tag{4.3}$$

We wish to show that $\alpha_j = 0$ for $j = 1, \ldots, k$. But now as

$$(T - \lambda_i I)\mathbf{u}_j = T\mathbf{u}_j - \lambda_i\mathbf{u}_j = (\lambda_j - \lambda_i)\mathbf{u}_j, \quad i, j = 1, \ldots, k\,,$$

it is readily checked that

$$(T - \lambda_2 I)\cdots(T - \lambda_k I)\mathbf{u}_j = \begin{cases} \mathbf{0} & \text{if} \quad j = 2, \ldots, k \\ (\lambda_1 - \lambda_2)\cdots(\lambda_1 - \lambda_k)\mathbf{u}_1 & \text{if} \quad j = 1. \end{cases}$$

Therefore, upon applying the product $(T - \lambda_2 I)\cdots(T - \lambda_k I)$ to both sides of (4.3), it is easily seen that the left-hand side

$$(T - \lambda_2 I)\cdots(T - \lambda_k I)(\alpha_1\mathbf{u}_1 + \cdots + \alpha_k\mathbf{u}_k) = (\lambda_1 - \lambda_2)\cdots(\lambda_1 - \lambda_n)\alpha_1\mathbf{u}_1\,,$$

whereas the right-hand side

$$(T - \lambda_2 I)\cdots(T - \lambda_k I)\mathbf{0} = \mathbf{0}\,.$$

Thus, $\alpha_1 = 0$, since these two sides must match. Similar considerations serve to show that $\alpha_2 = \cdots = \alpha_k = 0$. Consequently the vectors $\mathbf{u}_1, \ldots, \mathbf{u}_k$ are linearly independent and $k \leq n$, as claimed. □

4.4. Eigenvalues for matrices

The operation of multiplying vectors in $\mathbb{F}^n$ by a matrix $A \in \mathbb{F}^{n \times n}$ is a linear transformation of $\mathbb{F}^n$ into itself. Consequently, the definitions that were introduced earlier for the eigenvalue and eigenvector of a linear transformation can be reformulated directly in terms of matrices:

- A point $\lambda \in \mathbb{F}$ is said to be an **eigenvalue** of $A \in \mathbb{F}^{n\times n}$ if there exists a nonzero vector $\mathbf{u} \in \mathbb{F}^n$ such that $A\mathbf{u} = \lambda\mathbf{u}$, i.e., if
$$\mathcal{N}_{(A-\lambda I_n)} \neq \{\mathbf{0}\}.$$
Every nonzero vector $\mathbf{u} \in \mathcal{N}_{(A-\lambda I_n)}$ is said to be an **eigenvector** of A corresponding to the eigenvalue λ.
- A nonzero vector $\mathbf{u} \in \mathbb{F}^n$ is said to be a **generalized eigenvector** of the matrix $A \in \mathbb{F}^{n\times n}$ corresponding to the eigenvalue $\lambda \in \mathbb{F}$ if $\mathbf{u} \in \mathcal{N}_{(A-\lambda I_n)^n}$.
- A vector $\mathbf{u} \in \mathbb{F}^n$ is said to be a **generalized eigenvector of order** k of the matrix $A \in \mathbb{F}^{n\times n}$ corresponding to the eigenvalue $\lambda \in \mathbb{F}$ if $(A-\lambda I_n)^k\mathbf{u} = \mathbf{0}$, but $(A-\lambda I_n)^{k-1}\mathbf{u} \neq \mathbf{0}$. In this instance, the set of vectors $\mathbf{u}_j = (A-\lambda I_n)^{(k-j)}\mathbf{u}$ for $j = 1, \ldots, k$ is said to form a **Jordan chain of length** k; they satisfy the following chain of equalities:
$$\begin{aligned}(A-\lambda I_n)\mathbf{u}_1 &= \mathbf{0}\\ (A-\lambda I_n)\mathbf{u}_2 &= \mathbf{u}_1\\ &\vdots\\ (A-\lambda I_n)\mathbf{u}_k &= \mathbf{u}_{k-1}.\end{aligned}$$
This is equivalent to the formula

(4.4)
$$(A-\lambda I_n)\begin{bmatrix}\mathbf{u}_1 & \cdots & \mathbf{u}_k\end{bmatrix} = \begin{bmatrix}\mathbf{u}_1 & \cdots & \mathbf{u}_k\end{bmatrix} N, \quad \text{where} \quad N = \sum_{j=1}^{k-1} \mathbf{e}_j\mathbf{e}_{j+1}^T$$
and $\mathbf{e}_j$ denotes the j'th column of I_k. Thus, for example, if $k = 4$, then (4.4) reduces to the identity

$$(4.5)\quad (A-\lambda I_n)\begin{bmatrix}\mathbf{u}_1 & \mathbf{u}_2 & \mathbf{u}_3 & \mathbf{u}_4\end{bmatrix} = \begin{bmatrix}\mathbf{u}_1 & \mathbf{u}_2 & \mathbf{u}_3 & \mathbf{u}_4\end{bmatrix}\begin{bmatrix}0&1&0&0\\0&0&1&0\\0&0&0&1\\0&0&0&0\end{bmatrix}.$$

Exercise 4.6. Show that the vectors $\mathbf{u}_1, \ldots, \mathbf{u}_k$ in a Jordan chain of length k are linearly independent.

If $\lambda_1, \ldots, \lambda_k$ are distinct eigenvalues of a matrix $A \in \mathbb{F}^{n\times n}$, then:

- The number
$$\gamma_j = \dim \mathcal{N}_{(A-\lambda_j I_n)}, \; j = 1, \ldots, k,$$
is termed the **geometric multiplicity** of the eigenvalue λ_j. It is equal to the number of linearly independent eigenvectors associated with the eigenvalue λ_j.

- The number

$$\alpha_j = \dim \mathcal{N}_{(A-\lambda_j I_n)^n}, \; j = 1, \ldots, k,$$

is termed the **algebraic multiplicity** of the eigenvalue λ_j. It is equal to the number of linearly independent generalized eigenvectors associated with the eigenvalue λ_j.

- The inclusions

$$\mathcal{N}_{(A-\lambda_j I_n)} \subseteq \mathcal{N}_{(A-\lambda_j I_n)^2} \subseteq \cdots \subseteq \mathcal{N}_{(A-\lambda_j I_n)^n} \tag{4.6}$$

guarantee that

$$\gamma_j \leq \alpha_j \quad \text{for} \quad j = 1, \ldots, k\,, \tag{4.7}$$

and hence (as will follow in part from Theorem 4.12) that

$$\gamma_1 + \cdots + \gamma_k \leq \alpha_1 + \cdots + \alpha_k = n\,. \tag{4.8}$$

- The set

$$\sigma(A) = \{\lambda \in \mathbb{C} : \mathcal{N}_{(A-\lambda I_n)} \neq \{\mathbf{0}\}\} \tag{4.9}$$

is called the **spectrum** of A. Clearly, $\sigma(A)$ is equal to the set $\{\lambda_1, \ldots, \lambda_k\}$ of all the distinct eigenvalues of the matrix A in $\mathbb{C}$.

Theorems 4.2 and 4.3 imply that

(1) *Every matrix $A \in \mathbb{C}^{n\times n}$ has at least one eigenvalue $\lambda \in \mathbb{C}$.*

(2) *Every matrix $A \in \mathbb{C}^{n\times n}$ has at most n distinct eigenvalues in $\mathbb{C}$.*

(3) *Eigenvectors corresponding to distinct eigenvalues are automatically linearly independent.*

Even though (1) implies that $\sigma(A) \neq \emptyset$ for every $A \in \mathbb{C}^{n\times n}$, it does not guarantee that $\sigma(A) \cap \mathbb{R} \neq \emptyset$ if $A \in \mathbb{R}^{n\times n}$.

Exercise 4.7. Verify the inclusions (4.6).

Exercise 4.8. Show that the matrices

$$A = \begin{bmatrix} 1 & -1 \\ 1 & 1 \end{bmatrix} \quad \text{and} \quad A = \begin{bmatrix} 2 & -1 \\ 3 & -1 \end{bmatrix}$$

have no real eigenvalues, i.e., $\sigma(A) \cap \mathbb{R} = \emptyset$ in both cases.

Exercise 4.9. Show that although the following upper triangular matrices

$$\begin{bmatrix} 2 & 0 & 0 \\ 0 & 2 & 0 \\ 0 & 0 & 2 \end{bmatrix}, \quad \begin{bmatrix} 2 & 1 & 0 \\ 0 & 2 & 0 \\ 0 & 0 & 2 \end{bmatrix}, \quad \begin{bmatrix} 2 & 1 & 0 \\ 0 & 2 & 1 \\ 0 & 0 & 2 \end{bmatrix}$$

have the same diagonal, $\dim \mathcal{N}_{(A-2I_3)}$ is equal to three for the first, two for the second and one for the third. Calculate $\mathcal{N}_{(A-2I_3)^j}$ for $j = 1, 2, 3, 4$ for each of the three choices of A.

Exercise 4.10. Show that if $A \in \mathbb{F}^{n \times n}$ is a triangular matrix with entries a_{ij}, then $\sigma(A) = \cup_{i=1}^{n} \{a_{ii}\}$.

The cited theorems actually imply a little more:

Theorem 4.4. *Let $A \in \mathbb{C}^{n \times n}$ and let $\mathcal{U}$ be a nonzero subspace of $\mathbb{C}^n$ that is invariant under A, i.e., $\mathbf{u} \in \mathcal{U} \Longrightarrow A\mathbf{u} \in \mathcal{U}$. Then:*

(1) *There exists a nonzero vector $\mathbf{u} \in \mathcal{U}$ and a number $\lambda \in \mathbb{C}$ such that $A\mathbf{u} = \lambda\mathbf{u}$.*

(2) *If $\mathbf{u}_1, \ldots, \mathbf{u}_k \in \mathcal{U}$ are eigenvectors of A corresponding to distinct eigenvalues $\lambda_1, \ldots, \lambda_k$, then $k \leq \dim \mathcal{U}$.*

Exercise 4.11. Verify Theorem 4.4.

4.5. Direct sums

Let $\mathcal{U}$ and $\mathcal{V}$ be subspaces of a vector space $\mathcal{Y}$ over $\mathbb{F}$ and recall that

$$\mathcal{U} + \mathcal{V} = \{\mathbf{u} + \mathbf{v} : \mathbf{u} \in \mathcal{U} \text{ and } \mathbf{v} \in \mathcal{V}\}.$$

Clearly, $\mathcal{U} + \mathcal{V}$ is a subspace of $\mathcal{Y}$ with respect to the rules of vector addition and scalar multiplication that are inherited from the vector space $\mathcal{Y}$, since it is closed under vector addition and scalar multiplication.

The sum $\mathcal{U} + \mathcal{V}$ is said to be a **direct sum** if $\mathcal{U} \cap \mathcal{V} = \{\mathbf{0}\}$. Direct sums are denoted by the symbol $\dot{+}$, i.e., $\mathcal{U} \dot{+} \mathcal{V}$ rather than $\mathcal{U} + \mathcal{V}$.

The vector space $\mathcal{Y}$ is said to admit a **sum decomposition** if there exists a pair of subspaces $\mathcal{U}$ and $\mathcal{V}$ of $\mathcal{Y}$ such that

$$\mathcal{U} + \mathcal{V} = \mathcal{Y}.$$

In this instance, every vector $\mathbf{y} \in \mathcal{Y}$ can be expressed as a sum of the form $\mathbf{y} = \mathbf{u} + \mathbf{v}$ for **at least one** pair of vectors $\mathbf{u} \in \mathcal{U}$ and $\mathbf{v} \in \mathcal{V}$.

The vector space $\mathcal{Y}$ is said to admit a **direct sum decomposition** if there exist a pair of subspaces $\mathcal{U}$ and $\mathcal{V}$ of $\mathcal{Y}$ such that $\mathcal{U} \dot{+} \mathcal{V} = \mathcal{Y}$, i.e., if $\mathcal{U} + \mathcal{V} = \mathcal{Y}$ and $\mathcal{U} \cap \mathcal{V} = \{\mathbf{0}\}$. If this happens, then $\mathcal{V}$ is said to be a **complementary space** to $\mathcal{U}$ and $\mathcal{U}$ is said to be a complementary space to $\mathcal{V}$.

Lemma 4.5. *Let $\mathcal{Y}$ be a vector space over $\mathbb{F}$ and let $\mathcal{U}$ and $\mathcal{V}$ be subspaces of $\mathcal{Y}$ such that $\mathcal{U} \dot{+} \mathcal{V} = \mathcal{Y}$. Then every vector $\mathbf{y} \in \mathcal{Y}$ can be expressed as a sum of the form $\mathbf{y} = \mathbf{u} + \mathbf{v}$ for exactly one pair of vectors $\mathbf{u} \in \mathcal{U}$ and $\mathbf{v} \in \mathcal{V}$.*

Exercise 4.12. Verify Lemma 4.5.

Exercise 4.13. Let T be a linear transformation from a vector space $\mathcal{V}$ over $\mathbb{R}$ into itself and let $\mathcal{U}$ be a two dimensional subspace of $\mathcal{V}$ with basis $\{\mathbf{u}_1, \mathbf{u}_2\}$. Show that if $T\mathbf{u}_1 = \mathbf{u}_1 + 2\mathbf{u}_2$ and $T\mathbf{u}_2 = 2\mathbf{u}_1 + \mathbf{u}_2$, then $\mathcal{U}$ is the direct sum of two one dimensional spaces that are each invariant under T.

Lemma 4.6. *Let $\mathcal{U}, \mathcal{V}$ and $\mathcal{W}$ be subspaces of a vector space $\mathcal{Y}$ over $\mathbb{F}$ such that*

$$\mathcal{U}\dot{+}\mathcal{V} = \mathcal{Y} \quad \text{and} \quad \mathcal{U} \subseteq \mathcal{W}.$$

Then

$$\mathcal{W} = (\mathcal{W} \cap \mathcal{U})\dot{+}(\mathcal{W} \cap \mathcal{V}).$$

Proof. Clearly,

$$(\mathcal{W} \cap \mathcal{U}) + (\mathcal{W} \cap \mathcal{V}) \subseteq \mathcal{W} + \mathcal{W} = \mathcal{W}.$$

To establish the opposite inclusion, let $\mathbf{w} \in \mathcal{W}$. Then, since $\mathcal{Y} = \mathcal{U}\dot{+}\mathcal{V}$, $\mathbf{w} = \mathbf{u}+\mathbf{v}$ for exactly one pair of vectors $\mathbf{u} \in \mathcal{U}$ and $\mathbf{v} \in \mathcal{V}$. Moreover, under the added assumption that $\mathcal{U} \subseteq \mathcal{W}$, it follows that both $\mathbf{u}$ and $\mathbf{v} = \mathbf{w} - \mathbf{u}$ belong to $\mathcal{W}$. Therefore, $\mathbf{u} \in \mathcal{W} \cap \mathcal{U}$ and $\mathbf{v} \in \mathcal{W} \cap \mathcal{V}$, and hence

$$\mathcal{W} \subseteq (\mathcal{W} \cap \mathcal{U}) + (\mathcal{W} \cap \mathcal{V}).$$

□

Exercise 4.14. Provide an example of three subspaces $\mathcal{U}$, $\mathcal{V}$ and $\mathcal{W}$ of a vector space $\mathcal{Y}$ over $\mathbb{F}$ such that $\mathcal{U}\dot{+}\mathcal{V} = \mathcal{Y}$, but $\mathcal{W} \neq (\mathcal{W} \cap \mathcal{U})\dot{+}(\mathcal{W} \cap \mathcal{V})$. [HINT: Simple examples exist with $\mathcal{Y} = \mathbb{R}^2$.]

If $\mathcal{U}_j$, $j = 1, \ldots, k$, are finite dimensional subspaces of a vector space $\mathcal{Y}$ over $\mathbb{F}$, then the sum

$$\mathcal{U}_1 + \cdots + \mathcal{U}_k = \{\mathbf{u}_1 + \cdots + \mathbf{u}_k : \mathbf{u}_i \in \mathcal{U}_i \quad \text{for} \quad i = 1, \ldots, k\} \tag{4.10}$$

is said to be **direct** if

$$\dim \mathcal{U}_1 + \cdots + \dim \mathcal{U}_k = \dim \{\mathcal{U}_1 + \cdots + \mathcal{U}_k\}. \tag{4.11}$$

If $\mathcal{U} = \mathcal{U}_1 + \cdots + \mathcal{U}_k$ and the sum is direct, then we write

$$\mathcal{U} = \mathcal{U}_1\dot{+} \cdots \dot{+}\mathcal{U}_k.$$

If $k = 2$, then formula (2.16) implies that the sum $\mathcal{U}_1 + \mathcal{U}_2$ is direct if and only if $\mathcal{U}_1 \cap \mathcal{U}_2 = \{\mathbf{0}\}$. Therefore, the characterization (4.11) is consistent with the definition of the direct sum of two subspaces given earlier.

Exercise 4.15. Give an example of three subspaces $\mathcal{U}$, $\mathcal{V}$ and $\mathcal{W}$ of $\mathbb{R}^3$ such that $\mathcal{U} \cap \mathcal{V} = \{\mathbf{0}\}$, $\mathcal{U} \cap \mathcal{W} = \{\mathbf{0}\}$ and $\mathcal{V} \cap \mathcal{W} = \{\mathbf{0}\}$ yet the sum $\mathcal{U} + \mathcal{V} + \mathcal{W}$ is not direct.

Exercise 4.16. Let $\mathcal{Y}$ be a finite dimensional vector space over $\mathbb{F}$. Show that if $\mathcal{Y} = \mathcal{U}\dot{+}\mathcal{V}$ and $\mathcal{V} = \mathcal{X}\dot{+}\mathcal{W}$, then $\mathcal{Y} = \mathcal{U}\dot{+}\mathcal{X}\dot{+}\mathcal{W}$.

Lemma 4.7. *Let $\mathcal{U}_j$, $j = 1, \ldots, k$, be finite dimensional nonzero subspaces of a vector space $\mathcal{Y}$ over $\mathbb{F}$. Then the sum (4.10) is direct if and only if every set of nonzero vectors $\{\mathbf{u}_1, \ldots, \mathbf{u}_k\}$ with $\mathbf{u}_i \in \mathcal{U}_i$ for $i = 1, \ldots, k$ is a linearly independent set of vectors.*

Discussion. To ease the exposition, suppose that $k = 3$ and let $\{\mathbf{a}_1, \dots, \mathbf{a}_\ell\}$ be a basis for $\mathcal{U}_1$, $\{\mathbf{b}_1, \dots, \mathbf{b}_m\}$ be a basis for $\mathcal{U}_2$ and $\{\mathbf{c}_1, \dots, \mathbf{c}_n\}$ be a basis for $\mathcal{U}_3$. Clearly

$$\text{span}\,\{\mathbf{a}_1, \dots, \mathbf{a}_\ell,\ \mathbf{b}_1, \dots, \mathbf{b}_m,\ \mathbf{c}_1, \dots, \mathbf{c}_n\} = \mathcal{U}_1 + \mathcal{U}_2 + \mathcal{U}_3\,.$$

It is easily checked that if the sum is direct, then the $\ell + m + n$ vectors indicated above are linearly independent and hence if $\mathbf{u} = \sum \alpha_i \mathbf{a}_i$, $\mathbf{v} = \sum \beta_j \mathbf{b}_j$ and $\mathbf{w} = \sum \gamma_k \mathbf{a}_k$ are nonzero vectors in $\mathcal{U}_1$, $\mathcal{U}_2$ and $\mathcal{U}_3$, respectively, then they are linearly independent.

Suppose next that every set of nonzero vectors $\mathbf{u} \in \mathcal{U}_1$, $\mathbf{v} \in \mathcal{U}_2$ and $\mathbf{w} \in \mathcal{U}_3$ is linearly independent. Then $\{\mathbf{a}_1, \dots, \mathbf{a}_\ell,\ \mathbf{b}_1, \dots, \mathbf{b}_m,\ \mathbf{c}_1, \dots, \mathbf{c}_n\}$ must be a linearly independent set of vectors because if

$$\alpha_1 \mathbf{a}_1 + \cdots + \alpha_\ell \mathbf{a}_\ell + \beta_1 \mathbf{b}_1 + \cdots + \beta_m \mathbf{b}_m + \gamma_1 \mathbf{c}_1 + \cdots + \gamma_n \mathbf{c}_n = \mathbf{0}\,,$$

and if, say, $\alpha_1 \neq 0$, $\beta_1 \neq 0$ and $\gamma_1 \neq 0$, then

$$\alpha_1(\mathbf{a}_1 + \cdots + \alpha_1^{-1}\alpha_\ell \mathbf{a}_\ell) + \beta_1(\mathbf{b}_1 + \cdots + \beta_1^{-1}\beta_m \mathbf{b}_m) + \gamma_1(\mathbf{c}_1 + \cdots + \gamma_1^{-1}\gamma_n \mathbf{c}_n) = \mathbf{0}\,,$$

which implies that $\alpha_1 = \beta_1 = \gamma_1 = 0$, contrary to assumption. The same argument shows that all the remaining coefficients must be zero too. □

Exercise 4.17. Let $\mathcal{U} = \text{span}\{\mathbf{u}_1, \dots, \mathbf{u}_k\}$ over $\mathbb{F}$ and let $\mathcal{U}_j = \{\alpha \mathbf{u}_j : \alpha \in \mathbb{F}\}$. Show that the set of vectors $\{\mathbf{u}_1, \dots, \mathbf{u}_k\}$ is a basis for the vector space $\mathcal{U}$ over $\mathbb{F}$ if and only if $\mathcal{U}_1 \dot{+} \cdots \dot{+} \mathcal{U}_k = \mathcal{U}$.

4.6. Diagonalizable matrices

A matrix $A \in \mathbb{F}^{n \times n}$ is said to be **diagonalizable** if it is similar to a diagonal matrix, i.e., if there exists an invertible matrix $U \in \mathbb{F}^{n \times n}$ and a diagonal matrix $D \in \mathbb{F}^{n \times n}$ such that

$$A = UDU^{-1}\,. \tag{4.12}$$

Theorem 4.8. *Let $A \in \mathbb{C}^{n \times n}$ and suppose that A has exactly k distinct eigenvalues $\lambda_1, \dots, \lambda_k \in \mathbb{C}$. Then the sum*

$$\mathcal{N}_{(A-\lambda_1 I_n)} + \cdots + \mathcal{N}_{(A-\lambda_k I_n)}$$

is direct. Moreover, the following statements are equivalent:

(1) *A is diagonalizable.*

(2) $\dim \mathcal{N}_{(A-\lambda_1 I_n)} + \cdots + \dim \mathcal{N}_{(A-\lambda_k I_n)} = n$.

(3) $\mathcal{N}_{(A-\lambda_1 I_n)} \dot{+} \cdots \dot{+} \mathcal{N}_{(A-\lambda_k I_n)} = \mathbb{C}^n$.

Proof. Suppose first that A is diagonalizable. Then, the formula $A = UDU^{-1}$ implies that

$$A - \lambda I_n = UDU^{-1} - \lambda U I_n U^{-1} = U(D - \lambda I_n)U^{-1}$$

and hence that

$$\dim \mathcal{N}_{(A-\lambda I_n)} = \dim \mathcal{N}_{(D-\lambda I_n)} \quad \text{for every point} \quad \lambda \in \mathbb{C}.$$

In particular, if $\lambda = \lambda_j$ is an eigenvalue of A, then

$$\gamma_j = \dim \mathcal{N}_{(A-\lambda_j I_n)} = \dim \mathcal{N}_{(D-\lambda_j I_n)}$$

is equal to the number of times the number λ_j is repeated in the diagonal matrix D. Thus, $\gamma_1 + \cdots + \gamma_k = n$, i.e., (1) $\Longrightarrow$ (2) and, by Lemma 4.7, (2) $\Longleftrightarrow$ (3).

It remains to prove that (2) $\Longrightarrow$ (1). Take γ_j linearly independent vectors from $\mathcal{N}_{(A-\lambda_j I_n)}$ and array them as the column vectors of an $n \times \gamma_j$ matrix U_j. Then

$$AU_j = U_j \Lambda_j \quad \text{for} \quad j = 1, \ldots, k,$$

where Λ_j is the $\gamma_j \times \gamma_j$ diagonal matrix with λ_j on the diagonal. Thus, upon setting

$$U = [U_1 \cdots U_k] \quad \text{and} \quad D = \operatorname{diag}\{\Lambda_1, \ldots, \Lambda_k\},$$

it is readily seen that

$$AU = UD$$

and, with the help of Theorem 4.3, that the $\gamma_1 + \cdots + \gamma_k$ columns of U are linearly independent, i.e.,

$$\operatorname{rank} U = \gamma_1 + \cdots + \gamma_k.$$

The formula $AU = UD$ is valid even if $\gamma_1 + \cdots + \gamma_k < n$. However, if (2) is in force, then U is invertible and $A = UDU^{-1}$. $\square$

Corollary 4.9. Let $A \in \mathbb{C}^{n \times n}$ and suppose that A has n distinct eigenvalues in $\mathbb{C}$. Then A is diagonalizable.

Exercise 4.18. Verify the corollary.

Formula (4.12) is extremely useful. In particular, it implies that

$$A^2 = (UDU^{-1})(UDU^{-1}) = UD^2U^{-1},$$
$$A^3 = UD^3U^{-1}$$

etc.

The advantage is that the powers $D^2, D^3, \ldots, D^k$ are easy to compute:

$$D = \begin{bmatrix} \mu_1 & & \\ & \ddots & \\ & & \mu_n \end{bmatrix} \Longrightarrow D^k = \begin{bmatrix} \mu_1^k & & \\ & \ddots & \\ & & \mu_n^k \end{bmatrix}. \tag{4.13}$$

Moreover, this suggests that the matrix exponential (which will be introduced later)

$$e^A = \sum_{k=0}^{\infty} \frac{A^k}{k!} = U\left(\sum_{k=0}^{\infty} \frac{D^k}{k!}\right) U^{-1} = U \begin{bmatrix} e^{\mu_1} & & \\ & \ddots & \\ & & e^{\mu_n} \end{bmatrix} U^{-1},$$

all of which can be justified.

Exercise 4.19. Show that if a matrix $A \in \mathbb{F}^{n\times n}$ is diagonalizable, i.e., if $A = UDU^{-1}$ with $D = \mathrm{diag}\{\lambda_1, \dots, \lambda_n\}$, and if

$$U = \begin{bmatrix} \mathbf{u}_1 & \cdots & \mathbf{u}_n \end{bmatrix} \quad \text{and} \quad U^{-1} = \begin{bmatrix} \vec{\mathbf{v}}_1 \\ \vdots \\ \vec{\mathbf{v}}_n \end{bmatrix}, \text{ then :}$$

(1) $A^k = UD^kU^{-1} = \sum_{j=1}^{n} \lambda_j^k \mathbf{u}_j \, \vec{\mathbf{v}}_j$.

(2) $(A-\lambda I_n)^{-1} = U(D-\lambda I_n)^{-1}U^{-1} = \sum_{j=1}^{n}(\lambda_j-\lambda)^{-1}\mathbf{u}_j \, \vec{\mathbf{v}}_j$, if $\lambda \notin \sigma(A)$.

4.7. An algorithm for diagonalizing matrices

The verification of (2) $\Longrightarrow$ (1) in the proof of the Theorem 4.8 contains a recipe for constructing a pair of matrices U and D so that $A = UDU^{-1}$ for a matrix $A \in \mathbb{C}^{n\times n}$ with exactly k distinct eigenvalues $\lambda_1, \dots, \lambda_k$ when the geometric multiplicities meet the constraint $\gamma_1 + \cdots + \gamma_k = n$:

(1) Calculate the geometric multiplicity $\gamma_j = \dim \mathcal{N}_{(A-\lambda_j I_n)}$ for each eigenvalue λ_j of A.

(2) Obtain a basis for each of the spaces $\mathcal{N}_{(A-\lambda_j I_n)}$ for $j = 1, \dots, k$ and let U_j denote the $n \times \gamma_j$ matrix with columns equal to the vectors in this basis.

(3) Let $U = [U_1 \quad \cdots \quad U_k]$.

Then

$$AU = [AU_1 \quad \cdots \quad AU_k] = [\lambda_1 U_1 \quad \cdots \quad \lambda_k U_k] = UD\,,$$

where

$$D = \mathrm{diag}\,\{D_1, \dots, D_k\}$$

and D_j is a $\gamma_j \times \gamma_j$ diagonal matrix with λ_j on its diagonal. If $\gamma_1 + \cdots + \gamma_k = n$, then U will be invertible.

The next example illustrates the algorithm.

Example 4.10. Let $A \in \mathbb{C}^{6\times 6}$ and suppose that A has exactly 3 distinct eigenvalues $\lambda_1, \lambda_2, \lambda_3 \in \mathbb{C}$ with geometric multiplicities $\gamma_1 = 3$, $\gamma_2 = 1$ and $\gamma_3 = 2$, respectively. Let $\{\mathbf{u}_1, \mathbf{u}_2, \mathbf{u}_3\}$ be **any** basis for $\mathcal{N}_{(A-\lambda_1 I_n)}$, $\{\mathbf{u}_4\}$ be **any** basis for $\mathcal{N}_{(A-\lambda_2 I_n)}$ and $\{\mathbf{u}_5, \mathbf{u}_6\}$ be **any** basis for $\mathcal{N}_{(A-\lambda_3 I_n)}$. Then it is readily checked that

$$A[\mathbf{u}_1\ \mathbf{u}_2\ \mathbf{u}_3\ \mathbf{u}_4\ \mathbf{u}_5\ \mathbf{u}_6] = [\mathbf{u}_1\ \mathbf{u}_2\ \mathbf{u}_3\ \mathbf{u}_4\ \mathbf{u}_5\ \mathbf{u}_6]\left[\begin{array}{ccc:c:cc} \lambda_1 & 0 & 0 & 0 & 0 & 0 \\ 0 & \lambda_1 & 0 & 0 & 0 & 0 \\ 0 & 0 & \lambda_1 & 0 & 0 & 0 \\ \hdashline 0 & 0 & 0 & \lambda_2 & 0 & 0 \\ \hdashline 0 & 0 & 0 & 0 & \lambda_3 & 0 \\ 0 & 0 & 0 & 0 & 0 & \lambda_3 \end{array}\right].$$

But, upon setting

$$U = [\mathbf{u}_1\ \mathbf{u}_2\ \mathbf{u}_3\ \mathbf{u}_4\ \mathbf{u}_5\ \mathbf{u}_6] \quad \text{and} \quad D = \operatorname{diag}\{\lambda_1, \lambda_1, \lambda_1, \lambda_2, \lambda_3, \lambda_3\},$$

the preceding formula can be rewritten as

$$AU = UD \quad \text{or equivalently as} \quad A = UDU^{-1},$$

since U is invertible, because it is a 6×6 matrix with six linearly independent column vectors, thanks to Theorem 4.3.

Notice that in the notation used in Theorem 4.8 and its proof, $k = 3$, $U_1 = [\mathbf{u}_1\ \mathbf{u}_2\ \mathbf{u}_3]$, $U_2 = \mathbf{u}_4$, $U_5 = [\mathbf{u}_5\ \mathbf{u}_6]$, $\Lambda_1 = \operatorname{diag}\{\lambda_1, \lambda_1, \lambda_1\}$, $\Lambda_2 = \lambda_4$ and $\Lambda_3 = \operatorname{diag}\{\lambda_5, \lambda_6\}$.

4.8. Computing eigenvalues at this point

The eigenvalues of a matrix $A \in \mathbb{C}^{n\times n}$ are precisely those points $\lambda \in \mathbb{C}$ at which $\mathcal{N}_{(A-\lambda I_n)} \neq \{\mathbf{0}\}$. In Chapter 5, we shall identify these points with the values of $\lambda \in \mathbb{C}$ at which the determinant $\det(A - \lambda I_n)$ is equal to zero. However, as we have not introduced determinants yet, we shall discuss another method that uses Gaussian elimination to find those points $\lambda \in \mathbb{C}$ for which the equation $A\mathbf{x} - \lambda\mathbf{x} = \mathbf{0}$ has nonzero solutions $\mathbf{x} \in \mathbb{C}^n$. In particular, it is necessary to find those points λ for which the upper echelon matrix U corresponding to $A - \lambda I_n$ has less than n pivots.

Example 4.11. Let

$$A = \begin{bmatrix} 3 & 1 & 1 \\ 2 & 2 & 1 \\ 1 & 3 & 1 \end{bmatrix}.$$

Then

$$A - \lambda I_3 = \begin{bmatrix} 3-\lambda & 1 & 1 \\ 2 & 2-\lambda & 1 \\ 1 & 3 & 1-\lambda \end{bmatrix}.$$

Thus, permuting the first and third row of A for convenience, we obtain

$$\begin{bmatrix} 0 & 0 & 1 \\ 0 & 1 & 0 \\ 1 & 0 & 0 \end{bmatrix} A = \begin{bmatrix} 1 & 3 & 1-\lambda \\ 2 & 2-\lambda & 1 \\ 3-\lambda & 1 & 1 \end{bmatrix}.$$

Next, adding -2 times the first row to the second and $\lambda - 3$ times the first row to the third, yields

$$\begin{bmatrix} 1 & 0 & 0 \\ -2 & 1 & 0 \\ \lambda-3 & 0 & 1 \end{bmatrix} \begin{bmatrix} 1 & 3 & 1-\lambda \\ 2 & 2-\lambda & 1 \\ 3-\lambda & 1 & 1 \end{bmatrix} = \begin{bmatrix} 1 & 3 & 1-\lambda \\ 0 & -4-\lambda & 2\lambda-1 \\ 0 & 3\lambda-8 & x \end{bmatrix},$$

where

$$x = 1 + (\lambda - 3)(1 - \lambda).$$

Since the last matrix on the right is invertible when $\lambda = -4$, the vector space $\mathcal{N}_{A+4I_3} = \{\mathbf{0}\}$. Thus, we can assume that $\lambda + 4 \neq 0$ and add $(3\lambda - 8)/(\lambda + 4)$ times the second row to the third row to get

$$\begin{bmatrix} 1 & 0 & 0 \\ 0 & 1 & 0 \\ 0 & \dfrac{(3\lambda-8)}{(\lambda+4)} & 1 \end{bmatrix} \begin{bmatrix} 1 & 3 & 1-\lambda \\ 0 & -4-\lambda & 2\lambda-1 \\ 0 & 3\lambda-8 & x \end{bmatrix} = \begin{bmatrix} 1 & 3 & 1-\lambda \\ 0 & -4-\lambda & 2\lambda-1 \\ 0 & 0 & y \end{bmatrix},$$

where

$$\begin{aligned} y &= \frac{(3\lambda-8)(2\lambda-1) + \lambda + 4 + (\lambda+4)(\lambda-3)(1-\lambda)}{\lambda+4} \\ &= \frac{-\lambda(\lambda-5)(\lambda-1)}{\lambda+4}. \end{aligned}$$

Therefore, $\mathcal{N}_{(A-\lambda I_n)} \neq \{\mathbf{0}\}$ if and only if $\lambda = 0$, or $\lambda = 5$, or $\lambda = 1$.

Exercise 4.20. Find an invertible matrix $U \in \mathbb{C}^{3\times 3}$ and a diagonal matrix $D \in \mathbb{C}^{3\times 3}$ so that $A = UDU^{-1}$ when A is chosen equal to the matrix in the preceding example. [HINT: Follow the steps in the algorithm presented in the previous section.]

Exercise 4.21. Find an invertible matrix U such that $U^{-1}AU$ is equal to a diagonal matrix D for each of the following two choices of A:

$$\begin{bmatrix} 1 & 1 & 1 \\ 0 & 2 & 2 \\ 0 & 0 & 3 \end{bmatrix}, \quad \begin{bmatrix} 1 & 1 & 2 \\ 0 & 2 & 2 \\ 0 & 0 & 1 \end{bmatrix}.$$

Exercise 4.22. Repeat Exercise 4.21 for

$$A = \begin{bmatrix} 1 & 1 & 1 \\ 1 & 0 & 1 \\ 1 & 1 & 1 \end{bmatrix}.$$

[REMARK: This is a little harder than the previous exercise, but not much.]

4.9. Not all matrices are diagonalizable

Not all matrices are diagonalizable, even if complex eigenvalues are allowed. The problem is that a matrix may not have enough linearly independent eigenvectors; i.e., the criterion $\gamma_1 + \cdots + \gamma_k = n$ established in Theorem 4.8 may not be satisfied. Thus, for example, if

$$A = \begin{bmatrix} 2 & 1 \\ 0 & 2 \end{bmatrix}, \quad \text{then} \quad A - 2I_2 = \begin{bmatrix} 0 & 1 \\ 0 & 0 \end{bmatrix},$$

$\dim \mathcal{N}_{(A-2I_2)} = 1$ and $\dim \mathcal{N}_{(A-\lambda I_2)} = 0$ if $\lambda \neq 2$. Similarly, if

$$A = \begin{bmatrix} 2 & 1 & 0 \\ 0 & 2 & 1 \\ 0 & 0 & 2 \end{bmatrix}, \quad \text{then} \quad A - 2I_2 = \begin{bmatrix} 0 & 1 & 0 \\ 0 & 0 & 1 \\ 0 & 0 & 0 \end{bmatrix},$$

$\dim \mathcal{N}_{(A-2I_3)} = 1$ and $\dim \mathcal{N}_{(A-\lambda I_3)} = 0$ if $\lambda \neq 2$. More elaborate examples may be constructed by taking larger matrices of the same form or by putting such blocks together as in Exercise 4.23.

Exercise 4.23. Calculate $\dim \mathcal{N}_{(B_{\lambda_1} - \lambda_1 I_{13})^j}$ for $j = 1, 2, \ldots$ when

$$B_{\lambda_1} = \left[\begin{array}{ccccc|cccc|cc|c|c}
\lambda_1 & 1 & 0 & 0 & 0 & 0 & 0 & 0 & 0 & 0 & 0 & 0 & 0 \\
0 & \lambda_1 & 1 & 0 & 0 & 0 & 0 & 0 & 0 & 0 & 0 & 0 & 0 \\
0 & 0 & \lambda_1 & 1 & 0 & 0 & 0 & 0 & 0 & 0 & 0 & 0 & 0 \\
0 & 0 & 0 & \lambda_1 & 1 & 0 & 0 & 0 & 0 & 0 & 0 & 0 & 0 \\
0 & 0 & 0 & 0 & \lambda_1 & 0 & 0 & 0 & 0 & 0 & 0 & 0 & 0 \\
\hline
0 & 0 & 0 & 0 & 0 & \lambda_1 & 1 & 0 & 0 & 0 & 0 & 0 & 0 \\
0 & 0 & 0 & 0 & 0 & 0 & \lambda_1 & 1 & 0 & 0 & 0 & 0 & 0 \\
0 & 0 & 0 & 0 & 0 & 0 & 0 & \lambda_1 & 1 & 0 & 0 & 0 & 0 \\
0 & 0 & 0 & 0 & 0 & 0 & 0 & 0 & \lambda_1 & 0 & 0 & 0 & 0 \\
\hline
0 & 0 & 0 & 0 & 0 & 0 & 0 & 0 & 0 & \lambda_1 & 1 & 0 & 0 \\
0 & 0 & 0 & 0 & 0 & 0 & 0 & 0 & 0 & 0 & \lambda_1 & 0 & 0 \\
\hline
0 & 0 & 0 & 0 & 0 & 0 & 0 & 0 & 0 & 0 & 0 & \lambda_1 & 0 \\
\hline
0 & 0 & 0 & 0 & 0 & 0 & 0 & 0 & 0 & 0 & 0 & 0 & \lambda_1
\end{array}\right]$$

and build an array of symbols $\times$ with $\dim \mathcal{N}_{(B_{\lambda_1} - \lambda_1 I_{13})^i} - \dim \mathcal{N}_{(B_{\lambda_1} - \lambda_1 I_{13})^{i-1}}$ symbols $\times$ in the i'th row for $i = 1, 2, \ldots$. Check that the number of fundamental Jordan cells in B_{λ_1} of size $i \times i$ is equal to the number of columns of height i in the array corresponding to λ_j.

The notation

$$B_{\lambda_1} = \operatorname{diag}\{C_{\lambda_1}^{(5)}, C_{\lambda_1}^{(4)}, C_{\lambda_1}^{(2)}, C_{\lambda_1}^{(1)}, C_{\lambda_1}^{(1)}\}$$

is a convenient way to describe the matrix B_{λ_1} of Exercise 4.23 in terms of its fundamental **Jordan cells** $C_\alpha^{(\nu)}$, where

$$(4.14) \qquad C_\alpha^{(\nu)} = \begin{bmatrix} \alpha & 1 & 0 & & \cdots & 0 & 0 \\ 0 & \alpha & 1 & & \cdots & 0 & 0 \\ \vdots & \vdots & & \ddots & & \vdots & \vdots \\ 0 & 0 & & \cdots & & \alpha & 1 \\ 0 & 0 & & \cdots & & 0 & \alpha \end{bmatrix} = \alpha I_\nu + C_0^{(\nu)}$$

denotes the $\nu \times \nu$ matrix with α on the main diagonal, one on the diagonal line just above the main diagonal and zeros elsewhere. This helps to avoid such huge displays. Moreover, such block diagonal representations are convenient for calculation, because

$$(4.15) \quad B = \operatorname{diag}\{B_1, \ldots, B_k\} \Longrightarrow \dim \mathcal{N}_B = \dim \mathcal{N}_{B_1} + \cdots + \dim \mathcal{N}_{B_k}.$$

Nevertheless, the news is not all bad. There is a more general factorization formula than (4.12) in which the matrix D is replaced by a block diagonal matrix $J = \operatorname{diag}\{B_{\lambda_1}, \ldots, B_{\lambda_k}\}$, where B_{λ_j} is an $\alpha_j \times \alpha_j$ upper triangular matrix that is also a block diagonal matrix with γ_j Jordan cells (of assorted sizes) as blocks that is based on the following fact:

Theorem 4.12. *Let $A \in \mathbb{C}^{n\times n}$ and suppose that A has exactly k distinct eigenvalues, $\lambda_1, \ldots, \lambda_k \in \mathbb{C}$. Then*

$$\mathbb{C}^n = \mathcal{N}_{(A-\lambda_1 I_n)^n} \dotplus \cdots \dotplus \mathcal{N}_{(A-\lambda_k I_n)^n}.$$

The proof of this theorem will be carried out in the next few sections. At this point let us focus instead on its implications.

Example 4.13. Let $A \in \mathbb{C}^{9\times 9}$ and suppose that A has exactly three distinct eigenvalues λ_1, λ_2 and λ_3 with algebraic multiplicities $\alpha_1 = 4$, $\alpha_2 = 2$ and $\alpha_3 = 3$, respectively. Let $\{\mathbf{v}_1, \mathbf{v}_2, \mathbf{v}_3, \mathbf{v}_4\}$ be any basis for $\mathcal{N}_{(A-\lambda_1 I_9)^9}$, let $\{\mathbf{w}_1, \mathbf{w}_2\}$ be any basis for $\mathcal{N}_{(A-\lambda_2 I_9)^9}$ and let $\{\mathbf{x}_1, \mathbf{x}_2, \mathbf{x}_3\}$ be any basis for $\mathcal{N}_{(A-\lambda_3 I_9)^9}$. Then, since each of the spaces $\mathcal{N}_{(A-\lambda_j I_9)^9}$, $j = 1, 2, 3$ is invariant under multiplication by the matrix A,

$$\begin{aligned} A[\mathbf{v}_1\ \mathbf{v}_2\ \mathbf{v}_3\ \mathbf{v}_4] &= [\mathbf{v}_1\ \mathbf{v}_2\ \mathbf{v}_3\ \mathbf{v}_4]G_1 \\ A[\mathbf{w}_1\ \mathbf{w}_2] &= [\mathbf{w}_1\ \mathbf{w}_2]G_2 \\ A[\mathbf{x}_1\ \mathbf{x}_2\ \mathbf{x}_3] &= [\mathbf{x}_1\ \mathbf{x}_2\ \mathbf{x}_3]G_3 \end{aligned}$$

for some choice of $G_1 \in \mathbb{C}^{4\times 4}$, $G_2 \in \mathbb{C}^{2\times 2}$ and $G_3 \in \mathbb{C}^{3\times 3}$. In other notation, upon setting

$$V = [\mathbf{v}_1\ \mathbf{v}_2\ \mathbf{v}_3\ \mathbf{v}_4], \ W = [\mathbf{w}_1\ \mathbf{w}_2] \quad \text{and} \quad X = [\mathbf{x}_1\ \mathbf{x}_2\ \mathbf{x}_3],$$

one can write the preceding three sets of equations together as

$$A[V\ W\ X] = [V\ W\ X] \begin{bmatrix} G_1 & O & O \\ O & G_2 & O \\ O & O & G_3 \end{bmatrix}$$

or, equivalently, upon setting $U = [V\ W\ X]$, as

$$A = U \begin{bmatrix} G_1 & O & O \\ O & G_2 & O \\ O & O & G_3 \end{bmatrix} U^{-1}, \tag{4.16}$$

since the matrix $U = [V\ W\ X]$ is invertible, thanks to Theorem 4.12.

Formula (4.16) is the best that can be achieved with the given information. To say more, one needs to know more about the subspaces $\mathcal{N}_{(A-\lambda_i I_n)^j}$ for $j = 1, \ldots, \alpha_i$ and $i = 1, 2, 3$. Thus, for example, if $\dim \mathcal{N}_{(A-\lambda_i I_9)} = 1$ for $i = 1, 2, 3$, then the vectors in $\mathcal{N}_{(A-\lambda_i I_9)^{\alpha_i}}$ may be chosen so that

$$\operatorname{diag}\{G_1, G_2, G_3\} = \operatorname{diag}\{C_{\lambda_1}^{(4)}, C_{\lambda_2}^{(2)}, C_{\lambda_3}^{(3)}\}.$$

On the other hand, if $\dim \mathcal{N}_{(A-\lambda_i I_9)} = 2$ for $i = 1, 2, 3$ and $\dim \mathcal{N}_{(A-\lambda_1 I_9)^2} = 4$, then the vectors in $\mathcal{N}_{(A-\lambda_i I_9)^{\alpha_j}}$ may be chosen so that

$$\operatorname{diag}\{G_1, G_2, G_3\} = \operatorname{diag}\{C_{\lambda_1}^{(2)}, C_{\lambda_1}^{(2)}, C_{\lambda_2}^{(1)}, C_{\lambda_2}^{(1)}, C_{\lambda_3}^{(2)}, C_{\lambda_3}^{(1)}\}.$$

There are still more possibilities. The main facts are summarized in the statement of Theorem 4.14 in the next section.

4.10. The Jordan decomposition theorem

Theorem 4.14. *Let $A \in \mathbb{C}^{n\times n}$ and suppose that A has exactly k distinct eigenvalues $\lambda_1, \ldots, \lambda_k$ in $\mathbb{C}$ with geometric multiplicities $\gamma_1, \ldots, \gamma_k$ and algebraic multiplicities $\alpha_1, \ldots, \alpha_k$, respectively. Then there exists an invertible matrix $U \in \mathbb{C}^{n\times n}$ such that*

$$AU = UJ,$$

where:

(1) *$J = \operatorname{diag}\{B_{\lambda_1}, \ldots, B_{\lambda_k}\}$.*

(2) *B_{λ_j} is an $\alpha_j \times \alpha_j$ block diagonal matrix that is built out of γ_j Jordan cells $C_{\lambda_j}^{(\cdot)}$ of the form (4.14).*

(3) *The number of Jordan cells $C_{\lambda_j}^{(i)}$ in B_{λ_j} with $i \geq \ell$ is equal to*

$$\dim \mathcal{N}_{(A-\lambda_j I_n)^\ell} - \dim \mathcal{N}_{(A-\lambda_j I_n)^{\ell-1}}, \quad \ell = 2, \ldots, \alpha_j, \tag{4.17}$$

or, in friendlier terms, the number of Jordan cells $C_{\lambda_j}^{(i)}$ *in* B_{λ_j} *is equal to the number of columns of height* i *in the array of symbols* $\times$

$$\begin{array}{lll} \times & \cdots & \times \quad \text{with } \gamma_1 \text{ symbols in the first row} \\ \times & \cdots & \quad\quad \text{with } \dim \mathcal{N}_{(A-\lambda_j I_n)^2} - \dim \mathcal{N}_{(A-\lambda_j I_n)} \text{ symbols in row 2} \\ \times & \cdots & \quad\quad \text{with } \dim \mathcal{N}_{(A-\lambda_j I_n)^3} - \dim \mathcal{N}_{(A-\lambda_j I_n)^2} \text{ symbols in row 3} \\ \vdots & & \quad\quad \vdots \end{array}$$

(4) *The columns of* U *are generalized eigenvectors of the matrix* A.

(5) $(A - \lambda_1 I_n)^{\alpha_1} \cdots (A - \lambda_k I_n)^{\alpha_k} = O$.

(6) *If* $\nu_j = \min\{i : \dim \mathcal{N}_{(A-\lambda_j I_n)^i} = \dim \mathcal{N}_{(A-\lambda_j I_n)^n}\}$, *then* $\nu_j \leq \alpha_j$ *for* $j = 1, \ldots, k$ *and* $(A - \lambda_1 I_n)^{\nu_1} \cdots (A - \lambda_k I_n)^{\nu_k} = O$.

The verification of this theorem rests on Theorem 4.12. It amounts to showing that the basis of each of the spaces $\mathcal{N}_{(A-\lambda_j I_n)^n}$ can be organized in a suitable way. It turns out that the array constructed in (3) is a **Young diagram**, since the number of symbols in row $i + 1$ is less than or equal to the number of symbols in row i; see Corollary 6.5. Item (5) is the **Cayley-Hamilton theorem**. In view of (6), the polynomial $p(\lambda) = (\lambda - \lambda_1)^{\nu_1} \cdots (\lambda - \lambda_k)^{\nu_k}$ is referred to as the **minimal polynomial** for A. Moreover, the number ν_j is the "size" of the largest Jordan cell in B_{λ_j}. A detailed proof of the Jordan decomposition theorem is deferred to Chapter 6, though an illustrative example, which previews some of the key ideas, is furnished in the next section.

Exercise 4.24. Show that if

$$A = UC_\alpha^{(n)}U^{-1}, \quad \text{then} \quad \dim \mathcal{N}_{(A-\lambda I_n)} = \begin{cases} 0 & \text{if} \quad \lambda \neq \alpha \\ 1 & \text{if} \quad \lambda = \alpha \end{cases}.$$

Exercise 4.25. Calculate $\dim \mathcal{N}_{(A-\lambda I_p)^t}$ for every $\lambda \in \mathbb{C}$ and $t = 1, 2, \ldots$ for the 26×26 matrix $A = UJU^{-1}$ when $J = \operatorname{diag}\{B_{\lambda_1}, B_{\lambda_2}, B_{\lambda_3}\}$, the points $\lambda_1, \lambda_2, \lambda_3$ are distinct, B_{λ_1} is as in Exercise 4.23, $B_{\lambda_2} = \operatorname{diag}\{C_{\lambda_2}^{(3)}, C_{\lambda_2}^{(3)}\}$ and $B_{\lambda_3} = \operatorname{diag}\{C_{\lambda_3}^{(4)}, C_{\lambda_3}^{(2)}, C_{\lambda_3}^{(1)}\}$. Build an array of symbols $\times$ for each eigenvalue λ_j with $\dim \mathcal{N}_{(A-\lambda_j I_p)^i} - \dim \mathcal{N}_{(A-\lambda_j I_p)^{i-1}}$ symbols $\times$ in the i'th row for $i = 1, 2, \ldots$ and check that the number of fundamental Jordan cells in B_{λ_j} of size $i \times i$ is equal to the number of columns in the array of height i.

4.11. An instructive example

To develop some feeling for Theorem 4.14, we shall first investigate the implications of the factorization $A = UJU^{-1}$ on the matrix A when

$$U = \begin{bmatrix} \mathbf{u}_1 & \cdots & \mathbf{u}_5 \end{bmatrix}$$

is any 5×5 invertible matrix with columns $\mathbf{u}_1, \ldots, \mathbf{u}_5$ and

$$J = \left[\begin{array}{ccc:cc} \lambda_1 & 1 & 0 & 0 & 0 \\ 0 & \lambda_1 & 1 & 0 & 0 \\ 0 & 0 & \lambda_1 & 0 & 0 \\ \hdashline 0 & 0 & 0 & \lambda_2 & 1 \\ 0 & 0 & 0 & 0 & \lambda_2 \end{array}\right] = \operatorname{diag}\left\{C_{\lambda_1}^{(3)}, C_{\lambda_2}^{(2)}\right\}.$$

Then the matrix equation $AU = UJ$ can be replaced by five vector equations, one for each column of U:

$$\begin{aligned} A\mathbf{u}_1 &= \lambda_1 \mathbf{u}_1. \\ A\mathbf{u}_2 &= \mathbf{u}_1 + \lambda_1 \mathbf{u}_2. \\ A\mathbf{u}_3 &= \mathbf{u}_2 + \lambda_1 \mathbf{u}_3. \\ A\mathbf{u}_4 &= \lambda_2 \mathbf{u}_4. \\ A\mathbf{u}_5 &= \mathbf{u}_4 + \lambda_2 \mathbf{u}_5. \end{aligned}$$

The first three formulas imply in turn that

$$\begin{aligned} \mathbf{u}_1 &\in \mathcal{N}_{(A-\lambda_1 I_5)}\,; \text{ i.e., } \mathbf{u}_1 \text{ is an eigenvector corresponding to } \lambda_1. \\ \mathbf{u}_2 &\notin \mathcal{N}_{(A-\lambda_1 I_5)} \text{ but } \mathbf{u}_2 \in \mathcal{N}_{(A-\lambda_1 I_5)^2}. \\ \mathbf{u}_3 &\notin \mathcal{N}_{(A-\lambda_1 I_5)^2} \text{ but } \mathbf{u}_3 \in \mathcal{N}_{(A-\lambda_1 I_5)^3}\,; \end{aligned}$$

i.e., $\mathbf{u}_1, \mathbf{u}_2, \mathbf{u}_3$ is a Jordan chain of length 3. Similarly, $\mathbf{u}_4, \mathbf{u}_5$ is a Jordan chain of length 2. This calculation exhibits $\mathbf{u}_1$ and $\mathbf{u}_4$ as eigenvectors. In fact,

(1) if $\lambda_1 \neq \lambda_2$, then

$$\dim \mathcal{N}_{(A-\lambda I_5)} = \begin{cases} 1 & \text{if } \lambda = \lambda_1 \\ 1 & \text{if } \lambda = \lambda_2 \\ 0 & \text{otherwise} \end{cases}, \quad \dim \mathcal{N}_{(A-\lambda I_5)^2} = \begin{cases} 2 & \text{if } \lambda = \lambda_1 \\ 2 & \text{if } \lambda = \lambda_2 \\ 0 & \text{otherwise} \end{cases}$$

and

$$\dim \mathcal{N}_{(A-\lambda I_5)^k} = \begin{cases} 3 & \text{if } \lambda = \lambda_1 \\ 2 & \text{if } \lambda = \lambda_2 \\ 0 & \text{otherwise} \end{cases} \quad \text{for every integer} \quad k \geq 3;$$

(2) if $\lambda_1 = \lambda_2$, then

$$\dim \mathcal{N}_{(A-\lambda I_5)} = \begin{cases} 2 & \text{if } \lambda = \lambda_1 \\ 0 & \text{otherwise} \end{cases}, \quad \dim \mathcal{N}_{(A-\lambda I_5)^2} = \begin{cases} 4 & \text{if } \lambda = \lambda_1 \\ 0 & \text{otherwise} \end{cases}$$

and

$$\dim \mathcal{N}_{(A-\lambda I_5)^k} = \begin{cases} 5 & \text{if } \lambda = \lambda_1 \\ 0 & \text{otherwise} \end{cases} \quad \text{for every integer} \quad k \geq 3\,.$$

The key to these calculations is in the fact that

$$\dim \mathcal{N}_{(A-\lambda I_5)^k} = \dim \mathcal{N}_{(J-\lambda I_5)^k} \tag{4.18}$$

for $k = 1, 2, \ldots$, and the special structure of J. Formula (4.18) follows from the identity

$$(A - \lambda I_5)^k = U(J - \lambda I_5)^k U^{-1}. \tag{4.19}$$

Because of the block diagonal structure of J,

$$\operatorname{rank} J = \operatorname{rank} C_{\lambda_1}^{(3)} + \operatorname{rank} C_{\lambda_2}^{(2)}$$

and

$$\begin{aligned} \operatorname{rank}\,(J - \lambda I_5)^k &= \operatorname{rank}\left(C_{\lambda_1}^{(3)} - \lambda I_3\right)^k + \operatorname{rank}\left(C_{\lambda_2}^{(2)} - \lambda I_2\right)^k \\ &= \operatorname{rank}\left(C_{\lambda_1-\lambda}^{(3)}\right) + \operatorname{rank}\left(C_{\lambda_2-\lambda}^{(2)}\right). \end{aligned}$$

Moreover, it is easy to compute the indicated ranks, because a $\nu \times \nu$ Jordan cell $C_\beta^{(\nu)}$ is invertible if $\beta \neq 0$ and

$$\operatorname{rank}\left(C_0^{(\nu)}\right)^k = \nu - k \text{ for } k = 1, \ldots, \nu.$$

To illustrate even more graphically, observe that if $\beta = \lambda_1 - \lambda_2 \neq 0$, then

$$J - \lambda_1 I_5 = \left[\begin{array}{ccc:cc} 0 & 1 & 0 & 0 & 0 \\ 0 & 0 & 1 & 0 & 0 \\ 0 & 0 & 0 & 0 & 0 \\ \hdashline 0 & 0 & 0 & \beta & 1 \\ 0 & 0 & 0 & 0 & \beta \end{array}\right], \quad (J - \lambda_1 I_5)^2 = \left[\begin{array}{ccc:cc} 0 & 0 & 1 & 0 & 0 \\ 0 & 0 & 0 & 0 & 0 \\ 0 & 0 & 0 & 0 & 0 \\ \hdashline 0 & 0 & 0 & \beta^2 & 2\beta \\ 0 & 0 & 0 & 0 & \beta^2 \end{array}\right]$$

and

$$(J - \lambda_1 I_5)^3 = \left[\begin{array}{ccc:cc} 0 & 0 & 0 & 0 & 0 \\ 0 & 0 & 0 & 0 & 0 \\ 0 & 0 & 0 & 0 & 0 \\ \hdashline 0 & 0 & 0 & \beta^3 & 3\beta^2 \\ 0 & 0 & 0 & 0 & \beta^3 \end{array}\right].$$

Clearly one can construct more elaborate examples of nondiagonalizable matrices $A = UJU^{-1}$ by adding more diagonal block "cells" $C_{\lambda_j}^{(\nu)}$ to J.

4.12. The binomial formula

The familiar binomial identity

$$(a+b)^m = \sum_{k=0}^{m} \binom{m}{k} a^k b^{m-k}$$

for numbers a and b remains valid for square matrices A and B of the same size if they commute:

$$(A+B)^m = \sum_{k=0}^{m} \binom{m}{k} A^k B^{m-k} \quad \text{if } AB = BA. \tag{4.20}$$

If this is unfamiliar, try writing out $(A+B)^2$ and $(A+B)^3$. In particular,

$$(\lambda I + B)^m = \sum_{k=0}^{m} \binom{m}{k} \lambda^k B^{m-k}. \tag{4.21}$$

Exercise 4.26. Find a pair of matrices A and B for which the formula (4.20) fails.

4.13. More direct sum decompositions

Lemma 4.15. *Let $A \in \mathbb{F}^{n\times n}$, $\lambda \in \mathbb{F}$ and suppose that $(A-\lambda I_n)^j \mathbf{u} = \mathbf{0}$ for some $j \geq 1$ and $\mathbf{u} \in \mathbb{F}^n$. Then $(A-\lambda I_n)^n \mathbf{u} = \mathbf{0}$.*

Proof. Let $B = A - \lambda I_n$. Then, since the assertion is self-evident if $\mathbf{u} = \mathbf{0}$, it suffices to focus attention on the case when $\mathbf{u} \neq \mathbf{0}$ and k is equal to the smallest positive integer j such that $B^j\mathbf{u} = \mathbf{0}$ and to consider the set of nonzero vectors

$$\mathbf{u}, B\mathbf{u}, \ldots, B^{k-1}\mathbf{u}.$$

This set of vectors is linearly independent because if

$$c_0\mathbf{u} + c_1 B\mathbf{u} + \cdots + c_{k-1}B^{k-1}\mathbf{u} = \mathbf{0},$$

then the self-evident identity

$$B^{k-1}\left(c_0\mathbf{u} + \cdots + c_{k-1}B^{k-1}\mathbf{u}\right) = \mathbf{0}$$

implies that $c_0 = 0$. Similarly,

$$B^{k-2}\left(c_1 B\mathbf{u} + \cdots + c_{k-1}B^{k-1}\mathbf{u}\right) = \mathbf{0}$$

implies that $c_1 = 0$. After $k-1$ such steps we are left with

$$c_{k-1}B^{k-1}\mathbf{u} = \mathbf{0},$$

which implies that $c_{k-1} = 0$. This completes the proof of the asserted linear independence. But if k vectors in $\mathbb{F}^n$ are linearly independent, then $k \leq n$, and hence $B^n\mathbf{u} = \mathbf{0}$, as claimed. $\square$

Lemma 4.16. *Let $A \in \mathbb{F}^{n\times n}$ and $\lambda \in \mathbb{F}$. Then*

$$\mathbb{F}^n = \mathcal{N}_{(A-\lambda I_n)^n} \dot{+} \mathcal{R}_{(A-\lambda I_n)^n} . \tag{4.22}$$

Proof. Let $B = A - \lambda I_n$ and suppose first that

$$\mathbf{u} \in \mathcal{N}_{B^n} \cap \mathcal{R}_{B^n}.$$

Then $B^n\mathbf{u} = \mathbf{0}$ and $\mathbf{u} = B^n\mathbf{v}$ for some vector $\mathbf{v} \in \mathbb{F}^n$. Therefore,

$$\mathbf{0} = B^n\mathbf{u} = B^{2n}\mathbf{v}.$$

But, by the last lemma, this in fact implies that

$$\mathbf{u} = B^n\mathbf{v} = \mathbf{0}.$$

Thus, the sum is direct. It is all of $\mathbb{F}^n$ by the principle of conservation of dimension:

$$n = \dim \mathcal{R}_{B^n} + \dim \mathcal{N}_{B^n}.$$

□

Lemma 4.17. *Let $A \in \mathbb{F}^{n\times n}$, let $\lambda_1, \lambda_2 \in \mathbb{F}$ and suppose that $\lambda_1 \neq \lambda_2$. Then*

$$\mathcal{N}_{(A-\lambda_2 I_n)^n} \subseteq \mathcal{R}_{(A-\lambda_1 I_n)^n}.$$

Proof. Let $\mathbf{u} \in \mathcal{N}_{(A-\lambda_2 I_n)^n}$. Then, by formula (4.20),

$$\begin{aligned}
\mathbf{0} &= (A - \lambda_1 I_n + (\lambda_1 - \lambda_2) I_n)^n \,\mathbf{u} \\
&= \sum_{j=0}^{n} \binom{n}{j} (A - \lambda_1 I_n)^j (\lambda_1 - \lambda_2)^{n-j} \mathbf{u} \\
&= (\lambda_1 - \lambda_2)^n \mathbf{u} + \sum_{j=1}^{n} \binom{n}{j} (A - \lambda_1 I_n)^j (\lambda_1 - \lambda_2)^{n-j} \mathbf{u} \\
&= (\lambda_1 - \lambda_2)^n \mathbf{u} + (A - \lambda_1 I_n) \sum_{j=1}^{n} \binom{n}{j} (A - \lambda_1 I_n)^{j-1} (\lambda_1 - \lambda_2)^{n-j} \mathbf{u}\,.
\end{aligned}$$

Therefore,

$$\mathbf{u} = (A - \lambda_1 I_n) p(A) \mathbf{u},$$

for some polynomial

$$p(A) = c_0 I_n + c_1 A + \cdots + c_{n-1} A^{n-1}$$

in the matrix A.

Iterating the last identity for $\mathbf{u}$, we obtain

$$\begin{aligned}
\mathbf{u} &= (A - \lambda_1 I_n) p(A) (A - \lambda_1 I_n) p(A) \mathbf{u} \\
&= (A - \lambda_1 I_n)^2 p(A)^2 \mathbf{u},
\end{aligned}$$

since

$$(A - \lambda_1 I_n)p(A) = p(A)(A - \lambda_1 I_n).$$

Iterating $n-2$ more times we see that

$$\mathbf{u} = (A - \lambda_1 I_n)^n p(A)^n \mathbf{u},$$

which is to say that

$$\mathbf{u} \in \mathcal{R}_{(A-\lambda_1 I_n)^n},$$

as claimed. □

Remark 4.18. The last lemma may be exploited to give a quick proof of the fact that *generalized eigenvectors corresponding to distinct eigenvalues are automatically linearly independent.* To verify this, let

$$(A - \lambda_j I_n)^n \mathbf{u}_j = 0, \ j = 1, \ldots, k,$$

for some distinct set of eigenvalues $\lambda_1, \ldots, \lambda_k$ and suppose that

$$c_1\mathbf{u}_1 + \cdots + c_k\mathbf{u}_k = \mathbf{0}.$$

Then

$$-c_1\mathbf{u}_1 = c_2\mathbf{u}_2 + \cdots + c_k\mathbf{u}_k$$

and, since

$-c_1\mathbf{u}_1 \in \mathcal{N}_{(A-\lambda_1 I_n)^n}$ and, by Lemma 4.17, $c_2\mathbf{u}_2 + \cdots + c_k\mathbf{u}_k \in \mathcal{R}_{(A-\lambda_1 I_n)^n}$,

both sides of the last equality must equal zero, thanks to Lemma 4.16. Therefore $c_1 = 0$ and

$$c_2\mathbf{u}_2 + \cdots + c_k\mathbf{u}_k = \mathbf{0}.$$

To complete the verification, just keep on going.

Exercise 4.27. Complete the proof of the assertion in the preceding remark when $k = 3$.

4.14. Verification of Theorem 4.12

Lemma 4.16 guarantees that

$$\mathbb{F}^n = \mathcal{N}_{(A-\lambda I_n)^n} \dot{+} \mathcal{R}_{(A-\lambda I_n)^n} \tag{4.23}$$

for every point $\lambda \in \mathbb{F}$. The next step is to obtain an analogous direct sum decomposition for $\mathcal{R}_{(A-\lambda I_n)^n}$.

Lemma 4.19. *Let $A \in \mathbb{F}^{n\times n}$, let $\lambda_1, \lambda_2 \in \mathbb{F}$ and suppose that $\lambda_1 \neq \lambda_2$. Then*

$$\mathcal{R}_{(A-\lambda_1 I_n)^n} = \mathcal{N}_{(A-\lambda_2 I_n)^n} \dot{+} \{\mathcal{R}_{(A-\lambda_1 I_n)^n} \cap \mathcal{R}_{(A-\lambda_2 I_n)^n}\}. \tag{4.24}$$

Proof. The sum in (4.24) is direct, thanks to Lemma 4.16. Moreover, if

$$\mathbf{x} \in \mathcal{R}_{(A-\lambda_1 I_n)^n},$$

then, we can write

$$\mathbf{x} = \mathbf{u} + \mathbf{v}$$

where $\mathbf{u} \in \mathcal{N}_{(A-\lambda_2 I_n)^n}$, $\mathbf{v} \in \mathcal{R}_{(A-\lambda_2 I_n)^n}$ and, since the sum in (4.22) is direct, the vectors $\mathbf{u}$ and $\mathbf{v}$ are linearly independent. Lemma 4.17 guarantees that

$$\mathbf{u} \in \mathcal{R}_{(A-\lambda_1 I_n)^n}.$$

Therefore, the same holds true for $\mathbf{v}$. Thus, in view of Lemma 4.6,

$$\mathcal{R}_{(A-\lambda_1 I_n)^n} = \mathcal{R}_{(A-\lambda_1 I_n)^n} \cap \mathcal{N}_{(A-\lambda_2 I_n)^n} + \mathcal{R}_{(A-\lambda_1 I_n)^n} \cap \mathcal{R}_{(A-\lambda_2 I_n)^n},$$

which coincides with formula (4.24). □

There is a subtle point in the last proof that should not be overlooked; see Exercise 4.14.

Lemma 4.20. *Let $A \in \mathbb{C}^{n\times n}$ and suppose that A has exactly k distinct eigenvalues, $\lambda_1, \ldots, \lambda_k \in \mathbb{C}$. Then*

$$\mathcal{R}_{(A-\lambda_1 I_n)^n} \cap \mathcal{R}_{(A-\lambda_2 I_n)^n} \cap \cdots \cap \mathcal{R}_{(A-\lambda_k I_n)^n} = \{\mathbf{0}\}\,. \tag{4.25}$$

Proof. Let $\mathcal{M}$ denote the intersection of the k sets on the left-hand side of the asserted identity (4.25). Then it is readily checked that $\mathcal{M}$ is invariant under A; i.e, if $\mathbf{u} \in \mathcal{M}$, then $A\mathbf{u} \in \mathcal{M}$, because each of the sets $\mathcal{R}_{(A-\lambda_j I_n)^n}$ is invariant under A: if $\mathbf{u} \in \mathcal{R}_{(A-\lambda_j I_n)^n}$, then $\mathbf{u} = (A - \lambda_j I_n)^n \mathbf{v}_j$ and hence $A\mathbf{u} = (A - \lambda_j I_n)^n A\mathbf{v}_j$, for $j = 1, \ldots, k$. Consequently, if $\mathcal{M} \neq \{0\}$, then, by Theorem 4.4, there exists a complex number λ and a nonzero vector $\mathbf{v} \in \mathcal{M}$ such that $A\mathbf{v} - \lambda\mathbf{v} = 0$. But this means that λ is equal to one of the eigenvalues, say λ_t. Hence $\mathbf{v} \in \mathcal{N}_{(A-\lambda_t I_n)}$. But this in turn implies that

$$\mathbf{v} \in \mathcal{N}_{(A-\lambda_t I_n)^n} \cap \mathcal{R}_{(A-\lambda_t I_n)^n} = \{\mathbf{0}\}\,.$$

□

We are now ready to prove Theorem 4.12, which gives the theoretical justification for the decomposition of J into blocks B_{λ_j}, one for each distinct eigenvalue. It states that every vector $\mathbf{v} \in \mathbb{C}^n$ can be expressed as a linear combination of generalized eigenvectors of A. Moreover, generalized eigenvectors corresponding to distinct eigenvalues are linearly independent. The theorem is established by iterating Lemma 4.19.

Proof of Theorem 4.12. Let us suppose that $k \geq 3$. Then, by Lemmas 4.16 and 4.19,

$$\mathbb{C}^n = \mathcal{N}_{(A-\lambda_1 I_n)^n} \dot{+} \mathcal{R}_{(A-\lambda_1 I_n)^n}$$

and

$$\mathcal{R}_{(A-\lambda_1 I_n)^n} = \mathcal{N}_{(A-\lambda_2 I_n)^n} \dot{+} \mathcal{R}_{(A-\lambda_1 I_n)^n} \cap \mathcal{R}_{(A-\lambda_2 I_n)^n}.$$

Therefore,

$$\mathbb{C}^n = \mathcal{N}_{(A-\lambda_1 I_n)^n} \dot{+} \mathcal{N}_{(A-\lambda_2 I_n)^n} \dot{+} \mathcal{R}_{(A-\lambda_1 I_n)^n} \cap \mathcal{R}_{(A-\lambda_2 I_n)^n}\,.$$

Moreover, since

$$\mathcal{N}_{(A-\lambda_3 I_n)^n} \subseteq \mathcal{R}_{(A-\lambda_1 I_n)^n} \cap \mathcal{R}_{(A-\lambda_2 I_n)^n},$$

by Lemma 4.17, the supplementary formula

$$\mathcal{R}_{(A-\lambda_1 I_n)^n} \cap \mathcal{R}_{(A-\lambda_2 I_n)^n} = \mathcal{N}_{(A-\lambda_3 I_n)^n} \dot{+} \mathcal{R}_{(A-\lambda_1 I_n)^n} \cap \mathcal{R}_{(A-\lambda_2 I_n)^n} \cap \mathcal{R}_{(A-\lambda_3 I_n)^n}$$

may be verified just as in the proof of Lemma 4.19 and then substituted into the last formula for $\mathbb{C}^n$. To complete the proof, just keep on going until you run out of eigenvalues and then invoke Lemma 4.20. □

The point of Theorem 4.14 is that for every matrix $A \in \mathbb{C}^{n\times n}$ it is possible to find a set of n linearly independent generalized eigenvectors $\mathbf{u}_1, \ldots, \mathbf{u}_n$ such that

$$A[\mathbf{u}_1 \cdots \mathbf{u}_n] = [\mathbf{u}_1 \cdots \mathbf{u}_n]J\,.$$

The vectors have to be chosen properly. Details will be furnished in Chapter 6.

Exercise 4.28. If $B \in \mathbb{F}^{n\times n}$, then $\mathcal{R}_{B^n} \cap \mathcal{N}_{B^n} = \{\mathbf{0}\}$. Show by example that the vector space

$$\mathcal{R}_B \cap \mathcal{N}_B$$

may contain nonzero vectors.

Exercise 4.29. Show that if $A \in \mathbb{C}^{n\times n}$ has exactly two distinct eigenvalues in $\mathbb{C}$, then

$$\mathcal{R}_{(A-\lambda_1 I_n)^n} \cap \mathcal{R}_{(A-\lambda_2 I_n)^n} = \{\mathbf{0}\}\,.$$

Exercise 4.30. Show that if $A \in \mathbb{C}^{n\times n}$ has exactly k distinct eigenvalues $\lambda_1, \ldots, \lambda_k$ in $\mathbb{C}$ with algebraic multiplicities $\alpha_1, \ldots, \alpha_k$, then

$$\mathcal{N}_{(A-\lambda_1 I_n)^{\alpha_1}} \dot{+} \cdots \dot{+} \mathcal{N}_{(A-\lambda_k I_n)^{\alpha_k}} = \mathbb{C}^n\,.$$

Is it possible to reduce the powers further? Explain your answer.

Exercise 4.31. Verify formula (4.15). [HINT: In case of difficulty, start modestly by showing that if $B = \operatorname{diag}\{B_1, B_2, B_3\}$, then

$$\dim \mathcal{N}_B = \dim \mathcal{N}_{B_1} + \dim \mathcal{N}_{B_2} + \dim \mathcal{N}_{B_3}.]$$

Exercise 4.32. Let A be an $n \times n$ matrix.

- **(a):** Show that if $\mathbf{u}_1, \ldots, \mathbf{u}_k$ are eigenvectors corresponding to distinct eigenvalues $\lambda_1, \ldots, \lambda_k$, then the vectors $\mathbf{u}_1, \ldots, \mathbf{u}_k$ are linearly independent. (Try to give a simple direct proof that exploits the fact that $(A - \lambda_1 I_n) \cdots (A - \lambda_j I_n)\mathbf{u}_i = (\lambda_i - \lambda_1) \cdots (\lambda_i - \lambda_j)\mathbf{u}_i$.)
- **(b):** Use the conclusions of part (a) to show that if A has n distinct eigenvalues, then A is diagonalizable.

Exercise 4.33. Let $\mathbf{u} \in \mathbb{C}^n$, $\mathbf{v} \in \mathbb{C}^n$ and $B \in \mathbb{C}^{n\times n}$ be such that $B^4\mathbf{u} = 0$, $B^4\mathbf{v} = 0$ and the pair of vectors $B^3\mathbf{u}$ and $B^3\mathbf{v}$ are linearly independent in $\mathbb{C}^n$. Show that the eight vectors $\mathbf{u}$, $B\mathbf{u}$, $B^2\mathbf{u}$, $B^3\mathbf{u}$, $\mathbf{v}$, $B\mathbf{v}$, $B^2\mathbf{v}$ and $B^3\mathbf{v}$ are linearly independent in $\mathbb{C}^n$.

Exercise 4.34. Let $\mathbf{u} \in \mathbb{C}^n$, $\mathbf{v} \in \mathbb{C}^n$ and $B \in \mathbb{C}^{n\times n}$ be such that $B^4\mathbf{u} = 0$, $B^3\mathbf{v} = 0$ and the pair of vectors $B^3\mathbf{u}$ and $B^2\mathbf{v}$ are linearly independent in $\mathbb{C}^n$. Show that the seven vectors $\mathbf{u}$, $B\mathbf{u}$, $B^2\mathbf{u}$, $B^3\mathbf{u}$, $\mathbf{v}$, $B\mathbf{v}$ and $B^2\mathbf{v}$ are linearly independent in $\mathbb{C}^n$.

Exercise 4.35. Let $B \in \mathbb{C}^{n\times n}$. Show that $\mathcal{N}_B \subseteq \mathcal{N}_{B^2} \subseteq \mathcal{N}_{B^3} \subseteq \cdots$ and that if $\mathcal{N}_{B^j} = \mathcal{N}_{B^{j+1}}$ for $j = k$, then the equality prevails for every integer $j > k$ also.

Exercise 4.36. Show that if $B \in \mathbb{C}^{n\times n}$, then $\dim \mathcal{N}_{B^2} \leq 2 \dim \mathcal{N}_B$. [REMARK: The correct way to interpret this is: $\dim \mathcal{N}_{B^2} - \dim \mathcal{N}_B \leq \dim \mathcal{N}_B$.]

Exercise 4.37. Calculate $\begin{bmatrix} a & 1 \\ 0 & a \end{bmatrix}^{100}$. [HINT: To see the pattern, write the given matrix as $aI_2 + F$ and note that since $F^2 = 0$, $(aI_2 + F)^2$, $(aI_2 + F)^3, \ldots$, have a simple form.]

4.15. Bibliographical notes

Earlier versions of this chapter defined the eigenvalues of a matrix $A \in \mathbb{C}^{n\times n}$ in terms of the roots of the polynomial $\det(\lambda I_n - A)$. The present version, which was influenced by a conversation with Sheldon Axler at the Holomorphic Functions Session at MSRI, Berkeley, in 1995 and his paper [**5**] in the *American Math. Monthly*, avoids the use of determinants. They appear for the first time in the next chapter, and although they are extremely useful for calculating eigenvalues, they are not needed to establish the Jordan decomposition theorem.

To counter balance the title of [**5**] (which presumably was chosen for dramatic effect) it is perhaps appropriate to add the following words of the distinguished mathematical physicist L. D. Faddeev [**30**]: *If I had to choose a single term to characterize the technical tools used in my research, it would be determinants.*

Chapter 5

Determinants

Look at him, he doesn't drink, he doesn't smoke, he doesn't chew, he doesn't stay out late, and he still can't hit.

Casey Stengel

In this chapter we shall develop the theory of determinants. There are several ways to do this, many of which depend upon introducing unnatural looking formulas and/or recipes with little or no motivation and then showing that "they work". The approach adopted here is axiomatic. In particular we shall show that the determinant can be characterized as the one and only one **multilinear functional** $d(A)$ from $\mathbb{C}^{n\times n}$ to $\mathbb{C}$ that meets the two additional constraints $d(I_n) = 1$ and $d(PA) = -d(A)$ for every simple $n \times n$ permutation P. Later on, in Chapter 9, we shall also give a geometric interpretation of the determinant of a matrix $A \in \mathbb{R}^{n\times n}$ in terms of the volume of the parallelopiped generated by its column vectors.

5.1. Functionals

A function f from a vector space $\mathcal{V}$ over $\mathbb{F}$ into $\mathbb{F}$ is called a **functional**. A functional f on a vector space $\mathcal{V}$ over $\mathbb{F}$ is said to be a **linear functional** if it is a linear mapping from $\mathcal{V}$ into $\mathbb{F}$, i.e., if

$$f(\alpha\mathbf{u} + \beta\mathbf{v}) = \alpha f(\mathbf{u}) + \beta f(\mathbf{v})$$

for every choice of $\mathbf{u}, \mathbf{v} \in \mathcal{V}$ and $\alpha, \beta \in \mathbb{F}$. A linear functional on an n-dimensional vector space is completely determined by its action on the elements of a basis for the space: if $\{\mathbf{v}_1, \dots, \mathbf{v}_n\}$ is a basis for a vector space $\mathcal{V}$ over $\mathbb{F}$, and if $\mathbf{v} = \alpha_1\mathbf{v}_1 + \cdots + \alpha_n\mathbf{v}_n$, for some set of coefficients

$\{\alpha_1, \dots, \alpha_n\} \in \mathbb{F}$, then

$$f(\mathbf{v}) = f\left(\sum_{j=1}^{n} \alpha_j \mathbf{v}_j\right) = \sum_{j=1}^{n} \alpha_j f(\mathbf{v}_j) \tag{5.1}$$

is prescribed by the n numbers $f(\mathbf{v}_1), \dots, f(\mathbf{v}_n)$.

A functional $f(\mathbf{v}_1, \dots, \mathbf{v}_k)$ on an ordered set of vectors $\{\mathbf{v}_1, \dots, \mathbf{v}_k\}$ belonging to a vector space $\mathcal{V}$ is said to be **a multilinear functional** if it is linear in each entry separately; i.e.,

$$f(\mathbf{v}_1, \dots, \mathbf{v}_i + \mathbf{w}, \dots, \mathbf{v}_k) = f(\mathbf{v}_1, \dots, \mathbf{v}_i, \dots, \mathbf{v}_k) + f(\mathbf{v}_1, \dots, \mathbf{w}, \dots, \mathbf{v}_k)$$

for every integer i, $1 \le i \le k$, and

$$f(\mathbf{v}_1, \dots, \alpha\mathbf{v}_i, \dots, \mathbf{v}_k) = \alpha f(\mathbf{v}_1, \dots, \mathbf{v}_i, \dots, \mathbf{v}_k)$$

for every $\alpha \in \mathbb{F}$. Notice that if, say, $k = 3$, then this implies that

$$\begin{aligned} f(\mathbf{v}_1 + \mathbf{w}_1, \mathbf{v}_2 + \mathbf{w}_2, \mathbf{v}_3) &= f(\mathbf{v}_1, \mathbf{v}_2 + \mathbf{w}_2, \mathbf{v}_3) + f(\mathbf{w}_1, \mathbf{v}_2 + \mathbf{w}_2, \mathbf{v}_3) \\ &= f(\mathbf{v}_1, \mathbf{v}_2, \mathbf{v}_3) + f(\mathbf{v}_1, \mathbf{w}_2, \mathbf{v}_3) \\ &+ f(\mathbf{w}_1, \mathbf{v}_2, \mathbf{v}_3) + f(\mathbf{w}_1, \mathbf{w}_2, \mathbf{v}_3) \end{aligned}$$

and

$$f(\alpha\mathbf{v}_1, \alpha\mathbf{v}_2, \mathbf{v}_3) = \alpha^2 f(\mathbf{v}_1, \mathbf{v}_2, \mathbf{v}_3) \,.$$

5.2. Determinants

Let Σ_n denote the set of all the $n!$ one to one mappings σ of the set of integers $\{1, \dots, n\}$ onto itself and let $\mathbf{e}_i$ denote the i'th column of the identity matrix I_n. Then the formula

$$P_\sigma = \sum_{i=1}^{n} \mathbf{e}_i \mathbf{e}_{\sigma(i)}^T = \begin{bmatrix} \mathbf{e}_{\sigma(1)}^T \\ \vdots \\ \mathbf{e}_{\sigma(n)}^T \end{bmatrix}$$

that was introduced earlier defines a one to one correspondence between the set of all $n \times n$ permutation matrices P_σ and the set Σ_n. A permutation $P_\sigma \in \mathbb{R}^{n \times n}$ with $n \ge 2$ is said to be **simple** if σ interchanges exactly two of the integers in the set $\{1, \dots, n\}$ and leaves the rest alone; i.e., an $n \times n$ permutation matrix P is simple if and only if it can be expressed as

$$P = \sum_{j \in \Lambda} \mathbf{e}_j \mathbf{e}_j^T + \mathbf{e}_{i_1} \mathbf{e}_{i_2}^T + \mathbf{e}_{i_2} \mathbf{e}_{i_1}^T \,,$$

where $\Lambda = \{1, \dots, n\} \setminus \{i_1, i_2\}$ and i_1 and i_2 are distinct integers between 1 and n.

Exercise 5.1. Show that if P is a simple permutation, then $P = P^T$.

Theorem 5.1. *There is exactly one way of assigning a complex number $d(A)$ to each complex $n \times n$ matrix A that meets the following three requirements:*

1° $d(I_n) = 1$.

2° $d(PA) = -d(A)$ *for every simple permutation matrix P.*

3° $d(A)$ *is a multilinear functional of the rows of A.*

Discussion. The first two of these requirements are easily understood. The third is perhaps best visualized by example. Thus, if

$$A = \begin{bmatrix} a_{11} & a_{12} & a_{13} \\ a_{21} & a_{22} & a_{23} \\ a_{31} & a_{32} & a_{33} \end{bmatrix},$$

then, since

$$\begin{bmatrix} a_{11} & a_{12} & a_{13} \end{bmatrix} = \begin{bmatrix} a_{11} & 0 & 0 \end{bmatrix} + \begin{bmatrix} 0 & a_{12} & 0 \end{bmatrix} + \begin{bmatrix} 0 & 0 & a_{13} \end{bmatrix},$$

rule 3° applied to the top row of A implies that

$$\begin{aligned} d(A) &= d\left(\begin{bmatrix} a_{11} & 0 & 0 \\ a_{21} & a_{22} & a_{23} \\ a_{31} & a_{32} & a_{33} \end{bmatrix}\right) + d\left(\begin{bmatrix} 0 & a_{12} & 0 \\ a_{21} & a_{22} & a_{23} \\ a_{31} & a_{32} & a_{33} \end{bmatrix}\right) \\ &\quad + d\left(\begin{bmatrix} 0 & 0 & a_{13} \\ a_{21} & a_{22} & a_{23} \\ a_{31} & a_{32} & a_{33} \end{bmatrix}\right) \\ &= a_{11} d\left(\begin{bmatrix} 1 & 0 & 0 \\ a_{21} & a_{22} & a_{23} \\ a_{31} & a_{32} & a_{33} \end{bmatrix}\right) + a_{12} d\left(\begin{bmatrix} 0 & 1 & 0 \\ a_{21} & a_{22} & a_{23} \\ a_{31} & a_{32} & a_{33} \end{bmatrix}\right) \\ &\quad + a_{13} d\left(\begin{bmatrix} 0 & 0 & 1 \\ a_{21} & a_{22} & a_{23} \\ a_{31} & a_{32} & a_{33} \end{bmatrix}\right). \end{aligned}$$

This last formula can be rewritten more efficiently by invoking the notation $\vec{\mathbf{a}}_i$ and $\vec{\mathbf{e}}_i = \mathbf{e}_i^T$, for the i'th row of the matrix A and the i'th row of the identity matrix I_3, respectively, as

$$d(A) = \sum_{i=1}^{3} a_{1i} d\left(\begin{bmatrix} \vec{\mathbf{e}}_i \\ \vec{\mathbf{a}}_2 \\ \vec{\mathbf{a}}_3 \end{bmatrix}\right).$$

Moreover, since

$$\vec{\mathbf{a}}_2 = \sum_{j=1}^{3} a_{2j} \vec{\mathbf{e}}_j \quad \text{and} \quad \vec{\mathbf{a}}_3 = \sum_{k=1}^{3} a_{3k} \vec{\mathbf{e}}_k,$$

another two applications of rule 3° lead to the formula

$$\begin{aligned} d(A) &= \sum_{i=1}^{3} a_{1i} \left\{ \sum_{j=1}^{3} a_{2j} d \left(\begin{bmatrix} \vec{\mathbf{e}}_i \\ \vec{\mathbf{e}}_j \\ \vec{\mathbf{a}}_3 \end{bmatrix} \right) \right\} \\ &= \sum_{i=1}^{3} a_{1i} \left\{ \sum_{j=1}^{3} a_{2j} \left[\sum_{k=1}^{3} a_{3k} d \left(\begin{bmatrix} \vec{\mathbf{e}}_i \\ \vec{\mathbf{e}}_j \\ \vec{\mathbf{e}}_k \end{bmatrix} \right) \right] \right\}, \end{aligned}$$

which is an explicit formula for $d(A)$ in terms of the entries a_{st} in the matrix A and the numbers $d\left(\begin{bmatrix} \vec{\mathbf{e}}_i \\ \vec{\mathbf{e}}_j \\ \vec{\mathbf{e}}_k \end{bmatrix} \right)$, which in fact are equal to 0 if one or more of the rows coincide, thanks to the next lemma. Granting this fact for the moment, the last expression simplifies to

$$d(A) = \sum_{\sigma \in \Sigma_3} a_{1\sigma(1)} a_{2\sigma(2)} a_{3\sigma(3)} d \left(\begin{bmatrix} \mathbf{e}_{\sigma(1)}^T \\ \mathbf{e}_{\sigma(2)}^T \\ \mathbf{e}_{\sigma(3)}^T \end{bmatrix} \right),$$

where, as noted earlier, Σ_n denotes the set of all the $n!$ one to one mappings of the set $\{1, \ldots, n\}$ onto itself.

It is pretty clear that analogous formulas hold for $A \in \mathbb{C}^{n \times n}$ for every positive integer n:

(5.2)

$$d(A) = \sum_{\sigma \in \Sigma_n} a_{1\sigma(1)} \cdots a_{n\sigma(n)} d \left(\begin{bmatrix} \mathbf{e}_{\sigma(1)}^T \\ \vdots \\ \mathbf{e}_{\sigma(n)}^T \end{bmatrix} \right) = \sum_{\sigma \in \Sigma_n} a_{1\sigma(1)} \cdots a_{n\sigma(n)} d(P_\sigma).$$

Moreover, if P_σ is equal to the product of k simple permutations, then

$$d(P_\sigma) = (-1)^k d(I_n) = (-1)^k.$$

The unique number $d(A)$ that is determined by the three conditions in Theorem 5.1 is called the **determinant** of A and will be denoted $\det(A)$ or $\det A$ from now on.

Exercise 5.2. Use the three rules in Theorem 5.1 to show that if $A \in \mathbb{C}^{2 \times 2}$, then $\det A = a_{11}a_{22} - a_{12}a_{21}$.

Exercise 5.3. Use the three rules in Theorem 5.1 to show that if $A \in \mathbb{C}^{3 \times 3}$, then

$$\det A = a_{11}a_{22}a_{33} - a_{11}a_{23}a_{32} + a_{12}a_{23}a_{31} - a_{12}a_{21}a_{33} + a_{13}a_{21}a_{32} - a_{13}a_{22}a_{31}.$$

5.3. Useful rules for calculating determinants

Lemma 5.2. *The determinant of a matrix* $A \in \mathbb{C}^{n\times n}$ *satisfies the following rules:*

4° *If two rows of* A *are identical, then* $\det A = 0$.

5° *If* B *is the matrix that is obtained by adding a multiple of one row of* A *to another row of* A, *then* $\det B = \det A$.

6° *If* A *has a row in which all the entries are equal to zero, then* $\det A = 0$.

7° *If two rows of* A *are linearly dependent, then* $\det A = 0$.

Discussion. Rules 4°–7° are fairly easy consequences of 1°–3°, especially if you tackle them in the order that they are listed. Thus, for example, if two rows of A match and if P denotes the permutation that interchanges these two rows, then $A = PA$ and hence

$$\det A = \det(PA) = -\det A\,,$$

which clearly justifies 4°.

Rule 5° is most easily understood by example: If, say,

$$A = \begin{bmatrix} \vec{\mathbf{a}}_1 \\ \vec{\mathbf{a}}_2 \\ \vec{\mathbf{a}}_3 \\ \vec{\mathbf{a}}_4 \end{bmatrix} \quad \text{and} \quad B = A + \alpha \mathbf{e}_2\, \vec{\mathbf{a}}_4 = \begin{bmatrix} \vec{\mathbf{a}}_1 \\ \vec{\mathbf{a}}_2 + \alpha\, \vec{\mathbf{a}}_4 \\ \vec{\mathbf{a}}_3 \\ \vec{\mathbf{a}}_4 \end{bmatrix},$$

then

$$\det B = \det \begin{bmatrix} \vec{\mathbf{a}}_1 \\ \vec{\mathbf{a}}_2 \\ \vec{\mathbf{a}}_3 \\ \vec{\mathbf{a}}_4 \end{bmatrix} + \alpha \det \begin{bmatrix} \vec{\mathbf{a}}_1 \\ \vec{\mathbf{a}}_4 \\ \vec{\mathbf{a}}_3 \\ \vec{\mathbf{a}}_4 \end{bmatrix} = \det A + 0\,.$$

Rule 6° is left to the reader. Rule 7° follows from 6° and if, say, $n = 4$, $\alpha\mathbf{a}_3 + \beta\mathbf{a}_4 = \mathbf{0}$ and $\alpha \neq \mathbf{0}$, the observation that

$$\alpha \det A = \det \begin{bmatrix} \vec{\mathbf{a}}_1 \\ \vec{\mathbf{a}}_2 \\ \alpha\, \vec{\mathbf{a}}_3 \\ \vec{\mathbf{a}}_4 \end{bmatrix} = \det \begin{bmatrix} \vec{\mathbf{a}}_1 \\ \vec{\mathbf{a}}_2 \\ \alpha\, \vec{\mathbf{a}}_3 + \beta\, \vec{\mathbf{a}}_4 \\ \vec{\mathbf{a}}_4 \end{bmatrix} = 0\,.$$

A number of supplementary rules that are useful to calculate determinants will now be itemized in numbers running from 8° to 13°, interspersed with discussion.

8° If $A \in \mathbb{C}^{n\times n}$ is either upper triangular or lower triangular, then

$$\det A = a_{11}\cdots a_{nn}.$$

Discussion. To clarify 8°, suppose for example that

$$A = \begin{bmatrix} a_{11} & a_{12} & a_{13} \\ 0 & a_{22} & a_{23} \\ 0 & 0 & a_{33} \end{bmatrix}.$$

Then by successive applications of rules 3° and 5°, we obtain

$$\begin{aligned} \det A &= a_{33} \det \begin{bmatrix} a_{11} & a_{12} & a_{13} \\ 0 & a_{22} & a_{23} \\ 0 & 0 & 1 \end{bmatrix} = a_{33} \det \begin{bmatrix} a_{11} & a_{12} & 0 \\ 0 & a_{22} & 0 \\ 0 & 0 & 1 \end{bmatrix} \\ &= a_{33}a_{22} \det \begin{bmatrix} a_{11} & a_{12} & 0 \\ 0 & 1 & 0 \\ 0 & 0 & 1 \end{bmatrix} = a_{33}a_{22} \det \begin{bmatrix} a_{11} & 0 & 0 \\ 0 & 1 & 0 \\ 0 & 0 & 1 \end{bmatrix} \\ &= a_{33}a_{22}a_{11} \det I_3. \end{aligned}$$

Thus, in view of rule 1°,

$$\det A = a_{11}a_{22}a_{33},$$

as claimed. Much the same sort of argument works for lower triangular matrices, except then it is more convenient to work from the top row down rather than from the bottom row up.

Lemma 5.3. *If $E \in \mathbb{C}^{n\times n}$ is a lower triangular matrix with ones on the diagonal, then*

$$\det(EA) = \det A \tag{5.3}$$

for every $A \in \mathbb{C}^{n\times n}$.

Discussion. Rule 5° implies that $\det(EA) = \det A$ if E is a lower triangular matrix with ones on the diagonal and exactly one nonzero entry below the diagonal. But this is enough, since a general lower triangular matrix E with ones on the diagonal can be expressed as the product $E = E_1 \cdots E_k$ of k matrices with ones on the diagonal and exactly one nonzero entry below the diagonal, as in the next exercise. □

Exercise 5.4. Let

$$E = \begin{bmatrix} 1 & 0 & 0 & 0 \\ \alpha_{21} & 1 & 0 & 0 \\ \alpha_{31} & \alpha_{32} & 1 & 0 \\ \alpha_{41} & \alpha_{42} & \alpha_{43} & 1 \end{bmatrix}$$

and let $\mathbf{e}_i$, $i = 1, \ldots, 4$ denote the standard basis for $\mathbb{C}^4$. Show that

$$\begin{aligned} E &= (I_4 + \alpha_{21}\mathbf{e}_2\mathbf{e}_1^T)(I_4 + \alpha_{31}\mathbf{e}_3\mathbf{e}_1^T)(I_4 + \alpha_{41}\mathbf{e}_4\mathbf{e}_1^T) \\ &\times (I_4 + \alpha_{32}\mathbf{e}_3\mathbf{e}_2^T)(I_4 + \alpha_{42}\mathbf{e}_4\mathbf{e}_2^T)(I_4 + \alpha_{43}\mathbf{e}_4\mathbf{e}_3^T). \end{aligned}$$

9° If $A \in \mathbb{C}^{n\times n}$, then A is invertible if and only if $\det(A) \neq 0$.

Proof. In the usual notation, let

$$U = EPA \tag{5.4}$$

be in upper echelon form. Then U is automatically upper triangular (since it is square in this application) and, by the preceding rules,

$$\det(EPA) = \det(PA) = \pm \det A\,.$$

Therefore,

$$|\det A| = |\det U| = |u_{11} \cdots u_{nn}|\,.$$

But this serves to establish the assertion, since

$$A \text{ is invertible } \iff U \text{ is invertible}$$

and

$$U \text{ is invertible } \iff u_{11} \cdots u_{nn} \neq 0.$$

□

10° If $A, B \in \mathbb{C}^{n\times n}$, then $\det(AB) = \det\, A \det\, B = \det(BA)$.

Proof. If $\det\, B = 0$, then the asserted identities are immediate from rule 9°, since B, AB and BA are then all noninvertible matrices.

If $\det\, B \neq 0$, set

$$\varphi(A) = \frac{\det(AB)}{\det\, B}$$

and check that $\varphi(A)$ meets rules 1°– 3°. Then

$$\varphi(A) = \det\, A\,,$$

since there is only one functional that meets these three conditions, i.e.,

$$\det(AB) = \det\, A \det\, B\,,$$

as claimed. Now, having this last formula for every choice of A and B, invertible or not, we can interchange the roles of A and B to obtain

$$\det(BA) = \det\, B \det\, A = \det\, A \det\, B = \det(AB)\,.$$

□

Exercise 5.5. Show that if $\det\, B \neq 0$, then the functional $\varphi(A) = \dfrac{\det(AB)}{\det\, B}$ meets conditions 1°–3°. [HINT: To verify 3°, observe that if $\vec{a}_1, \ldots, \vec{a}_n$ designate the rows of A, then the rows of AB are $\vec{a}_1 B, \ldots, \vec{a}_n B$.]

11° If $A \in \mathbb{C}^{n\times n}$ and A is invertible, then $\det(A^{-1}) = \{\det\, A\}^{-1}$.

Proof. Invoke rule 10° and the formula $\det(AA^{-1}) = \det(I_n) = 1$. □

12° If $A \in \mathbb{C}^{n\times n}$, then $\det(A) = \det(A^T)$.

Proof. Invoking the formula $EPA = U$ and rules 10° and 8°, we see that

$$\det(P)\det(A) = \det(U) = u_{11}\cdots u_{nn}.$$

Next, another application of these rules to the transposed formula

$$A^T P^T E^T = U^T$$

leads to the formulas

$$\det(A^T)\det(P^T) = \det(U^T) = u_{11}\cdots u_{nn}.$$

But now as P can be written as the product

$$P = P_1 \cdots P_k$$

of simple permutations, it follows that

$$P^T = P_k^T \cdots P_1^T = P_k \cdots P_1$$

is again the product of k simple permutations. Therefore,

$$\det(P) = (-1)^k = \det(P^T)$$

and hence

$$\det(A) = (-1)^k u_{11}\cdots u_{nn} = \det(A^T),$$

as claimed. □

13° If $A \in \mathbb{C}^{n\times n}$, then rules 3° to 7° remain valid if the word *rows* is replaced by the word *columns* and the row interchange in rule 2° is replaced by a column interchange.

Proof. This is an easy consequence of 12°. The details are left to the reader. □

Exercise 5.6. Complete the proof of 13°.

Exercise 5.7. Calculate the determinants of the following matrices by Gaussian elimination:

$$\begin{bmatrix} 1 & 3 & 2 & 1 \\ 0 & 4 & 1 & 6 \\ 0 & 0 & 2 & 1 \\ 1 & 1 & 0 & 4 \end{bmatrix}, \quad \begin{bmatrix} 1 & 0 & 1 & 0 \\ 0 & 1 & 0 & 1 \\ 1 & 0 & 0 & 1 \\ 0 & 1 & 1 & 0 \end{bmatrix}, \quad \begin{bmatrix} 1 & 3 & 2 & 4 \\ 0 & 2 & 1 & 6 \\ 0 & 0 & 3 & 0 \\ 0 & 0 & 1 & 2 \end{bmatrix}, \quad \begin{bmatrix} 0 & 0 & 0 & 4 \\ 1 & 2 & 3 & 1 \\ 0 & 0 & 1 & 1 \\ 0 & 1 & 2 & 6 \end{bmatrix}.$$

[HINT: If, in the usual notation, $EPA = U$, then $|\det A| = |\det U|$.]

Exercise 5.8. Calculate the determinants of the matrices in the previous exercise by rules 1° to 13°.

5.4. Eigenvalues

Determinants play a useful role in calculating the eigenvalues of a matrix $A \in \mathbb{F}^{n\times n}$. In particular, if $A = UJU^{-1}$, where J is in Jordan form, then

$$\det(\lambda I_n - A) = \det(\lambda I_n - UJU^{-1}) = \det\left\{U(\lambda I_n - J)U^{-1}\right\}.$$

Therefore, by rules 10°, 11° and 8°, applied in that order,

$$\det(\lambda I_n - A) = \det(\lambda I_n - J) = (\lambda - j_{11})(\lambda - j_{22})\cdots(\lambda - j_{nn}),$$

where j_{ii}, $i = 1, \dots, n$, are the diagonal entries of J. The polynomial

$$p(\lambda) = \det(\lambda I_n - A) \tag{5.5}$$

is termed the **characteristic polynomial** of A. In particular, **a number λ is an eigenvalue of the matrix A if and only if** $p(\lambda) = 0$. Thus, for example, to find the eigenvalues of the matrix

$$A = \begin{bmatrix} 1 & 2 \\ 2 & 1 \end{bmatrix},$$

look for the roots of the polynomial

$$\det(\lambda I_2 - A) = (\lambda - 1)^2 - 2^2 = \lambda^2 - 2\lambda - 3.$$

This leads readily to the conclusion that the eigenvalues of the given matrix A are $\lambda_1 = 3$ and $\lambda_2 = -1$. Moreover, if $J = \text{diag}\{3, -1\}$, then

$$\begin{aligned} A^2 - 2A - 3I_2 &= (A - 3I_2)(A + I_2) = U(J - 3I_2)U^{-1}U(J + I_2)U^{-1} \\ &= U\begin{bmatrix} 0 & 0 \\ 0 & -4 \end{bmatrix}\begin{bmatrix} 4 & 0 \\ 0 & 0 \end{bmatrix}U^{-1} = U\begin{bmatrix} 0 & 0 \\ 0 & 0 \end{bmatrix}U^{-1}, \end{aligned}$$

which yields the far from obvious conclusion

$$A^2 - 2A - 3I_2 = O.$$

The argument propogates: If $\lambda_1, \dots, \lambda_k$ denote the distinct eigenvalues of A, and if α_i denotes the algebraic multiplicity of the eigenvalue λ_i, $i = 1, \dots, k$, then the characteristic polynomial can be written in the more revealing form

$$p(\lambda) = (\lambda - \lambda_1)^{\alpha_1}(\lambda - \lambda_2)^{\alpha_2}\cdots(\lambda - \lambda_k)^{\alpha_k}. \tag{5.6}$$

It is readily checked that

$$\begin{aligned} p(A) &= (A - \lambda_1 I_n)^{\alpha_1}(A - \lambda_2 I_n)^{\alpha_2}\cdots(A - \lambda_k I_n)^{\alpha_k} \\ &= U(J - \lambda_1 I_n)^{\alpha_1}(J - \lambda_2 I_n)^{\alpha_2}\cdots(J - \lambda_k I_n)^{\alpha_k}U^{-1} \\ &= O. \end{aligned}$$

This serves to justify the Cayley-Hamilton theorem that was referred to in the discussion of Theorem 4.14. In more striking terms, the **Cayley-Hamilton** theorem states that

$$\begin{aligned} \det(\lambda I_n - A) &= a_0 + \cdots + a_{n-1}\lambda^{n-1} + \lambda^n \\ &\Longrightarrow a_0 I_n + \cdots + a_{n-1}A^{n-1} + A^n = O \end{aligned} \tag{5.7}$$

Exercise 5.9. Show that if $J = \operatorname{diag}\{C_{\lambda_1}^{(5)}, C_{\lambda_2}^{(3)}, C_{\lambda_3}^{(2)}\}$, then $(J - \lambda_1 I_{10})^5 (J - \lambda_2 I_{10})^3 (J - \lambda_3 I_{10})^2 = O$.

Exercise 5.10. Show that if $\lambda_1 = \lambda_2$ in Exercise 5.9, then $(J - \lambda_1 I_{10})^5 (J - \lambda_3 I_{10})^2 = O$.

Exercise 5.10 illustrates the fact that if ν_j, $j = 1, \ldots, k$, denotes the size of the largest Jordan cell in the matrix J with λ_j on its diagonal, then $p(A) = 0$ holds for the possibly lower degree polynomial

$$p_{min}(\lambda) = (\lambda - \lambda_1)^{\nu_1} (\lambda - \lambda_2)^{\nu_2} \cdots (\lambda - \lambda_k)^{\nu_k},$$

which is the **minimal polynomial** referred to in the discussion of Theorem 4.14:

$$\begin{aligned} p_{min}(A) &= (A - \lambda_1 I_n)^{\nu_1} (A - \lambda_2 I_n)^{\nu_2} \cdots (A - \lambda_k I_n)^{\nu_k} \\ &= U (J - \lambda_1 I_n)^{\nu_1} (J - \lambda_2 I_n)^{\nu_2} \cdots (J - \lambda_k I_n)^{\nu_k} U^{-1} \\ &= O. \end{aligned}$$

Two more useful formulas that emerge from this analysis are:

$$\det A = \lambda_1^{\alpha_1} \lambda_2^{\alpha_2} \cdots \lambda_k^{\alpha_k} \tag{5.8}$$

and

$$\operatorname{trace} A = \alpha_1 \lambda_1 + \alpha_2 \lambda_2 + \cdots + \alpha_k \lambda_k, \tag{5.9}$$

where the **trace** of an $n \times n$ matrix A is defined as the sum of its diagonal elements:

$$\operatorname{trace} A = a_{11} + a_{22} + \cdots + a_{nn}. \tag{5.10}$$

The verification of the last formula depends upon the fact that

$$\operatorname{trace}(AB) = \sum_{i=1}^{n} \sum_{j=1}^{n} a_{ij} b_{ji} = \operatorname{trace}(BA). \tag{5.11}$$

Thus, in particular,

$$\operatorname{trace} A = \operatorname{trace}(UJU^{-1}) = \operatorname{trace}(JU^{-1}U) = \operatorname{trace} J, \tag{5.12}$$

which leads easily to the stated result.

5.5. Exploiting block structure

The calculation of determinants is often simplified by taking advantage of block structure.

Lemma 5.4. *If $A \in \mathbb{C}^{n\times n}$ is block diagonal, i.e., if A is of the form*

$$A = \begin{bmatrix} A_{11} & O \\ O & A_{22} \end{bmatrix}$$

with square blocks $A_{11} \in \mathbb{C}^{p\times p}$ and $A_{22} \in \mathbb{C}^{q\times q}$ on the diagonal, then

$$\det A = \det A_{11} \det A_{22} .$$

Proof. The basic factorization formula of Gaussian elimination guarantees the existence of a pair of lower triangular matrices $E_1 \in \mathbb{C}^{p\times p}$, $E_2 \in \mathbb{C}^{q\times q}$ with ones on the diagonal and a pair of permutation matrices $P_1 \in \mathbb{C}^{p\times p}$, $P_2 \in \mathbb{C}^{q\times q}$ such that

$$E_1P_1A_{11} = U_{11} \quad \text{and} \quad E_2P_2A_{22} = U_{22}$$

are in upper echelon form and hence automatically upper triangular. Thus,

$$\det U_{11} = \det E_1P_1A_{11} = \det P_1A_{11} \quad \text{and} \quad \det U_{22} = \det P_2A_{22} .$$

Moreover, since

$$\begin{bmatrix} E_1 & O \\ O & E_2 \end{bmatrix}\begin{bmatrix} P_1 & O \\ O & P_2 \end{bmatrix}\begin{bmatrix} A_{11} & O \\ O & A_{22} \end{bmatrix} = \begin{bmatrix} U_{11} & O \\ O & U_{22} \end{bmatrix}$$

and

$$\begin{bmatrix} E_1 & O \\ O & E_2 \end{bmatrix} \quad \text{and} \quad \begin{bmatrix} U_{11} & O \\ O & U_{22} \end{bmatrix}$$

are both triangular, it follows that

$$\det \begin{bmatrix} E_1 & O \\ O & E_2 \end{bmatrix} = \det E_1 \det E_2 = 1$$

and

$$\det \begin{bmatrix} U_{11} & O \\ O & U_{22} \end{bmatrix} = \det U_{11} \det U_{22} .$$

Therefore,

$$\begin{aligned} \det \left\{ \begin{bmatrix} P_1 & O \\ O & P_2 \end{bmatrix}\begin{bmatrix} A_{11} & O \\ O & A_{22} \end{bmatrix} \right\} &= \det \begin{bmatrix} U_{11} & O \\ O & U_{22} \end{bmatrix} \\ &= \det U_{11} \det U_{22} \\ &= \det P_1 \det A_{11} \det P_2 \det A_{22} . \end{aligned}$$

The proof is completed by checking that

$$\det \begin{bmatrix} P_1 & O \\ O & P_2 \end{bmatrix} = \det \begin{bmatrix} P_1 & O \\ O & I_q \end{bmatrix} \det \begin{bmatrix} I_p & O \\ O & P_2 \end{bmatrix} = \det P_1 \det P_2 .$$

□

Theorem 5.5. *Let $A \in \mathbb{C}^{n\times n}$ be expressed in block form as*

$$A = \begin{bmatrix} A_{11} & A_{12} \\ A_{21} & A_{22} \end{bmatrix},$$

with square blocks $A_{11} \in \mathbb{C}^{k\times k}$ and $A_{22} \in \mathbb{C}^{(n-k)\times(n-k)}$ on the diagonal.

(1) *If A_{22} is invertible, then*

$$\det A = \det(A_{11} - A_{12}A_{22}^{-1}A_{21}) \det A_{22} \,. \tag{5.13}$$

(2) *If A_{11} is invertible, then*

$$\det A = \det(A_{22} - A_{21}A_{11}^{-1}A_{12}) \det A_{11} \,. \tag{5.14}$$

Proof. The first assertion follows easily from Lemma 5.4 and the identity for the Schur complement of A_{22} with respect to A:

$$A = \begin{bmatrix} I_k & A_{12}A_{22}^{-1} \\ O & I_{n-k} \end{bmatrix} \begin{bmatrix} A_{11} - A_{12}A_{22}^{-1}A_{21} & O \\ O & A_{22} \end{bmatrix} \begin{bmatrix} I_k & O \\ A_{22}^{-1}A_{21} & I_{n-k} \end{bmatrix},$$

which is valid when A_{22} is invertible; the second rests on the identity

$$A = \begin{bmatrix} I_k & O \\ A_{21}A_{11}^{-1} & I_{n-k} \end{bmatrix} \begin{bmatrix} A_{11} & O \\ O & A_{22} - A_{21}A_{11}^{-1}A_{12} \end{bmatrix} \begin{bmatrix} I_k & A_{11}^{-1}A_{12} \\ O & I_{n-k} \end{bmatrix},$$

which is valid when A_{11} is invertible. □

Corollary 5.6. If A is block triangular, i.e., if

$$A = \begin{bmatrix} A_{11} & O \\ A_{21} & A_{22} \end{bmatrix} \text{ or } A = \begin{bmatrix} A_{11} & A_{12} \\ O & A_{22} \end{bmatrix}$$

with square diagonal blocks, then

$$\det A = \det A_{11} \det A_{22} \,.$$

Proof. Observe first that

$$\begin{bmatrix} A_{11} & A_{12} \\ O & A_{22} \end{bmatrix} = \begin{bmatrix} I_k & A_{12} \\ O & A_{22} \end{bmatrix} \begin{bmatrix} A_{11} & O \\ O & I_{n-k} \end{bmatrix}$$

and then invoke (2) of Theorem 5.5 to calculate the determinant of the first matrix on the right and (1) of the same theorem to calculate the determinant of the second matrix on the right. It is important to keep in mind that (1) and (2) cannot be invoked directly, because the diagonal blocks A_{11} and A_{22} are not assumed to be invertible.

The formula for the determinant of the second block matrix can be verified in much the same way or, what is even easier, by noting that the transpose of the second block matrix has the same block form as the first and then invoking the fact that $\det A = \det A^T$. □

Exercise 5.11. Show that if $A, B \in \mathbb{C}^{n\times n}$ are expressed in compatible four block form with $A_{11}, B_{11} \in \mathbb{C}^{k\times k}$, $A_{22}, B_{22} \in \mathbb{C}^{(n-k)\times(n-k)}$ and if $AB = I_n$, then

(5.15)
$$\det A_{11} \neq 0 \Longleftrightarrow \det B_{22} \neq 0 \quad \text{and} \quad \det A_{22} \neq 0 \Longleftrightarrow \det B_{11} \neq 0\,.$$

[HINT: $A_{11}\mathbf{x} = \mathbf{0} \Longrightarrow B_{22}A_{21}\mathbf{x} = \mathbf{0}$, $B_{22}\mathbf{y} = \mathbf{0} \Longrightarrow A_{11}B_{12}\mathbf{y} = \mathbf{0}$ etc.]

Exercise 5.12. Show that if $A, B \in \mathbb{C}^{n\times n}$ are expressed in compatible four block form with $A_{11}, B_{11} \in \mathbb{C}^{k\times k}$, $A_{22}, B_{22} \in \mathbb{C}^{(n-k)\times(n-k)}$ and if $AB = I_n$, then

$$\det B_{22} = \frac{\det A_{11}}{\det A} \quad \text{and} \quad \det B_{22} = \frac{\det A_{22}}{\det A}\,. \tag{5.16}$$

[HINT: In view of Exercise 5.11, it only remains to consider the cases $\det A_{11} \neq 0$ and $\det A_{22} \neq 0$, respectively.]

Remark 5.7. The identities in (5.16) are special cases of a more general formula due to Jacobi that expresses the determinant of an arbitrary square submatrix of A^{-1} in terms of the determinant of the complementary submatrix of A and $\det A$; see e.g., Exercise 5.21.

Exercise 5.13. Show that if B is a $p \times q$ matrix and C is a $q \times p$ matrix, then

$$\det\{I_p - BC\} = \det\{I_q - CB\}$$

and that

$$q + \operatorname{rank}\{I_p - BC\} = p + \operatorname{rank}\{I_q - CB\}\,.$$

[HINT: Imbed B and C appropriately in a $(p+q)\times(p+q)$ matrix and then exploit Theorem 5.5.]

Exercise 5.14. Show that if $A \in \mathbb{C}^{p\times q}$ and $B \in \mathbb{C}^{q\times p}$, then

$$\det(\lambda I_p - AB) = \lambda^{p-q}\det(\lambda I_q - BA) \quad \text{if} \quad \lambda \neq 0\,. \tag{5.17}$$

Exercise 5.15. Show that if $\mathbf{u} \in \mathbb{C}^p$, then $\det(I_p - \mathbf{u}\mathbf{u}^H) \neq 0$ if and only if $\mathbf{u}^H\mathbf{u} \neq 1$.

Exercise 5.16. Let $A \in \mathbb{C}^{n\times n}$ be invertible and let $\mathbf{u}$, $\mathbf{v} \in \mathbb{C}^n$. Show that the matrix $A + \mathbf{u}\mathbf{v}^T$ is invertible if and only if $1 + \mathbf{v}^T A^{-1}\mathbf{u} \neq 0$.

Exercise 5.17. Calculate the determinant of the matrix

$$A = \begin{bmatrix} 1 & 1 & 1 & 1 & 0 & 0 & 0 \\ 0 & 2 & 2 & 2 & 0 & 0 & 0 \\ 0 & 0 & 3 & 3 & 0 & 0 & 0 \\ 0 & 0 & 0 & 4 & 0 & 0 & 0 \\ 9 & 8 & 7 & 6 & 1 & 2 & 3 \\ 1 & 5 & 9 & 3 & 0 & 4 & 1 \\ 8 & 8 & 8 & 6 & 0 & 2 & 2 \end{bmatrix}$$

[HINT: This is easy if you exploit the block triangular structure of A.]

Exercise 5.18. Calculate the determinant of the matrix $\begin{bmatrix} O & I_n \\ I_k & O \end{bmatrix}$.

5.6. The Binet-Cauchy formula

The Binet-Cauchy formula is a useful tool for calculating the determinant of the product AB of a matrix $A \in \mathbb{C}^{k\times n}$ and a matrix $B \in \mathbb{C}^{n\times k}$ in terms of the determinants of certain square subblocks of A and B when $n \geq k$. The notation

$$C\begin{pmatrix} i_1, \ldots, i_k \\ j_1, \ldots, j_k \end{pmatrix} = \det \begin{bmatrix} c_{i_1j_1} & c_{i_1j_2} & \cdots & c_{i_1j_k} \\ \vdots & \vdots & & \vdots \\ c_{i_kj_1} & c_{i_kj_2} & \cdots & c_{i_kj_k} \end{bmatrix}$$

for the determinant of the indicated $k \times k$ subblock of the matrix C with entries taken from rows $i_1, \ldots, i_k$ and columns $j_1, \ldots, j_k$ of the matrix C will be needed. To amplify further, if, say, C is a 7×5 matrix, then

$$C\begin{pmatrix} 2,3,6 \\ 1,4,5 \end{pmatrix} = \det \begin{bmatrix} c_{21} & c_{24} & c_{25} \\ c_{31} & c_{34} & c_{35} \\ c_{61} & c_{64} & c_{65} \end{bmatrix}.$$

Theorem 5.8. *If $A \in \mathbb{C}^{k\times n}$, $B \in \mathbb{C}^{n\times k}$ and $k \leq n$, then*

$$\det AB = \sum_{1\leq j_1<j_2<\cdots<j_k\leq n} A\begin{pmatrix} 1, \ldots, k \\ j_1, \ldots, j_k \end{pmatrix} B\begin{pmatrix} j_1, \ldots, j_k \\ 1, \ldots, k \end{pmatrix}.$$

Discussion. The proof consists of two steps. The first is to exploit the identities

$$\begin{aligned} \begin{bmatrix} A & O_{k\times k} \\ -I_n & B \end{bmatrix} \begin{bmatrix} I_n & B \\ O_{k\times n} & I_k \end{bmatrix} &= \begin{bmatrix} A & AB \\ -I_n & O_{n\times k} \end{bmatrix} \\ &= \begin{bmatrix} AB & A \\ O_{n\times k} & -I_n \end{bmatrix} \begin{bmatrix} O & I_n \\ I_k & O \end{bmatrix} \end{aligned}$$

to obtain the formula

$$\begin{aligned} \det \begin{bmatrix} A & O_{k\times k} \\ -I_n & B \end{bmatrix} &= \det \begin{bmatrix} AB & A \\ O_{n\times k} & -I_n \end{bmatrix} \det \begin{bmatrix} O & I_n \\ I_k & O \end{bmatrix} \\ &= (-1)^{nk} \det \begin{bmatrix} AB & A \\ O_{n\times k} & -I_n \end{bmatrix} \\ &= (-1)^{nk} \det AB \, \det(-I_n) \\ &= (-1)^{n(k+1)} \det AB \end{aligned}$$

or, equivalently,

$$\det AB = (-1)^{n(k+1)} \det \begin{bmatrix} A & O_{k\times k} \\ -I_n & B \end{bmatrix} . \tag{5.18}$$

The second step is to use the multilinearity of determinants to help evaluate the determinant of the matrix on the right. Thus, for example, if

$A = [\mathbf{a}_1 \quad \mathbf{a}_2 \quad \mathbf{a}_3]$ is a 2×3 matrix, with columns $\mathbf{a}_1, \mathbf{a}_2, \mathbf{a}_3$ in $\mathbb{C}^2$,

$C = [\mathbf{c}_1 \quad \mathbf{c}_2 \quad \mathbf{c}_3]$ is a 3×3 matrix, with columns $\mathbf{c}_1, \mathbf{c}_2, \mathbf{c}_3$ in $\mathbb{C}^3$

and

$B = [\mathbf{b}_1 \quad \mathbf{b}_2]$ is a 3×2 matrix, with columns $\mathbf{b}_1, \mathbf{b}_2$ in $\mathbb{C}^3$,

then, since the determinant of the full matrix is a multilinear functional of its columns,

$$\begin{aligned} \det \begin{bmatrix} A & O_{2\times 2} \\ C & B \end{bmatrix} &= \det \begin{bmatrix} \mathbf{a}_1 & \mathbf{a}_2 & \mathbf{a}_3 & \mathbf{0} & \mathbf{0} \\ \mathbf{c}_1 & \mathbf{c}_2 & \mathbf{c}_3 & \mathbf{b}_1 & \mathbf{b}_2 \end{bmatrix} \\ &= ① + ② , \end{aligned}$$

where

$$① = \det \begin{bmatrix} \mathbf{a}_1 & \mathbf{a}_2 & \mathbf{0} & \mathbf{0} & \mathbf{0} \\ \mathbf{c}_1 & \mathbf{c}_2 & \mathbf{c}_3 & \mathbf{b}_1 & \mathbf{b}_2 \end{bmatrix} = \det \begin{bmatrix} \mathbf{a}_1 & \mathbf{a}_2 \end{bmatrix} \det \begin{bmatrix} \mathbf{c}_3 & \mathbf{b}_1 & \mathbf{b}_2 \end{bmatrix} ,$$

in view of Corollary 5.6, and

$$\begin{aligned} ② &= \det \begin{bmatrix} \mathbf{a}_1 & \mathbf{a}_2 & \mathbf{a}_3 & \mathbf{0} & \mathbf{0} \\ \mathbf{c}_1 & \mathbf{c}_2 & \mathbf{0} & \mathbf{b}_1 & \mathbf{b}_2 \end{bmatrix} \\ &= \det \begin{bmatrix} \mathbf{a}_1 & \mathbf{a}_2 & \mathbf{a}_3 & \mathbf{0} & \mathbf{0} \\ \mathbf{0} & \mathbf{c}_2 & \mathbf{0} & \mathbf{b}_1 & \mathbf{b}_2 \end{bmatrix} + \det \begin{bmatrix} \mathbf{0} & \mathbf{a}_2 & \mathbf{a}_3 & \mathbf{0} & \mathbf{0} \\ \mathbf{c}_1 & \mathbf{c}_2 & \mathbf{0} & \mathbf{b}_1 & \mathbf{b}_2 \end{bmatrix} \\ &= -\det \begin{bmatrix} \mathbf{a}_1 & \mathbf{a}_3 & \mathbf{a}_2 & \mathbf{0} & \mathbf{0} \\ \mathbf{0} & \mathbf{0} & \mathbf{c}_2 & \mathbf{b}_1 & \mathbf{b}_2 \end{bmatrix} + \det \begin{bmatrix} \mathbf{a}_2 & \mathbf{a}_3 & \mathbf{0} & \mathbf{0} & \mathbf{0} \\ \mathbf{c}_2 & \mathbf{0} & \mathbf{c}_1 & \mathbf{b}_1 & \mathbf{b}_2 \end{bmatrix} \\ &= -\det \begin{bmatrix} \mathbf{a}_1 & \mathbf{a}_3 \end{bmatrix} \det \begin{bmatrix} \mathbf{c}_2 & \mathbf{b}_1 & \mathbf{b}_2 \end{bmatrix} \\ &\qquad + \det \begin{bmatrix} \mathbf{a}_2 & \mathbf{a}_3 \end{bmatrix} \det \begin{bmatrix} \mathbf{c}_1 & \mathbf{b}_1 & \mathbf{b}_2 \end{bmatrix} . \end{aligned}$$

The next step is to show that if $C_3 = -I_3$, then

$$① = -A\begin{pmatrix} 1,2 \\ 1,2 \end{pmatrix} B\begin{pmatrix} 1,2 \\ 1,2 \end{pmatrix} \tag{5.19}$$

and

$$② = -A\begin{pmatrix} 1,2 \\ 1,3 \end{pmatrix} B\begin{pmatrix} 1,3 \\ 1,2 \end{pmatrix} - A\begin{pmatrix} 1,2 \\ 2,3 \end{pmatrix} B\begin{pmatrix} 2,3 \\ 1,2 \end{pmatrix} . \tag{5.20}$$

The Binet-Cauchy formula for this example is now easily completed by combining formulas (5.18)–(5.20).

Exercise 5.19. Verify formulas (5.19) and (5.20).

Exercise 5.20. Write the Binet–Cauchy formula for det AB for $A \in \mathbb{C}^{2\times 3}$ and $B \in \mathbb{C}^{2\times 3}$ in terms of the entries a_{ij} of A and b_{ij} of B. **Do not compute the relevant determinants.**

Exercise 5.21. Show that if $A, B \in \mathbb{C}^{5\times 5}$ and $AB = I_5$, then

(5.21)

$$B\begin{pmatrix} 2 & 3 & 5 \\ 1 & 4 & 5 \end{pmatrix} = \frac{A\begin{pmatrix} 1 & 4 \\ 2 & 3 \end{pmatrix}}{\det A} \quad \text{and} \quad B\begin{pmatrix} 2 & 5 \\ 1 & 3 \end{pmatrix} = -\frac{A\begin{pmatrix} 1 & 3 & 4 \\ 2 & 4 & 5 \end{pmatrix}}{\det A}.$$

5.7. Minors

The ij **minor** $A_{\{ij\}}$ of a matrix $A \in \mathbb{C}^{n\times n}$ is defined as the determinant of the $(n-1)\times(n-1)$ matrix that is obtained by deleting the i'th row and the j'th column of A. Thus, for example, if

$$A = \begin{bmatrix} 1 & 3 & 1 \\ 2 & 0 & 4 \\ 1 & 1 & 2 \end{bmatrix}, \text{ then } A_{\{12\}} = \det \begin{bmatrix} 2 & 4 \\ 1 & 2 \end{bmatrix}.$$

Theorem 5.9. *If A is an $n \times n$ matrix, then* det A *can be expressed as an* **expansion along the i'th row***:*

$$\det A = \sum_{j=1}^{n} a_{ij} A_{\{ij\}} (-1)^{i+j} \tag{5.22}$$

for each choice of i, $i = 1, \dots, n$, or as an **expansion along the j'th column***:*

$$\det A = \sum_{i=1}^{n} a_{ij} A_{\{ij\}} (-1)^{i+j} \tag{5.23}$$

for each choice of j, $j = 1, \dots, n$.

Discussion. The conclusion depends heavily upon the fact that the determinant is linear in each row separately and each column separately. Thus, for example, if A is a 4×4 matrix, then we can write the third row of A as

$$\begin{bmatrix} a_{31} & a_{32} & a_{33} & a_{34} \end{bmatrix} = a_{31}\begin{bmatrix} 1 & 0 & 0 & 0 \end{bmatrix} + a_{32}\begin{bmatrix} 0 & 1 & 0 & 0 \end{bmatrix} + a_{33}\begin{bmatrix} 0 & 0 & 1 & 0 \end{bmatrix} + a_{34}\begin{bmatrix} 0 & 0 & 0 & 1 \end{bmatrix}$$

and invoke linearity to obtain

$$\det A = a_{31} \det \begin{bmatrix} a_{11} & a_{12} & a_{13} & a_{14} \\ a_{21} & a_{22} & a_{23} & a_{24} \\ 1 & 0 & 0 & 0 \\ a_{41} & a_{42} & a_{43} & a_{44} \end{bmatrix} + a_{32} \det \begin{bmatrix} a_{11} & a_{12} & a_{13} & a_{14} \\ a_{21} & a_{22} & a_{23} & a_{24} \\ 0 & 1 & 0 & 0 \\ a_{41} & a_{42} & a_{43} & a_{44} \end{bmatrix}$$

$$+ a_{33} \det \begin{bmatrix} a_{11} & a_{12} & a_{13} & a_{14} \\ a_{21} & a_{22} & a_{23} & a_{24} \\ 0 & 0 & 1 & 0 \\ a_{41} & a_{42} & a_{43} & a_{44} \end{bmatrix} + a_{34} \det \begin{bmatrix} a_{11} & a_{12} & a_{13} & a_{14} \\ a_{21} & a_{22} & a_{23} & a_{24} \\ 0 & 0 & 0 & 1 \\ a_{41} & a_{42} & a_{43} & a_{44} \end{bmatrix}.$$

Next, by invoking rule 5°, we can knock out all the entries in the same column as the one in each of the last four determinants. Thus, for example, the determinant multiplying a_{32} is seen to be equal to

$$\det \begin{bmatrix} a_{11} & 0 & a_{13} & a_{14} \\ a_{21} & 0 & a_{23} & a_{24} \\ 0 & 1 & 0 & 0 \\ a_{41} & 0 & a_{43} & a_{44} \end{bmatrix} = (-1)^3 \det \begin{bmatrix} 1 & 0 & 0 & 0 \\ 0 & a_{11} & a_{13} & a_{14} \\ 0 & a_{21} & a_{23} & a_{24} \\ 0 & a_{41} & a_{43} & a_{44} \end{bmatrix}$$
$$= (-1)^3 A_{\{32\}},$$

thanks to rules 2°, 13° and Lemma 5.4. Similar considerations lead to the formula

$$\det A = a_{31} A_{\{31\}} - a_{32} A_{\{32\}} + a_{33} A_{\{33\}} - a_{34} A_{\{34\}}.$$

All the other formulas for computing the determinant of a matrix $A \in \mathbb{C}^{4\times 4}$ can be established in much the same way.

The verification of formula (5.22) for $A \in \mathbb{C}^{n\times n}$ is similar. It rests on the observation that the rows $\vec{\mathbf{a}}_i$ of the matrix A can be written as $\sum_{j=0}^{n} a_{ij}\mathbf{e}_j^T$ and hence, for example, the expansion in minors along the first row of A is obtained from the development

$$\det A = \sum_{j=0}^{n} a_{1j} \det \begin{bmatrix} \mathbf{e}_j^T \\ \vec{\mathbf{a}}_2 \\ \vdots \\ \vec{\mathbf{a}}_n \end{bmatrix}.$$

The verification of formula (5.23) rests on analogous decompositions for the columns of A.

MORAL: Formulas (5.22) and (5.23) yield $2n$ different ways of calculating the determinant of an $n \times n$ matrix, one for each row and and one for each column, respectively. It is usually advantageous to expand along the row or column with the most zeros.

Exercise 5.22. Evaluate the determinant of the 4×4 matrix

$$A = \begin{bmatrix} 5 & 2 & 3 & 1 \\ 3 & 0 & 0 & 2 \\ 1 & 1 & 0 & 1 \\ 0 & 2 & 0 & 1 \end{bmatrix}$$

twice, first begin by expanding in minors along the third column and then begin by expanding in minors along the fourth column.

Exercise 5.23. Show that if $A \in \mathbb{C}^{n\times n}$, then the ij minor $A_{\{ij\}}$ is equal to $(-1)^{i+j} \det \widetilde{A}$, where $\widetilde{A}$ denotes the matrix A with its i'th row replaced by $\mathbf{e}_j^T$.

Theorem 5.10. *If $A \in \mathbb{C}^{n\times n}$ and if $C \in \mathbb{C}^{n\times n}$ denotes the matrix with entries*

$$c_{ij} = (-1)^{i+j} A_{\{ji\}}\,, \quad i, j = 1, \ldots, n\,,$$

then:

(1) $AC = CA = \det A \cdot I_n$.

(2) *If* $\det A \neq 0$, *then A is invertible and* $A^{-1} = \dfrac{1}{\det A} C$.

Discussion. If A is a 3×3 matrix, then this theorem states that

(5.24)
$$\begin{bmatrix} a_{11} & a_{12} & a_{13} \\ a_{21} & a_{22} & a_{23} \\ a_{31} & a_{32} & a_{33} \end{bmatrix} \begin{bmatrix} A_{\{11\}} & -A_{\{21\}} & A_{\{31\}} \\ -A_{\{12\}} & A_{\{22\}} & -A_{\{32\}} \\ A_{\{13\}} & -A_{\{23\}} & A_{\{33\}} \end{bmatrix} = \begin{bmatrix} \det A & 0 & 0 \\ 0 & \det A & 0 \\ 0 & 0 & \det A \end{bmatrix}.$$

This formula may be verified by three simple sets of calculations. The first set is based on the formula

$$\det \begin{bmatrix} x & y & z \\ a_{21} & a_{22} & a_{23} \\ a_{31} & a_{32} & a_{33} \end{bmatrix} = xA_{\{11\}} - yA_{\{12\}} + zA_{\{13\}} \tag{5.25}$$

and the observation that:

- if $(x, y, z) = (a_{11}, a_{12}, a_{13})$, then the left-hand side of (5.25) is equal to $\det A$;
- if $(x, y, z) = (a_{21}, a_{22}, a_{23})$, then, by rule 4°, the left-hand side of (5.25) is equal to 0;
- if $(x, y, z) = (a_{31}, a_{32}, a_{33})$, then, by rule 4°, the left-hand side of (5.25) is equal to 0.

These three evaluations can be recorded in the following more revealing way:

$$\begin{bmatrix} a_{11} & a_{12} & a_{13} \\ a_{21} & a_{22} & a_{23} \\ a_{31} & a_{32} & a_{33} \end{bmatrix} \begin{bmatrix} A_{\{11\}} \\ -A_{\{12\}} \\ A_{\{13\}} \end{bmatrix} = \det A \begin{bmatrix} 1 \\ 0 \\ 0 \end{bmatrix}.$$

The next set of calculations uses the formula

$$\det \begin{bmatrix} a_{11} & a_{12} & a_{13} \\ x & y & z \\ a_{31} & a_{32} & a_{33} \end{bmatrix} = -xA_{\{21\}} + yA_{\{22\}} - zA_{\{23\}}$$

to verify that

$$\begin{bmatrix} a_{11} & a_{12} & a_{13} \\ a_{21} & a_{22} & a_{23} \\ a_{31} & a_{32} & a_{33} \end{bmatrix} \begin{bmatrix} -A_{\{21\}} \\ A_{\{22\}} \\ -A_{\{23\}} \end{bmatrix} = \det A \begin{bmatrix} 0 \\ 1 \\ 0 \end{bmatrix} .$$

The rest should be clear. □

Exercise 5.24. Formulate and verify the analogue of formula (5.24) for 4×4 matrices.

Exercise 5.25. Let A be an $n \times n$ matrix and let $f(x) = \det(I_n + xA)$. Show that $f'(0) = \operatorname{trace} A$. [HINT: First write the determinant in terms of the minors of either the first row or the first column and then differentiate. Lots of things drop out when $x = 0$. Try a 2×2 or a 3×3 first to get oriented.]

Exercise 5.26. Give a second proof of the formula in Exercise 5.25 on the basis of the Jordan decomposition $A = UJU^{-1}$ that is described in Theorem 4.14.

Exercise 5.27. Show that if $A \in \mathbb{C}^{3\times 3}$ is invertible and $A\mathbf{x} = \mathbf{b}$, then

$$x_1 = \frac{\det \begin{bmatrix} b_1 & a_{12} & a_{13} \\ b_2 & a_{22} & a_{23} \\ b_3 & a_{32} & a_{33} \end{bmatrix}}{\det A}, \quad x_2 = \frac{\det \begin{bmatrix} a_{11} & b_1 & a_{13} \\ a_{21} & b_2 & a_{23} \\ a_{31} & b_3 & a_{33} \end{bmatrix}}{\det A}$$

and state and verify the analogous formula for x_3. [REMARK: This is an example of **Cramer's rule.**]

Exercise 5.28. Show that if $A, B \in \mathbb{C}^{n\times n}$ and $AB = I_n$, then for every $\lambda \in \mathbb{C}$

$$(5.26) \qquad b_{11} + b_{21}\lambda + \cdots + b_{n1}\lambda^{n-1} = \frac{\det \begin{bmatrix} 1 & \lambda & \cdots & \lambda^{n-1} \\ a_{21} & a_{22} & \cdots & a_{2n} \\ \vdots & & & \vdots \\ a_{n1} & a_{n2} & \cdots & a_{2nn} \end{bmatrix}}{\det A} .$$

Exercise 5.29. Compute the inverse of the matrix $A = \begin{bmatrix} 1 & 2 & 2 \\ 2 & 1 & 2 \\ 1 & x & 0 \end{bmatrix}$ for those values of x for which A is invertible. [HINT: Exploit formula (5.24).]

Exercise 5.30. Show that if

$$\det \begin{bmatrix} a_{11} & a_{12} \\ a_{21} & a_{22} \end{bmatrix} \alpha + \det \begin{bmatrix} a_{11} & a_{13} \\ a_{21} & a_{23} \end{bmatrix} \beta + \det \begin{bmatrix} a_{12} & a_{13} \\ a_{22} & a_{23} \end{bmatrix} \gamma = 1\,,$$

then there exists a matrix $C \in \mathbb{C}^{3\times 2}$ such that $\begin{bmatrix} a_{11} & a_{12} & a_{13} \\ a_{21} & a_{22} & a_{23} \end{bmatrix} C = I_2$. [HINT: Imbed the given a_{ij} in an appropriately chosen 3×3 invertible matrix.]

Exercise 5.31. Let $\mathbf{a}_i \in \mathbb{C}^2$ and $\alpha_{ij} = \det \begin{bmatrix} \mathbf{a}_i & \mathbf{a}_j \end{bmatrix}$ for $i, j = 1, \ldots, 4$. Show that

$$\alpha_{12}\alpha_{34} - \alpha_{13}\alpha_{24} + \alpha_{14}\alpha_{23} = 0\,. \tag{5.27}$$

[HINT: Consider the matrix $\begin{bmatrix} \alpha_{14} & \alpha_{24} & \alpha_{34} \\ \mathbf{a}_1 & \mathbf{a}_2 & \mathbf{a}_3 \end{bmatrix}$.]

Exercise 5.32. Show that if the six numbers α_{12}, α_{13}, α_{14}, α_{34}, α_{24} and α_{23} satisfy the identity (5.27), then there exists a matrix $A \in \mathbb{C}^{2\times 4}$ such that $\alpha_{ij} = \det \begin{bmatrix} \mathbf{a}_i & \mathbf{a}_j \end{bmatrix}$, where $\mathbf{a}_i$ designates the i'th column of A. [HINT: Try $A = \begin{bmatrix} 1 & x & y & z \\ 0 & \alpha_{12} & \alpha_{13} & \alpha_{14} \end{bmatrix}$ for a start.]

5.8. Uses of determinants

In looking back over the results of this chapter, you should keep in mind that determinants play a useful role in:

(1) The calculation of the eigenvalues of a matrix A.

(2) Checking whether or not a matrix $A \in \mathbb{C}^{n\times n}$ is invertible.

(3) The calculation of the inverse of an invertible matrix $A \in \mathbb{C}^{n\times n}$.

(4) Solving equations of the form $A\mathbf{x} = \mathbf{b}$ when $A \in \mathbb{C}^{n\times n}$ is invertible.

Exercise 5.33. Let $A \in \mathbb{C}^{p\times q}$. Show that $\operatorname{rank} A = r$ if and only if the largest square invertible submatrix of A is of size $r \times r$. [REMARK: A $k \times k$ submatrix of A is obtained by deleting $p - k$ rows and $q - k$ columns.]

5.9. Companion matrices

A matrix $A \in \mathbb{C}^{n\times n}$ of the form

$$A = \begin{bmatrix} 0 & 1 & 0 & \cdots & 0 \\ 0 & 0 & 1 & \cdots & 0 \\ \vdots & & & \ddots & \vdots \\ 0 & 0 & 0 & \cdots & 1 \\ -a_0 & -a_1 & -a_2 & \cdots & -a_{n-1} \end{bmatrix} \tag{5.28}$$

is called a **companion matrix**. Companion matrices play a significant role in the theory of differential equations and the theory of Bezoutians, as will be discussed in Chapters 13 and 21.

Theorem 5.11. *Let $A \in \mathbb{C}^{n\times n}$ be a companion matrix of the form* (5.28) *with distinct eigenvalues $\lambda_1, \ldots, \lambda_k$ having geometric multiplicities $\gamma_1, \ldots, \gamma_k$ and algebraic multiplicities $\alpha_1, \ldots, \alpha_k$, respectively. Then:*

(1) $\det(\lambda I_n - A) = a_0 + a_1\lambda + \cdots + a_{n-1}\lambda^{n-1} + \lambda^n$.

(2) $\gamma_j = 1$ *for* $j = 1, \ldots, k$.

(3) *A is similar to the Jordan matrix* $J = \operatorname{diag}\{C_{\lambda_1}^{(\alpha_1)}, \ldots, C_{\lambda_k}^{(\alpha_k)}\}$.

(4) *A is invertible if and only if* $a_0 \neq 0$.

Proof. The formula in (1) is obtained by expanding in minors along the first column and taking advantage of the structure. The second assertion follows from the fact that $\dim \mathcal{R}_{(A-\lambda I_n)} \geq n-1$ for every point $\lambda \in \mathbb{C}$, because it implies that $\dim \mathcal{N}_{(A-\lambda_j I_n)} = 1$ for $j = 1, \ldots, k$. Therefore (3) follows: there is exactly one Jordan cell $C_{\lambda_j}^{(\alpha_j)}$ for each distinct eigenvalue λ_j. The last assertion (4) is left to the reader as an exercise. □

Exercise 5.34. Verify the formula in (1) for $n = 2$ and $n = 3$.

Exercise 5.35. Verify the formula in (1) for an arbitrary positive integer n by induction. [HINT: Expand in minors along the last column.]

Exercise 5.36. Verify assertion (4) in Theorem 5.11.

Exercise 5.37. Show that if $p(\lambda) = a_0 + a_1\lambda + a_2\lambda^2 + \lambda^3 = (\lambda - \mu)^3$, then

$$\begin{bmatrix} 0 & 1 & 0 \\ 0 & 0 & 1 \\ -a_0 & -a_1 & -a_2 \end{bmatrix} V = V \begin{bmatrix} \mu & 1 & 0 \\ 0 & \mu & 1 \\ 0 & 0 & \mu \end{bmatrix}, \quad \text{where} \quad V = \begin{bmatrix} 1 & 0 & 0 \\ \mu & 1 & 0 \\ \mu^2 & 2\mu & 1 \end{bmatrix}.$$

[HINT: Exploit the formulas $p(\mu) = 0$, $p'(\mu) = 0$ and $p''(\mu) = 0$, where p' denotes the derivative of p with respect to λ.]

5.10. Circulants and Vandermonde matrices

A matrix $A \in \mathbb{C}^{n\times n}$ of the form

$$A = a_0 I_n + a_1 P + \cdots + a_{n-1}P^{n-1} \tag{5.29}$$

based on the $n \times n$ permutation

$$P = \sum_{j=1}^{n-1} \mathbf{e}_j \mathbf{e}_{j+1}^T + \mathbf{e}_n \mathbf{e}_1^T \tag{5.30}$$

is termed a **circulant**. To illustrate more graphically, if $n = 5$, then

$$P = \begin{bmatrix} 0 & 1 & 0 & 0 & 0 \\ 0 & 0 & 1 & 0 & 0 \\ 0 & 0 & 0 & 1 & 0 \\ 0 & 0 & 0 & 0 & 1 \\ 1 & 0 & 0 & 0 & 0 \end{bmatrix} \quad \text{and} \quad A = \begin{bmatrix} a_0 & a_1 & a_2 & a_3 & a_4 \\ a_4 & a_0 & a_1 & a_2 & a_3 \\ a_3 & a_4 & a_0 & a_1 & a_2 \\ a_2 & a_3 & a_4 & a_0 & a_1 \\ a_1 & a_2 & a_3 & a_4 & a_0 \end{bmatrix}.$$

Because of the special form of A, it is reasonable to look for eigenvectors $\mathbf{u}$ of the special form $\mathbf{u}^T = [1\ x\ x^2\ x^3\ x^4]$. This will indeed be the case for appropriate choices of x. A matrix $V \in \mathbb{C}^{n\times n}$ with columns of this form:

$$V = \begin{bmatrix} 1 & \cdots & 1 \\ \lambda_1 & \cdots & \lambda_n \\ \vdots & & \vdots \\ \lambda_1^{n-1} & \cdots & \lambda_n^{n-1} \end{bmatrix}, \tag{5.31}$$

is termed a **Vandermonde** matrix.

Exercise 5.38. Show that if P denotes the permutation P defined by formula (5.30), then $\det(\lambda I_n - P) = -1 + \lambda^n$ and $P^n = I_n$. [HINT: Invoke Theorem 5.11 and the Cayley-Hamilton theorem.]

Exercise 5.39. Let $A \in \mathbb{C}^{n\times n}$ denote the circulant matrix defined by formula (5.29) and let $p(\lambda) = a_0 + a_1\lambda + \cdots + a_{n-1}\lambda^{n-1}$. Show that

$$AV = VD, \quad \text{where} \quad D = \operatorname{diag}\{p(\zeta_1), \ldots, p(\zeta_n)\},$$

$\zeta_j = \exp(2\pi i j/n)$ for $j = 1, \ldots, n$ and V is a Vandermonde matrix with $\lambda_j = \zeta_j$ for $j = 1, \ldots, n$. Conclude that

$$\det A = p(\zeta_1)\cdots p(\zeta_n). \tag{5.32}$$

Exercise 5.40. Show that if $A \in \mathbb{C}^{n\times n}$ and $B \in \mathbb{C}^{n\times n}$ are circulants, then $AB = BA$.

Exercise 5.41. Find the determinant of the (Vandermonde) matrix V given by formula (5.31) when $n = 3$. [REMARK: You can calculate this determinant by brute force. But a better way is to let $f(x)$ denote the value of the determinant when λ_1 is replaced by x and observe that $f(x)$ is a polynomial of degree two such that $f(\lambda_2) = f(\lambda_3) = 0$.]

Exercise 5.42. Let $\{\alpha_0, \ldots, \alpha_n\}$ and $\{\beta_0, \ldots, \beta_n\}$ be two sets of points in $\mathbb{C}$. Show that if $\alpha_i \neq \alpha_j$ when $i \neq j$, then there exists a unique polynomial $p(\lambda) = c_0 + c_1\lambda + \cdots + c_n\lambda^n$ of degree n such that $p(\alpha_i) = \beta_i$ for $i = 0, \ldots, n$ and find a formula for the coefficients c_j.

Chapter 6

Calculating Jordan forms

Some people believe that football is a matter of life or death. I'm very disappointed with that attitude. I can assure you its much, much more important than that.

Bill Shankly, former manager of Liverpool

The first (and main) part of this chapter is devoted to calculating the Jordan forms of a given matrix $A \in \mathbb{C}^{n\times n}$, i.e., to finding representations of the form

$$A = UJU^{-1}, \tag{6.1}$$

where $U \in \mathbb{C}^{n\times n}$ is invertible and $J \in \mathbb{C}^{n\times n}$ is a block diagonal matrix with Jordan cells as blocks. Subsequently, analogous representations for $A \in \mathbb{R}^{n\times n}$ when the matrices on the right are also constrained to have real entries are considered. The last section furnishes additional information on companion matrices. In particular, it is shown that a companion matrix S_f that is based on the polynomial

$$f(\lambda) = f_0 + f_1\lambda + \cdots + f_n\lambda^n = f_n(\lambda - \lambda_1)^{\alpha_1} \cdots (\lambda - \lambda_k)^{\alpha_k}$$

with k distinct roots $\lambda_1, \ldots, \lambda_k$ is similar to $J = \operatorname{diag}\{C_{\lambda_1}^{(\alpha_1)}, \ldots, C_{\lambda_k}^{(\alpha_k)}\}$. Thus, every matrix $A \in \mathbb{C}^{n\times n}$ is similar to a block diagonal matrix $\operatorname{diag}\{S_{g_1}, \ldots, S_{g_\ell}\}$ based on one or more companion matrices. The symbol S_f (S for significant other) is used for the companion matrix to avoid overburdening the letter C.

6.1. Overview

The calculation of the matrices in a representation of the form (6.1) can be conveniently divided into three parts (the notation corresponds to Theorem 4.14):

(1) Calculate the eigenvalues of the matrix A from the distinct roots $\lambda_1, \ldots, \lambda_k$ of the characteristic polynomial
$$p(\lambda) = \det(\lambda I_n - A)$$
by writing it in factored form as
$$p(\lambda) = (\lambda - \lambda_1)^{\alpha_1} \cdots (\lambda - \lambda_k)^{\alpha_k} .$$
The distinct roots $\lambda_1, \ldots, \lambda_k$ of $p(\lambda)$ are the distinct eigenvalues of A, and the numbers $\alpha_1, \ldots, \alpha_k$ are their algebraic multiplicities.

(2) Compute $J = \operatorname{diag}\{B_{\lambda_1} \ldots, B_{\lambda_k}\}$ by calculating
$$\dim \mathcal{N}_{(A-\lambda_j I_n)^i} \quad \text{for} \quad i = 1, \ldots, \alpha_j$$
for each of the distinct eigenvalues $\lambda_1, \ldots, \lambda_k$, in order to obtain the sizes of the Jordan cells in B_{λ_j} from either the diagram discussed in Section 6.4, or formula (4.17).

(3) Organize the vectors in $\mathcal{N}_{(A-\lambda_j I_n)^{\alpha_j}}$ into **Jordan chains**, one Jordan chain for each Jordan cell, in order to calculate the matrices U.

Remark 6.1. The information in (1) is enough to guarantee a factorization of A of the form $A = UGU^{-1}$, where $G = \operatorname{diag}\{G_1, \ldots, G_k\}$ for some choice of $G_j \in \mathbb{C}^{\alpha_j \times \alpha_j}$, $j = 1, \ldots, k$. The remaining two steps are to show that if the vectors in U are chosen appropriately, then each of the blocks G_j, $j = 1, \ldots, k$, can be expressed as a block diagonal matrix with γ_j Jordan cells as diagonal blocks and to determine the sizes of these cells.

6.2. Structure of the nullspaces $\mathcal{N}_{B^j}$

Lemma 6.2. *Let $B \in \mathbb{F}^{n \times n}$ and let $\mathbf{u} \in \mathbb{F}^n$ belong to $\mathcal{N}_{B^k}$ for some positive integer k. Then the vectors*
$$\{B^{k-1}\mathbf{u}, B^{k-2}\mathbf{u}, \ldots, \mathbf{u}\} \text{ are linearly independent over } \mathbb{F} \iff B^{k-1}\mathbf{u} \neq \mathbf{0} .$$

Proof. Suppose first that $B^{k-1}\mathbf{u} \neq \mathbf{0}$ and that
$$\alpha_0\mathbf{u} + \alpha_1 B\mathbf{u} + \cdots + \alpha_{k-1}B^{k-1}\mathbf{u} = \mathbf{0}$$
for some choice of coefficients $\alpha_0, \ldots, \alpha_k \in \mathbb{F}$. Then, since $\mathbf{u} \in \mathcal{N}_{B^k}$,
$$B^{k-1}(\alpha_0\mathbf{u} + \alpha_1 B\mathbf{u} + \cdots + \alpha_{k-1}B^{k-1}\mathbf{u}) = \alpha_0 B^{k-1}\mathbf{u} = \mathbf{0} ,$$
which clearly implies that $\alpha_0 = 0$. Similarly, the identity
$$B^{k-2}(\alpha_1 B\mathbf{u} + \cdots + \alpha_{k-1}B^{k-1}\mathbf{u}) = \alpha_1 B^{k-1}\mathbf{u} = \mathbf{0}$$

implies that $\alpha_1 = 0$. Continuing in this vein it is readily seen that $B^{k-1}\mathbf{u} \neq \mathbf{0} \Longrightarrow$ that the vectors $\{B^{k-1}\mathbf{u}, B^{k-2}\mathbf{u}, \ldots, \mathbf{u}\}$ are linearly independent over $\mathbb{F}$. Thus, as the converse is self-evident, the proof is complete. □

Lemma 6.3. *If $B \in \mathbb{F}^{n\times n}$, then:*

(1) *The null spaces $\mathcal{N}_{B^j}$ are ordered by inclusion:*

$$\mathcal{N}_B \subseteq \mathcal{N}_{B^2} \subseteq \mathcal{N}_{B^3} \subseteq \cdots .$$

(2) *If $\mathcal{N}_{B^j} = \mathcal{N}_{B^{j-1}}$ for some integer $j \geq 2$, then $\mathcal{N}_{B^{j+1}} = \mathcal{N}_{B^j}$.*

(3) *If $j \geq 1$ is an integer, then $\mathcal{N}_{B^j} \neq \{\mathbf{0}\} \Longleftrightarrow \mathcal{N}_{B^{j+1}} \neq \{\mathbf{0}\}$.*

Proof. If $\mathbf{u} \in \mathcal{N}_{B^j}$, then $B^{j+1}\mathbf{u} = B(B^j\mathbf{u}) = B\mathbf{0} = \mathbf{0}$, which clearly implies that $\mathbf{u} \in \mathcal{N}_{B^{j+1}}$ and hence justifies (1).

Suppose next that $\mathbf{u} \in \mathcal{N}_{B^{j+1}}$ and $\mathcal{N}_{B^j} = \mathcal{N}_{B^{j-1}}$. Then $B\mathbf{u} \in \mathcal{N}_{B^j}$, since $B^{j+1}\mathbf{u} = B^j(B\mathbf{u}) = \mathbf{0}$. Therefore, $B\mathbf{u} \in \mathcal{N}_{B^{j-1}}$, which implies in turn that $\mathbf{u} \in \mathcal{N}_{B^j}$. Thus,

$$\mathcal{N}_{B^j} = \mathcal{N}_{B^{j-1}} \Longrightarrow \mathcal{N}_{B^{j+1}} \subseteq \mathcal{N}_{B^j} ,$$

which, in view of (1), serves to prove (2). Finally, (3) is left to the reader as an exercise. □

Exercise 6.1. Verify assertion (3) in Lemma 6.3.

Lemma 6.4. *Let $B \in \mathbb{C}^{n\times n}$ and assume that $\mathcal{N}_{B^{j-1}}$ is a nonzero proper subspace of $\mathcal{N}_{B^j}$ for some integer $j \geq 2$.*

(1) *If $\{\mathbf{u}_1, \ldots, \mathbf{u}_k\}$ is a basis for $\mathcal{N}_{B^{j-1}}$, and $\{\mathbf{u}_1, \ldots, \mathbf{u}_k; \mathbf{v}_1, \ldots, \mathbf{v}_\ell\}$ is a basis for $\mathcal{N}_{B^j}$, then $\ell \leq k$ and span$\{B^{j-1}\mathbf{v}_1, \ldots, B^{j-1}\mathbf{v}_\ell\}$ is an ℓ-dimensional subspace of span$\{\mathbf{u}_1, \ldots, \mathbf{u}_k\}$.*

(2) *If also $\mathcal{N}_{B^j}$ is a proper subspace of $\mathcal{N}_{B^{j+1}}$ and $\{\mathbf{u}_1, \ldots, \mathbf{u}_k; \mathbf{v}_1, \ldots, \mathbf{v}_\ell; \mathbf{w}_1, \ldots, \mathbf{w}_m\}$ is a basis for $\mathcal{N}_{B^{j+1}}$, then $m \leq \ell \leq k$ and span$\{B^j\mathbf{w}_1, \ldots, B^j\mathbf{w}_m\}$ is an m-dimensional subspace of the ℓ-dimensional space span$\{B^{j-1}\mathbf{v}_1, \ldots, B^{j-1}\mathbf{v}_\ell\}$.*

Proof. To verify (2), observe first that the identity $B^j(B\mathbf{w}_i) = \mathbf{0}$ clearly implies that the vectors $B\mathbf{w}_i \in \mathcal{N}_{B^j}$ and hence that

$$\text{span}\,\{B\mathbf{w}_1, \ldots, B\mathbf{w}_m\} \subseteq \text{span}\,\{\mathbf{u}_1, \ldots, \mathbf{u}_k, \mathbf{v}_1, \ldots, \mathbf{v}_\ell\} .$$

But this in turn implies that

$$\text{span}\,\{B^j\mathbf{w}_1, \ldots, B^j\mathbf{w}_m\} \subseteq \text{span}\,\{B^{j-1}\mathbf{v}_1, \ldots, B^{j-1}\mathbf{v}_\ell\} .$$

The next step is to check that the vectors $B^j\mathbf{w}_1, \ldots, B^j\mathbf{w}_m$ are linearly independent. To this end, suppose that

$$\gamma_1 B^j\mathbf{w}_1 + \cdots + \gamma_m B^j\mathbf{w}_m = \mathbf{0}$$

for some set of constants $\gamma_1, \ldots, \gamma_m \in \mathbb{C}$. Then $\gamma_1 \mathbf{w}_1 + \cdots + \gamma_m \mathbf{w}_m \in \mathcal{N}_{B^j}$ and, consequently,

$$\gamma_1 \mathbf{w}_1 + \cdots + \gamma_m \mathbf{w}_m = \alpha_1 \mathbf{u}_1 + \cdots + \alpha_k \mathbf{u}_k + \beta_1 \mathbf{v}_1 + \cdots + \beta_\ell \mathbf{v}_\ell$$

for some choice of constants $\alpha_1, \ldots, \alpha_k, \beta_1, \ldots, \beta_\ell \in \mathbb{C}$. However, since the three sets of vectors are linearly independent, this is viable only if all the constants are zero. The same argument serves to show that the vectors $B^{j-1}\mathbf{v}_1, \ldots, B^{j-1}\mathbf{v}_\ell$ are linearly independent. This completes the proof of (2). The proof of (1) is similar. □

Corollary 6.5. If $B \in \mathbb{C}^{n\times n}$, then $\dim \mathcal{N}_{B^0} = 0$ and

$$\dim \mathcal{N}_{B^{j+1}} - \dim \mathcal{N}_{B^j} \le \dim \mathcal{N}_{B^j} - \dim \mathcal{N}_{B^{j-1}} \text{ for } j = 1, 2, \ldots . \tag{6.2}$$

In future applications of Lemma 6.4, we shall also need the following result, which will enable us to select basis elements in an appropriate way.

Lemma 6.6. *Let* $\text{span}\{\mathbf{u}_1, \ldots, \mathbf{u}_k\}$ *be a* k*-dimensional subspace of an* ℓ*-dimensional subspace* $\text{span}\{\mathbf{v}_1, \ldots, \mathbf{v}_\ell\}$ *of* $\mathbb{C}^n$, *where* $k < \ell$. *Then there exists an* $\ell \times \ell$ *permutation matrix* P *such that if*

$$[\widetilde{\mathbf{v}}_1 \quad \cdots \quad \widetilde{\mathbf{v}}_\ell] = [\mathbf{v}_1 \quad \cdots \quad \mathbf{v}_\ell]\, P,$$

then

$$\text{span}\{\mathbf{u}_1, \ldots, \mathbf{u}_k, \widetilde{\mathbf{v}}_1, \ldots, \widetilde{\mathbf{v}}_{\ell-k}\} = \text{span}\{\mathbf{v}_1, \ldots, \mathbf{v}_\ell\}.$$

Proof. Under the given assumptions there exists a matrix $A \in \mathbb{C}^{\ell\times k}$ with $\text{rank}\, A = k$ such that

$$[\mathbf{u}_1 \quad \cdots \quad \mathbf{u}_k] = [\mathbf{v}_1 \quad \cdots \quad \mathbf{v}_\ell]\, A.$$

Let P be an $\ell \times \ell$ permutation matrix such that the bottom $k \times k$ block of $P^T A$ is invertible. Thus, in terms of the block decomposition

$$P^T A = \begin{bmatrix} A_{12} \\ A_{22} \end{bmatrix}$$

with blocks $A_{12} \in \mathbb{C}^{(\ell-k)\times k}$ and $A_{22} \in \mathbb{C}^{k\times k}$, A_{22} is invertible and

$$\begin{aligned} [\mathbf{u}_1 \quad \cdots \quad \mathbf{u}_k] &= [\mathbf{v}_1 \quad \cdots \quad \mathbf{v}_\ell]\, P P^T A \\ &= [\mathbf{v}_1 \quad \cdots \quad \mathbf{v}_\ell]\, P \begin{bmatrix} A_{12} \\ A_{22} \end{bmatrix}. \end{aligned}$$

Consequently, the $\ell - k$ vectors defined by the formula

$$[\widetilde{\mathbf{v}}_1 \quad \cdots \quad \widetilde{\mathbf{v}}_{\ell-k}] = [\mathbf{v}_1 \quad \cdots \quad \mathbf{v}_\ell]\, P \begin{bmatrix} I_{\ell-k} \\ O \end{bmatrix}$$

meet the asserted conditions, since

$$[\widetilde{\mathbf{v}}_1 \quad \cdots \quad \widetilde{\mathbf{v}}_{\ell-k} \quad \mathbf{u}_1 \quad \cdots \quad \mathbf{u}_k] = [\mathbf{v}_1 \quad \cdots \quad \mathbf{v}_\ell]\, P \begin{bmatrix} I_{\ell-k} & A_{12} \\ O & A_{22} \end{bmatrix}$$

and the two $\ell \times \ell$ matrices on the far right in the last formula are invertible. $\square$

6.3. Chains and cells

Recall that a set of vectors $\mathbf{u}_1, \ldots, \mathbf{u}_k \in \mathbb{C}^n$ is said to form a **Jordan chain** of length k corresponding to an eigenvalue λ_0 of $A \in \mathbb{C}^{n \times n}$ if

$$\begin{aligned} (A - \lambda_0 I_n)\mathbf{u}_1 &= \mathbf{0} \\ (A - \lambda_0 I_n)\mathbf{u}_2 &= \mathbf{u}_1 \\ (A - \lambda_0 I_n)\mathbf{u}_3 &= \mathbf{u}_2 \\ &\vdots \\ (A - \lambda_0 I_n)\mathbf{u}_k &= \mathbf{u}_{k-1} \end{aligned}$$

and $\mathbf{u}_1 \neq \mathbf{0}$. In other words, the $n \times k$ matrix

$$U = [\mathbf{u}_1 \quad \cdots \quad \mathbf{u}_k]$$

with these vectors as its columns satisfies the identity

$$\begin{aligned} (A - \lambda_0 I_n)U &= [\mathbf{0} \quad \mathbf{u}_1 \quad \cdots \quad \mathbf{u}_{k-1}] \\ &= [\mathbf{u}_1 \quad \cdots \quad \mathbf{u}_k] \begin{bmatrix} 0 & 1 & 0 & \cdots & 0 \\ 0 & 0 & 1 & \cdots & 0 \\ \vdots & & & \ddots & \vdots \\ 0 & 0 & 0 & & 1 \\ 0 & 0 & 0 & \cdots & 0 \end{bmatrix} \\ &= [\mathbf{u}_1 \quad \cdots \quad \mathbf{u}_k]\, C_0^{(k)} \\ &= U(C_{\lambda_0}^{(k)} - \lambda_0 I_k)\,, \end{aligned}$$

i.e.,

$$AU = UC_{\lambda_0}^{(k)}\,.$$

To illustrate the computation of the number of Jordan cells in J, suppose for the sake of definiteness that B_{λ_1} is a block diagonal matrix with exactly k_1 Jordan cells $C_{\lambda_1}^{(1)}$, k_2 Jordan cells $C_{\lambda_1}^{(2)}$, k_3 Jordan cells $C_{\lambda_1}^{(3)}$ and k_4 Jordan cells $C_{\lambda_1}^{(4)}$ and let $B = B_{\lambda_1} - \lambda_1 I_{\alpha_1}$. Then

$$\begin{aligned} \dim \mathcal{N}_B &= k_1 + k_2 + k_3 + k_4 \\ \dim \mathcal{N}_{B^2} &= k_1 + 2k_2 + 2k_3 + 2k_4 \\ \dim \mathcal{N}_{B^3} &= k_1 + 2k_2 + 3k_3 + 3k_4 \\ \dim \mathcal{N}_{B^4} &= k_1 + 2k_2 + 3k_3 + 4k_4 = \alpha_1\,. \end{aligned}$$

Thus,

$$\begin{aligned}\dim \mathcal{N}_{B^2} - \dim \mathcal{N}_B &= k_2 + k_3 + k_4 \\ \dim \mathcal{N}_{B^3} - \dim \mathcal{N}_{B^2} &= k_3 + k_4 \\ \dim \mathcal{N}_{B^4} - \dim \mathcal{N}_{B^3} &= k_4\,,\end{aligned}$$

and hence

$$\begin{aligned}k_1 &= 2\dim \mathcal{N}_B - \dim \mathcal{N}_{B^2} \\ k_2 &= 2\dim \mathcal{N}_{B^2} - \dim \mathcal{N}_{B^3} - \dim \mathcal{N}_B \\ k_3 &= 2\dim \mathcal{N}_{B^3} - \dim \mathcal{N}_{B^4} - \dim \mathcal{N}_{B^2} \\ k_4 &= \dim \mathcal{N}_{B^4} - \dim \mathcal{N}_{B^3}\,.\end{aligned}$$

The last set of formulas can be written in the uniform pattern

(6.3)
$$k_j = 2\dim \mathcal{N}_{B^j} - \dim \mathcal{N}_{B^{j+1}} - \dim \mathcal{N}_{B^{j-1}} \quad \text{for} \quad j = 1, 2, \ldots, n-1\,,$$

since $\dim \mathcal{N}_{B^0} = 0$ and, for this choice of numbers, $\dim \mathcal{N}_{B^j} = \dim \mathcal{N}_{B^4}$ for $j = 4, \ldots, n$.

Exercise 6.2. Let $A \in \mathbb{C}^{n\times n}$ be similar to a Jordan matrix J that contains exactly k_j Jordan cells $C_\mu^{(j)}$ of size $j\times j$ with μ on the diagonal for $j = 1, \ldots, \ell$ and let $B = A - \mu I_n$. Show that formula (6.3) for k_j is still valid.

Exercise 6.3. Calculate $\dim \mathcal{N}_{B^j}$ for $j = 1, \ldots, 15$, when $B = B_{\lambda_1} - \lambda_1 I_{15}$ and

$$B_{\lambda_1} = \operatorname{diag}\{C_{\lambda_1}^{(5)}, C_{\lambda_1}^{(3)}, C_{\lambda_1}^{(3)}, C_{\lambda_1}^{(2)}, C_{\lambda_1}^{(1)}, C_{\lambda_1}^{(1)}\}\,.$$

Exercise 6.4. Find an 11×11 matrix B such that $\dim \mathcal{N}_B = 4$, $\dim \mathcal{N}_{B^2} = 7$, $\dim \mathcal{N}_{B^3} = 9$, $\dim \mathcal{N}_{B^4} = 10$ and $\dim \mathcal{N}_{B^5} = 11$.

Exercise 6.5. Let $A \in \mathbb{C}^{n\times n}$ be similar to a Jordan matrix J that contains exactly k_1 Jordan cells $C_\mu^{(1)}$, k_2 Jordan cells $C_\mu^{(2)}, \ldots$, k_ℓ Jordan cells $C_\mu^{(\ell)}$ with μ on the diagonal and let $B = A - \mu I_n$. Show that

(6.4)
$$\dim \mathcal{N}_{B^j} = \begin{cases} k_1 + k_2 + \cdots + k_\ell & \text{if} \quad j = 1 \\ k_1 + 2k_2 + \cdots + (j-1)k_{j-1} + j\sum_{i=j}^{\ell} k_i & \text{if} \quad j \geq 2\,. \end{cases}$$

Exercise 6.6. Show that in the setting of Exercise 6.5

(6.5)
$$\dim \mathcal{N}_{B^{j+1}} - \dim \mathcal{N}_{B^j} = k_{j+1} + \cdots + k_\ell \quad \text{for} \quad j = 1, \ldots, n-1\,.$$

6.4. Computing J

To illustrate the construction of J, let A be an $n \times n$ matrix with k distinct eigenvalues $\lambda_1, \ldots, \lambda_k$ having geometric multiplicities $\gamma_1, \ldots, \gamma_k$ and algebraic multiplicities $\alpha_1, \ldots, \alpha_k$, respectively. To construct the Jordan blocks associated with λ_1, let $B = A - \lambda_1 I_n$ for short and suppose for the sake of definiteness that $\gamma_1 = 6$, $\alpha_1 = 15$, and, to be more concrete, suppose that:

$$\dim \mathcal{N}_B = 6\,,\ \dim \mathcal{N}_{B^2} = 10\,,\ \dim \mathcal{N}_{B^3} = 13 \quad \text{and} \quad \dim \mathcal{N}_{B^4} = 15\,.$$

These numbers are chosen to meet the two constraints imposed by (1) of Lemma 6.3 and the inequalities in (6.2), but are otherwise completely arbitrary.

To see what to expect, construct an array of $\times$ symbols with 6 in the first row, $10 - 6 = 4$ in the second row, $13 - 10 = 3$ in the third row and $15 - 13 = 2$ in the fourth row:

$$\begin{array}{cccccc} \times & \times & \times & \times & \times & \times \\ \times & \times & \times & \times & & \\ \times & \times & \times & & & \\ \times & \times & & & & \end{array}$$

The Jordan cells will correspond in size to the number of $\times$ symbols in each column: two cells of size 4, one cell of size 3, one cell of size 2 and two cells of size 1.

The same construction works in general:

Theorem 6.7. *Let $A \in \mathbb{C}^{n\times n}$, $\mu \in \sigma(A)$, $B = A - \mu I_n$ and $d_j = \dim \mathcal{N}_{B^j}$ for $j = 0, \dots, n$. Now construct an array of $\times$ symbols with $d_j - d_{j-1}$ $\times$ symbols in the j'th row, stacked as in the example just above, and suppose that exactly ℓ rows contain at least one $\times$ symbol. Then the number k_j of Jordan cells $C_\mu^{(j)}$ in J is equal to the number of columns in the array that contain exactly j $\times$ symbols.*

Proof. In view of formula (6.5) and the fact that $d_0 = 0$,

$$\begin{aligned} d_1 - d_0 &= k_\ell + \cdots + k_2 + k_1 \\ d_2 - d_1 &= k_\ell + \cdots + k_2 \\ &\ \ \vdots \\ d_\ell - d_{\ell-1} &= k_\ell \end{aligned}$$

Therefore, there are k_j columns with exactly j $\times$ symbols in them for $j = 1, \dots, \ell$. $\square$

6.5. An algorithm for U

In this section we shall present an algorithm for choosing a basis of $\mathcal{N}_{(A-\lambda_j)^n}$ that serves to build the matrix U in the Jordan decomposition $A = UJU^{-1}$.

Let λ_i be an eigenvalue of $A \in \mathbb{C}^{n\times n}$, let $B = A - \lambda_i I_n$ and let

$\mathbf{a}_1, \ldots, \mathbf{a}_{\ell_1}$ be a basis for $\mathcal{N}_B$,

$\mathbf{a}_1, \ldots, \mathbf{a}_{\ell_1}; \mathbf{b}_1, \ldots, \mathbf{b}_{\ell_2}$ be a basis for $\mathcal{N}_{B^2}$,

$\mathbf{a}_1, \ldots, \mathbf{a}_{\ell_1}; \mathbf{b}_1, \ldots, \mathbf{b}_{\ell_2}; \mathbf{c}_1, \ldots, \mathbf{c}_{\ell_3}$ be a basis for $\mathcal{N}_{B^3}$,

$\mathbf{a}_1, \ldots, \mathbf{a}_{\ell_1}; \mathbf{b}_1, \ldots, \mathbf{b}_{\ell_2}; \mathbf{c}_1, \ldots, \mathbf{c}_{\ell_3}; \mathbf{d}_1, \ldots, \mathbf{d}_{\ell_4}$ be a basis for $\mathcal{N}_{B^4}$,

and suppose that $\mathcal{N}_{B^4} = \mathcal{N}_{B^n}$. Then, in view of Lemma 6.4, $\ell_1 \geq \ell_2 \geq \ell_3 \geq \ell_4$ and

span $\{B\mathbf{b}_1, \ldots, B\mathbf{b}_{\ell_2}\}$ is an ℓ_2-dimensional subspace of span$\{\mathbf{a}_1, \ldots, \mathbf{a}_{\ell_1}\}$,

span $\{B^2\mathbf{c}_1, \ldots, B^2\mathbf{c}_{\ell_3}\}$ is an ℓ_3-dimensional subspace of span$\{B\mathbf{b}_1, \ldots, B\mathbf{b}_{\ell_2}\}$,

span $\{B^3\mathbf{d}_1, \ldots, B^3\mathbf{d}_{\ell_4}\}$ is an ℓ_4-dimensional subspace of span$\{B^2\mathbf{c}_1, \ldots, B^2\mathbf{c}_{\ell_3}\}$.

Moreover, in view of Lemma 6.6, there exists a set of

$\ell_3 - \ell_4$ vectors $\widetilde{\mathbf{c}}_1, \ldots, \widetilde{\mathbf{c}}_{\ell_3-\ell_4}$ in $\{\mathbf{c}_1, \ldots, \mathbf{c}_{\ell_3}\}$

$\ell_2 - \ell_3$ vectors $\widetilde{\mathbf{b}}_1, \ldots, \widetilde{\mathbf{b}}_{\ell_2-\ell_3}$ in $\{\mathbf{b}_1, \ldots, \mathbf{b}_{\ell_2}\}$

$\ell_1 - \ell_2$ vectors $\widetilde{\mathbf{a}}_1, \ldots, \widetilde{\mathbf{a}}_{\ell_1-\ell_2}$ in $\{\mathbf{a}_1, \ldots, \mathbf{a}_{\ell_1}\}$,

such that the set of ℓ_1 vectors

$$\{B^3\mathbf{d}_1, \ldots, B^3\mathbf{d}_{\ell_4}; B^2\widetilde{\mathbf{c}}_1, \ldots, B^2\widetilde{\mathbf{c}}_{\ell_3-\ell_4}; B\widetilde{\mathbf{b}}_1, \ldots, B\widetilde{\mathbf{b}}_{\ell_2-\ell_3}; \widetilde{\mathbf{a}}_1, \ldots, \widetilde{\mathbf{a}}_{\ell_1-\ell_2}\}$$

is a basis for $\mathcal{N}_B$.

The next step is to supplement these vectors with the chains that they generate:

$$\begin{array}{ccccccc}
\ell_4 \text{ clmns} & & \ell_3 - \ell_4 \text{ clmns} & & \ell_2 - \ell_3 \text{ clmns} & & \ell_1 - \ell_2 \text{ clmns} \\
B^3\mathbf{d}_1 \cdots & \vdots & B^2\widetilde{\mathbf{c}}_1 \cdots & \vdots & B\widetilde{\mathbf{b}}_1 \cdots & \vdots & \widetilde{\mathbf{a}}_1 \cdots \\
B^2\mathbf{d}_1 \cdots & \vdots & B\widetilde{\mathbf{c}}_1 \cdots & \vdots & \widetilde{\mathbf{b}}_1 \cdots & \vdots & \\
B\mathbf{d}_1 \cdots & \vdots & \widetilde{\mathbf{c}}_1 \cdots & \vdots & & & \\
\mathbf{d}_1 \cdots & \vdots & & & & &
\end{array}$$

The algorithm produces a Jordan chain for each column, i.e.,

ℓ_4 Jordan chains of length 4,

$\ell_3 - \ell_4$ Jordan chains of length 3,

$\ell_2 - \ell_3$ Jordan chains of length 2,

$\ell_1 - \ell_2$ Jordan chains of length 1.

The total number of vectors in these chains is equal to

$$\begin{aligned} 4\ell_4 + 3(\ell_3 - \ell_4) + 2(\ell_2 - \ell_3) + (\ell_1 - \ell_2) &= \ell_4 + \ell_3 + \ell_2 + \ell_1 \\ &= \dim \mathcal{N}_{B^n} . \end{aligned}$$

Therefore, since this set of $\ell_4 + \ell_3 + \ell_2 + \ell_1$ vectors is linearly independent, it is a basis for $\mathcal{N}_{B^n}$.

Exercise 6.7. Verify that the $\ell_4+\ell_3+\ell_2+\ell_1$ vectors exhibited in the array just above are linearly independent.

Exercise 6.8. Show that in the array exhibited just above the set of vectors in the first k rows is a basis for $\mathcal{N}_{B^k}$ for $k = 1, 2, 3, 4$; i.e., the set of vectors in the first row is a basis for $\mathcal{N}_B$, the set of vectors in the first two rows is a basis for $\mathcal{N}_{B^2}$, etc.

To complete the construction, let

$$\begin{aligned} V_s &= \begin{bmatrix} B^3\mathbf{d}_s & B^2\mathbf{d}_s & B\mathbf{d}_s & \mathbf{d}_s \end{bmatrix} \quad \text{for} \quad s = 1, \ldots, \ell_4 , \\ W_s &= \begin{bmatrix} B^2\widetilde{\mathbf{c}}_s & B\widetilde{\mathbf{c}}_s & \widetilde{\mathbf{c}}_s \end{bmatrix} \quad \text{for} \quad s = 1, \ldots, \ell_3 - \ell_4 , \\ X_s &= \begin{bmatrix} B\widetilde{\mathbf{b}}_s & \widetilde{\mathbf{b}}_s \end{bmatrix} \quad \text{for} \quad s = 1, \ldots, \ell_2 - \ell_3 , \\ Y_s &= \begin{bmatrix} \widetilde{\mathbf{a}}_s \end{bmatrix} \quad \text{for} \quad s = 1, \ldots, \ell_1 - \ell_2 . \end{aligned}$$

Then it is readily checked that

$$BV_s = V_s C_0^{(4)} , \ BW_s = W_s C_0^{(3)} , \ BX_s = X_s C_0^{(2)} , \ BY_s = Y_s C_0^{(1)} ,$$

and hence that if

$$U_i = [V_1 \ \cdots \ V_{\ell_4} \ W_1 \ \cdots \ W_{\ell_3-\ell_4} \ X_1 \ \cdots \ X_{\ell_2-\ell_3} \ Y_1 \ \cdots \ Y_{\ell_1-\ell_2}] ,$$

then

$$BU_i = (A - \lambda_i I_n)U_i = U_i(B_{\lambda_i} - \lambda_i I_{\alpha_i}) ,$$

where B_{λ_i} is a block diagonal matrix with ℓ_4 Jordan cells $C_{\lambda_i}^{(4)}$, $\ell_3-\ell_4$ Jordan cells $C_{\lambda_i}^{(3)}$, $\ell_2 - \ell_3$ Jordan cells $C_{\lambda_i}^{(2)}$ and $\ell_1 - \ell_2$ Jordan cells $C_{\lambda_i}^{(1)}$ as blocks. The last identity is equivalent to the identity

$$AU_i = U_i B_{\lambda_i} .$$

This yields the vectors associated with λ_i and hence, upon setting

$$U = [U_1 \quad \cdots \quad U_k] \quad \text{and} \quad J = \operatorname{diag}\{B_{\lambda_1}, \ldots, B_{\lambda_k}\} ,$$

that

$$AU = UJ .$$

This completes the construction, since U is invertible, by Remark 4.18 and Exercise 6.7.

6.6. An example

Let $A \in \mathbb{C}^{n\times n}$, let $\lambda_1 \in \sigma(A)$, let $B = A - \lambda_1 I_n$ for short and suppose that

$$\dim \mathcal{N}_B = 2\ ,\ \dim \mathcal{N}_{B^2} = 4 \text{ and } \dim \mathcal{N}_{B^j} = 5 \text{ for } j = 3, \ldots, n\ .$$

The given information guarantees the existence of five linearly independent vectors $\mathbf{a}_1$, $\mathbf{a}_2$, $\mathbf{b}_1$, $\mathbf{b}_2$ and $\mathbf{c}_1$ such that $\{\mathbf{a}_1, \mathbf{a}_2\}$ is a basis for $\mathcal{N}_B$, $\{\mathbf{a}_1, \mathbf{a}_2, \mathbf{b}_1, \mathbf{b}_2\}$ is a basis for $\mathcal{N}_{B^2}$ and $\{\mathbf{a}_1, \mathbf{a}_2, \mathbf{b}_1, \mathbf{b}_2, \mathbf{c}_1\}$ is a basis for $\mathcal{N}_{B^3}$. Thus, upon supplementing these vectors with the chains that they generate, we obtain the array

$$\begin{array}{ccccc} 1 & 2 & 3 & 4 & 5 \\ B^2\mathbf{c}_1 & B\mathbf{b}_1 & B\mathbf{b}_2 & \mathbf{a}_1 & \mathbf{a}_2 \\ B\mathbf{c}_1 & \mathbf{b}_1 & \mathbf{b}_2 & & \\ \mathbf{c}_1 & & & & \end{array}$$

The five vectors in the first row of this array belong to the two-dimensional space $\mathcal{N}_B$. The analysis in Section 6.5 guarantees that at least one of the two sets of vectors $\{B^2\mathbf{c}_1, B\mathbf{b}_1\}$, $\{B^2\mathbf{c}_1, B\mathbf{b}_2\}$ is a set of linearly independent vectors. Suppose for the sake of definiteness that $B^2\mathbf{c}_1$ and $B\mathbf{b}_1$ are linearly independent. Then the earlier analysis also implies that $\{B^2\mathbf{c}_1, B\mathbf{c}_1, \mathbf{c}_1, B\mathbf{b}_1, \mathbf{b}_1\}$ is a basis for $\mathcal{N}_{B^n}$. Nevertheless, we shall redo the analysis of this special example from scratch in order to reenforce the underlying ideas. There are six main steps:

(1) Every vector in this array of nine vectors is nonzero.

(2) $\operatorname{span}\{B^2\mathbf{c}_1\} \subseteq \operatorname{span}\{B\mathbf{b}_1, B\mathbf{b}_2\} \subseteq \operatorname{span}\{\mathbf{a}_1, \mathbf{a}_2\}$.

(3) The vectors $B\mathbf{b}_1$ and $B\mathbf{b}_2$ are linearly independent.

(4) If the vectors $B^2\mathbf{c}_1$ and $B\mathbf{b}_1$ are linearly dependent, then the vectors $B^2\mathbf{c}_1$ and $B\mathbf{b}_2$ are linearly independent.

(5) If the vectors $B^2\mathbf{c}_1$ and $B\mathbf{b}_1$ are linearly independent, then the vectors in columns 1 and 2 are linearly independent.

(6) If the vectors $B^2\mathbf{c}_1$ and $B\mathbf{b}_2$ are linearly independent, then the vectors in columns 1 and 3 are linearly independent.

To verify (1), suppose first that $B^2\mathbf{c}_1 = \mathbf{0}$. Then $\mathbf{c}_1 \in \mathcal{N}_{B^2}$, which implies that $\mathbf{c}_1 \in \operatorname{span}\{\mathbf{a}_1, \mathbf{a}_2, \mathbf{b}_1, \mathbf{b}_2\}$. But this contradicts the presumed linear independence of the 5 vectors involved. Therefore, $B^2\mathbf{c}_1 \neq \mathbf{0}$ and hence, $B\mathbf{c}_1 \neq \mathbf{0}$ and $\mathbf{c}_1 \neq \mathbf{0}$. Similar reasons insure that $B\mathbf{b}_1 \neq \mathbf{0}$ and $B\mathbf{b}_2 \neq \mathbf{0}$. The vectors in the first row of the array are nonzero by choice.

To verify the first inclusion in (2), observe that $B\mathbf{c}_1 \in \mathcal{N}_{B^2}$, and hence it can be expressed as a linear combination of the basis vectors of that space:

$$B\mathbf{c}_1 = \alpha_1\mathbf{a}_1 + \alpha_2\mathbf{a}_2 + \beta_1\mathbf{b}_1 + \beta_2\mathbf{b}_2\,.$$

Therefore,

$$B^2\mathbf{c}_1 = \beta_1 B\mathbf{b}_1 + \beta_2 B\mathbf{b}_2 \,.$$

The second inclusion in (2) is self-evident, since $B\mathbf{b}_1, B\mathbf{b}_2 \in \mathcal{N}_B$ and $\{\mathbf{a}_1, \mathbf{a}_2\}$ is a basis for $\mathcal{N}_B$.

Next, to verify (3), suppose that

$$\beta_1 B\mathbf{b}_1 + \beta_2 B\mathbf{b}_2 = \mathbf{0}$$

for some choice of constants $\beta_1, \beta_2 \in \mathbb{C}$. Then the subsequent formula

$$B(\beta_1 \mathbf{b}_1 + \beta_2 \mathbf{b}_2) = \mathbf{0}$$

implies that $\beta_1 \mathbf{b}_1 + \beta_2 \mathbf{b}_2 \in \mathcal{N}_B$ and hence that

$$\beta_1 \mathbf{b}_1 + \beta_2 \mathbf{b}_2 = \alpha_1 \mathbf{a}_1 + \alpha_2 \mathbf{a}_2$$

for some choice of constants $\alpha_1, \alpha_2 \in \mathbb{C}$. However, since the four vectors in the last line are linearly independent, this means that $\alpha_1 = \alpha_2 = \beta_1 = \beta_2 = 0$. Therefore, (3) follows.

If $B^2\mathbf{c}_1$ and $B\mathbf{b}_1$ are linearly dependent, then $\gamma_1 B\mathbf{c}_1 + \beta_1 \mathbf{b}_1 \in \mathcal{N}_B$ for some choice of $\gamma_1, \beta_1 \in \mathbb{C}$ which are not both equal to zero. In fact, since $B\mathbf{c}_1 \notin \mathcal{N}_B$ and $\mathbf{b}_1 \notin \mathcal{N}_B$, both of these constants are different from zero. Similarly, if $B^2\mathbf{c}_1$ and $B\mathbf{b}_2$ are linearly dependent, then $\gamma_2 B\mathbf{c}_1 + \beta_2 \mathbf{b}_2 \in \mathcal{N}_B$ for some choice of constants $\gamma_2, \beta_2 \in \mathbb{C}$ which both differ from zero. Therefore,

$$\gamma_2(\gamma_1 B\mathbf{c}_1 + \beta_1 \mathbf{b}_1) - \gamma_1(\gamma_2 B\mathbf{c}_1 + \beta_2 \mathbf{b}_2) = \gamma_2\beta_1 \mathbf{b}_1 - \gamma_1\beta_2 \mathbf{b}_2$$

also belongs to $\mathcal{N}_B$, contrary to assumption, unless $\gamma_2\beta_1 = 0$ and $\gamma_1\beta_2 = 0$. This justifies (4).

Suppose next that the vectors $B^2\mathbf{c}_1$ and $B\mathbf{b}_1$ are linearly independent and that there exist constants such that

$$\gamma_1 \mathbf{c}_1 + \gamma_2 B\mathbf{c}_1 + \gamma_3 B^2\mathbf{c}_1 + \beta_1 \mathbf{b}_1 + \beta_2 B\mathbf{b}_1 = \mathbf{0} \,.$$

Then, upon multiplying both sides on the left by B^2, it is readily seen that $\gamma_1 = 0$. Next, upon multiplying both sides on the left by B, it follows that

$$\gamma_2 B^2\mathbf{c}_1 + \beta_1 B\mathbf{b}_1 = \mathbf{0} \,,$$

which, in view of the conditions imposed in (5), implies that $\gamma_2 = \beta_1 = 0$. Thus, the original linear combination of 5 vectors reduces to

$$\gamma_3 B^2\mathbf{c}_1 + \beta_2 B\mathbf{b}_1 = \mathbf{0} \,,$$

which forces $\gamma_3 = \beta_2 = 0$. This completes the proof of (5); the proof of (6) is similar.

In case (5), the set of vectors $\{B^2\mathbf{c}_1, B\mathbf{c}_1, \mathbf{c}_1, B\mathbf{b}_1, \mathbf{b}_1\}$ is a basis for $\mathcal{N}_{B^3}$. Moreover, since

$$\begin{aligned} B[B^2\mathbf{c}_1 \quad B\mathbf{c}_1 \quad \mathbf{c}_1 \quad B\mathbf{b}_1 \quad \mathbf{b}_1] &= [B^3\mathbf{c}_1 \quad B^2\mathbf{c}_1 \quad B\mathbf{c}_1 \quad B^2\mathbf{b}_1 \quad B\mathbf{b}_1] \\ &= [\mathbf{0} \quad B^2\mathbf{c}_1 \quad B\mathbf{c}_1 \quad \mathbf{0} \quad B\mathbf{b}_1] \\ &= [B^2\mathbf{c}_1 \quad B\mathbf{c}_1 \quad \mathbf{c}_1 \quad B\mathbf{b}_1 \quad \mathbf{b}_1]N, \end{aligned}$$

where

$$N = \operatorname{diag}\left\{C_0^{(3)}, C_0^{(2)}\right\},$$

it is now readily seen that the vectors

$$\mathbf{u}_1 = B^2\mathbf{c}_1,\ \mathbf{u}_2 = B\mathbf{c}_1,\ \mathbf{u}_3 = \mathbf{c}_1,\ \mathbf{u}_4 = B\mathbf{b}_1 \quad \text{and} \quad \mathbf{u}_5 = \mathbf{b}_1$$

are linearly independent and satisfy the equation

$$A[\mathbf{u}_1 \cdots \mathbf{u}_5] = [\mathbf{u}_1 \cdots \mathbf{u}_5] \begin{bmatrix} \lambda_1 & 1 & 0 & 0 & 0 \\ 0 & \lambda_1 & 1 & 0 & 0 \\ 0 & 0 & \lambda_1 & 0 & 0 \\ 0 & 0 & 0 & \lambda_1 & 1 \\ 0 & 0 & 0 & 0 & \lambda_1 \end{bmatrix}. \tag{6.6}$$

Similar conclusions prevail for case (6), but with $\mathbf{b}_2$ in place of $\mathbf{b}_1$.

6.7. Another example

In this section we shall present a second example to help clarify the general algorithm that was introduced in Section 6.5. To this end, assume for the sake of definiteness that

$$\begin{aligned} \dim \mathcal{N}_B &= 6,\ \dim \mathcal{N}_{B^2} = 10,\ \dim \mathcal{N}_{B^3} = 13 \quad \text{and} \\ \dim \mathcal{N}_{B^j} &= 15 \quad \text{for} \quad j = 4, \ldots, n. \end{aligned}$$

These numbers must meet the constraints imposed by (1) of Lemma 6.3 and the inequalities (6.2), but are otherwise completely arbitrary.

The eigenvectors and generalized eigenvectors corresponding to each Jordan block may be constructed as follows:

1. Construct a basis for $\mathcal{N}_{B^n}$ according to the following scheme:

$\mathbf{a}_1, \ldots, \mathbf{a}_6$ is a basis for $\mathcal{N}_B$,

$\mathbf{a}_1, \ldots, \mathbf{a}_6;\ \mathbf{b}_1, \ldots, \mathbf{b}_4$ is a basis for $\mathcal{N}_{B^2}$,

$\mathbf{a}_1, \ldots, \mathbf{a}_6;\ \mathbf{b}_1, \ldots, \mathbf{b}_4;\ \mathbf{c}_1, \mathbf{c}_2, \mathbf{c}_3$ is a basis for $\mathcal{N}_{B^3}$,

$\mathbf{a}_1, \ldots, \mathbf{a}_6;\ \mathbf{b}_1, \ldots, \mathbf{b}_4;\ \mathbf{c}_1, \mathbf{c}_2, \mathbf{c}_3;\ \mathbf{d}_1, \mathbf{d}_2$ is a basis for $\mathcal{N}_{B^4}$.

2. Construct **chains** of powers of B applied to each vector in the basis and display them in columns **of nonzero vectors** labeled 1-15:

1	2	3	4	5	6	7	8	9	10	$\cdots$	15
$B^3\mathbf{d}_1$	$B^3\mathbf{d}_2$	$B^2\mathbf{c}_1$	$B^2\mathbf{c}_2$	$B^2\mathbf{c}_3$	$B\mathbf{b}_1$	$B\mathbf{b}_2$	$B\mathbf{b}_3$	$B\mathbf{b}_4$	$\mathbf{a}_1$	$\cdots$	$\mathbf{a}_6$
$B^2\mathbf{d}_1$	$B^2\mathbf{d}_2$	$B\mathbf{c}_1$	$B\mathbf{c}_2$	$B\mathbf{c}_3$	$\mathbf{b}_1$	$\mathbf{b}_2$	$\mathbf{b}_3$	$\mathbf{b}_4$			
$B\mathbf{d}_1$	$B\mathbf{d}_2$	$\mathbf{c}_1$	$\mathbf{c}_2$	$\mathbf{c}_3$							
$\mathbf{d}_1$	$\mathbf{d}_2$										

3. Observe that the vectors in the first row of the preceding array belong to $\mathcal{N}_B$ and that

span $\{B^3\mathbf{d}_1, B^3\mathbf{d}_2\}$ is a 2-dimensional subspace of span $\{B^2\mathbf{c}_1, B^2\mathbf{c}_2, B^2\mathbf{c}_3\}$.

span $\{B^2\mathbf{c}_1, B^2\mathbf{c}_2, B^2\mathbf{c}_3\}$ is a 3-dimensional subspace of span $\{B\mathbf{b}_1, B\mathbf{b}_2, B\mathbf{b}_3, B\mathbf{b}_4\}$.

span $\{B\mathbf{b}_1, B\mathbf{b}_2, B\mathbf{b}_3, B\mathbf{b}_4\}$ is a 4-dimensional subspace of span $\{\mathbf{a}_1, \ldots, \mathbf{a}_6\}$.

Thus, for example, since $B\mathbf{d}_1 \in \mathcal{N}_{B^3}$, it follows that

$$B\mathbf{d}_1 = \sum_{j=1}^{6} \alpha_j \mathbf{a}_j + \sum_{j=1}^{4} \beta_j \mathbf{b}_j + \sum_{j=1}^{3} \gamma_j \mathbf{c}_j$$

for some choice of the 13 coefficients $\alpha_1, \ldots, \alpha_6$, $\beta_1, \ldots, \beta_4$, $\gamma_1, \gamma_2, \gamma_3$ and hence that

$$B^3\mathbf{d}_1 = \sum_{j=1}^{3} \gamma_j B^2 \mathbf{c}_j \,.$$

Moreover, if, say, $\alpha B^3\mathbf{d}_1 + \beta B^3\mathbf{d}_2 = \mathbf{0}$, then $\alpha\mathbf{d}_1 + \beta\mathbf{d}_2 \in \mathcal{N}_{B^3}$ and consequently, $\alpha\mathbf{d}_1 + \beta\mathbf{d}_2$ can be expressed as a linear combination of the vectors

$$\{\mathbf{a}_1, \ldots, \mathbf{a}_6, \mathbf{b}_1 \ldots, \mathbf{b}_4, \mathbf{c}_1 \ldots, \mathbf{c}_3\}\,.$$

However since all these vectors are linearly independent, this forces $\alpha = \beta = 0$. The remaining assertions may be verified in much the same way.

4. Build a basis for $\mathcal{N}_B$ by moving from left to right in the ordering

$$\begin{aligned} \operatorname{span}\{B^3\mathbf{d}_1, B^3\mathbf{d}_2\} &\subseteq \operatorname{span}\{B^2\mathbf{c}_1, B^2\mathbf{c}_2, B^2\mathbf{c}_3\} \\ &\subseteq \operatorname{span}\{B\mathbf{b}_1, B\mathbf{b}_2, B\mathbf{b}_3, B\mathbf{b}_4\} \subseteq \operatorname{span}\{\mathbf{a}_1, \ldots, \mathbf{a}_6\}\,, \end{aligned}$$

by adding vectors that increase the dimension of the space spanned by those selected earlier, starting with the set $\{B^3\mathbf{d}_1, B^3\mathbf{d}_2\}$. Thus, in the present setting, this is done as follows:

(i) Choose a vector $\mathbf{c}_j$ such that $B^2\mathbf{c}_j$ is linearly independent of $\{B^3\mathbf{d}_1, B^3\mathbf{d}_2\}$. There exists at least one such, say $\mathbf{c}_2$. Then span $\{B^3\mathbf{d}_1, B^3\mathbf{d}_2, B^2\mathbf{c}_2\}$ is a three-dimensional subspace of the three-dimensional space span $\{B^2\mathbf{c}_1, B^2\mathbf{c}_2, B^2\mathbf{c}_3\}$. Thus, these two spaces are equal and one moves to the next set of vectors on the right.

(ii) Choose a vector $\mathbf{b}_i$ such that $B\mathbf{b}_i$ is linearly independent of the vectors $\{B^3\mathbf{d}_1, B^3\mathbf{d}_2, B^2\mathbf{c}_2\}$. There exists at least one such, say $\mathbf{b}_1$. Then span $\{B^3\mathbf{d}_1, B^3\mathbf{d}_2, B^2\mathbf{c}_2, B\mathbf{b}_1\}$ is a four-dimensional subspace of the four-dimensional space span $\{B\mathbf{b}_1, B\mathbf{b}_2, B\mathbf{b}_3, B\mathbf{b}_4\}$. Thus, these two spaces are equal.

(iii) Choose a vector $\mathbf{a}_i$ that is linearly independent of $\{B^3\mathbf{d}_1, B^3\mathbf{d}_2, B^2\mathbf{c}_2, B\mathbf{b}_1\}$, say $\mathbf{a}_3$. Then span $\{B^3\mathbf{d}_1, B^3\mathbf{d}_2, B^2\mathbf{c}_2, B\mathbf{b}_1, \mathbf{a}_3\}$ is a five-dimensional subspace of the six-dimensional space span $\{\mathbf{a}_1, \ldots, \mathbf{a}_6\}$. Therefore another selection should be made from this set.

(iv) Choose a vector $\mathbf{a}_j$ that is linearly independent of $\{B^3\mathbf{d}_1, B^3\mathbf{d}_2, B^2\mathbf{c}_2, B\mathbf{b}_1, \mathbf{a}_3\}$, say $\mathbf{a}_5$. Then span $\{B^3\mathbf{d}_1, B^3\mathbf{d}_2, B^2\mathbf{c}_2, B\mathbf{b}_1, \mathbf{a}_3, \mathbf{a}_5\}$ is a six-dimensional subspace of the six-dimensional space span $\{\mathbf{a}_1, \ldots, \mathbf{a}_6\}$. Therefore the two spaces are equal and the selection procedure is complete.

5. The 15 vectors in the columns corresponding to $\{B^3\mathbf{d}_1, B^3\mathbf{d}_2, B^2\mathbf{c}_2, B\mathbf{b}_1, \mathbf{a}_3, \mathbf{a}_5\}$, i.e., columns 1, 2, 4, 6, 12, 14, are linearly independent. Since $\dim \mathcal{N}_{B^n} = 15$, these 15 linearly independent vectors in $\mathcal{N}_{B^n}$ form a basis for that space. Moreover, if $B = A - \lambda_1 I_n$, then each of the specified columns generates a Jordan cell $C_{\lambda_1}^{(\cdot)}$ of height equal to the height of the column.

Consider, for example, the cell corresponding to the first column. The four vectors in that column are stacked in order of decreasing powers of B to form an $n \times 4$ matrix:

$$\begin{aligned} B\begin{bmatrix} B^3\mathbf{d}_1 & B^2\mathbf{d}_1 & B\mathbf{d}_1 & \mathbf{d}_1 \end{bmatrix} &= \begin{bmatrix} \mathbf{0} & B^3\mathbf{d}_1 & B^2\mathbf{d}_1 & B\mathbf{d}_1 \end{bmatrix} \\ &= \begin{bmatrix} B^3\mathbf{d}_1 & B^2\mathbf{d}_1 & B\mathbf{d}_1 & \mathbf{d}_1 \end{bmatrix} C_0^{(4)} . \end{aligned}$$

Thus, upon writing $\mathbf{u}_1 = B^3\mathbf{d}_1$, $\mathbf{u}_2 = B^2\mathbf{d}_1$, $\mathbf{u}_3 = B\mathbf{d}_1$, $\mathbf{u}_4 = \mathbf{d}_1$ and setting $B = A - \lambda_1 I_n$, the last formula can be rewritten as

$$(A - \lambda_1 I_n)\begin{bmatrix} \mathbf{u}_1 & \mathbf{u}_2 & \mathbf{u}_3 & \mathbf{u}_4 \end{bmatrix} = \begin{bmatrix} \mathbf{u}_1 & \mathbf{u}_2 & \mathbf{u}_3 & \mathbf{u}_4 \end{bmatrix} \begin{bmatrix} 0 & 1 & 0 & 0 \\ 0 & 0 & 1 & 0 \\ 0 & 0 & 0 & 1 \\ 0 & 0 & 0 & 0 \end{bmatrix}$$

or, equivalently, as

$$A\begin{bmatrix}\mathbf{u}_1 & \mathbf{u}_2 & \mathbf{u}_3 & \mathbf{u}_4\end{bmatrix} = \begin{bmatrix}\mathbf{u}_1 & \mathbf{u}_2 & \mathbf{u}_3 & \mathbf{u}_4\end{bmatrix}\begin{bmatrix}\lambda_1 & 1 & 0 & 0\\ 0 & \lambda_1 & 1 & 0\\ 0 & 0 & \lambda_1 & 1\\ 0 & 0 & 0 & \lambda_1\end{bmatrix}.$$

Continuing in this fashion, set $\mathbf{u}_5 = B^3\mathbf{d}_2$, $\mathbf{u}_6 = B^2\mathbf{d}_2$, $\mathbf{u}_7 = B\mathbf{d}_2$, $\mathbf{u}_8 = \mathbf{d}_2$, $\mathbf{u}_9 = B^2\mathbf{c}_2$, $\mathbf{u}_{10} = B\mathbf{c}_2$, $\mathbf{u}_{11} = \mathbf{c}_2$, $\mathbf{u}_{12} = B\mathbf{b}_1$, $\mathbf{u}_{13} = \mathbf{b}_1$, $\mathbf{u}_{14} = \mathbf{a}_3$, $\mathbf{u}_{15} = \mathbf{a}_2$ and

$$U = [\mathbf{u}_1 \quad \mathbf{u}_2 \quad \cdots \quad \mathbf{u}_{15}].$$

It is readily seen that

$$AU = UB_{\lambda_1} \quad \text{where} \quad B_{\lambda_1} = \operatorname{diag}\{C_{\lambda_1}^{(4)}, C_{\lambda_1}^{(4)}, C_{\lambda_1}^{(3)}, C_{\lambda_1}^{(2)}, C_{\lambda_1}^{(1)}, C_{\lambda_1}^{(1)}\}.$$

Exercise 6.9. Find a Jordan form J and an invertible matrix U such that

$$\begin{bmatrix}2 & 0 & 0 & 0 & 2\\ 0 & 2 & 0 & 0 & 0\\ 0 & 2 & 2 & 0 & 0\\ 0 & 0 & 0 & 2 & 0\\ 0 & 0 & 0 & 0 & 2\end{bmatrix} = UJU^{-1}.$$

Exercise 6.10. Find a Jordan form J and an invertible matrix U such that

$$\begin{bmatrix}1 & 2 & 0 & 0\\ 0 & 1 & 2 & 0\\ 0 & 0 & 1 & 2\\ 0 & 0 & 0 & 1\end{bmatrix} = UJU^{-1}.$$

Exercise 6.11. Find a Jordan form J and an invertible matrix U such that

$$\begin{bmatrix}1 & 2 & 0 & 0\\ 0 & 1 & 0 & 0\\ x & 0 & 1 & 0\\ 0 & 0 & 2 & 1\end{bmatrix} = UJU^{-1},$$

first for $x = 0$ and then for $x = 1$.

Exercise 6.12. Find a Jordan form J and an invertible matrix U such that

$$A = \begin{bmatrix}2 & 0 & 0 & 0 & 0 & 0 & 1\\ 0 & 3 & 0 & 0 & 1 & 0 & 0\\ 0 & 0 & 3 & 0 & 0 & 0 & 0\\ 0 & 0 & 1 & 3 & 0 & 0 & 0\\ 0 & 0 & 0 & 0 & 3 & 0 & 0\\ 0 & 0 & 0 & 1 & 0 & 3 & 0\\ 0 & 0 & 0 & 0 & 0 & 0 & 2\end{bmatrix} = UJU^{-1}.$$

Exercise 6.13. Find a Jordan form J and an invertible matrix U such that

$$A = \begin{bmatrix} 1 & 1 & 0 & 0 & 0 \\ 1 & 2 & 1 & 0 & 0 \\ 13 & -8 & 6 & 0 & 0 \\ -11 & 8 & -5 & 1 & 1 \\ -22 & 17 & -12 & -4 & 5 \end{bmatrix} = UJU^{-1}.$$

[HINT: The first step is to compute $\det(\lambda I_5 - A)$. This is easier than it looks at first glance if you take advantage of the block triangular structure of A.]

6.8. Jordan decompositions for real matrices

The preceding analysis guarantees the existence of a Jordan decomposition $A = UJU^{-1}$ with $J, U \in \mathbb{C}^{n\times n}$ for every $A \in \mathbb{C}^{n\times n}$ and hence also for every $A \in \mathbb{R}^{n\times n}$. However, even if $A \in \mathbb{R}^{n\times n}$, J and U may have entries that are not real. Our next objective is to deduce analogous decompositions for $A \in \mathbb{R}^{n\times n}$, but with both J and U in $\mathbb{R}^{n\times n}$.

It suffices to focus on the Jordan chains that correspond to each of the Jordan cells that appear in J. Thus, if $C_\lambda^{(k)}$ is one of the diagonal blocks in J, the preceding analysis guarantees the existence of a set of linearly independent vectors $\mathbf{u}_1 \ldots, \mathbf{u}_k \in \mathbb{C}^n$ such that

$$A[\mathbf{u}_1 \quad \cdots \quad \mathbf{u}_k] = [\mathbf{u}_1 \quad \cdots \quad \mathbf{u}_k]C_\lambda^{(k)} \tag{6.7}$$

is in force or, equivalently, upon setting $B = A - \lambda I_n$, such that

$$[\mathbf{u}_1 \quad \cdots \quad \mathbf{u}_k] = [B^{k-1}\mathbf{u}_k \quad B^{k-2}\mathbf{u}_k \quad \cdots \quad \mathbf{u}_k] \quad \text{and} \quad B^k\mathbf{u}_k = \mathbf{0}. \tag{6.8}$$

Exercise 6.14. Show that the two conditions (6.7) and (6.8) are equivalent.

There are two cases to consider for $\lambda \in \sigma(A)$: $\lambda \in \mathbb{R}$ and $\lambda \notin \mathbb{R}$.

Case 1: $A \in \mathbb{R}^{n\times n}$ and $\lambda \in \sigma(A) \cap \mathbb{R}$.

Lemma 6.8. *Let $A \in \mathbb{R}^{n\times n}$, let $\lambda \in \sigma(A)\cap\mathbb{R}$ and let $\mathbf{u}_1 \ldots, \mathbf{u}_k$ be a Jordan chain in $\mathbb{C}^n$ corresponding to a Jordan cell $C_\lambda^{(k)}$. Then there exists a Jordan chain $\mathbf{v}_1 \ldots, \mathbf{v}_k$ in $\mathbb{R}^n$ corresponding to $C_\lambda^{(k)}$.*

Proof. Let $B = A - \lambda I_n$, let $\mathbf{u}_1 \ldots, \mathbf{u}_k$ be a Jordan chain in $\mathbb{C}^n$ corresponding to $C_\lambda^{(k)}$ and let $\mathbf{u}_j = \mathbf{x}_j + i\mathbf{y}_j$, where $\mathbf{x}_j$ and $\mathbf{y}_j$ denote the real and imaginary parts of the vector $\mathbf{u}_j$, respectively, for $j = 1, \ldots, k$. Then, by assumption, the given set of vectors satisfy the constraint (6.7) and are linearly independent over $\mathbb{C}$. Moreover, in view of the equivalence between (6.7) and (6.8), this means that

$$[\mathbf{u}_1 \quad \cdots \quad \mathbf{u}_k] = [B^{k-1}\mathbf{u}_k \quad B^{k-2}\mathbf{u}_k \quad \cdots \quad \mathbf{u}_k] \quad \text{and} \quad B^k\mathbf{u}_k = \mathbf{0}$$

or, equivalently, that

$$\begin{aligned} [\mathbf{x}_1 \quad \cdots \quad \mathbf{x}_k] &= [B^{k-1}\mathbf{x}_k \quad B^{k-2}\mathbf{x}_k \quad \cdots \quad \mathbf{x}_k], \quad B^k\mathbf{x}_k = \mathbf{0}, \\ [\mathbf{y}_1 \quad \cdots \quad \mathbf{y}_k] &= [B^{k-1}\mathbf{y}_k \quad B^{k-2}\mathbf{y}_k \quad \cdots \quad \mathbf{y}_k] \quad \text{and} \quad B^k\mathbf{y}_k = \mathbf{0}. \end{aligned}$$

It remains to check that at least one of the two sets of vectors

$$\{B^{k-1}\mathbf{x}_k, B^{k-2}\mathbf{x}_k, \ldots, \mathbf{x}_k\}, \quad \{B^{k-1}\mathbf{y}_k, B^{k-2}\mathbf{y}_k, \ldots, \mathbf{y}_k\}$$

is linearly independent over $\mathbb{R}$. In view of Lemma 6.2, it suffices to show that at least one of the two conditions $B^{k-1}\mathbf{x}_k \neq \mathbf{0}$ and $B^{k-1}\mathbf{y}_k \neq \mathbf{0}$ is in force. But this is clearly the case, since $B^{k-1}\mathbf{u}_k \neq \mathbf{0}$. □

Case 2: $A \in \mathbb{R}^{n\times n}$ and $\lambda \in \sigma(A) \cap (\mathbb{C} \setminus \mathbb{R})$.

If $A \in \mathbb{R}^{n\times n}$, then the characteristic polynomial $p(\lambda) = \det(\lambda I_n - A)$ has real coefficients. Therefore, the nonreal roots of $p(\lambda)$ come in conjugate pairs. Thus, for example, if

$$A\begin{bmatrix} \mathbf{u}_1 & \mathbf{u}_2 & \mathbf{u}_3 \end{bmatrix} = \begin{bmatrix} \mathbf{u}_1 & \mathbf{u}_2 & \mathbf{u}_3 \end{bmatrix}\begin{bmatrix} \lambda_1 & 1 & 0 \\ 0 & \lambda_1 & 1 \\ 0 & 0 & \lambda_1 \end{bmatrix} \quad \text{and} \quad \lambda_1 \notin \mathbb{R},$$

then, taking the complex conjugate of both sides,

$$A\begin{bmatrix} \overline{\mathbf{u}_1} & \overline{\mathbf{u}_2} & \overline{\mathbf{u}_3} \end{bmatrix} = \begin{bmatrix} \overline{\mathbf{u}_1} & \overline{\mathbf{u}_2} & \overline{\mathbf{u}_3} \end{bmatrix}\begin{bmatrix} \overline{\lambda_1} & 1 & 0 \\ 0 & \overline{\lambda_1} & 1 \\ 0 & 0 & \overline{\lambda_1} \end{bmatrix}$$

and, since $\lambda_1 \neq \overline{\lambda_1}$,

$$\operatorname{span}\{\mathbf{u}_1, \mathbf{u}_2, \mathbf{u}_3\} \cap \operatorname{span}\{\overline{\mathbf{u}_1}, \overline{\mathbf{u}_2}, \overline{\mathbf{u}_3}\} = \{\mathbf{0}\}.$$

Thus, the rank of the $n \times 6$ matrix $\begin{bmatrix} \mathbf{u}_1 & \mathbf{u}_2 & \mathbf{u}_3 & \overline{\mathbf{u}_1} & \overline{\mathbf{u}_2} & \overline{\mathbf{u}_3} \end{bmatrix}$ is equal to 6. Therefore, the same holds true for the $n \times 6$ real matrix

$$\begin{bmatrix} \mathbf{x}_1 & \mathbf{y}_1 & \mathbf{x}_2 & \mathbf{y}_2 & \mathbf{x}_3 & \mathbf{y}_3 \end{bmatrix} = \frac{1}{2}\begin{bmatrix} \mathbf{u}_1 & \mathbf{u}_2 & \mathbf{u}_3 & \overline{\mathbf{u}_1} & \overline{\mathbf{u}_2} & \overline{\mathbf{u}_3} \end{bmatrix} Q,$$

since the matrix

$$Q = \begin{bmatrix} 1 & -i & 0 & 0 & 0 & 0 \\ 0 & 0 & 1 & -i & 0 & 0 \\ 0 & 0 & 0 & 0 & 1 & -i \\ 1 & i & 0 & 0 & 0 & 0 \\ 0 & 0 & 1 & i & 0 & 0 \\ 0 & 0 & 0 & 0 & 1 & i \end{bmatrix}$$

is invertible. Moreover, upon writing λ_1 in polar coordinates as $\lambda_1 = re^{i\theta} = r\cos\theta + ir\sin\theta$, it is readily checked that

$$A\begin{bmatrix} \mathbf{x}_1 & \mathbf{y}_1 & \mathbf{x}_2 & \mathbf{y}_2 & \mathbf{x}_3 & \mathbf{y}_3 \end{bmatrix} = \begin{bmatrix} \mathbf{x}_1 & \mathbf{y}_1 & \mathbf{x}_2 & \mathbf{y}_2 & \mathbf{x}_3 & \mathbf{y}_3 \end{bmatrix}\Lambda,$$

where

$$\Lambda = \begin{bmatrix} r\cos\theta & r\sin\theta & 1 & 0 & 0 & 0 \\ -r\sin\theta & r\cos\theta & 0 & 1 & 0 & 0 \\ 0 & 0 & r\cos\theta & r\sin\theta & 1 & 0 \\ 0 & 0 & -r\sin\theta & r\cos\theta & 0 & 1 \\ 0 & 0 & 0 & 0 & r\cos\theta & r\sin\theta \\ 0 & 0 & 0 & 0 & -r\sin\theta & r\cos\theta \end{bmatrix}.$$

Analogous decompositions hold for other Jordan blocks.

Exercise 6.15. Let $A \in \mathbb{R}^{n\times n}$ and suppose that $n \geq 2$. Show that:

(1) There exists a one-dimensional subspace $\mathcal{U}$ of $\mathbb{C}^n$ that is invariant under A.

(2) There exists a subspace $\mathcal{V}$ of $\mathbb{R}^n$ of dimension less than or equal to two that is invariant under A.

6.9. Companion and generalized Vandermonde matrices

Lemma 6.9. *Let*

$$f(\lambda) = f_0 + f_1\lambda + \cdots + f_n\lambda^n, \; f_n \neq 0,$$

be a polynomial of degree n, let

$$S_f = \begin{bmatrix} 0 & 1 & \cdots & 0 & 0 \\ 0 & 0 & \cdots & 0 & 0 \\ \vdots & & & & \\ 0 & 0 & \cdots & 0 & 1 \\ -a_0 & -a_1 & \cdots & -a_{n-2} & -a_{n-1} \end{bmatrix}, \quad \textit{where} \quad a_j = f_j/f_n,$$

denote the companion matrix based on $f(\lambda)$ and let

$$\mathbf{v}(\lambda) = \begin{bmatrix} 1 \\ \lambda \\ \vdots \\ \lambda^{n-1} \end{bmatrix} \quad \textit{and} \quad \mathbf{f}(\lambda) = \begin{bmatrix} 0 \\ \vdots \\ 0 \\ f(\lambda) \end{bmatrix}.$$

Then

$$S_f\,\mathbf{v}(\lambda) = \lambda\,\mathbf{v}(\lambda) - \frac{1}{f_n}\mathbf{f}(\lambda) \tag{6.9}$$

and

$$S_f\,\mathbf{v}^{(j)}(\lambda) = \lambda\,\mathbf{v}^{(j)}(\lambda) + j\,\mathbf{v}^{(j-1)}(\lambda) - \frac{1}{f_n}\mathbf{f}^{(j)}(\lambda) \tag{6.10}$$

for $j = 1, \dots, n-1$.

Proof. By direct computation

$$\begin{aligned} S_f\,\mathbf{v}(\lambda) &= \begin{bmatrix} \lambda \\ \vdots \\ \lambda^{n-1} \\ -(a_0 + a_1\lambda + \cdots + a_{n-1}\lambda^{n-1}) \end{bmatrix} \\ &= \begin{bmatrix} \lambda \\ \vdots \\ \lambda^{n-1} \\ \lambda^n \end{bmatrix} - \frac{1}{f_n}\begin{bmatrix} 0 \\ \vdots \\ 0 \\ f(\lambda) \end{bmatrix}, \end{aligned}$$

which coincides with (6.9). The formulas in (6.10) are obtained by differentiating both sides of (6.9) j times with respect to λ. □

Corollary 6.10. In the setting of Lemma 6.9, assume that the polynomial $f(\lambda)$ admits a factorization of the form

$$f(\lambda) = f_n(\lambda - \lambda_1)^{m_1} \cdots (\lambda - \lambda_k)^{m_k}$$

with k distinct roots $\lambda_1, \cdots, \lambda_k$, and let

$$V_j = \begin{bmatrix} \frac{\mathbf{v}(\lambda_j)}{0!} & \frac{\mathbf{v}^{(1)}(\lambda_j)}{1!} & \cdots & \frac{\mathbf{v}^{(m_j-1)}(\lambda_j)}{(m_j-1)!} \end{bmatrix} \tag{6.11}$$

for $j = 1, \dots, k$. Then

$$S_f V_j = V_j(\lambda_j I_{m_j} + C_0^{(m_j)}) = V_j C_{\lambda_j}^{(m_j)} \quad \text{for} \quad j = 1, \dots, k\,. \tag{6.12}$$

Exercise 6.16. Verify formula (6.12) when $m_j = 4$.

A matrix of the form $V = [V_1 \quad \cdots \quad V_k]$, with V_j as in (6.11) is called a **generalized Vandermonde** matrix.

Corollary 6.11. The vectors in a generalized Vandermonde matrix are linearly independent.

Exercise 6.17. Verify Corollary 6.11.

Example 6.12. If

$$f(\lambda) = (\lambda - \alpha)^3(\lambda - \beta)^2\,,$$

then

$$\begin{bmatrix} 0 & 1 & 0 & 0 & 0 \\ 0 & 0 & 1 & 0 & 0 \\ 0 & 0 & 0 & 1 & 0 \\ 0 & 0 & 0 & 0 & 1 \\ -f_0 & -f_1 & -f_2 & -f_3 & -f_4 \end{bmatrix} \begin{bmatrix} 1 & 0 & 0 & 1 & 0 \\ \alpha & 1 & 0 & \beta & 1 \\ \alpha^2 & 2\alpha & 1 & \beta^2 & 2\beta \\ \alpha^3 & 3\alpha^2 & 3\alpha & \beta^3 & 3\beta^2 \\ \alpha^4 & 4\alpha^3 & 6\alpha^2 & \beta^4 & 4\beta^3 \end{bmatrix}$$

$$= \begin{bmatrix} 1 & 0 & 0 & 1 & 0 \\ \alpha & 1 & 0 & \beta & 1 \\ \alpha^2 & 2\alpha & 1 & \beta^2 & 2\beta \\ \alpha^3 & 3\alpha^2 & 3\alpha & \beta^3 & 3\beta^2 \\ \alpha^4 & 4\alpha^3 & 6\alpha^2 & \beta^4 & 4\beta^3 \end{bmatrix} \begin{bmatrix} \alpha & 1 & 0 & 0 & 0 \\ 0 & \alpha & 1 & 0 & 0 \\ 0 & 0 & \alpha & 0 & 0 \\ 0 & 0 & 0 & \beta & 1 \\ 0 & 0 & 0 & 0 & \beta \end{bmatrix}.$$

Exercise 6.18. Verify that the matrix identity in Example 6.12 is correct.

Theorem 6.13. *Let $f(\lambda)$ be a polynomial of degree n that admits a factorization of the form*

$$f(\lambda) = f_n(\lambda - \lambda_1)^{\alpha_1} \cdots (\lambda - \lambda_k)^{\alpha_k}$$

with k distinct roots $\lambda_1, \cdots, \lambda_k$. Then the companion matrix S_f is similar to a Jordan matrix J with one Jordan cell for each root:

$$S_f = VJV^{-1},$$

where V is a generalized Vandermonde matrix and

$$J = \operatorname{diag}\{C_{\lambda_1}^{(\alpha_1)}, \ldots, C_{\lambda_k}^{(\alpha_k)}\}.$$

Proof. This is an easy consequence of Corollary 6.10. □

This circle of ideas can also be run in the other direction, as indicated by the following two exercises.

Exercise 6.19. Let $A \in \mathbb{C}^{n\times n}$ have k distinct eigenvalues $\lambda_1, \ldots, \lambda_k$ with geometric multiplicities $\gamma_1, \ldots, \gamma_k$ and algebraic multiplicities $\alpha_1, \ldots, \alpha_k$, respectively. Show that if $\gamma_j = 1$ for $j = 1, \ldots, k$, then A is similar to a companion matrix S_f based on a polynomial $f(\lambda)$ and find $f(\lambda)$.

Exercise 6.20. Show that if $A \in \mathbb{C}^{n\times n}$ is similar to the Jordan matrix

$$J = \operatorname{diag}\{C_{\lambda_1}^{(4)}, C_{\lambda_1}^{(2)}, C_{\lambda_2}^{(3)}, C_{\lambda_2}^{(1)}, C_{\lambda_3}^{(3)}\},$$

then A is also similar to the block diagonal matrix $\operatorname{diag}\{S_{g_1}, S_{g_2}\}$ based on a pair of polynomials $g_1(\lambda)$ and $g_2(\lambda)$ and find the polynomials.

Exercise 6.21. Find a Jordan form for $\mu(I_n - \mu C_0^{(n)})^{-1}$ when $\mu \neq 0$.

Exercise 6.22. Find a Jordan form J for the matrix $A = \begin{bmatrix} 2 & 2 & 1 & 0 & 0 \\ 0 & 2 & 3 & 0 & 0 \\ 0 & 0 & 2 & 0 & 0 \\ 0 & 3 & 1 & 2 & 0 \\ 0 & 1 & 1 & 1 & 2 \end{bmatrix}$.

Exercise 6.23. Find an invertible matrix $U \in \mathbb{C}^{5\times 5}$ such that $A = UJU^{-1}$ for the matrix A with Jordan form J that was considered in Exercise 6.22.

Exercise 6.24. Let $B \in \mathbb{C}^{n\times n}$; let $\mathbf{u}_1 \in \mathcal{N}_B$; $\mathbf{v}_1, \mathbf{v}_2 \in \mathcal{N}_{B^2}$; $\mathbf{w}_1, \mathbf{w}_2 \in \mathcal{N}_{B^3}$; and assume that the 5 vectors $B^2\mathbf{w}_1$, $B^2\mathbf{w}_2$, $B\mathbf{v}_1$, $B\mathbf{v}_2$, $\mathbf{u}_1$ are linearly independent over $\mathbb{C}$. Show that the 11 vectors $B^2\mathbf{w}_1$, $B\mathbf{w}_1$, $\mathbf{w}_1$, $B^2\mathbf{w}_2$, $B\mathbf{w}_2$, $\mathbf{w}_2$, $B\mathbf{v}_1$, $\mathbf{v}_1$, $B\mathbf{v}_2$, $\mathbf{v}_2$, $\mathbf{u}_1$ are also linearly independent over $\mathbb{C}$.

Exercise 6.25. Find an invertible matrix U such that $U^{-1}AU$ is in Jordan form when $A = \begin{bmatrix} 1 & 0 & i \\ 0 & 2 & 0 \\ -i & 0 & 1 \end{bmatrix}$. [NOTE: $i = \sqrt{-1}$.]

Exercise 6.26. Find a Jordan form J for the matrix

$$A = \begin{bmatrix} 0 & 1 & 0 & 0 & 0 \\ 0 & 0 & 1 & 0 & 0 \\ 8 & -12 & 6 & 0 & 0 \\ -1 & 1 & 0 & 0 & 1 \\ -4 & 1 & 0 & -4 & 4 \end{bmatrix}.$$

[HINT: You may find the formula $x^3 - 6x^2 + 12x - 8 = (x-2)^3$ useful.]

Exercise 6.27. Find an invertible matrix U such that $AU = UJ$ for the matrices A and J considered in Exercise 6.26.

The next three exercises are adapted from [**58**].

Exercise 6.28. Show that if $n \geq 2$, then the matrix $(C_\mu^{(n)})^2$ is similar to the matrix $C_{\mu^2}^{(n)}$ if and only if $\mu \neq 0$.

Exercise 6.29. Let $B \in \mathbb{C}^{p\times p}$ be a triangular matrix with diagonal entries $b_{ii} = \lambda \neq 0$ for $i = 1, \ldots, p$ and let $V \in \mathbb{C}^{p\times p}$. Show that $B^2V = VB^2 \iff BV = VB$. [HINT: Separate even and odd powers of B in the binomial expansion of $(B - \lambda I_p)^p = O$ to obtain a pair of invertible matrices $P_1 = a_0 I_p + a_2 B^2 + \cdots$ and $P_2 = b_0 I_p + b_2 B^2 + \cdots$ such that $P_1 = BP_2$.]

Exercise 6.30. Let A, $B \in \mathbb{C}^{n\times n}$ and suppose that the eigenvalues of A and B are nonnegative and that $\mathcal{N}_A = \mathcal{N}_{A^2}$ and $\mathcal{N}_B = \mathcal{N}_{B^2}$. Show that $A^2 = B^2 \iff A = B$. [HINT: Invoke Exercise 6.28 to show that the Jordan decompositions $A = U_1 J_1 U_1^{-1}$ and $B = U_2 J_2 U_2^{-1}$ may be chosen so that $J_1 = J_2$ and hence that $J_1^2 V = V J_1^2$ for $V = U_1^{-1} U_2$. Then apply Exercise 6.29, block by block, to finish.]

Exercise 6.31. Let $A \in \mathbb{C}^{n\times n}$ be a companion matrix and let $p(\lambda) = \det(\lambda I_n - A)$. Show that

$$(6.13)\quad A\begin{bmatrix} 1 \\ \lambda \\ \vdots \\ \lambda^{n-1} \end{bmatrix} = \begin{bmatrix} \lambda \\ \lambda^2 \\ \vdots \\ \lambda^n - p(\lambda) \end{bmatrix}, \quad A\begin{bmatrix} 0 \\ 1 \\ \vdots \\ (n-1)\lambda^{n-2} \end{bmatrix} = \begin{bmatrix} 1 \\ 2\lambda \\ \vdots \\ n\lambda^{n-1} - p'(\lambda) \end{bmatrix}$$

and differentiate once again with respect to λ to obtain the next term in the indicated sequence of formulas.

Exercise 6.32. Find an invertible matrix U and a matrix J in Jordan form such that $A = UJU^{-1}$ if $A \in \mathbb{C}^{6\times 6}$ is a companion matrix, $\det(\lambda I_6 - A) = (\lambda - \lambda_1)^4(\lambda - \lambda_2)^2$ and $\lambda_1 \neq \lambda_2$. [HINT: Exploit the sequence of formulas indicated in (6.13), first with $\lambda = \lambda_1$ and then with $\lambda = \lambda_2$.]

Exercise 6.33. Let $A \in \mathbb{C}^{n\times n}$ with k distinct eigenvalues $\lambda_1, \ldots, \lambda_k$. Show that if the geometric multiplicity γ_j of λ_j is equal to one for $j = 1, \ldots, k$, then A is similar to a companion matrix.

Chapter 7

Normed linear spaces

I give you now Professor Twist, A conscientious scientist,

. . . . . .

Camped on a tropic riverside, One day he missed his loving bride.
She had, the guide informed him later, Been eaten by an alligator.
Professor Twist could not but smile. "You mean," he said, "a crocodile."

The Purist, by Ogden Nash

In this chapter we shall consider a number of different ways of assigning a number to each vector in a vector space $\mathcal{U}$ over $\mathbb{C}$ that gives some indication of its size. Ultimately, our main interest will be in the vector spaces $\mathbb{C}^n$ and $\mathbb{R}^n$. But at this stage of the game it is useful to develop the material in a more general framework, because the extra effort is small and the dividends are significant. We shall also show that if a matrix $B \in \mathbb{C}^{n\times n}$ is sufficiently close to an invertible matrix $A \in \mathbb{C}^{n\times n}$, then B is invertible too. In other words, the invertibility of a square matrix is preserved under small perturbations of its entries. This suggests the following question: Is the rank of a matrix $A \in \mathbb{C}^{p\times q}$ preserved under small perturbations of its entries? The answer, as we shall see later in Chapter 17, is yes if the rank of A is equal to either p or q, but not otherwise.

7.1. Four inequalities

Throughout this subsection $s > 1$ and $t > 1$ will be two fixed numbers that are connected by the formula

$$\frac{1}{s} + \frac{1}{t} = 1\,. \tag{7.1}$$

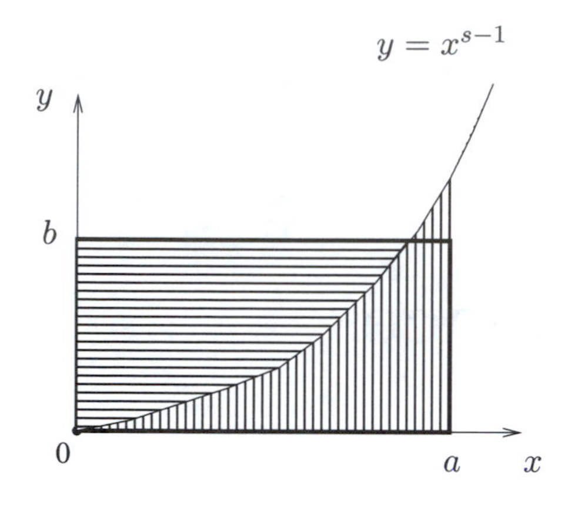

Figure 1

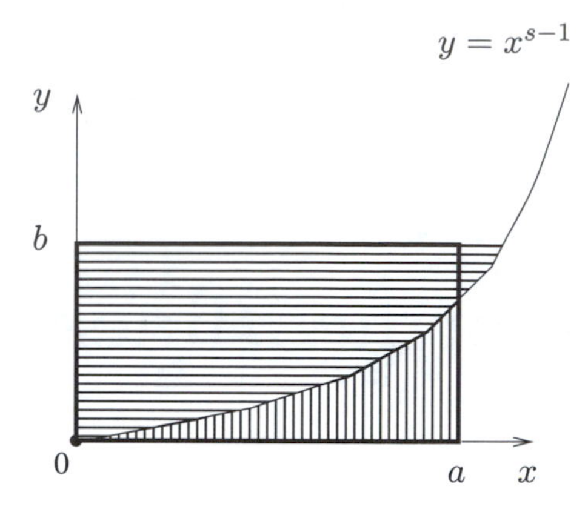

Figure 2

It is readily checked that

(7.2)
$$\frac{1}{s} + \frac{1}{t} = 1 \iff (s-1)(t-1) = 1 \iff (s-1)t = s \iff (t-1)s = t\ .$$

Lemma 7.1. *Let $a > 0$, $b > 0$, $s > 1$, $t > 1$ and $(s-1)(t-1) = 1$. Then*

$$ab \le \frac{a^s}{s} + \frac{b^t}{t}, \quad \text{with equality if and only if} \quad a^s = b^t\ . \tag{7.3}$$

Proof. The inequality will be obtained by comparing the areas of a rectangle R with horizontal sides of length a and vertical sides of length b with the area of the shaded regions that are formed between the x-axis and the curve $y = x^{s-1}$, $0 \le x \le a$, and the y axis and the same curve, now written as $x = y^{t-1}$, for $0 \le y \le b$, as sketched in the two figures.

The first figure corresponds to the case $a^{s-1} > b$; the second to the case $a^{s-1} < b$, and (7.2) guarantees that

$$y = x^{s-1} \iff x = y^{t-1}\ .$$

The rest is straightforward. It is clear from the figures that in each setting the area of the rectangle is less than or equal to the sum of the area of the vertically shaded piece and the area of the horizontally shaded piece:

$$\begin{aligned} ab &\le \int_0^a x^{s-1}dx + \int_0^b y^{t-1}dy \\ &= \frac{x^s}{s}\Big|_{x=0}^{x=a} + \frac{y^t}{t}\Big|_{y=0}^{y=b} \\ &= \frac{a^s}{s} + \frac{b^t}{t}\ . \end{aligned}$$

The figures also make it clear that equality will prevail in formula (7.3) if and only if $a^{s-1} = b$ or, equivalently, if and only if $a^{(s-1)t} = b^t$. But this is the same as the stated condition, since $a^s = a^{(s-1)t}$. □

Lemma 7.2. (Hölder's inequality) *Let $s > 1$, $t > 1$ and $(s-1)(t-1) = 1$. Then*

$$\sum_{k=1}^{n} |a_k b_k| \leq \left\{ \sum_{k=1}^{n} |a_k|^s \right\}^{1/s} \left\{ \sum_{k=1}^{n} |b_k|^t \right\}^{1/t} . \tag{7.4}$$

Moreover, equality will prevail in (7.4) if and only if the vectors $\mathbf{u}$ with components $u_j = |a_j|^s$ and $\mathbf{v}$ with components $v_j = |b_j|^t$ are linearly dependent.

Proof. We may assume that the right-hand side of the asserted inequality is not equal to zero, because otherwise the inequality is self-evident. [Why?] Let

$$\alpha_k = \frac{a_k}{\left\{\sum_{j=1}^n |a_j|^s\right\}^{1/s}} \quad \text{and} \quad \beta_k = \frac{b_k}{\left\{\sum_{j=1}^n |b_j|^t\right\}^{1/t}} .$$

Then

$$\sum_{k=1}^{n} |\alpha_k|^s = 1 \quad \text{and} \quad \sum_{k=1}^{n} |\beta_k|^t = 1 ,$$

and hence, in view of Lemma 7.1,

$$\sum_{k=1}^{n} |\alpha_k \beta_k| \leq \sum_{k=1}^{n} \frac{|\alpha_k|^s}{s} + \sum_{k=1}^{n} \frac{|\beta_k|^t}{t} = \frac{1}{s} + \frac{1}{t} = 1 .$$

This yields the desired inequality because

$$\sum_{k=1}^{n} |\alpha_k \beta_k| = \frac{\sum_{k=1}^n |a_k b_k|}{\left(\sum_{j=1}^n |a_j|^s\right)^{1/s} \left(\sum_{kj=1}^n |b_j|^t\right)^{1/t}} .$$

Finally, equality will prevail in (7.4) if and only if either (1) the right-hand side is equal to zero or (2) the right-hand side is not equal to zero and

$$|\alpha_i \beta_i| = \frac{|\alpha_i|^s}{s} + \frac{|\beta_i|^t}{t} \quad \text{for} \quad i = 1, \ldots, n .$$

Lemma 7.1 implies that the latter condition holds if and only if

$$|\alpha_i|^s = |\beta_i|^t \quad \text{for} \quad i = 1, \ldots, n ,$$

i.e., if and only if

$$\frac{|a_i|^s}{\sum_{j=1}^n |a_j|^s} = \frac{|b_i|^t}{\sum_{j=1}^n |b_j|^t} \quad \text{for} \quad i = 1, \ldots, n .$$

This completes the proof, since (1) and (2) are equivalent to the linear dependence of the vectors $\mathbf{u}$ and $\mathbf{v}$. □

The case $s = 2$ is of special interest because then $t = 2$ and the inequality (7.4) assumes a more symmetric form and gets a special name:

Lemma 7.3. (The Cauchy-Schwarz inequality) *Let* $\mathbf{a}, \mathbf{b} \in \mathbb{C}^n$ *with components* $a_1, \ldots, a_n$ *and* $b_1, \ldots, b_n$, *respectively. Then*

$$\sum_{k=1}^{n} |a_k b_k| \leq \left(\sum_{k=1}^{n} |a_k|^2\right)^{1/2} \left(\sum_{k=1}^{n} |b_k|^2\right)^{1/2}$$

with equality if and only if

$$\dim \operatorname{span}\{\mathbf{a}, \mathbf{b}\} \leq 1 .$$

Proof. The inequality is immediate from (7.4) by choosing $s = 2$ (which, as already remarked, then forces $t = 2$). ☐

Exercise 7.1. Show that if $\alpha, \beta \in \mathbb{R}$ and $\theta \in [0, 2\pi)$, then $\alpha \cos\theta + \beta \sin\theta \leq \sqrt{\alpha^2 + \beta^2}$ and that the upper bound is achieved for some choice of θ.

Lemma 7.4. (Minkowski's inequality) *Let* $1 \leq s < \infty$. *Then*

$$\left\{\sum_{k=1}^{n} |a_k + b_k|^s\right\}^{1/s} \leq \left\{\sum_{k=1}^{n} |a_k|^s\right\}^{1/s} + \left\{\sum_{k=1}^{n} |b_k|^s\right\}^{1/s} . \tag{7.5}$$

Proof. The case $s = 1$ is an immediate consequence of the fact that for every pair of complex numbers a and b, $|a + b| \leq |a| + |b|$. On the other hand, if $s > 1$, then

$$\begin{aligned} \sum_{k=1}^{n} |a_k + b_k|^s &= \sum_{k=1}^{n} |a_k + b_k|^{s-1} |a_k + b_k| \\ &\leq \sum_{k=1}^{n} |a_k + b_k|^{s-1} (|a_k| + |b_k|) . \end{aligned}$$

By Hölder's inequality,

$$\sum_{k=1}^{n} |a_k + b_k|^{s-1} |a_k| \leq \left\{\sum_{k=1}^{n} |a_k + b_k|^{(s-1)t}\right\}^{1/t} \left\{\sum_{k=1}^{n} |a_k|^s\right\}^{1/s}$$

and

$$\sum_{k=1}^{n} |a_k + b_k|^{s-1} |b_k| \leq \left\{\sum_{k=1}^{n} |a_k + b_k|^{(s-1)t}\right\}^{1/t} \left\{\sum_{k=1}^{n} |b_k|^s\right\}^{1/s}$$

Combining the last three inequalities, we obtain

$$\sum_{k=1}^{n} |a_k + b_k|^s \leq \left\{\sum_{k=1}^{n} |a_k + b_k|^s\right\}^{1/t} \left\{\left(\sum_{k=1}^{n} |a_k|^s\right)^{1/s} + \left(\sum_{k=1}^{n} |b_k|^s\right)^{1/s}\right\} .$$

Now, if

$$\sum_{k=1}^{n} |a_k + b_k|^s > 0 ,$$

then we can divide both sides of the last inequality by $\{\sum_{k=1}^{n} |a_k + b_k|^s\}^{1/t}$ to obtain the desired inequality (7.5).

It remains to consider the case $\sum_{k=1}^{n} |a_k + b_k|^s = 0$. But then the inequality (7.5) is self-evident. □

Exercise 7.2. Let $a_1, \ldots, a_n$ and $b_1, \ldots, b_n$ be nonnegative numbers and let $1 < s < \infty$. Show that

$$\left\{\sum_{k=1}^{n} |a_k + b_k|^s\right\}^{1/s} = \left\{\sum_{k=1}^{n} |a_k|^s\right\}^{1/s} + \left\{\sum_{k=1}^{n} |b_k|^s\right\}^{1/s} \tag{7.6}$$

if and only if the vectors **a** and **b** with components $a_1, \ldots, a_n$ and $b_1, \ldots, b_n$, respectively, are linearly dependent. [HINT: See how to change the inequalities in the proof of Minkowski's inequality to equalities.]

Remark 7.5. The inequality (7.3) is a special case of a more general statement that is usually referred to as **Young's inequality**:
If $a_1, \ldots, a_n$ *and* $p_1, \ldots, p_n$ *are positive numbers such that* $\frac{1}{p_1} + \cdots + \frac{1}{p_n} = 1$, *then*

$$a_1 \cdots a_n \leq \frac{a_1^{p_1}}{p_1} + \cdots + \frac{a_n^{p_n}}{p_n} . \tag{7.7}$$

A proof is spelled out in the following three exercises.

Exercise 7.3. Let $a_1, \ldots, a_n$ and $c_1, \ldots, c_n$ be positive numbers such that $c_1 + \cdots + c_n = 1$ and let $p > 1$. Show that

$$\left(\sum_{j=1}^{n} c_j a_j\right)^p \leq \sum_{j=1}^{n} c_j a_j^p . \tag{7.8}$$

[HINT: Write $c_j a_j = c_j^{1/q}(c_j^{1/p} a_j)$ and then invoke Hölder's inequality.]

Exercise 7.4. Verify Young's inequality when $n = 3$ by exploiting the inequality (7.3) to show that if

$$\frac{1}{p} = \frac{1}{p_1} + \frac{1}{p_2} \quad \text{and} \quad \frac{1}{q} = \frac{1}{p_3},$$

then:

(1) $a_1 a_2 a_3 \leq \dfrac{(a_1 a_2)^p}{p} + \dfrac{a_3^q}{q}$.

(2) $a_1 a_2 \leq \frac{p}{p_1} a_1^{p_1/p} + \frac{p}{p_2} a_2^{p_2/p}$.

(3) $\frac{p}{p_1}a_1^{p_1/p} + \frac{p}{p_2}a_2^{p_2/p} \le \left(\frac{p}{p_1}a_1^{p_1} + \frac{p}{p_2}a_2^{p_2}\right)^{1/p}$.

(4) Verify Young's inequality for $n = 3$.

[HINT: The inequality (7.8) is useful for (3).]

Exercise 7.5. Verify Young's inequality. [HINT: Use the steps in the preceding exercise as a guide.]

Exercise 7.6. Use Young's inequality to show that the geometric mean of a given set of positive numbers $b_1, \dots, b_n$ is less than or equal to its arithmetic mean, i.e.,

$$(b_1 b_2 \cdots b_n)^{1/n} \le \frac{b_1 + b_2 + \cdots + b_n}{n}. \tag{7.9}$$

7.2. Normed linear spaces

A vector space $\mathcal{U}$ over $\mathbb{F}$ is said to be a **normed linear space** if there exists a number $\varphi(\mathbf{x})$ assigned to each vector $\mathbf{x} \in \mathcal{U}$ such that for every choice of $\mathbf{x}, \mathbf{y} \in \mathcal{U}$ and every $\alpha \in \mathbb{F}$ the following four conditions are met:

(1) $\varphi(\mathbf{x}) \ge 0$.

(2) $\varphi(\mathbf{x}) = 0$ if and only if $\mathbf{x} = 0$.

(3) $\varphi(\alpha\mathbf{x}) = |\alpha|\varphi(\mathbf{x})$.

(4) $\varphi(\mathbf{x} + \mathbf{y}) \le \varphi(\mathbf{x}) + \varphi(\mathbf{y})$.

Any such function $\varphi(\mathbf{x})$ is said to be a **norm** and is usually denoted by the symbol $\|\mathbf{x}\|$, or by the symbol $\|\mathbf{x}\|_{\mathcal{U}}$, if it is desired to clarify the space under consideration. The inequality in (4) is called the **triangle inequality**.

Lemma 7.6. *Let $\mathcal{U}$ be a normed linear space over $\mathbb{F}$ with norm $\varphi(\mathbf{x})$. Then*

$$|\varphi(\mathbf{x}) - \varphi(\mathbf{y})| \le \varphi(\mathbf{x} - \mathbf{y})$$

for every choice of $\mathbf{x}$ and $\mathbf{y}$ in $\mathcal{U}$.

Proof. Item (4) in the property list of norms, i.e., the triangle inequality, implies that

$$\varphi(\mathbf{x}) = \varphi(\mathbf{x} - \mathbf{y} + \mathbf{y}) \le \varphi(\mathbf{x} - \mathbf{y}) + \varphi(\mathbf{y}).$$

Therefore,

$$\varphi(\mathbf{x}) - \varphi(\mathbf{y}) \le \varphi(\mathbf{x} - \mathbf{y}).$$

However, upon interchanging $\mathbf{x}$ and $\mathbf{y}$ in the last inequality, we obtain

$$\varphi(\mathbf{y}) - \varphi(\mathbf{x}) \le \varphi(\mathbf{y} - \mathbf{x}) = \varphi(\mathbf{x} - \mathbf{y}).$$

The last two inequalities imply that

$$-\varphi(\mathbf{x} - \mathbf{y}) \le \varphi(\mathbf{x}) - \varphi(\mathbf{y}) \le \varphi(\mathbf{x} - \mathbf{y}),$$

which is equivalent to the stated inequality. □

The simplest example of a norm on vectors $\mathbf{x} = \sum_{j=1}^n x_j\mathbf{u}_j$ in a finite dimensional vector space $\mathcal{U}$ over $\mathbb{F}$ with basis $\mathbf{u}_1, \ldots, \mathbf{u}_n$ is

$$\varphi(\mathbf{x}) = \max\{|x_j| : 1 \leq j \leq n\}.$$

To verify the triangle inequality, note that if $\mathbf{y} = \sum_{j=1}^n y_j\mathbf{u}_j$, then

$$|x_j + y_j| \leq |x_j| + |y_j| \leq \max_j |x_j| + \max_j |y_j|.$$

Exercise 7.7. Let $\mathcal{U}$ be a vector space over $\mathbb{C}$ with basis $\mathbf{u}_1, \ldots, \mathbf{u}_n$. Show that for each choice of s in the interval $1 \leq s < \infty$ the formula

$$\varphi\left(\sum_{j=1}^n x_j\mathbf{u}_j\right) = \left\{\sum_{j=1}^n |x_j|^s\right\}^{1/s}$$

also defines a norm on $\mathcal{U}$. [HINT: Minkowski's inequality (7.5) is useful.]

In the special case that $\mathcal{U} = \mathbb{F}^n$ and $\mathbf{u}_j = \mathbf{e}_j$, the j'th column of I_n, the norms considered above are commonly denoted by the symbols $\|\mathbf{x}\|_\infty$ and $\|\mathbf{x}\|_s$, respectively:

$$\|\mathbf{x}\|_\infty = \max\{|x_j| : 1 \leq j \leq n\} \tag{7.10}$$

and

$$\|\mathbf{x}\|_s = \left\{\sum_{j=1}^n |x_j|^s\right\}^{1/s} \quad \text{for } 1 \leq s < \infty. \tag{7.11}$$

This notation is also sometimes adopted in general normed linear spaces $\mathcal{U}$, but then care must be taken, because the numbers in formulas (7.10) and (7.11) depend upon the choice of the basis.

The **most important norms** in $\mathbb{C}^n$ are: $\|\mathbf{x}\|_1$, $\|\mathbf{x}\|_2$ and $\|\mathbf{x}\|_\infty$; the choice $s = 2$ yields the familiar Euclidean norm:

$$\|\mathbf{x}\|_2 = \left\{\sum_{j=1}^n |x_j|^2\right\}^{1/2}. \tag{7.12}$$

Exercise 7.8. Show that if $s \geq 1$ and $t \geq 0$, then

$$\|\mathbf{x}\|_s \geq \|\mathbf{x}\|_{s+t} \geq \|\mathbf{x}\|_\infty \quad \text{for each vector } \mathbf{x} \in \mathbb{C}^n. \tag{7.13}$$

[HINT: If $y_j = (\|\mathbf{x}\|_s)^{-1}|x_j|$, then $0 \leq y_j \leq 1$ and $\sum_{j=0}^n y_j^{s+t} \leq \sum_{j=0}^n y_j^s = 1$.]

Exercise 7.9. Show that $\lim_{s\uparrow\infty} \|\mathbf{x}\|_s = \|\mathbf{x}\|_\infty$ for each vector $\mathbf{x} \in \mathbb{C}^n$.

Exercise 7.10. Sketch the sets $\{\mathbf{x} \in \mathbb{R}^2 : \|\mathbf{x}\|_t \leq 1\}$ for $t = 1, 2$ and ∞.

Exercise 7.11. Show that $\mathbb{C}^{p\times q}$ is a normed linear space over $\mathbb{C}$ with respect to each of the norms

$$(7.14)\quad \|A\|_s = \begin{cases} \left\{\sum_{i=1}^{p}\sum_{j=1}^{q}|a_{ij}|^s\right\}^{1/s} & \text{if} \quad 1\le s<\infty \\ \max\{|a_{ij}| : i=1,\ldots,p;\, j=1,\ldots,q\} & \text{if} \quad s=\infty, \end{cases}$$

in which a_{ij} denotes the ij entry of the matrix A.

Exercise 7.12. Show that if $A = \begin{bmatrix} a & a \\ a & a \end{bmatrix}$ with $a > 0$, then $\|A^2\|_s > \|A\|_s^2$ when $2 < s \le \infty$.

Exercise 7.13. Show that the matrix A defined in Exercise 7.12 satisfies the inequality if $4\|A^2\|_s > \|A\|_s^2$ for each choice of s in the interval $1 < s \le \infty$.

A subset Q of a normed linear space $\mathcal{U}$ over $\mathbb{F}$ is said to be **convex** if

$$\mathbf{x},\mathbf{y}\in Q \Longrightarrow t\mathbf{x}+(1-t)\mathbf{y}\in Q \quad \text{for every} \quad 0\le t\le 1.$$

The balls of radius $r > 0$

(7.15)

$$B_r(\mathbf{a}) = \{\mathbf{x}\in\mathcal{U} : \|\mathbf{a}-\mathbf{x}\| < r\} \text{ and } \overline{B_r(\mathbf{a})} = \{\mathbf{x}\in\mathcal{U} : \|\mathbf{a}-\mathbf{x}\| \le r\}$$

are both convex sets.

Exercise 7.14. Verify the claim that the open and closed balls defined in (7.15) are both convex.

7.3. Equivalence of norms

The proof of the next theorem depends upon some elementary concepts from analysis. A brief survey of the facts needed here and further on is furnished in Appendix I. However, the proof may be skipped without loss of continuity.

Theorem 7.7. *On a finite dimensional vector space all norms are equivalent; i.e., if $\varphi(\mathbf{x})$ and $\psi(\mathbf{x})$ are norms on a finite dimensional vector space $\mathcal{U}$ over $\mathbb{F}$, then there exists a pair of positive constants γ_1 and γ_2 such that*

$$\gamma_1\varphi(\mathbf{x}) \le \psi(\mathbf{x}) \le \gamma_2\varphi(\mathbf{x}).$$

Proof. Let $\mathbf{u}_1,\ldots,\mathbf{u}_n$ be a basis for $\mathcal{U}$, let $\mathbf{x} = \sum_{j=1}^n x_j\mathbf{u}_j$ and $\mathbf{y} = \sum_{j=1}^n y_j\mathbf{u}_j$ and let $\mathbf{e}_i$ denote the i'th column of the identity matrix I_n. Then

$$\begin{aligned} |\varphi(\mathbf{x})-\varphi(\mathbf{y})| &\le \varphi(\mathbf{x}-\mathbf{y}) = \varphi\left(\sum_{i=1}^n (x_i-y_i)\mathbf{u}_i\right) \\ &\le \sum_{i=1}^n |x_i-y_i|\varphi(\mathbf{u}_i) \le \left\{\sum_{i=1}^n |x_i-y_i|^2\right\}^{1/2} \left\{\sum_{i=1}^n \varphi(\mathbf{u}_i)^2\right\}^{1/2}, \end{aligned}$$

by the Cauchy-Schwarz inequality. But this proves that

$$(7.16) \qquad |\varphi(\mathbf{x}) - \varphi(\mathbf{y})| \leq \beta \|S\mathbf{x} - S\mathbf{y}\|_2 \quad \text{and} \quad |\varphi(\mathbf{x})| \leq \beta \|S\mathbf{x}\|_2 ,$$

where

$$\beta = \left\{ \sum_{i=1}^{n} \varphi(\mathbf{u}_i)^2 \right\}^{1/2}$$

and S denotes the linear transformation from $\mathcal{U}$ **onto** $\mathbb{F}^n$ that is defined by the formula

$$S\left(\sum_{j=1}^{n} x_j \mathbf{u}_j \right) = \sum_{j=1}^{n} x_j \mathbf{e}_j .$$

The first inequality in (7.16) guarantees that $\varphi(\mathbf{x})$ is continuous. Moreover, by the properties of a norm, $\varphi(\mathbf{x}) > 0$ for every vector $\mathbf{x}$ in the set $\{\mathbf{x} : \|S\mathbf{x}\|_2 = 1\}$. Now let

$$\alpha = \inf \{\varphi(\mathbf{x}) : \|S\mathbf{x}\|_2 = 1\} .$$

By the definition of infimum, there exists a sequence of vectors $\mathbf{x}_1, \mathbf{x}_2, \ldots$ in $\mathcal{U}$ with $\|S\mathbf{x}_j\|_2 = 1$ such that $\varphi(\mathbf{x}_j) \to \alpha$ as $j \uparrow \infty$. Since $\{\mathbf{v} \in \mathbb{F}^n : \|\mathbf{v}\|_2 = 1\}$ is a closed bounded set in $\mathbb{F}^n$, it is a compact subset of $\mathbb{F}^n$ and hence there exist a subsequence of vectors $\mathbf{x}_{k_1}, \mathbf{x}_{k_2} \ldots$ in $\mathcal{U}$ and a vector $\mathbf{y} \in \mathcal{U}$ such that $\|S\mathbf{x}_{k_j} - S\mathbf{y}\|_2 \to 0$ as $j \uparrow \infty$. But this in turn implies that $\varphi(\mathbf{x}_{k_j}) \to \varphi(\mathbf{y})$ as $j \uparrow \infty$ and hence that $\alpha = \varphi(\mathbf{y}) > 0$. Therefore,

$$(7.17) \qquad \alpha \leq \varphi(\mathbf{u}) \leq \beta \quad \text{when} \quad \|S\mathbf{u}\|_2 = 1 .$$

Now take any vector $\mathbf{u} \in \mathcal{U}$ with $\mathbf{u} \neq \mathbf{0}$. Then the inequality (7.17) is applicable to $\mathbf{u}/\|S\mathbf{u}\|_2$ and implies that

$$\alpha \leq \varphi(\mathbf{u}/\|S\mathbf{u}\|_2) \leq \beta$$

or, equivalently, that

$$(7.18) \qquad \alpha \|S\mathbf{u}\|_2 \leq \varphi(\mathbf{u}) \leq \beta \|S\mathbf{u}\|_2$$

for every nonzero vector $\mathbf{u} \in \mathcal{U}$. But this last inequality is clearly valid for $\mathbf{u} = \mathbf{0}$ also. A similar pair of inequalities holds for $\psi(\mathbf{u})$:

$$(7.19) \qquad \alpha_1 \|S\mathbf{u}\|_2 \leq \psi(\mathbf{u}) \leq \beta_1 \|S\mathbf{u}\|_2$$

with $0 < \alpha_1 < \beta_1$. The statement of the theorem now follows easily by combining (7.18) and (7.19):

$$\frac{\alpha}{\beta_1} \psi(\mathbf{u}) \leq \varphi(\mathbf{u}) \leq \frac{\beta}{\alpha_1} \psi(\mathbf{u}) .$$

□

Remark 7.8. Even though all norms on a finite dimensional normed linear space are equivalent in the sense established above, particular choices may be most appropriate for certain applications. Thus, for example, if the entries u_i in a vector $\mathbf{u} \in \mathbb{R}^n$ denote deviations from a navigational path, such as a channel through shallow waters, it's important to keep $\|\mathbf{u}\|_\infty$ small. If $\mathbf{a}, \mathbf{b} \in \mathbb{R}^2$, then although $\|\mathbf{a} - \mathbf{b}\|_2$ is equal to the usual Euclidean distance between the points $\mathbf{a}$ and $\mathbf{b}$, the norm $\|\mathbf{a}-\mathbf{b}\|_1$ might give a better indication of the driving distance.

7.4. Norms of linear transformations

The set of linear transformations from a finite dimensional normed linear space $\mathcal{U}$ over $\mathbb{F}$ into a normed linear space $\mathcal{V}$ over $\mathbb{F}$ is clearly a vector space over $\mathbb{F}$ with respect to the natural rules of vector addition:

$$(S_1 + S_2)\mathbf{u} = S_1\mathbf{u} + S_2\mathbf{u}$$

and scalar multiplication:

$$(\alpha S)\mathbf{u} = \alpha(S\mathbf{u}) .$$

A particularly useful norm on a linear transformation S from a finite dimensional normed linear space $\mathcal{U}$ over $\mathbb{F}$ into a normed linear space $\mathcal{V}$ over $\mathbb{F}$ is defined by the following recipe:

$$\|S\|_{\mathcal{U},\mathcal{V}} = \max\{\|S\mathbf{u}\|_\mathcal{V} : \ \|\mathbf{u}\|_\mathcal{U} \le 1\} . \tag{7.20}$$

This norm will be referred to as the **operator norm** of the linear transformation. The usefulness of this number rests on the fact that it is **multiplicative** in the sense that is spelled out in Theorem 7.10. But the first order of business is to verify that this number defines a norm and to itemize some of its properties.

Theorem 7.9. *The number $\|S\|_{\mathcal{U},\mathcal{V}}$ that is defined by formula* (7.20) *defines a norm on the set of linear transformations S from a finite dimensional normed linear space $\mathcal{U}$ over $\mathbb{F}$ into a normed linear space $\mathcal{V}$ over $\mathbb{F}$. Moreover,*

(1) $\|S\mathbf{u}\|_\mathcal{U} \le \|S\|_{\mathcal{U},\mathcal{V}}\|\mathbf{u}\|_\mathcal{U}$ *for every vector* $\mathbf{u} \in \mathcal{U}$.

(2) $\|S\|_{\mathcal{U},\mathcal{V}} = \max\{\|S\mathbf{u}\|_\mathcal{V} : \mathbf{u} \in \mathcal{U} \quad \text{and} \quad \|\mathbf{u}\|_\mathcal{U} = 1\}$.

(3) $\|S\|_{\mathcal{U},\mathcal{V}} = \max\left\{\frac{\|S\mathbf{u}\|_\mathcal{V}}{\|\mathbf{u}\|_\mathcal{U}} : \mathbf{u} \in \mathcal{U} \quad \text{and} \quad \mathbf{u} \neq \mathbf{0}\right\}$.

Proof. It is readily checked that the number $\|S\|_{\mathcal{U},\mathcal{V}}$ meets the first three of the four stated requirements for a norm. To verify the fourth, let S_1 and S_2 be linear transformations from $\mathcal{U}$ into $\mathcal{V}$ and let $\mathbf{u} \in \mathcal{U}$ with $\|\mathbf{u}\|_\mathcal{U} \le 1$. Then, by the triangle inequality,

$$\|(S_1 + S_2)\mathbf{u}\|_\mathcal{V} = \|S_1\mathbf{u} + S_2\mathbf{u}\|_\mathcal{V} \le \|S_1\mathbf{u}\|_\mathcal{V} + \|S_2\mathbf{u}\|_\mathcal{V} \le \|S_1\|_{\mathcal{U},\mathcal{V}} + \|S_2\|_{\mathcal{U},\mathcal{V}} .$$

Thus, as this inequality holds for every choice of $\mathbf{u} \in \mathcal{U}$ with $\|\mathbf{u}\|_{\mathcal{U}} \leq 1$, it follows that

$$\|S_1 + S_2\|_{\mathcal{U},\mathcal{V}} \leq \|S_1\|_{\mathcal{U},\mathcal{V}} + \|S_2\|_{\mathcal{U},\mathcal{V}}.$$

Next, to verify (1), choose $\mathbf{u} \in \mathcal{U}$ with $\mathbf{u} \neq \mathbf{0}$ and let

$$\mathbf{w} = \frac{\mathbf{u}}{\|\mathbf{u}\|_{\mathcal{U}}}.$$

Then,

$$\|S\mathbf{u}\|_{\mathcal{V}} = \|S\mathbf{w}\|_{\mathcal{V}}\, \|\mathbf{u}\|_{\mathcal{U}} \leq \|S\|_{\mathcal{U},\mathcal{V}}\, \|\mathbf{u}\|_{\mathcal{U}},$$

since $\|\mathbf{w}\|_{\mathcal{U}} = 1$. This verifies (1). Moreover, the formula

$$\|S\mathbf{u}\|_{\mathcal{V}} = \|S\mathbf{w}\|_{\mathcal{V}}\, \|\mathbf{u}\|_{\mathcal{U}} \quad \text{with} \quad \|\mathbf{u}\|_{\mathcal{U}} \leq 1$$

implies that

$$\|S\mathbf{u}\|_{\mathcal{V}} \leq \max\{\|S\mathbf{y}\|_{\mathcal{V}} : \mathbf{y} \in \mathcal{U} \quad \text{and} \quad \|\mathbf{y}\|_{\mathcal{U}} = 1\}$$

and hence that

$$\|S\|_{\mathcal{U},\mathcal{V}} \leq \max\{\|S\mathbf{y}\|_{\mathcal{V}} : \mathbf{y} \in \mathcal{U} \text{ and } \|\mathbf{y}\|_{\mathcal{U}} = 1\}.$$

This serves to verify (2), since the opposite inequality is self-evident.

The verification of (3) rests on similar arguments and is left to the reader. □

Theorem 7.10. *Let $\mathcal{U}$, $\mathcal{V}$ and $\mathcal{W}$ be finite dimensional inner product spaces over $\mathbb{F}$ and let S_1 be a linear transformation from $\mathcal{U}$ into $\mathcal{V}$ and let S_2 be a linear transformation from $\mathcal{V}$ into $\mathcal{W}$. Then*

$$\|S_2S_1\|_{\mathcal{U},\mathcal{W}} \leq \|S_1\|_{\mathcal{U},\mathcal{V}}\, \|S_2\|_{\mathcal{V},\mathcal{W}}. \tag{7.21}$$

Proof. By (1) of Theorem 7.9,

$$\|S_2S_1\mathbf{u}\|_{\mathcal{W}} \leq \|S_2\|_{\mathcal{V},\mathcal{W}}\, \|S_1\mathbf{u}\|_{\mathcal{V}} \leq \|S_2\|_{\mathcal{V},\mathcal{W}}\, \|S_1\|_{\mathcal{U},\mathcal{V}}\, \|\mathbf{u}\|_{\mathcal{U}},$$

which leads easily to (7.21). □

Exercise 7.15. Let $A = \begin{bmatrix} 1 & 1 \\ 0 & 0 \end{bmatrix}$ and $B = \begin{bmatrix} 0 & 1 \\ 0 & 1 \end{bmatrix}$. Calculate $\|A\|_s$, $\|B\|_s$, $\|AB\|_s$ and $\|BA\|_s$ for $1 \leq s \leq \infty$, and determine for which values of s, if any, $\|AB\|_s \leq \|A\|_s\, \|B\|_s$.

7.5. Multiplicative norms

The ideas developed in the previous section are now applied to matrices. We shall say that a norm $\|\cdot\|$ on $\mathbb{C}^{n\times n}$ is **multiplicative** if

$$\|AB\| \leq \|A\|\|B\| \quad \text{for every} \quad A, B \in \mathbb{C}^{n\times n}. \tag{7.22}$$

In particular, we shall show that the norm of A as a linear transformation that sends $\mathbf{x} \in \mathbb{C}^n$ endowed with the norm $\|\mathbf{x}\|_s$ into $\mathbf{y} = A\mathbf{x} \in \mathbb{C}^n$ endowed with the norm $\|\mathbf{y}\|_s$ that is defined by the rule

$$\|A\|_{s,s} = \max\{\|A\mathbf{x}\|_s : \mathbf{x} \in \mathbb{C}^n \text{ and } \|\mathbf{x}\|_s \leq 1\} \tag{7.23}$$

for some choice of s in the interval $1 \leq s \leq \infty$ is a multiplicative norm. In fact this definition is easily extended to nonsquare matrices $A \in \mathbb{C}^{p\times q}$ with different norms on $\mathbb{C}^q$ and $\mathbb{C}^p$ by setting

$$\|A\|_{s,t} = \max\{\|A\mathbf{x}\|_t : \mathbf{x} \in \mathbb{C}^q \text{ and } \|\mathbf{x}\|_s \leq 1\} \quad \text{for} \quad 1 \leq s, t \leq \infty\,. \tag{7.24}$$

The norm $\|A\|_{s,t}$ is an operator norm; its value depends upon the choice of s and t. In this definition the numbers s and t are not linked by formula (7.1). Explicit evaluations of the number $\|A\|_{s,t}$ for some specific choices of s and t will be discussed in the next section. The remainder of this section is devoted to checking that formula (7.24) defines a norm on $\mathbb{C}^{p\times q}$ and that this norm is multiplicative in the sense that $\|AB\| \leq \|A\|\|B\|$ when the product of the two matrices is well defined. The first step is to explore the number $\|A\|_{s,t}$.

Lemma 7.11. *Let $A \in \mathbb{C}^{p\times q}$ and let $1 \leq s, t \leq \infty$. Then*

(1) $\|A\mathbf{x}\|_t \leq \|A\|_{s,t}\|\mathbf{x}\|_s$ *for every* $\mathbf{x} \in \mathbb{C}^q$.

(2) $\|A\|_{s,t} = \max\{\|A\mathbf{x}\|_t : \mathbf{x} \in \mathbb{C}^q$ *and* $\|\mathbf{x}\|_s = 1\}$.

(3) $\|A\|_{s,t} = \max\left\{\dfrac{\|A\mathbf{x}\|_t}{\|\mathbf{x}\|_s} : \mathbf{x} \neq \mathbf{0}\right\}$.

Proof. The inequality advertised in (1) is self-evident if $\mathbf{x} = \mathbf{0}$. Suppose, therefore, that $\mathbf{x} \neq \mathbf{0}$ and let $\mathbf{u} = (\|\mathbf{x}\|_s)^{-1}\mathbf{x}$. Then $\|\mathbf{u}\|_s = 1$ and

$$\|A\mathbf{x}\|_t = \|A\mathbf{u}\|_t\|\mathbf{x}\|_s \leq \|A\|_{s,t}\|\mathbf{x}\|_s\,.$$

Thus, the proof of (1) is complete.

Next, to establish (2), let

$$\mu_{s,t}(A) = \max\{\|A\mathbf{x}\|_t : \mathbf{x} \in \mathbb{C}^q \text{ and } \|\mathbf{x}\|_s = 1\}\,.$$

Then

$$\begin{aligned}
\|A\|_{s,t} &= \max\{\|A\mathbf{x}\|_t : \mathbf{x} \in \mathbb{C}^q \text{ and } \|\mathbf{x}\|_s \leq 1\} \\
&= \max\{\|A\mathbf{x}\|_t : \mathbf{x} \in \mathbb{C}^q,\ \|\mathbf{x}\|_s \leq 1 \text{ and } \mathbf{x} \neq \mathbf{0}\} \\
&= \max\{\|A\frac{\mathbf{x}}{\|\mathbf{x}\|_s}\|_t\|\mathbf{x}\|_s : \mathbf{x} \in \mathbb{C}^q,\ \|\mathbf{x}\|_s \leq 1 \text{ and } \mathbf{x} \neq \mathbf{0}\} \\
&\leq \mu_{s,t}(A)\max\{\|\mathbf{x}\|_s : \mathbf{x} \in \mathbb{C}^q \text{ and } \|\mathbf{x}\|_s \leq 1\}\,,
\end{aligned}$$

i.e.,

$$\|A\|_{s,t} \leq \mu_{s,t}(A)\,.$$

However, at the same time, $\mu_{s,t}(A) \le \|A\|_{s,t}$, since the maximization in the definition of the latter is taken over a larger set of vectors. Thus, (2) is in force. Much the same argument serves to justify (3). □

Exercise 7.16. Justify the formula for computing $\|A\|_{s,t}$ that is given in (3) of Lemma 7.11.

Lemma 7.12. *The vector space $\mathbb{C}^{p\times q}$ of $p \times q$ matrices A over $\mathbb{C}$ equipped with the norm $\|A\|_{s,t}$ is a normed linear space over $\mathbb{C}$.*

Proof. It is readily checked that $\|A\|_{s,t} \ge 0$, with equality if and only if $A = O_{p\times q}$, and that $\|\alpha A\|_{s,t} = |\alpha|\|A\|_{s,t}$. Therefore, it remains only to verify that if A and B are $p \times q$ matrices, then $\|A+B\|_{s,t} \le \|A\|_{s,t} + \|B\|_{s,t}$. But this is immediate from the definition of $\|A+B\|_{s,t}$ and the inequality

$$\|(A+B)\mathbf{x}\|_t \le \|A\mathbf{x}\|_t + \|B\mathbf{x}\|_t \le \|A\|_{s,t}\|\mathbf{x}\|_s + \|B\|_{s,t}\|\mathbf{x}\|_s.$$

□

Lemma 7.13. *Let $A \in \mathbb{C}^{p\times k}$, $B \in \mathbb{C}^{k\times q}$ and $1 \le r, s, t \le \infty$. Then*

$$\|AB\|_{r,t} \le \|B\|_{r,s}\|A\|_{s,t}.$$

Proof. The proof rests on the observation that

$$\begin{aligned} \|AB\mathbf{x}\|_t &\le \|A\|_{s,t}\|B\mathbf{x}\|_s \\ &\le \|A\|_{s,t}\|B\|_{r,s}\|\mathbf{x}\|_r \end{aligned}$$

and (3) of Lemma 7.11. □

The most important application of the last lemma is when $r = s = t = 2$:

$$\|AB\|_{2,2} \le \|B\|_{2,2}\|A\|_{2,2}. \tag{7.25}$$

Exercise 7.17. Show that if $A = \begin{bmatrix} a & b \\ 0 & c \end{bmatrix} \in \mathbb{R}^{2\times 2}$ and $d^2 = a^2 + b^2 + c^2$, then

$$\max\{\|A\mathbf{x}\|_2^2 : \mathbf{x} \in \mathbb{R}^2 \text{ and } \|\mathbf{x}\|_2 = 1\} = \frac{d^2 + \sqrt{d^4 - 4a^2c^2}}{2}.$$

[HINT: If $\mathbf{u} \in \mathbb{R}^2$ and $\|\mathbf{u}\|_2 = 1$, then $\mathbf{u}^T = [\cos\theta \quad \sin\theta]$, and to finish, refer to Exercise 7.1.]

7.6. Evaluating some operator norms

Lemma 7.14. *If $A \in \mathbb{C}^{p\times q}$, then:*

(1) $\|A\|_{1,1} = \max\limits_j \{\sum_{i=1}^p |a_{ij}|\}$.

(2) $\|A\|_{\infty,\infty} = \max\limits_i \left\{\sum_{j=1}^q |a_{ij}|\right\}$.

(3) $\|A\|_{2,2} = s_1$, *where s_1^2 is the maximum eigenvalue of the matrix $A^H A$.*

(4) $\|A\|_{1,\infty} = \|A\|_{\infty,1} = \max_{i,j} |a_{ij}|$.

(5) $\|A\|_{2,\infty} = \max_i \left\{ (\sum_{j=1}^q |a_{ij}|^2)^{1/2} \right\}$.

(6) $\|A\|_{1,2} = \max_j \left\{ (\sum_{i=1}^p |a_{ij}|^2)^{1/2} \right\}$.

Discussion. To obtain the first formula, observe that

$$\|A\mathbf{x}\|_1 = \sum_{i=1}^p \left| \sum_{j=1}^q a_{ij} x_j \right| \leq \sum_{j=1}^q \left(\sum_{i=1}^p |a_{ij}| \right) |x_j|$$

and hence that

$$\|A\mathbf{x}\|_1 \leq \max_j \left\{ \sum_{i=1}^p |a_{ij}| \right\} \|\mathbf{x}\|_1 . \tag{7.26}$$

This establishes the inequality

$$\|A\|_{1,1} \leq \max_j \left\{ \sum_{i=1}^p |a_{ij}| \right\} . \tag{7.27}$$

To obtain equality, it suffices to exhibit a vector $\mathbf{x} \in \mathbb{C}^q$ such that $\mathbf{x} \neq 0$ and equality prevails in formula (7.26). Suppose for the sake of definiteness that the maximum in (7.27) is achieved when $j = q$. Then for the vector $\mathbf{u}$ with $u_q = 1$ and all other coordinates equal to zero, we obtain $\|\mathbf{u}\|_1 = 1$ and

$$\|A\|_{1,1} \geq \frac{\|A\mathbf{u}\|_1}{\|\mathbf{u}\|_1} = \sum_{i=1}^p \left| \sum_{j=1}^q a_{ij} u_j \right| = \sum_{i=1}^p |a_{iq}| = \max_j \left\{ \sum_{i=1}^p |a_{ij}| \right\} .$$

This completes the proof of the first formula.

Next, to obtain the second formula, observe that

$$\left| \sum_{j=1}^q a_{ij} x_j \right| \leq \sum_{j=1}^q |a_{ij}||x_j| \leq \sum_{j=1}^q |a_{ij}| \|\mathbf{x}\|_\infty$$

and hence that

$$\|A\mathbf{x}\|_\infty = \max_i \left\{ \left| \sum_{j=1}^q a_{ij} x_j \right| \right\} \leq \max_i \left\{ \sum_{j=1}^q |a_{ij}| \right\} \|\mathbf{x}\|_\infty , \tag{7.28}$$

i.e.,

$$\|A\|_{\infty,\infty} \leq \max_i \left\{ \sum_{j=1}^q |a_{ij}| \right\} . \tag{7.29}$$

To obtain equality in (7.29), it suffices to exhibit a vector $\mathbf{x} \in \mathbb{C}^n$ such that $\mathbf{x} \neq 0$ and equality prevails in (7.28). Suppose for the sake of definiteness that the maximum in (7.29) is attained at $i = 1$ and that it is not equal to zero, and let $\mathbf{u}$ be the vector in $\mathbb{C}^n$ with entries

$$u_j = \begin{cases} \overline{a_{1j}}/|a_{1j}| & \text{if} \quad a_{1j} \neq 0 \\ 0 \quad \text{if} \quad a_{1j} = 0. \end{cases}$$

Then $\|\mathbf{u}\|_\infty = 1$ and

$$\|A\|_{\infty,\infty} \geq \frac{\|A\mathbf{u}\|_\infty}{\|\mathbf{u}\|_\infty} \geq \left|\sum_{j=1}^{q} a_{1j}u_j\right| = \sum_{j=1}^{q} |a_{1j}| = \max_i \left\{\sum_{j=1}^{q} |a_{ij}|\right\}.$$

This completes the proof of the second assertion if $A \neq O_{p\times q}$. However, if $A = O_{p\times q}$, then the asserted formula is self-evident.

We shall postpone the proof of the third assertion to Lemma 10.3 and leave the remaining assertions to the reader. □

Exercise 7.18. Compute the maximum eigenvalue of the matrix $A^H A$ when $A = \begin{bmatrix} a & b \\ 0 & c \end{bmatrix} \in \mathbb{R}^{2\times 2}$ and show that it is equal to the maximum that was calculated in Exercise 7.17.

7.7. Small perturbations

The central idea of this section is that if $A \in \mathbb{C}^{n\times n}$ is an invertible matrix and if $B \in \mathbb{C}^{n\times n}$ is close to A in the sense that $\|A - B\|$ is small enough with respect to some multiplicative norm, then B is also invertible. The main conclusion is based on the following lemma, which is important in its own right. Convergence of infinite sums and Cauchy sequences, which enter into the proof, are discussed briefly in Appendix I.

Lemma 7.15. *If $X \in \mathbb{C}^{n\times n}$ and $\|X\| < 1$ with respect to some multiplicative norm, then:*

(1) *$I_n - X$ is invertible.*

(2) *$(I_n - X)^{-1} = \sum_{j=0}^{\infty} X^j$; i.e., the sum converges.*

(3) $\|(I_n - X)^{-1}\| \leq \dfrac{1}{1 - \|X\|}$.

Proof. To verify (1) it suffices to show that 1 is not an eigenvalue of X. But if $\mathbf{u} = X\mathbf{u}$, then the self-evident inequality

$$\|\mathbf{u}\| = \|X\mathbf{u}\| \leq \|X\|\|\mathbf{u}\|$$

implies that

$$0 \leq (1 - \|X\|)\|\mathbf{u}\| \leq 0$$

and hence that $\mathbf{u} = \mathbf{0}$ and $I_n - X$ is invertible. Next, let

$$S_k = \sum_{i=0}^{k} X^i .$$

Then

$$\begin{aligned} \|S_{k+j} - S_k\| &= \|\sum_{i=k+1}^{k+j} X^i\| \le \sum_{i=k+1}^{k+j} \|X^i\| \\ &\le \sum_{i=k+1}^{k+j} \|X\|^i \le \sum_{i=k+1}^{\infty} \|X\|^i \\ &= \|X\|^{k+1}(1 - \|X\|)^{-1}, \end{aligned}$$

which can be made as small as you like by choosing k large enough, since $\|X\| < 1$. Consequently, the sequence of matrices $S_0, S_1, \ldots$ is a Cauchy sequence in $\mathbb{C}^{n\times n}$ and therefore must tend to a limit S_∞. (That is the meaning of the infinite sum in (2).)

The proof of (3) is a small variation of the proof of (2). The details are left to the reader. □

Exercise 7.19. Verify the bound in (3) of Lemma 7.15.

Exercise 7.20. Let $A \in \mathbb{C}^{n\times n}$ and $\lambda \in \mathbb{C}$. Show that if $|\lambda| > \|A\|$, then A is invertible and $\|(\lambda I_n - A)^{-1}\| \le (|\lambda| - \|A\|)^{-1}$.

Theorem 7.16. *Let $A, B \in \mathbb{C}^{n\times n}$ and suppose that A is invertible, that $\|A^{-1}\| \le \gamma$ and that $\|B - A\| < 1/\gamma$ with respect to some multiplicative norm for some number $\gamma > 0$. Then:*

(1) *B is invertible.*

(2) $\|B^{-1}\| \le \dfrac{\gamma}{1 - \gamma\|B - A\|}$.

Proof. Let

$$B = A - (A - B) = A(I_n - A^{-1}(A - B))$$

and set

$$X = A^{-1}(A - B) .$$

Then, since $B = A(I - X)$, the desired results are immediate from Lemma 7.15 and the estimate

$$\|X\| = \|A^{-1}(A - B)\| \le \|A^{-1}\|\|(A - B)\| \le \gamma\|(A - B)\| .$$

□

- The **spectral radius** $r_\sigma(A)$ of a matrix $A \in \mathbb{C}^{n\times n}$ is defined by the formula
$$r_\sigma(A) = \max\{|\lambda| : \lambda \in \sigma(A)\}.$$
Thus, if $\lambda_1, \dots, \lambda_k$ denote the distinct eigenvalues of A, then $r_\sigma(A) = \max\{|\lambda_1|, \dots, |\lambda_k|\}$.

Remark 7.17. Parts (1) and (2) of Lemma 7.15 hold under the assumption that $r_\sigma(X) < 1$, which is less restrictive than the constraint $\|X\| \le 1$, since

$$r_\sigma(X) \le \|X\| \tag{7.30}$$

for any multiplicative norm. The next exercise should help to clarify this point.

Exercise 7.21. Calculate $\|X\|$, $r_\sigma(X)$, $(I_2 - X)^{-1}$ and $\|(I_2 - X)^{-1}\|$ for the matrix
$$X = \begin{bmatrix} 1/2 & 2 \\ 0 & 1/2 \end{bmatrix},$$
using the formula in Exercise 7.17 to evaluate the norms.

Exercise 7.22. Show that if $X \in \mathbb{C}^{n\times n}$ and the spectral radius $r_\sigma(X) < 1$, then $I_n - X$ is invertible and the sequence of matrices $\{S_k\}$, $k = 0, 1, \dots$, defined in the proof of Lemma 7.15 is still a Cauchy sequence in $\mathbb{C}^{n\times n}$. However, the inequality $\|(I_n - X)^{-1}\| \le (1 - r_\sigma(X))^{-1}$ may fail. [HINT: You may find Exercise 7.21 helpful.]

Exercise 7.23. Calculate $\max\{\|(I_3 - X)^{-1}\mathbf{x}\|_2 : \mathbf{x} \in \mathbb{R}^3 \text{ and } \|\mathbf{x}\|_2 = 1\}$ for the matrix
$$X = \begin{bmatrix} 1/2 & 1 & 0 \\ 0 & 1/2 & 0 \\ 0 & 0 & 3 \end{bmatrix}.$$

7.8. Another estimate

The information in the next lemma will be useful in Chapter 17.

Lemma 7.18. *Let $A, B \in \mathbb{C}^{n\times n}$ and $\lambda \in \mathbb{C}$, and suppose that $\lambda I_n - A$ is invertible, that $\|(\lambda I_n - A)^{-1}\| \le \gamma$ and that $\|B - A\| < 1/\gamma$ with respect to some multiplicative norm for some number $\gamma > 0$. Then:*

(1) *$\lambda I_n - B$ is invertible.*

(2) $\|(\lambda I_n - B)^{-1}\| \le \dfrac{\gamma}{1 - \gamma\|B - A\|}$.

(3) $\|(\lambda I_n - A)^{-1} - (\lambda I_n - B)^{-1}\| \le \dfrac{\gamma^2\|A - B\|}{1 - \gamma\|A - B\|}$.

Proof. Clearly,

$$\begin{aligned} \lambda I_n - B &= \lambda I_n - A - (B - A) \\ &= (\lambda I_n - A)\{I_n - (\lambda I_n - A)^{-1}(B - A)\} \\ &= (\lambda I_n - A)(I_n - X), \end{aligned}$$

with

$$X = (\lambda I_n - A)^{-1}(B - A),$$

for short. Moreover,

$$\begin{aligned} \|X\| &= \|(\lambda I_n - A)^{-1}(B - A)\| \\ &\le \|(\lambda I_n - A)^{-1}\|\|B - A\| \\ &\le \gamma\|B - A\| < 1, \end{aligned}$$

by assumption. Therefore, by Lemma 7.15, the matrix $I_n - X$ is invertible and

$$\|(I_n - X)^{-1}\| \le (1 - \|X\|)^{-1}.$$

Thus,

$$\begin{aligned} \|(\lambda I_n - B)^{-1}\| &= \|(I_n - X)^{-1}(\lambda I_n - A)^{-1}\| \\ &\le \|(I_n - X)^{-1}\|\|(\lambda I_n - A)^{-1}\| \\ &\le \gamma(1 - \|X\|)^{-1} \le \gamma(1 - \gamma\|B - A\|)^{-1}, \end{aligned}$$

which justifies (2).

Finally, the bound furnished in (3) is an easy consequence of the formula

$$(\lambda I_n - A)^{-1} - (\lambda I_n - B)^{-1} = (\lambda I_n - A)^{-1}(A - B)(\lambda I_n - B)^{-1}$$

and the preceding estimates. □

7.9. Bounded linear functionals

A function $f(\mathbf{x})$ from a normed linear space $\mathcal{X}$ over $\mathbb{F}$ into $\mathbb{F}$ is said to be a **bounded linear functional** if

(1) $f(\alpha\mathbf{x}+\beta\mathbf{y}) = \alpha f(\mathbf{x})+\beta f(\mathbf{y})$ for every choice of $\mathbf{x}, \mathbf{y} \in \mathcal{X}$ and $\alpha, \beta \in \mathbb{F}$.

(2) There exists a constant c_f such that

$$|f(\mathbf{x})| \le c_f\|\mathbf{x}\| \text{ for every } \mathbf{x} \in \mathcal{X}.$$

The least constant c_f for which (2) holds is termed the norm of f and is designated by the symbol $\|f\|$. Thus,

$$\begin{aligned} \|f\| &= \sup\left\{\frac{|f(\mathbf{x})|}{\|\mathbf{x}\|} : \mathbf{x} \in \mathcal{X} \text{ and } \mathbf{x} \ne \mathbf{0}\right\} \\ &= \sup\{|f(\mathbf{x})| : \mathbf{x} \in \mathcal{X} \text{ and } \|\mathbf{x}\| = 1\}. \end{aligned}$$

Theorem 7.19. *Let f be a bounded linear functional from a normed linear space $\mathcal{X}$ over $\mathbb{F}$ into $\mathbb{F}$, let*

$$\mathcal{N}_f = \{\mathbf{x} \in \mathcal{X} : f(\mathbf{x}) = 0\}$$

and suppose that $\mathcal{N}_f \neq \mathcal{X}$. Then there exists a vector $\mathbf{x}_0 \in \mathcal{X}$ such that

$$\mathcal{X} = \mathcal{N}_f \dotplus \{\alpha \mathbf{x}_0 : \alpha \in \mathbb{F}\}.$$

Proof. Choose $\mathbf{x}_0 \in \mathcal{X}$ such that $f(\mathbf{x}_0) \neq 0$. Then the formula

$$\mathbf{x} = \left(\mathbf{x} - \frac{f(\mathbf{x})}{f(\mathbf{x}_0)}\mathbf{x}_0\right) + \left(\frac{f(\mathbf{x})}{f(\mathbf{x}_0)}\mathbf{x}_0\right)$$

clearly displays the fact that

$$\mathcal{N}_f + \operatorname{span}\{\mathbf{x}_0\} = \mathcal{X}.$$

It is also easy to see that this sum is direct, because if $\mathbf{y} = \alpha\mathbf{x}_0$ belongs to $\mathcal{N}_f$, then $f(\mathbf{y}) = \alpha f(\mathbf{x}_0) = 0$, which forces $\alpha = 0$ and hence $\mathbf{y} = \mathbf{0}$. □

Exercise 7.24. Let $f(\mathbf{x})$ be a linear functional on $\mathbb{C}^n$ and let $f(\mathbf{e}_j) = \alpha_j$ for $j = 1, \dots, n$, where $\mathbf{e}_j$ denotes the j'th column of I_n. Show that if $s > 1$ and $t = s/(s-1)$, then

$$\max\{|f(\mathbf{x})| : \|\mathbf{x}\|_s \leq 1\} = (\sum_{j=1}^{n} |\alpha_j|^t)^{1/t}.$$

[HINT: It's easy to show that $|f(\mathbf{x})| \leq (\sum_{j=1}^n |\alpha_j|^t)^{1/t}\|\mathbf{x}\|_s$. The particular vector $\mathbf{x}$ with coordinates $x_j = \overline{\alpha_j}|\alpha_j|^{t-2}$ when $\alpha_j \neq 0$ is useful for showing that the maximum is attained.]

Exercise 7.25. Let $f(\mathbf{x})$ be a linear functional on $\mathbb{C}^n$ and let $f(\mathbf{e}_j) = \alpha_j$ for $j = 1, \dots, n$, where $\mathbf{e}_j$ denotes the j'th column of I_n. Show that

$$\max\{|f(\mathbf{x})| : \|\mathbf{x}\|_\infty \leq 1\} = \sum_{j=1}^{n} |\alpha_j|.$$

[HINT: See the hint in Exercise 7.24.]

Exercise 7.26. Let $f(\mathbf{x})$ be a linear functional on $\mathbb{C}^n$ and let $f(\mathbf{e}_j) = \alpha_j$ for $j = 1, \dots, n$, where $\mathbf{e}_j$ denotes the j'th column of I_n. Show that

$$\max\{|f(\mathbf{x})| : \|\mathbf{x}\|_1 \leq 1\} = \max\{|\alpha_j| : j = 1, \dots, n\}.$$

[HINT: See the hint in Exercise 7.24.]

7.10. Extensions of bounded linear functionals

In this section we consider the problem of extending a bounded linear functional f that is specified on a proper subspace $\mathcal{U}$ of a normed linear space $\mathcal{X}$ over $\mathbb{F}$ to the full space $\mathcal{X}$ in such a way that the norm of the extension F is the same as the norm of f, i.e.,

$$\sup\{|F(\mathbf{x})| : \mathbf{x} \in \mathcal{X} \text{ and } \|\mathbf{x}\| \leq 1\} = \sup\{|f(\mathbf{u})| : \mathbf{u} \in \mathcal{U} \text{ and } \|\mathbf{u}\| \leq 1\}.$$

The fact that this is possible lies at the heart of the **Hahn-Banach** theorem.

It turns out to be more useful to phrase the extension problem in a slightly more general form that requires a new definition:

A real-valued function $p(\mathbf{x})$ on a vector space $\mathcal{X}$ over $\mathbb{F}$ is said to be a **seminorm** if for every choice of $\mathbf{x}, \mathbf{y} \in \mathcal{X}$ and $\alpha \in \mathbb{F}$ the following two conditions are met:

(1) $p(\mathbf{x}+\mathbf{y}) \leq p(\mathbf{x}) + p(\mathbf{y})$.

(2) $p(\alpha\mathbf{x}) = |\alpha| p(\mathbf{x})$.

Exercise 7.27. Show that if $p(\mathbf{x})$ is a seminorm on a vector space $\mathcal{X}$ over $\mathbb{F}$, then p also automatically satisfies the following additional three conditions:

(3) $p(\mathbf{0}) = 0$.

(4) $p(\mathbf{x}) \geq 0$ for every $\mathbf{x} \in \mathcal{X}$.

(5) $p(\mathbf{x}-\mathbf{y}) \geq |p(\mathbf{x}) - p(\mathbf{y})|$.

Theorem 7.20. *Let p be a seminorm on a finite dimensional vector space $\mathcal{X}$ over $\mathbb{F}$, and let f be a linear functional on a proper subspace $\mathcal{U}$ of $\mathcal{X}$ such that*

$$f(\mathbf{u}) \in \mathbb{F} \quad \textit{and} \quad |f(\mathbf{u})| \leq p(\mathbf{u}) \textit{ for every } \mathbf{u} \in \mathcal{U}.$$

Then there exists a linear functional F on the full space $\mathcal{X}$ such that

(1) *$F(\mathbf{u}) \in \mathbb{F}$ and $F(\mathbf{u}) = f(\mathbf{u})$ for every $\mathbf{u} \in \mathcal{U}$.*

(2) *$|F(\mathbf{x})| \leq p(\mathbf{x})$ for every $\mathbf{x} \in \mathcal{X}$.*

Proof. Suppose first that $\mathbb{F} = \mathbb{C}$ and let

$$g(\mathbf{u}) = \frac{f(\mathbf{u}) + \overline{f(\mathbf{u})}}{2} \quad \text{and} \quad h(\mathbf{u}) = \frac{f(\mathbf{u}) - \overline{f(\mathbf{u})}}{2i}$$

denote the real and imaginary parts of $f(\mathbf{u})$, respectively. Then

$$f(\mathbf{u}) = g(\mathbf{u}) + ih(\mathbf{u}),$$

and it is readily checked that

$$g(\alpha\mathbf{u}_1 + \beta\mathbf{u}_2) = \alpha g(\mathbf{u}_1) + \beta g(\mathbf{u}_2) \text{ and } h(\alpha\mathbf{u}_1 + \beta\mathbf{u}_2) = \alpha h(\mathbf{u}_1) + \beta h(\mathbf{u}_2)$$

for every choice of $\mathbf{u}_1, \mathbf{u}_2 \in \mathcal{U}$ and $\alpha, \beta \in \mathbb{R}$. Moreover,

$$g(\mathbf{u}) \leq |f(\mathbf{u})| \leq p(\mathbf{u}) \text{ for every } \mathbf{u} \in \mathcal{U}.$$

Let $\mathbf{v}_1 \in \mathcal{X} \setminus \mathcal{U}$ and suppose that there exists a real-valued function $G(\mathbf{u})$ such that

$$\begin{aligned} G(\mathbf{u} + \alpha\mathbf{v}_1) &= G(\mathbf{u}) + \alpha G(\mathbf{v}_1) \\ &= g(\mathbf{u}) + \alpha G(\mathbf{v}_1) \end{aligned}$$

and

$$G(\mathbf{u} + \alpha\mathbf{v}_1) \le p(\mathbf{u} + \alpha\mathbf{v}_1)$$

for every choice of $\mathbf{u} \in \mathcal{U}$ and $\alpha \in \mathbb{R}$. Then

$$g(\mathbf{u}) + \alpha G(\mathbf{v}_1) \le p(\mathbf{u} + \alpha\mathbf{v}_1)\,.$$

Thus, if $\alpha > 0$, then

$$\begin{aligned} g(\mathbf{u}) + \alpha G(\mathbf{v}_1) &\le p\left(\alpha\left\{\frac{\mathbf{u}}{\alpha} + \mathbf{v}_1\right\}\right) \\ &= \alpha p\left(\frac{\mathbf{u}}{\alpha} + \mathbf{v}_1\right) \end{aligned}$$

and

$$G(\mathbf{v}_1) \le p\left(\frac{\mathbf{u}}{\alpha} + \mathbf{v}_1\right) - g\left(\frac{\mathbf{u}}{\alpha}\right)$$

for every $\mathbf{u} \in \mathcal{U}$; i.e.,

$$G(\mathbf{v}_1) \le p(\mathbf{y} + \mathbf{v}_1) - g(\mathbf{y}) \tag{7.31}$$

for every $\mathbf{y} \in \mathcal{U}$.

On the other hand, if $\alpha < 0$, then

$$\begin{aligned} g(\mathbf{u}) + \alpha G(\mathbf{v}_1) &= g(\mathbf{u}) - |\alpha|G(\mathbf{v}_1) = G(\mathbf{u} - |\alpha|\mathbf{v}_1) \\ &\le p(\mathbf{u} - |\alpha|\mathbf{v}_1) = |\alpha|p\left(\frac{\mathbf{u}}{|\alpha|} - \mathbf{v}_1\right), \end{aligned}$$

and hence

$$G(\mathbf{v}_1) \ge g\left(\frac{\mathbf{u}}{|\alpha|}\right) - p\left(\frac{\mathbf{u}}{|\alpha|} - \mathbf{v}_1\right)$$

for every $\mathbf{u} \in \mathcal{U}$; i.e.,

$$G(\mathbf{v}_1) \ge g(\mathbf{x}) - p(\mathbf{x} - \mathbf{v}_1) \tag{7.32}$$

for every $\mathbf{x} \in \mathcal{U}$. Thus, in order for such an extension to exist, the number $G(\mathbf{v}_1)$ must meet the two sets of inequalities (7.31) and (7.32). Fortunately, this is possible: The inequality

$$\begin{aligned} g(\mathbf{x}) + g(\mathbf{y}) &= g(\mathbf{x} + \mathbf{y}) \le p(\mathbf{x} + \mathbf{y}) \\ &= p(\mathbf{x} - \mathbf{v}_1 + \mathbf{y} + \mathbf{v}_1) \\ &\le p(\mathbf{x} - \mathbf{v}_1) + p(\mathbf{y} + \mathbf{v}_1) \end{aligned}$$

implies that

$$g(\mathbf{x}) - p(\mathbf{x} - \mathbf{v}_1) \le p(\mathbf{y} + \mathbf{v}_1) - g(\mathbf{y})$$

and hence that

$$\text{(7.33)} \quad \sup\{g(\mathbf{x}) - p(\mathbf{x} - \mathbf{v}_1) : \mathbf{x} \in \mathcal{U}\} \leq \inf\{p(\mathbf{y} + \mathbf{v}_1) - g(\mathbf{y}) : \mathbf{y} \in \mathcal{U}\}\,.$$

The next step is to extend $f(\mathbf{u})$. This is facilitated by the observation that

$$g(i\mathbf{u}) + ih(i\mathbf{u}) = f(i\mathbf{u}) = if(\mathbf{u}) = ig(\mathbf{u}) - h(\mathbf{u})\,,$$

which implies that $h(\mathbf{u}) = -g(i\mathbf{u})$ and hence that

$$f(\mathbf{u}) = g(\mathbf{u}) - ig(i\mathbf{u})\,.$$

This suggests that

$$F(\mathbf{x}) = G(\mathbf{x}) - iG(i\mathbf{x})$$

might be a reasonable choice for the extension of $f(\mathbf{u})$, **if it's a linear functional** with respect to $\mathbb{C}$ that meets the requisite bound. Its clear that $F(\mathbf{x} + \mathbf{y}) = F(\mathbf{x}) + F(\mathbf{y})$. However, its not clear that $F(\alpha\mathbf{x}) = \alpha F(\mathbf{x})$ for every point $\alpha \in \mathbb{C}$, since $G(\alpha\mathbf{x}) = \alpha G(\mathbf{x})$ only for $\alpha \in \mathbb{R}$. To verify this, let $\alpha = a + ib$ with $a, b \in \mathbb{R}$. Then

$$\begin{aligned} F((a+ib)\mathbf{y}) &= G((a+ib)\mathbf{y}) - iG(i(a+ib)\mathbf{y}) \\ &= aG(\mathbf{y}) + bG(i\mathbf{y}) - iaG(i\mathbf{y}) + ibG(\mathbf{y}) \\ &= (a+ib)G(\mathbf{y}) - i(a+ib)G(i\mathbf{y}) \\ &= (a+ib)F(\mathbf{y}). \end{aligned}$$

Moreover, if $F(\mathbf{y}) \neq 0$, then upon writing the complex number $F(\mathbf{y})$ in polar coordinates as $F(\mathbf{y}) = e^{i\theta}|F(\mathbf{y})|$, it follows that

$$\begin{aligned} |F(\mathbf{y})| &= F(e^{-i\theta}\mathbf{y}) = G(e^{-i\theta}\mathbf{y}) \\ &\leq p(e^{-i\theta}\mathbf{y}) = |e^{-i\theta}|p(\mathbf{y}) = p(\mathbf{y}). \end{aligned}$$

Therefore, since $F(\mathbf{0}) = 0 = p(\mathbf{0})$, the inequality

$$|F(\mathbf{y})| \leq p(\mathbf{y})$$

holds for every vector $\mathbf{y} \in \mathcal{U} + \{\alpha\mathbf{v}_1 : \alpha \in \mathbb{C}\}$. This completes the proof for a one dimensional extension. The procedure can be repeated until the extension is defined on the full finite dimensional space $\mathcal{X}$ over $\mathbb{C}$.

The proof for the case $\mathbb{F} = \mathbb{R}$ is easily extracted from the preceding analysis and is left to the reader as an exercise. □

Exercise 7.28. Verify Theorem 7.20 for the case $\mathbb{F} = \mathbb{R}$. [HINT: The key is to verify (7.33) with $f = g$.]

Exercise 7.29. Show that if $\mathcal{X}$ is a finite dimensional normed linear space over $\mathbb{R}$, then Theorem 7.20 remains valid if $p(\mathbf{x})$ is only assumed to be a **sublinear functional** on $\mathcal{X}$; i.e., if for every choice of $\mathbf{x}, \mathbf{y} \in \mathcal{X}$, $p(\mathbf{x})$ satisfies the constraints (1) $\infty > p(\mathbf{x}) \geq 0$; (2) $p(\mathbf{x} + \mathbf{y}) \leq p(\mathbf{x}) + p(\mathbf{y})$; (3) $p(\alpha\mathbf{x} = \alpha p(\mathbf{x})$ for all $\alpha > 0$.

7.11. Banach spaces

A normed linear space $\mathcal{U}$ over $\mathbb{F}$ is said to be a **Banach space** over $\mathbb{F}$ if every Cauchy sequence $\mathbf{v}_1, \mathbf{v}_2, \ldots$ in $\mathcal{U}$ tends to a limit $\mathbf{v} \in \mathcal{U}$, i.e., if there exists a vector $\mathbf{v} \in \mathcal{U}$ such that $\lim_{n\uparrow\infty} \|\mathbf{v}_n - \mathbf{v}\|_{\mathcal{U}} = 0$. In this section we shall show that finite dimensional normed linear spaces are automatically Banach spaces. Not all normed linear spaces are Banach spaces.

Exercise 7.30. Let $\mathcal{U}$ be the space of continuous real-valued functions $f(x)$ on the interval $0 \le x \le 1$ equipped with the norm

$$\|f\|_{\mathcal{U}} = \int_0^1 |f(x)|dx.$$

Show that $\mathcal{U}$ is not a Banach space.

Remark 7.21. The vector space $\mathcal{U}$ considered in Exercise 7.30 is a Banach space with respect to the norm $\|f\|_{\mathcal{U}} = \max\{|f(x)| : 0 \le x \le 1\}$. This is a consequence of the Ascoli-Arzela theorem; see e.g., [**73**]. Thus, norms in infinite dimensional normed linear spaces are not necessarily equivalent.

Exercise 7.31. Let $\mathbf{u}_1, \ldots, \mathbf{u}_\ell$ be a basis for a normed linear space $\mathcal{U}$ over $\mathbb{F}$. Show that the functional

$$\varphi\left(\sum_{j=1}^{\ell} c_j \mathbf{u}_j\right) = \sum_{j=1}^{\ell} |c_j|$$

defines a norm on $\mathcal{U}$.

Theorem 7.22. *Every finite dimensional normed linear space $\mathcal{U}$ over $\mathbb{F}$ is automatically a Banach space over $\mathbb{F}$.*

Proof. Let $\mathbf{u}_1, \ldots, \mathbf{u}_\ell$ be a basis for $\mathcal{U}$ and let $\{\mathbf{v}_j\}_{j=1}^{\infty}$ be a Cauchy sequence in $\mathcal{U}$. Then, for every $\varepsilon > 0$ there exists a positive integer N such that

$$\|\mathbf{v}_{n+k} - \mathbf{v}_n\|_{\mathcal{U}} < \varepsilon \quad \text{for} \quad n \ge N \quad \text{and} \quad k \ge 1\,.$$

Let $\varphi(\mathbf{v})$ denote the norm introduced in Exercise 7.31. Then, in view of Theorem 7.7,

$$\alpha\|\mathbf{v}\|_{\mathcal{U}} \le \varphi(\mathbf{v}) \le \beta\|\mathbf{v}\|_{\mathcal{U}}$$

for some constants $0 < \alpha < \beta$. Thus the i'th coefficient c_{in} of $\mathbf{v}_n$ with respect to any basis $\{\mathbf{u}_1, \ldots, \mathbf{u}_\ell\}$ of $\mathcal{U}$ is subject to the bound

$$\begin{aligned} |c_{i,n+k} - c_{in}| &\le \sum_{j=1}^{\ell} |c_{j,n+k} - c_{jn}| \\ &= \varphi(\mathbf{v}_{n+k} - \mathbf{v}_n) \\ &\le \beta\|\mathbf{v}_{n+k} - \mathbf{v}_n\|_{\mathcal{U}}\,. \end{aligned}$$

But this means that $c_{in} \to d_i$ as $n \uparrow \infty$. Moreover, if

$$\mathbf{v} = \sum_{i=1}^{\ell} d_i \mathbf{u}_i \,,$$

then the inequalities

$$\begin{aligned} \alpha \|\mathbf{v} - \mathbf{v}_n\|_{\mathcal{U}} &\leq \varphi(\mathbf{v} - \mathbf{v}_n) \\ &= \varphi\left(\sum_{i=1}^{\ell} (c_{in} - d_i)\mathbf{u}_i\right) \\ &\leq \sum_{i=1}^{\ell} |c_{in} - d_i| \end{aligned}$$

clearly imply that $\|\mathbf{v} - \mathbf{v}_n\|_{\mathcal{U}} \to 0$ as $n \uparrow \infty$. □

Chapter 8

Inner product spaces and orthogonality

A proof should be as simple as possible, but no simpler.

Paraphrase of Albert Einstein's remark on deep truths

In this chapter we shall first introduce the notion of an inner product space and characterize its essential features. We then define orthogonality and study projections, orthogonal projections and related applications, including methods of orthogonalization and Gaussian quadrature.

8.1. Inner product spaces

A vector space $\mathcal{U}$ over $\mathbb{F}$ is said to be an **inner product space** if there is a number $\langle \mathbf{u}, \mathbf{v} \rangle_{\mathcal{U}} \in \mathbb{F}$ associated with every pair of vectors $\mathbf{u}, \mathbf{v} \in \mathcal{U}$ such that:

(1) $\langle \mathbf{u}+\mathbf{w}, \mathbf{v} \rangle_{\mathcal{U}} = \langle \mathbf{u}, \mathbf{v} \rangle_{\mathcal{U}} + \langle \mathbf{w}, \mathbf{v} \rangle_{\mathcal{U}}$ for every $\mathbf{w} \in \mathcal{U}$.

(2) $\langle \alpha\mathbf{u}, \mathbf{v} \rangle_{\mathcal{U}} = \alpha\langle \mathbf{u}, \mathbf{v} \rangle_{\mathcal{U}}$ for every $\alpha \in \mathbb{F}$.

(3) $\langle \mathbf{u}, \mathbf{v} \rangle_{\mathcal{U}} = \overline{\langle \mathbf{v}, \mathbf{u} \rangle_{\mathcal{U}}}$.

(4) $\langle \mathbf{u}, \mathbf{u} \rangle_{\mathcal{U}} \geq 0$ with equality if and only if $\mathbf{u} = \mathbf{0}$.

The number $\langle \mathbf{u}, \mathbf{v} \rangle_{\mathcal{U}}$ is termed the **inner product**. Items (1) and (2) imply that the inner product is linear in the first entry and hence, in particular, that

$$2\langle \mathbf{0}, \mathbf{v} \rangle_{\mathcal{U}} = \langle 2\mathbf{0}, \mathbf{v} \rangle_{\mathcal{U}} = \langle \mathbf{0}, \mathbf{v} \rangle_{\mathcal{U}},$$

which implies that $\langle \mathbf{0}, \mathbf{v}\rangle_{\mathcal{U}} = 0$. Item (3) then serves to guarantee that the inner product is additive in the second entry, i.e.,

$$\langle \mathbf{u}, \mathbf{v}+\mathbf{w}\rangle_{\mathcal{U}} = \langle \mathbf{u}, \mathbf{v}\rangle_{\mathcal{U}} + \langle \mathbf{u}, \mathbf{w}\rangle_{\mathcal{U}}\,; \quad \text{however,} \quad \langle \mathbf{u}, \beta\mathbf{v}\rangle_{\mathcal{U}} = \bar{\beta}\langle \mathbf{u}, \mathbf{v}\rangle_{\mathcal{U}}\,.$$

Usually we drop the subscript $\mathcal{U}$ from the symbol $\langle \mathbf{u}, \mathbf{v}\rangle_{\mathcal{U}}$ and simply write $\langle \mathbf{u}, \mathbf{v}\rangle$.

Exercise 8.1. Let $\mathcal{U}$ be an inner product space over $\mathbb{F}$ and let $\mathbf{u} \in \mathcal{U}$. Show that

$$\langle \mathbf{u}, \mathbf{v}\rangle = 0 \quad \text{for every} \quad \mathbf{v} \in \mathcal{U} \Longleftrightarrow \mathbf{u} = \mathbf{0}$$

and (consequently)

$$\langle \mathbf{u}_1, \mathbf{v}\rangle = \langle \mathbf{u}_2, \mathbf{v}\rangle \quad \text{for every} \quad \mathbf{v} \in \mathcal{U} \Longleftrightarrow \mathbf{u}_1 = \mathbf{u}_2\,.$$

The symbol $\langle \mathbf{x}, \mathbf{y}\rangle_{st}$, which is defined for $\mathbf{x}, \mathbf{y} \in \mathbb{F}^n$ by the formula

$$(8.1) \qquad \langle \mathbf{x}, \mathbf{y}\rangle_{st} = \mathbf{y}^H\mathbf{x} = \sum_{i=1}^{n} \overline{y_i} x_i\,,$$

will be used on occasion to denote the **standard inner product** on $\mathbb{F}^n$. The conjugation in this formula can be dropped if $\mathbf{x}$, $\mathbf{y} \in \mathbb{R}^n$. It is important to bear in mind that there are many other inner products that can be imposed on $\mathbb{F}^n$:

Exercise 8.2. Show that if $B \in \mathbb{C}^{n\times n}$ is invertible, then the formula

$$(8.2) \qquad \langle \mathbf{x}, \mathbf{y}\rangle = (B\mathbf{y})^H B\mathbf{x}$$

defines an inner product on $\mathbb{C}^n$.

Lemma 8.1. (The Cauchy-Schwarz inequality for inner products) *Let $\mathcal{U}$ be an inner product space over $\mathbb{C}$ with inner product $\langle \mathbf{u}, \mathbf{v}\rangle$ for every pair of vectors $\mathbf{u}, \mathbf{v} \in \mathcal{U}$. Then*

$$(8.3) \qquad |\langle \mathbf{u}, \mathbf{v}\rangle| \leq \{\langle \mathbf{u}, \mathbf{u}\rangle\}^{1/2} \{\langle \mathbf{v}, \mathbf{v}\rangle\}^{1/2}\,,$$

with equality if and only if dim span$\{\mathbf{u}, \mathbf{v}\} \leq 1$.

Proof. There are two cases to consider:

(1) If $\langle \mathbf{u}, \mathbf{v}\rangle = 0$, then the inequality is clear. Moreover, if equality holds in this case, then

$$0 = \{\langle \mathbf{u}, \mathbf{u}\rangle\}^{1/2} \{\langle \mathbf{v}, \mathbf{v}\rangle\}^{1/2}\,,$$

which forces at least one of the vectors $\mathbf{u}$,$\mathbf{v}$ to be equal to the vector $\mathbf{0}$, and hence the two vectors are linearly dependent.

(2) If $\langle \mathbf{u}, \mathbf{v} \rangle \neq 0$, then, in polar coordinates,

$$\langle \mathbf{u}, \mathbf{v} \rangle = |\langle \mathbf{u}, \mathbf{v} \rangle| e^{i\theta} = re^{i\theta}.$$

Set $\lambda = xe^{i\theta}$, $x \in \mathbb{R}$, $a = \langle \mathbf{u}, \mathbf{u} \rangle$, $b = \langle \mathbf{v}, \mathbf{v} \rangle$ and observe that

$$\begin{aligned} \langle \mathbf{u} + \lambda \mathbf{v}, \mathbf{u} + \lambda \mathbf{v} \rangle &= \langle \mathbf{u}, \mathbf{u} \rangle + \langle \mathbf{u}, \lambda \mathbf{v} \rangle + \langle \lambda \mathbf{v}, \mathbf{u} \rangle + \langle \lambda \mathbf{v}, \lambda \mathbf{v} \rangle \\ &= a + 2xr + x^2 b \\ &= b\left(x + \frac{r}{b}\right)^2 + a - \frac{r^2}{b} \\ &\geq 0 \end{aligned}$$

for every choice of $x \in \mathbb{R}$. (The condition $\langle \mathbf{u}, \mathbf{v} \rangle \neq 0$ insures that $\mathbf{v} \neq \mathbf{0}$ and hence permits us to divide by b.) Thus, upon choosing

$$x = -\frac{r}{b},$$

we conclude that

$$a - \frac{r^2}{b} = \langle \mathbf{u} + \lambda \mathbf{v}, \mathbf{u} + \lambda \mathbf{v} \rangle \geq 0,$$

which justifies the asserted inequality. Finally, if this last inequality is an equality, then

$$\langle \mathbf{u} + \lambda \mathbf{v}, \mathbf{u} + \lambda \mathbf{v} \rangle = 0$$

when $\lambda = xe^{i\theta}$ and x is chosen as above. But this implies that

$$\mathbf{u} + \lambda \mathbf{v} = \mathbf{0}$$

for this choice of λ and hence that $\mathbf{u}$ and $\mathbf{v}$ are linearly dependent. □

Exercise 8.3. Let $\mathcal{U}$ denote the set of continuous complex valued functions $f(t)$ on the finite closed interval $[a, b]$.

(a) Show that $\mathcal{U}$ is a vector space over $\mathbb{C}$ with respect to the natural rules of addition and multiplication by constants. Identify the zero element.

(b) Show that $\mathcal{U}$ is a normed linear space with respect to the norm $\|f\| = \left\{\int_a^b |f(t)|^2 dt\right\}^{1/2}$.

(c) Show that $\mathcal{U}$ is an inner product space with respect to the inner product $\langle f, g \rangle = \int_a^b f(t)\overline{g(t)}dt$.

Exercise 8.4. Show that if $f(t)$ and $g(t)$ are continuous complex valued functions $f(t)$ on the finite closed interval $[a, b]$, then

$$\left| \int_a^b f(t)\overline{g(t)}dt \right|^2 \leq \int_a^b |f(t)|^2 dt \int_a^b |g(t)|^2 dt$$

with equality if and only if there exists a pair of constants α, $\beta \in \mathbb{C}$ such that $\alpha f(t) + \beta g(t) = 0$ for every point $t \in [a, b]$.

Exercise 8.5. Show that the space $\mathcal{U} = \mathbb{C}^{p\times q}$ endowed with the inner product $\langle A, B\rangle = \text{trace}\{B^H A\}$ is a pq-dimensional inner product space.

8.2. A characterization of inner product spaces

An inner product space $\mathcal{U}$ over $\mathbb{F}$ is automatically a normed linear space with respect to the norm $\|\mathbf{u}\| = \{\langle \mathbf{u}, \mathbf{u}\rangle\}^{1/2}$:

$$\|\alpha\mathbf{u}\| = \{\langle \alpha\mathbf{u}, \alpha\mathbf{u}\rangle\}^{1/2} = \{\alpha\overline{\alpha}\langle \mathbf{u}, \mathbf{u}\rangle\}^{1/2} = |\alpha|\|\mathbf{u}\|,$$

$$\|\mathbf{u}\| \geq 0 \quad \text{with equality if and only if} \quad \mathbf{u} = 0,$$

and, by the Cauchy-Schwarz inequality for inner products,

$$\begin{aligned} \|\mathbf{u}+\mathbf{v}\|^2 &= \langle \mathbf{u}+\mathbf{v}, \mathbf{u}+\mathbf{v}\rangle = \|\mathbf{u}\|^2 + \langle \mathbf{u}, \mathbf{v}\rangle + \langle \mathbf{v}, \mathbf{u}\rangle + \|\mathbf{v}\|^2 \\ &\leq \|\mathbf{u}\|^2 + \|\mathbf{u}\|\|\mathbf{v}\| + \|\mathbf{v}\|\|\mathbf{u}\| + \|\mathbf{v}\|^2 \\ &= (\|\mathbf{u}\| + \|\mathbf{v}\|)^2. \end{aligned}$$

It is natural to ask whether or not the converse is true: Is every normed linear space automatically an inner product space? The answer is no, because the norm induced by the inner product has an extra property:

Lemma 8.2. *Let $\mathcal{U}$ be an inner product space over $\mathbb{F}$. Then the norm $\|\mathbf{u}\| = \{\langle \mathbf{u}, \mathbf{u}\rangle\}^{1/2}$ induced by the inner product satisfies the* **parallelogram law**

$$\|\mathbf{u}+\mathbf{v}\|^2 + \|\mathbf{u}-\mathbf{v}\|^2 = 2\|\mathbf{u}\|^2 + 2\|\mathbf{v}\|^2 . \tag{8.4}$$

Moreover, the inner product can be recovered from the norm by the formula

$$\langle \mathbf{u}, \mathbf{v}\rangle = \begin{cases} \frac{1}{4}\sum_{k=1}^{4} i^k\|\mathbf{u}+i^k\mathbf{v}\|^2 & \text{if} \quad \mathbb{F} = \mathbb{C} \\ \\ \frac{1}{2}\sum_{k=1}^{2}(-1)^k\|\mathbf{u}+(-1)^k\mathbf{v}\|^2 & \text{if} \quad \mathbb{F} = \mathbb{R} \end{cases} . \tag{8.5}$$

Proof. Both formulas are straightforward computations. □

Exercise 8.6. Verify formula (8.4) in the setting of Lemma 8.2.

Exercise 8.7. Verify formula (8.5) in the setting of Lemma 8.2.

Exercise 8.8. Let $\mathcal{U}$ be a normed linear space, and for $\mathbf{u}, \mathbf{v} \in \mathcal{U}$, let $f(\mathbf{u}) = \|\mathbf{u}+\mathbf{v}\|$. Show that $|f(\mathbf{u}_2) - f(\mathbf{u}_1)| \leq \|\mathbf{u}_2 - \mathbf{u}_1\|$ for any two elements $\mathbf{u}_1, \mathbf{u}_2 \in \mathcal{U}$.

Theorem 8.3. *Let $\mathcal{U}$ be a normed linear space over $\mathbb{C}$ in which the parallelogram law (8.4) holds. Then formula (8.5) defines an inner product in $\mathcal{U}$; i.e., the four defining characteristics of an inner product that were enumerated in the previous section are all met when $\langle \mathbf{u}, \mathbf{v}\rangle$ is defined by formula (8.5).*

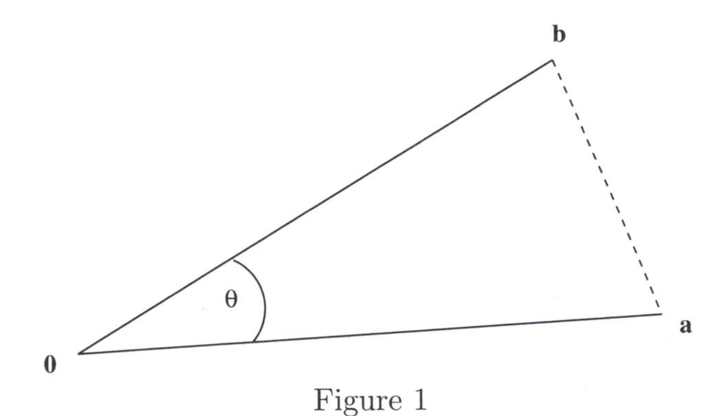

Figure 1

Discussion. We shall not give the details of the proof, but rather shall list a number of steps, each one of which is relatively simple to verify and which taken together yield the asserted result.

(a) $\langle \mathbf{u}, \mathbf{u} \rangle = \|\mathbf{u}\|^2$.

(b) $\langle \mathbf{u}, \mathbf{v} \rangle = \overline{\langle \mathbf{v}, \mathbf{u} \rangle}$.

(c) $\langle \mathbf{u}, \mathbf{0} \rangle = 0$.

(d) $\langle \mathbf{x}, \mathbf{y} \rangle + \langle \mathbf{u}, \mathbf{v} \rangle = 2\langle (\mathbf{x}+\mathbf{u})/2, (\mathbf{y}+\mathbf{v})/2 \rangle + 2\langle (\mathbf{x}-\mathbf{u})/2, (\mathbf{y}-\mathbf{v})/2 \rangle$.

(e) $\langle \mathbf{x}+\mathbf{u}, \mathbf{y} \rangle = \langle \mathbf{x}, \mathbf{y} \rangle + \langle \mathbf{u}, \mathbf{y} \rangle$.

(f) $\langle m\mathbf{x}, \mathbf{y} \rangle = m\langle \mathbf{x}, \mathbf{y} \rangle$ for any positive integer m.

(g) $\langle (m/n)\mathbf{x}, \mathbf{y} \rangle = (m/n)\langle \mathbf{x}, \mathbf{y} \rangle$ for any two positive integers m, n.

(h) $\langle \alpha\mathbf{x}, \mathbf{y} \rangle = \alpha\langle \mathbf{x}, \mathbf{y} \rangle$ for any real number α.

(i) $\langle \alpha\mathbf{x}, \mathbf{y} \rangle = \alpha\langle \mathbf{x}, \mathbf{y} \rangle$ for any complex number α.

HINT: Exercise 8.8 helps in the transition from (g) to (h). □

Exercise 8.9. Let $\mathcal{U} = \mathbb{C}^n$ endowed with the standard inner product. Show that if $n \geq 2$ and if $\mathbf{u} = \mathbf{e}_1$ and $\mathbf{v} = \mathbf{e}_2$, where $\mathbf{e}_j$ denotes the j'th column of I_n, and if $1 \leq s \leq \infty$, then

$$\|\mathbf{u}+\mathbf{v}\|_s^2 + \|\mathbf{u}-\mathbf{v}\|_s^2 = 2\|\mathbf{u}\|_s^2 + 2\|\mathbf{v}\|_s^2 \iff s = 2\,.$$

8.3. Orthogonality

The role of the inner product is perhaps best motivated by considering the law of cosines in $\mathbb{R}^3$ equipped with $\|\mathbf{x}\| = \|\mathbf{x}\|_2$. Then the cosine of the angle θ between the line segment running from $\mathbf{0}$ to $\mathbf{a} = (a_1, a_2, a_3)$ and the line segment running from $\mathbf{0}$ to $\mathbf{b} = (b_1, b_2, b_3)$ in Figure 1 is

$$\cos\theta = \frac{\|\mathbf{a}\|^2 + \|\mathbf{b}\|^2 - \|\mathbf{b}-\mathbf{a}\|^2}{2\|\mathbf{a}\|\|\mathbf{b}\|} = \frac{\sum_{i=1}^3 a_i b_i}{\sqrt{\sum_{i=1}^3 a_i^2}\sqrt{\sum_{i=1}^3 b_i^2}}\,.$$

In particular,

$$\sum_{i=1}^{3} a_i b_i = 0 \iff \cos\theta = 0 \iff \mathbf{a} \perp \mathbf{b}.$$

The law of cosines serves to motivate the following definitions in an inner product space $\mathcal{U}$ over $\mathbb{F}$ with inner product $\langle \mathbf{u}, \mathbf{v} \rangle_{\mathcal{U}}$:

- **Orthogonal vectors**: A pair of vectors $\mathbf{u}$ and $\mathbf{v}$ in $\mathcal{U}$ is said to be **orthogonal** if $\langle \mathbf{u}, \mathbf{v} \rangle_{\mathcal{U}} = 0$.
- **Orthogonal family**: A set of nonzero vectors $\{\mathbf{u}_1, \ldots, \mathbf{u}_k\}$ in $\mathcal{U}$ is said to be an **orthogonal family** if

$$\langle \mathbf{u}_i, \mathbf{u}_j \rangle_{\mathcal{U}} = 0 \text{ for } i \neq j.$$

  The assumption that none of the vectors $\mathbf{u}_1, \ldots, \mathbf{u}_k$ are equal to $\mathbf{0}$ serves to guarantee that they are automatically linearly independent.
- **Orthonormal family**: A set of vectors $\mathbf{u}_1, \ldots, \mathbf{u}_k$ in $\mathcal{U}$ is said to be an **orthonormal family** if
  (1) it is an orthogonal family and
  (2) the vectors $\mathbf{u}_i$, $i = 1, \ldots, k$, are all normalized to have unit length, i.e.,

$$\|\mathbf{u}_i\|_{\mathcal{U}}^2 = \langle \mathbf{u}_i, \mathbf{u}_i \rangle_{\mathcal{U}} = 1, \; i = 1, \ldots, k.$$

- **Orthogonal decomposition**: A pair of subspaces $\mathcal{V}$ and $\mathcal{W}$ of $\mathcal{U}$ is said to form an **orthogonal decomposition** of $\mathcal{U}$ if
  (1) $\mathcal{V} + \mathcal{W} = \mathcal{U}$,
  (2) $\langle \mathbf{v}, \mathbf{w} \rangle_{\mathcal{U}} = 0$ for every $\mathbf{v} \in \mathcal{V}$ and $\mathbf{w} \in \mathcal{W}$.
  Orthogonal decompositions will be indicated by the symbol

$$\mathcal{U} = \mathcal{V} \oplus \mathcal{W}.$$

- **Orthogonal complement**: If $\mathcal{V}$ is a subspace of an inner product space $\mathcal{U}$ over $\mathbb{F}$, then the set

$$\mathcal{V}^{\perp} = \{\mathbf{u} \in \mathcal{U} : \langle \mathbf{u}, \mathbf{v} \rangle_{\mathcal{U}} = 0 \quad \text{for every} \quad \mathbf{v} \in \mathcal{V}\}$$

  is referred to as the **orthogonal complement** of $\mathcal{V}$ in $\mathcal{U}$. It is a subspace of $\mathcal{U}$.

Exercise 8.10. Show that every orthogonal sum decomposition is a direct sum decomposition and give an example of a direct sum decomposition that is not an orthogonal decomposition.

Exercise 8.11. Show that if $\{\mathbf{u}_1, \ldots, \mathbf{u}_k\}$ is an orthogonal family of nonzero vectors in an inner product space $\mathcal{U}$ over $\mathbb{F}$, then $\mathbf{u}_1, \ldots, \mathbf{u}_k$ are linearly independent.

8.4. Gram matrices

Let $\mathbf{v}_1, \ldots, \mathbf{v}_k$ be a set of vectors in an inner product space $\mathcal{U}$ over $\mathbb{F}$. Then the $k \times k$ matrix G with entries

$$g_{ij} = \langle \mathbf{v}_j, \mathbf{v}_i \rangle_{\mathcal{U}} \quad \text{for} \quad i, j = 1, \ldots, k, \tag{8.6}$$

is called the **Gram matrix** of the given set of vectors. Note that $G = G^H$.

Lemma 8.4. *Let $\mathcal{U}$ be an inner product space over $\mathbb{F}$ and let G denote the Gram matrix of a set of vectors $\mathbf{v}_1, \ldots, \mathbf{v}_k$ in $\mathcal{U}$. Then G is invertible if and only if the vectors $\mathbf{v}_1, \ldots, \mathbf{v}_k$ are linearly independent over $\mathbb{F}$.*

Proof. Let $\mathbf{c}, \mathbf{d} \in \mathbb{F}^k$ with components $c_1, \ldots, c_k$ and $d_1, \ldots, d_k$, respectively, and let $\mathbf{v} = \sum_{j=1}^k c_j \mathbf{v}_j$ and $\mathbf{w} = \sum_{i=1}^k d_i \mathbf{v}_i$. Then it is readily checked that

$$\langle \mathbf{v}, \mathbf{w} \rangle_{\mathcal{U}} = \mathbf{d}^H G \mathbf{c}. \tag{8.7}$$

Suppose first that G is invertible and that $\sum_{j=1}^k c_j \mathbf{v}_j = \mathbf{0}$ for some choice of $c_1, \ldots, c_k \in \mathbb{F}$. Then, in view of formula (8.7),

$$0 = \langle \sum_{j=1}^k c_j \mathbf{v}_j, \sum_{i=1}^k d_i \mathbf{v}_i \rangle_{\mathcal{U}} = \mathbf{d}^H G \mathbf{c}$$

for every choice of $d_1, \ldots, d_k \in \mathbb{F}$. Therefore, $G\mathbf{c} = \mathbf{0}$, which in turn implies that $\mathbf{c} = \mathbf{0}$, since G is invertible. Thus, the vectors $\mathbf{v}_1, \ldots, \mathbf{v}_k$ are linearly independent.

Suppose next that the vectors $\mathbf{v}_1, \ldots, \mathbf{v}_k$ are linearly independent and that $\mathbf{c} \in \mathcal{N}_G$. Then, by formula (8.7),

$$\langle \sum_{j=1}^k c_j \mathbf{v}_j, \sum_{i=1}^k c_i \mathbf{v}_i \rangle_{\mathcal{U}} = \mathbf{c}^H G \mathbf{c} = 0.$$

Therefore, $\sum_{j=1}^k c_j \mathbf{v}_j = \mathbf{0}$ and hence, in view of the presumed linear independence, $c_1 = \cdots = c_k = 0$. Thus, G is invertible. $\square$

Exercise 8.12. Verify formula (8.7).

8.5. Adjoints

In this section we introduce the notion of the **adjoint** S^* of a linear transformation S from one inner product space into another. It is important to keep in mind that **the adjoint depends upon the inner products** of the two spaces.

Theorem 8.5. *Let $\mathcal{U}$ and $\mathcal{V}$ be a pair of finite dimensional inner product spaces over $\mathbb{F}$ and let S be a linear transformation from $\mathcal{U}$ into $\mathcal{V}$. Then there exists exactly one linear transformation S^* from $\mathcal{V}$ into $\mathcal{U}$ such that*

$$\langle S\mathbf{u}, \mathbf{v}\rangle_{\mathcal{V}} = \langle \mathbf{u}, S^*\mathbf{v}\rangle_{\mathcal{U}} \tag{8.8}$$

for every choice of $\mathbf{u} \in \mathcal{U}$ and $\mathbf{v} \in \mathcal{V}$.

Proof. It is easy to verify that there is at most one linear transformation from $\mathcal{V}$ into $\mathcal{U}$ for which (8.8) holds. If there were two such linear transformations, S_1^* and S_2^*, then

$$\langle S\mathbf{u}, \mathbf{v}\rangle_{\mathcal{V}} = \langle \mathbf{u}, S_1^*\mathbf{v}\rangle_{\mathcal{U}} = \langle \mathbf{u}, S_2^*\mathbf{v}\rangle_{\mathcal{U}}$$

and hence

$$\langle \mathbf{u}, (S_1^* - S_2^*)\mathbf{v}\rangle_{\mathcal{U}} = 0$$

for every choice of $\mathbf{u} \in \mathcal{U}$ and $\mathbf{v} \in \mathcal{V}$. Thus, upon choosing $\mathbf{u} = (S_1^* - S_2^*)\mathbf{v}$, it follows that

$$\langle (S_1^* - S_2^*)\mathbf{v}, (S_1^* - S_2^*)\mathbf{v}\rangle_{\mathcal{U}} = 0\,,$$

which implies that $(S_1^* - S_2^*)\mathbf{v} = \mathbf{0}$ for every vector $\mathbf{v} \in \mathcal{V}$, i.e., $S_1^* = S_2^*$.

The verification of the existence of S^* in the present setting is by computation (though a more elegant approach that is applicable in more general settings is via the Riesz representation theorem, which is discussed in the next section).

Let $\mathbf{u}_1, \ldots, \mathbf{u}_q$ be a basis for $\mathcal{U}$, let $\mathbf{v}_1, \ldots, \mathbf{v}_p$ be a basis for $\mathcal{V}$ and let U and V denote the corresponding Gram matrices, with entries

$$u_{ti} = \langle \mathbf{u}_i, \mathbf{u}_t\rangle_{\mathcal{U}} \quad \text{and} \quad v_{jk} = \langle \mathbf{v}_k, \mathbf{v}_j\rangle_{\mathcal{V}}\,,$$

respectively. It suffices to show that there exists a linear transformation S^* from $\mathcal{V}$ to $\mathcal{U}$ such that

$$\langle S\mathbf{u}_i, \mathbf{v}_j\rangle_{\mathcal{V}} = \langle \mathbf{u}_i, S^*\mathbf{v}_j\rangle_{\mathcal{U}} \quad \text{for} \quad i = 1, \ldots, q \quad \text{and} \quad j = 1, \ldots, p\,.$$

Let $S\mathbf{u}_i = \sum_{k=1}^p a_{ki}\mathbf{v}_k$ and suppose for the moment that S^* exists and that $S^*\mathbf{v}_j = \sum_{t=1}^q b_{tj}\mathbf{u}_t$. Then

$$\langle S\mathbf{u}_i, \mathbf{v}_j\rangle_{\mathcal{V}} = \sum_{k=1}^p a_{ki}\langle \mathbf{v}_k, \mathbf{v}_j\rangle_{\mathcal{V}} = \sum_{k=1}^p v_{jk}a_{ki}$$

and

$$\langle \mathbf{u}_i, S^*\mathbf{v}_j\rangle_{\mathcal{V}} = \sum_{t=1}^q \overline{b_{tj}}\langle \mathbf{u}_i, \mathbf{u}_t\rangle_{\mathcal{U}} = \sum_{t=1}^q u_{ti}\overline{b_{tj}}\,.$$

Thus, upon setting $A = [a_{ki}]$ and $B = [b_{tj}]$, it follows that the last two formulas will match if and only if $VA = B^HU$ and hence, since the Gram matrices V and U are invertible, if and only if

$$B = (U^H)^{-1}A^HV^H\,.$$

Thus, the linear transformation S^* from $\mathcal{V}$ into $\mathcal{U}$ that is defined by the formula

$$S^*\mathbf{v}_j = \sum_{t=1}^{q} b_{tj}\mathbf{u}_t \text{ (with } B = (U^H)^{-1}A^H V^H = [b_{tj}])$$

is the one and only linear transformation from $\mathcal{V}$ to $\mathcal{U}$ that meets the stated requirements. □

Corollary 8.6. If $\mathcal{U} = \mathbb{F}^q$ and $\mathcal{V} = \mathbb{F}^p$ are both endowed with the standard inner product and if the linear transformation is multiplication by $A \in \mathbb{C}^{p\times q}$, then $A^* = A^H$.

Exercise 8.13. Verify Corollary 8.6.

Example 8.7. Let $\mathcal{U} = \mathbb{C}^{p\times q}$ equipped with the inner product $\langle A, B\rangle_{\mathcal{U}} = \operatorname{trace} B^H A$, let $\mathcal{V} = \mathbb{C}^p$ equipped with the standard inner product, let $\mathbf{u} \in \mathbb{C}^q$ and let S denote the linear transformation from $\mathcal{U}$ into $\mathcal{V}$ that is defined by the formula $SA = A\mathbf{u}$ for every $A \in \mathcal{U}$. Then the adjoint S^* must satisfy the identity

$$\langle SA, \mathbf{v}\rangle_{\mathcal{V}} = \langle A, S^*\mathbf{v}\rangle_{\mathcal{U}}$$

for every choice of $A \in \mathbb{C}^{p\times q}$ and $\mathbf{v} \in \mathbb{C}^p$. Thus,

$$\begin{aligned} \langle SA, \mathbf{v}\rangle_{\mathcal{V}} &= \langle A\mathbf{u}, \mathbf{v}\rangle_{\mathcal{V}} = \mathbf{v}^H A\mathbf{u} \\ &= \operatorname{trace}\{\mathbf{v}^H A\mathbf{u}\} = \operatorname{trace}\{\mathbf{u}\mathbf{v}^H A\} \\ &= \langle A, \mathbf{v}\mathbf{u}^H\rangle_{\mathcal{U}}, \end{aligned}$$

i.e.,

$$\langle A, S^*\mathbf{v}\rangle_{\mathcal{U}} = \langle A, \mathbf{v}\mathbf{u}^H\rangle_{\mathcal{U}}$$

for every $\mathbf{v} \in \mathbb{C}^p$. Therefore, in view of Exercise 8.1,

$$S^*\mathbf{v} = \mathbf{v}\mathbf{u}^H \quad \text{for every} \quad \mathbf{v} \in \mathcal{V}. \tag{8.9}$$

Exercise 8.14. Verify the identification (8.9) in the preceding example by checking that for all rank one matrices $A = \mathbf{x}\mathbf{y}^H$, with $\mathbf{x} \in \mathbb{C}^p$ and $\mathbf{y} \in \mathbb{C}^q$,

$$\langle SA, \mathbf{v}\rangle_{\mathcal{V}} = \mathbf{y}^H\mathbf{u}\mathbf{v}^H\mathbf{x} \quad \text{and} \quad \langle A, S^*\mathbf{v}\rangle_{\mathcal{U}} = \mathbf{y}^H(S^*\mathbf{v})^H\mathbf{x}.$$

Exercise 8.15. Let $\mathcal{U} = \mathbb{C}^n$ equipped with the inner product $\langle \mathbf{u}, \mathbf{v}\rangle_{\mathcal{U}} = \sum_{j=1}^n j\overline{v_j}u_j$ for vectors $\mathbf{u}$, $\mathbf{v} \in \mathbb{C}^n$ with components $u_1, \dots, u_n$ and $v_1, \dots, v_n$, respectively. Find the adjoint A^* of a matrix $A \in \mathbb{C}^{n\times n}$ with respect to this inner product.

Lemma 8.8. *Let $\mathcal{U}$, $\mathcal{V}$ and $\mathcal{W}$ be finite dimensional inner product spaces over $\mathbb{F}$ and let S and S_1 be linear transformations from $\mathcal{U}$ into $\mathcal{V}$ and let T be a linear transformation from $\mathcal{V}$ into $\mathcal{W}$. Then:*

(1) $(\alpha S + \beta S_1)^* = \overline{\alpha}S^* + \overline{\beta}S_1^*$ *for every choice of α and β in $\mathbb{F}$.*

(2) $(S^*)^* = S$.

(3) $(TS)^* = S^*T^*$.

(4) *If* $\mathcal{U} = \mathcal{V}$ *and* I *denotes the identity in* $\mathcal{U}$, *then* $I^* = I$.

Proof. The formulas

$$\begin{aligned} \langle(\alpha S + \beta S_1)\mathbf{u}, \mathbf{v}\rangle_{\mathcal{V}} &= \alpha\langle S\mathbf{u}, \mathbf{v}\rangle_{\mathcal{V}} + \beta\langle S_1\mathbf{u}, \mathbf{v}\rangle_{\mathcal{V}} \\ &= \alpha\langle \mathbf{u}, S^*\mathbf{v}\rangle_{\mathcal{U}} + \beta\langle \mathbf{u}, S_1^*\mathbf{v}\rangle_{\mathcal{U}} \\ &= \langle \mathbf{u}, \overline{\alpha}S^*\mathbf{v}\rangle_{\mathcal{U}} + \langle \mathbf{u}, \overline{\beta}S_1^*\mathbf{v}\rangle_{\mathcal{U}} \\ &= \langle \mathbf{u}, (\overline{\alpha}S^* + \overline{\beta}S_1^*)\mathbf{v}\rangle_{\mathcal{U}} \end{aligned}$$

serve to verify (1). The proof of the remaining assertions is similar and is left to the reader. □

Exercise 8.16. Complete the proof of Lemma 8.8.

Lemma 8.9. *Let* $\mathcal{U}$ *and* $\mathcal{V}$ *be finite dimensional inner product spaces over* $\mathbb{F}$ *and let* S *be a linear transformation from* $\mathcal{U}$ *into* $\mathcal{V}$. *Then* $\mathcal{N}_S = \mathcal{N}_{S^*S}$, *i.e.,*

$$S\mathbf{u} = \mathbf{0} \iff S^*S\mathbf{u} = \mathbf{0}\,. \tag{8.10}$$

Proof. Suppose first that $S^*S\mathbf{u} = \mathbf{0}$ for some vector $\mathbf{u} \in \mathcal{U}$. Then the formulas

$$\langle S\mathbf{u}, S\mathbf{u}\rangle_{\mathcal{V}} = \langle S^*S\mathbf{u}, \mathbf{u}\rangle_{\mathcal{U}} = \langle \mathbf{0}, \mathbf{u}\rangle_{\mathcal{U}} = 0$$

imply that $S\mathbf{u} = \mathbf{0}$. Therefore, $\mathcal{N}_{S^*S} \subseteq \mathcal{N}_S$. To complete the proof, it remains to check that the opposite inclusion holds too. But that is self-evident. □

Exercise 8.17. Let T be a linear transformation from a finite dimensional inner product space $\mathcal{U}$ into itself, and let $\mathcal{V}$ be a subspace of $\mathcal{U}$. Show that

$$T\mathcal{V} \subseteq \mathcal{V} \iff T^*\mathcal{V}^\perp \subseteq \mathcal{V}^\perp\,. \tag{8.11}$$

8.6. The Riesz representation theorem

It is convenient to begin with an elementary exercise to help set the scene.

Exercise 8.18. Let $\mathcal{U}$ be an inner product space over $\mathbb{F}$, let $\mathbf{y} \in \mathcal{U}$ and let

$$f(\mathbf{x}) = \langle \mathbf{x}, \mathbf{y}\rangle_{\mathcal{U}} \quad \text{for every} \quad \mathbf{x} \in \mathcal{U}\,. \tag{8.12}$$

Show that f is a linear functional on $\mathcal{U}$ and that

$$\|\mathbf{y}\|_{\mathcal{U}} = \max\{|f(\mathbf{x})| : \mathbf{x} \in \mathcal{U} \quad \text{and} \quad \|\mathbf{x}\|_{\mathcal{U}} = 1\}\,. \tag{8.13}$$

A natural question is: Does every linear functional on $\mathcal{U}$ admit a representation of the form (8.12) for some vector $\mathbf{y} \in \mathcal{U}$? In view of Exercise 8.1, there is at most one such vector $\mathbf{y}$. The solution of the next exercise guarantees that there is at least one (and hence exactly one) such vector $\mathbf{y}$ if $\mathcal{U}$ is finite dimensional.

Exercise 8.19. Let $\mathcal{U}$ be a finite dimensional inner product space over $\mathbb{F}$, with basis $\{\mathbf{u}_1, \ldots, \mathbf{u}_n\}$ and Gram matrix G, and let f be a linear functional on $\mathcal{U}$. Show that

$$f(\mathbf{x}) = \langle \mathbf{x}, \mathbf{y} \rangle_{\mathcal{U}} \quad \text{for every} \quad \mathbf{x} \in \mathcal{U},$$

where $\mathbf{y} = \sum_{i=1}^{n} d_i \mathbf{u}_i$ and d_i is the i'th component of the vector

$$\mathbf{d} = G^{-1} \begin{bmatrix} f(\mathbf{u}_1) & \cdots & f(\mathbf{u}_n) \end{bmatrix}^H .$$

Exercise 8.19 is a good exercise in the use of Gram matrices. A more elegant approach that works equally well in infinite dimensional Hilbert spaces is based on the Riesz representation theorem; see Theorem 8.10, below.

An inner product space $\mathcal{U}$ over $\mathbb{F}$ is said to be a **Hilbert space** over $\mathbb{F}$ if every Cauchy sequence $\mathbf{x}_1, \mathbf{x}_2, \ldots$ in $\mathcal{U}$ tends to a limit $\mathbf{y} \in \mathcal{U}$ in the norm induced by the inner product. A Hilbert space may also be characterized as a Banach space in which the norm satisfies the parallelogram law, just as in Section 8.2. A linear transformation S from a Hilbert space $\mathcal{U}$ over $\mathbb{F}$ into a Hilbert space $\mathcal{V}$ over $\mathbb{F}$ is said to be **bounded** if there exists a number $\gamma > 0$ such that

$$\|S\mathbf{u}\|_{\mathcal{V}} \leq \gamma \|\mathbf{u}\|_{\mathcal{U}} \quad \text{for every vector} \quad \mathbf{u} \in \mathcal{U}. \tag{8.14}$$

Moreover, $\|S\|_{\mathcal{U},\mathcal{V}}$ is the smallest γ for which (8.14) holds; see Appendix I for additional discussion of this terminology. **Finite dimensional inner product spaces are automatically Hilbert spaces, and linear transformations from a finite dimensional inner product space $\mathcal{U}$ into an inner product space $\mathcal{V}$ are automatically bounded.**

Theorem 8.10. (Riesz representation) *Let $f(\mathbf{x})$ be a bounded linear functional on a Hilbert space $\mathcal{U}$ over $\mathbb{F}$. Then there exists a unique vector $\mathbf{y} \in \mathcal{U}$ such that*

$$f(\mathbf{x}) = \langle \mathbf{x}, \mathbf{y} \rangle_{\mathcal{U}} \quad \textit{for every} \quad \mathbf{x} \in \mathcal{U}. \tag{8.15}$$

Moreover,

$$\|\mathbf{y}\|_{\mathcal{U}} = \max\{|f(\mathbf{x})| : \mathbf{x} \in \mathcal{U} \quad \textit{and} \quad \|\mathbf{x}\|_{\mathcal{U}} = 1\}. \tag{8.16}$$

Proof. Let $\mathcal{N}_f = \{\mathbf{x} \in \mathcal{U} : f(\mathbf{x}) = 0\}$. Then, referring to Appendix I, if need be, it is readily checked that

(1) $\mathcal{N}_f$ is a closed subspace of $\mathcal{U}$.

(2) $\mathcal{N}_f = \mathcal{U}$ if and only if $f(\mathbf{x}) = 0$ for every $\mathbf{x} \in \mathcal{U}$.

(3) $f(\mathbf{u})\mathbf{v} - f(\mathbf{v})\mathbf{u} \in \mathcal{N}_f$ for every choice of $\mathbf{u}$ and $\mathbf{v}$ in $\mathcal{U}$.

(4) If $\mathcal{N}_f$ is a proper subspace of $\mathcal{U}$, then the orthogonal complement of $\mathcal{N}_f$ in $\mathcal{U}$ is a one dimensional subspace of $\mathcal{U}$.

Consequently, if $\mathcal{N}_f$ is a proper subspace of $\mathcal{U}$, then there exists a nonzero vector $\mathbf{x}_0 \in \mathcal{U}$ that is orthogonal to $\mathcal{N}_f$. Thus, as $f(\mathbf{x})\mathbf{x}_0 - f(\mathbf{x}_0)\mathbf{x} \in \mathcal{N}_f$, it follows that

$$f(\mathbf{x})\langle \mathbf{x}_0, \mathbf{x}_0\rangle_{\mathcal{U}} = f(\mathbf{x}_0)\langle \mathbf{x}, \mathbf{x}_0\rangle_{\mathcal{U}} \quad \text{for every vector} \quad \mathbf{x} \in \mathcal{U}\,.$$

This serves to establish (8.15) with

$$\mathbf{y} = \frac{\overline{f(\mathbf{x}_0)}}{\langle \mathbf{x}_0, \mathbf{x}_0\rangle_{\mathcal{U}}}\mathbf{x}_0\,.$$

If $\mathcal{N}_f = \mathcal{U}$, then $f(\mathbf{x}) = 0$ for every vector $\mathbf{x} \in \mathcal{U}$ and hence $\mathbf{y} = \mathbf{0}$. Assertion (8.16) has already been established in Exercise 8.18. □

Theorem 8.11. *Let $\mathcal{U}$ and $\mathcal{V}$ be Hilbert spaces over $\mathbb{F}$ and let S be a bounded linear transformation from $\mathcal{U}$ into $\mathcal{V}$. Then there exists exactly one bounded linear transformation S^* from $\mathcal{V}$ into $\mathcal{U}$ such that*

$$\langle S\mathbf{u}, \mathbf{v}\rangle_{\mathcal{V}} = \langle \mathbf{u}, S^*\mathbf{v}\rangle_{\mathcal{U}} \quad \textit{for every choice of} \quad \mathbf{u} \in \mathcal{U} \quad \textit{and} \quad \mathbf{v} \in \mathcal{V}\,.$$

Moreover, $\|S\| = \|S^*\|$.

Discussion. The formula

$$f_{\mathbf{v}}(\mathbf{u}) = \langle S\mathbf{u}, \mathbf{v}\rangle_{\mathcal{V}}$$

defines a bounded linear functional on $\mathcal{U}$ for each given $\mathbf{v} \in \mathcal{V}$. Therefore, by the Riesz representation theorem, there exists a unique vector $\mathbf{w} \in \mathcal{U}$ such that

$$f_{\mathbf{v}}(\mathbf{u}) = \langle \mathbf{u}, \mathbf{w}\rangle_{\mathcal{U}}\,.$$

But this in turn implies that

$$\langle S\mathbf{u}, \mathbf{v}\rangle_{\mathcal{V}} = \langle \mathbf{u}, \mathbf{w}\rangle_{\mathcal{U}} \quad \text{for every choice of} \quad \mathbf{u} \in \mathcal{U}\,.$$

Since there is only one such $\mathbf{w}$ for each $\mathbf{v} \in \mathcal{V}$, we may define $S^*\mathbf{v} = \mathbf{w}$. It remains to check that S^* is a bounded linear transformation from $\mathcal{V}$ into $\mathcal{U}$. The details are left to the reader. □

8.7. Normal, selfadjoint and unitary transformations

A linear transformation T from an inner product space $\mathcal{U}$ over $\mathbb{F}$ into itself is said to be

- **normal** if $T^*T = TT^*$, i.e., if

$$\langle T\mathbf{u}, T\mathbf{v}\rangle_{\mathcal{U}} = \langle T^*\mathbf{u}, T^*\mathbf{v}\rangle_{\mathcal{U}} \quad \text{for every choice of} \quad \mathbf{u}, \mathbf{v} \in \mathcal{U}\,.$$

- **selfadjoint** if $T = T^*$, i.e., if

$$\langle T\mathbf{u}, \mathbf{v}\rangle_{\mathcal{U}} = \langle \mathbf{u}, T\mathbf{v}\rangle_{\mathcal{U}} \quad \text{for every choice of} \quad \mathbf{u}, \mathbf{v} \in \mathcal{U}\,.$$

- **unitary** if $T^*T = TT^* = I$, i.e., if

$$\langle T\mathbf{u}, T\mathbf{v}\rangle_{\mathcal{U}} = \langle \mathbf{u}, \mathbf{v}\rangle_{\mathcal{U}} \quad \text{for every choice of} \quad \mathbf{u}, \mathbf{v} \in \mathcal{U}\,.$$

It is important to bear in mind that **each of these three classes of transformations depends upon the inner product** and that selfadjoint and unitary transformations are automatically normal.

Theorem 8.12. *Let T be a normal transformation from an n-dimensional inner product space $\mathcal{U}$ over $\mathbb{C}$ into itself. Then*

(1) $T\mathbf{u} = \lambda\mathbf{u} \iff T^*\mathbf{u} = \overline{\lambda}\mathbf{u}$.

(2) *There exists an orthonormal basis $\{\mathbf{u}_1, \ldots, \mathbf{u}_n\}$ of $\mathcal{U}$ and a set of complex numbers $\{\lambda_1, \ldots, \lambda_n\}$ (not necessarily distinct) such that*

$$T\mathbf{u}_j = \lambda_j\mathbf{u}_j \quad for \quad j = 1, \ldots, n\,.$$

(3) *If also $T = T^*$, then $\lambda_j \in \mathbb{R}$ for $j = 1, \ldots, n$.*

(4) *If also $T^*T = I$, then $|\lambda_j| = 1$ for $j = 1, \ldots, n$.*

Proof. The first assertion follows from Lemma 8.9 and the observation that

$$T\mathbf{u} = \mathbf{0} \iff T^*T\mathbf{u} = \mathbf{0} \iff TT^*\mathbf{u} = \mathbf{0} \iff T^*\mathbf{u} = \mathbf{0}\,,$$

with T replaced by $\lambda I - T$. This does the trick, since T is normal if and only if $\lambda I - T$ is normal and $(\lambda I - T)^* = \overline{\lambda}I - T^*$, i.e.,

$$(\lambda I - T)\mathbf{u} = \mathbf{0} \iff (\overline{\lambda}I - T^*)\mathbf{u} = \mathbf{0}\,.$$

Since $\mathcal{U}$ is invariant under T, Theorem 4.2 guarantees the existence of a vector $\mathbf{u}_1 \in \mathcal{U}$ such that $\|\mathbf{u}_1\|_{\mathcal{U}} = 1$ and $T\mathbf{u}_1 = \lambda_1\mathbf{u}_1$.

Now let us suppose that we have established the existence of an orthonormal family $\{\mathbf{u}_1, \ldots, \mathbf{u}_k\}$ of eigenvectors of T for some positive integer k and let

$$\mathcal{L}_k = \text{span}\,\{\mathbf{u}_1, \ldots, \mathbf{u}_k\} \text{ and } \mathcal{L}_k^{\perp} = \{\mathbf{u} \in \mathcal{U} : \langle \mathbf{v}, \mathbf{u}\rangle_{\mathcal{U}} = 0 \text{ for every } \mathbf{v} \in \mathcal{L}_k\}\,.$$

Then, $\mathcal{L}_k^{\perp}$ is invariant under T:

$$\langle \mathbf{v}, \mathbf{u}_j\rangle_{\mathcal{U}} = 0 \Longrightarrow \langle T\mathbf{v}, \mathbf{u}_j\rangle_{\mathcal{U}} = \langle \mathbf{v}, T^*\mathbf{u}_j\rangle_{\mathcal{U}} = \langle \mathbf{v}, \overline{\lambda_j}\mathbf{u}_j\rangle_{\mathcal{U}} = \lambda_j\langle \mathbf{v}, \mathbf{u}_j\rangle_{\mathcal{U}} = 0$$

for $j = 1, \ldots, k$. Therefore, if $k < n$, there is a vector $\mathbf{u}_{k+1}$ of norm one in $\mathcal{L}_k^{\perp}$ such that $T\mathbf{u}_{k+1} = \lambda_{k+1}\mathbf{u}_{k+1}$. Thus, if $\mathcal{L}_k$ exists for $k < n$, then $\mathcal{L}_{k+1}$ exists and the construction continues until $k = n$. This completes the proof of (2).

If T is selfadjoint, then the string of equalities

$$\begin{aligned} \lambda_j &= \lambda_j\langle \mathbf{u}_j, \mathbf{u}_j\rangle_{\mathcal{U}} = \langle T\mathbf{u}_j, \mathbf{u}_j\rangle_{\mathcal{U}} = \langle \mathbf{u}_j, T^*\mathbf{u}_j\rangle_{\mathcal{U}} \\ &= \langle \mathbf{u}_j, T\mathbf{u}_j\rangle_{\mathcal{U}} = \langle \mathbf{u}_j, \lambda_j\mathbf{u}_j\rangle_{\mathcal{U}} = \overline{\lambda_j}\langle \mathbf{u}_j, \mathbf{u}_j\rangle_{\mathcal{U}} \\ &= \overline{\lambda_j} \end{aligned}$$

implies that $\lambda_j \in \mathbb{R}$, i.e., the eigenvalues of T are real. This completes the proof of (3). The proof of (4) is left to the reader as an exercise. □

Exercise 8.20. Show that $|\lambda_j| = 1$ for each eigenvalue λ_j of the unitary transformation T considered in Theorem 8.12.

8.8. Projections and direct sum decompositions

- **Projections:** A linear transformation P of a vector space $\mathcal{U}$ over $\mathbb{F}$ into itself is said to be a **projection** if P is **idempotent**, i.e., if $P^2 = P$.
- **Orthogonal projections:** A linear transformation P of an inner product space $\mathcal{U}$ over $\mathbb{F}$ into itself is said to be an **orthogonal projection** if P is **idempotent and selfadjoint** with respect to the given inner product, i.e., if

$$P^2 = P \quad \text{and} \quad \langle P\mathbf{u}, \mathbf{v}\rangle_{\mathcal{U}} = \langle \mathbf{u}, P\mathbf{v}\rangle_{\mathcal{U}}$$

for every pair of vectors $\mathbf{u}, \mathbf{v} \in \mathcal{U}$.

Exercise 8.21. Let $\mathbf{u}_1$ and $\mathbf{u}_2$ be a pair of orthonormal vectors in an inner product space $\mathcal{U}$ over $\mathbb{F}$ and let $\alpha \in \mathbb{F}$. Show that the transformation P that is defined by the formula $P\mathbf{u} = \langle \mathbf{u}, \mathbf{u}_1 + \alpha\mathbf{u}_2\rangle_{\mathcal{U}}\mathbf{u}_1$ is a projection but is not an orthogonal projection unless $\alpha = 0$.

Lemma 8.13. *Let P be a projection in a vector space $\mathcal{U}$ over $\mathbb{F}$, and let*

$$\mathcal{R}_P = \{P\mathbf{x} : \mathbf{x} \in \mathcal{U}\} \quad \textit{and let} \quad \mathcal{N}_P = \{\mathbf{x} \in \mathcal{U} : P\mathbf{x} = \mathbf{0}\}.$$

Then

$$\mathcal{U} = \mathcal{R}_P \dot{+} \mathcal{N}_P. \tag{8.17}$$

Proof. Let $\mathbf{x} \in \mathcal{U}$. Then clearly

$$\mathbf{x} = P\mathbf{x} + (I - P)\mathbf{x}$$

and $P\mathbf{x} \in \mathcal{R}_P$. Moreover, $(I - P)\mathbf{x} \in \mathcal{N}_P$, since

$$P(I - P)\mathbf{x} = (P - P^2)\mathbf{x} = (P - P)\mathbf{x} = \mathbf{0}.$$

Thus,

$$\mathcal{U} = \mathcal{R}_P + \mathcal{N}_P.$$

The sum is direct because

$$\mathbf{y} \in \mathcal{R}_P \Longleftrightarrow \mathbf{y} = P\mathbf{y} \text{ and } \mathbf{y} \in \mathcal{N}_P \Longleftrightarrow P\mathbf{y} = \mathbf{0}.$$

□

Lemma 8.13 exhibits $\mathcal{U}$ as the direct sum of the spaces $\mathcal{V} = \mathcal{R}_P$ and $\mathcal{W} = \mathcal{N}_P$ that are defined in terms of a given projection P. Conversely, every direct sum decomposition $\mathcal{U} = \mathcal{V} \dot{+} \mathcal{W}$ defines a projection P on $\mathcal{U}$ with $\mathcal{V} = \mathcal{R}_P$ and $\mathcal{W} = \mathcal{N}_P$.

Lemma 8.14. *Let $\mathcal{V}$ and $\mathcal{W}$ be subspaces of a vector space $\mathcal{U}$ over $\mathbb{F}$ and suppose that $\mathcal{U} = \mathcal{V} \dot{+} \mathcal{W}$. Then:*

(1) *For every vector $\mathbf{u} \in \mathcal{U}$ there exists exactly one vector $P_{\mathcal{V}}\mathbf{u} \in \mathcal{V}$ such that $\mathbf{u} - P_{\mathcal{V}}\mathbf{u} \in \mathcal{W}$.*

(2) *$P_{\mathcal{V}}\mathbf{v} = \mathbf{v}$ for every $\mathbf{v} \in \mathcal{V}$.*

(3) *$P_{\mathcal{V}}\mathbf{w} = \mathbf{0}$ for every $\mathbf{w} \in \mathcal{W}$.*

(4) *$P_{\mathcal{V}}$ is linear on $\mathcal{U}$ and $P_{\mathcal{V}}^2 = P_{\mathcal{V}}$.*

(5) *$\mathcal{V} = \mathcal{R}_{P_{\mathcal{V}}}$ and $\mathcal{W} = \mathcal{N}_{P_{\mathcal{V}}}$.*

(6) *If $\mathcal{U}$ is an inner product space with inner product $\langle \cdot, \cdot \rangle_{\mathcal{U}}$, then*

$$\langle \mathbf{v}, \mathbf{w} \rangle_{\mathcal{U}} = 0 \quad \text{for every choice of} \quad \mathbf{v} \in \mathcal{V} \text{ and } \mathbf{w} \in \mathcal{W} \tag{8.18}$$

if and only if

$$\langle P_{\mathcal{V}}\mathbf{x}, \mathbf{y} \rangle_{\mathcal{U}} = \langle \mathbf{x}, P_{\mathcal{V}}\mathbf{y} \rangle_{\mathcal{U}} \quad \text{for every choice of} \quad \mathbf{x}, \mathbf{y} \in \mathcal{U}\,. \tag{8.19}$$

Proof. Item (1) is immediate from the definition of a direct sum decomposition. Items (2) and (3) then follow from the decompositions $\mathbf{v} = \mathbf{v} + \mathbf{0}$ and $\mathbf{w} = \mathbf{0} + \mathbf{w}$, respectively. Items (4) and (5) are left to the reader as exercises.

To verify (6), suppose first that (8.18) is in force. Then, since $P_{\mathcal{V}}\mathbf{u} \in \mathcal{V}$ and $(I - P_{\mathcal{V}})\mathbf{u} \in \mathcal{W}$ for every vector $\mathbf{u} \in \mathcal{U}$,

$$\langle P_{\mathcal{V}}\mathbf{x}, \mathbf{y} \rangle_{\mathcal{U}} = \langle P_{\mathcal{V}}\mathbf{x}, P_{\mathcal{V}}\mathbf{y} \rangle_{\mathcal{U}} + \langle P_{\mathcal{V}}\mathbf{x}, (I - P_{\mathcal{V}})\mathbf{y} \rangle_{\mathcal{U}} = \langle P_{\mathcal{V}}\mathbf{x}, P_{\mathcal{V}}\mathbf{y} \rangle_{\mathcal{U}}$$

and

$$\langle \mathbf{x}, P_{\mathcal{V}}\mathbf{y} \rangle_{\mathcal{U}} = \langle P_{\mathcal{V}}\mathbf{x}, P_{\mathcal{V}}\mathbf{y} \rangle_{\mathcal{U}} + \langle (I - P_{\mathcal{V}})\mathbf{x}, P_{\mathcal{V}}\mathbf{y} \rangle_{\mathcal{U}} = \langle P_{\mathcal{V}}\mathbf{x}, P_{\mathcal{V}}\mathbf{y} \rangle_{\mathcal{U}}$$

for every choice of $\mathbf{x}, \mathbf{y} \in \mathcal{U}$, i.e., (8.18) $\Longrightarrow$ (8.19). Conversely, if (8.19) is in force and $\mathbf{v} \in \mathcal{V}$ and $\mathbf{w} \in \mathcal{W}$, then

$$\langle \mathbf{v}, \mathbf{w} \rangle = \langle P_{\mathcal{V}}\mathbf{v}, \mathbf{w} \rangle = \langle \mathbf{v}, P_{\mathcal{V}}\mathbf{w} \rangle = \langle \mathbf{v}, \mathbf{0} \rangle = 0\,.$$

□

Exercise 8.22. Verify assertions (4) and (5) in Lemma 8.14.

Exercise 8.23. Let $\{\mathbf{v}, \mathbf{w}\}$ be a basis for a vector space $\mathcal{U}$ over $\mathbb{F}$. Find the projection $P_{\mathcal{V}}\mathbf{u}$ of the vector $\mathbf{u} = 2\mathbf{v} + 3\mathbf{w}$ onto the space $\mathcal{V}$ with respect to each of the following direct sum decompositions: $\mathcal{U} = \mathcal{V} \dot{+} \mathcal{W}$ and $\mathcal{U} = \mathcal{V} \dot{+} \mathcal{W}_1$, when $\mathcal{V} = \operatorname{span}\{\mathbf{v}\}$, $\mathcal{W} = \operatorname{span}\{\mathbf{w}\}$ and $\mathcal{W}_1 = \operatorname{span}\{\mathbf{w} + \mathbf{v}\}$.

It is important to keep in mind that $P_{\mathcal{V}}$ **depends upon both $\mathcal{V}$ and the complementary space $\mathcal{W}$.** However, if $P_{\mathcal{V}}$ is an orthogonal projection, then $\mathcal{W}$ is taken equal to the **orthogonal complement** $\mathcal{V}^{\perp}$ of $\mathcal{V}$, i.e.,

$$\mathcal{W} = \mathcal{V}^{\perp} = \{\mathbf{u} \in \mathcal{U} : \langle \mathbf{v}, \mathbf{u} \rangle_{\mathcal{U}} = 0 \quad \text{for every} \quad \mathbf{v} \in \mathcal{V}\}\,. \tag{8.20}$$

8.9. Orthogonal projections

The next result is an analogue of Lemma 8.14 for orthogonal projections that is formulated in terms of one subspace $\mathcal{V}$ of $\mathcal{U}$ rather than in terms of a pair of complementary subspaces $\mathcal{V}$ and $\mathcal{W}$. This is possible because, as noted at the end of the previous section, the second space $\mathcal{W}$ is implicitly specified as the **orthogonal complement** $\mathcal{V}^\perp$ of $\mathcal{V}$.

Lemma 8.15. *Let $\mathcal{U}$ be an inner product space over $\mathbb{F}$, let $\mathcal{V}$ be a subspace of $\mathcal{U}$ with basis $\{\mathbf{v}_1, \ldots, \mathbf{v}_k\}$ and Gram matrix G with entries $g_{ij} = \langle \mathbf{v}_j, \mathbf{v}_i\rangle_\mathcal{U}$. Then:*

(1) *For every vector $\mathbf{u} \in \mathcal{U}$, there exists exactly one vector $P_\mathcal{V}\mathbf{u} \in \mathcal{V}$ such that $\mathbf{u} - P_\mathcal{V}\mathbf{u} \in \mathcal{V}^\perp$; it is given by the formula*

$$P_\mathcal{V}\mathbf{u} = \sum_{j=1}^{k} (G^{-1}\mathbf{b})_j \mathbf{v}_j\,, \quad \textit{where} \quad \mathbf{b} = \begin{bmatrix} \langle \mathbf{u}, \mathbf{v}_1\rangle_\mathcal{U} \\ \vdots \\ \langle \mathbf{u}, \mathbf{v}_k\rangle_\mathcal{U} \end{bmatrix}; \tag{8.21}$$

i.e.,

$$\mathcal{U} = \mathcal{V} \oplus \mathcal{V}^\perp\,. \tag{8.22}$$

(2) *$P_\mathcal{V}$ is a linear transformation of $\mathcal{U}$ into $\mathcal{U}$ that maps $\mathcal{U}$ onto $\mathcal{V}$. Moreover, $P_\mathcal{V} = P_\mathcal{V}^2 = P_\mathcal{V}^*$; i.e., $P_\mathcal{V}$ is an orthogonal projection.*

(3) *$\|\mathbf{u} - \mathbf{v}\|_\mathcal{U}^2 \geq \|\mathbf{u} - P_\mathcal{V}\mathbf{u}\|_\mathcal{U}^2$ for every vector $\mathbf{v} \in \mathcal{V}$, with equality if and only if $\mathbf{v} = P_\mathcal{V}\mathbf{u}$.*

(4) *$\|P_\mathcal{V}\mathbf{u}\|_\mathcal{U}^2 \leq \|\mathbf{u}\|_\mathcal{U}^2$, with equality if and only if $\mathbf{u} \in \mathcal{V}$. Moreover,*

$$\|P_\mathcal{V}\mathbf{u}\|_\mathcal{U}^2 = \mathbf{b}^H G^{-1}\mathbf{b}\,.$$

(5) *If $\mathcal{U} = \mathbb{C}^n$ is endowed with the standard inner product and $V = [\mathbf{v}_1 \ \cdots \ \mathbf{v}_k]$, then $G = V^H V$ and*

$$P_\mathcal{V} = V(V^H V)^{-1} V^H\,. \tag{8.23}$$

Proof. The first assertion is equivalent to the claim that there exists exactly one choice of coefficients $c_1, \ldots, c_k \in \mathbb{C}$ such that

$$\left\langle \left(\mathbf{u} - \sum_{j=1}^{k} c_j \mathbf{v}_j\right), \mathbf{v}_i \right\rangle_\mathcal{U} = 0 \quad \text{for} \quad i = 1, \ldots, k\,,$$

or, equivalently in terms of the entries in the Gram matrix, that

$$\langle \mathbf{u}, \mathbf{v}_i\rangle_\mathcal{U} = \sum_{j=1}^{k} g_{ij} c_j \quad \text{for} \quad i = 1, \ldots, k\,.$$

But this in turn is the same as to say that the vector equation $\mathbf{b} = G\mathbf{c}$ has a unique solution $\mathbf{c} \in \mathbb{C}^k$ with components $c_1, \ldots, c_k$ for each choice of the vector $\mathbf{b}$. However, since G is invertible by Lemma 8.4, this is the case and $P_{\mathcal{V}}\mathbf{u}$ is uniquely specified by formula (8.21). Moreover, this formula clearly displays the fact that $P_{\mathcal{V}}$ is a linear transformation of $\mathcal{U}$ into $\mathcal{V}$, since the vector $\mathbf{b}$ depends linearly on $\mathbf{u}$. The rest of (2) is left to the reader as an exercise.

Next, since

$$\langle \mathbf{u} - P_{\mathcal{V}}\mathbf{u}, P_{\mathcal{V}}\mathbf{u} - \mathbf{v}\rangle_{\mathcal{U}} = 0 \quad \text{for} \quad \mathbf{v} \in \mathcal{V},$$

it is readily seen that

$$\|\mathbf{u} - \mathbf{v}\|_{\mathcal{U}}^2 = \|\mathbf{u} - P_{\mathcal{V}}\mathbf{u} + P_{\mathcal{V}}\mathbf{u} - \mathbf{v}\|_{\mathcal{U}}^2 = \|\mathbf{u} - P_{\mathcal{V}}\mathbf{u}\|_{\mathcal{U}}^2 + \|P_{\mathcal{V}}\mathbf{u} - \mathbf{v}\|_{\mathcal{U}}^2, \tag{8.24}$$

which serves to justify (3).

The first part of (4) follows from formula (8.24) with $\mathbf{v} = \mathbf{0}$; the second part is a straightforward calculation (it's a special case of (8.7)).

Finally, (5) follows from (1), since $G = V^H V$ and $\mathbf{b} = V^H\mathbf{u}$ in the given setting. □

Exercise 8.24. Show that in the setting of Lemma 8.15,

$$\|\mathbf{u} - \sum_{j=1}^{k} c_j\mathbf{v}_j\|^2 \geq \|\mathbf{u}\|^2 - \mathbf{b}^H G^{-1}\mathbf{b} \tag{8.25}$$

with equality if and only if $c_j = (G^{-1}\mathbf{b})_j$ for $j = 1, \ldots, k$.

Exercise 8.25. Verify directly that the transformation $P_{\mathcal{V}}$ defined by formula (8.23) for $\mathcal{U} = \mathbb{C}^n$ endowed with the standard inner product meets the following conditions:

(1) $(P_{\mathcal{V}})^2 = P_{\mathcal{V}}$.

(2) $(P_{\mathcal{V}})^H = P_{\mathcal{V}}$.

(3) $P_{\mathcal{V}}\mathbf{v}_j = \mathbf{v}_j$ for $j = 1, \ldots, k$.

(4) $P_{\mathcal{V}}\mathbf{u} = 0$ if $\mathbf{u} \in \mathcal{V}^\perp$, computed with respect to the standard inner product.

[HINT: Items (1), (2) and (4) are easy, as is (3), if you take advantage of the fact that $\mathbf{v}_j = V\mathbf{e}_j$, where $\mathbf{e}_j$ is the j'th column vector of I_n.]

Exercise 8.26. Calculate the norm of the projection P that is defined in Exercise 8.21.

Exercise 8.27. Show that if P is a nonzero projection matrix, then:

(a) $\|P\| = 1$ if P is an orthogonal projection matrix.

(b) $\|P\|$ can be very large if P is not an orthogonal projection matrix.

Exercise 8.28. Let

$$\begin{bmatrix}\mathbf{u}_1 & \mathbf{u}_2 & \mathbf{u}_3 & \mathbf{u}_4 & \mathbf{u}_5 & \mathbf{u}_6\end{bmatrix} = \begin{bmatrix} 1 & 1 & 1 & 2 & 3 & 4 \\ 2 & 0 & 4 & 1 & 5 & 0 \\ 1 & 1 & 1 & 0 & 1 & 0 \\ 0 & 1 & -1 & 0 & -1 & 1 \end{bmatrix},$$

and let $\mathcal{U} = \text{span}\{\mathbf{u}_1, \mathbf{u}_2, \mathbf{u}_3, \mathbf{u}_4\}$, $\mathcal{V} = \text{span}\{\mathbf{u}_1, \mathbf{u}_2, \mathbf{u}_3\}$, $\mathcal{W}_1 = \text{span}\{\mathbf{u}_4\}$ and $\mathcal{W}_2 = \text{span}\{\mathbf{u}_5\}$.

(a) Find a basis for the vector space $\mathcal{V}$.

(b) Show that $\mathcal{U} = \mathcal{V} \dot{+} \mathcal{W}_1$ and $\mathcal{U} = \mathcal{V} \dot{+} \mathcal{W}_2$.

(c) Find the projection of the vector $\mathbf{u}_6$ onto the space $\mathcal{V}$ with respect to the first direct sum decomposition.

(d) Find the projection of the vector $\mathbf{u}_6$ onto the space $\mathcal{V}$ with respect to the second direct sum decomposition.

(e) Find the orthogonal projection of the vector $\mathbf{u}_6$ onto the space $\mathcal{V}$.

8.10. Orthogonal expansions

If the vectors $\mathbf{v}_1, \ldots, \mathbf{v}_k$ that are specified in the Lemma 8.15 are orthonormal in $\mathcal{U}$, then the formulas simplify, because $G = I_k$, and the conclusions can be reformulated as follows:

Lemma 8.16. *Let* $\mathbf{v}_1, \ldots, \mathbf{v}_k$ *be an orthonormal set of vectors in an inner product space* $\mathcal{U}$ *over* $\mathbb{F}$ *and let* $\mathcal{V} = \text{span}\{\mathbf{v}_1, \ldots, \mathbf{v}_k\}$. *Then:*

(1) *The vectors* $\mathbf{v}_1, \ldots, \mathbf{v}_k$ *are linearly independent.*

(2) $\mathcal{V}^\perp = \{\mathbf{u} \in \mathcal{U} : \langle \mathbf{v}_j, \mathbf{u} \rangle_\mathcal{U} = 0 \text{ for } j = 1, \ldots, k\}$.

(3) *The orthogonal projection* $P_\mathcal{V}\mathbf{u}$ *of a vector* $\mathbf{u} \in \mathcal{U}$ *onto* $\mathcal{V}$ *is given by the formula*

$$P_\mathcal{V}\mathbf{u} = \langle \mathbf{u}, \mathbf{v}_1 \rangle_\mathcal{U}\mathbf{v}_1 + \cdots + \langle \mathbf{u}, \mathbf{v}_k \rangle_\mathcal{U}\mathbf{v}_k.$$

(4) $\|P_\mathcal{V}\mathbf{u}\|_\mathcal{U}^2 = \sum_{j=1}^k |\langle \mathbf{u}, \mathbf{v}_j \rangle_\mathcal{U}|^2$ *for every vector* $\mathbf{u} \in \mathcal{U}$.

(5) **(Bessel's inequality)** $\sum_{j=1}^k |\langle \mathbf{u}, \mathbf{v}_j \rangle_\mathcal{U}|^2 \leq \|\mathbf{u}\|_\mathcal{U}^2$, *with equality if and only if* $\mathbf{u} \in \mathcal{V}$.

Notice that:

(1) It is easy to calculate the coefficients of a vector $\mathbf{v}$ in the span of $\mathbf{v}_1, \ldots, \mathbf{v}_k$ and its norm in terms of these coefficients:

$$\mathbf{v} = \sum_{j=1}^k c_j\mathbf{v}_j \Longrightarrow c_j = \langle \mathbf{v}, \mathbf{v}_j \rangle_\mathcal{U} \quad \text{and} \quad \|\mathbf{v}\|_\mathcal{U}^2 = \sum_{j=1}^k |c_j|^2.$$

(2) It is easy to calculate the coefficients of the projection $P_\mathcal{V}$ of a vector $\mathbf{u}$ onto $\mathcal{V} = \text{span}\{\mathbf{v}_1, \ldots, \mathbf{v}_k\}$:

$$P_\mathcal{V}\mathbf{u} = \sum_{j=1}^{k} c_j\mathbf{v}_j \Longrightarrow c_j = \langle \mathbf{u}, \mathbf{v}_j \rangle_\mathcal{U} \quad \text{and} \quad \|P_\mathcal{V}\mathbf{u}\|_\mathcal{U}^2 = \sum_{j=1}^{k} |c_j|^2 .$$

(3) The coefficients c_j, $j = 1, \ldots, k$, computed in (2) do not change if the space $\mathcal{V}$ is enlarged by adding more orthonormal vectors.

It is important to note that to this point the analysis in this section is applicable to any inner product space. Thus, for example, we may choose $\mathcal{U}$ equal to the set of continuous complex valued functions on the interval $[0, 1]$, with inner product

$$\langle f, g \rangle_\mathcal{U} = \int_0^1 f(t)\overline{g(t)}dt.$$

Then it is readily checked that the set of functions

$$\varphi_j(t) = e^{j2\pi it}, \quad j = 1, \ldots, k,$$

is an orthonormal family in $\mathcal{U}$ for any choice of the integer k. Consequently,

$$\sum_{j=1}^{k} \left| \int_0^1 f(t)\overline{\varphi_j(t)}dt \right|^2 \leq \int_0^1 |f(t)|^2 dt,$$

by the last item in Lemma 8.16.

Exercise 8.29. Show that no matter how large you choose k, the family $\varphi_j(t) = e^{j2\pi it}$, $j = 1, \ldots, k$, is not a basis for the space $\mathcal{U}$ considered just above.

Lemma 8.17. *Let $\mathcal{U}$ be a k-dimensional inner product space over $\mathbb{F}$ with inner product $\langle \cdot, \cdot \rangle_\mathcal{U}$, and let $\mathbf{u}_1, \ldots, \mathbf{u}_k$ be an orthonormal basis for $\mathcal{U}$. Then*

$$\text{(8.26)} \quad \langle \mathbf{u}, \mathbf{w} \rangle_\mathcal{U} = \sum_{j=1}^{k} \langle \mathbf{u}, \mathbf{u}_j \rangle_\mathcal{U} \overline{\langle \mathbf{w}, \mathbf{u}_j \rangle_\mathcal{U}} \quad \textit{and} \quad \langle \mathbf{u}, \mathbf{u} \rangle_\mathcal{U} = \sum_{j=1}^{k} |\langle \mathbf{u}, \mathbf{u}_j \rangle_\mathcal{U}|^2$$

for every choice of $\mathbf{u}$ and $\mathbf{w}$ in $\mathcal{U}$.

Proof. Since the given basis for $\mathcal{U}$ is orthonormal,

$$\mathbf{u} = \sum_{j=1}^{k} c_j\mathbf{u}_j \quad \text{and} \quad \mathbf{w} = \sum_{i=1}^{k} d_i\mathbf{u}_j ,$$

where

$$c_j = \langle \mathbf{u}, \mathbf{u}_j \rangle_\mathcal{U} \quad \text{and} \quad d_j = \langle \mathbf{w}, \mathbf{v}_j \rangle_\mathcal{U} , \quad \text{for} \quad j = 1, \ldots, k .$$

Therefore,

$$\langle \mathbf{u}, \mathbf{w}\rangle_{\mathcal{U}} = \langle \sum_{j=1}^{k} c_j \mathbf{u}_j, \sum_{i=1}^{k} d_i \mathbf{u}_j\rangle = \sum_{i,j=1}^{k} c_j \overline{d_i} \langle \mathbf{u}_j, \mathbf{u}_i\rangle_{\mathcal{U}} = \sum_{i=1}^{k} c_i \overline{d_i}\,,$$

by the presumed orthonormality of the basis. The rest is plain. □

Lemma 8.18. *Let $\mathcal{U}$ and $\mathcal{V}$ be inner product spaces over $\mathbb{F}$ with orthonormal bases $\mathbf{u}_1, \ldots, \mathbf{u}_q$ and $\mathbf{v}_1, \ldots, \mathbf{v}_p$, respectively, and let S be a linear transformation from $\mathcal{U}$ into $\mathcal{V}$. Then*

$$\sum_{j=1}^{q} \|S\mathbf{u}_j\|_{\mathcal{V}}^2 = \sum_{i=1}^{p} \|S^*\mathbf{v}_i\|_{\mathcal{U}}^2. \tag{8.27}$$

Proof. By Lemma 8.17,

$$\|S\mathbf{u}_j\|_{\mathcal{V}}^2 = \sum_{i=1}^{p} |\langle S\mathbf{u}_j, \mathbf{v}_i\rangle_{\mathcal{V}}|^2 = \sum_{i=1}^{p} |\langle \mathbf{u}_j, S^*\mathbf{v}_i\rangle_{\mathcal{U}}|^2\,.$$

Therefore,

$$\sum_{j=1}^{q} \|S\mathbf{u}_j\|_{\mathcal{V}}^2 = \sum_{i=1}^{p}\sum_{j=1}^{q} |\langle \mathbf{u}_j, S^*\mathbf{v}_i\rangle_{\mathcal{U}}|^2 = \sum_{i=1}^{p} \|S^*\mathbf{v}_i\|_{\mathcal{U}}^2\,.$$

□

Lemma 8.19. *Let S be a linear transformation from an inner product space $\mathcal{U}$ over $\mathbb{F}$ into itself and let $\{\mathbf{u}_1, \ldots, \mathbf{u}_n\}$ and $\{\mathbf{w}_1, \ldots, \mathbf{w}_n\}$ be any two orthonormal bases for $\mathcal{U}$. Then:*

$$\sum_{j=1}^{n} \langle S\mathbf{u}_j, \mathbf{u}_j\rangle_{\mathcal{U}} = \sum_{j=1}^{n} \langle S\mathbf{w}_j, \mathbf{w}_j\rangle_{\mathcal{U}}\,. \tag{8.28}$$

Proof. By Lemma 8.17,

$$\langle S\mathbf{u}_j, \mathbf{u}_j\rangle_{\mathcal{U}} = \sum_{i=1}^{n} \langle S\mathbf{u}_j, \mathbf{w}_i\rangle_{\mathcal{U}} \overline{\langle \mathbf{u}_j, \mathbf{w}_i\rangle_{\mathcal{U}}} = \sum_{i=1}^{n} \langle S\mathbf{u}_j, \mathbf{w}_i\rangle_{\mathcal{U}} \langle \mathbf{w}_i, \mathbf{u}_j\rangle_{\mathcal{U}}$$

and

$$\langle \mathbf{w}_i, S^*\mathbf{w}_i\rangle_{\mathcal{U}} = \sum_{j=1}^{n} \langle \mathbf{w}_i, \mathbf{u}_j\rangle_{\mathcal{U}} \overline{\langle S^*\mathbf{w}_i, \mathbf{u}_j\rangle_{\mathcal{U}}} = \sum_{j=1}^{n} \langle \mathbf{w}_i, \mathbf{u}_j\rangle_{\mathcal{U}} \langle S\mathbf{u}_j, \mathbf{w}_i\rangle_{\mathcal{U}}\,.$$

Therefore,

$$\sum_{j=1}^{n} \langle S\mathbf{u}_j, \mathbf{u}_j\rangle_{\mathcal{U}} = \sum_{i=1}^{n} \langle \mathbf{w}_i, S^*\mathbf{w}_i\rangle_{\mathcal{U}} = \sum_{i=1}^{n} \langle S\mathbf{w}_i, \mathbf{w}_i\rangle_{\mathcal{U}}\,,$$

as claimed. □

Exercise 8.30. Show that in the setting of Lemma 8.19

$$\sum_{j=1}^{n} \|S\mathbf{u}_j\|_{\mathcal{U}}^2 = \sum_{j=1}^{n} \|S\mathbf{w}_j\|_{\mathcal{U}}^2 = \sum_{j=1}^{n} \|S^*\mathbf{u}_j\|_{\mathcal{U}}^2 . \tag{8.29}$$

8.11. The Gram-Schmidt method

Let $\{\mathbf{u}_1, \ldots, \mathbf{u}_k\}$ be a set of linearly independent vectors in an inner product space $\mathcal{U}$ over $\mathbb{F}$. The Gram-Schmidt procedure is a method for finding a set of orthonormal vectors $\{\mathbf{v}_1, \ldots, \mathbf{v}_k\}$ such that

$$\mathcal{V}_j = \text{span}\{\mathbf{v}_1, \ldots, \mathbf{v}_j\} = \text{span}\{\mathbf{u}_1, \ldots, \mathbf{u}_j\} \text{ for } j = 1, \ldots, k.$$

The steps are as follows: Let $P_{\mathcal{V}_j}$ denote the orthogonal projection onto $\mathcal{V}_j$ and then:

(1) Set $\mathbf{v}_1 = \mathbf{u}_1/\|\mathbf{u}_1\|_{\mathcal{U}}$. Then $\|\mathbf{v}_1\|_{\mathcal{U}} = 1$.

(2) Set

$$\mathbf{w}_2 = \mathbf{u}_2 - P_{\mathcal{V}_1}\mathbf{u}_2 = \mathbf{u}_2 - \langle \mathbf{u}_2, \mathbf{v}_1\rangle_{\mathcal{U}}\mathbf{v}_1$$

and check that

$$\langle \mathbf{w}_2, \mathbf{v}_1\rangle_{\mathcal{U}} = \langle \mathbf{u}_2, \mathbf{v}_1\rangle_{\mathcal{U}} - \langle \mathbf{u}_2, \mathbf{v}_1\rangle_{\mathcal{U}} \langle \mathbf{v}_1, \mathbf{v}_1\rangle_{\mathcal{U}} = 0$$

and, since $\mathbf{u}_2$ and $\mathbf{v}_1$ are linearly independent, $\|\mathbf{w}_2\|_{\mathcal{U}} \neq 0$. Therefore, we can set $\mathbf{v}_2 = \mathbf{w}_2/\|\mathbf{w}_2\|_{\mathcal{U}}$. Notice that

$$\text{span}\{\mathbf{u}_1, \mathbf{u}_2\} = \text{span}\{\mathbf{v}_1, \mathbf{v}_2\} .$$

(3) Set

$$\mathbf{w}_3 = \mathbf{u}_3 - P_{\mathcal{V}_2}\mathbf{u}_3 = \mathbf{u}_3 - \langle \mathbf{u}_3, \mathbf{v}_1\rangle_{\mathcal{U}}\mathbf{v}_1 - \langle \mathbf{u}_3, \mathbf{v}_2\rangle_{\mathcal{U}}\mathbf{v}_2$$

and check that

$$\langle \mathbf{w}_3, \mathbf{v}_1\rangle_{\mathcal{U}} = 0 \ , \ \langle \mathbf{w}_3, \mathbf{v}_2\rangle_{\mathcal{U}} = 0 \ , \ \|\mathbf{w}_3\|_{\mathcal{U}} \neq 0 \ .$$

Therefore we can set

$$\mathbf{v}_3 = \mathbf{w}_3/\|\mathbf{w}_3\|_{\mathcal{U}}$$

and verify that $\mathbf{v}_1, \mathbf{v}_2, \mathbf{v}_3$ is an orthonormal set of vectors such that

$$\text{span}\{\mathbf{v}_1, \mathbf{v}_2, \mathbf{v}_3\} = \text{span}\{\mathbf{u}_1, \mathbf{u}_2, \mathbf{u}_3\} .$$

The first three steps should suffice to transmit the general idea. A complete formal proof depends upon induction, i.e., showing that if the procedure works for the first j vectors (and $j < k$), then it works for the first $j+1$ vectors: Thus suppose $\mathbf{v}_1, \ldots, \mathbf{v}_j$ is an orthonormal family of vectors such that

$$\text{span}\{\mathbf{v}_1, \ldots, \mathbf{v}_j\} = \text{span}\{\mathbf{u}_1, \ldots, \mathbf{u}_j\} .$$

Then, set

$$\mathbf{w}_{j+1} = \mathbf{u}_{j+1} - P_{\mathcal{V}_j}\mathbf{u}_{j+1} = \mathbf{u}_{j+1} - \langle \mathbf{u}_{j+1}, \mathbf{v}_1\rangle_{\mathcal{U}}\mathbf{v}_1 - \cdots - \langle \mathbf{u}_{j+1}, \mathbf{v}_j\rangle_{\mathcal{U}}\mathbf{v}_j$$

and observe that

$$\langle \mathbf{w}_{j+1}, \mathbf{v}_1\rangle_{\mathcal{U}} = \cdots = \langle \mathbf{w}_{j+1}, \mathbf{v}_j\rangle_{\mathcal{U}} = 0 \quad \text{and} \quad \|\mathbf{w}_{j+1}\|_{\mathcal{U}} \neq 0\ .$$

Therefore, we can set $\mathbf{v}_{j+1} = \mathbf{w}_{j+1}/\|\mathbf{w}_{j+1}\|_{\mathcal{U}}$ and check that $\mathbf{v}_1, \ldots, \mathbf{v}_{j+1}$ is an orthonormal set of vectors such that

$$\mathrm{span}\{\mathbf{v}_1, \ldots, \mathbf{v}_{j+1}\} = \mathrm{span}\{\mathbf{u}_1, \ldots, \mathbf{u}_{j+1}\}$$

to finish. □

Exercise 8.31. Show that if $\{\mathbf{u}_1, \ldots, \mathbf{u}_k\}$ is a set of k linearly independent vectors in $\mathbb{F}^n$, then there exists an invertible upper triangular matrix $B \in \mathbb{F}^{k\times k}$ such that the matrix $V = \begin{bmatrix}\mathbf{u}_1 & \cdots & \mathbf{u}_k\end{bmatrix} B$ has orthonormal columns. [HINT: This is a byproduct of the Gram-Schmidt procedure.]

Exercise 8.32. Show that if $A \in \mathbb{C}^{n\times n}$ is invertible, then there exists an invertible upper triangular matrix B such that AB is unitary and an invertible lower triangular matrix C such that CA is unitary. [HINT: Exploit Exercise 8.31.]

Exercise 8.33. Show that if $A \in \mathbb{R}^{n\times n}$ is invertible, then there exist an invertible upper triangular matrix B such that AB is an orthogonal matrix and an invertible lower triangular matrix C such that CA is an orthogonal matrix. [HINT: Exploit Exercise 8.31.]

Exercise 8.34. Find a set of three polynomials $p_0(t) = a$, $p_1(t) = b+ct$, and $p_3(t) = d+et+ft^2$ with real coefficients a, b, c, d, e, f so that they form an orthonormal set with respect to the real inner product $\langle f, g\rangle = \int_0^2 f(t)g(t)dt$.

8.12. Toeplitz and Hankel matrices

The structure of the Gram matrix G of a set of vectors $\{\mathbf{v}_1, \ldots, \mathbf{v}_k\}$ in an inner product space $\mathcal{U}$ over $\mathbb{C}$ depends upon the choice of the vectors and upon the inner product. In many applications the Gram matrix that comes into play is either a Toeplitz matrix or a Hankel matrix. A matrix $A \in \mathbb{C}^{n\times n}$ is said to be a **Toeplitz** matrix if its entries a_{ij}, $i, j = 1, \ldots, n$, depend only upon $i - j$, i.e., if

$$a_{ij} = \alpha_{i-j} \text{ for } i, j = 1, \ldots, n,$$

for some set of $2n - 1$ numbers $\alpha_{-(n-1)}, \ldots, \alpha_0, \ldots, \alpha_{n-1}$. Toeplitz matrices occur naturally in the theory of stationary (and weakly stationary) sequences and in approximation and extension problems involving trigonometric polynomials.

A matrix $A \in \mathbb{C}^{n\times n}$ is said to be a **Hankel** matrix if its entries a_{ij}, $i,j = 1,\ldots,n$, depend only upon $i+j$, i.e., if

$$a_{ij} = \beta_{i+j-1} \text{ for } i,j = 1,\ldots,n,$$

for some choice of the $2n-1$ numbers $\beta_1,\ldots,\beta_{2n-1}$. Hankel matrices occur in problems involving polynomial (and rational) approximation on $\mathbb{R}$ or subsets of $\mathbb{R}$.

If $\mathcal{U}$ is the space of continuous functions $f(e^{i\theta})$ on the unit circle and an inner product is defined in terms of a function $w(e^{i\theta})$ by the formula

$$\langle f,g\rangle_{\mathcal{U}} = \frac{1}{2\pi}\int_0^{2\pi} \overline{g(e^{i\theta})}w(e^{i\theta})f(e^{i\theta})d\theta\,,$$

where $w(e^{i\theta}) > 0$ for $0 \le \theta < 2\pi$, and if $\varphi_j(e^{i\theta}) = e^{ij\theta}$ for $j = 1,\ldots,n$, then

$$g_{jk} = \langle \varphi_k,\varphi_j\rangle_{\mathcal{U}} = \frac{1}{2\pi}\int_0^{2\pi} e^{-ij\theta}w(e^{i\theta})e^{ik\theta}d\theta = a_{j-k}\,,$$

where

$$a_j = \frac{1}{2\pi}\int_0^{2\pi} w(e^{i\theta})e^{-ij\theta}d\theta$$

is the j'th Fourier coefficient of $w(e^{i\theta})$; i.e., G is a Toeplitz matrix.

On the other hand, if $\mathcal{U}$ is the space of continuous functions $f(x)$ on a subinterval of $\mathbb{R}$ and an inner product is defined in terms of a function $w(x)$ by the formula

$$\langle f,g\rangle_{\mathcal{U}} = \int_c^d \overline{g(x)}w(x)f(x)dx\,,$$

where $w(x) > 0$ on the interval $c < x < d$, and if $\varphi_j(x) = x^j$, then

$$g_{jk} = \langle \varphi_k,\varphi_j\rangle_{\mathcal{U}} = \int_c^d x^j w(x)x^k dx = b_{j+k}\,,$$

where

$$b_j = \int_c^d x^j w(x)dx\,;$$

i.e., G is a Hankel matrix.

These simple examples help to illustrate the great interest in developing efficient schemes for solving matrix equations of the form $G\mathbf{x} = \mathbf{b}$ and calculating G^{-1} when G is either a Toeplitz or a Hankel matrix.

Let

$$(8.30)\quad Z_n = \sum_{j=1}^n \mathbf{e}_j\mathbf{e}_{n-j+1}^T = \begin{bmatrix} 0 & 0 & \cdots & 0 & 1\\ 0 & 0 & \cdots & 1 & 0\\ \vdots & & & \vdots & \\ 1 & 0 & \cdots & 0 & 0\end{bmatrix} \quad\text{and}\quad N_n = \sum_{j=1}^{n-1} \mathbf{e}_j\mathbf{e}_{j+1}^T\,.$$

Exercise 8.35. Show that $A \in \mathbb{C}^{n\times n}$ is a Toeplitz matrix if and only if Z_nA is a Hankel matrix.

Exercise 8.36. Show that if $A \in \mathbb{C}^{n\times n}$ is a Hankel matrix with $a_{ij} = \beta_{i+j-1}$ for $i, j = 1, \ldots, n$, then, in terms of the matrices $Z = Z_n$ and $N = N_n$ defined in formula (8.30),

$$A = \sum_{j=1}^{n} \beta_j Z(N^T)^{n-j} + \sum_{j=1}^{n-1} \beta_{n+j} Z N^j \,.$$

Exercise 8.37. Show that if $A \in \mathbb{C}^{n\times n}$ is a Toeplitz matrix with $a_{ij} = \alpha_{i-j}$, then, in terms of the matrices $Z = Z_n$ and $N = N_n$ defined in formula (8.30),

$$A = \sum_{i=0}^{n-1} \alpha_{-i} N^i + \sum_{i=1}^{n-1} \alpha_i (N^T)^i \,.$$

Exercise 8.38. The $n\times n$ Hankel matrix H_n with entries $h_{ij} = 1/(i+j+1)$ for $i, j = 0, \ldots, n-1$ is known as the **Hilbert matrix**. Show that the Hilbert matrix is invertible. [HINT: $\int_0^1 x^i x^j dx = 1/(i+j+1)$.]

Exercise 8.39. Show that if H_n denotes the $n\times n$ Hankel matrix introduced in Exercise 8.38 and if $\mathbf{a} \in \mathbb{C}^n$ and $\mathbf{b} \in \mathbb{C}^n$ are vectors with components $a_0, \ldots, a_{n-1}$ and $b_0, \ldots, b_{n-1}$, respectively, then

(8.31)

$$\langle H_n\mathbf{a}, \mathbf{b}\rangle_{st} = \frac{1}{2\pi}\int_0^{2\pi} \left(\sum_{k=0}^{n-1} \overline{b_k} e^{-ikt}\right) \left(ie^{-it}(\pi - t)\right) \left(\sum_{j=0}^{n-1} a_j e^{-ijt}\right) dt \,.$$

Exercise 8.40. Show that if H_n denotes the $n\times n$ Hankel matrix introduced in Exercise 8.38, then $\|H_n\|_{2,2} < \pi$. [HINT: First use formula (8.31) to prove that $|\langle H_n\mathbf{a}, \mathbf{b}\rangle_{st}| \le \pi\|\mathbf{a}\|_2\|\mathbf{b}\|_2$.]

8.13. Gaussian quadrature

Let $w(x)$ denote a positive continuous function on a finite interval $a \le x \le b$ and let $\mathcal{U}$ denote the inner product space over $\mathbb{R}$ of continuous complex valued functions on this interval, equipped with the inner product

$$\langle f, g\rangle_{\mathcal{U}} = \int_a^b \overline{g(x)} w(x) f(x) dx \,.$$

Let $\mathcal{P}_k$, $k = 0, 1, \ldots$, denote the $k+1$-dimensional subspace of polynomials of degree less than or equal to k (with complex coefficients), let $\mathfrak{P}_n$ denote the orthogonal projection of $\mathcal{U}$ onto $\mathcal{P}_n$, let M_x denote the linear transformation on $\mathcal{U}$ of multiplication by the independent variable x and let

$$S_n = \mathfrak{P}_n M_x|_{\mathcal{P}_n} \quad \text{for} \quad n = 0, 1 \ldots ;$$

i.e., S_n maps $f \in \mathcal{P}_n \longrightarrow \mathfrak{P}_n M_x f$. Then clearly $\mathcal{P}_n$ is invariant under S_n and $S_n = S_n^*$; i.e.,

$$\langle S_n f, g\rangle_{\mathcal{U}} = \langle f, S_n g\rangle_{\mathcal{U}} \quad \text{for every choice of} \quad f, g \in \mathcal{P}_n\,.$$

Consequently, there exists an orthonormal set of vectors $\varphi_j \in \mathcal{P}_n$, $j = 0, \ldots, n$, such that

$$S_n\varphi_j = \lambda_j\varphi_j \quad \text{and} \quad \lambda_j \in \mathbb{R} \quad \text{for} \quad j = 0, \ldots, n\,.$$

Thus, if

$$\pi_k(x) = x^k \quad \text{for} \quad k = 0, 1, \ldots,$$

then, for $1 \le k \le n+1$ and $j = 0, \ldots, n$,

$$\begin{aligned} \langle \varphi_j, \pi_k\rangle_{\mathcal{U}} &= \langle \varphi_j, S_n\pi_{k-1}\rangle_{\mathcal{U}} \\ &= \langle S_n\varphi_j, \pi_{k-1}\rangle_{\mathcal{U}} \\ &= \lambda_j\langle \varphi_j, \pi_{k-1}\rangle_{\mathcal{U}}\,, \end{aligned}$$

and hence, upon iterating this formula, we obtain

(8.32)

$$\langle \varphi_j, \pi_k\rangle_{\mathcal{U}} = \lambda_j^k\langle \varphi_j, \pi_0\rangle_{\mathcal{U}} \quad \text{for} \quad j = 0, \ldots, n \quad \text{and} \quad k = 1, \ldots, n+1\,.$$

Lemma 8.20. *If $p(x)$ is any polynomial of degree less than or equal to $n+1$ with complex coefficients, then:*

(1) $\langle \varphi_j, p\rangle_{\mathcal{U}} = p(\lambda_j)\langle \varphi_j, \pi_0\rangle_{\mathcal{U}}$.

(2) $\varphi_j(\lambda_j)\langle \varphi_j, \pi_0\rangle_{\mathcal{U}} = 1$ *and* $\varphi_k(\lambda_j) = 0$ *if* $j \neq k$.

(3) *If $p(x)$ is a polynomial of degree $n+1$ such that*

$$\langle \varphi_j, p\rangle = 0 \quad \textit{for} \quad j = 0, \ldots, n\,,$$

then $p(\lambda_j) = 0$ for $j = 0, \ldots, n$.

Proof. Let $p(x) = \sum_{i=0}^{n+1} c_i x^i$. Then, in view of formula (8.32),

$$\begin{aligned} \langle \varphi_j, p\rangle_{\mathcal{U}} &= \langle \varphi_j, c_0\rangle_{\mathcal{U}} + \sum_{i=1}^{n+1} c_i\langle \varphi_j, \pi_i\rangle_{\mathcal{U}} \\ &= c_0\langle \varphi_j, 1\rangle_{\mathcal{U}} + \sum_{i=1}^{n+1} c_i\lambda_j^i\langle \varphi_j, \pi_0\rangle_{\mathcal{U}} \\ &= p(\lambda_j)\langle \varphi_j, \pi_0\rangle_{\mathcal{U}}\,, \end{aligned}$$

which verifies (1) and, upon choosing $p = \varphi_k$, yields the formula

$$\varphi_k(\lambda_j)\langle \varphi_j, \pi_0\rangle_{\mathcal{U}} = \langle \varphi_j, \varphi_k\rangle_{\mathcal{U}}\,,$$

which leads easily to (2), since the φ_j are orthonormal.

Finally, (3) is an easy consequence of the formula in (1) and the fact that $\langle \varphi_j, \pi_0\rangle_{\mathcal{U}} \neq 0$ for $j = 0, \ldots, n$, which is immediate from (2). □

Theorem 8.21. *Let $W_j = |\langle \pi_0, \varphi_j\rangle_{\mathcal{U}}|^2$ for $j = 0, \ldots, n$. Then the formula*

$$\int_a^b w(x)f(x)dx = \sum_{i=0}^{n} W_i f(\lambda_i) \tag{8.33}$$

is valid for every polynomial $f(x)$ of degree less than or equal to $2n+1$ with complex coefficients.

Proof. It suffices to verify this formula for $f(x) = \pi_k(x)$, $k = 0, \ldots, 2n+1$. Consider first the case $k = i + j$ with $i, j = 0, \ldots, n$. Then

$$\begin{aligned}
\int_a^b \pi_i(x)w(x)\pi_j(x)dx &= \langle \pi_j, \pi_i\rangle_{\mathcal{U}} \\
&= \left\langle \sum_{s=0}^{n} \langle \pi_j, \varphi_s\rangle_{\mathcal{U}}\, \varphi_s, \sum_{t=0}^{n} \langle \pi_i, \varphi_t\rangle_{\mathcal{U}}\, \varphi_t \right\rangle_{\mathcal{U}} \\
&= \sum_{s,t=0}^{n} \langle \pi_j, \varphi_s\rangle_{\mathcal{U}}\, \langle \varphi_t, \pi_i\rangle_{\mathcal{U}}\, \langle \varphi_s, \varphi_t\rangle_{\mathcal{U}} \\
&= \sum_{s=0}^{n} \overline{\pi_j(\lambda_s)}\pi_i(\lambda_s)\langle \varphi_s, \pi_0\rangle_{\mathcal{U}}\langle \pi_0, \varphi_s\rangle_{\mathcal{U}} \\
&= \sum_{s=0}^{n} \lambda_s^{i+j} |\langle \pi_0, \varphi_s\rangle_{\mathcal{U}}|^2 .
\end{aligned}$$

To complete the proof, it remains only to check that

$$\int_a^b w(x)x^{2n+1}dx = \sum_{t=0}^{2n+1} \lambda_s^{2n+1} |\langle \pi_0, \varphi_s\rangle_{\mathcal{U}}|^2 . \tag{8.34}$$

The details are left to the reader. □

Exercise 8.41. Justify formula (8.34). [HINT: First check that the integral is equal to $\langle \pi_n, S_n\pi_n\rangle_{\mathcal{U}}$.]

Finite sums like (8.33) that serve to approximate definite integrals, with equality for a reasonable class of functions, are termed **quadrature formulas**.

The notation

(8.35)

$$h_j = \langle x^j, 1\rangle_{\mathcal{U}}\,, \quad H_n = \begin{bmatrix} h_0 & \cdots & h_n \\ \vdots & & \vdots \\ h_n & \cdots & h_{2n} \end{bmatrix} \quad \text{and} \quad K_n = \begin{bmatrix} h_1 & \cdots & h_{n+1} \\ \vdots & & \vdots \\ h_{n+1} & \cdots & h_{2n+1} \end{bmatrix}$$

will prove useful in the next three exercises.

Exercise 8.42. Show that in the setting of this section,

$$\mathfrak{P}_n x^{n+1} = \sum_{j=0}^{n} b_j x^j\,, \qquad \text{where} \quad \begin{bmatrix} b_0 \\ \vdots \\ b_n \end{bmatrix} = H_n^{-1} \begin{bmatrix} h_{n+1} \\ \vdots \\ h_{2n+1} \end{bmatrix},$$

h_j and the $(n+1) \times (n+1)$ Hankel matrix H_n are defined in (8.35).

Exercise 8.43. Show that in the setting of this section,

$$S_n \sum_{j=0}^{n} a_j x^j = \sum_{j=0}^{n} b_j x^j \Longleftrightarrow \begin{bmatrix} b_0 \\ \vdots \\ b_n \end{bmatrix} = H_n^{-1} K_n \begin{bmatrix} a_0 \\ \vdots \\ a_n \end{bmatrix},$$

where the $(n+1) \times (n+1)$ Hankel matrices H_n and K_n are defined in (8.35).

Exercise 8.44. Show that if $\mathbf{a}$ and $\mathbf{b}$ are vectors with components $a_0, \ldots, a_n$ and $b_0, \ldots, b_n$, respectively, then

(8.36)

$$\langle \sum_{j=0}^{n} a_j x^j, \sum_{k=0}^{n} b_k x^k \rangle_{\mathcal{U}} = \mathbf{b}^H H_n \mathbf{a} \quad \text{and} \quad \langle S_n \sum_{j=0}^{n} a_j x^j, \sum_{k=0}^{n} b_k x^k \rangle_{\mathcal{U}} = \mathbf{b}^H K_n \mathbf{a}\,,$$

where the $(n+1) \times (n+1)$ Hankel matrices H_n and K_n are defined in (8.35).

8.14. Bibliographical notes

Example 8.7 is adapted from the monograph by Borwein and Lewis [**10**]. The bound on the norm of the Hilbert matrix that is developed in Exercises 8.39 and 8.40 is adapted from the discussion of the Hilbert matrix in Peller [**56**]. Another way to compute this bound is discussed in Chapter 16. The treatment of Gaussian quadrature that is presented in this chapter, including the proofs of Lemma 8.20 and Theorem 8.21, is adapted from the PhD thesis of Ilan Degani [**19**].

Chapter 9

Symmetric, Hermitian and normal matrices

Not everything that one thinks, should one say; not everything that one says, should one write, and not everything that one writes, should one publish.

Dictum of the Soloveitchik family, cited in [**12**], p. 135

In this chapter we shall focus primarily on the inner product space $\mathbb{C}^n$ equipped with the standard inner product

$$\langle \mathbf{u}, \mathbf{v} \rangle_{st} = \mathbf{v}^H \mathbf{u} \,. \tag{9.1}$$

The subscript st will be used initially to emphasize the dependence of the result under discussion on the inner product, especially in the formulation of the result, but may be dropped when the intent is clear from the surrounding content. A number of facts that are easily available from the analysis in Chapter 8 will be re-proved in this setting by different arguments because of the importance of both the facts and the methods. Moreover, in many respects this chapter is a natural continuation of Chapter 6, since it focuses on matrices and the implications of extra structure of a matrix A on the corresponding Jordan forms J.

Recall that a matrix $A \in \mathbb{F}^{n\times n}$ is said to be **symmetric** if $A = A^T$. It is said to be **Hermitian** if $A = A^H$. If $A \in \mathbb{R}^{n\times n}$, then A is Hermitian if and only if it is symmetric. But this is not true if $A \in \mathbb{C}^{n\times n}$. Thus, for example, the matrix

$$A = \begin{bmatrix} 1 & i \\ i & 1 \end{bmatrix} \text{ is symmetric but not Hermitian,}$$

whereas the matrix

$$B = \begin{bmatrix} 1 & i \\ -i & 1 \end{bmatrix} \text{ is Hermitian but not symmetric.}$$

A matrix $A \in \mathbb{F}^{p\times q}$ is said to be **isometric** if $A^H A = I_q$.
A matrix $A \in \mathbb{R}^{n\times n}$ is said to be an **orthogonal matrix** if

$$A^T A = I_n .$$

A matrix $A \in \mathbb{C}^{n\times n}$ is said to be a **unitary matrix** if

$$A^H A = I_n .$$

The preceding three definitions are linked to the standard inner product; see Exercises 9.1, 9.2 and 9.13.

Exercise 9.1. Show that the columns of an $n \times n$ orthogonal matrix form an orthonormal family in $\mathbb{R}^n$ with respect to the standard inner product (9.1).

Exercise 9.2. Show that the columns of an $n \times n$ unitary matrix form an orthonormal family in $\mathbb{C}^n$ with respect to the standard inner product (9.1).

Exercise 9.3. Let $A \in \mathbb{C}^{n\times n}$. Show that $A^H A = I_n$ if and only if

$$\langle A\mathbf{x}, A\mathbf{x}\rangle_{st} = \langle \mathbf{x}, \mathbf{x}\rangle_{st} \quad \text{for every} \quad \mathbf{x} \in \mathbb{C}^n .$$

Our first main objective is to show that every Hermitian matrix is diagonalizable; i.e., $A = UDU^{-1}$, where D is a diagonal matrix with **real** entries and U may be chosen to be unitary.

9.1. Hermitian matrices are diagonalizable

Lemma 9.1. *Let $A \in \mathbb{C}^{n\times n}$ be a Hermitian matrix. Then:*

(1) *The eigenvalues of A are real (even if A is complex).*

(2) *Eigenvectors corresponding to distinct eigenvalues of A are orthogonal with respect to the standard inner product* (9.1).

Proof. Let $\mathbf{u}, \mathbf{v} \in \mathbb{C}^n$ be such that $A\mathbf{u} = \alpha\mathbf{u}$ and $A\mathbf{v} = \beta\mathbf{v}$ for some pair of nonzero vectors $\mathbf{u}$ and $\mathbf{v}$. Then

$$\begin{aligned} \alpha\langle \mathbf{u}, \mathbf{v}\rangle_{st} &= \langle \alpha\mathbf{u}, \mathbf{v}\rangle_{st} = \langle A\mathbf{u}, \mathbf{v}\rangle_{st} \\ &= \langle \mathbf{u}, A^H\mathbf{v}\rangle_{st} = \langle \mathbf{u}, A\mathbf{v}\rangle_{st} \\ &= \langle \mathbf{u}, \beta\mathbf{v}\rangle_{st} = \overline{\beta}\langle \mathbf{u}, \mathbf{v}\rangle_{st} . \end{aligned}$$

Therefore,

$$(\alpha - \overline{\beta})\langle \mathbf{u}, \mathbf{v}\rangle_{st} = 0 \quad \text{and} \quad (\alpha - \overline{\alpha})\langle \mathbf{u}, \mathbf{u}\rangle_{st} = 0 . \tag{9.2}$$

The second equality, which follows from the first by choosing $\mathbf{v} = \mathbf{u}$ and $\alpha = \beta$, implies that

$$\alpha = \overline{\alpha},$$

and hence that the eigenvalues of A are automatically real. Thus, $\overline{\beta} = \beta$, and if $\alpha \neq \beta$, then the first formula in (9.2) implies that $\mathbf{u}$ is orthogonal to $\mathbf{v}$. □

Exercise 9.4. Let $A \in \mathbb{C}^{p\times q}$.

(a) Show that $\mathcal{R}_A \cap \mathcal{N}_{A^H} = \{\mathbf{0}\}$.

(b) Show that if $A = A^H$, then $p = q$ and $\mathcal{R}_A \cap \mathcal{N}_A = \{\mathbf{0}\}$.

Lemma 9.2. *If $A \in \mathbb{C}^{n\times n}$ is Hermitian, then*

$$\mathcal{N}_{(A-\lambda I_n)^k} = \mathcal{N}_{(A-\lambda I_n)}$$

for every positive integer k and every complex number λ.

Proof. If λ is not an eigenvalue of A, then $A - \lambda I_n$ and $(A - \lambda I_n)^k$ are both invertible matrices and hence $\mathcal{N}_{(A-\lambda I_n)} = \mathcal{N}_{(A-\lambda I_n)^k} = \{\mathbf{0}\}$.

On the other hand, if λ is an eigenvalue of A and $\mathbf{u} \in \mathcal{N}_{(A-\lambda I_n)^{k+1}}$ for some positive integer k, then the vector

$$\mathbf{v} = (A - \lambda I_n)^k \mathbf{u}$$

belongs to $\mathcal{R}_{(A-\lambda I_n)} \cap \mathcal{N}_{(A-\lambda I_n)}$. Moreover, since $\lambda \in \mathbb{R}$, the matrix $A - \lambda I_n$ is also Hermitian; i.e., $A - \lambda I_n = (A - \lambda I_n)^H$. Therefore, $\mathbf{v} = \mathbf{0}$, because $\mathcal{R}_{(A-\lambda I_n)} \cap \mathcal{N}_{(A-\lambda I_n)^H} = \{\mathbf{0}\}$, thanks to Exercise 9.4. Thus,

$$(A - \lambda I_n)^{k+1}\mathbf{u} = 0 \Longrightarrow (A - \lambda I_n)^k \mathbf{u} = 0$$

for every positive integer k, which justifies the asserted identity. □

Theorem 9.3. *If $A \in \mathbb{C}^{n\times n}$ is Hermitian, then A is unitarily equivalent to a diagonal matrix $D \in \mathbb{R}^{n\times n}$; i.e., there exists a unitary matrix $U \in \mathbb{C}^{n\times n}$ and a diagonal matrix $D \in \mathbb{R}^{n\times n}$ such that*

$$A = UDU^H. \tag{9.3}$$

Proof. Let $\lambda_1, \dots, \lambda_k$ denote the distinct eigenvalues of A. Then, in view of Lemma 9.2, the algebraic multiplicity α_j of each eigenvalue λ_j is equal to the geometric multiplicity γ_j. Therefore, each of the Jordan cells in the Jordan decomposition of A is 1×1; that is to say, the Jordan matrix J in the Jordan decomposition

$$A = UJU^{-1}$$

must be of the form

$$J = \text{diag}\{B_{\lambda_1}, \dots, B_{\lambda_k}\},$$

where B_{λ_j} is an $\alpha_j \times \alpha_j$ diagonal matrix with λ_j on the diagonal. In particular, J is a diagonal matrix. Consequently each column in the matrix U is an eigenvector of A.

By Lemma 9.1, the eigenvectors corresponding to distinct eigenvalues are automatically orthogonal. Moreover, the columns in U corresponding to the same eigenvalue can be chosen orthonormal (by the Gram-Schmidt procedure). Thus, by choosing all the columns in U to have norm one, we end up with a unitary matrix U. □

Example 9.4. Let A be a 5×5 Hermitian matrix with characteristic polynomial $p(\lambda) = (\lambda - \lambda_1)^3(\lambda - \lambda_2)^2$, where $\lambda_1 \neq \lambda_2$. Then, by Theorem 9.3, $\dim \mathcal{N}_{(\lambda_1 I_5 - A)} = 3$ and $\dim \mathcal{N}_{(\lambda_2 I_5 - A)} = 2$. Let $\mathbf{u}_1, \mathbf{u}_2, \mathbf{u}_3$ be an orthonormal basis for $\mathcal{N}_{(\lambda_1 I_5 - A)}$ and let $\mathbf{u}_4, \mathbf{u}_5$ be an orthonormal basis for $\mathcal{N}_{(\lambda_2 I_5 - A)}$. This can always be achieved by invoking the Gram-Schmidt method, in each nullspace separately, if need be. Therefore, since the eigenvectors of a Hermitian matrix that correspond to distinct eigenvalues are automatically orthogonal, the full set $\mathbf{u}_1, \ldots, \mathbf{u}_5$ is an orthonormal basis for $\mathbb{C}^5$. Thus, upon setting

$$U = [\mathbf{u}_1 \cdots \mathbf{u}_5] \text{ and } D = \text{diagonal}\{\lambda_1, \lambda_1, \lambda_1, \lambda_2, \lambda_2\},$$

one can readily check that

$$AU = UD$$

and that U is unitary.

Remark 9.5. Since

$$U \text{ is unitary} \iff U \text{ is invertible and } U^{-1} = U^H,$$

the computation of the inverse of a unitary matrix is remarkably simple. Moreover,

$$U \quad \text{unitary} \quad \Longrightarrow \langle U\mathbf{u}, U\mathbf{v}\rangle_{st} = \langle \mathbf{u}, \mathbf{v}\rangle_{st}$$

for every choice of $\mathbf{u}, \mathbf{v} \in \mathbb{C}^n$.

Exercise 9.5. Show that if $E = \begin{bmatrix} A & B \\ C & D \end{bmatrix}$ is a $2p \times 2p$ matrix with $p \times p$ blocks $A = A^H$, $B = B^H$, $C = C^H$ and $D = D^H$, then $\lambda \in \sigma(E) \iff \overline{\lambda} \in \sigma(E)$. [HINT: It suffices to focus on $\lambda \notin \mathbb{R}$.]

9.2. Commuting Hermitian matrices

Theorem 9.6. *Let $A \in \mathbb{C}^{n \times n}$ and $B \in \mathbb{C}^{n \times n}$ be Hermitian matrices. Then $AB = BA$ if and only if there exists a single unitary matrix $U \in \mathbb{C}^{n \times n}$ that diagonalizes both A and B.*

Proof. Suppose first that there exists a unitary matrix $U \in \mathbb{C}^{n\times n}$ such that $D_A = U^HAU$ and $D_B = U^HBU$ are both diagonal matrices. Then, since $D_AD_B = D_BD_A$,

$$\begin{aligned} AB &= UD_AU^HUD_BU^H = UD_AD_BU^H \\ &= UD_BD_AU^H = UD_BU^HUD_AU^H = BA\,. \end{aligned}$$

The proof of the converse is equally simple in the special case that A has n distinct eigenvalues $\lambda_1,\ldots,\lambda_n$, because then the formulas

$$Au_j = \lambda_j u_j,\ j = 1,\ldots,n,$$

and

$$ABu_j = BAu_j = \lambda_j Bu_j,\ j = 1,\ldots,n,$$

imply that

$$B\mathbf{u}_j = \beta_j \mathbf{u}_j \quad \text{for} \quad j = 1,\ldots,n$$

and some choice of $\beta_1,\ldots,\beta_n \in \mathbb{C}$. But, since $\mathbf{u}_1,\ldots,\mathbf{u}_n$ may be chosen orthonormal, this is the same as to say that

$$AU = U\begin{bmatrix} \lambda_1 & & \\ & \ddots & \\ & & \lambda_n \end{bmatrix} \text{ and } BU = U\begin{bmatrix} \beta_1 & & \\ & \ddots & \\ & & \beta_n \end{bmatrix}$$

for some unitary matrix $U \in \mathbb{C}^{n\times n}$.

Suppose next that $AB = BA$ and A has k distinct eigenvalues $\lambda_1,\ldots,\lambda_k$ with geometric multiplicities $\gamma_1,\ldots,\gamma_k$, respectively. Then there exists a set of k isometric matrices $U_1 \in \mathbb{C}^{n\times\gamma_1},\ldots,U_k \in \mathbb{C}^{n\times\gamma_k}$ such that $U = \begin{bmatrix} U_1 & \cdots & U_k \end{bmatrix}$ is unitary and $AU_j = \lambda_j U_j$. Therefore,

$$ABU_j = BAU_j = \lambda_j BU_j \quad \text{for} \quad j = 1,\ldots,k\,,$$

which implies that the columns of the matrix BU_j belong to $\mathcal{N}_{(A-\lambda_j I_n)}$ and hence, since the columns of U_j form a basis for that space, that there exist matrices C_j such that

$$BU_j = U_jC_j \quad \text{for} \quad j = 1,\ldots,k\,.$$

The supplementary formulas

$$C_j = U_j^HU_jC_j = U_j^HBU_j = C_j^H \quad \text{for} \quad j = 1,\ldots,k$$

exhibit C_j as a $\gamma_j \times \gamma_j$ Hermitian matrix. Therefore, upon writing

$$C_j = W_jD_jW_j^H \quad \text{for} \quad j = 1,\ldots,k\,,$$

with W_j unitary and D_j diagonal, and setting

$$V_j = U_jW_j \quad \text{for} \quad j = 1,\ldots,k \quad \text{and} \quad W = \text{diag}\,\{W_1,\ldots,W_k\}\,,$$

one can readily check that

$$AV_j = AU_jW_j = \lambda_j U_jW_j = \lambda_j V_j \quad \text{for} \quad j = 1,\ldots,k$$

and

$$BV_j = BU_jW_j = U_jC_jW_j = U_jW_jD_j = V_jD_j \quad \text{for} \quad j = 1, \ldots, k.$$

Thus, the matrix $V = \begin{bmatrix} V_1 & \cdots & V_k \end{bmatrix} = UW$ is a unitary matrix that serves to diagonalize both A and B. □

9.3. Real Hermitian matrices

In this section we shall show that if $A = A^H$ and $A \in \mathbb{R}^{n\times n}$, then the unitary matrix U in formula (9.3) may also be chosen in $\mathbb{R}^{n\times n}$.

Theorem 9.7. *If $A = A^H$ and $A \in \mathbb{R}^{n\times n}$, then there exist an orthogonal matrix $Q \in \mathbb{R}^{n\times n}$ and a real diagonal matrix $D \in \mathbb{R}^{n\times n}$ such that*

$$A = QDQ^T . \tag{9.4}$$

Proof. Let $\mu \in \sigma(A)$ and let $\mathbf{u}_1, \ldots, \mathbf{u}_\ell$ be a basis for the nullspace of the matrix $B = A - \mu I_n$. Then, since $B \in \mathbb{R}^{n\times n}$, the real and imaginary parts of the vectors $\mathbf{u}_j$ also belong to $\mathcal{N}_B$: If $\mathbf{u}_j = \mathbf{x}_j + i\mathbf{y}_j$ with $\mathbf{x}_j$ and $\mathbf{y}_j$ in $\mathbb{R}^n$ for $j = 1, \ldots, \ell$, then

$$A(\mathbf{x}_j + i\mathbf{y}_j) = \mu(\mathbf{x}_j + i\mathbf{y}_j) \Longrightarrow A\mathbf{x}_j = \mu\mathbf{x}_j \text{ and } A\mathbf{y}_j = \mu\mathbf{y}_j \text{ for } j = 1, \ldots, \ell;$$

i.e., the vectors

$$\mathbf{x}_j = \frac{\mathbf{u}_j + \overline{\mathbf{u}_j}}{2} \quad \text{and} \quad \mathbf{y}_j = \frac{\mathbf{u}_j - \overline{\mathbf{u}_j}}{2i}$$

also belong to $\mathcal{N}_B$. Moreover, since

$$\text{span}\,\{\mathbf{x}_1, \ldots, \mathbf{x}_\ell, \mathbf{y}_1, \ldots, \mathbf{y}_\ell\} = \mathcal{N}_B\,,$$

ℓ of these vectors form a basis for $\mathcal{N}_B$. Next, by invoking the Gram-Schmidt procedure, we can find an orthonormal basis of ℓ vectors in $\mathbb{R}^n$ for $\mathcal{N}_B$.

If A has k distinct eigenvalues $\lambda_1, \ldots, \lambda_k$, let Q_i, $i = 1, \ldots, k$, denote the $n \times \gamma_i$ matrix that is obtained by stacking the vectors that are obtained by applying the procedure described above to $B_i = A - \lambda_i I_n$ for $i = 1, \ldots, k$. Then, $AQ_i = \lambda_i Q_i$ and

$$A\begin{bmatrix} Q_1 & \cdots & Q_k \end{bmatrix} = \begin{bmatrix} Q_1 & \cdots & Q_k \end{bmatrix} D \text{ where } D = \text{diag}\,\{\lambda_1 I_{\gamma_1}, \ldots, \lambda_k I_{\gamma_k}\}\,.$$

Moreover, the matrix $Q = \begin{bmatrix} Q_1 & \cdots & Q_k \end{bmatrix}$ is an orthogonal matrix, since all the columns in Q have norm one and, by Lemma 9.1, the columns in Q_i are orthogonal to the columns in Q_j if $i \neq j$. □

Lemma 9.8. *If $A \in \mathbb{R}^{p\times q}$, then*

$$\begin{aligned} &\max\,\{\|A\mathbf{x}\|_{st} : \mathbf{x} \in \mathbb{C}^q \quad \textit{and} \quad \|\mathbf{x}\|_{st} = 1\} \\ &\qquad = \max\,\{\|A\mathbf{x}\|_{st} : \mathbf{x} \in \mathbb{R}^q \quad \textit{and} \quad \|\mathbf{x}\|_{st} = 1\}\,. \end{aligned}$$

Proof. Since $A \in \mathbb{R}^{p\times q}$, A^HA is a real $q\times q$ Hermitian matrix. Therefore, $A^HA = QDQ^T$, where $Q \in \mathbb{R}^{q\times q}$ is orthogonal and $D \in \mathbb{R}^{q\times q}$ is diagonal. Let $\delta = \max\{\lambda : \lambda \in \sigma(A^HA)\}$, let $\mathbf{x} \in \mathbb{C}^q$ and let $\mathbf{y} = Q^T\mathbf{x}$. Then $\delta \geq 0$ and

$$\begin{aligned} \|A\mathbf{x}\|_{st}^2 &= \langle A^HA\mathbf{x}, \mathbf{x}\rangle_{st} = \langle QDQ^T\mathbf{x}, \mathbf{x}\rangle_{st} \\ &= \langle DQ^T\mathbf{x}, Q^T\mathbf{x}\rangle_{st} = \langle D\mathbf{y}, \mathbf{y}\rangle_{st} \\ &= \sum_{j=1}^n d_{jj}\overline{y_j}y_j \leq \delta\sum_{j=1}^n \overline{y_j}y_j \\ &= \delta\|\mathbf{y}\|_{st}^2 = \delta\|Q^T\mathbf{x}\|_{st}^2 = \delta\|\mathbf{x}\|_{st}^2 . \end{aligned}$$

Thus,

$$\max\{\|A\mathbf{x}\|_{st} : \mathbf{x} \in \mathbb{C}^n \quad \text{and} \quad \|\mathbf{x}\|_{st} = 1\} = \sqrt{\delta}\,.$$

However, it is readily seen that this maximum can be attained by choosing $\mathbf{x} = Q\mathbf{e}_1$, the first column of Q. But this proves the claim, since $Q\mathbf{e}_1 \in \mathbb{R}^q$. □

9.4. Projections and direct sums in $\mathbb{F}^n$

- **Projections:** A matrix $P \in \mathbb{F}^{n\times n}$ is said to be a projection if $P^2 = P$.
- **Orthogonal projections:** A matrix $P \in \mathbb{F}^{n\times n}$ is said to be an orthogonal projection (with respect to the standard inner product (9.1)) if $P^2 = P$ and $P^H = P$. Thus, for example,

$$P = \begin{bmatrix} 1 & 0 \\ \alpha & 0 \end{bmatrix}$$

is a projection, but it is not an orthogonal projection with respect to the standard inner product unless $\alpha = 0$.

Lemma 9.9. *Let $P \in \mathbb{F}^{n\times n}$ be a projection, let $\mathcal{R}_P = \{P\mathbf{x} : \mathbf{x} \in \mathbb{F}^n\}$ and let $\mathcal{N}_P = \{\mathbf{x} \in \mathbb{F}^n : P\mathbf{x} = \mathbf{0}\}$. Then*

$$\mathbb{F}^n = \mathcal{R}_P \dot{+} \mathcal{N}_P\,. \tag{9.5}$$

Proof. Clearly

$$\mathbf{x} = P\mathbf{x} + (I_n - P)\mathbf{x} \quad \text{for every vector} \quad \mathbf{x} \in \mathbb{F}^n\,.$$

Therefore, since $P\mathbf{x} \in \mathcal{R}_P$ and $(I_n - P)\mathbf{x} \in \mathcal{N}_P$, it follows that $\mathbb{F}^n = \mathcal{R}_P + \mathcal{N}_P$. It remains only to show that the indicated sum is direct. This is left to the reader as an exercise. □

Exercise 9.6. Show that if P is a projection on a vector space $\mathcal{U}$ over $\mathbb{F}$, then $\mathcal{R}_P \cap \mathcal{N}_P = \{\mathbf{0}\}$.

Exercise 9.7. Let $\mathbf{e}_j$ denote the j'th column of the identity matrix I_4 for $j = 1, \ldots, 4$, and let

$$\mathbf{u} = \begin{bmatrix} 1 \\ 3 \\ 4 \\ 1 \end{bmatrix}, \mathbf{w}_1 = \begin{bmatrix} 1 \\ 1 \\ 1 \\ 0 \end{bmatrix}, \mathbf{w}_2 = \begin{bmatrix} 1 \\ 1 \\ 1 \\ 1 \end{bmatrix}, \mathbf{w}_3 = \begin{bmatrix} 1 \\ 0 \\ 1 \\ 0 \end{bmatrix}, \mathbf{w}_4 = \begin{bmatrix} 1 \\ 1 \\ 0 \\ 1 \end{bmatrix}.$$

Compute the projection of the vector $\mathbf{u}$ onto the subspace $\mathcal{V}$ with respect to the direct sum decomposition $\mathbb{F}4 = \mathcal{V}\dot{+}\mathcal{W}$ when:

(a) $\mathcal{V} = \text{span}\{\mathbf{e}_1, \mathbf{e}_2\}$ and $\mathcal{W} = \text{span}\{\mathbf{w}_1, \mathbf{w}_2\}$.

(b) $\mathcal{V} = \text{span}\{\mathbf{e}_1, \mathbf{e}_2\}$ and $\mathcal{W} = \text{span}\{\mathbf{w}_3, \mathbf{w}_4\}$.

(c) $\mathcal{V} = \text{span}\{\mathbf{e}_1, \mathbf{e}_2, \mathbf{w}_1\}$ and $\mathcal{W} = \text{span}\{\mathbf{w}_4\}$.

[REMARK: The point of this exercise is that the coefficients of $\mathbf{e}_1$ and $\mathbf{e}_2$ are different in all three settings.]

Lemma 9.10. *Let $\mathbb{F}^n = \mathcal{V}\dot{+}\mathcal{W}$ and let $V = \begin{bmatrix} \mathbf{v}_1 & \cdots & \mathbf{v}_k \end{bmatrix}$, where $\{\mathbf{v}_1, \ldots, \mathbf{v}_k\}$ is a basis for $\mathcal{V}$, and let $W = \begin{bmatrix} \mathbf{w}_1 & \cdots & \mathbf{w}_\ell \end{bmatrix}$, where $\{\mathbf{w}_1, \ldots, \mathbf{w}_\ell\}$ is a basis for $\mathcal{W}$. Then:*

(1) *The matrix $[V \quad W]$ is invertible.*

(2) *The projection $P_\mathcal{V}$ of $\mathbb{F}^n$ onto $\mathcal{V}$ with respect to the decomposition $\mathbb{F}^n = \mathcal{V}\dot{+}\mathcal{W}$ is given by the formula*

$$P_\mathcal{V} = V[I_k \quad O][V \quad W]^{-1}. \tag{9.6}$$

Proof. Let $\mathbf{u} \in \mathbb{F}^n$. Then there exists a unique vector $\mathbf{c} \in \mathbb{F}^n$ with entries $c_1, \ldots, c_n$ such that $\mathbf{u} = c_1\mathbf{v}_1 + \cdots + c_k\mathbf{v}_k + c_{k+1}\mathbf{w}_1 + \cdots + c_n\mathbf{w}_\ell$ or, equivalently,

$$\mathbf{u} = [V \quad W]\mathbf{c}.$$

Consequently, (1) holds and

$$P_\mathcal{V}\mathbf{u} = c_1\mathbf{v}_1 + \cdots + c_k\mathbf{v}_k = V[I_k \quad O]\mathbf{c}, \quad \text{where} \quad \mathbf{c} = [V \quad W]^{-1}\mathbf{u}.$$

□

If P is an orthogonal projection, then formula (9.6) simplifies with the help of the following simple observation:

Lemma 9.11. *Let $A \in \mathbb{F}^{p\times q}$, let $\mathbf{u} \in \mathbb{F}^q$ and $\mathbf{v} \in \mathbb{F}^p$. Then*

$$\langle A\mathbf{u}, \mathbf{v}\rangle_{st} = \langle \mathbf{u}, A^H\mathbf{v}\rangle_{st}. \tag{9.7}$$

Proof. This is a pure computation:

$$\begin{aligned}\langle A\mathbf{u}, \mathbf{v}\rangle_{st} &= \mathbf{v}^H(A\mathbf{u}) = (\mathbf{v}^H A)\mathbf{u} \\ &= (A^H\mathbf{v})^H\mathbf{u} = \langle \mathbf{u}, A^H\mathbf{v}\rangle_{st}.\end{aligned}$$

□

Lemma 9.12. *In the setting of Lemma 9.10, $P_{\mathcal{V}}$ is an orthogonal projection with respect to the standard inner product (9.1) if and only if*

$$\langle V\mathbf{x}, W\mathbf{y}\rangle_{st} = 0 \quad \text{for every choice of} \quad \mathbf{x} \in \mathbb{F}^k \quad \text{and} \quad \mathbf{y} \in \mathbb{F}^\ell . \tag{9.8}$$

Moreover, if (9.8) is in force, then

$$P_{\mathcal{V}} = V(V^H V)^{-1} V^H . \tag{9.9}$$

Proof. Let $P = P_{\mathcal{V}}$. If $P = P^H$, then

$$\begin{aligned} \langle \mathbf{v}_i, \mathbf{w}_j\rangle &= \langle P\mathbf{v}_i, \mathbf{w}_j\rangle = \langle \mathbf{v}_i, P^H\mathbf{w}_j\rangle \\ &= \langle \mathbf{v}_i, P\mathbf{w}_j\rangle = \langle \mathbf{v}_i, \mathbf{0}\rangle \\ &= 0 . \end{aligned}$$

Thus, the constraint (9.8) is in force. Conversely, if (9.8) is in force, then, since $V^H W = O_{k\times\ell}$ and $W^H V = O_{\ell\times k}$, it is readily checked that

$$\begin{bmatrix} V & W \end{bmatrix}^{-1} = \begin{bmatrix} (V^H V)^{-1} V^H \\ (W^H W)^{-1} W^H \end{bmatrix} \tag{9.10}$$

and hence that formula (9.6) simplifies to

$$P_{\mathcal{V}} = V[I_k \quad O] \begin{bmatrix} (V^H V)^{-1} V^H \\ (W^H W)^{-1} W^H \end{bmatrix} = V(V^H V)^{-1} V^H ,$$

as claimed. □

Exercise 9.8. Double check the validity of formula (9.10) by computing $\begin{bmatrix} V & W \end{bmatrix}^{-1} [V \quad W]$ and $[V \quad W] \begin{bmatrix} V & W \end{bmatrix}^{-1}$.

Lemma 9.13. *Let $P \in \mathbb{F}^{n\times n}$ be an orthogonal projection (with respect to the standard inner product (9.1)). Then*

(1) *$\mathcal{N}_P$ is orthogonal to $\mathcal{R}_P$ (with respect to the standard inner product).*

(2) *$\mathbb{F}^n = \mathcal{R}_P \oplus \mathcal{N}_P$.*

Proof. Let $\mathbf{u} \in \mathcal{R}_P$ and $\mathbf{v} \in \mathcal{N}_P$. Then

$$\begin{aligned} \langle \mathbf{u}, \mathbf{v}\rangle_{st} &= \langle P\mathbf{u}, \mathbf{v}\rangle_{st} = \langle \mathbf{u}, P^H\mathbf{v}\rangle_{st} \\ &= \langle \mathbf{u}, P\mathbf{v}\rangle_{st} = \langle \mathbf{u}, \mathbf{0}\rangle_{st} = 0 , \end{aligned}$$

since $P = P^H$ and $\mathbf{v} \in \mathcal{N}_P$. This completes the proof of (1). The second assertion is then immediate from Lemma 9.9. □

The next result includes a more general version of (2) of the last lemma that is often useful.

Lemma 9.14. *Let $A \in \mathbb{F}^{p\times q}$, $\mathcal{R}_A = \{A\mathbf{x} : \mathbf{x} \in \mathbb{F}^q\}$, $\mathcal{N}_A = \{\mathbf{x} \in \mathbb{F}^q : A\mathbf{x} = \mathbf{0}\}$, etc. Then, with respect to the standard inner product:*

(1) *$\mathbb{F}^p = \mathcal{R}_A \oplus \mathcal{N}_{A^H}$.*

(2) $\mathbb{F}^q = \mathcal{R}_{A^H} \oplus \mathcal{N}_A$.

(3) $\mathcal{R}_A = \mathcal{R}_{AA^H}$ *and* $\mathcal{R}_{A^H} = \mathcal{R}_{A^H A}$.

(4) $\mathcal{N}_A = \mathcal{N}_{A^H A}$ *and* $\mathcal{N}_{A^H} = \mathcal{N}_{AA^H}$.

Proof. Since $\operatorname{rank} A = \operatorname{rank} A^T = \operatorname{rank} A^H$, the principle of conservation of dimension implies that

$$\begin{aligned} p &= \dim \mathcal{R}_{A^H} + \dim \mathcal{N}_{A^H} \\ &= \dim \mathcal{R}_A + \dim \mathcal{N}_{A^H} \ . \end{aligned}$$

Therefore, to complete the proof of the first assertion, it suffices to show that $\mathcal{R}_A$ is orthogonal to $\mathcal{N}_{A^H}$. To this end, let $\mathbf{u} \in \mathcal{R}_A$ and $\mathbf{v} \in \mathcal{N}_{A^H}$. Then, since $\mathbf{u} = A\mathbf{x}$ for some vector $\mathbf{x} \in \mathbb{C}^q$,

$$\begin{aligned} \langle \mathbf{u}, \mathbf{v} \rangle_{st} &= \langle A\mathbf{x}, \mathbf{v} \rangle_{st} = \langle \mathbf{x}, A^H \mathbf{v} \rangle_{st} \\ &= \langle \mathbf{x}, \mathbf{0} \rangle_{st} = 0. \end{aligned}$$

This completes the proof of the first assertion. The second then follows immediately by replacing A by A^H in the first.

In particular, (2) implies that every vector $\mathbf{u} \in \mathbb{F}^q$ can be expressed as a sum of the form $\mathbf{u} = A^H\mathbf{v} + \mathbf{w}$ for some choice of $\mathbf{v} \in \mathbb{F}^p$ and $\mathbf{w} \in \mathcal{N}_A$. Thus,

$$A\mathbf{u} = A(A^H\mathbf{v} + \mathbf{w}) = AA^H\mathbf{v}.$$

This shows that $\mathcal{R}_A \subseteq \mathcal{R}_{AA^H}$. Therefore, since the opposite inclusion is self-evident, equality must prevail, which proves the first formula in (3).

Next, the implications

$$A^H A\mathbf{u} = \mathbf{0} \Longrightarrow \mathbf{u}^H A^H A\mathbf{u} = \mathbf{0} \Longrightarrow \|A\mathbf{u}\|_{st} = 0 \Longrightarrow A\mathbf{u} = \mathbf{0}$$

yield the inclusion $\mathcal{N}_{A^H A} \subseteq \mathcal{N}_A$. Therefore, since the opposite inclusion is self-evident, equality must prevail. This justifies the first assertion in (4). The second assertions in (3) and (4) follow by interchanging A and A^H. $\square$

Exercise 9.9. Show directly that $\mathcal{R}_A \cap \mathcal{N}_{A^H} = \mathbf{0}$ for every $p \times q$ matrix A. [HINT: It suffices to show that $\langle \mathbf{u}, \mathbf{u} \rangle_{st} = 0$ for vectors $\mathbf{u}$ that belong to both of these spaces.]

The next lemma will be useful in the sequel, particularly in the development of singular value decompositions, in the next chapter.

Lemma 9.15. *Let $V \in \mathbb{F}^{n\times r}$ be a matrix with r columns that are orthonormal in $\mathbb{F}^n$ with respect to the standard inner product. Then $r \leq n$. Moreover, if $r < n$, then we can add $n - r$ columns to V to obtain a unitary matrix.*

Proof. By Lemma 9.14,

$$\mathbb{F}^n = \mathcal{R}_V \oplus \mathcal{N}_{V^H}.$$

By assumption, the columns $\mathbf{v}_1, \ldots, \mathbf{v}_r$ span $\mathcal{R}_V$. If $n = r$, then V is unitary and there is nothing left to do. If $r < n$, let $\mathbf{w}_{r+1}, \ldots, \mathbf{w}_n$ be a basis for $\mathcal{N}_{V^H}$. By the Gram-Schmidt algorithm, there exists an orthonormal family $\mathbf{v}_{r+1}, \ldots, \mathbf{v}_n$ that also spans $\mathcal{N}_{V^H}$. The matrix $[V \ \mathbf{v}_{r+1} \cdots \mathbf{v}_n]$ with columns $\mathbf{v}_1, \ldots, \mathbf{v}_n$ is unitary. □

Exercise 9.10. Let $A \in \mathbb{C}^{p\times q}$ and $B \in \mathbb{C}^{p\times r}$. Show that

$$\operatorname{rank}\begin{bmatrix} A & B \end{bmatrix} = p \Longleftrightarrow \mathcal{N}_{A^H} \cap \mathcal{N}_{B^H} = \{\mathbf{0}\}.$$

9.5. Projections and rank

Lemma 9.16. *Let P and Q be projection matrices in $\mathbb{F}^{n\times n}$ such that $\|P - Q\| < 1$. Then* $\operatorname{rank} P = \operatorname{rank} Q$.

Proof. The inequality $\|P - Q\| < 1$ implies that the matrix $I_n - (P - Q)$ is invertible and hence that

$$\begin{aligned} \operatorname{rank} P &= \operatorname{rank}\{P(I_n - (P - Q)\} = \operatorname{rank}\{PQ\} \\ &\leq \min\{\operatorname{rank} P, \operatorname{rank} Q\}. \end{aligned}$$

Therefore,

$$\operatorname{rank} P \leq \operatorname{rank} Q.$$

On the other hand, since Q and P can be interchanged in the preceding analysis, the inequality $\operatorname{rank} P \leq \operatorname{rank} Q$ must also be in force. Therefore, $\operatorname{rank} P = \operatorname{rank} Q$, as claimed. □

9.6. Normal matrices

In Section 9.1, we showed that every Hermitian matrix A can be diagonalized "by" a unitary matrix U; i.e., $AU = UD$, where D is a diagonal matrix. This is such a useful result that it would be nice if it held true for other classes of matrices. If so, then a natural question is: What is the largest class of matrices which can be diagonalized "by" a unitary matrix? The answer is the class of normal matrices.

- **normal matrices**: A matrix $A \in \mathbb{C}^{n\times n}$ is said to be **normal** if $A^H A = AA^H$. Notice that in addition to the class of $n \times n$ Hermitian matrices, the class of $n\times n$ normal matrices includes the class of $n \times n$ unitary matrices.

Lemma 9.17. *If $A \in \mathbb{C}^{n\times n}$ is normal, then*

$$\mathcal{N}_A = \mathcal{N}_{A^H}.$$

Proof. This is a consequence of the following sequence of implications:

$$\begin{aligned} A\mathbf{u} = \mathbf{0} &\iff \|A\mathbf{u}\|_{st} = 0 \\ &\iff \langle A^H A\mathbf{u}, \mathbf{u}\rangle_{st} = 0 \\ &\iff \langle AA^H\mathbf{u}, \mathbf{u}\rangle_{st} = 0 \\ &\iff A^H\mathbf{u} = 0. \end{aligned}$$

□

Lemma 9.18. *If $A \in \mathbb{C}^{n\times n}$ is a normal matrix, then*

$$\mathcal{N}_{A^k} = \mathcal{N}_A$$

for every positive integer k.

Proof. Let k be a positive integer and let $\mathbf{u} \in \mathcal{N}_{A^{k+1}}$. Then the vector $\mathbf{v} = A^k\mathbf{u}$ belongs to $\mathcal{R}_A \cap \mathcal{N}_A$. Therefore, since $\mathcal{N}_A = \mathcal{N}_{A^H}$ by Lemma 9.17 and $\mathcal{R}_A \cap \mathcal{N}_{A^H} = \{\mathbf{0}\}$ by Lemma 9.14, it follows that $\mathbf{v} = \mathbf{0}$. This proves that $\mathcal{N}_{A^{k+1}} \subseteq \mathcal{N}_{A^k}$ and hence, as the opposite inclusion is self-evident, that $\mathcal{N}_{A^{k+1}} = \mathcal{N}_{A^k}$ for every positive integer k. Thus, $\mathcal{N}_{A^k} = \mathcal{N}_A$ for every positive integer k, as advertised. □

Exercise 9.11. Show that if $A \in \mathbb{C}^{n\times n}$ is normal, then $(A^k)^H A = A(A^H)^k$ for every nonnegative integer k.

Exercise 9.12. Give a second proof of Lemma 9.18 by justifying the following sequence of implications:

$$\begin{aligned} A^k\mathbf{u} = \mathbf{0} &\iff (A^k)^H A^k\mathbf{u} = \mathbf{0} \iff (A^H A)^k\mathbf{u} = \mathbf{0} \\ &\iff A^H A\mathbf{u} = \mathbf{0} \iff A\mathbf{u} = \mathbf{0}\,. \end{aligned}$$

[HINT: Lemma 9.2 is applicable to $A^H A$.]

Lemma 9.19. *Let $A \in \mathbb{C}^{n\times n}$ be a normal matrix. Then*

$$\mathcal{N}_{(A-\lambda I_n)^k} = \mathcal{N}_{(A-\lambda I_n)}$$

for every complex number $\lambda \in \mathbb{C}$ and every positive integer k.

Proof. In view of Lemma 9.18, it suffices to observe that if A is normal, then $A - \lambda I_n$ is also normal for every complex number $\lambda \in \mathbb{C}$. □

The preceding result guarantees that normal matrices are diagonalizable. It remains to check that the diagonalization can be effected by unitary matrices.

Lemma 9.20. *The eigenvectors of a normal matrix corresponding to distinct eigenvalues are automatically orthogonal.*

Proof. Let $A \in \mathbb{C}^{n\times n}$ be a normal matrix and suppose that

$$A\mathbf{u} = \alpha\mathbf{u} \text{ and } A\mathbf{v} = \beta\mathbf{v}$$

for some pair of nonzero vectors $\mathbf{u}$ and $\mathbf{v}$. Then, by Lemma 9.17,

$$A^H\mathbf{v} = \overline{\beta}\mathbf{v}$$

and hence

$$\begin{aligned}\alpha\langle\mathbf{u},\mathbf{v}\rangle_{st} &= \langle A\mathbf{u},\mathbf{v}\rangle_{st} \\ &= \langle\mathbf{u},A^H\mathbf{v}\rangle_{st} \\ &= \langle\mathbf{u},\overline{\beta}\mathbf{v}\rangle_{st} \\ &= \beta\langle\mathbf{u},\mathbf{v}\rangle_{st}\,.\end{aligned}$$

Therefore,

$$(\alpha-\beta)\langle\mathbf{u},\mathbf{v}\rangle_{st} = 0,$$

which clearly implies that $\mathbf{u}$ is orthogonal to $\mathbf{v}$ if $\alpha \neq \beta$. □

The preceding analysis shows that normal matrices can be diagonalized by unitary matrices. The next theorem shows that this is the end of the line:

Theorem 9.21. *Let $A \in \mathbb{C}^{n\times n}$. Then there exists an orthonormal basis of $\mathbb{C}^n$ consisting of eigenvectors of A if and only if A is normal.*

Proof. Suppose that $\mathbf{u}_1,\ldots,\mathbf{u}_n$ is an orthonormal family of eigenvectors of A, let $\lambda_1,\ldots,\lambda_n$ denote the corresponding eigenvalues and let

$$U = \begin{bmatrix}\mathbf{u}_1 & \cdots & \mathbf{u}_n\end{bmatrix}$$

be the $n\times n$ matrix with columns $\mathbf{u}_1,\ldots,\mathbf{u}_n$ and $D = \operatorname{diag}\{\lambda_1\ldots,\lambda_n\}$. Then

$$AU = UD$$

and, since U is unitary,

$$A = UDU^H \quad \text{and} \quad A^H = UD^HU^H\,.$$

Therefore, since

$$DD^H = D^HD = \operatorname{diag}\{|\lambda_1|^2,\ldots,|\lambda_n|^2\}\,,$$

one can readily see that

$$AA^H = UDD^HU^H = UD^HDU^H = A^HA;$$

i.e., A is normal.

Conversely, if A is normal, then there exists an orthonormal basis of $\mathbb{C}^n$ made up of eigenvectors of A, since the eigenvectors corresponding to distinct eigenvalues are orthogonal by Lemma 9.20 and a linearly independent set of eigenvectors corresponding to the same eigenvalue can be replaced by an orthonormal set via the Gram–Schmidt orthogonalization method. □

Exercise 9.13. Show that if $A \in \mathbb{F}^{p\times q}$ is an isometric marix, then

$$\langle A\mathbf{u}, A\mathbf{v}\rangle_{st} = \langle \mathbf{u}, \mathbf{v}\rangle_{st}$$

for every choice of $\mathbf{u}, \mathbf{v} \in \mathbb{F}^q$.

Exercise 9.14. Let U be an $n \times n$ unitary matrix. Show directly that the eigenvectors corresponding to distinct eigenvalues are orthogonal with respect to the standard inner product in $\mathbb{C}^n$ and that if μ is an eigenvalue of U then $|\mu| = 1$.

9.7. Schur's theorem

We have observed that every normal matrix is unitarily equivalent to a diagonal matrix. A rather useful theorem of Issai Schur states that every square matrix is unitarily equivalent to a triangular matrix:

Theorem 9.22. *Let $A \in \mathbb{C}^{n\times n}$. Then there exists a unitary matrix $V \in \mathbb{C}^{n\times n}$ such that*

$$V^H A V = S \tag{9.11}$$

is upper triangular. Moreover, if A is similar to an upper triangular matrix B, then V can be chosen so that the diagonal entries of S coincide with the diagonal entries of B; i.e., $s_{jj} = b_{jj}$ for $j = 1, \dots, n$.

Proof. Let $A \in \mathbb{C}^{n\times n}$ and let $U = \begin{bmatrix}\mathbf{u}_1 & \cdots & \mathbf{u}_n\end{bmatrix}$ be an invertible matrix in $\mathbb{C}^{n\times n}$ with columns $\mathbf{u}_1, \dots, \mathbf{u}_n$ such that

$$B = U^{-1}AU$$

is upper triangular. Such a matrix U always exists, because B may be taken equal to a Jordan form of A. Then it is readily checked that

$$\begin{aligned} A\mathbf{u}_1 &\in \operatorname{span}\{\mathbf{u}_1\} \\ A\mathbf{u}_2 &\in \operatorname{span}\{\mathbf{u}_1, \mathbf{u}_2\} \\ &\vdots \\ A\mathbf{u}_j &\in \operatorname{span}\{\mathbf{u}_1, \dots, \mathbf{u}_j\} \end{aligned}$$

and hence that

$$\mathcal{M}_j = \operatorname{span}\{\mathbf{u}_1, \dots, \mathbf{u}_j\}$$

is a j-dimensional subspace of $\mathbb{C}^n$ such that

$$\mathcal{M}_1 \subset \mathcal{M}_2 \subset \cdots \subset \mathcal{M}_n = \mathbb{C}^n$$

and

$$A\mathcal{M}_j \subseteq \mathcal{M}_j \ , \ \text{for} \ \ j = 1, \dots, n \ .$$

By the Gram-Schmidt procedure we can construct an orthonormal set of vectors $\mathbf{v}_1, \ldots, \mathbf{v}_n$ such that

$$\operatorname{span}\{\mathbf{v}_1, \ldots, \mathbf{v}_j\} = \operatorname{span}\{\mathbf{u}_1, \ldots, \mathbf{u}_j\} \quad \text{for} \quad j = 1, \ldots, n\,.$$

Therefore,

$$A\mathbf{v}_j \in \operatorname{span}\{\mathbf{v}_1, \ldots, \mathbf{v}_j\} \;\; \text{for} \;\; j = 1, \ldots, n\,.$$

But this means that the matrix V with columns $\mathbf{v}_1, \ldots, \mathbf{v}_n$ can be written as

$$V = US$$

for some invertible upper triangular matrix S and hence that

$$AV = AUS = UBS = VS^{-1}BS\,.$$

This proves the first assertion, since V is unitary and $S^{-1}BS$ is upper triangular. Moreover, upon writing

$$S = D_1 + X_1\,, \;\; B = D_0 + X_0 \;\; \text{and} \;\; S^{-1} = D_2 + X_2\,,$$

where D_j is diagonal and X_j is strictly upper triangular (i.e., upper triangular with zero entries on the diagonal), it is readily checked that

$$\begin{aligned} S^{-1}BS &= (D_2 + X_2)(D_0 + X_0)(D_1 + X_1) \\ &= D_2 D_0 D_1 + X_3\,, \end{aligned}$$

where X_3 is strictly upper triangular and hence as

$$D_2 D_0 D_1 = D_0 D_2 D_1 \;\; \text{and} \;\; D_2 = D_1^{-1}\,,$$

the diagonal component of S agrees with the diagonal component of B, as claimed. □

The usefulness of the decomposition (9.11) rests on the fact that the diagonal entries of S run through the eigenvalues of A repeated according to algebraic multiplicity. To verify this statement, recall that if A is an $n \times n$ matrix with eigenvalues $\mu_1, \ldots, \mu_n$, then

$$\det(\lambda I_n - A) = (\lambda - \mu_1) \cdots (\lambda - \mu_n)\,.$$

On the other hand, formula (9.11) implies that

$$\det(\lambda I_n - A) = \det(\lambda I_n - S) = (\lambda - s_{11}) \cdots (\lambda - s_{nn})\,,$$

since S is triangular. In particular,

$$\sum_{j=1}^{n} |\mu_j|^2 = \sum_{j=1}^{n} |s_{jj}|^2\,.$$

Corollary 9.23. Let A be an $n \times n$ matrix with eigenvalues $\mu_1, \ldots, \mu_n$, repeated according to algebraic multiplicity. Then

$$\sum_{j=1}^{n} |\mu_j|^2 \leq \sum_{i,j=1}^{n} |a_{ij}|^2 . \tag{9.12}$$

Proof. By Schur's theorem, there exists a unitary matrix U such that

$$U^H AU = S$$

is upper triangular and $s_{ii} = \mu_i$ for $i = 1, \ldots, n$. Therefore,

$$\begin{aligned} \operatorname{trace}\{A^H A\} &= \operatorname{trace}\{US^H U^H USU^H\} = \operatorname{trace}\{US^H SU^H\} \\ &= \operatorname{trace}\{S^H SU^H U\} = \operatorname{trace}\{S^H S\}, \end{aligned}$$

which supplies the identity

$$\sum_{i,j=1}^{n} |s_{ij}|^2 = \sum_{i,j=1}^{n} |a_{ij}|^2$$

for the entries s_{ij} of S and a_{ij} of A. Therefore,

$$\sum_{i=1}^{n} |\mu_i|^2 = \sum_{i=1}^{n} |s_{ii}|^2 \leq \sum_{i,j=1}^{n} |a_{ij}|^2 .$$

□

Schur used the estimate (9.12) to give a simple proof of one of Hadamard's inequalities (9.13):

Corollary 9.24. Let $A \in \mathbb{C}^{n\times n}$ and let $\gamma = \max\{|a_{ij}| : i, j = 1, \ldots, n\}$. Then

$$|\det A| \leq \gamma^n n^{n/2} . \tag{9.13}$$

Proof. Let $\mu_1, \ldots, \mu_n$ denote the eigenvalues of A repeated according to their algebraic multiplicity. Then $\det A = \mu_1\mu_2 \cdots \mu_n$ and the inequality between the geometric and arithmetic means followed by an application of the bound (9.12) implies that

$$\begin{aligned} |\mu_1|^2|\mu_2|^2 \cdots |\mu_n|^2 &\leq \left(\frac{|\mu_1|^2 + |\mu_2|^2 + \cdots + |\mu_n|^2}{n}\right)^n \\ &\leq \left(\frac{1}{n}\sum_{i,j=1}^{n} |a_{ij}|^2\right)^n \\ &\leq \left(\frac{n^2\gamma^2}{n}\right)^n . \end{aligned}$$

The rest is plain sailing. □

Exercise 9.15. Let $A, B \in \mathbb{C}^{n\times n}$. Show that

(a) $\det(\lambda I_n - AB) = \det(\lambda I_n - BA)$ for every point $\lambda \in \mathbb{C}$.

(b) The matrices AB and BA have the same set of eigenvalues with the same algebraic multiplicities.

Exercise 9.16. Let $A \in \mathbb{C}^{n\times n}$. Show that $\operatorname{trace} A = \sum_{j=1}^{n} \langle A\mathbf{u}_j, \mathbf{u}_j\rangle$ for every orthonormal basis $\{\mathbf{u}_1, \ldots, \mathbf{u}_n\}$ of $\mathbb{C}^n$. [HINT: $\operatorname{trace} A = \operatorname{trace}(U^H AU)$.]

Exercise 9.17. Let $\mu_1, \ldots, \mu_n$ denote the eigenvalues of $A \in \mathbb{C}^{n\times n}$, repeated according to their algebraic multiplicity and let $\delta_1(A), \ldots, \delta_n(A)$ denote the eigenvalues of $A^H A$. Show that $\sum_{j=1}^{n} |\mu_j|^2 \leq \sum_{j=1}^{n} \delta_j(A)$.

Exercise 9.18. Let $A, B \in \mathbb{C}^{n\times n}$ and assume that $AB = (AB)^H$. Show that $\operatorname{trace}\{(AB)^H AB\} \leq \operatorname{trace}\{(BA)^H BA\}$. [HINT: Exploit the preceding exercises.]

9.8. QR factorization

Lemma 9.25. *Let $A \in \mathbb{F}^{p\times q}$ and suppose that $\operatorname{rank} A = q$. Then there exist a unique isometric matrix $Q \in \mathbb{F}^{p\times q}$ and a unique upper triangular matrix $R \in \mathbb{F}^{q\times q}$ with positive entries on the diagonal such that $A = QR$.*

Proof. The existence of at least one factorization of the indicated form is a consequence of the Gram-Schmidt procedure. To verify the asserted uniqueness, suppose that there were two such factorizations: $A = Q_1R_1$ and $A = Q_2R_2$. Then

$$R_1^H R_1 = R_1^H Q_1^H Q_1 R_1 = A^H A = R_2^H Q_2^H Q_2 R_2 = R_2^H R_2$$

and hence,

$$R_1(R_2)^{-1} = (R_1^H)^{-1} R_2^H = \{(R_1(R_2)^{-1})^H\}^{-1}.$$

Therefore, since the left-hand side of the last equality is upper triangular while the right-hand side is lower triangular, it follows that

$$R_1(R_2)^{-1} = D = \{D^H\}^{-1}$$

is a diagonal matrix with positive diagonal entries. Therefore, in view of the relation $D = \{D^H\}^{-1}$, it follows that $D = I_q$. Thus, $Q_1 = Q_2$ and $R_1 = R_2$, as claimed. □

Exercise 9.19. Let $A = \begin{bmatrix} 1 & 1 & 2 \\ 1 & 2 & 0 \\ 0 & 0 & 1 \\ 1 & 1 & 0 \end{bmatrix}$. Find an isometric matrix $Q \in \mathbb{R}^{4\times 3}$ and an upper triangular matrix $R \in \mathbb{R}^{3\times 3}$ with positive entries on the diagonal such that $A = QR$.

9.9. Areas, volumes and determinants

To warm up, consider the following:

Exercise 9.20. Let $\mathbf{a}, \mathbf{b} \in \mathbb{R}^2$ and let $V = [\mathbf{a} \quad \mathbf{b}]$ denote the 2×2 matrix with columns $\mathbf{a}$ and $\mathbf{b}$. Then the area of the parallelogram generated by $\mathbf{a}$ and $\mathbf{b}$ is equal to $|\det V|$.

Lemma 9.26. *Let* $\mathbf{a}, \mathbf{b} \in \mathbb{R}^n$ *and let* $V = [\mathbf{a} \quad \mathbf{b}]$ *denote the* $n \times 2$ *matrix with columns* $\mathbf{a}$ *and* $\mathbf{b}$. *Then the area of the parallelogram generated by* $\mathbf{a}$ *and* $\mathbf{b}$ *is equal to* $\{\det(V^H V)\}^{1/2}$.

Proof. To begin with, let us assume that $\mathbf{a}$ and $\mathbf{b}$ are linearly independent. Then the area of the parallelogram of interest (as drawn in Figure 1) is equal to $h\|a\|$.

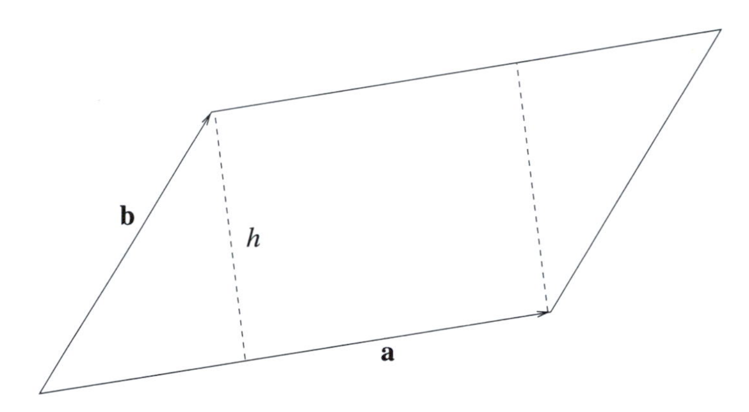

Figure 1. Parallelogram generated by $\mathbf{a}$ and $\mathbf{b}$

Thus, the main chore is to figure out h. To this end, let $\mathcal{A} = \text{span}\{\mathbf{a}\}$. Then, by formula (8.23),

$$P_{\mathcal{A}} = \mathbf{a}(\mathbf{a}^H\mathbf{a})^{-1}\mathbf{a}^H \, .$$

Therefore,

$$P_{\mathcal{A}}\mathbf{b} = \mathbf{a}(\mathbf{a}^H\mathbf{a})^{-1}\mathbf{a}^H\mathbf{b} = \langle \mathbf{b}, \mathbf{a}\rangle \|\mathbf{a}\|^{-2}\mathbf{a} \, ,$$
$$\mathbf{b} - P_{\mathcal{A}}\mathbf{b} = \mathbf{b} - \mathbf{a}(\mathbf{a}^H\mathbf{a})^{-1}\mathbf{a}^H\mathbf{b}$$

and

$$h = \|\mathbf{b} - P_{\mathcal{A}}\mathbf{b}\| \, .$$

Thus,

$$\begin{aligned} h^2 &= \langle \mathbf{b} - P_{\mathcal{A}}\mathbf{b}, \mathbf{b} - P_{\mathcal{A}}\mathbf{b}\rangle \\ &= \langle \mathbf{b} - P_{\mathcal{A}}\mathbf{b}, \mathbf{b}\rangle \\ &= \mathbf{b}^H\mathbf{b} - \mathbf{b}^H\mathbf{a}(\mathbf{a}^H\mathbf{a})^{-1}\mathbf{a}^H\mathbf{b} \\ &= \|\mathbf{b}\|^2 - |\langle \mathbf{a}, \mathbf{b}\rangle|^2\|\mathbf{a}\|^{-2} \, . \end{aligned}$$

Consequently,

$$(\text{area})^2 = \|\mathbf{a}\|^2\|\mathbf{b}\|^2 - |\langle \mathbf{a}, \mathbf{b}\rangle|^2 \,. \tag{9.14}$$

To complete the proof, observe that

$$\begin{aligned} V^H V &= \begin{bmatrix} \mathbf{a}^H \\ \mathbf{b}^H \end{bmatrix} [\mathbf{a} \quad \mathbf{b}] \\ &= \begin{bmatrix} \mathbf{a}^H\mathbf{a} & \mathbf{a}^H\mathbf{b} \\ \mathbf{b}^H\mathbf{a} & \mathbf{b}^H\mathbf{b} \end{bmatrix} = \begin{bmatrix} \|\mathbf{a}\|^2 & \langle \mathbf{b}, \mathbf{a}\rangle \\ \langle \mathbf{a}, \mathbf{b}\rangle & \|\mathbf{b}\|^2 \end{bmatrix} \end{aligned}$$

and hence that

$$\det(V^H V) = \|\mathbf{a}\|^2\|\mathbf{b}\|^2 - |\langle \mathbf{a}, \mathbf{b}\rangle|^2 \,. \tag{9.15}$$

Thus, we obtain

$$\text{area} = \{\det(V^H V)\}^{1/2} \,, \tag{9.16}$$

as claimed, at least when $\mathbf{a}$ and $\mathbf{b}$ are linearly independent. Formula (9.16) remains valid, however, even if $\mathbf{a}$ and $\mathbf{b}$ are linearly dependent because then both sides in formula (9.16) are equal to zero. □

As a byproduct of the proof of the last lemma we obtain the formula

$$|\langle \mathbf{a}, \mathbf{b}\rangle|^2 = \|\mathbf{a}\|^2\|\mathbf{b}\|^2 - (\text{area})^2 \,. \tag{9.17}$$

This yields another proof of the Cauchy-Schwarz inequality for vectors in $\mathbb{R}^n$:

$$|\langle \mathbf{a}, \mathbf{b}\rangle| \leq \|\mathbf{a}\|\|\mathbf{b}\|$$

with equality if and only if the area is equal to zero, i.e. if and only if $\mathbf{a}$ and $\mathbf{b}$ are colinear.

Lemma 9.27. *Let* $\operatorname{vol}\{\mathbf{a}, \mathbf{b}, \mathbf{c}\}$ *denote the volume of the parallelopiped generated by the vectors* $\mathbf{a}, \mathbf{b}, \mathbf{c} \in \mathbb{R}^3$ *and let*

$$W = [\mathbf{a} \quad \mathbf{b} \quad \mathbf{c}] \,.$$

Then

$$\operatorname{vol}\{\mathbf{a}, \mathbf{b}, \mathbf{c}\} = |\det W| \,. \tag{9.18}$$

Proof. Suppose first that the three given vectors are linearly independent. Then in view of Lemma 9.26, the volume we are after is given by the formula

$$\operatorname{vol}\{\mathbf{a}, \mathbf{b}, \mathbf{c}\} = h\{\det[V^H V]\}^{1/2} \,,$$

where V is the 3×2 matrix with columns $\mathbf{a}$ and $\mathbf{b}$ and h is the distance of the point (c_1, c_2, c_3) from the plane $\mathcal{V}$ generated by the vectors $\mathbf{a}$ and $\mathbf{b}$. Thus, h is equal to the length of the vector $\mathbf{c} - P_{\mathcal{V}}\mathbf{c} = \mathbf{c} - V(V^H V)^{-1}V^H\mathbf{c}$:

$$\begin{aligned} h^2 &= \langle \mathbf{c} - V(V^H V)^{-1}V^H\mathbf{c}, \mathbf{c} - V(V^H V)^{-1}V^H\mathbf{c}\rangle \\ &= \langle \mathbf{c} - V(V^H V)^{-1}V^H\mathbf{c}, \mathbf{c}\rangle \,, \end{aligned}$$

since $P_{\mathcal{V}} = V(V^HV)^{-1}V^H$ is an orthogonal projection, i.e.,

$$\langle \mathbf{c} - P_{\mathcal{V}}\mathbf{c}, P_{\mathcal{V}}\mathbf{c}\rangle = \langle P_{\mathcal{V}}(\mathbf{c} - P_{\mathcal{V}}\mathbf{c}), \mathbf{c}\rangle = \langle \mathbf{0}, \mathbf{c}\rangle = 0 .$$

Consequently

$$\begin{aligned} h^2 &= \|\mathbf{c}\|^2 - \langle V(V^HV)^{-1}V^H\mathbf{c}, \mathbf{c}\rangle \\ &= \mathbf{c}^H\mathbf{c} - \mathbf{c}^HV(V^HV)^{-1}V^H\mathbf{c} \end{aligned}$$

and thus,

$$(\mathrm{vol}\,\{\mathbf{a}, \mathbf{b}, \mathbf{c}\})^2 = \{\|\mathbf{c}\|^2 - \mathbf{c}^HV(V^HV)^{-1}V^H\mathbf{c}\}\det(V^HV) .$$

The next step is to observe that

$$W = [V \quad \mathbf{c}] \quad \text{and} \quad W^HW = \begin{bmatrix} V^HV & V^H\mathbf{c} \\ \mathbf{c}^HV & \mathbf{c}^H\mathbf{c} \end{bmatrix} .$$

Therefore, by formula (5.14),

$$\det W^HW = \det(V^HV)\{\mathbf{c}^H\mathbf{c} - \mathbf{c}^HV(V^HV)^{-1}V^H\mathbf{c}\} ,$$

which serves to complete the proof for the case of three linearly independent vectors, since

$$|\det W|^2 = \det W^HW .$$

However, the formula remains valid even if $\mathbf{a}$, $\mathbf{b}$ and $\mathbf{c}$ are linearly dependent because then both sides of formula (9.18) are equal to zero. □

In order to extend these formulas to higher dimensions, it is necessary to develop the definition of volume in $\mathbb{R}^n$ for $n > 3$, which we shall not do. The monograph [**71**] provides good introduction to this class of ideas.

Exercise 9.21. Let $\mathcal{A} = \{A \in \mathbb{R}^{n\times n} : A^T = A\}$ and let S denote the linear transformation from $\mathcal{A}$ into $\mathbb{R}^{kn}$ that is defined by the formula

$$S(A) = \begin{bmatrix} A\mathbf{e}_1 \\ \vdots \\ A\mathbf{e}_k \end{bmatrix} ,$$

where $\mathbf{e}_j$ denotes the j'th column of I_n for $j = 1, \dots, n$ and $k \le n$. Show that

(a) $\mathcal{A}$ is an $(n^2 + n)/2$ dimensional vector space over $\mathbb{R}$.

(b) $\dim \mathcal{N}_S = \dfrac{(n-k)^2 + n - k}{2}$.

(c) $nk - \dim \mathcal{R}_S = \dfrac{k(k-1)}{2}$.

Exercise 9.22. Show that the conclusions of Exercise 9.21 remain valid if $\mathbf{e}_1, \dots, \mathbf{e}_k$ is replaced by any set $\mathbf{v}_1, \dots, \mathbf{v}_k$ of k linearly independent vectors in $\mathbb{R}^n$.

Exercise 9.23. Let $\{\mathbf{u}_1, \ldots, \mathbf{u}_k\}$ and $\{\mathbf{v}_1, \ldots, \mathbf{v}_\ell\}$ be two sets of linearly independent vectors in $\mathbb{R}^n$ and let S denote the linear transformation from $\mathbb{R}^{n\times n}$ into $\mathbb{R}^{(k+\ell)n}$ that is defined by the formula

$$S(A) = \begin{bmatrix} A\mathbf{u}_1 \\ \vdots \\ A\mathbf{u}_k \\ A^T\mathbf{v}_1 \\ \vdots \\ A\mathbf{v}_\ell^T \end{bmatrix}.$$

Show that

(a) $\dim \mathcal{N}_S = (n-k)(n-\ell)$.

(b) $n(k+\ell) - \dim \mathcal{R}_S = k\ell$.

Exercise 9.24. Let $\mathbf{v} \in \mathbb{R}^n$, $A \in \mathbb{R}^{k\times n}$ and let $\mathbf{u}_1, \ldots, \mathbf{u}_q$ be linearly independent vectors in $\mathcal{N}_A \cap \mathbb{R}^n$ such that $\langle \mathbf{v}, \mathbf{u}_j \rangle_{st} = 0$ for $j = 1, \ldots, q$. Show that if $\operatorname{rank} A = n - q$, then $\mathbf{v}^T$ is a linear combination of the rows of A.

Exercise 9.25. Let $A = \begin{bmatrix} \alpha & \gamma \\ \beta & \delta \end{bmatrix}$ and $G(\alpha) = \begin{bmatrix} \alpha & \beta \\ \beta & -\overline{\alpha} \end{bmatrix}$ be unitary matrices with $\beta \geq 0$. Show that $\beta = \sqrt{1 - |\alpha|^2}$ and that $A = G(\alpha) \begin{bmatrix} 1 & 0 \\ 0 & \alpha_1 \end{bmatrix}$, where $|\alpha_1| = 1$. [HINT: Consider $G(\alpha)^H A$.]

A matrix $A \in \mathbb{C}^{n\times n}$ is said to be an **upper Hessenberg** matrix if $a_{ij} = 0$ for $i \geq j+2$. Thus, for example,

$$A = \begin{bmatrix} a_{11} & a_{12} & a_{13} & a_{14} \\ a_{21} & a_{22} & a_{23} & a_{24} \\ 0 & a_{32} & a_{33} & a_{34} \\ 0 & 0 & a_{43} & a_{44} \end{bmatrix}$$

is a 4×4 upper Hessenberg matrix.

Exercise 9.26. Show that if A is a unitary 4×4 upper Hessenberg matrix with nonnegative entries on the subdiagonal and $G(\alpha)$ is as in Exercise 9.25, then A admits a factorization of the form

$$A = \begin{bmatrix} G(\alpha_1) & O \\ O & I_2 \end{bmatrix} \begin{bmatrix} O & O & O \\ O & G(\alpha_2) & O \\ O & O & 1 \end{bmatrix} \begin{bmatrix} I_2 & O \\ O & G(\alpha_3) \end{bmatrix} \begin{bmatrix} I_3 & O \\ O & \alpha_4 \end{bmatrix}$$

for some choice of constants α_j with $|\alpha_j| \leq 1$ and $\beta_j = a_{j+1,j}$ for $j = 1, 2, 3$ and $|\alpha_4| = 1$. [HINT: Keep the conclusions of Exercise 9.25 in mind.]

Exercise 9.27. Formulate and solve an analogue of Exercise 9.26 for a 5×5 unitary upper Hessenberg matrix A with nonnegative entries on the subdiagonal.

9.10. Bibliographical notes

Exercise 9.18 is adapted from a statement in an expository article by Bhatia [**8**]. Exercises 9.21, 9.22 and 9.23 are adapted from Lemma 9.5 in Camino et al. [**14**]. Exercises 9.25–9.27 are adapted from a theorem in [**2**].

Chapter 10

Singular values and related inequalities

Let's throw everything away. Then there will be room for what's left.

Irene Dym

• **WARNING:** From now on, unless explicitly indicated otherwise, $\langle \mathbf{u}, \mathbf{v} \rangle = \mathbf{v}^H \mathbf{u}$ and $\|\mathbf{u}\| = \sqrt{\mathbf{u}^H \mathbf{u}}$ for vectors $\mathbf{u}, \mathbf{v} \in \mathbb{F}^k$ and $\|A\| = \|A\|_{2,2}$ for matrices A.

10.1. Singular value decompositions

Let $A \in \mathbb{C}^{p\times q}$. Then the matrices $A^H A$ and AA^H are Hermitian matrices of sizes $q\times q$ and $p\times p$, respectively. Moreover, if $A^H A\mathbf{u} = \alpha \mathbf{u}$ and $AA^H \mathbf{v} = \beta \mathbf{v}$, then the formulas

$$\alpha\langle \mathbf{u}, \mathbf{u}\rangle = \langle A^H A\mathbf{u}, \mathbf{u}\rangle = \langle A\mathbf{u}, A\mathbf{u}\rangle \geq 0$$

and

$$\beta\langle \mathbf{v}, \mathbf{v}\rangle = \langle AA^H \mathbf{v}, \mathbf{v}\rangle = \langle A^H \mathbf{v}, A^H \mathbf{v}\rangle \geq 0$$

clearly imply that the eigenvalues of $A^H A$ and AA^H are nonnegative. Therefore, by Theorem 9.3, there exists a unitary matrix $U \in \mathbb{C}^{q\times q}$ such that

$$A^H A = U \begin{bmatrix} s_1^2 & & \\ & \ddots & \\ & & s_q^2 \end{bmatrix} U^H, \tag{10.1}$$

where the numbers $s_1^2, \ldots, s_q^2$ designate the eigenvalues of $A^H A$, and, in keeping with the usual conventions, it is assumed that they are indexed so that

$$s_1 \geq s_2 \geq \cdots \geq s_q \geq 0\,.$$

The numbers $s_j, j = 1, \ldots, q$, are referred to as the **singular values** of A. If $\operatorname{rank} A = r$, then $s_r > 0$ and $s_j = 0$ for $j = r+1, \ldots, q$, if $r < q$.

Exercise 10.1. Show that if $A \in \mathbb{C}^{p\times q}$, then

$$\mathcal{R}_A = \mathcal{R}_{AA^H} \quad \text{and} \quad \operatorname{rank} A^H A = \operatorname{rank} A = \operatorname{rank} AA^H\,. \tag{10.2}$$

Exercise 10.2. Show that if $A \in \mathbb{C}^{p\times q}$ and $1 \leq r < q$, then

$$\operatorname{rank} A^H A = r \Longleftrightarrow s_r > 0 \text{ and } s_{r+1} = 0\,. \tag{10.3}$$

Theorem 10.1. *Let $A \in \mathbb{C}^{p\times q}$ be a matrix of rank r with singular values $s_1, \ldots, s_q$ and let*

$$D = \operatorname{diag}\{s_1, \ldots, s_r\} \in \mathbb{R}^{r\times r} \quad (\textit{with} \quad s_1 \geq \cdots \geq s_r > 0)\,.$$

Then there exists a unitary matrix $V \in \mathbb{C}^{p\times p}$ and a unitary matrix $U \in \mathbb{C}^{q\times q}$ such that

$$A = \begin{cases} V\begin{bmatrix} D & O_{r\times(q-r)} \\ O_{(p-r)\times r} & O_{(p-r)\times(q-r)}\end{bmatrix} U^H & \textit{if} \quad r < \min\{p,q\} \\[2ex] V\begin{bmatrix} D \\ O_{(p-r)\times r}\end{bmatrix} U^H & \textit{if} \quad r = q < p \\[2ex] V\begin{bmatrix} D & O_{r\times(q-r)}\end{bmatrix} U^H & \textit{if} \quad r = p < q \\[2ex] VDU^H & \textit{if} \quad r = p = q \end{cases}\,. \tag{10.4}$$

Moreover, if $A \in \mathbb{R}^{p\times q}$, then the unitary matrices U and V in (10.4) *may be chosen to have real entries, i.e., to be orthogonal matrices.*

Proof. Let $\mathbf{u}_j$ denote the j'th column of the unitary matrix U that appears in formula (10.1). Then

$$A^H A\mathbf{u}_j = s_j^2 \mathbf{u}_j \quad \text{for} \quad j = 1, \ldots, q\,.$$

Let

$$A\mathbf{u}_j = s_j \mathbf{v}_j \quad \text{for} \quad j = 1, \ldots, r\,. \tag{10.5}$$

Then

$$\begin{aligned} \langle s_j\mathbf{v}_j, s_k\mathbf{v}_k\rangle &= \langle A\mathbf{u}_j, A\mathbf{u}_k\rangle \\ &= \langle A^H A\mathbf{u}_j, \mathbf{u}_k\rangle \\ &= s_j^2\langle \mathbf{u}_j, \mathbf{u}_k\rangle \quad \text{for} \quad j,k = 1, \ldots, r\,. \end{aligned}$$

Therefore,

$$\langle \mathbf{v}_j, \mathbf{v}_k \rangle = \begin{cases} 0 & \text{if } j \neq k \\ 1 & \text{if } j = k \end{cases}$$

for $j, k = 1, \ldots, r$. Thus, in matrix notation,

(10.6)

$$A \begin{bmatrix} \mathbf{u}_1 & \cdots & \mathbf{u}_r \end{bmatrix} = \begin{bmatrix} \mathbf{v}_1 & \cdots & \mathbf{v}_r \end{bmatrix} \begin{bmatrix} s_1 & & \\ & \ddots & \\ & & s_r \end{bmatrix} = \begin{bmatrix} \mathbf{v}_1 & \cdots & \mathbf{v}_r \end{bmatrix} D\,.$$

If $r = p$, then the matrix $V = \begin{bmatrix} \mathbf{v}_1 & \cdots & \mathbf{v}_r \end{bmatrix}$ is unitary. If $r < p$, then, in view of Lemma 9.15, we can add columns $\mathbf{v}_{r+1}, \ldots, \mathbf{v}_p$ so that

$$V = \begin{bmatrix} \mathbf{v}_1 & \cdots & \mathbf{v}_p \end{bmatrix}$$

is a unitary matrix. Consequently, we can rewrite formula (10.6) as

$$A \begin{bmatrix} \mathbf{u}_1 & \cdots & \mathbf{u}_r \end{bmatrix} = \begin{cases} VD & \text{if} \quad r = p \\ V \begin{bmatrix} D \\ O_{(p-r)\times r} \end{bmatrix} & \text{if} \quad r < p \end{cases}. \tag{10.7}$$

If $r = q$, then (10.7) yields the second formula in (10.4) if $r < p$ and the fourth formula in (10.4) if $r = p$.

If $r < q$, then, since

$$A^H A\mathbf{u}_j = \mathbf{0} \quad \text{for} \quad j = r+1, \ldots, q \Longrightarrow A\mathbf{u}_j = \mathbf{0} \quad \text{for} \quad j = r+1, \ldots, q\,,$$

we can add the last $q - r$ columns of U on the left of (10.7), balanced by $r - q$ zero columns on the right of (10.7), to obtain

$$AU = \begin{cases} V \begin{bmatrix} D & O_{r\times(q-r)} \end{bmatrix} & \text{if} \quad r = p < q \\ V \begin{bmatrix} D & O_{r\times(q-r)} \\ O_{(p-r)\times r} & O_{(p-r)\times(q-r)} \end{bmatrix} & \text{if} \quad r < \min\{p, q\}\,, \end{cases}$$

which yields the remaining two formulas in (10.4).

Finally, if $A \in \mathbb{R}^{p\times q}$, then $A^H A$ is a real Hermitian matrix and so, in view of the analysis in Section 9.3, the unitary matrix U in formula (10.1) may be chosen in $\mathbb{R}^{q\times q}$ and hence the unitary matrix V may be chosen in $\mathbb{R}^{p\times p}$. □

Formula (10.4) is called the **singular value decomposition** of A.

Exercise 10.3. Show that if $A \in \mathbb{C}^{p\times q}$ and $\operatorname{rank} A = r$, then the nonzero singular values of A coincide with the nonzero singular values of A^H; i.e., $s_j(A) = s_j(A^H)$ for $j = 1, \ldots, r$.

Corollary 10.2. Let $A \in \mathbb{C}^{p\times q}$ be a matrix of rank r, $r \geq 1$, and let $\mathbf{v}_1, \dots, \mathbf{v}_p$ and $\mathbf{u}_1, \dots, \mathbf{u}_q$ denote the columns of the unitary matrices $V \in \mathbb{C}^{p\times p}$ and $U \in \mathbb{C}^{q\times q}$ that appear in the singular value decomposition (10.4) of A. Let

(10.8)

$$U_1 = \begin{bmatrix}\mathbf{u}_1 & \cdots & \mathbf{u}_r\end{bmatrix}, \quad D = \operatorname{diag}\{s_1, \dots, s_r\} \quad \text{and} \quad V_1 = \begin{bmatrix}\mathbf{v}_1 & \cdots & \mathbf{v}_r\end{bmatrix}.$$

Then

$$A = V_1 D U_1^H = \sum_{j=1}^{r} \mathbf{v}_j s_j \mathbf{u}_j^H \quad \text{and} \quad A^H = U_1 D V_1^H = \sum_{j=1}^{r} \mathbf{u}_j s_j \mathbf{v}_j^H. \tag{10.9}$$

Proof. The first formula is equivalent to formula (10.4), and the second follows easily from the first. □

Exercise 10.4. Show that if $A \in \mathbb{C}^{p\times q}$ is expressed in the form (10.9), then $\mathcal{R}_A = \operatorname{span}\{\mathbf{v}_1, \dots, \mathbf{v}_r\}$ and $\mathcal{N}_A = \operatorname{span}\{\mathbf{u}_{r+1}, \dots, \mathbf{u}_q\}$.

Exercise 10.5. Show that if $A \in \mathbb{C}^{p\times q}$ is expressed in the form (10.9), then $\mathcal{R}_{A^H} = \operatorname{span}\{\mathbf{u}_1, \dots, \mathbf{u}_r\}$ and $\mathcal{N}_{A^H} = \operatorname{span}\{\mathbf{v}_{r+1}, \dots, \mathbf{v}_p\}$.

Exercise 10.6. Let $A \in \mathbb{C}^{p\times q}$ be expressed in the form (10.9) and let

$$A^\dagger = U_1 D^{-1} V_1^H. \tag{10.10}$$

Show that $AA^\dagger A = A^\dagger$, $AA^\dagger A = A$, $(A^\dagger A)^H = A^\dagger A$ and $(AA^\dagger)^H = AA^\dagger$.

In the next chapter, we shall identify the matrix $A^\dagger$ as the **Moore-Penrose inverse** of A and shall show that it is the only matrix in $\mathbb{C}^{q\times p}$ that meets the four conditions in Exercise 10.6.

Lemma 10.3. *Let $A \in \mathbb{C}^{p\times q}$. Then $\|A\|_{2,2} = s_1$.*

Proof. Let M denote the middle term in the decomposition (10.4) so that $A = VMU^H$, and let $\mathbf{x} \in \mathbb{C}^q$ and $\mathbf{y} = U^H\mathbf{x}$. Then (10.4),

$$\begin{aligned}\|A\mathbf{x}\|_2^2 &= \|VMU^H\mathbf{x}\|_2^2 = \|MU^H\mathbf{x}\|_2^2 \\ &= \|M\mathbf{y}\|_2^2 = \sum_{j=1}^{r} |s_j y_j|^2 \\ &\leq s_1^2 \sum_{j=1}^{r} |y_j|^2 \leq s_1^2 \|\mathbf{y}\|_2^2.\end{aligned}$$

Therefore, since $\|\mathbf{y}\|_2 = \|\mathbf{x}\|_2$, it follows that

$$\|A\mathbf{x}\|_2 \leq s_1 \|\mathbf{x}\|_2$$

for every $\mathbf{x} \in \mathbb{C}^q$ and hence that $\|A\|_{2,2} \leq s_1$. On the other hand, since

$$\|A\mathbf{u}_1\|_2 = \|MU^H\mathbf{u}_1\|_2 = \|M\mathbf{e}_1\|_2 = s_1,$$

it follows that $\|A\|_{2,2} \geq s_1$ and hence that equality prevails. □

The next result extends Lemma 10.4 and serves to characterize all the singular values of $A \in \mathbb{C}^{p\times q}$ in terms of an approximation problem.

Lemma 10.4. *If $A \in \mathbb{C}^{p\times q}$, then its singular values $s_1, \ldots, s_q$ can be characterized as*

$$s_{j+1} = \min\left\{\|A - B\|_{2,2} : B \in \mathbb{C}^{p\times q} \quad \textit{and} \quad \operatorname{rank} B \leq j\right\}. \tag{10.11}$$

Proof. Suppose first that $\operatorname{rank} B = \operatorname{rank} A$. Then clearly the choice $B = A$ minimizes the norm in (10.11). Suppose next that $\operatorname{rank} A = r$ and $\operatorname{rank} B = k$ with $k < r$ and let $\mathbf{u}_1, \ldots, \mathbf{u}_q$ and $\mathbf{v}_1, \ldots, \mathbf{v}_p$ denote the columns of a pair of unitary matrices U and V in a singular value decomposition of A. Let

$$\mathbf{x} = \sum_{j=1}^{k+1} c_j \mathbf{u}_j \quad \text{with} \quad \sum_{j=1}^{k+1} |c_j|^2 = 1\,.$$

Then

$$A\mathbf{x} = \sum_{j=1}^{k+1} c_j s_j \mathbf{v}_j \quad \text{and} \quad B\mathbf{x} = \sum_{j=1}^{p} \langle B\mathbf{x}, \mathbf{v}_j\rangle \mathbf{v}_j\,.$$

The next step is to show that $\mathbf{x}$ may be chosen to be orthogonal to all the vectors $B^H\mathbf{v}_1, \ldots, B^H\mathbf{v}_{k+1}$. In view of the chosen form of $\mathbf{x}$, this is the same as to say that there exists a choice of coefficients $c_1, \ldots, c_{k+1}$ such that

$$\sum_{j=1}^{k+1} \mathbf{v}_i^H B\mathbf{u}_j c_j = \sum_{j=1}^{k+1} c_j \langle B\mathbf{u}_j, \mathbf{v}_i\rangle = 0 \quad \text{for} \quad i = 1, \ldots, k+1\,.$$

This last requirement can be written more transparently in terms of the matrices $U_1 = \begin{bmatrix}\mathbf{u}_1 & \cdots & \mathbf{u}_{k+1}\end{bmatrix}$ and $V_1 = \begin{bmatrix}\mathbf{v}_1 & \cdots & \mathbf{v}_{k+1}\end{bmatrix}$ and the vector $\mathbf{c} \in \mathbb{C}^{k+1}$ with components $c_1, \ldots, c_{k+1}$ as

$$V_1^H B U_1 \mathbf{c} = \mathbf{0}\,.$$

However, since $\operatorname{rank} V_1^H B U_1 \leq k$, there exists a vector $\mathbf{c}$ with $\|\mathbf{c}\| = 1$ that meets this requirement. For such a choice of $\mathbf{c}$,

$$\begin{aligned}
\|A\mathbf{x} - B\mathbf{x}\|_{2,2}^2 &= \left\|\sum_{j=1}^{k+1} c_j s_j \mathbf{v}_j - \sum_{j=k+2}^{p} \langle B\mathbf{x}, \mathbf{v}_j\rangle \mathbf{v}_j\right\|_{2,2}^2 \\
&= \sum_{j=1}^{k+1} |c_j|^2 s_j^2 + \sum_{j=k+2}^{p} |\langle B\mathbf{x}, \mathbf{v}_j\rangle|^2 \\
&\geq \sum_{j=1}^{k+1} |c_j|^2 s_j^2 \geq s_{k+1}^2 \sum_{j=1}^{k+1} |c_j|^2 = s_{k+1}^2\,.
\end{aligned}$$

Thus,

$$\|A - B\|_{2,2} \geq s_{k+1} \quad \text{for every } B \in \mathbb{C}^{p\times q} \text{ with } \quad \operatorname{rank} B \leq k\,. \tag{10.12}$$

It remains to show that there exists a choice of $B \in \mathbb{C}^{p\times q}$ with $\operatorname{rank} B \leq k$ that attains equality in (10.12). This is an easy consequence of the singular value decomposition and is left to the reader as an exercise. □

Exercise 10.7. Show that there exists a choice of $B \in \mathbb{C}^{p\times q}$ with $\operatorname{rank} B \leq k$ that attains equality in (10.12).

Exercise 10.8. Let $A \in \mathbb{C}^{p\times q}$. Show that

$$\|A\| = \max\{|\langle A\mathbf{x}, \mathbf{y}\rangle| : \mathbf{x} \in \mathbb{C}^q\,,\ \mathbf{y} \in \mathbb{C}^p\,,\ \text{and } \|\mathbf{x}\| = \|\mathbf{y}\| = 1\}\,. \tag{10.13}$$

Exercise 10.9. Let $A \in \mathbb{C}^{p\times q}$. Show that

$$\|A\| = \|A^H\| = \|A^H A\|^{1/2} = \|AA^H\|^{1/2}\,. \tag{10.14}$$

Exercise 10.10. Let $A \in \mathbb{C}^{p\times q}$ and suppose that $s_1(A) \leq 1$. Show that the matrix $I_p - AB$ is invertible for every choice of $B \in \mathbb{C}^{q\times p}$ with $s_1(B) \leq 1$ if and only if $s_1(A) < 1$.

Exercise 10.11. Let $A \in \mathbb{C}^{n\times n}$ and let $\lambda_1, \ldots, \lambda_{2n}$ denote the eigenvalues of the matrix

$$B = \begin{bmatrix} O & A \\ A^H & O \end{bmatrix}$$

repeated according to their mutiplicity and indexed so that $\lambda_1 \geq \cdots \geq \lambda_{2n}$. Express these eigenvalues in terms of the singular values of A.

Exercise 10.12. Show that if $A = A^H \in \mathbb{C}^{n\times n}$ and $B = B^H \in \mathbb{C}^{n\times n}$, then $\|AB\|^2 = \|BA^2B\|$.

Exercise 10.13. Redo Exercises 9.17 and 9.18 using singular value decompositions.

10.2. Complex symmetric matrices

If $A \in \mathbb{C}^{n\times n}$ is symmetric, then $A^H = \overline{A}$.

Theorem 10.5. (Takagi) *If $A \in \mathbb{C}^{n\times n}$ and $A = A^T$, then there exists a unitary matrix $W \in \mathbb{C}^{n\times n}$ and a diagonal matrix D such that*

$$A = WDW^T \quad \textit{and} \quad D^H D = \mathit{diag}\{s_1^2, \ldots, s_n^2\}\,, \tag{10.15}$$

where $s_1 \geq \cdots \geq s_n$ are the singular values of A.

Proof. Let $\sigma_1 > \cdots > \sigma_\ell$ denote the distinct singular values of A. Then the condition $A = A^T$ implies that the spaces $\mathcal{M}_i = \mathcal{N}_{(A^H A - \sigma_i^2 I_n)}$ meet the constraint $\overline{A\mathcal{M}_i} \subseteq \mathcal{M}_i$ for $i = 1, \ldots, \ell$; i.e., if $\mathbf{u} \in \mathcal{M}_i$, then $\overline{A\mathbf{u}} \in \mathcal{M}_i$:

$$A^H A(\overline{A\mathbf{u}}) = \overline{A}A(\overline{A\mathbf{u}}) = \overline{A(A^H A\mathbf{u})} = \overline{A(\sigma_i^2\mathbf{u})} = \sigma_i^2\overline{A\mathbf{u}}\,.$$

If $\dim \mathcal{M}_i = 1$ for $i = 1, \dots, \ell$, then $\ell = n$ and the matrix $U \in \mathbb{C}^{n\times n}$ with columns $\mathbf{u}_i \in \mathcal{M}_i$ of norm $\|\mathbf{u}_i\| = 1$ for $i = 1, \dots, n$ is unitary and meets the supplementary condition

$$AU = \overline{U}D\,, \text{ where } D \text{ is a diagonal matrix and } D^H D = \operatorname{diag}\{s_1^2, \dots, s_n^2\}\,.$$

Consequently, $W = \overline{U}$ satisfies the conditions in (10.15).

On the other hand, if $\dim \mathcal{M}_i = k_i > 1$ for some i, then the construction of the matrix W is a little more complicated: let $\mathbf{u}_1 \in \mathcal{M}_i$ with $\|\mathbf{u}_1\| = 1$ and let

$$\mathbf{v}_1 = \begin{cases} \mathbf{u}_1 & \text{if } \mathbf{u}_1 \text{ and } \overline{A\mathbf{u}_1} \text{ are linearly dependent} \\ \sigma_i \mathbf{u}_1 + \overline{A\mathbf{u}_1} & \text{if } \mathbf{u}_1 \text{ and } \overline{A\mathbf{u}_1} \text{ are linearly independent} \end{cases}.$$

Then it is readily checked that $\mathbf{v}_1 \in \mathcal{M}_i$ and $A\mathbf{v}_1 = \alpha_1 \overline{\mathbf{v}_1}$, where $|\alpha_1| = \sigma_i$.

Next, choose a nonzero vector $\mathbf{u}_2 \in \mathcal{M}_i$ that is orthogonal to $\mathbf{v}_1$ and define

$$\mathbf{v}_2 = \begin{cases} \mathbf{u}_2 & \text{if } \mathbf{u}_2 \text{ and } \overline{A\mathbf{u}_2} \text{ are linearly dependent} \\ \sigma_i \mathbf{u}_2 + \overline{A\mathbf{u}_2} & \text{if } \mathbf{u}_2 \text{ and } \overline{A\mathbf{u}_2} \text{ are linearly independent} \end{cases}$$

and check that $\mathbf{v}_2 \in \mathcal{M}_i$ and $A\mathbf{v}_2 = \alpha_2 \overline{\mathbf{v}_2}$, where $|\alpha_2| = \sigma_i$, and that

$$\langle \mathbf{v}_2, \mathbf{v}_1 \rangle = 0\,.$$

Continuing this way, generate an orthogonal basis $\mathbf{v}_1, \dots, \mathbf{v}_{k_i}$ of $\mathcal{M}_i$ with the property

$$A\mathbf{v}_j = \alpha_j \overline{\mathbf{v}}_j \quad \text{and} \quad |\alpha_j| = \sigma_i$$

for $j = 1, \dots, k_i$. Let W_i denote the $n \times k_i$ matrix with columns $\mathbf{w}_1, \dots, \mathbf{w}_{k_i}$ based on the normalized vectors

$$\mathbf{w}_j = \|\mathbf{v}_j\|^{-1} \overline{\mathbf{v}_j}\,.$$

Then, since

$$W_i^T \overline{W_i} = I_{k_i} \quad \text{and} \quad A\overline{W_i} = W_i D_i \quad \text{where} \quad D_i = \operatorname{diag}\{\alpha_1, \dots, \alpha_{k_i}\}\,,$$

it is readily checked that the matrix $W = \begin{bmatrix} W_1 & \cdots & W_\ell \end{bmatrix}$ with blocks W_i of size $n \times k_i$ is a unitary matrix that meets the conditions of (10.15). $\square$

10.3. Approximate solutions of linear equations

Let $A \in \mathbb{C}^{p\times q}$ and let $\mathbf{b} \in \mathbb{C}^p$. Then the equation

$$A\mathbf{x} = \mathbf{b}$$

has a solution $\mathbf{x} \in \mathbb{C}^q$ if and only if $\mathbf{b} \in \mathcal{R}_A$. However, if $\mathbf{b} \notin \mathcal{R}_A$, then a reasonable strategy is to

$$\text{minimize}\,\{\|A\mathbf{x} - \mathbf{b}\| \text{ over } \mathbf{x} \in \mathbb{C}^q\}$$

with respect to some norm $\|\cdot\|$. The most convenient norm is $\|\cdot\|_2$, because it fits naturally with the standard inner product. More often than not we shall, as warned earlier, drop the subscript.

Lemma 10.6. *If $A \in \mathbb{C}^{p\times q}$, $\mathbf{b} \in \mathbb{C}^p$ and $P_{\mathcal{R}_A}$ denotes the orthogonal projection of $\mathbb{C}^p$ onto $\mathcal{R}_A$, then*

$$\|A\mathbf{x} - \mathbf{b}\|^2 \geq \|(I_p - P_{\mathcal{R}_A})\mathbf{b}\|^2 \,, \tag{10.16}$$

with equality if and only if $A\mathbf{x} = P_{\mathcal{R}_A}\mathbf{b}$. Moreover, if $\operatorname{rank} A = r$ *and $V_1 = \begin{bmatrix} \mathbf{v}_1 & \cdots & \mathbf{v}_r \end{bmatrix}$ is built from the first r columns in the matrix V in (10.4) and $s_1, \ldots, s_r$ are the positive singular values of A, then*

$$P_{\mathcal{R}_A}\mathbf{b} = V_1V_1^H\mathbf{b} = \sum_{j=1}^{r}\langle \mathbf{b}, \mathbf{v}_j\rangle \mathbf{v}_j \tag{10.17}$$

and

$$\|(I_p - P_{\mathcal{R}_A})\mathbf{b}\|^2 = \|\mathbf{b}\|^2 - \sum_{j=1}^{r} |\langle \mathbf{b}, \mathbf{v}_j\rangle|^2 = \sum_{j=r+1}^{p} |\langle \mathbf{b}, \mathbf{v}_j\rangle|^2 \,. \tag{10.18}$$

Moreover, if $\mathbf{u}_1, \ldots, \mathbf{u}_q$ denote the columns in the matrix U in (10.4), $U_1 = \begin{bmatrix} \mathbf{u}_1 & \cdots & \mathbf{u}_r \end{bmatrix}$ and $U_2 = \begin{bmatrix} \mathbf{u}_{r+1} & \cdots & \mathbf{u}_q \end{bmatrix}$, then the vector

$$\mathbf{x} = U_1D^{-1}V_1^H\mathbf{b} + U_2\begin{bmatrix} c_{r+1} \\ \vdots \\ c_q \end{bmatrix} = \sum_{j=1}^{r} \frac{\langle \mathbf{b}, \mathbf{v}_j\rangle}{s_j}\mathbf{u}_j + \sum_{j=r+1}^{q} c_j\mathbf{u}_j \tag{10.19}$$

is a solution of the equation

$$A\mathbf{x} = P_{\mathcal{R}_A}\mathbf{b} \tag{10.20}$$

for every choice of the coefficients $c_{r+1}, \ldots, c_q$.

Proof. Formula (10.17) is an application of Exercise 10.4 and Lemma 9.12. Formula (10.18) then follows from the decomposition

$$\mathbf{b} - P_{\mathcal{R}_A}\mathbf{b} = \sum_{j=1}^{p}\langle \mathbf{b}, \mathbf{v}_j\rangle \mathbf{v}_j - \sum_{j=1}^{r}\langle \mathbf{b}, \mathbf{v}_j\rangle \mathbf{v}_j = \sum_{j=r+1}^{p}\langle \mathbf{b}, \mathbf{v}_j\rangle \mathbf{v}_j$$

and the fact that the vectors $\mathbf{v}_1, \ldots, \mathbf{v}_r$ form an orthonormal basis for $\mathcal{R}_A$.

Finally, since the vectors $\mathbf{u}_1, \ldots, \mathbf{u}_q$ form an orthonormal basis for $\mathbb{C}^q$, every vector $\mathbf{x} \in \mathbb{C}^q$ can be expressed as

$$\mathbf{x} = \sum_{j=1}^{q} c_j\mathbf{u}_j \,.$$

Thus, with the aid of Corollary 10.2, it follows that

$$Ax = \sum_{j=1}^{r} s_j c_j \mathbf{v}_j$$

and hence that $A\mathbf{x} = P_{\mathcal{R}_A}\mathbf{b}$ if and only if

$$c_j = \frac{\langle \mathbf{b}, \mathbf{v}_j \rangle}{s_j} \quad \text{for} \quad j = 1, \ldots, r.$$

Since there are no constraints on c_j for $j = r+1, \ldots, q$, formula (10.19) is indeed a solution of (10.20). □

Exercise 10.14. In the setting of Lemma 10.6, show that if $r < q$ and $A^\dagger$ denotes the Moore-Penrose inverse of A introduced in Exercise 10.6, then the vector

$$\mathbf{x} = A^\dagger \mathbf{b} = \sum_{j=1}^{r} \frac{\langle \mathbf{b}, \mathbf{v}_j \rangle}{s_j} \mathbf{u}_j \tag{10.21}$$

may be characterized as the solution of (10.20) with the smallest norm.

Exercise 10.15. In the setting of Lemma 10.6, show that if $r = q$, then $A^H A$ is invertible and the solution $\mathbf{x}$ of equation (10.20) given by formula (10.19) may be expressed as $\mathbf{x} = (A^H A)^{-1} A^H \mathbf{b}$.

10.4. The Courant-Fischer theorem

Let $\mathcal{S}_j$ denote the set of all j-dimensional subspaces of $\mathbb{C}^n$ for $j = 0, \ldots, n$, where it is to be understood that $\mathcal{S}_0 = \{\mathbf{0}\}$ and $\mathcal{S}_n = \mathbb{C}^n$.

Theorem 10.7. (Courant-Fischer) *Let $A \in \mathbb{C}^{n \times n}$ be a Hermitian matrix with eigenvalues $\lambda_1, \ldots, \lambda_n$.*

(1) *If $\lambda_1 \leq \cdots \leq \lambda_n$, then*

(10.22)
$$\lambda_j = \min_{\mathcal{X} \in \mathcal{S}_j} \max \left\{ \frac{\langle A\mathbf{x}, \mathbf{x} \rangle}{\langle \mathbf{x}, \mathbf{x} \rangle} : \mathbf{x} \in \mathcal{X} \quad \text{and} \quad \mathbf{x} \neq \mathbf{0} \right\}, \ j = 1, \ldots, n.$$

(10.23)
$$\lambda_{n+1-j} = \max_{\mathcal{X} \in \mathcal{S}_j} \min \left\{ \frac{\langle A\mathbf{x}, \mathbf{x} \rangle}{\langle \mathbf{x}, \mathbf{x} \rangle} : \mathbf{x} \in \mathcal{X} \quad \text{and} \quad \mathbf{x} \neq \mathbf{0} \right\}, \ j = 1, \ldots, n.$$

(2) *If $\lambda_1 \geq \cdots \geq \lambda_n$, then*

$$\lambda_j = \max_{\mathcal{X} \in \mathcal{S}_j} \min \left\{ \frac{\langle A\mathbf{x}, \mathbf{x} \rangle}{\langle \mathbf{x}, \mathbf{x} \rangle} : \mathbf{x} \in \mathcal{X} \quad \text{and} \quad \mathbf{x} \neq \mathbf{0} \right\}, \ j = 1, \ldots, n. \tag{10.24}$$

(10.25)

$$\lambda_{n+1-j} = \min_{\mathcal{X}\in\mathcal{S}_j} \max\left\{\frac{\langle A\mathbf{x},\mathbf{x}\rangle}{\langle \mathbf{x},\mathbf{x}\rangle} : \mathbf{x}\in\mathcal{X} \quad \textit{and} \quad \mathbf{x}\neq\mathbf{0}\right\}, \ j=1,\ldots,n.$$

Proof. Let $\mathbf{u}_1,\ldots,\mathbf{u}_n$ be an orthonormal set of eigenvectors of A corresponding to the eigenvalues $\lambda_1,\ldots,\lambda_n$.

To prove (10.22), let $\mathcal{U}_j = \operatorname{span}\{\mathbf{u}_j,\ldots,\mathbf{u}_n\}$. Then

$$\mathcal{X}\cap\mathcal{U}_j \neq \{\mathbf{0}\} \quad \text{for every choice of} \quad \mathcal{X}\in\mathcal{S}_j\,,$$

since $\dim\mathcal{X} = j$ and $\dim\mathcal{U}_j = n+1-j$. Choose $\mathbf{v}\in\mathcal{X}\cap\mathcal{U}_j$ with $\mathbf{v}\neq\mathbf{0}$. Then

$$\mathbf{v} = \sum_{i=j}^{n} c_i\mathbf{u}_i$$

and hence

$$\langle A\mathbf{v},\mathbf{v}\rangle = \sum_{i=j}^{n}\lambda_i|c_i|^2 \geq \lambda_j\sum_{i=j}^{n}|c_i|^2 = \lambda_j\langle\mathbf{v},\mathbf{v}\rangle\,.$$

Therefore,

$$\max\left\{\frac{\langle A\mathbf{x},\mathbf{x}\rangle}{\langle \mathbf{x},\mathbf{x}\rangle} : \mathbf{x}\in\mathcal{X} \quad \text{and} \quad \mathbf{x}\neq\mathbf{0}\right\} \geq \lambda_j\,,$$

for every $\mathcal{X}\in\mathcal{S}_j$, which in turn implies that

$$\min_{\mathcal{X}\in\mathcal{S}_j}\max\left\{\frac{\langle A\mathbf{x},\mathbf{x}\rangle}{\langle \mathbf{x},\mathbf{x}\rangle} : \mathbf{x}\in\mathcal{X} \quad \text{and} \quad \mathbf{x}\neq\mathbf{0}\right\} \geq \lambda_j\,.$$

On the other hand, as

$$\max\left\{\frac{\langle A\mathbf{x},\mathbf{x}\rangle}{\langle \mathbf{x},\mathbf{x}\rangle} : \mathbf{x}\in\operatorname{span}\{\mathbf{u}_1,\ldots,\mathbf{u}_j\} \quad \text{and} \quad \mathbf{x}\neq\mathbf{0}\right\} = \lambda_j\,,$$

it follows that

$$\min_{\mathcal{X}\in\mathcal{S}_j}\max\left\{\frac{\langle A\mathbf{x},\mathbf{x}\rangle}{\langle \mathbf{x},\mathbf{x}\rangle} : \mathbf{x}\in\mathcal{X} \quad \text{and} \quad \mathbf{x}\neq\mathbf{0}\right\} \leq \lambda_j\,,$$

and hence that equality prevails.

To prove (10.23), let

$$\mathcal{W}_j = \operatorname{span}\{\mathbf{u}_1,\ldots,\mathbf{u}_{n-j+1}\}\,.$$

Then

$$\mathcal{X}\cap\mathcal{W}_j\neq\{\mathbf{0}\}$$

for every $\mathcal{X}\in\mathcal{S}_j$. Thus, for every $\mathcal{X}\in\mathcal{S}_j$, we can find a nonzero vector $\mathbf{w}\in\mathcal{X}\cap\mathcal{W}_j$. But this implies that

$$\mathbf{w} = \sum_{i=1}^{n-j+1} c_i\mathbf{u}_i$$

and hence that

$$\langle A\mathbf{w}, \mathbf{w}\rangle = \sum_{i=1}^{n-j+1} \lambda_i |c_i|^2 \le \lambda_{n-j+1} \sum_{i=1}^{n-j+1} |c_i|^2 = \lambda_{n-j+1}\langle \mathbf{w}, \mathbf{w}\rangle\,.$$

Therefore,

$$\min\left\{\frac{\langle A\mathbf{x}, \mathbf{x}\rangle}{\langle \mathbf{x}, \mathbf{x}\rangle} : \mathbf{x} \in \mathcal{X} \quad \text{and} \quad \mathbf{x} \ne \mathbf{0}\right\} \le \frac{\langle A\mathbf{w}, \mathbf{w}\rangle}{\langle \mathbf{w}, \mathbf{w}\rangle} \le \lambda_{n-j+1}\,.$$

Thus, as the space $\mathcal{X}$ is an arbitrary member of $\mathcal{S}_j$, it follows that

$$\max_{\mathcal{X}\in\mathcal{S}_j} \min\left\{\frac{\langle A\mathbf{x}, \mathbf{x}\rangle}{\langle \mathbf{x}, \mathbf{x}\rangle} : \mathbf{x} \in \mathcal{X} \quad \text{and} \quad \mathbf{x} \ne \mathbf{0}\right\} \le \lambda_{n-j+1}\,.$$

To get the opposite inequality, it suffices to note that

$$\min\left\{\frac{\langle A\mathbf{y}, \mathbf{y}\rangle}{\langle \mathbf{y}, \mathbf{y}\rangle} : \mathbf{y} \in \text{span}\,\{\mathbf{u}_{n-j+1}, \ldots, \mathbf{u}_n\} \quad \text{and} \quad \mathbf{y} \ne \mathbf{0}\right\} = \lambda_{n-j+1}\,.$$

The verification of (10.24) and (10.25) is left to the reader. □

Exercise 10.16. Show that if a_{ij} denote the entries of the matrix

$$A = \begin{bmatrix} 0 & 1 & 2 \\ 3 & 1 & 0 \\ 0 & 2 & 1 \end{bmatrix}, \quad \text{then} \quad \max_i \left\{\min_j \,[a_{ij}]\right\} \ne \min_j \left\{\max_i \,[a_{ij}]\right\}.$$

Exercise 10.17. Show that if the eigenvalues of $A = A^H$ are ordered so that $\lambda_1 \ge \cdots \ge \lambda_n$, then formulas (10.24) and (10.25) hold.

Exercise 10.18. Show that if $A \in \mathbb{C}^{n\times n}$ is a Hermitian matrix with eigenvalues $\lambda_1 \le \cdots \le \lambda_n$ and $\mathcal{X}^\perp$ denotes the orthogonal complement of $\mathcal{X}$ in $\mathbb{C}^n$, then

$$\lambda_{n-j+1} = \min_{\mathcal{X}\in\mathcal{S}_j} \max\left\{\frac{\langle A\mathbf{x}, \mathbf{x}\rangle}{\langle \mathbf{x}, \mathbf{x}\rangle} : \mathbf{x} \in \mathcal{X}^\perp \quad \text{and} \quad \mathbf{x} \ne \mathbf{0}\right\} \text{ for } j = 1, \ldots, n\,.$$

Exercise 10.19. Show that if $A \in \mathbb{C}^{n\times n}$ is a Hermitian matrix with eigenvalues $\lambda_1 \le \cdots \le \lambda_n$ and $\mathcal{X}^\perp$ denotes the orthogonal complement of $\mathcal{X}$ in $\mathbb{C}^n$, then

$$\lambda_j = \max_{\mathcal{X}\in\mathcal{S}_j} \min\left\{\frac{\langle A\mathbf{x}, \mathbf{x}\rangle}{\langle \mathbf{x}, \mathbf{x}\rangle} : \mathbf{x} \in \mathcal{X}^\perp \quad \text{and} \quad \mathbf{x} \ne \mathbf{0}\right\} \quad \text{for} \quad j = 1, \ldots, n\,.$$

Lemma 10.8. *Let $A, B \in \mathbb{C}^{n\times n}$ and let $s_j(A)$ and $s_j(BA)$, $j = 1, \ldots, n$, denote the singular values of A, B and BA, respectively. Then:*

$$s_j(BA) \le \|B\| s_j(A)\,.$$

Proof. Since $s_1(A) \geq \cdots \geq s_n(A)$,

$$\begin{aligned} s_j^2(A) &= \max_{\mathcal{Y}\in\mathcal{S}_j} \min\left\{\langle A^H A\mathbf{y}, \mathbf{y}\rangle : \mathbf{y} \in \mathcal{Y} \text{ and } \|\mathbf{y}\| = 1\right\} \\ &= \max_{\mathcal{Y}\in\mathcal{S}_j} \min\left\{\|A\mathbf{y}\|^2 : y \in \mathcal{Y} \text{ and } \|\mathbf{y}\| = 1\right\} . \end{aligned}$$

Correspondingly,

$$s_j^2(BA) = \max_{\mathcal{Y}\in\mathcal{S}_j} \min\left\{\|BA\mathbf{y}\|^2 : y \in \mathcal{Y} \text{ and } \|\mathbf{y}\| = 1\right\} .$$

Therefore, as

$$\|BA\mathbf{y}\| \leq \|B\|\|A\mathbf{y}\| ,$$

it follows that

$$\begin{aligned} &\min\left\{\|BA\mathbf{y}\|^2 : \mathbf{y} \in \mathcal{Y} \text{ and } \|\mathbf{y}\| = 1\right\} \\ &\qquad \leq \|B\|^2 \min\left\{\|A\mathbf{y}\|^2 : \mathbf{y} \in \mathcal{Y} \text{ and } \|\mathbf{y}\| = 1\right\} \end{aligned}$$

and hence that

$$s_j(BA)^2 \leq \|B\|^2 s_j(A)^2 \text{ for } j = 1, \ldots, n .$$

This serves to prove the lemma, since $s_j(BA) \geq 0$ and $s_j(A) \geq 0$ by definition. □

Exercise 10.20. Let $A \in \mathbb{C}^{n\times n}$, $B \in \mathbb{C}^{n\times n}$ and let $s_j(AB)$ and $s_j(A)$ denote the singular values of the matrices AB and A, respectively. Show that $s_j(AB) \leq \|B\| s_j(A)$.

Exercise 10.21. Let $A \in \mathbb{C}^{n\times n}$ be a Hermitian matrix with eigenvalues $\lambda_1 \geq \cdots \geq \lambda_n$. Show that $\lambda_n \leq \min a_{ii} \leq \max a_{ii} \leq \lambda_1$.

10.5. Inequalities for singular values

Lemma 10.9. *Let $A \in \mathbb{C}^{n\times n}$, let $s_1 \geq \cdots \geq s_n$ denote the singular values of A and let $1 \leq k \leq n$. Then*

$$\det(W^H A^H A W) \leq s_1^2 \ldots s_k^2 \det(W^H W) \quad \textit{for every choice of} \quad W \in \mathbb{C}^{n\times k} .$$

Proof. Theorem 10.1 guarantees the existence of a unitary matrix $U \in \mathbb{C}^{n\times n}$ such that

$$U^H A^H A U = D^2 \quad \text{with} \quad D = \text{diag}\{s_1, \ldots, s_n\} .$$

Therefore,

$$\begin{aligned} W^H A^H A W &= W^H U D^2 U^H W \\ &= W^H U D D U^H W \\ &= B B^H , \end{aligned}$$

where $B = FD$ and $F = W^H U$. Let $C = B^H$. Then, by the Binet-Cauchy formula,

$$\det(BB^H) = \sum_{1 \le j_1 < \cdots < j_k \le n} B\begin{pmatrix} 1, \ldots, k \\ j_1, \ldots, j_k \end{pmatrix} C\begin{pmatrix} j_1, \ldots, j_k \\ 1, \ldots, k \end{pmatrix}.$$

Consequently, as

$$\overline{B\begin{pmatrix} 1, \ldots, k \\ j_1, \ldots, j_k \end{pmatrix}} = C\begin{pmatrix} j_1, \ldots, j_k \\ 1, \ldots, k \end{pmatrix}, \tag{10.26}$$

it follows that

$$\det(BB^H) = \sum_{1 \le j_1 < \cdots < j_k \le n} \left| B\begin{pmatrix} 1, \ldots, k \\ j_1, \ldots, j_k \end{pmatrix} \right|^2. \tag{10.27}$$

Moreover, since

$$\begin{aligned} B\begin{pmatrix} 1, \ldots, k \\ j_1, \ldots, j_k \end{pmatrix} &= \det \begin{bmatrix} b_{1j_1} & \cdots & b_{1j_k} \\ \vdots & & \vdots \\ b_{kj_1} & \cdots & b_{kj_k} \end{bmatrix} \\ &= \det \left\{ \begin{bmatrix} f_{1j_1} & \cdots & f_{1j_k} \\ \vdots & & \vdots \\ f_{kj_1} & \cdots & f_{kj_k} \end{bmatrix} \begin{bmatrix} s_{j_1} & & \\ & \ddots & \\ & & s_{j_k} \end{bmatrix} \right\} \\ &= s_{j_1} \cdots s_{j_k} F\begin{pmatrix} 1, \ldots, k \\ j_1, \ldots, j_k \end{pmatrix} \end{aligned}$$

and $0 \le s_{j_1} \cdots s_{j_k} \le s_1 \cdots s_k$, it follows that

$$\left| B\begin{pmatrix} 1, \ldots, k \\ j_1, \ldots, j_k \end{pmatrix} \right|^2 \le s_1^2 \cdots s_k^2 \left| F\begin{pmatrix} 1, \ldots, k \\ j_1, \ldots, j_k \end{pmatrix} \right|^2$$

and hence, by formula (10.27),

$$\det(BB^H) \le s_1^2 \cdots s_k^2 \det(FF^H).$$

Thus,

$$\begin{aligned} \det(W^H A^H A W) &\le s_1^2 \cdots s_k^2 \det(W^H U^H U W) \\ &= s_1^2 \cdots s_k^2 \det(W^H W), \end{aligned}$$

as claimed. □

Exercise 10.22. Verify formula (10.26).

Lemma 10.10. *Let $A \in \mathbb{C}^{n\times n}$, let $s_1 \geq \cdots \geq s_n$ denote the singular values of A and suppose that the eigenvalues of A are repeated according to algebraic multiplicity and are indexed so that $|\lambda_1| \geq |\lambda_2| \geq \cdots \geq |\lambda_n|$. Then*

$$|\lambda_1|\cdots|\lambda_k| \leq s_1\cdots s_k \quad for \quad k = 1,\ldots,n\,.$$

Proof. By Schur's theorem, there exists a unitary matrix U such that $T = U^H AU$ is upper triangular and $t_{jj} = \lambda_j$ for $j = 1,\ldots,n$. Thus, if $V^H = [I_k \quad O_{k\times(n-k)}]$, then

$$\lambda_1\cdots\lambda_k = \det\{V^H TV\} = \det T_{11}\,,$$

where T_{11} denotes the upper left-hand $k\times k$ corner of T. Moreover, since T is upper triangular,

$$V^H T^H TV = V^H \begin{bmatrix} T_{11}^H & O \\ T_{12}^H & T_{22}^H \end{bmatrix} \begin{bmatrix} T_{11} & T_{12} \\ O & T_{22} \end{bmatrix} \begin{bmatrix} I_k \\ O \end{bmatrix} = \begin{bmatrix} T_{11} \\ O \end{bmatrix} = T_{11}^H T_{11}\,.$$

Therefore,

$$\begin{aligned} |\lambda_1\cdots\lambda_k|^2 &= |\det\{V^H TV\}|^2 = |\det T_{11}|^2 = \det\{T_{11}^H T_{11}\} \\ &= \det\{V^H T^H TV\} = \det\{V^H U^H A^H U U^H AUV\} \\ &= \det\{V^H U^H A^H AUV\} \\ &\leq s_1^2\cdots s_k^2 \det\{V^H U^H UV\}\,, \end{aligned}$$

by Lemma 10.9. This is equivalent to the claimed inequality, since

$$\det\{V^H U^H UV\} = 1$$

for the given choice of V^H when U is unitary. □

Corollary 10.11. Let $A \in \mathbb{C}^{n\times n}$ with singular values $s_1 \geq \cdots \geq s_n$ and eigenvalues $\lambda_1,\ldots,\lambda_n$ repeated according to algebraic multiplicity and indexed so that $|\lambda_1| \geq \cdots \geq |\lambda_n|$. Then $s_k = 0 \Longrightarrow \lambda_k = 0$ (i.e., $|\lambda_k| > 0 \Longrightarrow s_k > 0$).

Lemma 10.12. *Let $\{a_1,\ldots,a_n\}$ and $\{b_1,\ldots,b_n\}$ be two sequences of real numbers such that $a_1 \geq a_2 \geq \cdots \geq a_n$; $b_1 \geq b_2 \geq \cdots \geq b_n$ and*

$$\sum_{j=1}^k a_j \leq \sum_{j=1}^k b_j\,, \quad for \quad k = 1,\ldots,n\,.$$

Then

$$\sum_{j=1}^k e^{a_j} \leq \sum_{j=1}^k e^{b_j}\,, \quad for \quad k = 1,\ldots,n\,.$$

Proof. It is readily checked that

$$e^x = \int_{-\infty}^x (x-s)e^s ds$$

or, equivalently, in terms of the notation

$$(x-s)_+ = \begin{cases} x-s & \text{for} \quad x-s>0 \\ 0 & \qquad\quad x-s \le 0 \,, \end{cases}$$

that

$$e^x = \int_{-\infty}^{\infty} (x-s)_+ e^s ds \,.$$

Consequently,

$$\sum_{j=1}^{k} e^{a_j} = \sum_{j=1}^{k} \int_{-\infty}^{\infty} (a_j - s)_+ e^s ds$$

and

$$\sum_{j=1}^{k} e^{b_j} = \sum_{j=1}^{k} \int_{-\infty}^{\infty} (b_j - s)_+ e^s ds \,.$$

Thus, in order to establish the stated inequality, it suffices to show that

$$\sum_{j=1}^{k} (a_j - s)_+ \le \sum_{j=1}^{k} (b_j - s)_+$$

for every $s \in \mathbb{R}$. To this end, let

$\alpha(s) = (a_1 - s)_+ + \cdots + (a_k - s)_+$ and $\beta(s) = (b_1 - s)_+ + \cdots + (b_k - s)_+$

and consider the following cases:

(1) If $s < a_k$, then

$$\begin{aligned} \alpha(s) &= (a_1 - s) + \cdots + (a_k - s) \\ &\le (b_1 - s) + \cdots + (b_k - s) \\ &\le (b_1 - s)_+ + \cdots + (b_k - s)_+ = \beta(s) \,. \end{aligned}$$

(2) If $a_j \le s < a_{j-1}$, for $j = 2, \ldots, k$, then

$$\begin{aligned} \alpha(s) &= (a_1 - s)_+ + \cdots + (a_j - s)_+ \\ &= (a_1 - s) + \cdots + (a_{j-1} - s) \\ &\le (b_1 - s) + \cdots + (b_{j-1} - s) \\ &\le (b_1 - s)_+ + \cdots + (b_k - s)_+ = \beta(s) \,. \end{aligned}$$

(3) If $s \ge a_1$, then $\alpha(s) = 0$ and so $\beta(s) \ge \alpha(s)$, since $\beta(s) \ge 0$. □

Theorem 10.13. *Let $A \in \mathbb{C}^{n\times n}$, let $s_1, \ldots, s_n$ denote the singular values of A and let $\lambda_1, \ldots, \lambda_n$ denote the eigenvalues of A repeated according to algebraic multiplicity and indexed so that $|\lambda_1| \ge \cdots \ge |\lambda_n|$. Then*

(1) $\displaystyle\sum_{j=1}^{k} |\lambda_j|^p \le \sum_{j=1}^{k} s_j^p$ *for $p > 0$ and $k = 1, \ldots, n$.*

(2) $\displaystyle\prod_{j=1}^{k}(1+r|\lambda_j|) \le \prod_{j=1}^{k}(1+rs_j)$ *for* $r > 0$ *and* $k = 1, \ldots, n$.

Proof. Lemma 10.10 guarantees that

$$|\lambda_1| \cdots |\lambda_k| \le s_1 \cdots s_k \, .$$

Suppose that $|\lambda_k| > 0$. Then

$$\ln |\lambda_1| + \cdots + \ln |\lambda_k| \le \ln s_1 + \cdots + \ln s_k$$

and hence, if $p > 0$,

$$p\{\ln |\lambda_1| + \cdots + \ln |\lambda_k|\} \le p\{\ln s_1 + \cdots + \ln s_k\}$$

or, equivalently,

$$\ln |\lambda_1|^p + \cdots + \ln |\lambda_k|^p \le \ln s_1^p + \cdots + \ln s_k^p \, .$$

Consequently, Lemma 10.12 is applicable to the numbers $a_j = \ln |\lambda_j|^p$, $b_j = \ln s_j^p$, $j = 1, \ldots, k$, and yields the inequality

$$e^{\ln |\lambda_1|^p} + \cdots + e^{\ln |\lambda_k|^p} \le e^{\ln s_1^p} + \cdots + e^{\ln s_k^p} \, ,$$

which is equivalent to

$$|\lambda_1|^p + \cdots + |\lambda_k|^p \le s_1^p + \cdots + s_k^p \, .$$

Thus we have established the inequality (1) for every integer $k \in \{1, \ldots, n\}$ for which $|\lambda_k| > 0$. However, this is really enough, because if $\lambda_\ell = 0$, then $|\lambda_j| \le s_j$ for $j = \ell, \ldots, n$. Thus, for example, if $n = 5$ and $|\lambda_3| > 0$ but $\lambda_4 = 0$, then the asserted inequality (1) holds for $k = 1, 2, 3$ by the preceding analysis. However, it must also hold for $k = 4$ and $k = 5$, since $\lambda_4 = 0 \Longrightarrow \lambda_5 = 0$ and thus $|\lambda_4| \le s_4$ and $|\lambda_5| \le s_5$.

The second inequality may be verified in much the same way by invoking the formula

$$\varphi(x) = \int_{-\infty}^{x} (x - s)\varphi''(s)ds$$

with

$$\varphi(x) = \ln(1 + re^x) \quad \text{and} \quad r > 0 \, .$$

This works because

$$\varphi''(x) = \frac{re^x}{(1 + re^x)^2} \ge 0 \quad \text{for every} \quad x \in \mathbb{R} \, .$$

The details are left to the reader. □

Exercise 10.23. Verify the integral representation for $\varphi(x)$, assuming that $\varphi(x)$, $\varphi'(x)$ and $\varphi''(x)$ are nice continuous functions that tend to zero quickly enough as $x \to -\infty$ so that the integrals referred to in the following hint converge. [HINT: Under the given assumptions on φ,

$$\begin{aligned}\varphi(x) &= \int_{-\infty}^{x} \varphi'(s)ds = \int_{-\infty}^{x} \left\{ \int_{-\infty}^{s} \varphi''(u)du \right\} ds \\ &= \int_{-\infty}^{x} \left(\int_{u}^{x} ds \right) \varphi''(u)du \,.]\end{aligned}$$

Lemma 10.14. *Let $A \in \mathbb{C}^{n\times n}$ with singular values $s_1 \geq \cdots \geq s_n$. Then*

$$\begin{aligned}&s_1 + \cdots + s_k \\ &\quad = \max\{|\mathrm{trace}(V^H UAV)| : UU^H = I_n\,,\ V \in \mathbb{C}^{n\times k} \text{ and } V^H V = I_k\}\,.\end{aligned}$$

Proof. Let $B = V^H UAV$ and let $\lambda_1(B), \ldots, \lambda_k(B)$ denote the eigenvalues of B repeated according to their algebraic multiplicity. Then, since

$$\mathrm{trace}\ B = \sum_{j=1}^{k} \lambda_j(B)\,,$$

Theorem 10.13 implies that

$$|\mathrm{trace}\ B| \leq \sum_{j=1}^{k} |\lambda_j(B)| \leq \sum_{j=1}^{k} s_j(B)\,.$$

Moreover, by Lemma 10.8,

$$\begin{aligned}s_j(B) &= s_j(V^H UAV) \\ &\leq \|V^H\| \|U\| s_j(A) \|V\| \\ &= s_j(A)\,.\end{aligned}$$

Therefore, for every choice of $U \in \mathbb{C}^{n\times n}$ and $V \in \mathbb{C}^{n\times k}$, with $U^H U = I_n$ and $V^H V = I_k$,

$$|\mathrm{trace}(V^H UAV)| \leq \sum_{j=1}^{k} s_j(A)\,.$$

The next step is to show that there exists a choice of U and V of the requisite form for which equality is attained. The key ingredient is the singular value decomposition

$$A = V_1 S U_1^H\,,$$

in which V_1 and U_1 are unitary,

$$S = \mathrm{diag}\{s_1, \ldots, s_n\} \quad \text{and} \quad s_j = s_j(A)\,.$$

In these terms,

$$\mathrm{trace}\,(V^H UAV) = \mathrm{trace}\,(V^H U V_1 S U_1^H V\}\,,$$

which, upon choosing

$$V^H = [I_k \quad O]U_1^H \text{ and } U = U_1 V_1^H,$$

simplifies to

$$\begin{aligned}\operatorname{trace}(V^H UAV) &= \operatorname{trace}\{[I_k \quad O]S[I_k \quad O]^H\} \\ &= s_1 + \cdots + s_k .\end{aligned}$$

□

The next theorem summarizes a number of important properties of singular values.

Theorem 10.15. *Let $A, B \in \mathbb{C}^{n\times n}$ and let $s_j(A)$ and $s_j(B)$, $j = 1, \ldots, n$, denote the singular values of A and B, respectively. Then:*

(1) $s_j(A) = s_j(A^H)$.

(2) $s_j(BA) \le \|B\| s_j(A)$.

(3) $s_j(AB) \le s_j(A)\|B\|$.

(4) $\prod_{j=1}^k s_j(AB) \le \prod_{j=1}^k s_j(A) \prod_{j=1}^k s_j(B)$.

(5) $\sum_{j=1}^k s_j(A+B) \le \sum_{j=1}^k s_j(A) + \sum_{j=1}^k s_j(B)$.

Proof. Items (1)–(3) are covered by Exercise 10.3, Lemma 10.8 and Exercise 10.20, respectively.

Next, a double application of Lemma 10.9 yields the inequalities

$$\begin{aligned}\det\{V^H B^H A^H ABV\} &\le s_1(A)^2 \cdots s_k(A)^2 \det\{V^H B^H BV\} \\ &\le s_1(A)^2 \cdots s_k(A)^2 s_1(B)^2 \cdots s_k(B)^2 \det\{V^H V\}\end{aligned}$$

for every matrix $V \in \mathbb{C}^{n\times k}$. Thus, if U is a unitary matrix such that

$$B^H A^H AB = U \begin{bmatrix} s_1(AB)^2 & & \\ & \ddots & \\ & & s_n(AB)^2 \end{bmatrix} U^H ,$$

the choice

$$V^H = [I_k \quad O]U^H$$

yields the formulas

$$\begin{aligned}V^H B^H A^H ABV &= [I_k \quad O] \begin{bmatrix} s_1(AB)^2 & & \\ & \ddots & \\ & & s_n(AB)^2 \end{bmatrix} \begin{bmatrix} I_k \\ O \end{bmatrix} \\ &= \begin{bmatrix} s_1(AB)^2 & & \\ & \ddots & \\ & & s_k(AB)^2 \end{bmatrix},\end{aligned}$$

which leads easily to the inequality (4).

Finally, the justification of (5) rests on Lemma 10.14 and the observation that for any unitary matrix $U \in \mathbb{C}^{n\times n}$ and any $V \in \mathbb{C}^{n\times k}$ with $V^HV = I_k$,

$$\text{trace}\{V^HU(A+B)V\} = \text{trace}\{V^HUAV\} + \text{trace}\{V^HUBV\}$$

and hence (by that lemma) that

$$\begin{aligned}|\text{trace}\{V^HU(A+B)V\}| &\leq |\text{trace}\{V^HUAV\}| + |\text{trace}\{V^HUBV\}| \\ &\leq \sum_{j=1}^{k} s_j(A) + \sum_{j=1}^{k} s_j(B)\ .\end{aligned}$$

The exhibited inequality is valid for every choice of U and V of the indicated form. Thus, upon maximizing the left-hand side, we obtain (5). □

Exercise 10.24. Let $A \in \mathbb{C}^{n\times n}$, let $\beta_1, \ldots, \beta_n$ and $\gamma_1, \ldots, \gamma_n$ denote the eigenvalues of the Hermitian matrices

$$B = (A + A^H)/2 \quad \text{and} \quad C = (A - A^H)/(2i)\,,$$

respectively, and let $\lambda \in \sigma(A)$. Show that

$$\beta_1 \leq \frac{\lambda + \overline{\lambda}}{2} \leq \beta_n \quad \text{and} \quad \gamma_1 \leq \frac{\lambda - \overline{\lambda}}{2i} \leq \gamma_n\,.$$

[HINT: If $A\mathbf{x} = \lambda\mathbf{x}$, then $\lambda + \overline{\lambda} = \langle A\mathbf{x}, \mathbf{x}\rangle + \langle \mathbf{x}, A\mathbf{x}\rangle$.]

Remark 10.16. A number of inequalities exist for the real and imaginary parts of eigenvalues. Thus, for example, if $A \in \mathbb{C}^{n\times n}$ and $\lambda_j = \beta_j + i\gamma_j$, $j = 1, \ldots, n$, denote the eigenvalues of A, repeated according to algebraic multiplicity and indexed so that $|\lambda_1| \geq \cdots \geq |\lambda_n|$ and $B = (A - A^H)/(2i)$, then

$$\sum_{j=1}^{k} |\gamma_j| \leq \sum_{j=1}^{k} s_j(B)\ .$$

See e.g., p. 57 of [**35**].

10.6. Bibliographical notes

Theorem 10.5 was adapted from an article by Takagi [**66**]. The last section was adapted from Gohberg-Krein [**35**], which contains Hilbert space versions of most of the cited results, and is an excellent source of supplementary information.

Chapter 11

Pseudoinverses

"How long have you been hearing confessions?" "About fifteen years." "What has confession taught you about men?" ... "the fundamental fact is that there is no such thing as a grown up person...."

Andre Malraux [**49**]

To set the stage, it is useful to recall that if $A \in \mathbb{F}^{p\times q}$, then

$$A \quad \text{is left invertible} \quad \Longleftrightarrow \mathcal{N}_A = \{\mathbf{0}\} \Longleftrightarrow \operatorname{rank} A = q$$

and

$$A \quad \text{is right invertible} \quad \Longleftrightarrow \mathcal{R}_A = \mathbb{F}^p \Longleftrightarrow \operatorname{rank} A = p\,.$$

Thus, if $\operatorname{rank} A < \min\{p, q\}$, then A is neither left invertible nor right invertible.

11.1. Pseudoinverses

A matrix $A^\circ \in \mathbb{F}^{q\times p}$ is said to be a **pseudoinverse** of a matrix $A \in \mathbb{F}^{p\times q}$ if

$$(11.1) \qquad AA^\circ A = A \quad \text{and} \quad A^\circ AA^\circ = A^\circ\,.$$

It is readily checked that if A is left invertible, then every left inverse of A is a pseudoinverse, i.e.,

$$(11.2) \qquad BA = I_q \Longrightarrow ABA = A \text{ and } BAB = B\,.$$

Similarly, if A is right invertible, then every right inverse of A is a pseudoinverse, i.e.,

$$(11.3) \qquad AC = I_p \Longrightarrow ACA = A \text{ and } CAC = C\,.$$

However, although there are matrices A which are neither left invertible nor right invertible, every matrix A has a pseudoinverse. Moreover, *A° is a pseudoinverse of A if and only if A is a pseudoinverse of A°.*

Exercise 11.1. Let A be a 4×5 matrix such that $EPA = U$ is an upper echelon matrix with pivots in the 11, 22 and 34 positions. Show that there exists an invertible 5×5 lower triangular matrix F and a 5×5 permutation matrix Π such that

$$\Pi F U^T = \begin{bmatrix} 1 & 0 & 0 & 0 \\ 0 & 1 & 0 & 0 \\ 0 & 0 & 1 & 0 \\ 0 & 0 & 0 & 0 \\ 0 & 0 & 0 & 0 \end{bmatrix} \quad \text{and} \quad A = (EP)^{-1} \begin{bmatrix} I_3 & O_{3\times 2} \\ O_{1\times 3} & O_{1\times 2} \end{bmatrix} \left(F^T \Pi^T\right)^{-1}.$$

Theorem 11.1. *Every matrix $A \in \mathbb{F}^{p\times q}$ admits a pseudoinverse $A^\circ \in \mathbb{F}^{q\times p}$. Moreover, if $\operatorname{rank} A = r > 0$ and the singular value decomposition (10.4) of A is expressed generically as*

$$A = V \begin{bmatrix} D & O_{r\times(q-r)} \\ O_{(p-r)\times r} & O_{(p-r)\times(q-r)} \end{bmatrix} U^H, \tag{11.4}$$

where $V \in \mathbb{F}^{p\times p}$ and $U \in \mathbb{F}^{q\times q}$ are unitary and $D = diag\{s_1, \dots, s_r\}$ is invertible, then

$$A^\circ = U \begin{bmatrix} D^{-1} & B_1 \\ B_2 & B_2 D B_1 \end{bmatrix} V^H \tag{11.5}$$

is a pseudoinverse of A for every choice of $B_1 \in \mathbb{F}^{r\times(p-r)}$ and $B_2 \in \mathbb{F}^{(q-r)\times r}$. Furthermore, every generalized inverse A° of A can be expressed this way.

Proof. If $A = O_{p\times q}$, then $A^\circ = O_{q\times p}$ is readily seen to be the one and only pseudoinverse of A.

Suppose next that $\operatorname{rank} A = r > 0$ and note that every matrix $\widetilde{A} \in \mathbb{F}^{q\times p}$ can be written in the form

$$\widetilde{A} = U \begin{bmatrix} R_{11} & R_{12} \\ R_{21} & R_{22} \end{bmatrix} V^H,$$

where U, V are as in (11.4), $R_{11} \in \mathbb{F}^{r\times r}$, $R_{12} \in \mathbb{F}^{r\times(p-r)}$, $R_{21} \in \mathbb{F}^{(q-r)\times r}$ and $R_{22} \in \mathbb{F}^{(q-r)\times(p-r)}$. The constraint $A\widetilde{A}A = A$ is met if and only if

$$\begin{bmatrix} D & O \\ O & O \end{bmatrix} \begin{bmatrix} R_{11} & R_{12} \\ R_{21} & R_{22} \end{bmatrix} \begin{bmatrix} D & O \\ O & O \end{bmatrix} = \begin{bmatrix} D & O \\ O & O \end{bmatrix},$$

i.e., if and only if

$$R_{11} = D^{-1}.$$

Next, fixing $R_{11} = D^{-1}$, we see that the second constraint $\widetilde{A}A\widetilde{A} = \widetilde{A}$ is met if and only if

$$\begin{bmatrix} D^{-1} & R_{12} \\ R_{21} & R_{22} \end{bmatrix} \begin{bmatrix} D & O \\ O & O \end{bmatrix} \begin{bmatrix} D^{-1} & R_{12} \\ R_{21} & R_{22} \end{bmatrix} = \begin{bmatrix} D^{-1} & R_{12} \\ R_{21} & R_{22} \end{bmatrix},$$

i.e., if and only if

$$R_{21}DR_{12} = R_{22}.$$

Thus, $\widetilde{A}$ is a pseudoinverse of A if and only if it can be expressed in the form

$$\widetilde{A} = U \begin{bmatrix} D^{-1} & R_{12} \\ R_{21} & R_{21}DR_{12} \end{bmatrix} V^H.$$

But this is exactly the assertion of the lemma (with $B_1 = R_{12}$ and $B_2 = R_{21}$). □

Lemma 11.2. *Let $A \in \mathbb{F}^{p\times q}$ and let A° be a pseudoinverse of A. Then:*

(1) $\mathcal{R}_{AA^\circ} = \mathcal{R}_A$.

(2) $\mathcal{N}_{AA^\circ} = \mathcal{N}_{A^\circ}$.

(3) $\dim \mathcal{R}_A = \dim \mathcal{R}_{A^\circ}$.

Proof. Clearly

$$\mathcal{R}_{AA^\circ} \subseteq \mathcal{R}_A = \mathcal{R}_{AA^\circ A} \subseteq \mathcal{R}_{AA^\circ}.$$

Therefore, equality (1) prevails.

On the other hand,

$$\mathcal{N}_{A^\circ} \subseteq \mathcal{N}_{AA^\circ} \subseteq \mathcal{N}_{A^\circ AA^\circ} = \mathcal{N}_{A^\circ},$$

which serves to establish (2).

Finally, by the principle of conservation of dimension (applied first to A° and then to AA°) and the preceding two formulas,

$$\begin{aligned} p &= \dim \mathcal{N}_{A^\circ} + \dim \mathcal{R}_{A^\circ} \\ &= \dim \mathcal{N}_{AA^\circ} + \dim \mathcal{R}_{A^\circ}, \end{aligned}$$

whereas

$$\begin{aligned} p &= \dim \mathcal{N}_{AA^\circ} + \dim \mathcal{R}_{AA^\circ} \\ &= \dim \mathcal{N}_{AA^\circ} + \dim \mathcal{R}_{A}. \end{aligned}$$

Assertion (3) drops out by comparing the two formulas for p. □

It is instructive to verify assertions (1)–(3) of the last lemma via the decompositions (11.4) and (11.5).

Exercise 11.2. Verify assertions (1)–(3) of Lemma 11.2 via the decompositions (11.4) and (11.5).

Lemma 11.3. *Let $A^\circ \in \mathbb{F}^{q\times p}$ be a pseudoinverse of $A \in \mathbb{F}^{p\times q}$. Then:*

(1) $\mathbb{F}^p = \mathcal{R}_A \dot{+} \mathcal{N}_{A^\circ}$.

(2) $\mathbb{F}^q = \mathcal{R}_{A^\circ} \dot{+} \mathcal{N}_A$.

Proof. First observe that AA° and $A^\circ A$ are both projections, since

$$(AA^\circ)(AA^\circ) = (AA^\circ A)A^\circ = AA^\circ$$

and

$$(A^\circ A)(A^\circ A) = (A^\circ AA^\circ)A = A^\circ A.$$

Thus, as

$$\mathcal{R}_P \dot{+} \mathcal{N}_P = \mathbb{F}^k \quad \text{for any projection} \quad P \in \mathbb{F}^{k\times k}, \tag{11.6}$$

$$\mathbb{F}^p = \mathcal{R}_{AA^\circ} \dot{+} \mathcal{N}_{AA^\circ} \quad \text{and} \quad \mathbb{F}^q = \mathcal{R}_{A^\circ A} \dot{+} \mathcal{N}_{A^\circ A}.$$

The first conclusion now drops out easily from Lemma 11.2. The second follows from the first since A° is a pseudoinverse of A if and only if A is a pseudoinverse of A°. □

Remark 11.4. Lemma 11.2 exhibits the fact that if A° is a generalized inverse for a matrix $A \in \mathbb{F}^{p\times q}$, then $\mathcal{N}_{A^\circ}$ is a complementary subspace for $\mathcal{R}_A$ in $\mathbb{F}^p$ and $\mathcal{R}_{A^\circ}$ is a complementary subspace for $\mathcal{N}_A$ in $\mathbb{F}^q$. Our next objective is to establish a converse statement. The proof will exploit the general form (11.5) for pseudoinverses and the following preliminary observation.

Lemma 11.5. *Let $A \in \mathbb{F}^{p\times q}$, let $A^\circ \in \mathbb{F}^{q\times p}$ be a pseudoinverse of A and suppose that $\operatorname{rank} A = r > 0$ and that A and A° are expressed in the forms* (11.4) *and* (11.5), *respectively. Then:*

$$\mathcal{R}_A = \left\{ V \begin{bmatrix} I_r \\ O_{(p-r)\times r} \end{bmatrix} \mathbf{u} : \mathbf{u} \in \mathbb{F}^r \right\}, \tag{11.7}$$

$$\mathcal{N}_{A^\circ} = \left\{ V \begin{bmatrix} -DB_1 \\ I_{p-r} \end{bmatrix} \mathbf{v} : \mathbf{v} \in \mathbb{F}^{p-r} \right\}, \tag{11.8}$$

$$\mathcal{R}_{A^\circ} = \left\{ U \begin{bmatrix} I_r \\ B_2 D \end{bmatrix} \mathbf{u} : \mathbf{u} \in \mathbb{F}^r \right\}, \tag{11.9}$$

$$\mathcal{N}_A = \left\{ U \begin{bmatrix} O_{r\times(q-r)} \\ I_{q-r} \end{bmatrix} \mathbf{v} : \mathbf{v} \in \mathbb{F}^{(q-r)} \right\}. \tag{11.10}$$

Proof. By definition,

$$\begin{aligned} \mathcal{R}_A &= \left\{ V \begin{bmatrix} D & O \\ O & O \end{bmatrix} U^H \mathbf{x} : \mathbf{x} \in \mathbb{F}^q \right\} \\ &= \left\{ V \begin{bmatrix} D \\ O \end{bmatrix} \begin{bmatrix} I_r & O \end{bmatrix} \mathbf{x} : \mathbf{x} \in \mathbb{F}^q \right\} \\ &= \left\{ V \begin{bmatrix} I_r \\ O \end{bmatrix} \mathbf{u} : \mathbf{u} \in \mathbb{F}^r \right\}. \end{aligned}$$

Similarly,

$$\begin{aligned}\mathcal{R}_{A^\circ} &= \left\{ U \begin{bmatrix} D^{-1} \\ B_2 \end{bmatrix} \begin{bmatrix} I_r & DB_1 \end{bmatrix} V^H \mathbf{x} : \mathbf{x} \in \mathbb{F}^p \right\} \\ &= \left\{ U \begin{bmatrix} D^{-1} \\ B_2 \end{bmatrix} \begin{bmatrix} I_r & DB_1 \end{bmatrix} \mathbf{x} : \mathbf{x} \in \mathbb{F}^p \right\} \\ &= \left\{ U \begin{bmatrix} I_r \\ B_2 D \end{bmatrix} \mathbf{u} : \mathbf{u} \in \mathbb{F}^r \right\}.\end{aligned}$$

Suppose next that $\mathbf{x} \in \mathcal{N}_{A^\circ}$. Then

$$U \begin{bmatrix} D^{-1} \\ B_2 \end{bmatrix} \begin{bmatrix} I_r & DB_1 \end{bmatrix} V^H \mathbf{x} = \mathbf{0}.$$

However, since $U \begin{bmatrix} D^{-1} \\ B_2 \end{bmatrix}$ is left invertible, this holds if and only if

$$\begin{bmatrix} I_r & DB_1 \end{bmatrix} V^H \mathbf{x} = \mathbf{0}.$$

Thus, upon writing

$$V^H \mathbf{x} = \begin{bmatrix} \mathbf{u} \\ \mathbf{v} \end{bmatrix},$$

with $\mathbf{u} \in \mathbb{F}^r$ and $\mathbf{v} \in \mathbb{F}^{p-r}$, we see that

$$\begin{bmatrix} I_r & DB_1 \end{bmatrix} \begin{bmatrix} \mathbf{u} \\ \mathbf{v} \end{bmatrix} = \mathbf{0}$$

or equivalently that

$$\mathbf{u} = -DB_1 \mathbf{v}.$$

Therefore,

$$\mathbf{x} = V \begin{bmatrix} -DB_1 \\ I_{p-r} \end{bmatrix} \mathbf{v}.$$

This proves that

$$\mathcal{N}_{A^\circ} \subseteq \left\{ V \begin{bmatrix} -DB_1 \\ I_{p-r} \end{bmatrix} \mathbf{v} : \mathbf{v} \in \mathbb{F}^{p-r} \right\}$$

and hence in fact serves to establish equality, since the opposite inclusion is self-evident.

The formula for $\mathcal{N}_A$ is established in much the same way. □

Exercise 11.3. Verify the formula for $\mathcal{N}_A$ that is given in Lemma 11.5.

Remark 11.6. Formulas (11.7) and (11.8) confirm the already established fact that

$$\begin{aligned}\mathcal{R}_A+\mathcal{N}_{A^\circ}&=\left\{V\begin{bmatrix}I_r & -DB_1\\ O_{(p-r)\times r} & I_{(p-r)\times(p-r)}\end{bmatrix}\begin{bmatrix}\mathbf{u}\\ \mathbf{v}\end{bmatrix}:\mathbf{u}\in\mathbb{F}^r \text{ and } \mathbf{v}\in\mathbb{F}^{p-r}\right\}\\ &=\left\{V\begin{bmatrix}I_r & -DB_1\\ O_{(p-r)\times r} & I_{(p-r)\times(p-r)}\end{bmatrix}\mathbf{x}:\mathbf{x}\in\mathbb{F}^p\right\}\\ &=\mathbb{F}^p,\end{aligned}$$

since both of the $p\times p$ matrices are invertible.

Similarly, formulas (11.9) and (11.10) confirm the already established fact that

$$\begin{aligned}\mathcal{R}_{A^\circ}+\mathcal{N}_A&=\left\{U\begin{bmatrix}I_r & O_{r\times(q-r)}\\ B_2D & I_{q-r}\end{bmatrix}\begin{bmatrix}\mathbf{u}\\ \mathbf{v}\end{bmatrix}:\mathbf{u}\in\mathbb{F}^r \text{ and } \mathbf{v}\in\mathbb{F}^{q-r}\right\}\\ &=\mathbb{F}^q.\end{aligned}$$

Exercise 11.4. Use formulas (11.7)–(11.10) to confirm that

$$\mathcal{R}_A\cap\mathcal{N}_{A^\circ}=\{\mathbf{0}\} \text{ and } \mathcal{R}_{A^\circ}\cap\mathcal{N}_A=\{\mathbf{0}\}.$$

Exercise 11.5. Show that, in the setting of Lemma 11.5,

$$\mathcal{R}_A \quad\text{is orthogonal to}\quad \mathcal{N}_{A^\circ}\Longleftrightarrow B_1=O_{r\times(p-r)} \tag{11.11}$$

and

$$\mathcal{N}_A \quad\text{is orthogonal to}\quad \mathcal{R}_{A^\circ}\Longleftrightarrow B_2=O_{(q-r)\times r}\,. \tag{11.12}$$

Theorem 11.7. *Let $A\in\mathbb{F}^{p\times q}$ and let $\mathcal{X}$ and $\mathcal{Y}$ be subspaces of $\mathbb{F}^p$ and $\mathbb{F}^q$ respectively, such that*

$$\mathcal{R}_A\dot{+}\mathcal{X}=\mathbb{F}^p \quad\text{and}\quad \mathcal{N}_A\dot{+}\mathcal{Y}=\mathbb{F}^q.$$

Then there exists a pseudoinverse A° of A such that

$$\mathcal{N}_{A^\circ}=\mathcal{X} \quad\text{and}\quad \mathcal{R}_{A^\circ}=\mathcal{Y}\,.$$

Proof. Suppose first that $\mathcal{X}$ and $\mathcal{Y}$ are proper nonzero subspaces of $\mathbb{F}^p$ and $\mathbb{F}^q$ respectively, let $r=\operatorname{rank}A$ and let $\{\mathbf{x}_1,\ldots,\mathbf{x}_{p-r}\}$ be a basis for $\mathcal{X}$. Then, in terms of the representation (11.4), we can write

$$\begin{bmatrix}\mathbf{x}_1 & \cdots & \mathbf{x}_{p-r}\end{bmatrix}=V\begin{bmatrix}C\\ E\end{bmatrix}$$

for some choice of $C\in\mathbb{F}^{r\times(p-r)}$ and $E\in\mathbb{F}^{(p-r)\times(p-r)}$. Thus, in view of (11.7),

$$\mathcal{R}_A+\mathcal{X}=\left\{V\begin{bmatrix}I_r & C\\ O_{(p-r)\times r} & E\end{bmatrix}\begin{bmatrix}\mathbf{u}\\ \mathbf{v}\end{bmatrix}:\mathbf{u}\in\mathbb{F}^r \text{ and } \mathbf{v}\in\mathbb{F}^{p-r}\right\}.$$

Moreover, since

$$\mathcal{R}_A + \mathcal{X} = \mathbb{F}^p \Longleftrightarrow E \quad \text{is invertible},$$

it follows that E is invertible and

$$\mathcal{R}_A + \mathcal{X} = \left\{ V \begin{bmatrix} I_r & CE^{-1} \\ O_{(p-r)\times r} & I_{p-r} \end{bmatrix} \begin{bmatrix} \mathbf{u} \\ E\mathbf{v} \end{bmatrix} : \mathbf{u} \in \mathbb{F}^r \text{ and } \mathbf{v} \in \mathbb{F}^{p-r} \right\}.$$

Choose

$$B_1 = -D^{-1}CE^{-1}.$$

Next, let $\mathbf{y}_1, \ldots, \mathbf{y}_r$ be a basis for $\mathcal{Y}$ and write

$$\begin{bmatrix} \mathbf{y}_1 & \cdots & \mathbf{y}_r \end{bmatrix} = U \begin{bmatrix} G \\ H \end{bmatrix},$$

where $G \in \mathbb{F}^{r\times r}$ and $H \in \mathbb{F}^{(q-r)\times r}$. Then

$$\mathcal{N}_A + \mathcal{Y} = \left\{ U \begin{bmatrix} O_{r\times(q-r)} & G \\ I_{q-r} & H \end{bmatrix} \begin{bmatrix} \mathbf{u} \\ \mathbf{v} \end{bmatrix} : \mathbf{u} \in \mathbb{F}^{q-r} \text{ and } \mathbf{v} \in \mathbb{F}^r \right\} = \mathbb{F}^q$$

if and only if G is invertible. Thus,

$$\begin{aligned} \mathcal{N}_A + \mathcal{Y} &= \left\{ U \begin{bmatrix} O_{r\times(q-r)} & I_r \\ I_{q-r} & HG^{-1} \end{bmatrix} \begin{bmatrix} \mathbf{u} \\ G\mathbf{v} \end{bmatrix} : \mathbf{u} \in \mathbb{F}^{q-r} \text{ and } \mathbf{v} \in \mathbb{F}^r \right\} \\ &= \left\{ U \begin{bmatrix} O_{r\times(q-r)} & I_r \\ I_{q-r} & HG^{-1} \end{bmatrix} \mathbf{x} : \mathbf{x} \in \mathbb{F}^q \right\}. \end{aligned}$$

Choose $B_2 = HG^{-1}D^{-1}$. Then for the specified choices of B_1 and B_2, the matrix A° defined by formula (11.5) is a pseudoinverse of the matrix A in (11.5) such that

$$\mathcal{X} = \left\{ V \begin{bmatrix} -DB_1 \\ I_{p-r} \end{bmatrix} \mathbf{v} : \mathbf{v} \in \mathbb{F}^{p-r} \right\} = \mathcal{N}_{A^\circ}$$

and

$$\mathcal{Y} = \left\{ U \begin{bmatrix} I_r \\ B_2 D \end{bmatrix} \mathbf{u} : \mathbf{u} \in \mathbb{F}^r \right\} = \mathcal{R}_{A^\circ},$$

as claimed; see Lemma 11.5. This completes the proof when $\mathcal{X}$ and $\mathcal{Y}$ are proper subspaces of $\mathbb{F}^p$ and $\mathbb{F}^q$, respectively. The remaining cases are left to the reader. □

Exercise 11.6. Let $A \in \mathbb{F}^{p\times q}$ and suppose that $\mathcal{R}_A = \mathbb{F}^p$ and $\mathcal{N}_A \dot{+} \mathcal{Y} = \mathbb{F}^q$ for some proper nonzero subspace $\mathcal{Y}$ of $\mathbb{F}^q$. Show that there exists a pseudoinverse A° of A such that $\mathcal{N}_{A^\circ} = \{\mathbf{0}\}$ and $\mathcal{R}_{A^\circ} = \mathcal{Y}$.

Exercise 11.7. Let

$$A = \begin{bmatrix} 1 & 1 & 0 \\ 1 & 0 & 0 \\ 0 & 0 & 0 \\ 0 & 0 & 0 \end{bmatrix}, \quad \mathcal{X} = \text{span}\left\{ \begin{bmatrix} 1 \\ 1 \\ 1 \end{bmatrix}, \begin{bmatrix} 1 \\ 0 \\ 1 \end{bmatrix} \right\} \text{ and } \mathcal{Y} = \text{span}\left\{ \begin{bmatrix} 1 \\ 0 \\ 1 \\ 1 \end{bmatrix}, \begin{bmatrix} 1 \\ 1 \\ 1 \\ 0 \end{bmatrix} \right\}.$$

Find a pseudoinverse A° of the matrix A such that $\mathcal{N}_{A^\circ} = \mathcal{Y}$ and $\mathcal{R}_{A^\circ} = \mathcal{X}$.

Exercise 11.8. Let $A \in \mathbb{C}^{p\times q}$ admit a singular value decomposition of the form $A = VSU^H$, where $V \in \mathbb{C}^{p\times p}$ and $U \in \mathbb{C}^{q\times q}$ are both unitary. Suppose further that $\operatorname{rank} A = r$, $S = \operatorname{diag}\{D, O_{(p-r)\times(q-r)}\}$ and that $1 \le r < \min\{p, q\}$.

(1) Find formulas for A^HA, AA^H, $AA^\dagger$ and $A^\dagger A$.

(2) Show that the ranges of A^HA and $A^\dagger A$ coincide.

(3) Show that the ranges of AA^H and $AA^\dagger$ coincide.

(4) Describe the null spaces of the four matrices considered in (1) in terms of appropriately chosen sub-blocks of U and V.

11.2. The Moore-Penrose inverse

Theorem 11.8. *Let $A \in \mathbb{F}^{p\times q}$. Then there exists exactly one matrix $A^\dagger \in \mathbb{F}^{q\times p}$ that meets the four conditions*

(11.13)
$$AA^\dagger A = A\,, \quad A^\dagger AA^\dagger = A^\dagger\,, \quad AA^\dagger = (AA^\dagger)^H \quad \textit{and} \quad A^\dagger A = (A^\dagger A)^H\,.$$

Proof. If $A = O_{p\times q}$, then the matrix $A^\dagger = O_{q\times p}$ clearly meets the four conditions in (11.13).

If $\operatorname{rank} A = r > 0$, and

$$A = V\begin{bmatrix} D & O_{r\times(q-r)} \\ O_{(p-r)\times r} & O_{(p-r)\times(q-r)} \end{bmatrix} U^H = V_1 D U_1^H, \tag{11.14}$$

where $V = \begin{bmatrix} V_1 & V_2 \end{bmatrix}$ and $U = \begin{bmatrix} U_1 & U_2 \end{bmatrix}$ are unitary matrices with first blocks V_1 and U_1 of sizes $p \times r$ and $q \times r$, respectively, and

$$D = \operatorname{diag}\{s_1, \ldots, s_r\} \tag{11.15}$$

is a diagonal matrix based on the nonzero singular values $s_1 \ge \cdots \ge s_r > 0$ of A, then the matrix $A^\dagger \in \mathbb{F}^{q\times p}$ defined by the formula

$$A^\dagger = U\begin{bmatrix} D^{-1} & O \\ O & O \end{bmatrix} V^H = U_1 D^{-1} V_1^H \tag{11.16}$$

meets the four conditions in (11.13).

It remains to check uniqueness. Let $B \in \mathbb{F}^{q\times p}$ and $C \in \mathbb{F}^{q\times p}$ both satisfy the four conditions in (11.13) and let $Y = B^H - C^H$. Then the formulas

$$A = ABA = A(BA)^H = AA^HB^H$$

and

$$A = ACA = A(CA)^H = AA^HC^H$$

imply that $O = AA^HY$ and hence that $\mathcal{R}_Y \subseteq \mathcal{N}_{AA^H} = \mathcal{N}_{A^H}$.

On the other hand, the formulas

$$B = BAB = B(AB)^H = BB^H A^H$$

and

$$C = CAC = C(AC)^H = CC^H A^H$$

imply that

$$Y = A(BB^H - CC^H)$$

and hence that $\mathcal{R}_Y \subseteq \mathcal{R}_A$. Thus, as $\mathcal{R}_A \cap \mathcal{N}_{A^H} = \{\mathbf{0}\}$ by Lemma 9.14, it follows that $Y = O$, as needed. □

The unique matrix $A^\dagger$ that satisfies the four conditions in (11.13) is called the **Moore-Penrose inverse** of A.

In view of the last two formulas in (11.13), $AA^\dagger$ and $A^\dagger A$ are both orthogonal projections with respect to the standard inner product. Correspondingly the direct sum decompositions exhibited in Lemma 11.3 become orthogonal decompositions if the Moore-Penrose inverse $A^\dagger$ is used in place of an arbitrary pseudoinverse A°.

Lemma 11.9. *Let $A^\dagger \in \mathbb{F}^{q\times p}$ be the Moore-Penrose inverse of $A \in \mathbb{F}^{p\times q}$. Then:*

(1) $\mathbb{F}^p = \mathcal{R}_A \oplus \mathcal{N}_{A^\dagger}$.

(2) $\mathbb{F}^q = \mathcal{R}_{A^\dagger} \oplus \mathcal{N}_A$.

Proof. Since $A^\dagger$ is a pseudoinverse, Lemma 11.3 guarantees the direct sum decompositions

$$\mathbb{F}^p = \mathcal{R}_A \dot{+} \mathcal{N}_{A^\dagger} \quad \text{and} \quad \mathbb{F}^q = \mathcal{N}_A \dot{+} \mathcal{R}_{A^\dagger}\,.$$

To complete the proof of (1), we need to show that the two spaces $\mathcal{R}_A$ and $\mathcal{N}_{A^\dagger}$ are orthogonal with respect to the standard inner product. To this end let $\mathbf{x} \in \mathcal{R}_A$ and $\mathbf{y} \in \mathcal{N}_{A^\dagger}$. Then, since $\mathcal{R}_A = \mathcal{R}_{AA^\dagger}$ and $AA^\dagger$ is a projection,

$$\mathbf{x} = AA^\dagger \mathbf{x}\,.$$

Therefore,

$$\begin{aligned} \langle \mathbf{x}, \mathbf{y} \rangle &= \langle AA^\dagger \mathbf{x}, \mathbf{y} \rangle = \langle \mathbf{x}, (AA^\dagger)^H \mathbf{y} \rangle \\ &= \langle \mathbf{x}, AA^\dagger \mathbf{y} \rangle = 0\,, \end{aligned}$$

as needed. The proof of (2) is immediate from (1) and the fact that $(A^\dagger)^\dagger = A$. □

Exercise 11.9. Show that if $A \in \mathbb{C}^{p\times p}$, $B \in \mathbb{C}^{p\times q}$ and $\mathcal{R}_B \subseteq \mathcal{R}_A$, then $AA^\dagger B = B$.

Lemma 11.10. *Let*

$$M = \begin{bmatrix} A & B \\ B^H & C \end{bmatrix}$$

be a Hermitian matrix with square diagonal blocks such that $\mathcal{R}_B \subseteq \mathcal{R}_A$. *Then* M *admits a factorization of the form*

$$M = \begin{bmatrix} I & O \\ B^H A^\dagger & I \end{bmatrix} \begin{bmatrix} A & O \\ O & C - B^H A^\dagger B \end{bmatrix} \begin{bmatrix} I & A^\dagger B \\ O & I \end{bmatrix}. \tag{11.17}$$

Proof. The formula is easily verified by direct calculation, since $B^H A^\dagger A = B^H$ and $AA^\dagger B = B$ when the presumed inclusion is in force. □

Exercise 11.10. Show that $\mathcal{R}_{A^\dagger} = \mathcal{R}_{A^H}$ and $\mathcal{N}_{A^\dagger} = \mathcal{N}_{A^H}$. [HINT: This is an easy consequence of the representations (11.14) and (11.16).]

Exercise 11.11. Show that if $A \in \mathbb{C}^{p\times q}$, then the matrix $AA^\dagger$ is an orthogonal projection from $\mathbb{C}^p$ onto $\mathcal{R}_A$.

Exercise 11.12. Use the representation formulas (11.14) and (11.16) to give a new proof of the following two formulas, for any matrix $A \in \mathbb{C}^{p\times q}$:

(1) $\mathbb{C}^p = \mathcal{R}_A \oplus \mathcal{N}_{A^H}$ (with respect to the standard inner product).

(2) $\mathbb{C}^q = \mathcal{R}_{A^H} \oplus \mathcal{N}_A$ (with respect to the standard inner product).

Exercise 11.13. Show that if $B, C \in \mathbb{C}^{p\times q}$, $A \in \mathbb{C}^{q\times q}$ and $\operatorname{rank} B = \operatorname{rank} C = \operatorname{rank} A = q$, then

$$(BAC^H)^\dagger = C(C^H C)^{-1} A^{-1} (B^H B)^{-1} B^H \tag{11.18}$$

and give explicit formulas for $(BAC^H)^\dagger(BAC^H)$ and $(BAC^H)(BAC^H)^\dagger$ in terms of B, B^H, C and C^H.

Exercise 11.14. Show that if $A \in \mathbb{C}^{p\times q}$, then $A^\dagger A A^H = A^H A A^\dagger = A^H$.

Exercise 11.15. Show that if

$$E = \begin{bmatrix} O & B \\ B^H & C \end{bmatrix}$$

is a Hermitian matrix such that $\mathcal{R}_C \subseteq \mathcal{R}_{B^H}$, then the Moore-Penrose inverse $E^\dagger$ of E is given by the formula

$$E^\dagger = \begin{bmatrix} -(B^\dagger)^H C B^\dagger & (B^\dagger)^H \\ B^\dagger & O \end{bmatrix}. \tag{11.19}$$

[HINT: Exploit Exercise 11.14 and the fact that $\mathcal{R}_C \subseteq \mathcal{R}_{B^H} \Longrightarrow B^\dagger B C = C$.]

Exercise 11.16. Let

$$C = B \begin{bmatrix} A & O \\ O & O \end{bmatrix} B^H,$$

where B is invertible, and let $A^\dagger$ denote the Moore-Penrose inverse of A. Show that the matrix

$$(B^{-1})^H \begin{bmatrix} A^\dagger & O \\ O & O \end{bmatrix} B^{-1}$$

is a pseudoinverse of C, but it is not a Moore-Penrose inverse.

Exercise 11.17. Show that the matrix

$$\begin{bmatrix} AA^\dagger & O \\ B^H A^\dagger & O \end{bmatrix}$$

is a projection but not an orthogonal projection with respect to the standard inner product (unless $B^H A^\dagger = O$).

Exercise 11.18. Let $A_1, A_2 \in \mathbb{C}^{p\times q}$ and $B_1, B_2 \in \mathbb{C}^{p\times r}$ and suppose that $\mathcal{R}_{B_1} \subseteq \mathcal{R}_{A_1}$ and $\mathcal{R}_{B_2} \subseteq \mathcal{R}_{A_2}$. Show by example that this **does not imply** that $\mathcal{R}_{B_1+B_2} \subseteq \mathcal{R}_{A_1+A_2}$. [HINT: Try $B_i = \mathbf{u}_i\mathbf{v}_i^H$ and $A_i = \mathbf{u}_i\mathbf{w}_i^H$ for $i = 1, 2$, with $\mathbf{v}_1$ orthogonal to $\mathbf{v}_2$ and $\mathbf{w}_1 = \mathbf{w}_2$.]

Exercise 11.19. Let $A \in \mathbb{C}^{p\times p}$, $B \in \mathbb{C}^{p\times q}$,

$$M = \begin{bmatrix} A & B \\ B^H & O \end{bmatrix}$$

and suppose that $BB^H = I_p$. Show that the Moore-Penrose inverse

$$M^\dagger = \begin{bmatrix} O & B \\ B^H & -B^H AB \end{bmatrix}.$$

Exercise 11.20. Let $B \in \mathbb{C}^{p\times q}$. Show that:

(1) $B^\dagger B$ is the orthogonal projection of $\mathbb{C}^q$ onto $\mathcal{R}_{B^H}$.

(2) $BB^\dagger$ is the orthogonal projection of $\mathbb{C}^p$ onto $\mathcal{R}_B$.

11.3. Best approximation in terms of Moore-Penrose inverses

The Moore-Penrose inverse of a matrix A with singular value decomposition (11.14) is given by the formula (11.16).

Lemma 11.11. *If $A \in \mathbb{C}^{p\times q}$ and $\mathbf{b} \in \mathbb{C}^p$, then*

$$\|A\mathbf{x} - \mathbf{b}\|^2 = \|A\mathbf{x} - AA^\dagger\mathbf{b}\|^2 + \|(I_p - AA^\dagger)\mathbf{b}\|^2 \tag{11.20}$$

for every $\mathbf{x} \in \mathbb{C}^q$.

Proof. The stated formula follows easily from the decomposition

$$\begin{aligned} A\mathbf{x} - \mathbf{b} &= (A\mathbf{x} - AA^\dagger\mathbf{b}) - (I_p - AA^\dagger)\mathbf{b} \\ &= AA^\dagger(A\mathbf{x} - \mathbf{b}) - (I_p - AA^\dagger)\mathbf{b}, \end{aligned}$$

since

$$\begin{aligned}\langle AA^\dagger(A\mathbf{x}-\mathbf{b}),(I_p-AA^\dagger)\mathbf{b}\rangle &= \langle (A\mathbf{x}-\mathbf{b}),(AA^\dagger)^H(I_p-AA^\dagger)\mathbf{b}\rangle\\ &= \langle A\mathbf{x}-\mathbf{b},(AA^\dagger)(I_p-AA^\dagger)\mathbf{b}\rangle\\ &= \langle A\mathbf{x}-\mathbf{b},\mathbf{0}\rangle\\ &= 0\,.\end{aligned}$$

□

Formula (11.20) exhibits the fact that the best approximation to $\mathbf{b}$ that we can hope to get by vectors of the form $A\mathbf{x}$ is obtained by choosing $\mathbf{x}$ so that

$$A\mathbf{x} = AA^\dagger\mathbf{b}\,.$$

This is eminently reasonable, since $AA^\dagger\mathbf{b}$ is equal to the orthogonal projection of $\mathbf{b}$ onto $\mathcal{R}_A$. The particular choice

$$\mathbf{x} = A^\dagger\mathbf{b}$$

has one more feature:

$$\|A^\dagger\mathbf{b}\| = \min\{\|\mathbf{y}\| : \mathbf{y}\in\mathbb{C}^q \quad\text{and}\quad A\mathbf{y}=AA^\dagger\mathbf{b}\}\,. \tag{11.21}$$

To verify this, observe that if $\mathbf{y}$ is any vector for which $A\mathbf{y} = AA^\dagger\mathbf{b}$, then

$$\mathbf{y} - A^\dagger\mathbf{b} \in \mathcal{N}_A\,;$$

i.e.,

$$\mathbf{y} = A^\dagger\mathbf{b} + \mathbf{u}$$

for some vector $\mathbf{u}\in\mathcal{N}_A$. Therefore, since

$$\langle A^\dagger\mathbf{b},\mathbf{u}\rangle = \langle A^\dagger AA^\dagger\mathbf{b},\mathbf{u}\rangle = \langle A^\dagger\mathbf{b},A^\dagger A\mathbf{u}\rangle = \langle A^\dagger\mathbf{b},\mathbf{0}\rangle = 0\,,$$

it follows that

$$\|\mathbf{y}\|^2 = \|A^\dagger\mathbf{b}\|^2 + \|\mathbf{u}\|^2.$$

Thus,

$$\|\mathbf{y}\| \ge \|A^\dagger\mathbf{b}\|$$

with equality if and only if

$$\mathbf{y} = A^\dagger\mathbf{b}.$$

Remark 11.12. If A is a $p\times q$ matrix of rank q, then A^HA is invertible and another recipe for obtaining an approximate solution to the equation $A\mathbf{x}=\mathbf{b}$ is based on the observation that if $\mathbf{x}$ is a solution, then

$$A^HA\mathbf{x} = A^H\mathbf{b}$$

and hence

$$\mathbf{x} = (A^HA)^{-1}A^H\mathbf{b}\,.$$

Exercise 11.21. Let $A\in\mathbb{F}^{p\times q}$. Show that if $\operatorname{rank} A = q$, then A^HA is invertible and

$$(A^HA)^{-1}A^H = A^\dagger\,.$$

Chapter 12

Triangular factorization and positive definite matrices

Half the harm that is done in this world
Is due to people who want to feel important.
They don't mean to do harm—but the harm does not interest them.
Or they do not see it, or they justify it
Because they are absorbed in the endless struggle
To think well of themselves.

T. S. Elliot, *The Cocktail Party*

This chapter is devoted primarily to positive definite and semidefinite matrices and related applications. To add perspective, however, it is convenient to begin with some general observations on the triangular factorization of matrices. In a sense this is not new, because the formula

$$EPA = U \quad \text{or, equivalently,} \quad A = P^{-1}E^{-1}U$$

that emerged from the discussion of Gaussian elimination is almost a triangular factorization. Under appropriate extra assumptions on the matrix $A \in \mathbb{C}^{n\times n}$, the formula $A = P^{-1}E^{-1}U$ holds with $P = I_n$.

• **WARNING:** We remind the reader that from now on $\langle \mathbf{u}, \mathbf{v}\rangle = \langle \mathbf{u}, \mathbf{v}\rangle_{st}$, the standard inner product, and $\|\mathbf{u}\| = \|\mathbf{u}\|_2$ for vectors $\mathbf{u}, \mathbf{v} \in \mathbb{F}^n$, unless indicated otherwise. Correspondingly, $\|A\| = \|A\|_{2,2}$ for matrices A.

12.1. A detour on triangular factorization

The notation

$$(12.1)\quad A_{[j,k]} = \begin{bmatrix} a_{jj} & \cdots & a_{jk} \\ \vdots & \cdots & \vdots \\ a_{kj} & \cdots & a_{kk} \end{bmatrix} \quad \text{for} \quad A \in \mathbb{C}^{n\times n} \quad \text{and} \quad 1 \le j \le k \le n$$

will be convenient.

Theorem 12.1. *A matrix $A \in \mathbb{C}^{n\times n}$ admits a factorization of the form*

$$(12.2)\qquad A = LDU\,,$$

where $L \in \mathbb{C}^{n\times n}$ is a lower triangular matrix with ones on the diagonal, $U \in \mathbb{C}^{n\times n}$ is an upper triangular matrix with ones on the diagonal and $D \in \mathbb{C}^{n\times n}$ is an invertible diagonal matrix, if and only if the submatrices

$$(12.3)\qquad A_{[1,k]} \quad \textit{are invertible for} \quad k = 1, \ldots, n\,.$$

Moreover, if the conditions in (12.3) *are met, then there is only one set of matrices, L, D and U, with the stated properties for which* (12.2) *holds.*

Proof. Suppose first that the condition (12.3) is in force. Then, upon expressing

$$A = \begin{bmatrix} A_{11} & A_{12} \\ A_{21} & A_{22} \end{bmatrix}$$

in block form with $A_{11} \in \mathbb{C}^{p\times p}$, $A_{22} \in \mathbb{C}^{q\times q}$ and $p + q = n$, we can invoke the first Schur complement formula

$$A = \begin{bmatrix} I_p & O \\ A_{21}A_{11}^{-1} & I_q \end{bmatrix} \begin{bmatrix} A_{11} & O \\ O & A_{22} - A_{21}A_{11}^{-1}A_{12} \end{bmatrix} \begin{bmatrix} I_p & A_{11}^{-1}A_{12} \\ O & I_q \end{bmatrix}$$

repeatedly to obtain the asserted factorization formula (12.2). Thus, if $A_{11} = A_{[1,n-1]}$, then $\alpha_n = A_{22} - A_{21}A_{11}^{-1}A_{12}$ is a nonzero number and the exhibited formula states that

$$A = L_n \begin{bmatrix} A_{[1,n-1]} & O \\ O & \alpha_n \end{bmatrix} U_n\,,$$

where $L_n \in \mathbb{C}^{n\times n}$ is a lower triangular matrix with ones on the diagonal and $U_n \in \mathbb{C}^{n\times n}$ is an upper triangular matrix with ones on the diagonal. The next step is to apply the same procedure to the $(n-1)\times(n-1)$ matrix $A_{[1,n-1]}$. This yields a factorization of the form

$$A_{[1,n-1]} = \widetilde{L}_{n-1} \begin{bmatrix} A_{[1,n-2]} & O \\ O & \alpha_{n-1} \end{bmatrix} \widetilde{U}_{n-1}\,,$$

where $\widetilde{L}_{n-1} \in \mathbb{C}^{(n-1)\times(n-1)}$ is a lower triangular matrix with ones on the diagonal and $\widetilde{U}_{n-1} \in \mathbb{C}^{(n-1)\times(n-1)}$ is an upper triangular matrix with ones on the diagonal. Therefore,

$$A = L_n \begin{bmatrix} \widetilde{L}_{n-1} & O \\ O & 1 \end{bmatrix} \begin{bmatrix} A_{[1,n-2]} & O & O \\ O & \alpha_{n-1} & O \\ O & O & \alpha_n \end{bmatrix} \begin{bmatrix} \widetilde{U}_{n-1} & O \\ O & 1 \end{bmatrix} U_n \,,$$

which is one step further down the line. The final formula is obtained by iterating this procedure $n-3$ more times.

Conversely, if A admits a factorization of the form (12.2) with the stated properties, then, upon writing the factorization in block form as

$$\begin{bmatrix} A_{11} & A_{12} \\ A_{21} & A_{22} \end{bmatrix} = \begin{bmatrix} L_{11} & O \\ L_{21} & L_{22} \end{bmatrix} \begin{bmatrix} D_{11} & O \\ O & D_{22} \end{bmatrix} \begin{bmatrix} U_{11} & U_{12} \\ O & U_{22} \end{bmatrix},$$

it is readily checked that

$$A_{11} = L_{11} D_{11} U_{11}$$

or, equivalently, that

$$A_{[1,k]} = L_{[1,k]} D_{[1,k]} U_{[1,k]} \quad \text{for} \quad k = 1, \ldots, n\,.$$

Thus, $A_{[1,k]}$ is invertible for $k = 1, \ldots, n$, as needed.

To verify uniqueness, suppose that $A = L_1 D_1 U_1 = L_2 D_2 U_2$. Then the identity $L_2^{-1} L_1 D_1 = D_2 U_2 U_1^{-1}$ implies that $L_2^{-1} L_1 D_1$ is both upper and lower triangular and hence must be a diagonal matrix, which is readily seen to be equal to D_1. Therefore, $L_1 = L_2$ and by an analogous argument $U_1 = U_2$, which then forces $D_1 = D_2$. □

Theorem 12.2. *A matrix $A \in \mathbb{C}^{n\times n}$ admits a factorization of the form*

$$A = UDL\,, \tag{12.4}$$

where $L \in \mathbb{C}^{n\times n}$ is a lower triangular matrix with ones on the diagonal, $U \in \mathbb{C}^{n\times n}$ is an upper triangular matrix with ones on the diagonal and $D \in \mathbb{C}^{n\times n}$ is an invertible diagonal matrix, if and only if the blocks

$$A_{[k,n]} \quad \textit{are invertible for} \quad k = 1, \ldots, n. \tag{12.5}$$

Moreover, if the conditions in (12.5) *are met, then there is only one set of matrices, L, D and U, with the stated properties for which* (12.4) *holds.*

Proof. The details are left to the reader. They are easily filled in with the proof of Theorem 12.1 as a guide. □

Exercise 12.1. Prove Theorem 12.2.

Exercise 12.2. Let $P_k = \text{diag}\,\{I_k, O_{(n-k)\times(n-k)}\}$. Show that

(a) $A \in \mathbb{C}^{n\times n}$ is upper triangular if and only if $AP_k = P_kAP_k$ for $k = 1, \dots, n$.

(b) $A \in \mathbb{C}^{n\times n}$ is lower triangular if and only if $P_kA = P_kAP_k$ for $k = 1, \dots, n$.

Exercise 12.3. Show that if $L \in \mathbb{C}^{n\times n}$ is lower triangular, $U \in \mathbb{C}^{n\times n}$ is upper triangular and $D \in \mathbb{C}^{n\times n}$ is diagonal, then

$$(12.6)\quad (LDU)_{[1,k]} = L_{[1,k]}D_{[1,k]}U_{[1,k]} \quad \text{and} \quad (UDL)_{[k,n]} = U_{[k,n]}D_{[k,n]}L_{[k,n]}$$

for $k = 1, \dots, n$.

12.2. Definite and semidefinite matrices

A matrix $A \in \mathbb{C}^{n\times n}$ is said to be **positive semidefinite over** $\mathbb{C}^n$ if

$$(12.7)\qquad \langle A\mathbf{x}, \mathbf{x}\rangle \geq 0 \text{ for every } \mathbf{x} \in \mathbb{C}^n;$$

it is said to be **positive definite over** $\mathbb{C}^n$ if

$$(12.8)\qquad \langle A\mathbf{x}, \mathbf{x}\rangle > 0 \text{ for every nonzero vector } \mathbf{x} \in \mathbb{C}^n.$$

The **notation**
$A \succeq O$ will be used to indicate that the matrix $A \in \mathbb{C}^{n\times n}$ is positive semidefinite over $\mathbb{C}^n$. Similarly, the notation
$A \succ O$ will be used to indicate that the matrix $A \in \mathbb{C}^{n\times n}$ is positive definite over $\mathbb{C}^n$. Moreover, if $A \in \mathbb{C}^{n\times n}$ and $B \in \mathbb{C}^{n\times n}$, then
$A \succeq B$ and $A \succ B$ means that $A - B \succeq O$ and $A - B \succ O$, respectively. Correspondingly, a matrix $A \in \mathbb{C}^{n\times n}$ is said to be **negative semidefinite** over $\mathbb{C}^n$ if $-A \succeq O$ and **negative definite** over $\mathbb{C}^n$ if $-A \succ O$.

Lemma 12.3. *If $A \in \mathbb{C}^{n\times n}$ and $A \succeq O$, then:*

(1) *A is automatically Hermitian.*

(2) *The eigenvalues of A are nonnegative numbers.*

(3) *$A \succ O \iff$ the eigenvalues of A are all positive* $\iff \det A > 0$.

Proof. If $A \succeq O$, then

$$\langle A\mathbf{x}, \mathbf{x}\rangle = \overline{\langle A\mathbf{x}, \mathbf{x}\rangle} = \langle \mathbf{x}, A\mathbf{x}\rangle$$

for every $\mathbf{x} \in \mathbb{C}^n$. Therefore, by a straightforward calculation,

$$\begin{aligned} 4\langle A\mathbf{x}, \mathbf{y}\rangle &= \sum_{k=1}^{4} i^k \langle A(\mathbf{x} + i^k\mathbf{y}), (\mathbf{x} + i^k\mathbf{y})\rangle \\ &= \sum_{k=1}^{4} i^k \langle (\mathbf{x} + i^k\mathbf{y}), A(\mathbf{x} + i^k\mathbf{y})\rangle = 4\langle \mathbf{x}, A\mathbf{y}\rangle\,; \end{aligned}$$

i.e., $\langle A\mathbf{x}, \mathbf{y}\rangle = \langle \mathbf{x}, A\mathbf{y}\rangle$ for every choice of $\mathbf{x}$, $\mathbf{y} \in \mathbb{C}^n$. Therefore, (1) holds.

Next, let $\mathbf{x}$ be an eigenvector of A corresponding to the eigenvalue λ. Then

$$\lambda\langle \mathbf{x}, \mathbf{x}\rangle = \langle A\mathbf{x}, \mathbf{x}\rangle \geq 0.$$

Therefore $\lambda \geq 0$, since $\langle \mathbf{x}, \mathbf{x}\rangle > 0$. This justifies assertion (2); the proof of (3) is left to the reader. □

WARNING: The conclusions of Lemma 12.3 are not true under the less restrictive constraint

$$\langle A\mathbf{x}, \mathbf{x}\rangle \geq 0 \text{ for every } \mathbf{x} \in \mathbb{R}^n.$$

Thus, for example, if

$$A = \begin{bmatrix} 2 & -2 \\ 0 & 2 \end{bmatrix} \text{ and } \mathbf{x} = \begin{bmatrix} x_1 \\ x_2 \end{bmatrix},$$

then

$$\langle A\mathbf{x}, \mathbf{x}\rangle = (x_1 - x_2)^2 + x_1^2 + x_2^2 > 0$$

for every nonzero vector $\mathbf{x} \in \mathbb{R}^n$. However, A is clearly not Hermitian.

Exercise 12.4. Let $A \in \mathbb{C}^{n\times n}$. Show that if $A \succeq O$, then

$$A \succ O \iff \text{all the eigenvalues of } A \text{ are positive} \iff \det A > 0\,.$$

Exercise 12.5. Show that if $V \in \mathbb{C}^{n\times n}$ is invertible, then

$$A \succ O \iff V^H A V \succ O\,.$$

Exercise 12.6. Show that if $V \in \mathbb{C}^{n\times k}$ and $\operatorname{rank} V = k$, then

$$A \succ O \Longrightarrow V^H A V \succ O\,,$$

but the converse implication is not true if $k < n$.

Exercise 12.7. Show that if the $n \times n$ matrix $A = [a_{ij}]$, $i, j = 1, \ldots, n$, is positive semidefinite over $\mathbb{C}^n$, then $|a_{ij}|^2 \leq a_{ii}a_{jj}$.

Exercise 12.8. Show that if $A \in \mathbb{C}^{n\times n}$, $n = p + q$ and

$$A = \begin{bmatrix} A_{11} & A_{12} \\ A_{21} & A_{22} \end{bmatrix},$$

where $A_{11} \in \mathbb{C}^{p\times p}$, $A_{22} \in \mathbb{C}^{q\times q}$, then

$$A \succ O \iff A_{11} \succ O\,, \quad A_{21} = A_{12}^H \quad \text{and} \quad A_{22} - A_{21}A_{11}^{-1}A_{12} \succ O\,.$$

Exercise 12.9. Show that if $A \in \mathbb{C}^{p\times q}$, then

$$\|A\| \leq 1 \iff I_q - A^H A \succeq O \iff I_p - AA^H \succeq O\,.$$

[HINT: Use the singular value decomposition of A.]

Exercise 12.10. Show that if $A \in \mathbb{C}^{n\times n}$ and $A = A^H$, then

$$\begin{bmatrix} A^2 & A \\ A & I_n \end{bmatrix} \succeq O\,.$$

Exercise 12.11. Show that if $A \in \mathbb{C}^{n\times n}$ and $A \succeq O$, then

$$\begin{bmatrix} A & A \\ A & A \end{bmatrix} \succeq O\,.$$

Exercise 12.12. Let $U \in \mathbb{C}^{n\times n}$ be unitary and let $A \in \mathbb{C}^{n\times n}$. Show that if $A \succ O$ and $AU \succ O$, then $U = I_n$. [HINT: Consider $\langle A\mathbf{x}, \mathbf{x}\rangle$ for eigenvectors $\mathbf{x}$ of U.]

12.3. Characterizations of positive definite matrices

A basic question of interest is to check when an $n \times n$ matrix $A = [a_{ij}], i, j = 1, \ldots, n$ is positive definite over $\mathbb{C}^n$. The next theorem supplies a number of equivalent characterizations.

Theorem 12.4. *If $A \in \mathbb{C}^{n\times n}$, then the following statements are equivalent:*

(1) $A \succ O$.

(2) $A = A^H$ *and the eigenvalues of A are all positive; i.e.* $\lambda_j > 0$ *for* $j = 1, \ldots, n$.

(3) $A = V^H V$ *for some $n \times n$ invertible matrix V.*

(4) $A = A^H$ *and* $\det A_{[1,k]} > 0 \quad \text{for} \quad k = 1, \ldots, n$.

(5) $A = LL^H$, *where L is a lower triangular invertible matrix.*

(6) $A = A^H$ *and* $\det A_{[k,n]} > 0 \quad \text{for} \quad k = 1, \ldots, n$.

(7) $A = UU^H$, *where U is an upper triangular invertible matrix.*

Proof. Let $\{\mathbf{u}_1, \ldots, \mathbf{u}_n\}$ denote an orthonormal set of eigenvectors corresponding to $\lambda_1, \ldots, \lambda_n$. Then, since $\langle \mathbf{u}_j, \mathbf{u}_j\rangle = 1$, the formula

$$\lambda_j = \lambda_j\langle \mathbf{u}_j, \mathbf{u}_j\rangle = \langle A\mathbf{u}_j, \mathbf{u}_j\rangle\,, \quad \text{for} \quad j = 1, \ldots, n\,,$$

clearly displays the fact that (1)$\Longrightarrow$(2). Next, if (2) is in force, then

$$D = \text{diag}\{\lambda_1, \ldots, \lambda_n\}$$

admits a square root

$$D^{1/2} = \text{diag}\{\sqrt{\lambda_1}, \ldots, \sqrt{\lambda_n}\}$$

and hence the diagonalization formula

$$A = UDU^H \quad \text{with} \quad U = \begin{bmatrix} \mathbf{u}_1 & \cdots & \mathbf{u}_n \end{bmatrix}$$

can be rewritten as

$$A = V^H V \quad \text{with} \quad V = D^{1/2} U^H \quad \text{invertible}\,.$$

Thus, (2) $\Longrightarrow$ (3), and, upon setting

$$\Pi_k = \begin{bmatrix} I_k \\ O_{(n-k)\times k} \end{bmatrix} \quad \text{and} \quad V_1 = V\Pi_k \,,$$

it is readily seen that

$$\begin{aligned} A_{[1,k]} &= \Pi_k^H A \Pi_k \\ &= \Pi_k^H V^H V \Pi_k \\ &= V_1^H V_1 \,. \end{aligned}$$

But this implies that

$$\begin{aligned} \langle \Pi_k^H A \Pi_k \mathbf{x}, \mathbf{x} \rangle &= \langle V_1^H V_1 \mathbf{x}, \mathbf{x} \rangle \\ &= \langle V_1 \mathbf{x}, V_1 \mathbf{x} \rangle \\ &> 0 \end{aligned}$$

for every nonzero vector $\mathbf{x} \in \mathbb{C}^k$, since V_1 has k linearly independent columns. Therefore, (3) implies (4). However, in view of Theorem 12.1, (4) implies that $A = L_1 D U_1$, where $L_1 \in \mathbb{C}^{n\times n}$ is a lower triangular matrix with ones on the diagonal, $U_1 \in \mathbb{C}^{n\times n}$ is an upper triangular matrix with ones on the diagonal and $D \in \mathbb{C}^{n\times n}$ is an invertible diagonal matrix. Thus, as $A = A^H$ in the present setting, it follows that

$$(U_1^H)^{-1} L_1 = D^H L_1^H U_1^{-1} D^{-1}$$

and therefore, since the left-hand side of the last identity is lower triangular and the right-hand side is upper triangular, the matrix $(U_1^H)^{-1}L_1$ must be a diagonal matrix. Moreover, since both U_1 and L_1 have ones on their diagonals, it follows that $(U_1^H)^{-1}L_1 = I_n$, i.e., $U_1^H = L_1$. Consequently,

$$A_{[1,k]} = \Pi_k^H A \Pi_k = \Pi_k^H U_1^H D U_1 \Pi_k = (\Pi_k^H U_1^H \Pi_k)(\Pi_k^H D \Pi_k)(\Pi_k^H U_1 \Pi_k)$$

and

$$\det A_{[1,k]} = \det\{(L_1)_{[1,k]}\} \det\{D_{[1,k]}\} \det\{(U_1)_{[1,k]}\} = d_{11} \cdots d_{kk}$$

for $k = 1, \ldots, n$. Therefore, D is positive definite over $\mathbb{C}^n$ as is $A = L_1 D L_1^H$. The formula advertised in (5) is obtained by setting $L = L_1 D^{1/2}$. It is also clear that (5) implies (1). Next, the matrix identity

$$\begin{bmatrix} O & I_{n-k} \\ I_k & O \end{bmatrix} \begin{bmatrix} A_{11} & A_{12} \\ A_{21} & A_{22} \end{bmatrix} \begin{bmatrix} O & I_k \\ I_{n-k} & O \end{bmatrix} = \begin{bmatrix} A_{22} & A_{21} \\ A_{12} & A_{11} \end{bmatrix}$$

clearly displays the fact that (4) holds if and only if (6) holds. Moreover, since (7) implies (1), it remains only to show that (6) implies (7) in order to complete the proof. This is left to the reader as an exercise.

Exercise 12.13. Verify the implication (6)$\Longrightarrow$(7) in Theorem 12.4.

Exercise 12.14. Let $A \in \mathbb{C}^{n\times n}$ and let $D_A = \text{diag}\{a_{11}\dots, a_{nn}\}$ denote the $n \times n$ diagonal matrix with diagonal entries equal to the diagonal entries of A. Show that D_A is multiplicative on upper triangular matrices in the sense that if A and B are both $n \times n$ upper triangular matrices, then $D_{AB} = D_A D_B$ and thus, if A is invertible, $D_{A^{-1}} = (D_A)^{-1}$.

Remark 12.5. The proof that a matrix A that is positive definite over $\mathbb{C}^n$ admits a factorization of the form $A = LL^H$ for some lower triangular invertible matrix L can also be based on the general factorization formula $EPA = U$ that was established as a byproduct of Gaussian elimination. The proof may be split into two parts. The first part is to check that, since $A \succ O$, there always exists a lower triangular matrix E with ones on the diagonal such that

$$EA = U$$

is in upper echelon form and hence upper triangular. Once this is verified, the second part is easy: The identity

$$UE^H = EAE^H = (EAE^H)^H = EU^H$$

implies that $D = UE^H$ is a positive definite matrix that is both lower triangular and upper triangular. Therefore,

$$D = \text{diag}\{d_{11}, \dots, d_{nn}\}$$

is a diagonal matrix with $d_{jj} > 0$ for $j = 1, \dots, n$. Thus, D has a positive square root:

$$D = F^2 ,$$

where

$$F = \text{diag}\{(d_{11})^{1/2}, \dots, (d_{nn})^{1/2}\} ,$$

and consequently

$$A = (E^{-1}F)(E^{-1}F)^H .$$

This is a representation of the desired form, since $L = E^{-1}F$ is lower triangular.

Notice that d_{jj} is the j'th pivot of U and $E^{-1} = (D^{-1}U)^H$.

Exercise 12.15. Show that if $A \in \mathbb{C}^{3\times 3}$ and $A \succ O$, then there exists a lower triangular matrix E with ones on the diagonal such that EA is upper triangular.

Exercise 12.16. Show that if $A \in \mathbb{C}^{3\times 3}$ and $A \succ O$, then there exists an upper triangular matrix F with ones on the diagonal such that FA is lower triangular. [HINT: This is very much like Gaussian elimination in spirit, except that now you work from the bottom row up instead of from the top row down.]

Exercise 12.17. Let $A = \begin{bmatrix} A_1 & A_2 \end{bmatrix}$, where $A_1 \in \mathbb{C}^{n\times s}$, $A_2 \in \mathbb{C}^{n\times t}$ and $s+t=r$. Show that if $\operatorname{rank} A = r$, then the matrices A^HA, $A_1^HA_1$, $A_2^HA_2$ and $A_2^HA_2 - A_2^HA_1(A_1^HA_1)^{-1}A_1^HA_2$ are all positive definite (over complex spaces of appropriate sizes).

Exercise 12.18. Show that if $x \in \mathbb{R}$, then the matrix $\begin{bmatrix} 3 & 2 & x \\ 2 & 2 & 1 \\ x & 1 & 1 \end{bmatrix}$ will be positive definite over $\mathbb{C}^3$ if and only if $(x-1)^2 < 1/2$.

12.4. An application of factorization

Lemma 12.6. *Let $A \in \mathbb{C}^{n\times n}$ and suppose that $A \succ O$ and that*

$$A = \begin{bmatrix} a_{11} & \mathbf{c}^H \\ \mathbf{c} & D \end{bmatrix} \quad \text{and} \quad A^{-1} = \begin{bmatrix} b_{11} & \mathbf{d}^H \\ \mathbf{d} & E \end{bmatrix},$$

where $\mathbf{c}, \mathbf{d} \in \mathbb{C}^{n-1}$ and $D, E \in \mathbb{C}^{(n-1)\times(n-1)}$. Then

(12.9)

$$\min_{x_2,\ldots,x_n} \langle A(\mathbf{e}_1 - \sum_{j=2}^n x_j\mathbf{e}_j), \mathbf{e}_1 - \sum_{j=2}^n x_j\mathbf{e}_j\rangle = \{a_{11} - \mathbf{c}^HD^{-1}\mathbf{c}\}^{1/2} = b_{11}^{-1/2}.$$

Proof. In view of Theorem 12.4, $A = LL^H$, where $L \in \mathbb{C}^{n\times n}$ is an invertible lower triangular matrix. Therefore,

$$\langle A(\mathbf{e}_1 - \sum_{j=2}^n x_j\mathbf{e}_j), \mathbf{e}_1 - \sum_{j=2}^n x_j\mathbf{e}_j\rangle = \|L^H(\mathbf{e}_1 - \sum_{j=2}^n x_j\mathbf{e}_j)\|^2.$$

Let $\mathbf{v}_j = L^H\mathbf{e}_j$ for $j = 1, \ldots, n$,

$$V = \begin{bmatrix} \mathbf{v}_2 & \cdots & \mathbf{v}_n \end{bmatrix} \quad \text{and} \quad \mathcal{V} = \operatorname{span}\{\mathbf{v}_2, \ldots, \mathbf{v}_n\}.$$

Then, since L is invertible, the vectors $\mathbf{v}_1, \ldots, \mathbf{v}_n$ are linearly independent and hence the orthogonal projection $P_\mathcal{V}$ of $\mathbb{C}^n$ onto $\mathcal{V}$ is given by the formula

$$P_\mathcal{V} = V(V^HV)^{-1}V^H.$$

Thus the minimum of interest is equal to

$$\|\mathbf{v}_1 - P_\mathcal{V}\mathbf{v}_1\|^2 = \langle \mathbf{v}_1 - P_\mathcal{V}\mathbf{v}_1, \mathbf{v}_1\rangle = \|\mathbf{v}_1\|^2 - \mathbf{v}_1^HV(V^HV)^{-1}V^H\mathbf{v}_1.$$

It remains to express this number in terms of the entries in the original matrix A by taking advantage of the formulas

$$\begin{bmatrix} a_{11} & \mathbf{c}^H \\ \mathbf{c} & D \end{bmatrix} = A = LL^H = \begin{bmatrix} \mathbf{v}_1^H \\ V^H \end{bmatrix} \begin{bmatrix} \mathbf{v}_1 & V \end{bmatrix} = \begin{bmatrix} \mathbf{v}_1^H\mathbf{v}_1 & \mathbf{v}_1^HV \\ V^H\mathbf{v}_1 & V^HV \end{bmatrix}.$$

The rest is left to the reader. □

Exercise 12.19. Complete the proof of Lemma 12.6.

Exercise 12.20. Let $A \in \mathbb{C}^{n\times n}$ and assume that $A \succ O$. Evaluate

$$\min_{x_1,\ldots,x_{n-1}} \langle A(\mathbf{e}_n - \sum_{j=1}^{n-1} x_j\mathbf{e}_j),\, \mathbf{e}_n - \sum_{j=1}^{n-1} x_j\mathbf{e}_j\rangle$$

in terms of the entries in A and the entries in A^{-1}.

12.5. Positive definite Toeplitz matrices

In this section we shall sketch some applications related to factorization in the special case that the given positive definite matrix is a Toeplitz matrix.

Theorem 12.7. *Let*

$$T_n = \begin{bmatrix} t_0 & t_{-1} & \cdots & t_{-n} \\ t_1 & t_0 & \cdots & t_{1-n} \\ \vdots & \ddots & \ddots & \vdots \\ t_n & \cdots & t_1 & t_0 \end{bmatrix} \succ O \quad and \quad \Gamma_n = \begin{bmatrix} \gamma_{00}^{(n)} & \cdots & \gamma_{0n}^{(n)} \\ \vdots & & \vdots \\ \gamma_{nn}^{(n)} & \cdots & \gamma_{nn}^{(n)} \end{bmatrix} = T_n^{-1}$$

and let

$$p_n(\lambda) = \sum_{j=0}^{n} \gamma_{j0}^{(n)}\lambda^j \quad and \quad q_n(\lambda) = \sum_{j=0}^{n} \gamma_{jn}^{(n)}\lambda^j .$$

Then:

(1) $\sum_{i,j=0}^{n} \lambda^i\gamma_{ij}^{(n)}\overline{\omega}^j$ *is related to the polynomials* $p_n(\lambda)$ *and* $q_n(\lambda)$ *by the formula*

$$\sum_{i,j=0}^{n} \lambda^i\gamma_{ij}^{(n)}\overline{\omega}^j = \frac{p_n(\lambda)\{\gamma_{00}^{(n)}\}^{-1}p_n(\omega)^* - \lambda\overline{\omega}q_n(\lambda)\{\gamma_{nn}^{(n)}\}^{-1}q_n(\omega)^*}{1-\lambda\overline{\omega}} . \tag{12.10}$$

(2) $\gamma_{ij}^{(n)} = \gamma_{n-j,n-i}^{(n)} = \overline{\gamma_{ji}^{(n)}}$ *for* $i,j = 0,\ldots,n$.

(3) $q_n(\lambda) = \lambda^n\overline{p_n(1/\overline{\lambda})}$.

(4) *The polynomial* $p_n(\lambda)$ *has no roots in the closed unit disc.*

(5) *If* $S_n \succ O$ *is an* $(n+1)\times(n+1)$ *Toeplitz matrix such that* $T_n^{-1}\mathbf{e}_1 = S_n^{-1}\mathbf{e}_1$, *then* $S_n = T_n$.

Proof. Since $T_n \succ O \Longrightarrow \Gamma_n \succ O$ we can invoke the Schur complement formulas to write

(12.11)

$$\begin{aligned} \Gamma_n &= \begin{bmatrix} \gamma_{00}^{(n)} & \mathbf{x}^H \\ \mathbf{x} & X \end{bmatrix} \\ &= \begin{bmatrix} 1 & O \\ \mathbf{x}\{\gamma_{00}^{(n)}\}^{-1} & I_n \end{bmatrix} \begin{bmatrix} \gamma_{00}^{(n)} & O \\ O & X - \mathbf{x}\{\gamma_{00}^{(n)}\}^{-1}\mathbf{x}^H \end{bmatrix} \begin{bmatrix} 1 & \{\gamma_{00}^{(n)}\}^{-1}\mathbf{x}^H \\ O & I_n \end{bmatrix} \end{aligned}$$

and

(12.12)

$$\begin{aligned}\Gamma_n &= \begin{bmatrix} Y & \mathbf{y} \\ \mathbf{y}^H & \gamma_{nn}^{(n)} \end{bmatrix} \\ &= \begin{bmatrix} I_n & \mathbf{y}\{\gamma_{nn}^{(n)}\}^{-1} \\ O & 1 \end{bmatrix} \begin{bmatrix} Y - \mathbf{y}\{\gamma_{nn}^{(n)}\}^{-1}\mathbf{y}^H & O \\ O & \gamma_{nn}^{(n)} \end{bmatrix} \begin{bmatrix} I_n & O \\ \{\gamma_{nn}^{(n)}\}^{-1}\mathbf{y}^H & 1 \end{bmatrix},\end{aligned}$$

where $\mathbf{x}^H = \begin{bmatrix} \gamma_{01}^{(n)} & \cdots & \gamma_{0n}^{(n)} \end{bmatrix}$, $\mathbf{y}^H = \begin{bmatrix} \gamma_{n0}^{(n)} & \cdots & \gamma_{n,n-1}^{(n)} \end{bmatrix}$, X denotes the lower right-hand $n \times n$ corner of Γ_n and Y denotes the upper left-hand $n \times n$ corner of Γ_n. Thus,

$$\begin{aligned}\begin{bmatrix} 1 & \lambda & \cdots & \lambda^n \end{bmatrix} \Gamma_n \begin{bmatrix} 1 \\ \overline{\omega} \\ \vdots \\ \overline{\omega}^n \end{bmatrix} &= p_n(\lambda)\{\gamma_{00}^{(n)}\}^{-1} p_n(\omega)^* \\ &\quad + \begin{bmatrix} \lambda & \cdots & \lambda^n \end{bmatrix} \begin{bmatrix} X - \mathbf{x}\{\gamma_{00}^{(n)}\}^{-1}\mathbf{x}^H \end{bmatrix} \begin{bmatrix} \overline{\omega} \\ \vdots \\ \overline{\omega}^n \end{bmatrix},\end{aligned}$$

and a second development based on the second Schur complement formula yields the identity

$$\begin{aligned}\begin{bmatrix} 1 & \lambda & \cdots & \lambda^n \end{bmatrix} \Gamma_n \begin{bmatrix} 1 \\ \overline{\omega} \\ \vdots \\ \overline{\omega}^n \end{bmatrix} &= q_n(\lambda)\{\gamma_{nn}^{(n)}\}^{-1} q_n(\omega)^* \\ &\quad + \begin{bmatrix} 1 & \cdots & \lambda^{n-1} \end{bmatrix} \begin{bmatrix} Y - \mathbf{y}\{\gamma_{nn}^{(n)}\}^{-1}\mathbf{y}^H \end{bmatrix} \begin{bmatrix} 1 \\ \vdots \\ \overline{\omega}^{n-1} \end{bmatrix}.\end{aligned}$$

The proof of formula (12.10) is now completed by verifying that

$$X - \mathbf{x}\{\gamma_{00}^{(n)}\}^{-1}\mathbf{x}^H = \Gamma_{n-1} = Y - \mathbf{y}\{\gamma_{nn}^{(n)}\}^{-1}\mathbf{y}^H, \tag{12.13}$$

where

(12.14)

$$\Gamma_k = \begin{bmatrix} \gamma_{00}^{(k)} & \cdots & \gamma_{0k}^{(k)} \\ \vdots & & \vdots \\ \gamma_{kk}^{(k)} & \cdots & \gamma_{kk}^{(k)} \end{bmatrix} = T_k^{-1} \quad \text{and} \quad T_k = \begin{bmatrix} t_0 & t_{-1} & \cdots & t_{-k} \\ \vdots & \vdots & & \vdots \\ t_k & t_{k-1} & \cdots & t_0 \end{bmatrix}$$

for $k = 0, \ldots, n$. The details are left to the reader as an exercise.

To verify (2), let δ_{ij} denote the Kronecker delta symbol, i.e.,

$$\delta_{ij} = \begin{cases} 1 & \text{if} \quad i = j \\ 0 & \text{if} \quad i \neq j \end{cases},$$

and write

$$\begin{aligned} \sum_{s=0}^{n} t_{i-s}\gamma_{sj}^{(n)} &= \delta_{ij} = \delta_{n-i,n-j} = \sum_{s=0}^{n} t_{n-i-s}\gamma_{s,n-j}^{(n)} \\ &= \sum_{s=0}^{n} t_{n-i-(n-s)}\gamma_{n-s,n-j}^{(n)} = \sum_{s=0}^{n} \gamma_{n-s,n-j}^{(n)} t_{s-i} \\ &= \sum_{s=0}^{n} \gamma_{js}^{(n)} t_{s-i}\,, \end{aligned}$$

which, upon comparing the last two sums, yields the first formula in (2); the second follows from the fact that T_n and Γ_n are Hermitian matrices.

Suppose next that $p_n(\omega) = 0$. Then formula (12.10) implies that

$$(12.15) \qquad -|\omega|^2 q_n(\omega)\left\{\gamma_{nn}^{(n)}\right\}^{-1} q_n(\omega)^* = (1 - |\omega|^2)\sum_{i,j=0}^{n} \omega^i \gamma_{ij}^{(n)} \overline{\omega}^j\,,$$

which is impossible if $|\omega| < 1$, because then the left-hand side of the identity (12.15) is less than or equal to zero, whereas the right-hand side is positive. Thus, $|p_n(\omega)| > 0$ if $|\omega| < 1$. Moreover, if $|\omega| = 1$, then formula (12.15) implies that $q_n(\omega) = 0$ also. Thus, formula (12.10) implies that

$$(12.16) \qquad 0 = \begin{bmatrix} 1 & \lambda & \cdots & \lambda^n \end{bmatrix} \Gamma_n \begin{bmatrix} 1 \\ \overline{\omega} \\ \vdots \\ \overline{\omega}^n \end{bmatrix}$$

for all $\lambda \in \mathbb{C}$, which is impossible.

Finally, in view of items (2) and (3), formula (12.10) can be rewritten as

(12.17)

$$\sum_{i,j=0}^{n} \lambda^i \gamma_{ij}^{(n)} \overline{\omega}^j = \frac{p_n(\lambda)\{\gamma_{00}^{(n)}\}^{-1} p_n(\omega)^* - \lambda^{n+1} p_n(1/\lambda)\{\gamma_{00}^{(n)}\}^{-1} \overline{\omega}^{n+1} p_n(1/\omega)^*}{1 - \lambda\overline{\omega}},$$

which exhibits the fact that if $T_n \succ O$, then all the entries $\gamma_{ij}^{(n)}$ are completely determined by the first column of Γ_n, and hence serves to verify (5). □

Exercise 12.21. Verify the identity (12.13) in the setting of Theorem 12.7. [HINT: Use the Schur complement formulas to calculate Γ_n^{-1} alias T_n from the two block decompositions (12.11) and (12.12).]

Theorem 12.7 is just the tip of the iceberg; it can be generalized in many directions. Some indications are sketched in the next several exercises and the next section, all of which can be skipped without loss of continuity.

Exercise 12.22. Show that if $T_n \succ O$, then Γ_n admits the triangular factorization

$$\Gamma_n = L_n D_n L_n^H \,, \tag{12.18}$$

where

(12.19)

$$L_n = \begin{bmatrix} \gamma_{00}^{(n)} & O & \cdots & O \\ \gamma_{10}^{(n)} & \gamma_{00}^{(n-1)} & \cdots & O \\ \vdots & & \ddots & \vdots \\ \gamma_{n0}^{(n)} & \gamma_{n-1,0}^{(n-1)} & \cdots & \gamma_{00}^{(0)} \end{bmatrix}, \quad L_n^H = \begin{bmatrix} \gamma_{00}^{(n)} & \gamma_{01}^{(n)} & \cdots & \gamma_{0n}^{(n)} \\ O & \gamma_{00}^{(n-1)} & \cdots & \gamma_{0,n-1}^{(n-1)} \\ \vdots & & \ddots & \vdots \\ O & O & \cdots & \gamma_{00}^{(0)} \end{bmatrix}$$

and

$$D_n = \operatorname{diag}\,\{\{\gamma_{00}^{(n)}\}^{-1}, \{\gamma_{00}^{(n-1)}\}^{-1}, \ldots, \{\gamma_{00}^{(0)}\}^{-1}\}\,. \tag{12.20}$$

Exercise 12.23. Find formulas in terms of $\gamma_{ij}^{(k)}$ analogous to those given in the preceding exercise for the factors in a triangular factorization of the form

$$\Gamma_n = U_n D_n U_n^H \,, \tag{12.21}$$

where $T_n \succ O$, U_n is an upper triangular matrix and D_n is a diagonal matrix.

Positive definite Toeplitz matrices play a significant role in the theory of prediction of stationary stochastic sequences, which, when recast in the language of trigonometric approximation, focuses on evaluations of the following sort:

(12.22)

$$\min\left\{\frac{1}{2\pi}\int_0^{2\pi} |e^{in\theta} - \sum_{j=0}^{n-1} c_j e^{ij\theta}|^2 f(e^{i\theta})d\theta \,:\, c_0,\ldots,c_{n-1}\in\mathbb{C}\right\} = \{\gamma_{nn}^{(n)}\}^{-1}$$

and

(12.23)

$$\min\left\{\frac{1}{2\pi}\int_0^{2\pi} |1 - \sum_{j=1}^{n} c_j e^{ij\theta}|^2 f(e^{i\theta})d\theta \,:\, c_1,\ldots,c_n\in\mathbb{C}\right\} = \{\gamma_{00}^{(n)}\}^{-1},$$

where, for ease of exposition, we assume that $f(e^{i\theta})$ is a continuous function of θ on the interval $0 \le \theta \le 2\pi$ such that $f(e^{i\theta}) > 0$ on this interval. Let

$T_n = T_n(f)$ denote the Toeplitz matrix with entries

$$t_j = \frac{1}{2\pi}\int_0^{2\pi} f(e^{i\theta})e^{-ij\theta}d\theta \quad \text{for} \quad j = 0, \pm 1, \pm 2, \ldots . \tag{12.24}$$

Exercise 12.24. Show that

$$\text{if} \quad \mathbf{b} = \begin{bmatrix} b_0 \\ \vdots \\ b_n \end{bmatrix}, \quad \text{then} \quad \frac{1}{2\pi}\int_0^{2\pi} |\sum_{j=0}^{n} b_j e^{ij\theta}|^2 f(e^{i\theta})d\theta = \mathbf{b}^H T_n \mathbf{b}. \tag{12.25}$$

Exercise 12.25. Show that if $T_n \succ O$ and $\mathbf{u}^H = \begin{bmatrix} t_n & \cdots & t_1 \end{bmatrix}$, then

$$T_n = \begin{bmatrix} I_n & \mathbf{0} \\ \mathbf{u}^H T_{n-1}^{-1} & 1 \end{bmatrix} \begin{bmatrix} T_{n-1} & \mathbf{0} \\ \mathbf{0}^H & \{\gamma_{nn}^{(n)}\}^{-1} \end{bmatrix} \begin{bmatrix} I_n & T_{n-1}^{-1}\mathbf{u} \\ \mathbf{0}^H & 1 \end{bmatrix}. \tag{12.26}$$

Exercise 12.26. Show that if $T_n \succ O$ and $\mathbf{v}^H = \begin{bmatrix} t_1 & \cdots & t_n \end{bmatrix}$, then

$$T_n = \begin{bmatrix} 1 & \mathbf{v}^H T_{n-1}^{-1} \\ \mathbf{0} & 1 \end{bmatrix} \begin{bmatrix} \{\gamma_{00}^{(n)}\}^{-1} & \mathbf{0}^H \\ \mathbf{0} & T_{n-1} \end{bmatrix} \begin{bmatrix} 1 & \mathbf{0}^H \\ T_{n-1}^{-1}\mathbf{v} & I_n \end{bmatrix}. \tag{12.27}$$

Exercise 12.27. Show that if $T_n \succ O$, then

$$\begin{aligned} \det T_n &= \{\gamma_{00}^{(n)}\}^{-1}\{\gamma_{00}^{(n-1)}\}^{-1}\cdots\{\gamma_{00}^{(0)}\}^{-1} \\ &= \{\gamma_{nn}^{(n)}\}^{-1}\{\gamma_{n-1,n-1}^{(n-1)}\}^{-1}\cdots\{\gamma_{00}^{(0)}\}^{-1}. \end{aligned} \tag{12.28}$$

Exercise 12.28. Verify formula (12.22). [HINT: Exploit formulas (12.25) and (12.26).]

Exercise 12.29. Verify formula (12.23). [HINT: Exploit formulas (12.25) and (12.27).]

Exercise 12.30. Use the formulas in Lemma 8.15 for calculating orthogonal projections to verify (12.22) and (12.23) another way.

Exercise 12.31. Show that if $T_n \succ O$, then $\gamma_{00}^{(n)} = \gamma_{nn}^{(n)}$ and

$$\gamma_{00}^{(n-1)} \geq \gamma_{00}^{(n)}. \tag{12.29}$$

[HINT: The monotonicity is an easy consequence of formula (12.23).]

Exercise 12.32. Show that if $T_n \succ O$, then the polynomials

$$q_k(\lambda) = \sum_{j=0}^{k} \gamma_{jk}^{(k)}\lambda^j \quad \text{for} \quad k = 0, \ldots, n \tag{12.30}$$

are orthogonal with respect to the inner product

$$\langle q_j, q_k\rangle_f = \frac{1}{2\pi}\int_0^{2\pi} \overline{q_k(e^{i\theta})} f(e^{i\theta}) q_j(e^{i\theta})d\theta, \quad \text{and} \quad \langle q_k, q_k\rangle_f = \gamma_{kk}^{(k)}. \tag{12.31}$$

Exercise 12.33. Use the orthogonal polynomials defined by formula (12.30) to give a new proof of formula (12.22). [HINT: Write $\zeta^n = \sum_{j=0}^n c_j q_j(\zeta)$.]

Exercise 12.34. Let $f(e^{i\theta}) = |h(e^{i\theta}|^2$, where $h(\zeta) = \sum_{j=0}^\infty h_j\zeta^j$, $\sum_{j=0}^\infty |h_j| < \infty$ and $|h(\zeta)| > 0$ for $|\zeta| \le 1$. Granting that $1/h$ has the same properties as h (which follows from a theorem of Norbert Wiener), show that

$$\lim_{n\uparrow\infty}\{\gamma_{nn}^{(n)}\}^{-1} = |h_0|^2 . \tag{12.32}$$

[HINT: $|1 - \sum_{j=1}^n c_j e^{ij\theta}|^2 f(e^{i\theta}) = |h(e^{i\theta}) - \sum_{j=1}^n c_j e^{ij\theta} h(e^{i\theta})|^2 = |h_0 + u(e^{i\theta})|^2$, where $u(e^{i\theta}) = \sum_{j=1}^\infty h_j e^{ij\theta} - \sum_{j=1}^n c_j e^{ij\theta} h(e^{i\theta})$ is orthogonal to h_0 with respect to the inner product of Exercise 8.3 adapted to $[0, 2\pi]$.]

The next lemma serves to guarantee that the conditions imposed on $f(e^{i\theta})$ in Exercise 12.34 are met if $f(\zeta) = a(\zeta)/b(\zeta)$, where $a(\zeta) = \sum_{j=-k}^k a_j\zeta^j$ and $b(\zeta) = \sum_{j=-\ell}^\ell b_j\zeta^j$ are trigonometric polynomials such that $a(\zeta) > 0$ and $b(\zeta) > 0$ when $|\zeta| = 1$.

Lemma 12.8. (Riesz-Fejér) *Let*

$$f(\zeta) = \sum_{j=-n}^n f_j\zeta^j \quad \textit{for} \quad |\zeta| = 1$$

be a trigonometric polynomial such that $|f(\zeta)| > 0$ for every point $\zeta \in \mathbb{C}$ with $|\zeta| = 1$ and $f_n \neq 0$. Then there exists a polynomial $\varphi_n(\zeta) = a(\zeta - \alpha_1)\cdots(\zeta - \alpha_n)$ such that

$$f(\zeta) = |\varphi_n(\zeta)|^2 \quad \textit{for} \quad |\zeta| = 1$$

and $|\alpha_j| > 1$ for $j = 1, \ldots, n$.

Proof. Under the given assumptions, it is readily checked that $f_{-j} = \overline{f_j}$ for $j = 0, \ldots, n$ and hence, that $f(\beta) = \overline{f(1/\overline{\beta})}$ for every point $\beta \in \mathbb{C} \setminus \{0\}$. Moreover, since $g(\zeta) = \zeta^n f(\zeta) = f_{-n} + f_{1-n}\zeta + \cdots + f_n\zeta^{2n}$ is a polynomial of degree $2n$ with $g(0) = f_{-n} \neq 0$,

$$g(\zeta) = a(\zeta - \beta_1)\cdots(\zeta - \beta_{2n})$$

for some choice of points $a, \beta_1, \ldots, \beta_{2n} \in \mathbb{C} \setminus \{0\}$. However, in view of the preceding discussion, these roots can be indexed so that $|\beta_j| > 1$ and $\beta_{j+n} = 1/\overline{\beta_j}$ for $j = 1, \ldots, n$. Therefore,

$$\begin{aligned} f(\zeta) &= \zeta^{-n} a \prod_{j=1}^n (\zeta - \beta_j)(\zeta - 1/\overline{\beta_j}) \\ &= (-1)^n a(\overline{\beta_1}\cdots\overline{\beta_n})^{-1} \prod_{j=1}^n (\zeta - \beta_j)(\overline{\zeta} - \overline{\beta_j}) \quad \text{if} \quad |\zeta| = 1 . \end{aligned}$$

The polynomial $\varphi_n(\zeta) = \sqrt{(-1)^n a(\overline{\beta_1}\cdots\overline{\beta_n})^{-1}}\prod_{j=1}^n(\zeta-\beta_j)$ meets the stated requirements of the lemma. □

12.6. Detour on block Toeplitz matrices

The interplay between the two Schur complements that was used to establish formula (12.10) is easily adapted to the more general setting of block Toeplitz matrices

$$(12.33)\qquad T_k = \begin{bmatrix} t_0 & \cdots & t_{-k} \\ \vdots & \ddots & \vdots \\ t_k & \cdots & t_0 \end{bmatrix} \text{ with blocks } t_i \in \mathbb{C}^{p\times p},\ i=0,\ldots,k,$$

and their inverses

$$(12.34)\ \Gamma_k = \begin{bmatrix} \gamma_{00}^{(k)} & \cdots & \gamma_{0k}^{(k)} \\ \vdots & & \vdots \\ \gamma_{k0}^{(k)} & \cdots & \gamma_{kk}^{(k)} \end{bmatrix} \text{ with blocks } \gamma_{ij}^{(k)} \in \mathbb{C}^{p\times p},\ i,j=0,\ldots,k,$$

when they exist.

Lemma 12.9. *Let T_k denote the block Toeplitz matrices defined by formula (12.33) and suppose that $n \geq 1$ and T_n is invertible. Let $\Gamma_n = T_n^{-1}$ be decomposed into blocks as in formula* (12.34). *Then the following are equivalent:*

(1) *T_n and $\gamma_{00}^{(n)}$ are invertible.*

(2) *T_n and T_{n-1} are invertible.*

(3) *T_n and $\gamma_{nn}^{(n)}$ are invertible.*

Proof. The proof is an easy consequence of the two Schur decompositions used in the proof of Theorem 12.7 except that now block Schur decompositions are used and the vectors $\mathbf{x}^H$ and $\mathbf{y}^H$ are replaced by the block rows $\begin{bmatrix}\gamma_{01}^{(n)} & \cdots & \gamma_{0n}^{(n)}\end{bmatrix}$ and $\begin{bmatrix}\gamma_{n0}^{(n)} & \cdots & \gamma_{n,n-1}^{(n)}\end{bmatrix}$, respectively. The details are left to the reader. □

Exercise 12.35. Show that in the setting of Lemma 12.9,

$$\det\, \gamma_{00}^{(n)} = \det\, \gamma_{nn}^{(n)}\,.$$

Theorem 12.10. *Let T_n and T_{n-1} be invertible block Toeplitz matrices of the form* (12.33) *and let $\Gamma_n = T_n^{-1}$. Then the matrix polynomials*

$$(12.35)\qquad P_n(\lambda) = \sum_{j=0}^n \lambda^j\gamma_{j0}^{(n)},\quad P_n^\circ(\lambda) = \sum_{j=0}^n \lambda^j\gamma_{0j}^{(n)}$$

$$(12.36)\qquad Q_n(\lambda) = \sum_{j=0}^n \lambda^j\gamma_{jn}^{(n)} \quad \textit{and} \quad Q_n^\circ(\lambda) = \sum_{j=0}^n \lambda^j\gamma_{nj}^{(n)}$$

are connected by the formula

$$\text{(12.37)} \quad \frac{P_n(\lambda)\{\gamma_{00}^{(n)}\}^{-1}P_n^\circ(\overline{\omega}) - \lambda\overline{\omega}Q_n(\lambda)\{\gamma_{nn}^{(n)}\}^{-1}Q_n^\circ(\overline{\omega})}{1-\lambda\overline{\omega}} = \sum_{i,j=0}^{n} \lambda^i \gamma_{ij}^{(n)} \overline{\omega}^j .$$

Proof. The proof is an almost exact paraphrase of the verification of formula (12.10), except that now block Schur decompositions are used and the vectors $\mathbf{x}^H$ and $\mathbf{y}^H$ are replaced by the block rows $\begin{bmatrix}\gamma_{01}^{(n)} & \cdots & \gamma_{0n}^{(n)}\end{bmatrix}$ and $\begin{bmatrix}\gamma_{n0}^{(n)} & \cdots & \gamma_{n,n-1}^{(n)}\end{bmatrix}$, respectively.

The well known Gohberg-Heinig formula for Γ_n can be obtained from formula (12.37) with the aid of the following evaluation:

Lemma 12.11. *Let*

$$X(\lambda) = \sum_{i=0}^{\ell} \lambda^i X_i \quad \textit{and} \quad Y(\lambda) = \sum_{i=0}^{\ell} \lambda^i Y_i$$

be matrix polynomials with coefficients $X_i \in \mathbb{C}^{p\times q}$ and $Y_i \in \mathbb{C}^{q\times p}$ for $i = 0, \ldots, \ell$ such that

$$\text{(12.38)} \qquad X(e^{i\theta})Y(e^{i\theta})^H = O_{p\times p} \quad \textit{for} \quad 0 \le \theta < 2\pi .$$

Then

$$\text{(12.39)} \quad \frac{X(\lambda)Y(\omega)^H}{1-\lambda\overline{\omega}}$$

$$= -\Psi(\lambda)\begin{bmatrix} X_\ell & X_{\ell-1} & \cdots & X_1 \\ O & X_\ell & \cdots & X_2 \\ \vdots & \ddots & \ddots & \vdots \\ O & \cdots & O & X_\ell \end{bmatrix}\begin{bmatrix} Y_\ell^H & O & \cdots & O \\ Y_{\ell-1}^H & Y_\ell^H & \ddots & \vdots \\ \vdots & & \ddots & O \\ Y_1^H & Y_2^H & \cdots & Y_\ell^H \end{bmatrix}\Psi(\omega)^H$$

$$= -\Psi(\lambda)\begin{bmatrix} X_1 & X_2 & \cdots & X_\ell \\ X_2 & X_3 & \cdots & O \\ \vdots & & & \vdots \\ X_\ell & O & \cdots & O \end{bmatrix}\begin{bmatrix} Y_1^H & Y_2^H & \cdots & Y_\ell^H \\ Y_2^H & Y_3^H & \cdots & O \\ \vdots & & & \vdots \\ Y_\ell^H & O & \cdots & O \end{bmatrix}\Psi(\omega)^H ,$$

where

$$\Psi(\lambda) = \Psi_{\ell-1}(\lambda) = \begin{bmatrix} I_p & \lambda I_p & \cdots & \lambda^{\ell-1} \end{bmatrix} .$$

Proof. The condition (12.38) is equivalent to the condition

$$X(\lambda)Y(1/\overline{\lambda})^H = O_{p\times p} \quad \text{for} \quad \lambda \in \mathbb{C} \setminus \{0\} .$$

Thus, if $\omega \neq 0$ and $\mu = 1/\overline{\omega}$, then

$$\begin{aligned}
\frac{X(\lambda)Y(\omega)^H}{1-\lambda\overline{\omega}} &= \frac{\{X(\lambda) - X(1/\overline{\omega})\}Y(\omega)^H}{1-\lambda\overline{\omega}} \\
&= -\mu\sum_{j=0}^{\ell}\frac{\lambda^j-\mu^j}{\lambda-\mu}X_jY(\omega)^H \\
&= -\mu\sum_{j=1}^{\ell}\frac{\lambda^j-\mu^j}{\lambda-\mu}X_jY(\omega)^H \\
&= -\mu\sum_{j=1}^{\ell}\sum_{i=0}^{j-1}\lambda^i\mu^{j-1-i}X_jY(\omega)^H \\
&= -\mu\sum_{s=0}^{\ell-1}\sum_{i=0}^{s}\lambda^i\mu^{s-i}X_{s+1}Y(\omega)^H \\
&= -\mu\sum_{i=0}^{\ell-1}\lambda^i\sum_{s=i}^{\ell-1}\mu^{s-i}X_{s+1}Y(\omega)^H \,,
\end{aligned}$$

which serves to identify the coefficient of λ^i in the first formula on the right-hand side of (12.39) with

$$-\mu\sum_{s=i}^{\ell-1}\mu^{s-i}X_{s+1}Y(\omega)^H = -\sum_{s=i}^{\ell-1}\sum_{t=0}^{\ell}\overline{\omega}^{t-s+i-1}X_{s+1}Y_t^H \,.$$

Moreover, since the right-hand side of the last formula is a polynomial in $\overline{\omega}$, it can be reexpressed as

$$\begin{aligned}
& -\left\{X_{i+1}\sum_{t=0}^{\ell}\overline{\omega}^{t-1}Y_t^H + X_{i+2}\sum_{t=0}^{\ell}\overline{\omega}^{t-2}Y_t^H + \cdots + X_\ell\sum_{t=0}^{\ell}\overline{\omega}^{t-\ell+i}Y_t^H\right\} \\
= & -\left\{X_{i+1}\sum_{t=1}^{\ell}\overline{\omega}^{t-1}Y_t^H + X_{i+2}\sum_{t=2}^{\ell}\overline{\omega}^{t-2}Y_t^H + \cdots + X_\ell\sum_{t=\ell-i}^{\ell}\overline{\omega}^{t-\ell+i}Y_t^H\right\} \\
= & -\left\{X_{i+1}\sum_{t=0}^{\ell-1}\overline{\omega}^{t}Y_{t+1}^H + X_{i+2}\sum_{t=0}^{\ell-2}\overline{\omega}^{t}Y_{t+2}^H + \cdots + X_\ell\sum_{t=0}^{i}\overline{\omega}^{t}Y_{t+\ell-i}^H\right\} .
\end{aligned}$$

Therefore, the coefficient of $\lambda^i\overline{\omega}^j$ in the first formula on the right-hand side of (12.39) is equal to

$$-\left\{X_{i+1}Y_{j+1}^H + X_{i+2}Y_{j+2}^H + \cdots + X_\ell Y_{\ell-i+j}^H\right\} \quad \text{if} \quad i \geq j$$

and to

$$-\left\{X_{i+1}Y_{j+1}^H + X_{i+2}Y_{j+2}^H + \cdots + X_{\ell-i+j}Y_\ell^H\right\} \quad \text{if} \quad i \leq j \,,$$

which yields the first formula in (12.39). The second formula follows easily from the first upon noting that

$$\begin{bmatrix} X_1 & X_2 & \cdots & X_\ell \\ X_2 & X_3 & \cdots & O \\ \vdots & & & \vdots \\ X_\ell & O & \cdots & O \end{bmatrix} = \begin{bmatrix} X_\ell & X_{\ell-1} & \cdots & X_1 \\ O & X_\ell & \cdots & X_2 \\ \vdots & & & \vdots \\ O & O & \cdots & X_\ell \end{bmatrix} \begin{bmatrix} O & \cdots & O & I_p \\ O & \cdots & I_p & O \\ \vdots & & & \vdots \\ I_p & \cdots & O & O \end{bmatrix}.$$

□

Theorem 12.12. (Gohberg-Heinig) *In the setting of Theorem 12.10,*

$$(12.40)\quad T_n^{-1} = \begin{bmatrix} \gamma_{nn}^{(n)} & \gamma_{n-1,n}^{(n)} & \cdots & \gamma_{0n}^{(n)} \\ O & \gamma_{nn}^{(n)} & \cdots & \gamma_{1n}^{(n)} \\ \vdots & & \ddots & \vdots \\ O & O & \cdots & \gamma_{nn}^{(n)} \end{bmatrix} D_n^{-1} \begin{bmatrix} \gamma_{nn}^{(n)} & O & \cdots & O \\ \gamma_{n,n-1}^{(n)} & \gamma_{nn}^{(n)} & \cdots & O \\ \vdots & & \ddots & \vdots \\ \gamma_{n0}^{(n)} & \gamma_{n1}^{(n)} & \cdots & \gamma_{nn}^{(n)} \end{bmatrix}$$

$$- \begin{bmatrix} O & \gamma_{n0}^{(n)} & \gamma_{n-1,0}^{(n)} & \cdots & \gamma_{10}^{(n)} \\ O & O & \gamma_{n0}^{(n)} & \cdots & \gamma_{20}^{(n)} \\ \vdots & & \cdots & \ddots & \vdots \\ O & O & O & \cdots & \gamma_{n0}^{(n)} \\ O & O & O & \cdots & O \end{bmatrix} D_0^{-1} \begin{bmatrix} O & O & \cdots & O \\ \gamma_{0n}^{(n)} & O & \cdots & O \\ \gamma_{0,n-1}^{(n)} & \gamma_{0n}^{(n)} & \cdots & O \\ \vdots & & \ddots & \vdots \\ \gamma_{01}^{(n)} & \gamma_{02}^{(n)} & \cdots & \gamma_{0n}^{(n)} \end{bmatrix},$$

where

$$D_0 = \operatorname{diag}\{\gamma_{00}^{(n)}, \ldots, \gamma_{00}^{(n)}\} \quad \textit{and} \quad \operatorname{diag}\{\gamma_{nn}^{(n)}, \ldots, \gamma_{nn}^{(n)}\}.$$

Proof. Let

$$X(\lambda) = \begin{bmatrix} P_n(\lambda)\{\gamma_{00}^{(n)}\}^{-1} & \lambda Q_n(\lambda)\{\gamma_{nn}^{(n)}\}^{-1} \end{bmatrix}$$

and

$$Y(\lambda) = \begin{bmatrix} P_n^\circ(\overline{\lambda})^H & -\lambda Q_n^\circ(\overline{\lambda})^H \end{bmatrix}.$$

Then

$$X_j = \begin{cases} \begin{bmatrix} \gamma_{j0}^{(n)}\{\gamma_{00}^{(n)}\}^{-1} & \gamma_{j-1,n}^{(n)}\{\gamma_{nn}^{(n)}\}^{-1} \end{bmatrix} & \text{for} \quad j = 1, \ldots, n \\ \begin{bmatrix} O & I_p \end{bmatrix} \quad \text{for} \quad j = n+1 \end{cases}$$

and

$$Y_j = \begin{cases} \begin{bmatrix} \{\gamma_{0j}^{(n)}\}^H & -\{\gamma_{n,j-1}^{(n)}\}^H \end{bmatrix} & \text{for} \quad j = 1, \ldots, n \\ \begin{bmatrix} O & -\{\gamma_{nn}^{(n)}\}^H \end{bmatrix} \quad \text{for} \quad j = n+1 \end{cases}$$

and the formula emerges from formula (12.39) upon making the requisite substitutions. □

12.7. A maximum entropy matrix completion problem

In this section we shall consider the problem of completing a matrix in $\mathbb{C}^{n\times n}$ that belongs to the class

$$\mathbb{C}_{\succ}^{n\times n} = \{A \in \mathbb{C}^{n\times n} : A \succ O\}$$

when only the entries in the $2m+1$ central diagonals of A, i.e., the entries a_{ij} with indices in the set

$$\Lambda_m = \{(i,j) : i,j = 1,\ldots,n \quad \text{and} \quad |i-j| \le m\}, \tag{12.41}$$

are given. It is tempting to set the unknown entries equal to zero. However, the matrix that is obtained this way is not necessarily positive definite; see Exercise 12.18 for a simple example. A remarkable fact is that there exists exactly one completion $\widetilde{A} \in \mathbb{C}_{\succ}^{n\times n}$ of the partially specified A such that $\mathbf{e}_i^T(\widetilde{A})^{-1}\mathbf{e}_j = 0$ for $(i,j) \notin \Lambda_m$. We shall sketch an algorithm for obtaining this particular completion that is based on factorization and shall show that $\widetilde{A}$ can also be characterized as the completion which maximizes the determinant. Because of this property $\widetilde{A}$ is commonly referred to as the **maximum entropy completion**.

Theorem 12.13. *Let m be an integer such that $0 \le m \le n-1$ and let*

$$\{b_{ij} : (i,j) \in \Lambda_m\}$$

be a given set of complex numbers. Then there exists a matrix $A \in \mathbb{C}_{\succ}^{n\times n}$ such that

$$a_{ij} = b_{ij} \quad \textit{for} \quad (i,j) \in \Lambda_m \tag{12.42}$$

if and only if

$$\begin{bmatrix} b_{jj} & \cdots & b_{j,j+m} \\ \vdots & & \vdots \\ b_{j+m,j} & \cdots & b_{j+m,j+m} \end{bmatrix} \succ O \quad \textit{for} \quad j = 1,\ldots,n-m. \tag{12.43}$$

Discussion. The proof of necessity is easy and is left to the reader as an exercise. The verification of sufficiency is by construction and is most easily understood by example. To this end, let $n=5$ and $m=2$, and let $X \in \mathbb{C}^{5\times 5}$ denote the lower triangular matrix with entries x_{ij} that are set equal to zero for $i > j+2$ (i.e., $x_{41} = x_{51} = x_{52} = 0$) and are determined by the following equations when $j \le i \le j+2$:

$$\begin{bmatrix} b_{11} & b_{12} & b_{13} \\ b_{21} & b_{22} & b_{23} \\ b_{31} & b_{32} & b_{33} \end{bmatrix} \begin{bmatrix} x_{11} \\ x_{21} \\ x_{31} \end{bmatrix} = \begin{bmatrix} 1 \\ 0 \\ 0 \end{bmatrix}, \quad \begin{bmatrix} b_{22} & b_{23} & b_{24} \\ b_{32} & b_{33} & b_{34} \\ b_{42} & b_{43} & b_{44} \end{bmatrix} \begin{bmatrix} x_{22} \\ x_{32} \\ x_{42} \end{bmatrix} = \begin{bmatrix} 1 \\ 0 \\ 0 \end{bmatrix}, \tag{12.44}$$

(12.45)

$$\begin{bmatrix} b_{33} & b_{34} & b_{35} \\ b_{43} & b_{44} & b_{45} \\ b_{53} & b_{54} & b_{55} \end{bmatrix} \begin{bmatrix} x_{33} \\ x_{43} \\ x_{53} \end{bmatrix} = \begin{bmatrix} 1 \\ 0 \\ 0 \end{bmatrix}, \quad \begin{bmatrix} b_{44} & b_{45} \\ b_{54} & b_{55} \end{bmatrix} \begin{bmatrix} x_{44} \\ x_{54} \end{bmatrix} = \begin{bmatrix} 1 \\ 0 \end{bmatrix}, \quad b_{55}x_{55} = 1 .$$

Next, let $D = \operatorname{diag}\{x_{11}, \ldots, x_{55}\}$. Since $x_{jj} > 0$ for $j = 1, \ldots, 5$, $D \succ O$ and the matrix $L = XD^{-1}$ is lower triangular with ones on the diagonal. Now set

$$A = (L^H)^{-1} D^{-1} L^{-1} . \tag{12.46}$$

Then clearly $A \in \mathbb{C}_{\succ}^{5\times 5}$ and equations (12.44) and (12.45) are in force, but with a_{ij} in place of b_{ij}. Therefore, the numbers $c_{ij} = b_{ij} - a_{ij}$ are solutions of the equations

$$\begin{bmatrix} c_{11} & c_{12} & c_{13} \\ c_{21} & c_{22} & c_{23} \\ c_{31} & c_{32} & c_{33} \end{bmatrix} \begin{bmatrix} x_{11} \\ x_{21} \\ x_{31} \end{bmatrix} = \begin{bmatrix} 0 \\ 0 \\ 0 \end{bmatrix}, \quad \begin{bmatrix} c_{22} & c_{23} & c_{24} \\ c_{32} & c_{33} & c_{34} \\ c_{42} & c_{43} & c_{44} \end{bmatrix} \begin{bmatrix} x_{22} \\ x_{32} \\ x_{42} \end{bmatrix} = \begin{bmatrix} 0 \\ 0 \\ 0 \end{bmatrix}, \tag{12.47}$$

(12.48)

$$\begin{bmatrix} c_{33} & c_{34} & c_{35} \\ c_{43} & c_{44} & c_{45} \\ c_{53} & c_{54} & c_{55} \end{bmatrix} \begin{bmatrix} x_{33} \\ x_{43} \\ x_{53} \end{bmatrix} = \begin{bmatrix} 0 \\ 0 \\ 0 \end{bmatrix}, \quad \begin{bmatrix} c_{44} & c_{45} \\ c_{54} & c_{55} \end{bmatrix} \begin{bmatrix} x_{44} \\ x_{54} \end{bmatrix} = \begin{bmatrix} 0 \\ 0 \end{bmatrix}, \quad c_{55}x_{55} = 0 .$$

But, since the x_{jj} are positive and each of the five submatrices are Hermitian, it is readily seen that $c_{ij} = 0$ for all the indicated entries; i.e., $a_{ij} = b_{ij}$ for $|i - j| \le 2$. Thus, the matrix A constructed above is a positive definite completion. □

At first glance it might seem that the missing entries in the partially specified matrix should be set equal to zero. However, as we have already noted, Exercise 12.18 shows that the matrix that arises this way is not necessarily positive definite over $\mathbb{C}^n$.

The matrix A that is constructed in Theorem 12.13 inherits special properties from the construction.

Lemma 12.14. *If $A = XX^H$ and $X \in \mathbb{C}^{n\times n}$ is a lower triangular invertible matrix, then*

$$a_{ij} = 0 \quad \textit{for} \quad i - j \ge k \iff x_{ij} = 0 \quad \textit{for} \quad i - j \ge k . \tag{12.49}$$

Discussion. The verification of (12.49) becomes transparent if the calculations are organized properly. The underlying ideas are best conveyed by example. Let $A \in \mathbb{C}^{7\times 7}$ and suppose that $k = 3$. Then the entries a_{ij} in A that meet the constraint $i - j \ge k$ with $k = 3$ can be expressed in terms of the corresponding entries x_{ij} in the lower triangular matrix X by means of the formulas

$$
\begin{bmatrix} a_{41} \\ a_{51} \\ a_{61} \\ a_{71} \end{bmatrix} = \begin{bmatrix} x_{41} \\ x_{51} \\ x_{61} \\ x_{71} \end{bmatrix} \overline{x_{11}}, \quad \begin{bmatrix} a_{52} \\ a_{62} \\ a_{72} \end{bmatrix} = \begin{bmatrix} x_{51} & x_{52} \\ x_{61} & x_{62} \\ x_{71} & x_{72} \end{bmatrix} \begin{bmatrix} \overline{x_{21}} \\ \overline{x_{22}} \end{bmatrix},
$$

(12.50)

$$
\begin{bmatrix} a_{63} \\ a_{73} \end{bmatrix} = \begin{bmatrix} x_{61} & x_{62} & x_{63} \\ x_{71} & x_{72} & x_{73} \end{bmatrix} \begin{bmatrix} \overline{x_{31}} \\ \overline{x_{32}} \\ \overline{x_{33}} \end{bmatrix} \quad \text{and} \quad a_{74} = \sum_{j=1}^{4} x_{7j}\overline{x_{4j}}\,.
$$

Thus, as the diagonal entries of X are all nonzero, assertion (12.71) is easily verified for this special case. The general case may be established in just the same way.

There is a companion result, which we state without proof:

Lemma 12.15. *If $A = YY^H$ and $Y \in \mathbb{C}^{n\times n}$ is an upper triangular invertible matrix, then*

$$
a_{ij} = 0 \quad \text{for} \quad i - j \le -k \Longleftrightarrow y_{ij} = 0 \quad \text{for} \quad i - j \le -k\,. \tag{12.51}
$$

Exercise 12.36. Verify Lemma 12.15 if $n = 7$ and $k = 3$.

Theorem 12.16. *Let m be an integer such that $0 \le m \le n-1$ and let*

$$
\{b_{ij} : (i,j) \in \Lambda_m\}
$$

be a given set of complex numbers such that the conditions (12.43) are in force. Then there exists exactly one matrix $A \in \mathbb{C}_{\succ}^{n\times n}$ such that

$$
a_{ij} = b_{ij} \quad \text{for} \quad (i,j) \in \Lambda_m \tag{12.52}
$$

and

$$
\mathbf{e}_i^T A^{-1} \mathbf{e}_j = 0 \quad \text{for} \quad (i,j) \notin \Lambda_m\,. \tag{12.53}
$$

Proof. In view of Lemma 12.14, the matrix $A = (L^H)^{-1}D^{-1}L^{-1}$ that was constructed in the discussion of Theorem 12.13 meets both of the stated conditions. Suppose next that $A_1 \in \mathbb{C}_{\succ}^{n\times n}$ is a second matrix that meets the conditions (12.52) and (12.53). Then, by Theorem 12.4, A_1 admits a factorization of the form

$$
A_1^{-1} = L_1 D_1 L_1^H
$$

for some lower triangular matrix $L_1 \in \mathbb{C}^{n\times n}$ with ones on the diagonal and some diagonal matrix $D_1 \in \mathbb{C}_{\succ}^{n\times n}$. Moreover, by Lemma 12.14 the entries z_{ij} in the lower triangular matrix $Z = L_1 D$ are equal to zero for $i > j + m$. Consequently, the entries z_{ij} with $j \le i \le j+m$ are determined by the same equations as the x_{ij} for $j \le i \le j+m$, i.e., by equations (12.44) and (12.45) if $n = 5$ and $m = 2$, or, in general, by the equations

$$B_{[j,j+m]} \begin{bmatrix} x_{jj} \\ \vdots \\ x_{j+m,j} \end{bmatrix} = \begin{bmatrix} 1 \\ 0 \\ \vdots \\ 0 \end{bmatrix} \quad \text{for} \quad j = 1, \dots, n-m \tag{12.54}$$

and

$$B_{[j,n]} \begin{bmatrix} x_{jj} \\ \vdots \\ x_{n,j} \end{bmatrix} = \begin{bmatrix} 1 \\ 0 \\ \vdots \end{bmatrix} \quad \text{for} \quad j = n-m+1, \dots, n\,. \tag{12.55}$$

Thus, $z_{ij} = x_{ij}$ for $i, j = 1, \dots, n$ and hence $A_1 = A$; i.e., the proof of uniqueness is complete. □

Theorem 12.17. *Let m be an integer such that $0 \le m \le n-1$ and let*

$$\{b_{ij} : (i,j) \in \Lambda_m\}$$

be a given set of complex numbers such that the conditions (12.43) *are in force. Let $A \in \mathbb{C}_{\succ}^{n\times n}$ meet conditions* (12.52) *and* (12.53) *and let $C \in \mathbb{C}_{\succ}^{n\times n}$ meet condition* (12.52)*. Then*

(1) $\det A \ge \det C$, *with equality if and only if $A = C$.*
(2) *If $A = L_A D_A L_A^H$ and $C = L_C D_C L_C^H$, in which L_A and L_C are lower triangular with ones on the diagonal and D_A and D_C are $n \times n$ diagonal matrices, then $D_A \succeq D_C$, with equality if and only if $A = C$.*

Proof. In view of Theorem 12.4,

$$A = (X^H)^{-1} D X^{-1} \quad \text{and} \quad C = (Y^H)^{-1} G Y^{-1}\,,$$

where $X \in \mathbb{C}^{n\times n}$ and $Y \in \mathbb{C}^{n\times n}$ are lower triangular matrices with ones on the diagonal, $D \in \mathbb{C}_{\succ}^{n\times n}$ and $G \in \mathbb{C}_{\succ}^{n\times n}$ are diagonal matrices and $x_{ij} = 0$ for $i \ge m+j$. Therefore, the formulas

$$C = A + (C - A) \quad \text{and} \quad Z = Y^{-1}X$$

imply that

$$Z^H G Z = D + X^H (C - A) X\,.$$

Thus, as Z is lower triangular with ones on the diagonal and the diagonal entries of $X^H(C-A)X$ are all equal to zero,

$$d_{jj} = \sum_{s=j}^{n} g_{ss} |z_{sj}|^2 = g_{jj} + \sum_{s>j} g_{ss}|z_{sj}|^2 \ge g_{jj}$$

with strict inequality unless $z_{sj} = 0$ for $s > j$, i.e., unless $Z = I_n$. This completes the proof of (1). Much the same argument serves to justify (2). □

Remark 12.18. Theorem 12.16 can also be expressed in terms of the orthogonal projection P_{Λ_m} that is defined by the formula

$$P_{\Lambda_m} A = \sum_{(i,j)\in\Lambda_m} \langle A, \mathbf{e}_i\mathbf{e}_j^T\rangle \mathbf{e}_i\mathbf{e}_j^T$$

on the inner product space $\mathbb{C}^{n\times n}$ with inner product $\langle A, B\rangle = \text{trace}\,\{B^H A\}$: If the conditions of Theorem 12.16 are met and if $Q \in \mathbb{C}^{n\times n}$ with $q_{ij} = b_{ij}$ for $(i,j) \in \Lambda_m$, then there exists exactly one matrix $A \in \mathbb{C}_{\succ}^{n\times n}$ such that

$$P_{\Lambda_m} A = Q \quad \text{and} \quad (I_n - P_{\Lambda_m})A^{-1} = O\,.$$

This formulation suggests that results analogous to those discussed above can be obtained in other algebras, which is indeed the case.

12.8. Schur complements for semidefinite matrices

In this section we shall show that if $E \succeq O$, then analogues of the Schur complement formulas hold even if neither of the block diagonal entries are invertible. (Similar formulas hold if $E \preceq O$.)

Lemma 12.19. *Let $A \in \mathbb{C}^{p\times p}$, $D \in \mathbb{C}^{q\times q}$, $n = p + q$, and let*

$$E = \begin{bmatrix} A & B \\ B^H & D \end{bmatrix}$$

be positive semidefinite over $\mathbb{C}^n$. Then:

1. $\mathcal{N}_A \subseteq \mathcal{N}_{B^H}$ *and* $\mathcal{N}_D \subseteq \mathcal{N}_B$.
2. $\mathcal{R}_B \subseteq \mathcal{R}_A$ *and* $\mathcal{R}_{B^H} \subseteq \mathcal{R}_D$.
3. $AA^\dagger B = B$ *and* $DD^\dagger B^H = B^H$.
4. *The matrix E admits the (lower-upper) factorization*

$$E = \begin{bmatrix} I_p & O \\ B^H A^\dagger & I_q \end{bmatrix} \begin{bmatrix} A & O \\ O & D - B^H A^\dagger B \end{bmatrix} \begin{bmatrix} I_p & A^\dagger B \\ O & I_q \end{bmatrix}, \tag{12.56}$$

where $A^\dagger$ denotes the Moore-Penrose inverse of A.

5. *The matrix E admits the (upper-lower) factorization*

$$E = \begin{bmatrix} I_p & BD^\dagger \\ O & I_q \end{bmatrix} \begin{bmatrix} A - BD^\dagger B^H & O \\ O & D \end{bmatrix} \begin{bmatrix} I_p & O \\ D^\dagger B^H & I_q \end{bmatrix}, \tag{12.57}$$

where $D^\dagger$ denotes the Moore-Penrose inverse of D.

Proof. Since E is presumed to be positive semidefinite, the inequality

$$\mathbf{x}^H(A\mathbf{x} + B\mathbf{y}) + \mathbf{y}^H(B^H\mathbf{x} + D\mathbf{y}) \geq 0$$

must be in force for every choice of $\mathbf{x} \in \mathbb{C}^p$ and $\mathbf{y} \in \mathbb{C}^q$. If, in particular, $\mathbf{x} \in \mathcal{N}_A$, then this reduces to

$$\mathbf{x}^H B\mathbf{y} + \mathbf{y}^H(B^H\mathbf{x} + D\mathbf{y}) \geq 0$$

for every choice of $\mathbf{y} \in \mathbb{C}^q$ and hence, upon replacing $\mathbf{y}$ by $\varepsilon\mathbf{y}$, to

$$\varepsilon\mathbf{x}^H B\mathbf{y} + \varepsilon\mathbf{y}^H B^H\mathbf{x} + \varepsilon^2\mathbf{y}^H D\mathbf{y} \geq 0$$

for every choice of $\varepsilon > 0$ as well. Consequently, upon dividing through by ε and then letting $\varepsilon \downarrow 0$, it follows that

$$\mathbf{x}^H B\mathbf{y} + \mathbf{y}^H B^H\mathbf{x} \geq 0$$

for every choice of $\mathbf{y} \in \mathbb{C}^q$. But if $\mathbf{y} = -B^H\mathbf{x}$, then the last inequality implies that

$$-2\|B^H\mathbf{x}\|^2 = -\mathbf{x}^H BB^H\mathbf{x} - \mathbf{x}^H BB^H\mathbf{x} \geq 0\,.$$

Therefore,

$$B^H\mathbf{x} = \mathbf{0}\,,$$

which serves to complete the proof of the first statement in (1) and, since the orthogonal complements of the indicated sets satisfy the opposite inclusion, implies that

$$\mathcal{R}_{A^H} = (\mathcal{N}_A)^\perp \supseteq (\mathcal{N}_{B^H})^\perp = \mathcal{R}_B\,.$$

Since $A = A^H$, this verifies the first assertion in (2); the proofs of the second assertions in (1) and (2) are similar. The fourth assertion is a straightforward consequence of the formula $AA^\dagger A = A$ and the fact that

$$A = A^H \Longrightarrow (A^\dagger)^H = A^\dagger\,.$$

Items (3) and (5) are left to the reader. □

Exercise 12.37. Verify items (3) and (5) of Lemma 12.19.

Theorem 12.20. *If $A \in \mathbb{C}^{n\times n}$ and $A \succeq O$, then A admits factorizations of the form*

$$A = LD_1L^H \quad \text{and} \quad A = UD_2U^H\,, \tag{12.58}$$

where L is lower triangular with ones on the diagonal, U is upper triangular with ones on the diagonal, and D_1 and D_2 are $n \times n$ diagonal matrices with nonnegative entries.

Proof. Since $A_{[1,k]} \succeq O$ for $k = 1, \ldots, n$, formula (12.56) implies that

$$A_{[1,k]} = \widetilde{L}_k \begin{bmatrix} A_{[1,k-1]} & O \\ O & \alpha_k \end{bmatrix} \widetilde{L}_k^H \quad \text{for} \quad k = 2, \ldots, n\,,$$

where $\widetilde{L}_k$ is a $k \times k$ lower triangular matrix with ones on the diagonal and $\alpha_k \geq 0$:

$$A = \widetilde{L}_n \begin{bmatrix} A_{[1,n-1]} & O \\ O & \alpha_n \end{bmatrix} \widetilde{L}_n^H\,, \ldots, A_{[1,2]} = \widetilde{L}_2 \begin{bmatrix} A_{[1,1]} & O \\ O & \alpha_2 \end{bmatrix} \widetilde{L}_2^H\,.$$

The first formula in (12.58) is obtained by setting $L_k = \operatorname{diag}\{\widetilde{L}_k, I_{n-k}\}$ and writing

$$A = A_{[1,n]} = L_n L_{n-1} \cdots L_2 \operatorname{diag}\{a_{11}, \alpha_2, \ldots, \alpha_n\} L_2^H \cdots L_{n-1}^H L_n^H .$$

The second formula in (12.57) is verified on the basis of (12.58) in much the same way. □

Exercise 12.38. Let

$$A = \begin{bmatrix} a & b \\ b & c \end{bmatrix} \quad \text{and} \quad B = \begin{bmatrix} c & b \\ b & a \end{bmatrix} .$$

Show that if $a > b > c > 0$ and $ac > b^2$, then:

(1) The matrices A and B are both positive definite over $\mathbb{C}^2$.

(2) The matrix AB is not positive definite over $\mathbb{C}^2$.

(3) The matrix $AB + BA$ is not positive definite over $\mathbb{C}^2$.

Exercise 12.39. Show that the matrix AB considered in Exercise 12.38 is not positive definite over $\mathbb{R}^2$.

Exercise 12.40. Let $A \in \mathbb{C}^{n\times n}$ and $B \in \mathbb{C}^{n\times n}$ both be positive semidefinite over $\mathbb{C}^n$. Show that $A^2B^2 + B^2A^2$ need not be positive semidefinite over $\mathbb{C}^n$. [HINT: See Exercise 12.38.]

Exercise 12.41. Let $A \in \mathbb{C}^{n\times n}$ be expressed in block form as

$$A = \begin{bmatrix} A_{11} & A_{12} \\ A_{21} & A_{22} \end{bmatrix}$$

with square blocks A_{11} and A_{22} and suppose that $A \succeq O$. Show that:

(1) There exists a matrix $K \in \mathbb{C}^{p\times q}$ such that $A_{12} = A_{11}K$.

(2) $A = \begin{bmatrix} I_p & O \\ K^H & I_q \end{bmatrix} \begin{bmatrix} A_{11} & O \\ O & A_{22} - K^H A_{11} K \end{bmatrix} \begin{bmatrix} I_p & K \\ O & I_q \end{bmatrix}$.

(3) $K^H A_{11} K = A_{21} A_{11}^\dagger A_{12}$.

Exercise 12.42. Show that in the setting of Exercise 12.41

(1) There exists a matrix $K \in \mathbb{C}^{q\times p}$ such that $A_{21} = A_{22}K$.

(2) $A = \begin{bmatrix} I_p & K^H \\ O & I_q \end{bmatrix} \begin{bmatrix} A_{11} - KA_{22}K^H & O \\ O & A_{22} \end{bmatrix} \begin{bmatrix} I_p & O \\ K & I_q \end{bmatrix}$.

(3) $KA_{22}K^H = A_{12}A_{22}^\dagger A_{21}$.

Exercise 12.43. Let $A = BB^H$, where $B \in \mathbb{C}^{n\times k}$ and $\operatorname{rank} B = k$; let $\mathbf{u}_1, \ldots, \mathbf{u}_k$ be an orthonormal basis for $\mathcal{R}_{B^H}$; and let $A_\ell = \sum_{j=1}^{\ell} B\mathbf{u}_j\mathbf{u}_j^H B^H$ for $\ell = 1, \ldots, k$. Show that $A = A_k$ and that $A - A_\ell$ is a positive semidefinite matrix of rank $k - \ell$ for $\ell = 1, \ldots, k-1$.

Exercise 12.44. Show that if $A \in \mathbb{C}^{n\times n}$, then

$$(12.59)\quad A \succeq O \Longrightarrow A = A^H \quad \text{and} \quad \det A_{[1,k]} \geq 0 \quad \text{for} \quad k = 1, \ldots, n\,,$$

but the converse implication is false.

Exercise 12.45. Let $A \in \mathbb{C}^{n\times n}$. Show that

$$(12.60)\qquad A = A^H \quad \text{and} \quad \sigma(A) \subset [0, \infty) \Longleftrightarrow A \succeq O\,.$$

12.9. Square roots

Theorem 12.21. *If $A \in \mathbb{C}^{n\times n}$ and $A \succeq O$, then there is exactly one matrix $B \in \mathbb{C}^{n\times n}$ such that $B \succeq O$ and $B^2 = A$.*

Proof. If $A \in \mathbb{C}^{n\times n}$ and $A \succeq O$, then there exists a unitary matrix U and a diagonal matrix

$$D = \operatorname{diag}\{d_{11}, \ldots, d_{nn}\}$$

with nonnegative entries such that $A = UDU^H$. Therefore, upon setting

$$D^{1/2} = \operatorname{diag}\{d_{11}^{1/2}, \ldots, d_{nn}^{1/2}\}\,,$$

it is readily checked that the matrix $B = UD^{1/2}U^H$ is again positive semidefinite and

$$B^2 = (UD^{1/2}U^H)UD^{1/2}U^H) = UDU^H = A\,.$$

This completes the proof of the existence of at least one positive semidefinite square root of A.

Suppose next that there are two positive semidefinite square roots of A, say B_1 and B_2. Then, since B_1 and B_2 are both positive semidefinite over $\mathbb{C}^n$ and hence Hermitian, there exist a pair of unitary matrices U_1 and U_2 and a pair of diagonal matrices $D_1 \succeq O$ and $D_2 \succeq O$ such that

$$B_1 = U_1D_1U_1^H \quad \text{and} \quad B_2 = U_2D_2U_2^H\,.$$

Thus, as

$$U_1D_1^2U_1^H = B_1^2 = A = B_2^2 = U_2D_2^2U_2^H\,,$$

it follows that

$$U_2^HU_1D_1^2 = D_2^2U_2^HU_1$$

and hence that

$$(U_2^HU_1D_1 - D_2U_2^HU_1)D_1 + D_2(U_2^HU_1D_1 - D_2U_2^HU_1) = O\,.$$

But this in turn implies that the matrix

$$X = U_2^HU_1D_1 - D_2U_2^HU_1$$

is a solution of the equation

$$XD_1 + D_2X = O.$$

The next step is to show that $X = O$ is the only solution of this equation. Upon writing

$$D_1 = \text{diag}\,\{d_{11}^{(1)}, \dots, d_{nn}^{(1)}\} \text{ and } D_2 = \text{diag}\,\{d_{11}^{(2)}, \dots, d_{nn}^{(2)}\},$$

one can readily check that x_{ij}, the ij entry of the matrix X, is a solution of the equation

$$x_{ij} d_{jj}^{(1)} + d_{ii}^{(2)} x_{ij} = 0.$$

Thus, if $d_{jj}^{(1)} + d_{ii}^{(2)} > 0$, then $x_{ij} = 0$. On the other hand, if $d_{jj}^{(1)} + d_{ii}^{(2)} = 0$, then $d_{jj}^{(1)} = d_{ii}^{(2)} = 0$ and, as follows from the definition of X, $x_{ij} = 0$ in this case too. Consequently,

$$U_2^H U_1 D_1 - D_2 U_2^H U_1 = X = O;$$

i.e.,

$$B_1 = U_1 D_1 U_1^H = U_2 D_2 U_2^H = B_2,$$

as claimed. □

If $A \succeq O$, the symbol $A^{1/2}$ will be used to denote the unique $n \times n$ matrix $B \succeq O$ with $B^2 = A$. Correspondingly, B will be referred to as the **square root** of A. The restriction that $B \succeq O$ is essential to insure uniqueness. Thus, for example, if A is Hermitian, then the formula

$$\begin{bmatrix} A & O \\ C & -A \end{bmatrix} \begin{bmatrix} A & O \\ C & -A \end{bmatrix} = \begin{bmatrix} A^2 & O \\ CA - AC & A^2 \end{bmatrix}$$

exhibits the matrix

$$\begin{bmatrix} A & O \\ C & -A \end{bmatrix} \quad \text{as a square root of} \quad \begin{bmatrix} A^2 & O \\ O & A^2 \end{bmatrix}$$

for every choice of C that commutes with A. In particular,

$$\begin{bmatrix} I_k & O \\ C & -I_k \end{bmatrix} \begin{bmatrix} I_k & O \\ C & -I_k \end{bmatrix} = \begin{bmatrix} I_k & O \\ O & I_k \end{bmatrix} \quad \text{for every} \quad C \in \mathbb{C}^{k \times k}.$$

Exercise 12.46. Show that if $A, B \in \mathbb{C}^{n \times n}$ and if $A \succ O$ and $B = B^H$, then there exists a matrix $V \in \mathbb{C}^{n \times n}$ such that

$$V^H A V = I_n \quad \text{and} \quad V^H B V = D = \text{diag}\,\{\lambda_1, \dots, \lambda_n\}.$$

[HINT: Reexpress the problem in terms of $U = A^{1/2} V$.]

Exercise 12.47. Show that if A, $B \in \mathbb{C}^{n \times n}$ and $A \succeq B \succ O$, then $B^{-1} \succeq A^{-1} \succ O$. [HINT: $A - B \succ O \Longrightarrow A^{-1/2} B A^{-1/2} \prec I_n$.]

Exercise 12.48. Show that if $A, B \in \mathbb{C}^{n \times n}$ and if $A \succeq O$ and $B \succeq O$, then trace $AB \geq 0$ (even if $AB \not\succeq O$).

12.10. Polar forms

If $A \in \mathbb{C}^{p\times q}$ and $r = \operatorname{rank} A \geq 1$, then the formula $A = V_1 D U_1^H$ that was obtained in Corollary 10.2 on the basis of the singular value decomposition of A can be reexpressed in **polar form**:

$$A = V_1 U_1^H (U_1 D U_1^H) \quad \text{and} \quad A = (V_1 D V_1^H) V_1 U_1^H\,, \tag{12.61}$$

where $V_1 U_1^H$ maps $\mathcal{R}_{A^H}$ isometrically onto $\mathcal{R}_A$, $U_1 D U_1^H = \{A^H A\}^{1/2}$ is positive definite on $\mathcal{R}_{A^H}$ and $V_1 D V_1^H = \{AA^H\}^{1/2}$ is positive definite on $\mathcal{R}_A$. These formulas are matrix analogues of the polar decomposition of a complex number.

Theorem 12.22. *Let $A \in \mathbb{C}^{p\times q}$. Then*

(1) $\operatorname{rank} A = q$ *if and only if A admits a factorization of the form $A = V_1 P_1$, where $V_1 \in \mathbb{C}^{p\times q}$ is* **isometric**; *i.e., $V_1^H V_1 = I_q$, and $P_1 \in \mathbb{C}^{q\times q}$ is positive definite over $\mathbb{C}^q$.*

(2) $\operatorname{rank} A = p$ *if and only if A admits a factorization of the form $A = P_2 U_2$, where $U_2 \in \mathbb{C}^{p\times q}$ is* **coisometric**; *i.e., $U_2 U_2^H = I_p$, and $P_2 \in \mathbb{C}^{p\times p}$ is positive definite over $\mathbb{C}^p$.*

Proof. If $\operatorname{rank} A = q$, then $p \geq q$ and, by Theorem 10.1, A admits a factorization of the form

$$A = V \begin{bmatrix} D \\ O \end{bmatrix} U = V \begin{bmatrix} I_q \\ O \end{bmatrix} DU\,,$$

where V and U are unitary matrices of sizes $p \times p$ and $q \times q$, respectively, and $D \in \mathbb{C}^{q\times q}$ is positive definite over $\mathbb{C}^q$. But this yields a factorization of the asserted form with $V_1 = V \begin{bmatrix} I_q \\ O \end{bmatrix} U$ and $P_1 = U^H D U$. Conversely, if A admits a factorization of this form, it is easily seen that $\operatorname{rank} A = q$. The details are left to the reader.

Assertion (2) may be established in much the same way or by invoking (1) and passing to transposes. The details are left to the reader. □

Exercise 12.49. Complete the proof of assertion (1) in Theorem 12.22.

Exercise 12.50. Verify assertion (2) in Theorem 12.22.

Exercise 12.51. Show that if $UU^H = VV^H$ for a pair of matrices $U, V \in \mathbb{C}^{n\times d}$ with $\operatorname{rank} U = \operatorname{rank} V = d$, then $U = VK$ for some unitary matrix $K \in \mathbb{C}^{d\times d}$.

Exercise 12.52. Find an isometric matrix V_1 and a matrix $P_1 \succ O$ such that $\begin{bmatrix} 1 & 0 \\ 1 & 1 \\ 0 & 1 \end{bmatrix} = V_1 P_1$.

12.11. Matrix inequalities

Lemma 12.23. *If $F \in \mathbb{C}^{p\times q}$, $G \in \mathbb{C}^{r\times q}$ and $F^HF - G^HG \succeq O$, then there exists exactly one matrix $K \in \mathbb{C}^{r\times p}$ such that*

$$G = KF \quad \text{and} \quad K\mathbf{u} = \mathbf{0} \quad \text{for every} \quad \mathbf{u} \in \mathcal{N}_{F^H}\,. \tag{12.62}$$

Moreover, this matrix K is contractive: $\|K\| \leq 1$.

Proof. The given conditions imply that

$$\langle F^HF\mathbf{x}, \mathbf{x}\rangle \geq \langle G^HG\mathbf{x}, \mathbf{x}\rangle \quad \text{for every} \quad \mathbf{x} \in \mathbb{C}^q\,.$$

Thus,

$$F\mathbf{x} = \mathbf{0} \Longrightarrow \|G\mathbf{x}\| = 0 \Longrightarrow G\mathbf{x} = \mathbf{0}\,;$$

i.e., $\mathcal{N}_F \subseteq \mathcal{N}_G$ and hence $\mathcal{R}_{G^H} \subseteq \mathcal{R}_{F^H}$. Therefore, there exists a matrix $K_1^H \in \mathbb{C}^{p\times r}$ such that $G^H = F^HK_1^H$.

If $\mathcal{N}_{F^H} = \{\mathbf{0}\}$, then the matrix $K = K_1$ meets both of the conditions in (12.62).

If $\mathcal{N}_{F^H} \neq \{\mathbf{0}\}$ and $V \in \mathbb{C}^{p\times \ell}$ is a matrix whose columns form a basis for $\mathcal{N}_{F^H}$, then

$$F^H(K_1^H + VL) = F^HK_1^H = G^H$$

for every choice of $L \in \mathbb{C}^{\ell\times r}$. Moreover,

$$\begin{aligned}(K_1 + L^HV^H)V = O \quad &\Longleftrightarrow \quad L^H = -K_1V(V^HV)^{-1} \\ &\Longleftrightarrow \quad K_1 + L^HV^H = K_1(I_p - V(V^HV)^{-1}V^H)\,.\end{aligned}$$

Thus, the matrix $K = K_1(I_p - V(V^HV)^{-1}V)$ meets the two conditions stated in (12.62). This is eminently reasonable, since $I_p - V(V^HV)^{-1}V^H$ is the formula for the orthogonal projection of $\mathbb{C}^p$ onto $\mathcal{R}_F$.

It is readily checked that if $\widetilde{K} \in \mathbb{C}^{r\times p}$ is a second matrix that meets the two conditions in (12.62), then $\widetilde{K} = K$. The details are left to the reader.

It remains to check that K is contractive. Since

$$\mathbb{C}^p = \mathcal{R}_F \oplus \mathcal{N}_{F^H}\,,$$

every vector $\mathbf{u} \in \mathbb{C}^p$ can be expressed as $\mathbf{u} = F\mathbf{x} + V\mathbf{y}$ for some choice of $\mathbf{x} \in \mathbb{C}^q$ and $\mathbf{y} \in \mathbb{C}^\ell$. Correspondingly,

$$\begin{aligned}\langle K\mathbf{u}, K\mathbf{u}\rangle &= \langle K(F\mathbf{x} + V\mathbf{y}), K(F\mathbf{x} + V\mathbf{y})\rangle = \langle KF\mathbf{x}, KF\mathbf{x}\rangle \\ &= \langle G\mathbf{x}, G\mathbf{x}\rangle \leq \langle F\mathbf{x}, F\mathbf{x}\rangle \\ &\leq \langle F\mathbf{x}, F\mathbf{x}\rangle + \langle V\mathbf{y}, V\mathbf{y}\rangle = \langle \mathbf{u}, \mathbf{u}\rangle\,.\end{aligned}$$

□

Exercise 12.53. Show that if $K \in \mathbb{C}^{r\times p}$ and $\widetilde{K} \in \mathbb{C}^{r\times p}$ both meet the two conditions in (12.62), then $K = \widetilde{K}$ and hence that K is uniquely specified in terms of the Moore-Penrose inverse $F^\dagger$ of F by the formula $K = GFF^\dagger$.

Corollary 12.24. If, in the setting of Lemma 12.23, $F^H F = G^H G$, then the unique matrix K that meets the two conditions in (12.62) is an isometry on $\mathcal{R}_F$.

Proof. This is immediate from the identity

$$\langle KF\mathbf{x}, KF\mathbf{x}\rangle = \langle G\mathbf{x}, G\mathbf{x}\rangle = \langle F\mathbf{x}, F\mathbf{x}\rangle ,$$

which is valid for every $\mathbf{x} \in \mathbb{C}^q$. $\square$

Lemma 12.25. *Let $A \in \mathbb{C}^{n\times n}$ and $B \in \mathbb{C}^{n\times n}$, and suppose that $A \succeq B \succeq O$. Then:*

(1) *There exists a matrix $K \in \mathbb{C}^{n\times n}$ such that $B = K^H AK$ and $I_n - K^H K \succeq O$.*

(2) $A^{1/2} \succeq B^{1/2}$.

(3) $\det A \geq \det B \geq O$.

Moreover, if $A \succ O$, then

(4) $\det A = \det B$ *if and only if* $A = B$.

Proof. Lemma 12.23 with $F = A^{1/2}$ and $G = B^{1/2}$ guarantees the existence of a contractive matrix $K \in \mathbb{C}^{n\times n}$ such that $KA^{1/2} = B^{1/2}$. Therefore, since $B^{1/2} = (B^{1/2})^H = A^{1/2}K^H$,

$$B = B^{1/2}B^{1/2} = KA^{1/2}(KA^{1/2})^H = KAK^H .$$

Next, in view of Exercise 20.1, it suffices to show that all the eigenvalues of the Hermitian matrix $A^{1/2} - B^{1/2}$ are nonnegative in order to verify (2). To this end, let $(A^{1/2} - B^{1/2})\mathbf{u} = \lambda\mathbf{u}$ for some nonzero vector $\mathbf{u}$. Then

$$\begin{aligned}
\lambda\langle(A^{1/2} + B^{1/2})\mathbf{u}, \mathbf{u}\rangle &= \langle(A^{1/2} + B^{1/2})(A^{1/2} - B^{1/2})\mathbf{u}, \mathbf{u}\rangle \\
&= \langle(A + B^{1/2}A^{1/2} - A^{1/2}B^{1/2} - B)\mathbf{u}, \mathbf{u}\rangle \\
&= \langle(A - B)\mathbf{u}, \mathbf{u}\rangle \geq 0 ,
\end{aligned}$$

since

$$\langle B^{1/2}A^{1/2}\mathbf{u}, \mathbf{u}\rangle = \langle B^{1/2}\mathbf{u}, B^{1/2}\mathbf{u}\rangle + \lambda\langle\mathbf{u}, \mathbf{u}\rangle = \langle A^{1/2}B^{1/2}\mathbf{u}, \mathbf{u}\rangle .$$

The last inequality implies that $\lambda \geq 0$ if $\langle(A^{1/2} + B^{1/2})\mathbf{u}, \mathbf{u}\rangle > 0$. On the other hand, if $\langle(A^{1/2} + B^{1/2})\mathbf{u}, \mathbf{u}\rangle = 0$, then $\langle A^{1/2}\mathbf{u}, \mathbf{u}\rangle = \langle B^{1/2}\mathbf{u}, \mathbf{u}\rangle = 0$ and hence $\lambda = 0$.

To obtain (3), observe first that in view of (1), the eigenvalues $\mu_1, \ldots , \mu_n$ of $K^H K$ are subject to the bounds $0 \leq \mu_j \leq 1$ for $j = 1, \ldots , n$. Therefore,

$$\det B = \det(KAK^H) = \det(K^H K) \det A = (\mu_1 \cdots \mu_n) \det A \leq \det A .$$

Moreover, if $\det B = \det A$ and A is invertible, then $\mu_1 = \cdots = \mu_n = 1$, i.e., $K^H K = I_n$. Therefore, since $KA^{1/2} = B^{1/2} = (B^{1/2})^H = A^{1/2}K^H$,

$$B = A^{1/2}K^H K A^{1/2} = A^{1/2} I_n A^{1/2} = A\,,$$

which justifies (4) and completes the proof. $\square$

Exercise 12.54. Let $A = \begin{bmatrix} 2 & 1 \\ 1 & 1 \end{bmatrix}$ and $B = \begin{bmatrix} 1 & 0 \\ 0 & 0 \end{bmatrix}$. Show that $A - B \succ O$, but $A^2 - B^2$ has one positive eigenvalue and one negative eigenvalue.

Theorem 12.26. *If $A_1 \in \mathbb{C}^{n\times s}$, $A_2 \in \mathbb{C}^{n\times t}$, $\operatorname{rank} A_1 = s$, $\operatorname{rank} A_2 = t$ and $A = \begin{bmatrix} A_1 & A_2 \end{bmatrix}$, then*

$$\det(A^H A) \le \det(A_1^H A_1)\,\det(A_2^H A_2)\,,$$

with equality if and only if $A_1^H A_2 = O$.

Proof. Clearly

$$A^H A = \begin{bmatrix} A_1^H A_1 & A_1^H A_2 \\ A_2^H A_1 & A_2^H A_2 \end{bmatrix}.$$

Therefore, since $A_1^H A_1$ is invertible by Exercise 12.17, it follows from the Schur complement formulas that

$$\det(A^H A) = \det(A_1^H A_1)\,\det(A_2^H A_2 - A_2^H A_1 (A_1^H A_1)^{-1} A_1^H A_2)\,.$$

Thus, as

$$A_2^H A_2 - A_2^H A_1 (A_1^H A_1)^{-1} A_1^H A_2 \preceq A_2^H A_2\,,$$

Lemma 12.25 guarantees that

$$\det(A_2^H A_2 - A_2^H A_1 (A_1^H A_1)^{-1} A_1^H A_2) \le \det(A_2^H A_2)\,,$$

with equality if and only if

$$A_2^H A_2 - A_2^H A_1 (A_1^H A_1)^{-1} A_1^H A_2 = A_2^H A_2\,.$$

This serves to complete the proof, since the last equality holds if and only if $A_1^H A_2 = O$. $\square$

The lemma leads to another inequality (12.63) that is also credited to Hadamard. This inequality is sharper than the inequality (9.13).

Corollary 12.27. Let $A = \begin{bmatrix} \mathbf{a}_1 & \cdots & \mathbf{a}_n \end{bmatrix}$ be an $n \times n$ matrix with columns $\mathbf{a}_j \in \mathbb{C}^n$ for $j = 1, \ldots, n$. Then

$$|\det A|^2 \le \prod_{j=1}^{n} \mathbf{a}_j^T \mathbf{a}_j\,. \tag{12.63}$$

Moreover, if A is invertible, then equality holds in (12.63) if and only if the columns of A are orthogonal.

Proof. The basic strategy is to iterate Theorem 12.26. The details are left to the reader. □

Exercise 12.55. Complete the proof of Corollary 12.27.

Exercise 12.56. Show that if $U, V \in \mathbb{C}^{n\times d}$ and $\operatorname{rank} U = \operatorname{rank} V = d$, then

$$UU^H = VV^H \Longleftrightarrow U = VK \quad \text{for some unitary matrix} \quad K \in \mathbb{C}^{d\times d}.$$

Exercise 12.57. Show that if $A \in \mathbb{C}^{n\times n}$ and $O \preceq A \preceq I_n$, then a vector $\mathbf{x} \in \mathcal{R}_{(I_n - A)}$ if and only if

$$\lim_{\delta\uparrow 1} \langle (I_n - \delta A)^{-1}\mathbf{x}, \mathbf{x}\rangle < \infty .$$

[HINT: The result is transparent if A is diagonal.]

Exercise 12.58. Show that if $A, B \in \mathbb{C}^{n\times n}$ and if $A \succ O$ and $B \succ O$, then

$$\sqrt{\det A \det B} \leq \det \frac{A+B}{2}. \tag{12.64}$$

Exercise 12.59. Show that if $A, B \in \mathbb{C}^{n\times n}$ and if $AB = O$ but $A+B \succ O$, then there exists a unitary matrix $U \in \mathbb{C}^{n\times n}$ such that

$$U^H AU = \begin{bmatrix} A_{11} & O \\ O & O \end{bmatrix} \quad \text{and} \quad U^H BU = \begin{bmatrix} O & O \\ O & B_{22} \end{bmatrix},$$

where $A_{11} \succ O$ and $B_{22} \succ O$.

12.12. A minimal norm completion problem

The next result, which is usually referred to as Parrott's lemma, is a nice application of the preceding circle of ideas.

Lemma 12.28. *Let $A \in \mathbb{C}^{p\times q}$, $B \in \mathbb{C}^{p\times r}$ and $C \in \mathbb{C}^{s\times q}$. Then*

$$\min\left\{\left\|\begin{bmatrix} A & B \\ C & D \end{bmatrix}\right\| : D \in \mathbb{C}^{s\times r}\right\} = \max\left\{\left\|\begin{bmatrix} A \\ C \end{bmatrix}\right\|, \|\begin{bmatrix} A & B \end{bmatrix}\|\right\}. \tag{12.65}$$

The proof will be developed in a sequence of auxiliary lemmas, most of which will be left to the reader to verify.

Lemma 12.29. *Let $A \in \mathbb{C}^{p\times q}$. Then*

$$\|A\| \leq \gamma \Longleftrightarrow \gamma^2 I_q - A^H A \succeq O \Longleftrightarrow \gamma^2 I_p - AA^H \succeq O .$$

Proof. The proof is easily extracted from the inequalities in Exercise 12.9. □

Lemma 12.30. *Let $A \in \mathbb{C}^{p\times q}$, $B \in \mathbb{C}^{p\times r}$ and $C \in \mathbb{C}^{s\times q}$. Then*

(1) $\gamma \geq \left\|\begin{bmatrix} A \\ C \end{bmatrix}\right\| \Longleftrightarrow \gamma^2 I_q - A^H A \succeq C^H C.$

(2) $\gamma \geq \| [\ A \quad B\ ] \| \Longleftrightarrow \gamma^2 I_p - AA^H \succeq BB^H$.

Proof. This is an easy consequence of the preceding lemma and the fact that $\|E\| = \|E^H\|$. □

Lemma 12.31. *If $A \in \mathbb{C}^{p\times q}$ and $\|A\| \leq \gamma$, then:*

$$(\gamma^2 I_q - A^H A)^{1/2} A^H = A^H (\gamma^2 I_p - AA^H)^{1/2} \tag{12.66}$$

and

$$(\gamma^2 I_p - AA^H)^{1/2} A = A(\gamma^2 I_q - A^H A)^{1/2}\,. \tag{12.67}$$

Proof. These formulas may also be established with the aid of the singular value decomposition of A. □

Lemma 12.32. *If $A \in \mathbb{C}^{p\times q}$, $p + q = n$ and $\|A\| \leq \gamma$, then the matrix*

$$E = \begin{bmatrix} A & (\gamma^2 I_p - AA^H)^{1/2} \\ (\gamma^2 I_q - A^H A)^{1/2} & -A^H \end{bmatrix} \tag{12.68}$$

satisfies the identity

$$EE^H = \begin{bmatrix} \gamma^2 I_p & O \\ O & \gamma^2 I_q \end{bmatrix} = \gamma^2 I_n\,.$$

Proof. This is a straightforward multiplication, thanks to Lemma 12.31. □

Lemma 12.33. *Let $A \in \mathbb{C}^{p\times q}$, $B \in \mathbb{C}^{p\times r}$ and $C \in \mathbb{C}^{s\times q}$ and suppose that*

$$\gamma \geq \max\left\{ \left\| \begin{bmatrix} A \\ C \end{bmatrix} \right\|, \ \| [\ A \quad B\ ] \| \right\}.$$

Then there exists a matrix $D \in \mathbb{C}^{s\times r}$ such that

$$\left\| \begin{bmatrix} A & B \\ C & D \end{bmatrix} \right\| \leq \gamma\,.$$

Proof. The given inequality implies that

$$\gamma^2 I_q - A^H A \succeq C^H C \quad \text{and} \quad \gamma^2 I_p - AA^H \succeq BB^H\,.$$

Therefore, by Lemma 12.23,

$$B = (\gamma^2 I_p - AA^H)^{1/2} X \quad \text{and} \quad C = Y(\gamma^2 I_q - A^H A)^{1/2} \tag{12.69}$$

for some choice of $X \in \mathbb{C}^{p\times r}$ and $Y \in \mathbb{C}^{s\times q}$ with $\|X\| \leq 1$ and $\|Y\| \leq 1$. Thus, upon setting $D = -YA^H X$, it is readily seen that

$$\begin{bmatrix} A & B \\ C & D \end{bmatrix} = \begin{bmatrix} I_p & O \\ O & Y \end{bmatrix} E \begin{bmatrix} I_q & O \\ O & X \end{bmatrix},$$

where E is given by formula (12.68). But this does the trick, since $EE^H = \gamma^2 I_n$ by Lemma 12.32 and the norm of each of the two outside factors on the right is equal to one. □

12.13. A description of all solutions to the minimal norm completion problem

Theorem 12.34. *A matrix $D \in \mathbb{C}^{s\times r}$ achieves the minimum in* (12.65) *if and only if it can be expressed in the form*

$$D = -YA^HX + (I_s - YY^H)^{1/2}Z(I_r - X^HX)^{1/2}, \tag{12.70}$$

where

$$X = \left\{\gamma^2 I_p - AA^H)^{1/2}\right\}^\dagger B, \quad Y = C\left\{\gamma^2 I_q - A^HA)^{1/2}\right\}^\dagger \tag{12.71}$$

and

$$Z \quad \text{is any matrix in } \mathbb{C}^{s\times r} \text{ such that} \quad Z^HZ \leq \gamma^2 I_r. \tag{12.72}$$

Discussion. We shall outline the main steps in the proof:

1.

$$\begin{bmatrix} A^H & C^H \\ B^H & D^H \end{bmatrix} \begin{bmatrix} A & B \\ C & D \end{bmatrix} \preceq \gamma^2 I_{q+r}$$

if and only if

$$\begin{bmatrix} C^H \\ D^H \end{bmatrix} \begin{bmatrix} C & D \end{bmatrix} \preceq \gamma^2 I_{q+r} - \begin{bmatrix} A^H \\ B^H \end{bmatrix} \begin{bmatrix} A & B \end{bmatrix}. \tag{12.73}$$

2. In view of Lemma 12.31 and the formulas in (12.69),

$$\gamma^2 I_{q+r} - \begin{bmatrix} A^H \\ B^H \end{bmatrix} \begin{bmatrix} A & B \end{bmatrix} = M^HM,$$

where

$$M = \begin{bmatrix} (\gamma^2 I_q - A^HA)^{1/2} & -A^HX \\ O & \gamma(I_r - X^HX)^{1/2} \end{bmatrix}.$$

3. In view of (12.73), the identity in Step 2 and Lemma 12.23, there exists a unique matrix $\begin{bmatrix} K_1 & K_2 \end{bmatrix}$ with components $K_1 \in \mathbb{C}^{s\times q}$ and $K_2 \in \mathbb{C}^{s\times r}$ such that

$$\begin{aligned} \begin{bmatrix} C & D \end{bmatrix} &= \begin{bmatrix} K_1 & K_2 \end{bmatrix} M \\ &= \begin{bmatrix} K_1(\gamma^2 I_q - A^HA)^{1/2} & -K_1A^HX + K_2\gamma(I_r - X^HX)^{1/2} \end{bmatrix} \end{aligned}$$

and

$$K_1\mathbf{u}_1 + K_2\mathbf{u}_2 = \mathbf{0} \quad \text{if} \quad M^H \begin{bmatrix} \mathbf{u}_1 \\ \mathbf{u}_2 \end{bmatrix} = \mathbf{0}.$$

4. $K_1 = Y$, since

$$\begin{aligned} M^H \begin{bmatrix} \mathbf{u}_1 \\ \mathbf{u}_2 \end{bmatrix} = \mathbf{0} \iff & (\gamma^2 I_q - A^HA)^{1/2}\mathbf{u}_1 = \mathbf{0} \quad \text{and} \\ & -X^HA\mathbf{u}_1 + \gamma(I_r - X^HX)^{1/2}\mathbf{u}_2 = \mathbf{0} \\ \iff & (\gamma^2 I_q - A^HA)^{1/2}\mathbf{u}_1 = \mathbf{0} \text{ and } (I_r - X^HX)^{1/2}\mathbf{u}_2 = \mathbf{0}, \end{aligned}$$

because

$$X^HA = B^H\left\{(\gamma^2I_q - A^HA)^{1/2}\right\}^\dagger A = B^HA\left\{(\gamma^2I_p - AA^H)^{1/2}\right\}^\dagger$$

and $\mathcal{N}_{W^H} = \mathcal{N}_{W^\dagger}$ for any matrix $W \in \mathbb{C}^{k\times k}$.

5. Extract the formula

$$D = -K_1A^HX + \gamma K_2(I_q - XX^H)^{1/2}$$

from Step 3 and then, taking note of the fact that $K_1K_1^H + K_2K_2^H \preceq I_s$, replace K_1 by Y and γK_2 by $(I_s - YY^H)^{1/2}Z$.

12.14. Bibliographical notes

The section on maximum entropy interpolants is adapted from the paper [**24**]. It is included here to illustrate the power of factorization methods. The underlying algebraic structure is clarified in [**25**]; see also [**34**] for further generalizations. A description of all completions of the problem considered in Section 12.7 may be found e.g., in Chapter 10 of [**21**]. Formulas (12.32) and (12.28) imply that

$$\lim_{n\uparrow\infty}\frac{\ln\det T_n(f)}{n} = \ln|h_0|^2 = \frac{1}{2\pi}\int_0^{2\pi}\ln f(e^{i\theta})d\theta \tag{12.74}$$

for the Toeplitz matrix $T_n(f)$ based on the Fourier coefficients of the considered function f. This is a special case of a theorem that was proved by Szegö in 1915 and is still the subject of active research today; see e.g., [**9**] and [**65**] for two recent expository articles on the subject; [**64**] for additional background material; and the references cited in all three. Lemma 12.8 is due to Fejér and Riesz. Formula (12.40) is one way of writing a formula due to Gohberg and Heinig. Other variants may be obtained by invoking appropriate generalizations of the observation

$$\begin{bmatrix}0&0&1\\0&1&0\\1&0&0\end{bmatrix}\begin{bmatrix}a&0&0\\b&a&0\\c&b&a\end{bmatrix}\begin{bmatrix}0&0&1\\0&1&0\\1&0&0\end{bmatrix} = \begin{bmatrix}a&b&c\\0&a&b\\0&0&a\end{bmatrix}.$$

The minimal norm completion problem is adapted from [**31**] and [**74**], both of which cite [**18**] as a basic reference for this problem. Exercises 12.58 and 12.59 are adapted from [**69**] and [**26**], respectively.

Chapter 13

Difference equations and differential equations

There are few vacancies in the Big Leagues for the man who is liable to steal second with the bases full.

Christy Mathewson, cited in [**40**], p. 136

In this chapter we shall focus primarily on four classes of equations:

(1) $\mathbf{x}_{k+1} = A\mathbf{x}_k$, $k = 0, 1, \ldots$, in which $A \in \mathbb{F}^{p\times p}$ and $\mathbf{x}_0 \in \mathbb{F}^p$ are specified.

(2) $\mathbf{x}'(t) = A\mathbf{x}(t)$ for $t \geq a$, in which $A \in \mathbb{F}^{p\times p}$ and $\mathbf{x}(a) \in \mathbb{F}^p$ are specified.

(3) $x_{k+p} = a_1x_{k+p-1}+\cdots+a_px_k$, for $k = p, p+1, \ldots$, in which $a_1, \ldots, a_p \in \mathbb{F}$, $a_p \neq 0$ and $x_0, \ldots, x_{p-1}$ are specified.

(4) $x^{(p)}(t) = a_1x^{(p-1)}(t)+a_2x^{(p-2)}(t)+\cdots+a_px(t)$, in which $a_1, \ldots, a_p \in \mathbb{F}$ and $x(a), \ldots, x^{(p-1)}(a)$ are specified.

It is easy to exhibit solutions to the first-order vector equations described in (1) and (2). The main effort is to understand the behavior of these solutions when k and t tend to ∞ with the help of the Jordan decomposition of the matrix A. The equations in (3) and (4) are then solved by imbedding them in first-order vector equations of the kind considered in (1) and (2), respectively. Two extra sections that deal with second-order equations with nonconstant coefficients have been added because of the importance of this material in applications.

13.1. Systems of difference equations

The easiest place to start is with the system of difference equations (or, in other terminology, the discrete dynamical system)

$$\mathbf{x}_{k+1} = A\mathbf{x}_k\,, \quad k = 0, 1, \ldots\,, \tag{13.1}$$

in which $A \in \mathbb{C}^{p\times p}$ and $\mathbf{x}_0 \in \mathbb{C}^p$ are specified and the objective is to understand the behavior of the solution $\mathbf{x}_n$ as n gets large. Clearly

$$\mathbf{x}_n = A^n\mathbf{x}_0\,.$$

However, this formula does not provide much insight into the behavior of $\mathbf{x}_n$. This is where the fact that A is similar to a Jordan matrix J comes into play:

$$A = VJV^{-1} \Longrightarrow \mathbf{x}_n = VJ^nV^{-1}\mathbf{x}_0\,, \quad \text{for} \quad n = 0, 1, \ldots\,. \tag{13.2}$$

The advantage of this new formulation is that J^n is relatively easy to compute: If A is diagonalizable, then

$$J = \operatorname{diag}\{\lambda_1, \ldots, \lambda_p\}\,, \quad J^n = \operatorname{diag}\{\lambda_1^n, \ldots, \lambda_p^n\}$$

and

$$\mathbf{x}_n = \sum_{j=1}^{p} d_j\lambda_j^n\mathbf{v}_j$$

is a linear combination of the eigenvectors $\mathbf{v}_j$ of A, alias the columns of V, with coefficients that are proportional to λ_j^n. If A is not diagonalizable, then

$$J = \operatorname{diag}\{J_1, \ldots, J_r\}\,,$$

where each block entry J_i is a Jordan cell, and

$$J^n = \operatorname{diag}\{J_1^n, \ldots, J_r^n\}\,.$$

Consequently the key issue reduces to understanding the behavior of the n'th power $(C_\lambda^{(m)})^n$ of the $m\times m$ Jordan cell $C_\lambda^{(m)}$ as n tends to ∞. Fortunately, this is still relatively easy:

Lemma 13.1. *If $N = C_\lambda^{(m)} - \lambda I_m = C_0^{(m)}$, then*

$$(C_\lambda^{(m)})^n = \sum_{j=0}^{m-1} \binom{n}{j} \lambda^{n-j}N^j \quad \textit{when} \quad n \geq m\,. \tag{13.3}$$

Proof. Since N commutes with λI_m, the binomial theorem is applicable and supplies the formula

$$(C_\lambda^{(m)})^n = (\lambda I_m + N)^n = \sum_{j=0}^{n} \binom{n}{j} \lambda^{n-j}N^j\,.$$

But this is the same as formula (13.3), since $N^j = 0$ for $j \geq m$. $\square$

Exercise 13.1. Show that if $J = \operatorname{diag}\{\lambda_1, \ldots, \lambda_p\}$, $V = \begin{bmatrix} \mathbf{v}_1 & \cdots & \mathbf{v}_p \end{bmatrix}$ and $(V^{-1})^T = \begin{bmatrix} \mathbf{w}_1 & \cdots & \mathbf{w}_p \end{bmatrix}$, then the solution (13.2) of the system (13.1) can be expressed in the form

$$\mathbf{x}_n = \sum_{j=1}^{p} \lambda_j^n \mathbf{v}_j \mathbf{w}_j^T \mathbf{x}_0 \,.$$

Exercise 13.2. Show that if, in the setting of Exercise 13.1, $|\lambda_1| > |\lambda_j|$ for $j = 2, \ldots, p$, then

$$\lim_{n \uparrow \infty} \frac{1}{\lambda_1^n} \mathbf{x}_n = \mathbf{v}_1 \mathbf{w}_1^T \mathbf{x}_0 \,.$$

Exercise 13.3. The output $\mathbf{u}_n$ of a chemical plant at time n, $n = 0, 1, \ldots,$ is modelled by a system of the form $\mathbf{u}_n = A^n \mathbf{u}_0$. Show that if

$$A = \begin{bmatrix} 1 & -3/2 & 0 \\ 0 & 1/2 & 0 \\ 0 & 0 & 1/4 \end{bmatrix} \quad \text{and } \mathbf{u}_0 = \begin{bmatrix} a \\ b \\ c \end{bmatrix}, \text{ then } \lim_{n \to \infty} \mathbf{u}_n = \begin{bmatrix} a - 3b \\ 0 \\ 0 \end{bmatrix}.$$

Exercise 13.4. Find an explicit formula for the solution $\mathbf{u}_n$ of the system $\mathbf{u}_n = A^n \mathbf{u}_0$ when

$$A = \begin{bmatrix} 1 & 2 & 0 \\ 0 & 1 & 3 \\ 0 & 0 & 1 \end{bmatrix} \quad \text{and} \quad \mathbf{u}_0 = \begin{bmatrix} 2 \\ 3 \\ 0 \end{bmatrix}.$$

Notice that it is **not necessary to compute** V^{-1} in the formula for the solution in (13.2). It is enough to compute $V^{-1}\mathbf{x}_0$, which is often much less work: set

$$\mathbf{y}_0 = V^{-1}\mathbf{x}_0 \quad \text{and solve the equation} \quad V\mathbf{y}_0 = \mathbf{x}_0 \,.$$

Exercise 13.5. Calculate $V^{-1}\mathbf{x}_0$ when $V = \begin{bmatrix} 6 & 2 & 2 \\ 0 & 3 & 1 \\ 0 & 0 & 1 \end{bmatrix}$ and $\mathbf{x}_0 = \begin{bmatrix} 6 \\ 0 \\ 0 \end{bmatrix}$ both directly (i.e., by first calculating V^{-1} and then calculating the product $V^{-1}\mathbf{x}_0$) and indirectly by solving the equation $V\mathbf{y}_0 = \mathbf{x}_0$, and compare the effort.

13.2. The exponential e^{tA}

Our next objective is to develop formulas analogous to (13.2) for the solution of a first-order vector differential equation. To do this, it is useful to first discuss the exponential e^{tA} of a matrix $A \in \mathbb{C}^{n \times n}$.

It is well known that for every complex number α the exponential e^{α} may be expressed as a power series

$$e^{\alpha} = \sum_{k=0}^{\infty} \frac{\alpha^k}{k!} \,,$$

which converges in the full complex plane $\mathbb{C}$. The same recipe may be used for square matrices A, thanks to the following lemma.

Lemma 13.2. *Let* $A = [a_{ij}]$, $i, j = 1, \ldots, p$, *be a* $p \times p$ *matrix and let*

$$\alpha = \max\{|a_{ij}| : i, j = 1, \ldots, p\}.$$

Then the ij *entry of* A^k *is subject to the bound*

$$|(A^k)_{ij}| \leq \frac{(\alpha p)^k}{p}, \text{ for } i, j = 1, \ldots, p \text{ and } k = 1, 2, \ldots. \tag{13.4}$$

Proof. The proof is by induction. The details are left to the reader. □

Thus, for $A \in \mathbb{C}^{p\times p}$, we may define

$$e^A = I_p + A + \frac{A^2}{2!} + \cdots. \tag{13.5}$$

Exercise 13.6. Verify the bound (13.4).

Exercise 13.7. Show that if $A \in \mathbb{C}^{p\times p}$, then the partial sums

$$S_k = \sum_{j=0}^{k} \frac{A^k}{k!}$$

form a Cauchy sequence in the normed linear space $\mathbb{C}^{p\times p}$ with respect to any multiplicative norm on that space.

Exercise 13.8. Show that if $A \in \mathbb{C}^{p\times p}$, then

$$\left\|\frac{e^{hA} - I_p - hA}{h}\right\| \leq \frac{e^{|h|\|A\|} - 1 - |h|\|A\|}{|h|} \leq (e^{|h|\|A\|} - 1)\|A\|. \tag{13.6}$$

Exercise 13.9. Show that if A, $B \in \mathbb{C}^{p\times p}$ and $AB = BA$, then

$$e^{A+B} = e^A e^B.$$

WARNING: In general, $e^{A+B} \neq e^A e^B$.

Exercise 13.10. Exhibit a pair of matrices A, $B \in \mathbb{C}^{p\times p}$ such that $e^{A+B} \neq e^A e^B$.

Exercise 13.11. Show that if A, $B \in \mathbb{C}^{p\times p}$, then

$$\lim_{(s,t)\to(0,0)} \frac{e^{tA}e^{sB}e^{-tA}e^{-sB} - I_p}{st} = AB - BA.$$

Let

$$F(t) = e^{tA} = I_p + tA + t^2\frac{A^2}{2!} + \cdots.$$

Then

$$F(0) = I_p$$

and

$$\frac{F(t+h)-F(t)}{h} = \frac{e^{(t+h)A}-e^{tA}}{h} = e^{tA}\left(\frac{e^{hA}-I_p}{h}\right),$$

which tends to

$$e^{tA}A = Ae^{tA}$$

as h tends to zero, thanks to the bound (13.6). Thus, the derivative

$$F'(t) = \lim_{h\to 0}\frac{F(t+h)-F(t)}{h} = AF(t)\,.$$

The same definition is used for the derivative of any suitably smooth matrix valued function $F(t) = [f_{ij}(t)]$ with entries $f_{ij}(t)$ and implies that

$$F'(t) = [f'_{ij}(t)]\,, \quad \text{and correspondingly} \quad \int_a^b F(s)ds = \left[\int_a^b f_{ij}(s)ds\right]\,;$$

i.e., **differentiation and integration of a matrix valued function is carried out on each entry in the matrix separately.**

Exercise 13.12. Show that if $F(t)$ is an invertible suitably smooth $p \times p$ matrix valued function on the interval $a < t < b$, then

$$\lim_{h\to 0}\frac{F(t+h)^{-1}-F(t)^{-1}}{h} = -F(t)^{-1}F'(t)F(t)^{-1} \text{ for } a < t < b\,. \tag{13.7}$$

[HINT: $F(t+h)^{-1} - F(t)^{-1} = F(t+h)^{-1}(F(t)-F(t+h))F(t)^{-1}$.]

Exercise 13.13. Calculate e^A when $A = \begin{bmatrix} 0 & b \\ c & 0 \end{bmatrix}$.

Exercise 13.14. Calculate e^A when $A = \begin{bmatrix} a & b \\ b & a \end{bmatrix}$. [HINT: aI_2 and $A - aI_2$ commute.]

Exercise 13.15. Calculate e^A when $A = \begin{bmatrix} a & b \\ -b & a \end{bmatrix}$. [HINT: aI_2 and $A - aI_2$ commute.]

13.3. Systems of differential equations

In view of the preceding analysis, it should be clear that for any vector $\mathbf{c} \in \mathbb{C}^p$, the vector function

$$\mathbf{x}(t) = e^{(t-a)A}\mathbf{c} \tag{13.8}$$

is a solution of the system

$$\mathbf{x}'(t) = A\mathbf{x}(t)\,, \quad t \geq a\,, \quad \text{with initial conditions} \quad \mathbf{x}(a) = \mathbf{c}\,. \tag{13.9}$$

The advantage of this formulation is its simplicity. The disadvantage is that it is hard to see what's going on. But this is where the Jordan decomposition theorem comes to the rescue, just as before: If

$$A = VJV^{-1} \quad \text{for some Jordan matrix } J \text{, then} \quad e^{tA} = Ve^{tJ}V^{-1}$$

and

$$\mathbf{x}(t) = Ve^{(t-a)J}\mathbf{d}\,, \quad \text{where} \quad \mathbf{d} = V^{-1}\mathbf{x}(a)\,. \tag{13.10}$$

Note that **it is not necessary to calculate** V^{-1}, since only $\mathbf{d}$ is needed. The advantage of this new formula is that it is easy to calculate e^{tJ}: If

$$J = \operatorname{diag}\{\lambda_1, \ldots, \lambda_p\}\,, \quad \text{then} \quad e^{tJ} = \operatorname{diag}\{e^{t\lambda_1}, \ldots, e^{t\lambda_p}\}$$

and hence, upon writing $V = \begin{bmatrix}\mathbf{v}_1 & \cdots & \mathbf{v}_p\end{bmatrix}$ and $\mathbf{d}^T = \begin{bmatrix}d_1 & \cdots & d_p\end{bmatrix}$,

$$\mathbf{x}(t) = \sum_{j=1}^{p} d_j e^{(t-a)\lambda_j}\mathbf{v}_j\,, \tag{13.11}$$

which exhibits the solution $\mathbf{x}(t)$ of the system (13.9) as a linear combination of the eigenvectors $\mathbf{v}_1, \ldots, \mathbf{v}_p$ of A with coefficients that depend upon the eigenvalues of A and vary with t. If A is not diagonalizable, then

$$J = \operatorname{diag}\{J_1, \ldots, J_r\} \quad \text{and} \quad e^{tJ} = \operatorname{diag}\{e^{tJ_1}, \ldots, e^{tJ_r}\}\,,$$

where each block entry J_i is a Jordan cell. Consequently, the solution $\mathbf{x}(t)$ of the system (13.9) is now a linear combination of generalized eigenvectors of A and it is important to understand the behavior of $e^{tC_\lambda^{(m)}}$ as t tends to ∞. Fortunately, this too is relatively easy. Thus, for example, if $m = 3$ and $N = C_\lambda^{(3)} - \lambda I_3$, then

$$\begin{aligned} e^{tC_\lambda^{(3)}} &= e^{t\lambda I_3}e^{tN} = e^{t\lambda}e^{tN} \\ &= e^{t\lambda}\left\{I_3 + tN + \frac{t^2N^2}{2!}\right\} \\ &= \begin{bmatrix} e^{t\lambda} & te^{t\lambda} & \frac{t^2}{2!}e^{t\lambda} \\ 0 & e^{t\lambda} & te^{t\lambda} \\ 0 & 0 & e^{t\lambda} \end{bmatrix}. \end{aligned}$$

The same pattern propagates for every Jordan cell:

Lemma 13.3. *If* $N = C_\lambda^{(m)} - \lambda I_m = C_0^{(m)}$, *then*

$$e^{tC_\lambda^{(m)}} = e^{t\lambda}e^{tN} = e^{t\lambda}\sum_{j=0}^{m-1}\frac{(tN)^j}{j!}\,. \tag{13.12}$$

Proof. The proof is easy and is left to the reader as an exercise. □

Exercise 13.16. Verify formula (13.12).

Exercise 13.17. Show that if $J = \mathrm{diag}\{\lambda_1, \ldots, \lambda_p\}$, $V = \begin{bmatrix}\mathbf{v}_1 & \cdots & \mathbf{v}_p\end{bmatrix}$ and $(V^{-1})^T = \begin{bmatrix}\mathbf{w}_1 & \cdots & \mathbf{w}_p\end{bmatrix}$, then the solution (13.8) of the system (13.9) can be expressed in the form

$$\mathbf{x}(t) = \sum_{j=1}^{p} e^{(t-a)\lambda_j}\mathbf{v}_j\mathbf{w}_j^T\mathbf{x}(a)\,.$$

Exercise 13.18. Show that if, in the setting of Exercise 13.17, $|\lambda_1| > |\lambda_j|$ for $j = 2, \ldots, p$, then

$$\lim_{t\uparrow\infty} e^{-t\lambda_1}\mathbf{x}(t) = e^{-a\lambda_1}\mathbf{v}_1\mathbf{w}_1^T\mathbf{x}(a)\,.$$

Exercise 13.19. Give an explicit formula for e^{tA} when

$$A = \begin{bmatrix} 0 & 1 & 0 \\ -1 & 0 & 1 \\ 0 & -1 & 0 \end{bmatrix}.$$

[HINT: You may use the fact that the eigenvalues of A are equal to 0, $i\sqrt{2}$ and $-i\sqrt{2}$.]

Exercise 13.20. Let $A = VJV^{-1}$, where $J = \begin{bmatrix} 2 & 1 & 0 \\ 0 & 2 & 0 \\ 0 & 0 & 3 \end{bmatrix}$, $V = [\mathbf{v}_1\ \mathbf{v}_2\ \mathbf{v}_3]$ and $(V^T)^{-1} = [\mathbf{w}_1\ \mathbf{w}_2\ \mathbf{w}_3]$. Evaluate the limit of the matrix valued function $e^{-3t}e^{tA}$ as $t \uparrow \infty$.

13.4. Uniqueness

Formula (13.8) provides a (smooth) solution to the first-order vector differential equation (13.9). However, it remains to check that there are no others.

Lemma 13.4. *The differential equation* (13.9) *has only one solution* $\mathbf{x}(t)$ *with continuous derivative* $\mathbf{x}'(t)$ *on the interval* $a \le t \le b$ *that meets the specified initial condition at* $t = a$.

Proof. Suppose to the contrary that there are two solutions $\mathbf{x}(t)$ and $\mathbf{y}(t)$. Then

$$\begin{aligned}\mathbf{x}(t) - \mathbf{y}(t) &= \int_a^t \{\mathbf{x}'(s) - \mathbf{y}'(s)\}ds \\ &= \int_a^t A\{\mathbf{x}(s) - \mathbf{y}(s)\}ds\,.\end{aligned}$$

Therefore, upon setting $\mathbf{u}(s) = \mathbf{x}(s) - \mathbf{y}(s)$ for $a \le s \le b$ and iterating the last equality, we obtain the formula

$$\mathbf{u}(t) = A^n \int_a^t \int_a^{s_1} \cdots \int_a^{s_{n-1}} \mathbf{u}(s_n) ds_n \cdots ds_1 ,$$

which in turn leads to the inequality

$$M \le M\|A^n\| \frac{(b-a)^n}{n!} \le M\|A\|^n \frac{(b-a)^n}{n!}$$

for

$$M = \max\{\|\mathbf{u}(t)\| : a \le t \le b\} .$$

If n is large enough, then $\|A\|^n(b-a)^n/n! < 1$ and hence,

$$0 \le M\left(1 - \|A\|^n \frac{(b-a)^n}{n!}\right) \le 0 .$$

Therefore, $M = 0$; i.e., there is only one smooth solution of the differential equation (13.9) that meets the given initial conditions. □

Much the same sort of analysis leads to Gronwall's inequality:

Exercise 13.21. Let $h(t)$ be a continuous real-valued function on the interval $a \le t \le b$. Show that

$$\int_a^t h(s_2)\left\{\int_a^{s_2} h(s_1)ds_1\right\} ds_2 = \left(\int_a^t h(s)ds\right)^2 /2! ,$$

$$\int_a^t h(s_3)\left[\int_a^{s_3} h(s_2)\left\{\int_a^{s_2} h(s_1)ds_1\right\} ds_2\right] ds_3 = \left(\int_a^t h(s)ds\right)^3 /3! ,$$

etc.

Exercise 13.22. (Gronwall's inequality) Let $\alpha > 0$ and let $u(t)$ and $h(t)$ be continuous real-valued functions on the interval $a \le t \le b$ such that

$$u(t) \le \alpha + \int_a^t h(s)u(s)ds \quad \text{and} \quad h(t) \ge 0 \quad \text{for} \quad a \le t \le b .$$

Show that

$$u(t) \le \alpha \exp\left(\int_a^t h(s)ds\right) \quad \text{for} \quad a \le t \le b .$$

[HINT: Iterate the inequality and exploit Exercise 13.21.]

13.5. Isometric and isospectral flows

A matrix $B \in \mathbb{R}^{p\times p}$ is said to be **skew-symmetric** if $B = -B^T$. Analogously, $B \in \mathbb{C}^{p\times p}$ is said to be **skew-Hermitian** if $B = -B^H$.

Exercise 13.23. Let $B \in \mathbb{C}^{p\times p}$. Show that if B is skew-Hermitian, then e^B is unitary.

Exercise 13.24. Let $F(t) = e^{tB}$, where $B \in \mathbb{R}^{p\times p}$. Show that $F(t)$ is an orthogonal matrix for every $t \in \mathbb{R}$ if and only if B is skew-symmetric. [HINT: If $F(t)$ is orthogonal, then the derivative $\{F(t)F(t)^T\}' = 0$.]

Exercise 13.25. Let $B \in \mathbb{R}^{p\times p}$ and let $\mathbf{x}(t)$, $t \geq 0$, denote the solution of the differential equation $\mathbf{x}'(t) = B\mathbf{x}(t)$ for $t \geq 0$ that meets the initial condition $\mathbf{x}(0) = \mathbf{c} \in \mathbb{R}^p$.

(a) Show that $\frac{d}{dt}\|\mathbf{x}(t)\|^2 = \mathbf{x}(t)^T(B + B^T)\mathbf{x}(t)$ for every $t \geq 0$.

(b) Show that if B is skew-symmetric, then $\|\mathbf{x}(t)\| = \|\mathbf{x}(0)\|$ for every $t \geq 0$.

Exercise 13.26. Let $A \in \mathbb{R}^{p\times p}$ and $U(t)$, $t \geq 0$, be a one-parameter family of $p \times p$ real matrices such that $U'(t) = B(t)U(t)$ for $t > 0$ and $U(0) = I_p$. Show that $F(t) = U(t)AU(t)^{-1}$ is a solution of the differential equation

$$F'(t) = B(t)F(t) - F(t)B(t) \qquad \text{for } t \geq 0\,. \tag{13.13}$$

Exercise 13.27. Show that if $F(t)$ is the only smooth solution of a differential equation of the form (13.13) with suitably smooth $B(t)$, then $F(t) = U(t)F(0)U(t)^{-1}$ for $t \geq 0$. [HINT: Consider $U(t)F(0)U(t)^{-1}$ when $U(t)$ is a solution of $U'(t) = B(t)U(t)$ with $U(0) = I_p$.]

A pair of matrix valued functions $F(t)$ and $B(t)$ that are related by equation (13.13) is said to be a **Lax pair**, and the solution $F(t) = U(t)F(0)U(t)^{-1}$ is said to be **isospectral** because its eigenvalues are independent of t.

13.6. Second-order differential systems

If $A = VJV^{-1}$ is a 2×2 matrix that is similar to a Jordan matrix J, then either

$$J = \begin{bmatrix} \lambda_1 & 0 \\ 0 & \lambda_2 \end{bmatrix} \quad \text{or} \quad J = \begin{bmatrix} \lambda_1 & 1 \\ 0 & \lambda_1 \end{bmatrix}.$$

In the first case, A has two linearly independent eigenvectors, $\mathbf{v}_1$ and $\mathbf{v}_2$:

$$A\begin{bmatrix} \mathbf{v}_1 & \mathbf{v}_2 \end{bmatrix} = \begin{bmatrix} \mathbf{v}_1 & \mathbf{v}_2 \end{bmatrix} \begin{bmatrix} \lambda_1 & 0 \\ 0 & \lambda_2 \end{bmatrix}$$

and

$$\begin{aligned} \mathbf{u}(t) &= e^{tA}\mathbf{c} \\ &= V\begin{bmatrix} e^{t\lambda_1} & 0 \\ 0 & e^{t\lambda_2} \end{bmatrix} V^{-1}\mathbf{c} \\ &= \begin{bmatrix} \mathbf{v}_1 & \mathbf{v}_2 \end{bmatrix} \begin{bmatrix} e^{t\lambda_1} & 0 \\ 0 & e^{t\lambda_2} \end{bmatrix} \begin{bmatrix} d_1 \\ d_2 \end{bmatrix} \\ &= e^{t\lambda_1}d_1\mathbf{v}_1 + e^{t\lambda_2}d_2\mathbf{v}_2\,, \end{aligned}$$

where we have set

$$V = [\mathbf{v}_1 \quad \mathbf{v}_2] \quad \text{and} \quad \begin{bmatrix} d_1 \\ d_2 \end{bmatrix} = V^{-1}\mathbf{c}.$$

In the second case

$$e^{tJ} = \begin{bmatrix} e^{t\lambda_1} & te^{t\lambda_1} \\ 0 & e^{t\lambda_1} \end{bmatrix}$$

and only the first column of

$$V = \begin{bmatrix} \mathbf{v}_1 & \mathbf{v}_2 \end{bmatrix}$$

is an eigenvector of A. Defining d_1 and d_2 as before, we now obtain the formula

$$\begin{aligned} \mathbf{u}(t) &= \begin{bmatrix} \mathbf{v}_1 & \mathbf{v}_2 \end{bmatrix} \begin{bmatrix} e^{t\lambda_1} & te^{t\lambda_1} \\ 0 & e^{t\lambda_1} \end{bmatrix} \begin{bmatrix} d_1 \\ d_2 \end{bmatrix} \\ &= d_1 e^{t\lambda_1}\mathbf{v}_1 + d_2(te^{t\lambda_1}\mathbf{v}_1 + e^{t\lambda_1}\mathbf{v}_2)\,. \end{aligned}$$

13.7. Stability

The formulas

$$\mathbf{u}_n = VJ^nV^{-1}\mathbf{u}_0 \quad \text{and} \quad \mathbf{x}(t) = Ve^{(t-a)J}V^{-1}\mathbf{x}(a)$$

express the solutions $\mathbf{u}_n$ and $\mathbf{x}(t)$ of equations (13.1) and (13.9) as linear combinations of the eigenvectors and generalized eigenvectors of A with coefficients that depend upon n and t, respectively, and the eigenvalues. Thus, the "dynamic" behavior depends essentially upon the magnitudes $|\lambda_j|$, $j = 1, \ldots, p$, in the first case and the real parts of λ_j, $j = 1, \ldots, p$, in the second:

$$\|\mathbf{u}_n\| = \|VJ^nV^{-1}\mathbf{u}_0\| \leq \|V\|\|J^n\|\|V^{-1}\mathbf{u}_0\|$$

and, similarly,

$$\|\mathbf{x}(t)\| \leq \|V\|\|e^{(t-a)J}\|\|V^{-1}\mathbf{x}(a)\|\,.$$

These bounds are particularly transparent when

$$J = \operatorname{diag}\{\lambda_1, \ldots, \lambda_p\}\,,$$

because then

$$\|J^n\| \leq \alpha^n \quad \text{and} \quad \|e^{tJ}\| \leq e^{t\beta}\,,$$

where

$$\alpha = \max\{|\lambda_j| : j = 1, \ldots, p\} \quad \text{and} \quad \beta = \max\{\lambda_j + \overline{\lambda_j} : j = 1, \ldots, p\}\,.$$

In particular,

(a) J diagonal (or not) and $|\alpha| < 1 \Longrightarrow \lim_{n\uparrow\infty} \|\mathbf{u}_n\| = 0$.

(b) J diagonal and $|\alpha| \leq 1 \Longrightarrow \|\mathbf{u}_n\|$ is bounded.

(c) J diagonal (or not) and $\beta < 0 \Longrightarrow \lim_{t\to\infty} \|\mathbf{x}(t)\| = 0$.

(d) J diagonal and $\beta \le 0 \Longrightarrow \|\mathbf{x}(t)\|$ is bounded for $t > 0$.

Exercise 13.28. Show by example that item (b) in the list just above is not necessarily correct if the assumption that J is a diagonal matrix is dropped.

Exercise 13.29. Show by example that item (d) in the list just above is not necessarily correct if the assumption that J is a diagonal matrix is dropped.

13.8. Nonhomogeneous differential systems

In this section we shall consider nonhomogeneous differential systems, i.e., systems of the form

$$\mathbf{x}'(t) = A\mathbf{x}(t) + \mathbf{g}(t)\,, \quad \alpha \le t < \beta\,,$$

where $A \in \mathbb{R}^{n\times n}$ and $\mathbf{g}(t)$ is a continuous $n \times 1$ real vector valued function on the interval $\alpha \le t < \beta$. Then, since

$$\mathbf{x}'(t) - A\mathbf{x}(t) = e^{tA}\left(e^{-tA}\mathbf{x}(t)\right)'\,,$$

it is readily seen that the given system can be reexpressed as

$$\left(e^{-sA}\mathbf{x}(s)\right)' = e^{-sA}\mathbf{g}(s)$$

and hence, upon integrating both sides from α to a point $t \in (\alpha, \beta)$, that

$$e^{-tA}\mathbf{x}(t) - e^{-\alpha A}\mathbf{x}(\alpha) = \int_\alpha^t \left(e^{-sA}\mathbf{x}(s)\right)' ds = \int_\alpha^t e^{-sA}\mathbf{g}(s)ds$$

or, equivalently, that

$$\mathbf{x}(t) = e^{(t-\alpha)A}\mathbf{x}(\alpha) + \int_\alpha^t e^{(t-s)A}\mathbf{g}(s)ds \quad \text{for} \quad \alpha \le t < \beta\,. \tag{13.14}$$

13.9. Strategy for equations

To this point we have shown how to exploit the Jordan decomposition of a matrix in order to study the solutions of a first-order vector difference equation and a first-order vector differential equation. The next item of business is to study higher order scalar difference equations and higher order scalar differential equations. In both cases the strategy is to identify the solution with a particular coordinate of the solution of a first-order vector equation. This will lead to vector equations of the form $\mathbf{u}_{k+1} = A\mathbf{u}_k$ and $\mathbf{x}'(t) = A\mathbf{x}(t)$, respectively. However, now A will be a **companion matrix** and hence Theorem 5.11 supplies an explicit formula for $\det(\lambda I_n - A)$, which is simply related to the scalar difference/differential equation under consideration. Moreover, A is similar to a Jordan matrix with only one Jordan cell for each distinct eigenvalue. Consequently, it is possible to develop an algorithm for writing down the solution, as will be noted in subsequent sections.

Exercise 13.30. Show that if A is a companion matrix, then, in the notation of Theorem 5.11,

$$A \quad \text{is invertible} \Longleftrightarrow \lambda_1 \cdots \lambda_k \neq 0 \Longleftrightarrow a_0 \neq 0\,. \tag{13.15}$$

13.10. Second-order difference equations

To warm up, we shall begin with the second-order difference equation

$$x_n = ax_{n-1} + bx_{n-2},\ n = 2, 3, \ldots\,, \ \text{with}\ b \neq 0\,, \tag{13.16}$$

where a and b are fixed and x_0 and x_1 are given. The objective is to obtain a formula for x_n and, if possible, to understand how x_n behaves as $n \uparrow \infty$.

We shall solve this second-order difference equation by embedding it into a first-order vector equation as follows:

First observe that

$$x_n = \begin{bmatrix} b & a \end{bmatrix} \begin{bmatrix} x_{n-2} \\ x_{n-1} \end{bmatrix}$$

and then, to fill out the left-hand side, add the row

$$x_{n-1} = \begin{bmatrix} 0 & 1 \end{bmatrix} \begin{bmatrix} x_{n-2} \\ x_{n-1} \end{bmatrix}$$

to get

$$\begin{bmatrix} x_{n-1} \\ x_n \end{bmatrix} = \begin{bmatrix} 0 & 1 \\ b & a \end{bmatrix} \begin{bmatrix} x_{n-2} \\ x_{n-1} \end{bmatrix}, \quad n = 2, 3, \ldots\,.$$

Thus, upon setting

$$\mathbf{u}_0 = \begin{bmatrix} x_0 \\ x_1 \end{bmatrix}, \ \mathbf{u}_1 = \begin{bmatrix} x_1 \\ x_2 \end{bmatrix}, \ \ldots\,, \ \mathbf{u}_n = \begin{bmatrix} x_n \\ x_{n+1} \end{bmatrix}$$

and

$$A = \begin{bmatrix} 0 & 1 \\ b & a \end{bmatrix}, \tag{13.17}$$

we obtain the sequence

$$\mathbf{u}_1 = A\mathbf{u}_0,\ \mathbf{u}_2 = A\mathbf{u}_1,\ \ldots\,,$$

i.e.,

$$\mathbf{u}_n = A^n \mathbf{u}_0\,.$$

Since A is a companion matrix, Theorem 5.11 implies that

$$\det(\lambda I_2 - A) = \lambda^2 - a\lambda - b$$

and hence the eigenvalues of A are

$$\lambda_1 = \frac{a + \sqrt{a^2 + 4b}}{2}, \quad \lambda_2 = \frac{a - \sqrt{a^2 + 4b}}{2}\,.$$

Therefore, A is similar to a Jordan matrix of the form

$$J = \begin{bmatrix} \lambda_1 & 0 \\ 0 & \lambda_2 \end{bmatrix} \quad \text{if} \quad \lambda_1 \neq \lambda_2 \quad \text{and} \quad J = \begin{bmatrix} \lambda_1 & 1 \\ 0 & \lambda_1 \end{bmatrix} \quad \text{if} \quad \lambda_1 = \lambda_2 .$$

Moreover, since $b \neq 0$ by assumption, the formula

$$(\lambda - \lambda_1)(\lambda - \lambda_2) = \lambda^2 - a\lambda - b \Longrightarrow \lambda_1\lambda_2 \neq 0 .$$

Case 1 ($\lambda_1 \neq \lambda_2$):

$$A = VJV^{-1} \Longrightarrow \mathbf{u}_n = A^n\mathbf{u}_0 = V \begin{bmatrix} \lambda_1^n & 0 \\ 0 & \lambda_2^n \end{bmatrix} V^{-1}\mathbf{u}_0. \tag{13.18}$$

Consequently,

$$x_n = \begin{bmatrix} 1 & 0 \end{bmatrix} V \begin{bmatrix} \lambda_1^n & 0 \\ 0 & \lambda_2^n \end{bmatrix} V^{-1}\mathbf{u}_0 . \tag{13.19}$$

However, it is not necessary to calculate V and V^{-1}. It suffices to note that formula (13.19) guarantees that x_n must be of the form

$$x_n = \alpha\lambda_1^n + \beta\lambda_2^n \quad (\lambda_1 \neq \lambda_2)$$

and then to solve for α and β from the given "initial conditions" x_0 and x_1.

Example 13.5.

$$x_n = 3x_{n-1} + 4x_{n-2},\ n = 2, 3, \dots ,$$

$$x_0 = 5 \text{ and } x_1 = 0.$$

Discussion. The roots of the equation $\lambda^2 - 3\lambda - 4$ are $\lambda_1 = 4$ and $\lambda_2 = -1$. Therefore, the solution x_n must be of the form

$$x_n = \alpha 4^n + \beta(-1)^n,\ n = 0, 1, \dots .$$

The initial condition

$$x_0 = 5 \Longrightarrow \alpha + \beta = 5,$$

whereas, the initial condition

$$x_1 = 0 \Longrightarrow 4\alpha - \beta = 0.$$

Thus, we see that $\alpha = 1, \beta = 4$, and hence the solution is

$$x_n = 4^n + 4(-1)^n \quad \text{for} \quad n = 0, 1, \dots .$$

Case 2 ($\lambda_1 = \lambda_2$):

$$\mathbf{u}_n = A^n\mathbf{u}_0 = V \begin{bmatrix} \lambda_1 & 1 \\ 0 & \lambda_1 \end{bmatrix}^n V^{-1}\mathbf{u}_0 = V \begin{bmatrix} \lambda_1^n & n\lambda_1^{n-1} \\ 0 & \lambda_1^n \end{bmatrix} V^{-1}\mathbf{u}_0 .$$

Consequently

$$x_n = \begin{bmatrix} 0 & 1 \end{bmatrix} V \begin{bmatrix} \lambda_1^n & n\lambda_1^{n-1} \\ 0 & \lambda_1^n \end{bmatrix} V^{-1}\mathbf{u}_0$$

must be of the form

$$x_n = \alpha\lambda_1^n + \beta n\lambda_1^n\,.$$

Notice that since $\lambda_1 \neq 0$, a (positive or negative) power of λ_1 can be absorbed into the constant β in the last formula for x_n.

Example 13.6.

$$x_n = 2x_{n-1} - x_{n-2} \quad \text{for} \quad n = 2, 3, \ldots$$

$$x_0 = 3 \quad \text{and} \quad x_1 = 5.$$

Discussion. The equation $\lambda^2 - 2\lambda + 1 = 0$ has two equal roots:

$$\lambda_1 = \lambda_2 = 1.$$

Therefore,

$$x_n = \alpha(1)^n + \beta n(1)^n = \alpha + \beta n\,.$$

Substituting the initial conditions

$$x_0 = \alpha = 3 \quad \text{and} \quad x_1 = 3 + \beta = 5\,,$$

we see that $\beta = 2$ and hence that

$$x_n = 3 + 2n \quad \text{for} \quad n = 0, 1, \ldots\,.$$

We are thus led to the following **recipe**: The solution of the second-order difference equation

$$x_n = ax_{n-1} + bx_{n-2},\ n = 2, 3, \ldots\,, \quad \text{with} \quad b \neq 0\,,$$

$$x_0 = c \quad \text{and} \quad x_1 = d$$

may be obtained as follows:

(1) Solve for the roots λ_1, λ_2 of the quadratic equation

$$\lambda^2 = a\lambda + b$$

and note that the factorization

$$(\lambda - \lambda_1)(\lambda - \lambda_2) = \lambda^2 - a\lambda - b$$

implies that $\lambda_1\lambda_2 = b \neq 0$.

(2) Express the solution as

$$x_n = \begin{cases} \alpha\lambda_1^n + \beta\lambda_2^n & \text{if} \quad \lambda_1 \neq \lambda_2 \\ \\ \alpha\lambda_1^n + \beta n\lambda_1^n & \text{if} \quad \lambda_1 = \lambda_2 \end{cases}$$

for some choice of α and β.

(3) Solve for α and β by invoking the initial conditions:

$$\begin{cases} c = x_0 = \alpha + \beta \quad \text{and} \quad d = x_1 = \alpha\lambda_1 + \beta\lambda_2 & \text{if} \quad \lambda_1 \neq \lambda_2 \\ \\ c = x_0 = \alpha \quad \text{and} \quad d = x_1 = \alpha\lambda_1 + \beta\lambda_1 & \text{if} \quad \lambda_1 = \lambda_2 \end{cases}\,.$$

Exercise 13.31. Find an explicit formula for x_n, for $n = 0, 1, \ldots$, given that $x_0 = -1$, $x_1 = 2$ and $x_{k+1} = 3x_k - 2x_{k-1}$ for $k = 1, 2, \ldots$.

Exercise 13.32. The **Fibonacci sequence** x_n, $n = 0, 1, \ldots$, is prescribed by the initial conditions $x_0 = 1$, $x_1 = 1$ and the difference equation $x_{n+1} = x_n + x_{n-1}$ for $n = 1, 2, \ldots$. Find an explicit formula for x_n and use it to calculate the **golden mean**, $\lim_{n\uparrow\infty} x_n/x_{n+1}$.

13.11. Higher order difference equations

Similar considerations apply to higher order difference equations. The solution to the p'th order equation

(13.20)

$$x_{n+p} = c_1 x_{n+p-1} + c_2 x_{n+p-2} + \cdots + c_p x_n \,, \quad n = 0, 1, \ldots \,, \quad \text{with} \quad c_p \neq 0$$

and given initial conditions $x_0, x_1, \ldots, x_{p-1}$, can be obtained from the solution to the first-order vector equation

$$\mathbf{u}_n = A\mathbf{u}_{n-1} \quad \text{for} \quad n = p, p+1, \ldots$$

where

$$\text{(13.21)} \qquad \mathbf{u}_n = \begin{bmatrix} x_{n-p+1} \\ \vdots \\ x_{n-1} \\ x_n \end{bmatrix} \quad \text{and} \quad A = \begin{bmatrix} 0 & 1 & 0 & \cdots & 0 \\ 0 & 0 & 1 & & 0 \\ \vdots & & & & \vdots \\ 0 & 0 & 0 & & 1 \\ c_p & c_{p-1} & c_{p-2} & \cdots & c_1 \end{bmatrix} .$$

The nature of the solution will depend on the eigenvalues of the matrix A.

A convenient **recipe** for obtaining the solution of equation (13.20) is:

(1) Find the roots of the polynomial $\lambda^p - c_1\lambda^{p-1} - \cdots - c_p$.

(2) If $\lambda^p - c_1\lambda^{p-1} - \cdots - c_p = (\lambda - \lambda_1)^{\alpha_1} \cdots (\lambda - \lambda_k)^{\alpha_k}$ with distinct roots $\lambda_1, \ldots, \lambda_k$, then the solution must be of the form

$$x_n = \sum_{j=1}^{k} p_j(n)\lambda_j^n \,, \quad \text{where} \quad p_j \quad \text{is a polynomial of degree} \quad \alpha_j - 1 \,.$$

(3) Invoke the initial conditions to solve for the coefficients of the polynomials p_j.

Discussion. The algorithm works because A is a companion matrix. Thus,

$$\det(\lambda I_p - A) = \lambda^p - c_1\lambda^{p-1} - \cdots - c_p$$

and hence, if

$$\det(\lambda I_p - A) = (\lambda - \lambda_1)^{\alpha_1} \cdots (\lambda - \lambda_k)^{\alpha_k}$$

with distinct roots $\lambda_1, \ldots, \lambda_k$, then A is similar to the Jordan matrix

$$J = \text{diag}\,\{C_{\lambda_1}^{(\alpha_1)}, \ldots, C_{\lambda_k}^{(\alpha_k)}\},$$

with one Jordan cell for each distinct eigenvalue. Therefore, the solution must be of the form indicated in (2).

Remark 13.7. The equation $\lambda^p - c_1\lambda^{p-1} - \cdots - c_p = 0$ may be obtained with minimum thought by letting $x_j = \lambda^j$ in equation (13.20) and then factoring out the highest common power of λ.

Exercise 13.33. Find the solution of the third-order difference equation

$$x_{n+3} = 3x_{n+2} - 3x_{n+1} + x_n\ ,\ \ n = 0, 1, \ldots$$

subject to the initial conditions $x_0 = 1$, $x_1 = 2$ and $x_2 = 8$. [HINT: $(x-1)^3 = x^3 - 3x^2 + 3x - 1$.]

13.12. Ordinary differential equations

Ordinary differential equations with constant coefficients can be solved by imbedding them in first-order vector differential equations and exploiting the theory developed in Section 13.3. Thus, for example, to solve the second-order differential equation

$$x''(t) = ax'(t) + bx(t),\ \ t \geq 0 \tag{13.22}$$

with initial conditions

$$x(0) = c \text{ and } x'(0) = d\,,$$

introduce the new variables

$$u_1(t) = x(t) \text{ and } u_2(t) = x'(t).$$

Then

$$u_1'(t) = u_2(t)$$

and

$$u_2'(t) = x''(t) = au_2(t) + bu_1(t).$$

Consequently, the vector

$$\mathbf{u}(t) = \begin{bmatrix} u_1(t) \\ u_2(t) \end{bmatrix}$$

is a solution of the first-order vector equation

$$\begin{aligned} \mathbf{u}'(t) = \begin{bmatrix} u_1'(t) \\ u_2'(t) \end{bmatrix} &= \begin{bmatrix} 0 & 1 \\ b & a \end{bmatrix} \begin{bmatrix} u_1(t) \\ u_2(t) \end{bmatrix} \\ &= A\mathbf{u}(t),\ \ t \geq 0\,, \end{aligned}$$

with

$$A = \begin{bmatrix} 0 & 1 \\ b & a \end{bmatrix} \quad \text{and} \quad \mathbf{u}(0) = \begin{bmatrix} c \\ d \end{bmatrix}.$$

Thus,

$$\mathbf{u}(t) = e^{tA} \begin{bmatrix} c \\ d \end{bmatrix}$$

and

$$x(t) = \begin{bmatrix} 1 & 0 \end{bmatrix} e^{tA} \begin{bmatrix} c \\ d \end{bmatrix}.$$

Let λ_1, λ_2 denote the roots of $\lambda^2 - a\lambda - b$. Then, since A is a companion matrix, there are only two possible Jordan forms:

$$J = \begin{bmatrix} \lambda_1 & 0 \\ 0 & \lambda_2 \end{bmatrix} \quad \text{if} \quad \lambda_1 \neq \lambda_2 \quad \text{and} \quad J = \begin{bmatrix} \lambda_1 & 1 \\ 0 & \lambda_1 \end{bmatrix} \quad \text{if} \quad \lambda_1 = \lambda_2\,.$$

Case 1 $(\lambda_1 \neq \lambda_2)$**:**

$$e^{tJ} = \begin{bmatrix} e^{\lambda_1 t} & 0 \\ 0 & e^{\lambda_2 t} \end{bmatrix}$$

and hence the solution $x(t)$ of equation (13.22) must be of the form

$$x(t) = \alpha e^{\lambda_1 t} + \beta e^{\lambda_2 t},$$

for some choice of the constants α and β.

Case 2 $(\lambda_1 = \lambda_2)$**:**

$$e^{tJ} = \begin{bmatrix} e^{\lambda_1 t} & te^{\lambda_1 t} \\ 0 & e^{\lambda_1 t} \end{bmatrix}$$

and hence the solution $x(t)$ of the equation must be of the form

$$x(t) = \alpha e^{\lambda_1 t} + \beta te^{\lambda_1 t}.$$

In both cases, the constants α and β are determined by the initial conditions.

The **recipe** for solving a p'th order differential equation

$$x^{(p)}(t) = a_1 x^{(p-1)}(t) + a_2 x^{(p-2)}(t) + \cdots + a_p x(t) \quad \text{for} \quad t \geq a \tag{13.23}$$

with constant coefficients that is subject to the constraint $a_p \neq 0$ and to the initial conditions

$$x(a) = c_1, \ldots, x^{(p-1)}(a) = c_p$$

is similar:

(1) Find the roots of the polynomial $\lambda^p - (a_1\lambda^{p-1} + \cdots + a_p)$.

(2) If $\lambda^p - (a_1\lambda^{p-1} + \cdots + a_p) = (\lambda - \lambda_1)^{\alpha_1} \cdots (\lambda - \lambda_k)^{\alpha_k}$ with k distinct roots $\lambda_1, \ldots, \lambda_k$, then the solution $x(t)$ to the given equation is of the form

$$x(t) = e^{(t-a)\lambda_1} p_1(t) + \cdots + e^{(t-a)\lambda_k} p_k(t)\,,$$

where $p_j(t)$ is a polynomial of degree $\alpha_j - 1$ for $j = 1, \ldots, k$.

(3) Find the coefficients of the polynomials $p_j(t)$ by imposing the initial conditions.

Discussion. Let

$$\mathbf{u}(t) = \begin{bmatrix} x(t) \\ x^{(1)}(t) \\ \vdots \\ x^{(p-1)}(t) \end{bmatrix}.$$

Then

$$\mathbf{u}'(t) = A\mathbf{u}(t) \quad \text{for} \quad t \geq a\,,$$

where

$$A = \begin{bmatrix} 0 & 1 & 0 & \cdots & 0 \\ 0 & 0 & 1 & \cdots & 0 \\ \vdots & & & \ddots & \vdots \\ 0 & 0 & 0 & \cdots & 1 \\ a_p & a_{p-1} & a_{p-2} & \cdots & a_1 \end{bmatrix} \quad \text{and} \quad \mathbf{u}(a) = \mathbf{c} = \begin{bmatrix} c_1 \\ \vdots \\ c_p \end{bmatrix}.$$

Thus,

$$\mathbf{u}(t) = e^{(t-a)A}\mathbf{c} \quad \text{and} \quad x(t) = \begin{bmatrix} 1 & 0 & \cdots & 0 \end{bmatrix} \mathbf{u}(t) \quad \text{for} \quad t \geq a\,.$$

The special form of the solution indicated in (2) follows from the fact that A is a companion matrix and hence is similar to the Jordan matrix

$$J = \operatorname{diag}\{C_{\lambda_1}^{(\alpha_1)}, \ldots, C_{\lambda_k}^{(\alpha_k)}\}\,.$$

□

Remark 13.8. The equation $\lambda^p - a_1\lambda^{p-1} - \cdots - a_p = 0$ may be obtained with minimum thought in this setting too by letting $x(t) = e^{\lambda t}$ in equation (13.23) and then factoring out the term $e^{\lambda t}$.

Example 13.9. The recipe for solving the third-order differential equation

$$x'''(t) = ax''(t) + bx'(t) + cx(t), \;\; t \geq 0 \text{ and } c \neq 0\,,$$

is:

(1) Solve for the roots $\lambda_1, \lambda_2, \lambda_3$ of the polynomial $\lambda^3 - a\lambda^2 - b\lambda - c$.

(2) The solution is

$$x(t) = \alpha e^{\lambda_1 t} + \beta e^{\lambda_2 t} + \gamma e^{\lambda_3 t} \text{ if } \lambda_1, \lambda_2, \lambda_3 \text{ are all different},$$
$$x(t) = \alpha e^{\lambda_1 t} + \beta t e^{\lambda_1 t} + \gamma e^{\lambda_3 t} \text{ if } \lambda_1 = \lambda_2 \neq \lambda_3,$$
$$x(t) = \alpha e^{\lambda_1 t} + \beta t e^{\lambda_1 t} + \gamma t^2 e^{\lambda_1 t} \text{ if } \lambda_1 = \lambda_2 = \lambda_3.$$

(3) Determine the constants α, β, γ from the initial conditions $x(0), x'(0)$ and $x''(0)$.

Exercise 13.34. Find the solution of the third-order differential equation

$$x^{(3)}(t) = 3x^{(2)}(t) - 3x^{(1)}(t) + x(t) \ , \ t \geq 0 \ ,$$

subject to the initial conditions

$$x(0) = 1 \ , \ x^{(1)}(0) = 2 \ , \ x^{(2)}(0) = 8 \ .$$

Exercise 13.35. Let $\mathbf{u}'(t) = \begin{bmatrix} 0 & \alpha \\ \alpha & 0 \end{bmatrix} \mathbf{u}(t)$ for $t \geq 0$. Show in two different ways that $\|\mathbf{u}(t)\|_2 = \|\mathbf{u}(0)\|_2$ if $\alpha + \overline{\alpha} = 0$: first by showing that the derivative of $\|\mathbf{u}(t)\|_2$ with respect to t is constant and then by invoking Exercise 13.23.

Exercise 13.36. In the setting of Exercise 13.35, describe $\|\mathbf{u}(t)\|_2$ as $t \uparrow \infty$ if $\alpha + \overline{\alpha} \neq 0$.

Exercise 13.37. Evaluate $\lim_{t \uparrow \infty} t^{-2} e^{-2t} \mathbf{y}(t)$ for the solution $\mathbf{y}(t)$ of the equation

$$\mathbf{y}'(t) = \begin{bmatrix} 0 & 1 & 0 \\ 0 & 0 & 1 \\ 8 & -12 & 6 \end{bmatrix} \mathbf{y}(t), \ t \geq 0, \text{ when } \mathbf{y}(0) = \begin{bmatrix} 8 \\ 8 \\ 8 \end{bmatrix}.$$

13.13. Wronskians

To this point we have considered only differential equations with constant coefficients. A significant number of applications involve differential equations with coefficients that also depend upon the independent variable, i.e., equations of the form

(13.24)
$$a_p(t)x^{(p)}(t) + a_{p-1}(t)x^{(p-1)}(t) + \cdots + a_1(t)x^{(1)}(t) + a_0(t)x(t) = g(t)$$

on either a finite or infinite subinterval of $\mathbb{R}$. Although we shall consider only second-order differential equations in the sequel, it is instructive to begin in the more general setting of p'th order differential equations.

Lemma 13.10. *Let $u_1(t), \ldots, u_p(t)$ be solutions of the homogeneous equation*

$$a_p(t)x^{(p)}(t)+a_{p-1}(t)x^{(p-1)}(t)+\cdots+a_1(t)x^{(1)}(t)+a_0(t)x(t)=0\,,\ \alpha\le t\le\beta\,,$$

in which the coefficients are assumed to be continuous real-valued functions on a finite interval $\alpha \le t \le \beta$ with $a_p(t) > 0$ on this interval. Let

$$\varphi(t) = \det\begin{bmatrix} u_1(t) & \cdots & u_p(t) \\ u_1^{(1)}(t) & \cdots & u_p^{(1)}(t) \\ \vdots & & \vdots \\ u_1^{(p-1)}(t) & \cdots & u_p^{(p-1)}(t) \end{bmatrix}. \tag{13.25}$$

Then

$$\varphi(t) = \exp\left\{-\int_\alpha^t \frac{a_{p-1}(s)}{a_p(s)}ds\right\}\varphi(\alpha)\,.$$

Discussion. Let $p = 3$. Then

$$\begin{aligned}\varphi'(t) &= \det\begin{bmatrix} u_1^{(1)}(t) & u_2^{(1)}(t) & u_3^{(1)}(t) \\ u_1^{(1)}(t) & u_2^{(1)}(t) & u_3^{(1)}(t) \\ u_1^{(2)}(t) & u_2^{(2)}(t) & u_3^{(2)}(t) \end{bmatrix} + \det\begin{bmatrix} u_1(t) & u_2(t) & u_3(t) \\ u_1^{(2)}(t) & u_2^{(2)}(t) & u_3^{(2)}(t) \\ u_1^{(2)}(t) & u_2^{(2)}(t) & u_3^{(2)}(t) \end{bmatrix} \\ &\quad + \begin{bmatrix} u_1(t) & u_2(t) & u_3(t) \\ u_1^{(1)}(t) & u_2^{(1)}(t) & u_3^{(1)}(t) \\ u_1^{(3)}(t) & u_2^{(3)}(t) & u_3^{(3)}(t) \end{bmatrix} \\ &= 0+0+\begin{bmatrix} u_1(t) & u_2(t) & u_3(t) \\ u_1^{(1)}(t) & u_2^{(1)}(t) & u_3^{(1)}(t) \\ u_1^{(3)}(t) & u_2^{(3)}(t) & u_3^{(3)}(t) \end{bmatrix} \\ &= -\frac{a_2(t)}{a_3(t)}\varphi(t)\,,\end{aligned}$$

since

$$a_3(t)u_j^{(3)}(t) + a_2(t)u_j^{(2)}(t) + a_1(t)u_j^{(1)}(t) + a_0(t)u_j(t) = 0 \quad \text{for} \quad j = 1, 2, 3\,.$$

But this leads easily to the stated conclusion. It is clear that the same argument is applicable for general p.

The function $\varphi(t)$ defined by formula (13.25) is called the **Wronskian** of the functions $u_1(t), \ldots, u_p(t)$.

Exercise 13.38. Let $u_1(t), \ldots, u_p(t)$ be solutions of the homogeneous differential equation considered in Lemma 13.10. Show that the vectors $\mathbf{u}_j(t)$ that are defined by the formulas $\mathbf{u}_j(t)^T = \begin{bmatrix} u_j(t) & u_j^{(1)}(t) & \cdots & u_j^{(p-1)}(t) \end{bmatrix}$ for $j = 1, \ldots, p$ are linearly independent at one point in the interval $\alpha \le t \le \beta$ if and only if they are linearly independent at every point in the interval.

13.14. Variation of parameters

The method of **variation of parameters** provides a solution to a nonhomogeneous equation of the form (13.24) in terms of linear combinations of the solutions to the corresponding homogeneous equation, in which the coefficients are permitted to depend on the independent variable t. We shall illustrate the method for the second-order differential equation

$$a(t)y''(t) + b(t)y'(t) + c(t)y(t) = g(t)\,. \tag{13.26}$$

Let

$$u(t) = d_1(t)u_1(t) + d_2(t)u_2(t)$$

be a linear combination of solutions $u_1(t)$ and $u_2(t)$ to the homogeneous equation

$$a(t)y''(t) + b(t)y'(t) + c(t)y(t) = 0\,, \tag{13.27}$$

with coefficients $d_1(t)$ and $d_2(t)$ that are allowed to vary with the independent variable t. To explore this idea, note that

$$u'(t) = d_1u_1' + d_2u_2' + d_1'u_1 + d_2'u_2\,,$$

and hence upon choosing $d_1(t)$ and $d_2(t)$ so that

$$d_1'u_1 + d_2'u_2 = 0$$

on the interval $\alpha \le t < \beta$, it follows that

$$u''(t) = d_1u_1'' + d_2u_2'' + d_1'u_1' + d_2'u_2'\,.$$

Therefore

$$\begin{aligned} au'' + bu' + cu &= d_1(au_1'' + bu_1' + cu_1) + d_2(au_2'' + bu_2' + cu_2) \\ &\quad + a(d_1'u_1' + d_2'u_2') \\ &= a(d_1'u_1' + d_2'u_2')\,. \end{aligned}$$

Thus, the problem of interest reduces to finding coefficients $d_1(t)$ and $d_2(t)$ such that

$$\begin{aligned} d_1'u_1 + d_2'u_2 &= 0 \\ a(d_1'u_1' + d_2'u_2') &= g(t) \end{aligned}$$

or, equivalently,

$$\begin{bmatrix} u_1 & u_2 \\ pu_1' & pu_2' \end{bmatrix} \begin{bmatrix} d_1' \\ d_2' \end{bmatrix} = \begin{bmatrix} 0 \\ a^{-1}pg \end{bmatrix} \quad \text{with} \quad p(t) = \exp\left\{\int_\alpha^t \frac{b(s)}{a(s)}ds\right\}.$$

The extra factor $p(t)$ has been introduced in the array of equations in order to take advantage of Lemma 13.10, which guarantees that the determinant of the 2×2 matrix on the left is equal to a constant:

$$\det \begin{bmatrix} u_1 & u_2 \\ pu_1' & pu_2' \end{bmatrix} = \varphi(t)p(t) = \varphi(\alpha)\,.$$

Let $\gamma = \varphi(\alpha)$, for short. If the constant $\gamma = p(u_1u_2' - u_2u_1')$ is nonzero, then

$$\begin{bmatrix} d_1' \\ d_2' \end{bmatrix} = \frac{\begin{bmatrix} pu_2' & -u_2 \\ -pu_1' & u_1 \end{bmatrix}\begin{bmatrix} 0 \\ pg/a \end{bmatrix}}{p(u_1u_2' - u_2u_1')} = \frac{\begin{bmatrix} -u_2pg/a \\ u_1pg/a \end{bmatrix}}{\gamma}.$$

Thus,

(13.28)

$$u(t) = \int_\alpha^t \frac{p(s)}{\gamma a(s)}\left(u_2(t)u_1(s) - u_1(t)u_2(s)\right)g(s)ds + d_1(\alpha)u_1(t) + d_2(\alpha)u_2(t)\,.$$

Exercise 13.39. Verify that the function $u(t)$ specified in formula (13.28) is a solution of the differential equation (13.26) for every choice of the constants $d_1(\alpha)$ and $d_2(\alpha)$. [HINT: The formulas

$$\frac{d}{dt}\int_\alpha^t f(s)ds = f(t) \quad \text{and} \quad \frac{d}{dt}\int_t^\beta f(s)ds = -f(t) \tag{13.29}$$

may be useful.]

Exercise 13.40. Show that if the arbitrary constants in formula (13.28) are specified as $d_2(\alpha) = 0$ and $d_1(\alpha) = \gamma^{-1}\int_\alpha^\beta a(s)^{-1}p(s)u_2(s)g(s)ds$, then the solution can be expressed as

$$u(t) = \int_\alpha^\beta G(t,s)\frac{p(s)}{\gamma a(s)}g(s)ds\,,$$

where

$$G(t,s) = \begin{cases} u_1(t)u_2(s) & \text{if} \quad \alpha \le t \le s \le \beta \\ u_1(s)u_2(t) & \text{if} \quad \alpha \le s \le t \le \beta \end{cases}.$$

(The kernel $G(t,s)$ is called the **Green function** of the problem.)

Exercise 13.41. Use the formulas in Exercise 13.40 to show that for any choice of $a \in \mathbb{R}$ and $b \in \mathbb{R}$, there exist a pair of constants κ_1 and κ_2 such that $au(\alpha)+bu'(\alpha) = \kappa_1(au_1(\alpha)+bu_1'(\alpha))$ and $au(\beta)+bu'(\beta) = \kappa_2(au_2(\beta)+bu_2'(\beta))$.

Chapter 14

Vector valued functions

In my experience, those people who think they know all the answers, don't know all the questions. A Chinese proverb puts it well: Trust only those who doubt.

In this chapter we shall discuss vector valued functions of one and many variables and some of their applications. We begin with some notation for classes of functions with different degrees of smoothness that will prove convenient.

Let Q be an open subset of $\mathbb{R}^n$ and let $\overline{Q}$ denote the closure of Q. A function f that maps Q into $\mathbb{R}$ is said to belong to the class

$\mathcal{C}(Q)$ if f is continuous on Q,

$\mathcal{C}(\overline{Q})$ if f is continuous on $\overline{Q}$,

$\mathcal{C}^k(Q)$ for some positive integer k if f and all its partial derivatives of order up to and including k are continuous on Q,

$\mathcal{C}^k(\overline{Q})$ for some positive integer k if $f \in \mathcal{C}(Q)$ and f and all its partial derivatives of order up to and including k extend continuously to $\overline{Q}$.

A vector valued function $\mathbf{f}$ from Q or $\overline{Q}$ into $\mathbb{R}^m$ is said to belong to one of the four classes listed above if all its components belong to that class. Moreover, on occasion, $\mathbf{f}$ is said to be **smooth** if it belongs to $\mathcal{C}^k(Q)$ for k large enough for the application at hand. The notation $B_r(\mathbf{a})$ and $\overline{B_r(\mathbf{a})}$ for balls of radius $r > 0$ that was introduced in (7.15) will be useful.

- The **warnings** posted in the preceding chapters are still in effect.

14.1. Mean value theorems

We begin with the classical mean value theorem for real-valued functions $f(x)$ of one variable x that are defined on the closed interval

$$[a,b] = \{x \in \mathbb{R} : a \le x \le b\}.$$

The proof can be found in many textbooks (see e.g. [**4**]) and will not be given here.

Theorem 14.1. *Let $f(x)$ be a continuous real-valued function on the finite closed interval $[a,b]$ and suppose that the derivative $f'(x)$ exists for each point x in the open interval*

$$(a,b) = \{x \in \mathbb{R} : a < x < b\}.$$

Then

$$f(b) - f(a) = f'(c)(b-a) \tag{14.1}$$

for some point $c \in (a,b)$.

We turn next to the generalized mean value theorem.

Theorem 14.2. *Let $f(x)$ and $g(x)$ be continuous real-valued functions on the finite closed interval $[a,b]$ and suppose that the derivatives $f'(x)$ and $g'(x)$ exist for each point x in the open interval (a,b). Then*

$$\{f(b) - f(a)\}g'(c) = \{g(b) - g(a)\}f'(c) \tag{14.2}$$

for some point $c \in (a,b)$.

Proof. Let

$$h(x) = f(x)\{g(b) - g(a)\} - g(x)\{f(b) - f(a)\}.$$

Then, by Theorem 14.1,

$$h(b) - h(a) = h'(c)(b-a)$$

for some point $c \in (a,b)$. However, since

$$h(b) - h(a) = 0 \text{ and } b > a,$$

this implies that

$$h'(c) = 0$$

for some point $c \in (a,b)$. But that is the same as the asserted statement. □

Exercise 14.1. Let $p(x) = a_0 + a_1x + \cdots + a_nx^n$ be a polynomial of degree n with n distinct real roots $\alpha_1 < \cdots < \alpha_n$, where $n \ge 2$. Show that $p'(x)$ has $n-1$ real roots $\beta_1 < \cdots < \beta_{n-1}$ such that $\alpha_j < \beta_j < \beta_{j+1}$ for $j = 1, \ldots, n-1$.

Exercise 14.2. Use the mean value theorem to show that if $b > a > 0$, then $\sqrt{ab} - a \le (b-a)/2$.

14.2. Taylor's formula with remainder

Theorem 14.3. *Let $f \in \mathcal{C}^{n-1}([a,b])$ on the finite closed interval $[a,b]$ and suppose that the n'th order derivative $f^{(n)}(x)$ exists for each point x in the open interval (a,b). Then*

$$f(b) = f(a) + \sum_{k=1}^{n-1} f^{(k)}(a)\frac{(b-a)^k}{k!} + f^{(n)}(c)\frac{(b-a)^n}{n!} \tag{14.3}$$

for some point $c \in (a,b)$.

Proof. Let $g \in \mathcal{C}^{n-1}([a,b])$ on the finite closed interval $[a,b]$ be such that the n'th order derivative $g^{(n)}(x)$ exists for each point x in the open interval (a,b) and let

$$\varphi(x) = f(x) + \sum_{k=1}^{n-1} \frac{f^{(k)}(x)}{k!}(b-x)^k$$

and

$$\psi(x) = g(x) + \sum_{k=1}^{n-1} \frac{g^{(k)}(x)}{k!}(b-x)^k.$$

Then $\varphi(x)$ and $\psi(x)$ meet the hypotheses of the generalized mean value theorem, Theorem 14.2. Therefore, by that theorem,

$$\{\varphi(b) - \varphi(a)\}\psi'(c) = \{\psi(b) - \psi(a)\}\varphi'(c)$$

for some point $c \in (a,b)$. Thus, as

$$\varphi(b) = f(b), \quad \psi(b) = g(b)$$

and, by a short calculation,

$$\varphi'(x) = \frac{f^{(n)}(x)}{(n-1)!}(b-x)^{n-1} \quad \text{and} \quad \psi'(x) = \frac{g^{(n)}(x)}{(n-1)!}(b-x)^{n-1},$$

we see that

$$\{f(b) - \varphi(a)\}g^{(n)}(c) = \{g(b) - \psi(a)\}f^{(n)}(c)$$

for some point $c \in (a,b)$. To complete the proof, let

$$g(x) = (x-a)^n.$$

Then, as

$$\begin{aligned} g^{(k)}(a) &= 0 \quad \text{for} \quad k = 0, \dots, n-1, \\ g^{(n)}(x) &= n! \quad \text{for every point} \quad x \in \mathbb{R} \end{aligned}$$

and

$$g(b) = (b-a)^n,$$

the last formula reduces to

$$\{f(b) - \varphi(a)\}n! = (b-a)^n f^{(n)}(c)\,,$$

which is equivalent to formula (14.3). □

14.3. Application of Taylor's formula with remainder

Formula (14.3) is useful for calculating $f(b)$ from $f(a)$ and its derivatives $f^{(1)}(a), f^{(2)}(a), \ldots$, when b is close to a. Thus, upon setting $b = a + h$, we can reexpress formula (14.3) as

$$f(a+h) - \left\{ f(a) + \sum_{k=1}^{n-1} f^{(k)}(a)\frac{h^k}{k!} \right\} = f^{(n)}(c)\frac{h^n}{n!} \tag{14.4}$$

and use the right-hand side to estimate the difference between the true value of $f(a+h)$ and the approximant:

$$f(a) + \sum_{k=1}^{n-1} f^{(k)}(a)\frac{h^k}{k!}$$

Thus, for example, in order to calculate $(27.1)^{5/3}$ to an accuracy of $1/100$, let

$$f(x) = x^{5/3}, a = 27 \text{ and } b = 27.1.$$

Then as

$$f'(x) = \frac{5}{3}x^{2/3} \text{ and } f''(x) = \frac{10}{9}x^{-1/3},$$

the formula

$$f(b) = f(a) + f'(a)(b-a) + f''(c)\frac{(b-a)^2}{2!}$$

translates to

$$(27.1)^{5/3} - \left\{ (27)^{5/3} + \frac{5}{3}(27)^{2/3}\frac{1}{10} \right\} = \frac{10}{9}c^{-1/3}\frac{1}{200};$$

that is

$$\left| (27.1)^{5/3} - \left\{ 3^5 + \frac{3}{2} \right\} \right| = \frac{c^{-1/3}}{180},$$

for some number c that lies between 27 and 27.1. In particular, this constraint implies that $c > 27$ and hence that $c^{-1/3} < \frac{1}{3}$. Consequently

$$\left| (27.1)^{5/3} - \left\{ 3^5 + \frac{3}{2} \right\} \right| \le \frac{(1/3)}{180} = \frac{1}{540}.$$

Thus, the error in approximating $(27.1)^{5/3}$ by $3^5 + \frac{3}{2}$ is less than $1/(540)$.

Exercise 14.3. Show that the error in approximating $(27.1)^{5/3}$ by $(27)^{5/3}$ is bigger than $3/2$.

14.4. Mean value theorem for functions of several variables

Let $f(\mathbf{x}) = f(x_1, \dots, x_n)$ be a real-valued function of the vector $\mathbf{x}$ with components $x_1, \dots, x_n$ and suppose that the partial derivatives $\frac{\partial f}{\partial x_j}(\mathbf{x})$ exist in some region, say $a_1 < x_1 < b_1, \dots, a_n < x_n < b_n$. Then we shall write

$$(\nabla f)(\mathbf{x}) = \begin{bmatrix} \frac{\partial f}{\partial x_1}(\mathbf{x}) & \cdots & \frac{\partial f}{\partial x_n}(\mathbf{x}) \end{bmatrix} \tag{14.5}$$

for the $1 \times n$ row vector with entries $\frac{\partial f}{\partial x_j}(\mathbf{x})$, $j = 1, \dots, n$. The vector $(\nabla f)(\mathbf{x})$ is termed the **gradient** of f.

Theorem 14.4. *Let $Q = \{\mathbf{x} \in \mathbb{R}^n : a_1 < x_1 < b_1, \dots, a_n < x_n < b_n\}$ and let $f(\mathbf{x}) = f(x_1, \dots, x_n)$ be a continuous real-valued function of the variables $x_1, \dots, x_n$ in the bounded closed region $\overline{Q}$ and suppose that the partial derivatives $\frac{\partial f}{\partial x_j}(x_1, \dots, x_n)$ exist for each point $(x_1, \dots, x_n)$ in the open region Q. Then*

$$f(\mathbf{b}) - f(\mathbf{a}) = (\nabla f)(\mathbf{c})(\mathbf{b} - \mathbf{a}) \tag{14.6}$$

for some point $\mathbf{c} = \mathbf{a} + t_0(\mathbf{b} - \mathbf{a})$, $0 < t_0 < 1$, on the open line segment between $\mathbf{a}$ and $\mathbf{b}$.

Proof. Let

$$\begin{aligned} h(t) &= f(\mathbf{a} + t(\mathbf{b} - \mathbf{a})) \\ &= f(x_1(t), \dots, x_n(t)), \quad \text{where} \quad x_j(t) = a_j + t(b_j - a_j) \end{aligned}$$

for $0 \leq t \leq 1$. Then clearly $h(t)$ is continuous on the interval $[0, 1]$, and the derivative

$$\begin{aligned} h'(t) &= \sum_{j=1}^{n} \frac{\partial f}{\partial x_j}(\mathbf{a} + t(\mathbf{b} - \mathbf{a}))(b_j - a_j) \\ &= (\nabla f)(\mathbf{a} + t(\mathbf{b} - \mathbf{a}))(\mathbf{b} - \mathbf{a}) \end{aligned}$$

exists for each point t in the open interval $(0, 1)$. Therefore, by Theorem 14.1, there exists a point $t_0 \in (0, 1)$ such that $h(1) - h(0) = h'(t_0)$. But, in view of the preceding calculation, this is easily seen to be the same as formula (14.6) with $\mathbf{c} = \mathbf{a} + t_0(\mathbf{b} - \mathbf{a})$. $\square$

14.5. Mean value theorems for vector valued functions of several variables

We turn now to vector valued functions

$$\mathbf{f}(\mathbf{x}) = \begin{bmatrix} f_1(x_1, \dots, x_q) \\ \vdots \\ f_p(x_1, \dots, x_q) \end{bmatrix}$$

of several variables. We assume that each of the components $f_i(\mathbf{x})$, $i = 1,\ldots,p$, of $\mathbf{f}(\mathbf{x})$ is real- valued. Thus $\mathbf{f}(\mathbf{x})$ defines a mapping from some subset of $\mathbb{R}^q$ into $\mathbb{R}^p$.

Theorem 14.5. *Assume that each of the components $f_i(\mathbf{x})$, $i = 1,\ldots,p$, of $\mathbf{f}(\mathbf{x})$ is a continuous real-valued function of the variables $x_1,\ldots,x_q$ in the bounded closed region $a_1 \le x_1 \le b_1,\ldots,a_q \le x_q \le b_q$ and that the partial derivatives*

$$\frac{\partial f_i}{\partial x_j}(\mathbf{x}) = \frac{\partial f_i}{\partial x_j}(x_1,\ldots,x_q)$$

exist for each point $(x_1,\ldots,x_q)$ in the open region $a_1 < x_1 < b_1,\ldots,a_q < x_q < b_q$. Then

$$\mathbf{f}(\mathbf{b}) - \mathbf{f}(\mathbf{a}) = \begin{bmatrix} (\nabla f_1)(\mathbf{c}_1) \\ \vdots \\ (\nabla f_p)(\mathbf{c}_p) \end{bmatrix} (\mathbf{b} - \mathbf{a}) \tag{14.7}$$

for some set of points

$$\mathbf{c}_i = \mathbf{a} + t_i(\mathbf{b} - \mathbf{a}),\ 0 < t_i < 1, i = 1,\ldots,p.$$

Proof. This is an immediate consequence of Theorem 14.4, applied to each component $f_i(\mathbf{x})$ of $\mathbf{f}(\mathbf{x})$ separately. □

Corollary 14.6. In the setting of Theorem 14.5,

$$\|\mathbf{f}(\mathbf{b}) - \mathbf{f}(\mathbf{a})\| \le \left\{ \sum_{i=1}^{p} \|\nabla f_i(\mathbf{c}_i)\|^2 \right\}^{1/2} \|\mathbf{b} - \mathbf{a}\|$$

for some set of points $\mathbf{c}_1,\ldots,\mathbf{c}_p$ in the open line segment between $\mathbf{a}$ and $\mathbf{b}$.

Proof. By definition,

$$\|\mathbf{f}(\mathbf{b}) - \mathbf{f}(\mathbf{a})\|^2 = \sum_{i=1}^{p} \{f_i(\mathbf{b}) - f_i(\mathbf{a})\}^2 .$$

Moreover, by Theorem 14.5 and the Cauchy–Schwarz inequality,

$$\begin{aligned} |f_i(\mathbf{b}) - f_i(\mathbf{a})| &= |\langle \mathbf{b} - \mathbf{a}, \nabla f_i(\mathbf{c}_i)^T \rangle| \\ &\le \|\mathbf{b} - \mathbf{a}\| \|\nabla f_i(\mathbf{c}_i)^T\| \\ &= \|\mathbf{b} - \mathbf{a}\| \|\nabla f_i(\mathbf{c}_i)\|. \end{aligned}$$

But this is easily seen to be equivalent to the asserted statement. □

Theorem 14.7. *Let*

$$\mathbf{f}(\mathbf{x}) = \begin{bmatrix} f_1(x_1, \ldots, x_q) \\ \vdots \\ f_p(x_1, \ldots, x_q) \end{bmatrix}$$

be a continuous map from the bounded region $a_1 \leq x_1 \leq b_1, \ldots, a_q \leq x_q \leq b_q$ *into* $\mathbb{R}^p$ *such that all the partial derivatives* $\frac{\partial f_j}{\partial x_k}(\mathbf{x})$, $j = 1, \ldots, p$, $k = 1, \ldots, q$, *exist for* $a_1 < x_1 < b_1, \ldots, a_k < x_q < b_k$ *and let*

$$J_{\mathbf{f}}(\mathbf{x}) = \begin{bmatrix} \frac{\partial f_1}{\partial x_1}(\mathbf{x}) & \cdots & \frac{\partial f_1}{\partial x_q}(\mathbf{x}) \\ \vdots & & \vdots \\ \frac{\partial f_p}{\partial x_1}(\mathbf{x}) & \cdots & \frac{\partial f_p}{\partial x_q}(\mathbf{x}) \end{bmatrix}$$

denote the **Jacobian matrix** *of the mapping* $\mathbf{f}$. *Then*

$$\|\mathbf{f}(\mathbf{b}) - \mathbf{f}(\mathbf{a})\| \leq \|J_{\mathbf{f}}(\mathbf{c})\| \|\mathbf{b} - \mathbf{a}\|$$

for some point $\mathbf{c}$ *on the open line segment between* $\mathbf{a}$ *and* $\mathbf{b}$.

Proof. Let

$$\mathbf{u} = \mathbf{f}(\mathbf{b}) - \mathbf{f}(\mathbf{a})$$

and let

$$h(t) = \mathbf{u}^T \mathbf{f}(\mathbf{a} + t(\mathbf{b} - \mathbf{a})).$$

Then, by the classical mean value theorem,

$$h(1) - h(0) = h'(t_0)(1 - 0)$$

for some point $t_0 \in (0, 1)$. But now as

$$h(1) - h(0) = \mathbf{u}^T \mathbf{f}(\mathbf{b}) - \mathbf{u}^T \mathbf{f}(\mathbf{a}) = \|\mathbf{u}\|^2$$

and

$$\begin{aligned} h'(t) &= \frac{d}{dt} \sum_{j=1}^{p} u_j f_j(\mathbf{a} + t(\mathbf{b} - \mathbf{a})) \\ &= \sum_{j=1}^{p} u_j \sum_{k=1}^{q} \frac{\partial f_j}{\partial x_k}(\mathbf{a} + t(\mathbf{b} - \mathbf{a}))(b_k - a_k) \\ &= \langle J_{\mathbf{f}}(\mathbf{a} + t(\mathbf{b} - \mathbf{a}))(\mathbf{b} - \mathbf{a}), \mathbf{u} \rangle, \end{aligned}$$

the mean value theorem yields the formula

$$\|\mathbf{u}\|^2 = \langle J_{\mathbf{f}}(\mathbf{c})(\mathbf{b} - \mathbf{a}), \mathbf{u} \rangle$$

for some point

$$\mathbf{c} = \mathbf{a} + t_0(\mathbf{b} - \mathbf{a})$$

on the open line segment between $\mathbf{a}$ and $\mathbf{b}$. Thus, by the Cauchy-Schwarz inequality,

$$\|\mathbf{u}\|^2 \leq \|J_{\mathbf{f}}(\mathbf{c})(\mathbf{b}-\mathbf{a})\|\|\mathbf{u}\|,$$

which leads easily to the advertised result. □

14.6. Newton's method

Newton's method is an iterative scheme for solving equations of the form $\mathbf{f}(\mathbf{x}) = \mathbf{0}$ for vector valued functions $\mathbf{f} \in \mathcal{C}^2(Q)$ that map an open subset Q of $\mathbb{R}^p$ into $\mathbb{R}^p$. The underlying idea is that if $\mathbf{x}_0$ is close to a point $\mathbf{u}^\circ$ at which $\mathbf{f}(\mathbf{u}^\circ) = 0$ and the Jacobian matrix $J_{\mathbf{f}}(\mathbf{u}^\circ)$ is invertible, then, since

$$\mathbf{0} = \mathbf{f}(\mathbf{u}^\circ) \approx \mathbf{f}(\mathbf{x}_0) + J_{\mathbf{f}}(\mathbf{x}_0)(\mathbf{u}^\circ - \mathbf{x}_0)$$

and $J_{\mathbf{f}}(\mathbf{x}_0)$ is invertible, the point $\mathbf{u}^\circ$ should be close to

$$\mathbf{x}_1 = \mathbf{x}_0 - J_{\mathbf{f}}(\mathbf{x}_0)^{-1}\mathbf{f}(\mathbf{x}_0)$$

and even closer to

$$\mathbf{x}_2 = \mathbf{x}_1 - J_{\mathbf{f}}(\mathbf{x}_1)^{-1}\mathbf{f}(\mathbf{x}_1)\,,$$

etc.

Theorem 14.8. *Let Q be a nonempty open subset of $\mathbb{R}^p$, let $\mathbf{f} \in \mathcal{C}^1(Q)$ map Q into $\mathbb{R}^p$ and suppose that there exists a point $\mathbf{u}^\circ \in Q$ such that*

(1) $\mathbf{f}(\mathbf{u}^\circ) = \mathbf{0}$.

(2) *The Jacobian matrix*

$$J_{\mathbf{f}}(\mathbf{x}) = \begin{bmatrix} \frac{\partial f_1}{\partial x_1}(\mathbf{x}) & \cdots & \frac{\partial f_1}{\partial x_p}(\mathbf{x}) \\ \vdots & & \vdots \\ \frac{\partial f_p}{\partial x_1}(\mathbf{x}) & \cdots & \frac{\partial f_p}{\partial x_p}(\mathbf{x}) \end{bmatrix}$$

is invertible at $\mathbf{u}^\circ$.

(3) *There exists a pair of numbers $\alpha > 0$ and $\rho > 0$ such that the open ball $B_\rho(\mathbf{u}^\circ) \subset Q$ and*

$$\|J_{\mathbf{f}}(\mathbf{a}) - J_{\mathbf{f}}(\mathbf{b})\| \leq \alpha\|\mathbf{a}-\mathbf{b}\| \quad \textit{for} \quad \mathbf{a}, \mathbf{b} \in B_\rho(\mathbf{u}^\circ)\,.$$

Then there exists a pair of numbers $\beta > 0$ and $\delta > 0$ such that

$$J_{\mathbf{f}}(\mathbf{x}) \quad \textit{is invertible and} \quad \|J_{\mathbf{f}}(\mathbf{x})^{-1}\| \leq \beta \quad \textit{for} \quad \mathbf{x} \in B_\delta(\mathbf{u}^\circ)\,. \tag{14.8}$$

Moreover, if

$$\mathbf{x}_{i+1} = \mathbf{x}_i - J_{\mathbf{f}}(\mathbf{x}_i)^{-1}\mathbf{f}(\mathbf{x}_i)\,, \quad i = 0, 1, \ldots\,, \quad \textit{and} \quad \mathbf{x}_i \in B_\delta(\mathbf{u}^\circ)\,, \tag{14.9}$$

then

$$\|\mathbf{x}_{i+1} - \mathbf{u}^\circ\| \le \frac{\alpha\beta}{2}\|\mathbf{x}_i - \mathbf{u}^\circ\|^2 . \tag{14.10}$$

Proof. Suppose that the vector $\mathbf{x}_i$ belongs to the open ball $B_\delta(\mathbf{u}^\circ)$ for some choice of δ in the interval $0 < \delta \le \rho$ such that the conditions in (14.8) are in force and let

$$\mathbf{x}_{i+1} = \mathbf{x}_i - J_{\mathbf{f}}(\mathbf{x}_i)^{-1}\mathbf{f}(\mathbf{x}_i) .$$

Then

$$\mathbf{x}_{i+1} - \mathbf{u}^\circ = \mathbf{x}_i - \mathbf{u}^\circ - J_{\mathbf{f}}(\mathbf{x}_i)^{-1}\{\mathbf{f}(\mathbf{x}_i) - \mathbf{f}(\mathbf{u}^\circ)\} ,$$

since $\mathbf{f}(\mathbf{u}^\circ) = \mathbf{0}$. Let

$$\mathbf{h}(s) = \mathbf{f}(\mathbf{x}_i + s(\mathbf{u}^\circ - \mathbf{x}_i)) \quad \text{for} \quad 0 \le s \le 1 .$$

Then

$$\begin{aligned} \mathbf{f}(\mathbf{u}^\circ) - \mathbf{f}(\mathbf{x}_i) &= \mathbf{h}(1) - \mathbf{h}(0) = \int_0^1 \mathbf{h}'(s)ds \\ &= \int_0^1 (J_{\mathbf{f}})(\mathbf{x}_i + s(\mathbf{u}^\circ - \mathbf{x}_i))ds(\mathbf{u}^\circ - \mathbf{x}_i) . \end{aligned}$$

Thus,

$$\mathbf{x}_{i+1} - \mathbf{u}^\circ = J_{\mathbf{f}}(\mathbf{x}_i)^{-1} \int_0^1 \{J_{\mathbf{f}}(\mathbf{x}_i + s(\mathbf{u}^\circ - \mathbf{x}_i)) - J_{\mathbf{f}}(\mathbf{x}_i)\}ds(\mathbf{u}^\circ - \mathbf{x}_i) \tag{14.11}$$

and hence by (3),

$$\begin{aligned} \|\mathbf{x}_{i+1} - \mathbf{u}^\circ\| &= \left\| J_{\mathbf{f}}(\mathbf{x}_i)^{-1} \int_0^1 \{J_{\mathbf{f}}(\mathbf{x}_i + s(\mathbf{u}^\circ - \mathbf{x}_i)) - J_{\mathbf{f}}(\mathbf{x}_i)\}ds(\mathbf{u}^\circ - \mathbf{x}_i) \right\| \\ &\le \alpha\beta \int_0^1 \|\mathbf{x}_i + s(\mathbf{u}^\circ - \mathbf{x}_i) - \mathbf{x}_i\|ds\|\mathbf{u}^\circ - \mathbf{x}_i\| \\ &\le \alpha\beta \int_0^1 sds\|\mathbf{u}^\circ - \mathbf{x}_i\|^2 , \end{aligned}$$

which coincides with (14.10). □

Corollary 14.9. If, in the setting of Theorem 14.8, $\delta_1 = \min\{\delta, 2/(\alpha\beta)\}$ and $\mathbf{x}_0 \in B_{\delta_1}(\mathbf{u}^\circ)$, then

$$\mathbf{x}_i \in B_{\delta_1}(\mathbf{u}^\circ) \quad \text{and} \quad \|\mathbf{u}^\circ - \mathbf{x}_{i+1}\| \le \|\mathbf{u}^\circ - \mathbf{x}_i\| \quad \text{for} \quad i = 1, 2, \ldots .$$

Proof. The proof is left to the reader. □

Exercise 14.4. Verify Corollary 14.9.

Exercise 14.5. Show that the **Newton step** (14.11) for solving the equation $x^2 - a = 0$ to find the square roots of $a > 0$ is

$$x_{n+1} = \frac{1}{2}\left(x_n + \frac{a}{x_n}\right) \quad \text{if} \quad x_n \neq 0\,,$$

and calculate x_1, x_2, x_3 when $a = 4$ and $x_0 = \pm 1$.

Exercise 14.6. Show that in the setting of Exercise 14.5

$$\|x_{n+1} - x_n\| \leq \frac{1}{2}\|x_n^{-1}\|\|x_n - x_{n-1}\|^2\,.$$

Exercise 14.7. Show that the Newton step (14.11) for solving the equation $x^3 - a = 0$ to find the cube roots of a is

$$x_{n+1} = \frac{1}{3}\left(2x_n + \frac{a}{x_n^2}\right) \quad \text{if} \quad x_n \neq 0\,,$$

and calculate x_1, x_2, x_3 when $a = 8$ and $x_0 = \pm 1$.

14.7. A contractive fixed point theorem

Theorem 14.10. *Let* $\mathbf{f}(\mathbf{x})$ *be a continuous map of a closed subset* E *of a normed linear space* $\mathcal{X}$ *over* $\mathbb{F}$ *into itself such that*

$$\|\mathbf{f}(\mathbf{b}) - \mathbf{f}(\mathbf{a})\| \leq K\|\mathbf{b} - \mathbf{a}\|$$

for some constant $K, 0 < K < 1$, *and every pair of points* $\mathbf{a}, \mathbf{b}$ *in the set* E. *Then:*

(1) *There is exactly one point* $\mathbf{x}_* \in E$ *such that* $\mathbf{f}(\mathbf{x}_*) = \mathbf{x}_*$.

(2) *If* $\mathbf{x}_0 \in E$ *and* $\mathbf{x}_{n+1} = \mathbf{f}(\mathbf{x}_n)$ *for* $n = 0, 1, \ldots$, *then*

$$\mathbf{x}_* = \lim_{n\uparrow\infty} \mathbf{x}_n\,;$$

i.e., the limit exists and is independent of how the initial point $\mathbf{x}_0$ *is chosen and*

(3) $\|\mathbf{x}_* - \mathbf{x}_n\| \leq \dfrac{K^n}{1-K}\|\mathbf{x}_1 - \mathbf{x}_0\|$.

Proof. Choose any point $\mathbf{x}_0 \in E$ and then define the sequence of points $\mathbf{x}_1, \mathbf{x}_2, \ldots$ by the rule

$$\mathbf{x}_{n+1} = f(\mathbf{x}_n).$$

Then clearly

$$\begin{aligned}
\|\mathbf{x}_2 - \mathbf{x}_1\| &= \|f(\mathbf{x}_1) - f(\mathbf{x}_0)\| \leq K\|\mathbf{x}_1 - \mathbf{x}_0\| \\
\|\mathbf{x}_3 - \mathbf{x}_2\| &= \|f(\mathbf{x}_2) - f(\mathbf{x}_1)\| \leq K\|\mathbf{x}_2 - \mathbf{x}_1\| \leq K^2\|\mathbf{x}_1 - \mathbf{x}_0\| \\
&\vdots \\
\|\mathbf{x}_{n+1} - \mathbf{x}_n\| &\leq K^n\|\mathbf{x}_1 - \mathbf{x}_0\|\,,
\end{aligned}$$

and hence

$$\begin{aligned}\|\mathbf{x}_{n+k}-\mathbf{x}_n\| &\le \|\mathbf{x}_{n+k}-\mathbf{x}_{n+k-1}\|+\cdots+\|\mathbf{x}_{n+1}-\mathbf{x}_n\| \\ &\le (K^{n+k-1}+\cdots+K^n)\|\mathbf{x}_1-\mathbf{x}_0\| \\ &\le \frac{K^n}{1-K}\|\mathbf{x}_1-\mathbf{x}_0\|.\end{aligned}$$

Therefore, since K^n tends to 0 as $n \uparrow \infty$, this last bound guarantees that the sequence $\{\mathbf{x}_n\}$ is a Cauchy sequence in the closed subset E of $\mathcal{X}$. Thus, x_{n+k} converges to a limit $\mathbf{x}_*$ in E as $k \uparrow \infty$, which justifies the inequality in (3). Moreover,

$$\begin{aligned}\|f(\mathbf{x}_*)-\mathbf{x}_*\| &= \|f(\mathbf{x}_*)-f(\mathbf{x}_n)+\mathbf{x}_{n+1}-\mathbf{x}_*\| \\ &\le \|f(\mathbf{x}_*)-f(\mathbf{x}_n)\|+\|\mathbf{x}_{n+1}-\mathbf{x}_*\| \\ &\le K\|\mathbf{x}_*-\mathbf{x}_n\|+\|\mathbf{x}_{n+1}-\mathbf{x}_*\| \\ &\le \frac{2K^n}{1-K}\|\mathbf{x}_1-\mathbf{x}_0\|.\end{aligned}$$

Thus, as this upper bound can be made arbitrarily small by choosing n large, we must have

$$\mathbf{x}_* = f(\mathbf{x}_*)\,;$$

i.e., $\mathbf{x}_*$ is a **fixed point** of f. This establishes the existence of a fixed point. The next step is to verify uniqueness. To this end, suppose that $\mathbf{x}_*$ and $\mathbf{y}_*$ are both fixed points of f in the set E. Then

$$\begin{aligned}0 \le \|\mathbf{x}_*-\mathbf{y}_*\| &= \|f(\mathbf{x}_*)-f(\mathbf{y}_*)\| \\ &\le K\|\mathbf{x}_*-\mathbf{y}_*\|.\end{aligned}$$

Therefore,

$$0 \le (1-K)\|\mathbf{x}_*-\mathbf{y}_*\| \le 0.$$

This proves that

$$\mathbf{x}_* = \mathbf{y}_*.$$

□

Example 14.11. Let $A \in \mathbb{C}^{p\times p}$, $B \in \mathbb{C}^{q\times q}$ and $C \in \mathbb{C}^{p\times q}$. Then the equation

$$X - AXB = C$$

has a unique solution $X \in \mathbb{C}^{p\times q}$ if $\|A\|\|B\| < 1$. (Much stronger results will be obtained in Chapter 18.)

Discussion. Let

$$f(X) = C + AXB\,.$$

Then clearly the function f maps $X \in \mathbb{C}^{p\times q}$ into $f(X) \in \mathbb{C}^{p\times q}$, and X is a solution of the given equation if and only if $f(X) = X$. The conclusion now follows from the last theorem (with $E = \mathbb{C}^{p\times q}$ identified as $\mathbb{C}^{pq}$) and the observation that

$$\|f(X) - f(Y)\| = \|A(X - Y)B\| \leq \|A\|\|X - Y\|\|B\|.$$

Exercise 14.8. Let $E = \{x \in \mathbb{R} : 0 \leq x \leq 1\}$ and let $f(x) = (1 + x^2)/2$. Show that:

(a) f maps E into E.

(b) There does not exist a positive constant $\gamma < 1$ such that $|f(b) - f(a)| \leq \gamma|b - a|$ for every choice of $a, b \in E$.

(c) f has exactly one fixed point $x_* \in E$.

Exercise 14.9. Show that the polynomial $p(x) = 1 - 4x + x^2 - x^3$ has at least one root in the interval $0 \leq x \leq 1$. [HINT: Use the fixed point theorem.]

Exercise 14.10. Show that the function $\mathbf{f}(x, y) = \begin{bmatrix} (1 - y)/2 \\ (1 + x^2)/3 \end{bmatrix}$ has a fixed point inside the set of points $(x, y) \in \mathbb{R}^2 : x^2 + y^2 \leq 1$.

Exercise 14.11. Show that if $A \in \mathbb{R}^{p\times p}$ and $\|I_p - A\| < 1$, then, for any choice of $\mathbf{b} \in \mathbb{R}^p$ and $\mathbf{u}_0 \in \mathbb{R}^p$, the vectors $\mathbf{u}_{n+1} = \mathbf{b} + (I_p - A)\mathbf{u}_n$ converge to a solution $\mathbf{x}$ of the equation $A\mathbf{x} = \mathbf{b}$ as n tends to infinity. [HINT: $A\mathbf{x} = \mathbf{b}$ if and only if $\mathbf{b} + (I_p - A)\mathbf{x} = \mathbf{x}$.]

14.8. A refined contractive fixed point theorem

The fixed point theorem that we proved above assumed that $\mathbf{f}(\mathbf{x})$ was a continuous mapping of a closed subset E of a normed linear space $\mathcal{V}$ over $\mathbb{F}$ into itself such that

$$\|\mathbf{f}(\mathbf{x}) - \mathbf{f}(\mathbf{y})\| \leq K\|\mathbf{x} - \mathbf{y}\|$$

for some constant K, $0 < K < 1$, for every pair of vectors $\mathbf{x}$ and $\mathbf{y}$ in E. The next theorem relaxes this constraint.

Theorem 14.12. *Let $\mathbf{f}(\mathbf{x})$ be a continuous map of a closed subset E of a normed linear space $\mathcal{V}$ over $\mathbb{F}$ into itself such that the j'th iterate*

$$\mathbf{f}^{[j]} = \mathbf{f} \circ \mathbf{f} \circ \cdots \circ \mathbf{f}$$

of f satisfies the constraint

$$\left\|\mathbf{f}^{[j]}(\mathbf{x}) - \mathbf{f}^{[j]}(\mathbf{y})\right\| \leq K\|\mathbf{x} - \mathbf{y}\|$$

for some constant K, $0 < K < 1$, with respect to any norm $\|\ \|$ on $\mathcal{V}$. Then $\mathbf{f}$ has a unique fixed point $\mathbf{x}_*$ in E.

Proof. Let $\mathbf{g}(\mathbf{x}) = \mathbf{f}^{[j]}(\mathbf{x})$. Then, by Theorem 14.7, $\mathbf{g}$ has a unique fixed point $\mathbf{x}_*$. Moreover,

$$\mathbf{f}(\mathbf{x}_*) = \mathbf{f}(\mathbf{g}(\mathbf{x}_*)) = \mathbf{g}(\mathbf{f}(\mathbf{x}_*)) .$$

But this exhibits $\mathbf{f}(\mathbf{x}_*)$ as a fixed point of $\mathbf{g}$. Thus, as $\mathbf{g}$ has only one fixed point, we must have

$$\mathbf{f}(\mathbf{x}_*) = \mathbf{x}_*;$$

i.e., $\mathbf{x}_*$ is a fixed point of $\mathbf{f}$. □

Example 14.13. Let $A \in \mathbb{C}^{p\times p}$, $B \in \mathbb{C}^{q\times q}$ and $C \in \mathbb{C}^{p\times q}$. Then the equation

$$X - AXB = C$$

has a unique solution $X \in \mathbb{C}^{p\times q}$ if $\|A^n\|\|B^n\| < 1$ for some positive integer n.

This conclusion is stronger than the one obtained in the last section. It rests on the observation that the j'th iterate of the function $f(X) = C + AXB$ satisfies the inequality

$$\|f^{[j]}(X) - f^{[j]}(Y)\| = \|A^j(X-Y)B^j\| \leq \|A^j\|\|X-Y\|\|B^j\| .$$

14.9. Spectral radius

The last example (and in fact earlier considerations on the growth of the solutions to the equations studied in Chapter 13 indicates the importance of estimates of the size of $\|A^n\|$. The next theorem provides a remarkable connection between these numbers and the **spectral radius**

(14.12) $$r_\sigma(A) = \max\{|\lambda| : \lambda \in \sigma(A)\} \quad \text{for} \quad A \in \mathbb{F}^{p\times p} .$$

It is easy to obtain a bound:

Lemma 14.14. *Let $A \in \mathbb{F}^{p\times p}$. Then*

(14.13) $$r_\sigma(A) \leq \|A^n\|^{1/n}$$

for every positive integer n.

Proof. Let $A\mathbf{x} = \lambda\mathbf{x}$ for some nonzero vector $\mathbf{x} \in \mathbb{C}^p$. Then, since $A^n\mathbf{x} = \lambda^n\mathbf{x}$, it is readily seen that

$$|\lambda^n|\|\mathbf{x}\| = \|A^n\mathbf{x}\| \leq \|A^n\|\|\mathbf{x}\|$$

and hence that

$$|\lambda^n| \leq \|A^n\|$$

for every eigenvalue λ of A. Therefore, the spectral radius $r_\sigma(A)$ of the matrix J is clearly subject to the bound (14.13) for every positive integer n. □

It is a little harder to obtain appropriate upper bounds on $\|A^n\|^{1/n}$. It is convenient to first establish the following result:

Lemma 14.15. *If* $A \in \mathbb{F}^{p\times p}$ *is similar to* $B \in \mathbb{F}^{p\times p}$ *and if* $\lim_{n\uparrow\infty} \|A^n\|^{1/n}$ *exists, then*

$$\lim_{n\uparrow\infty} \|A^n\|^{1/n} = \lim_{n\uparrow\infty} \|B^n\|^{1/n}.$$

Proof. By assumption, there exists an invertible matrix $U \in \mathbb{F}^{p\times p}$ such that $A = UBU^{-1}$. Therefore,

$$\|A^n\| = \|UB^nU^{-1}\| \leq \|U\|\|B^n\|\|U^{-1}\|$$

and

$$\|A^n\|^{1/n} \leq \|U\|^{1/n}\|B^n\|^{1/n}\|U^{-1}\|^{1/n} = \|B^n\|^{1/n}(1+\varepsilon_n),$$

where

$$(1+\varepsilon_n) = \{\|U\|\|U^{-1}\|\}^{1/n} \geq \|UU^{-1}\|^{1/n} = 1$$

and $\varepsilon_n \downarrow 0$ as $n \uparrow \infty$. The proof is now easily completed, since the formula $B = U^{-1}AU$ yields the supplementary bound

$$\|B^n\|^{1/n} \leq \|U^{-1}\|^{1/n}\|A^n\|^{1/n}\|U\|^{1/n} = \|A^n\|^{1/n}(1+\varepsilon_n).$$

□

Exercise 14.12. Show that the numbers ε_n that are defined in the proof of Lemma 14.15 tend monotonically to zero as $n \uparrow \infty$.

Theorem 14.16. *Let* $A \in \mathbb{F}^{p\times p}$. *Then*

$$\lim_{n\uparrow\infty} \|A^n\|^{1/n} = r_\sigma(A); \tag{14.14}$$

i.e., the indicated limit exists and is equal to the spectral radius of A.

Proof. Let A be similar to a Jordan matrix J. Then $r_\sigma(A) = r_\sigma(J)$, and hence, in view of Lemma 14.15, it suffices to show that

$$\lim_{n\uparrow\infty} \|J^n\|^{1/n} = r_\sigma(J). \tag{14.15}$$

Since (14.15) is clear if $r_\sigma(J) = 0$, it is necessary to consider only the case $r_\sigma(J) > 0$.

Suppose first that $J = C_\mu^{(p)}$ is a single Jordan cell with $\mu \neq 0$. Then $J = B + C$, with $B = \mu I_p$ and $C = C_\mu^{(p)} - \mu I_p = C_0^{(p)}$. Therefore, since

$$BC = CB \quad \text{and} \quad C^k = 0 \quad \text{for} \quad k \geq p,$$

the binomial theorem is applicable: If $n > p$, then

$$\begin{aligned} J^n &= \sum_{k=0}^{n} \binom{n}{k} B^{n-k} C^k \\ &= B^n + \binom{n}{1} B^{n-1} C + \cdots + \binom{n}{p-1} B^{n-p+1} C^{p-1} \\ &= B^{n-p+1} \left\{ B^{p-1} + \binom{n}{1} B^{p-2} C + \cdots + \binom{n}{p-1} C^{p-1} \right\} \end{aligned}$$

and, since $\|B\| = |\mu|$ and $\|C\| = 1$,

$$\begin{aligned} \|J^n\| &\leq \|B^{n-p+1}\| \left\{ \|B^{p-1}\| + n\|B^{p-2}C\| + \cdots + n^{p-1}\|C^{p-1}\| \right\} \\ &\leq \|B^{n-p+1}\| n^{p-1} (1 + \|B\|)^{p-1} \\ &= |\mu|^n n^{p-1} (1 + |\mu|^{-1})^{p-1} . \end{aligned}$$

Therefore,

$$\|J^n\|^{1/n} \leq |\mu|(1 + \delta_n) , \tag{14.16}$$

where

$$\begin{aligned} 1 + \delta_n &= \left\{ n(1 + |\mu|^{-1}) \right\}^{(p-1)/n} \\ &= \exp \left\{ \frac{p-1}{n} \left[\ln n + \ln(1 + |\mu|^{-1}) \right] \right\} \\ &\rightarrow 1 \quad \text{as} \quad n \uparrow \infty . \end{aligned}$$

Thus, the two bounds (14.13) and (14.16) imply that

$$r_\sigma(J) \leq \|J^n\|^{1/n} \leq r_\sigma(J)\{1 + \delta_n\} ,$$

which serves to complete the verification of (14.15) when $J = C_\mu^{(p)}$ is a single Jordan cell with $\mu \neq 0$, since $\delta_n \rightarrow 0$ as $n \uparrow \infty$.

The next step is to observe that formula (14.15) holds if the $p \times p$ matrix

$$J = \operatorname{diag}\{J_1, \ldots, J_r\} ,$$

where

$$J_i = C_{\mu_i}^{(\nu_i)}$$

is a Jordan cell of size $\nu_i \times \nu_i$ with $\mu_i \neq 0$ for $i = 1, \ldots, r$. Then

$$\begin{aligned} \|J^n\|^{1/n} &= \max\{\|J_1^n\|^{1/n}, \ldots, \|J_r^n\|^{1/n}\} \\ &\leq \max\{|\mu_1| n^p, \ldots, |\mu_r| n^p\} \\ &\leq r_\sigma(J) n^p \quad \text{for large enough} \quad n . \end{aligned}$$

□

Remark 14.17. Formula (14.14) is valid in a much wider context than was considered here; see e.g., Chapter 18 of W. Rudin [**60**].

Theorem 14.18. *Let A and B be $p \times p$ matrices that commute. Then*

$$\sigma(A+B) \subseteq \sigma(A) + \sigma(B).$$

Proof. Let $\mathbf{u}$ be an eigenvector of $A+B$ corresponding to the eigenvalue μ. Then

$$(A+B)\mathbf{u} = \mu\mathbf{u}$$

and hence, since $BA = AB$,

$$(A+B)B\mathbf{u} = B(A+B)\mathbf{u} = \mu B\mathbf{u}\,;$$

that is to say, $\mathcal{N}_{(A+B-\mu I_p)}$ is invariant under B. Therefore, by Theorem 4.2, there exists an eigenvector $\mathbf{v}$ of B in this null space. This is the same as to say that

$$(A+B)\mathbf{v} = \mu\mathbf{v} \text{ and } B\mathbf{v} = \beta\mathbf{v}$$

where β is an eigenvalue of B. But this in turn implies that

$$A\mathbf{v} = (\mu - \beta)\mathbf{v}\,;$$

i.e., the number $\alpha = \mu - \beta$ is an eigenvalue of A. Thus we have shown that

$$\mu \in \sigma(A+B) \Longrightarrow \mu = \alpha + \beta\,, \quad \text{where} \quad \alpha \in \sigma(A) \text{ and } \beta \in \sigma(B).$$

But that is exactly what we wanted to prove. □

Theorem 14.19. *If A and B are $p \times p$ matrices such that $AB = BA$, then*

(1) $r_\sigma(A+B) \le r_\sigma(A) + r_\sigma(B)$.

(2) $r_\sigma(AB) \le r_\sigma(A) r_\sigma(B)$.

Proof. The first assertion is an immediate consequence of Theorem 14.18 and the definition of spectral radius. The second is left to the reader as an exercise. □

Exercise 14.13. Verify the second assertion in Theorem 14.19.

Exercise 14.14. Verify the first assertion in Theorem 14.19 by estimating $\|(A+B)^n\|$ with the aid of the binomial theorem. [REMARK: This is not as easy as the proof furnished above, but has the advantage of being applicable in wider circumstances.]

Exercise 14.15. Show that if $A, B \in \mathbb{C}^{n\times n}$, then $r_\sigma(AB) = r_\sigma(BA)$, even if $AB \ne BA$. [HINT: Recall formula (5.17).]

Exercise 14.16. Show that if $A = \begin{bmatrix} 1 & 1 \\ 0 & 0 \end{bmatrix}$ and $B = \begin{bmatrix} 1 & 0 \\ 1 & 0 \end{bmatrix}$, then

$$r_\sigma(AB) > r_\sigma(A)\, r_\sigma(B) \quad \text{and} \quad r_\sigma(A+B) > r_\sigma(A) + r_\sigma(B)\,.$$

Exercise 14.17. Show that if A is a normal matrix, then $r_\sigma(A) = \|A\|$.

Exercise 14.18. Show that if $A, B \in \mathbb{C}^{n\times n}$, then $r_{A+B} \le r_\sigma(A) + \|B\|$, even if the two matrices do not commute.

14.10. The Brouwer fixed point theorem

A set K is said to have the **fixed point property** if for every continuous mapping T of K into K, there is an $\mathbf{x} \in K$ such that $T\mathbf{x} = \mathbf{x}$. This section is devoted to the Brouwer fixed point theorem, which states that the closed unit ball has the fixed point property. The proof rests on the following preliminary result:

Theorem 14.20. *Let $\overline{B} = \{x \in \mathbb{R}^n : \|x\| \leq 1\}$. There does not exist a function $\mathbf{f} \in \mathcal{C}^2(\overline{B})$ that maps $\overline{B}$ into its boundary $S = \{\mathbf{x} \in \mathbb{R}^n : \|\mathbf{x}\| = 1\}$ such that $f(\mathbf{x}) = \mathbf{x}$ for every point $\mathbf{x} \in S$.*

Discussion. Suppose to the contrary that there does exist a function $\mathbf{f} \in \mathcal{C}^2(\overline{B})$ that maps $\overline{B}$ into its boundary S such that $f(\mathbf{x}) = \mathbf{x}$ for every point $\mathbf{x} \in S$, and, to ease the exposition, let us focus on the case $n = 3$, so that

$$\mathbf{f}(\mathbf{x}) = \begin{bmatrix} f_1(\mathbf{x}) \\ f_2(\mathbf{x}) \\ f_3(\mathbf{x}) \end{bmatrix} = \begin{bmatrix} f_1(x_1, x_2, x_3) \\ f_2(x_1, x_2, x_3) \\ f_3(x_1, x_2, x_3) \end{bmatrix}.$$

Let

$$D_{\mathbf{f}}(\mathbf{x}) = \det J_{\mathbf{f}}(\mathbf{x}) = \det \begin{bmatrix} \frac{\partial f_1}{\partial x_1}(\mathbf{x}) & \frac{\partial f_1}{\partial x_2}(\mathbf{x}) & \frac{\partial f_1}{\partial x_3}(\mathbf{x}) \\ \frac{\partial f_2}{\partial x_1}(\mathbf{x}) & \frac{\partial f_2}{\partial x_2}(\mathbf{x}) & \frac{\partial f_2}{\partial x_3}(\mathbf{x}) \\ \frac{\partial f_3}{\partial x_1}(\mathbf{x}) & \frac{\partial f_3}{\partial x_2}(\mathbf{x}) & \frac{\partial f_3}{\partial x_3}(\mathbf{x}) \end{bmatrix}.$$

Then, since $\mathbf{f}$ maps $\overline{B}$ into the boundary S,

$$1 = \|\mathbf{f}(\mathbf{x})\|^2 = \sum_{i=1}^{3} f_i(\mathbf{x})^2 \quad \text{for every point} \quad \mathbf{x} \in \overline{B}.$$

Therefore,

$$0 = \frac{\partial}{\partial x_j}\left(\sum_{i=1}^{3} f_i(\mathbf{x})^2\right) = 2\sum_{i=1}^{3} f_i(\mathbf{x})\frac{\partial f_i}{\partial x_j}(\mathbf{x})$$

for $j = 1, 2, 3$ and $\mathbf{x} \in B$, and consequently

$$\begin{bmatrix} f_1(\mathbf{x}) & f_2(\mathbf{x}) & f_3(\mathbf{x}) \end{bmatrix} \begin{bmatrix} \frac{\partial f_1}{\partial x_1}(\mathbf{x}) & \frac{\partial f_1}{\partial x_2}(\mathbf{x}) & \frac{\partial f_1}{\partial x_3}(\mathbf{x}) \\ \frac{\partial f_2}{\partial x_1}(\mathbf{x}) & \frac{\partial f_2}{\partial x_2}(\mathbf{x}) & \frac{\partial f_2}{\partial x_3}(\mathbf{x}) \\ \frac{\partial f_3}{\partial x_1}(\mathbf{x}) & \frac{\partial f_3}{\partial x_2}(\mathbf{x}) & \frac{\partial f_3}{\partial x_3}(\mathbf{x}) \end{bmatrix} = \begin{bmatrix} 0 & 0 & 0 \end{bmatrix}$$

if $\mathbf{x} \in B$. Thus, $D_{\mathbf{f}}(\mathbf{x}) = 0$ for every point $\mathbf{x} \in B$. Moreover, if $M_{ij}(\mathbf{x})$ denotes the ij minor of the matrix under consideration, then

$$\begin{aligned} D_{\mathbf{f}}(\mathbf{x}) &= \sum_{j=1}^{3}(-1)^{1+j}\frac{\partial f_1}{\partial x_j}(\mathbf{x})M_{1j}(\mathbf{x}) \\ &= \sum_{j=1}^{3}(-1)^{1+j}\left\{\frac{\partial}{\partial x_j}(f_1M_{1j}) - f_1\frac{\partial M_{1j}}{\partial x_j}\right\}. \end{aligned}$$

Next, in order to evaluate the sum of the second terms on the right, it is convenient to let

$$\mathbf{g} = \begin{bmatrix} f_2 \\ f_3 \end{bmatrix}$$

and then to note that

$$\begin{aligned} &\frac{\partial M_{11}}{\partial x_1} - \frac{\partial M_{12}}{\partial x_2} + \frac{\partial M_{13}}{\partial x_3} \\ &= \frac{\partial}{\partial x_1}\det\left[\frac{\partial \mathbf{g}}{\partial x_2}\ \frac{\partial \mathbf{g}}{\partial x_3}\right] - \frac{\partial}{\partial x_2}\det\left[\frac{\partial \mathbf{g}}{\partial x_1}\ \frac{\partial \mathbf{g}}{\partial x_3}\right] + \frac{\partial}{\partial x_3}\det\left[\frac{\partial \mathbf{g}}{\partial x_1}\ \frac{\partial \mathbf{g}}{\partial x_2}\right] \\ &= \det\left[\frac{\partial^2 \mathbf{g}}{\partial x_1\partial x_2}\ \frac{\partial \mathbf{g}}{\partial x_3}\right] + \det\left[\frac{\partial \mathbf{g}}{\partial x_2}\ \frac{\partial^2 \mathbf{g}}{\partial x_1\partial x_3}\right] \\ &- \det\left[\frac{\partial^2 \mathbf{g}}{\partial x_2\partial x_1}\ \frac{\partial \mathbf{g}}{\partial x_3}\right] - \det\left[\frac{\partial \mathbf{g}}{\partial x_1}\ \frac{\partial^2 \mathbf{g}}{\partial x_2\partial x_3}\right] \\ &+ \det\left[\frac{\partial^2 \mathbf{g}}{\partial x_3\partial x_1}\ \frac{\partial \mathbf{g}}{\partial x_2}\right] + \det\left[\frac{\partial \mathbf{g}}{\partial x_1}\ \frac{\partial^2 \mathbf{g}}{\partial x_3\partial x_2}\right] = 0\,. \end{aligned}$$

Thus, to this point we know that

$$\begin{aligned} 0 &= \int\!\!\int\!\!\int_B D_{\mathbf{f}}(\mathbf{x})dx_1dx_2dx_3 \\ &= \int\!\!\int\!\!\int_B \sum_{j=1}^{3}(-1)^{1+j}\frac{\partial}{\partial x_j}(f_1M_{1j})dx_1dx_2dx_3\,. \end{aligned}$$

The next step is to evaluate the last integral another way with the aid of **Gauss' divergence theorem**, which serves to reexpress the volume integral of interest in terms of a surface integral over the boundary of B:

$$\int\!\!\int\!\!\int_B \left(\frac{\partial g_1}{\partial x_1} + \frac{\partial g_2}{\partial x_2} + \frac{\partial g_3}{\partial x_3}\right) dx_1dx_2dx_3 = \int\!\!\int_S (g_1x_1 + g_2x_2 + g_3x_3)\,d\sigma\,.$$

This implies that

$$\int\int\int_B D_{\mathbf{f}}(\mathbf{x})dx_1dx_2dx_3 = \int\int_S f_1(\mathbf{x})\sum_{j=1}^{3}(-1)^{1+j}x_jM_{1j}(\mathbf{x})d\sigma$$

and leads to the problem of evaluating

$$\sum_{j=1}^{3}(-1)^{1+j}x_jM_{1j} = \det\begin{bmatrix} x_1 & x_2 & x_3 \\ \frac{\partial f_2}{\partial x_1} & \frac{\partial f_2}{\partial x_2} & \frac{\partial f_2}{\partial x_3} \\ \frac{\partial f_3}{\partial x_1} & \frac{\partial f_3}{\partial x_2} & \frac{\partial f_3}{\partial x_3}\end{bmatrix}$$

on the boundary S of the ball B. To this end, let $\mathbf{x}(t)$, $-1 \leq t \leq 1$, be a smooth curve in S such that $\mathbf{x}(0) = \mathbf{u}$ and $\mathbf{x}'(0) = \mathbf{v}$. Then

$$\frac{df_i(\mathbf{x}(t))}{dt} = \frac{\partial f_i}{\partial x_1}x_1'(t) + \frac{\partial f_i}{\partial x_2}x_2'(t) + \frac{\partial f_i}{\partial x_3}x_3'(t)\,.$$

However, since $f_i(\mathbf{x}(t)) = x_i(t)$ this last expression is also equal to $x_i'(t)$. Thus, writing the gradient $\operatorname{grad} f_i(\mathbf{u}) = \nabla f_i(\mathbf{u})$ as a column vector,

$$\langle \operatorname{grad} f_i(\mathbf{u}) - \mathbf{e}_i, \mathbf{v}\rangle = 0 \quad \text{for} \quad \mathbf{u} \in S$$

for every choice of $\mathbf{v}$ that is tangent to S at the point $\mathbf{u}$. Therefore, $\operatorname{grad} f_i(\mathbf{u}) - \mathbf{e}_i = \lambda_i\mathbf{u}$ for some constant $\lambda_i \in \mathbb{R}$. In other notation,

$$\operatorname{grad} f_i(\mathbf{x}) - \mathbf{e}_i = \lambda_i\mathbf{x}\,.$$

Thus, the determinant of interest is equal to

$$\det\begin{bmatrix} x_1 & x_2 & x_3 \\ \lambda_2x_1 & \lambda_2x_2+1 & \lambda_2x_3 \\ \lambda_3x_1 & \lambda_3x_2 & \lambda_3x_3+1\end{bmatrix} = x_1\,,$$

which leads to the contradiction

$$0 = \int\int_S x_1^2d\sigma\,.$$

Therefore there does not exist a function $\mathbf{f} \in \mathcal{C}^2(\overline{B})$ that maps $\overline{B}$ into its boundary S such that $f(\mathbf{x}) = \mathbf{x}$ for every point $\mathbf{x} \in S$. □

Theorem 14.21. *Let $f(x)$ be a continuous mapping of the closed unit ball*

$$\overline{B} = \{\mathbf{x} \in \mathbb{R}^n : \|\mathbf{x}\| \leq 1\}$$

into itself. Then there is a point $\mathbf{x} \in \overline{B}$ such that $f(\mathbf{x}) = \mathbf{x}$.

Proof. If the theorem is false, then there exists a continuous function $\mathbf{f}(\mathbf{x})$ that maps $\overline{B}$ into itself such that $\|\mathbf{f}(\mathbf{x}) - \mathbf{x}\| > 0$ for every point $\mathbf{x} \in \overline{B}$. Therefore, since $\overline{B}$ is compact and $\mathbf{f}(\mathbf{x}) - \mathbf{x}$ is continuous on $\overline{B}$, there exists an $\varepsilon > 0$ such that $\|\mathbf{f}(\mathbf{x}) - \mathbf{x}\| \geqslant \varepsilon$ for every point $\mathbf{x} \in \overline{B}$.

Let $\mathbf{g} \in \mathcal{C}^2(\overline{B})$ be a mapping of $\overline{B}$ into itself such that $\|\mathbf{g}(\mathbf{x}) - \mathbf{f}(\mathbf{x})\| \leqslant \varepsilon/2$ for $\mathbf{x} \in \overline{B}$. Then $\|\mathbf{g}(\mathbf{x}) - \mathbf{x}\| \geqslant \varepsilon/2$ for $\mathbf{x} \in \overline{B}$.

Now choose a point $\mathbf{b}(\mathbf{x})$ on the line generated by $\mathbf{x}$ and $\mathbf{g}(\mathbf{x})$ such that $\|\mathbf{b}(\mathbf{x})\| = 1$ and $\mathbf{x}$ lies between $\mathbf{b}(\mathbf{x})$ and $\mathbf{g}(\mathbf{x})$ in the sense that

$$\mathbf{x} = t\mathbf{b}(\mathbf{x}) + (1-t)\mathbf{g}(\mathbf{x})$$

for some $0 < t \leq 1$; $t = 0$ is ruled out since $\|\mathbf{g}(\mathbf{x}) - \mathbf{x}\| \geqslant \varepsilon/2$. Then

$$\mathbf{b}(\mathbf{x}) = \mathbf{x} + \left(\frac{1-t}{t}\right)(\mathbf{x} - \mathbf{g}(\mathbf{x})) = \mathbf{x} + c(\mathbf{x})(\mathbf{x} - \mathbf{g}(\mathbf{x})),$$

where the coefficient $c(\mathbf{x}) = (1-t)/t$ is nonnegative and may be expressed as

$$c(\mathbf{x}) = \frac{-\langle \mathbf{x} - \mathbf{g}(\mathbf{x}), \mathbf{x}\rangle + \{\langle \mathbf{x} - \mathbf{g}(\mathbf{x}), \mathbf{x}\rangle^2 + \|\mathbf{x} - \mathbf{g}(\mathbf{x})\|^2(1 - \|\mathbf{x}\|^2)\}^{1/2}}{\|\mathbf{x} - \mathbf{g}(\mathbf{x})\|^2},$$

since $\mathbf{c}(\mathbf{x}) \geqslant 0$ and $\|\mathbf{b}(\mathbf{x})\| = 1$. But this exhibits $\mathbf{b}(\mathbf{x})$ as a function of class $\mathcal{C}^2(\overline{B})$ such that $\mathbf{b}(\mathbf{x}) = \mathbf{x}$ for points $\mathbf{x} \in \overline{B}$ with $\|\mathbf{x}\| = 1$, which is impossible in view of Theorem 14.20. □

The Brouwer fixed point theorem can be strengthened to: *Every closed bounded convex subset of* $\mathbb{R}^n$ *has the fixed point property*; see Chapter 22.

There are also more general versions in infinite dimensional spaces:

The Leray-Schauder Theorem: Every compact convex subset in a Banach space has the fixed point property; see e.g. [**62**], for a start.

14.11. Bibliographical notes

The discussion of Newton's method is adapted from [**57**]. A more sophisticated version due to Kantorovich may be found in the book [**62**] by Saaty and Bram. The discussion of Theorem 14.20 is adopted from an expository article by Yakar Kannai [**41**]. The proof of Theorem 14.21 is adapted from [**62**].

Chapter 15

The implicit function theorem

It seems that physicists do not object to rigorous proofs provided that they are short and simple. I have much sympathy with this point of view. Unfortunately it has not always been possible to provide proofs of this kind.

E. C. Titchmarsh [**67**]

This chapter is devoted primarily to the implicit function theorem and a few of its applications. The last two sections are devoted to an application of vector calculus to dynamical systems and a test for their stability.

15.1. Preliminary discussion

To warm up, consider first the problem of describing the set of solutions $\mathbf{u} \in \mathbb{R}^n$ to the equation $A\mathbf{u} = \mathbf{b}$, when $A \in \mathbb{R}^{p\times n}$, $\mathbf{b} \in \mathbb{R}^p$ and $\operatorname{rank} A = p$. The rank condition implies that there exists an $n \times n$ permutation matrix P such that the last p columns of the matrix AP are linearly independent. Thus, upon writing

$$AP = \begin{bmatrix} A_{11} & A_{12} \end{bmatrix} \quad \text{and} \quad \begin{bmatrix} \mathbf{x} \\ \mathbf{y} \end{bmatrix} = P^T\mathbf{u}$$

with $A_{12} \in \mathbb{R}^{p\times p}$ invertible, $\mathbf{x} \in \mathbb{R}^q$, $\mathbf{y} \in \mathbb{R}^p$ and $n = p + q$, the original equation can be rewritten as

$$\begin{aligned} \mathbf{0} &= \mathbf{b} - A\mathbf{u} = \mathbf{b} - APP^T\mathbf{u} \\ &= \mathbf{b} - \begin{bmatrix} A_{11} & A_{12} \end{bmatrix} \begin{bmatrix} \mathbf{x} \\ \mathbf{y} \end{bmatrix} = \mathbf{b} - A_{11}\mathbf{x} - A_{12}\mathbf{y}\,. \end{aligned}$$

Thus,

$$A\mathbf{u} - \mathbf{b} = \mathbf{0} \Longleftrightarrow \mathbf{y} = A_{12}^{-1}(\mathbf{b} - A_{11}\mathbf{x})\,; \tag{15.1}$$

i.e., the constraint $A\mathbf{u} - \mathbf{b} = \mathbf{0}$ implicitly prescribes some of the entries in bu in terms of the others. The implicit function theorem is based on the same circle of ideas applied to the problem of describing the set of vectors $\mathbf{u} \in \mathbb{R}^n$ of the vector equation

$$\mathbf{g}(\mathbf{u}) = \mathbf{0}$$

when $\mathbf{g}(\mathbf{u}) \in \mathbb{R}^p$. The idea is that if $\mathbf{g}(\mathbf{a}) = \mathbf{0}$ and $\mathbf{u}$ is close to $\mathbf{a}$, then $\mathbf{g}(\mathbf{u})$ should behave approximately the same as $\mathbf{g}(\mathbf{a}) + \nabla\mathbf{g}(\mathbf{a})\,(\mathbf{u} - \mathbf{a})$, where

$$\nabla\mathbf{g} = \nabla \begin{bmatrix} g_1 \\ \vdots \\ g_p \end{bmatrix} = \begin{bmatrix} \nabla g_1 \\ \vdots \\ \nabla g_p \end{bmatrix}.$$

To elaborate further, it is instructive to consider the example of a vector valued function

$$\mathbf{g}(\mathbf{x}, \mathbf{y}) = \begin{bmatrix} g_1(x_1, x_2, y_1, y_2, y_3) \\ g_2(x_1, x_2, y_1, y_2, y_3) \\ g_3(x_1, x_2, y_1, y_2, y_3) \end{bmatrix}$$

that maps the set

$$Q = \{(\mathbf{x}, \mathbf{y}) \in \mathbb{R}^2 \times \mathbb{R}^3 : \|\mathbf{x} - \mathbf{x}_0\| < \alpha \quad \text{and} \quad \|\mathbf{y} - \mathbf{y}_0\| < \beta\}$$

into $\mathbb{R}^3$ such that:

(1) $\mathbf{g} \in \mathcal{C}^1(Q)$.

(2) $\mathbf{g}(\mathbf{x}_0, \mathbf{y}_0) = \mathbf{0}$.

(3) The matrix

$$B_0 = \begin{bmatrix} \frac{\partial g_1}{\partial y_1}(\mathbf{x}_0, \mathbf{y}_0) & \cdots & \frac{\partial g_1}{\partial y_3}(\mathbf{x}_0, \mathbf{y}_0) \\ \vdots & & \vdots \\ \frac{\partial g_3}{\partial y_1}(\mathbf{x}_0, \mathbf{y}_0) & \cdots & \frac{\partial g_3}{\partial y_3}(\mathbf{x}_0, \mathbf{y}_0) \end{bmatrix}$$

is invertible.

The mean value theorem yields the formula

$$\mathbf{g}(\mathbf{x}, \mathbf{y}) - \mathbf{g}(\mathbf{x}_0, \mathbf{y}_0) = \begin{bmatrix} \nabla g_1(\mathbf{c}_1) \\ \nabla g_2(\mathbf{c}_2) \\ \nabla g_3(\mathbf{c}_3) \end{bmatrix} \begin{bmatrix} \mathbf{x} - \mathbf{x}_0 \\ \mathbf{y} - \mathbf{y}_0 \end{bmatrix}$$

for points $(\mathbf{x}, \mathbf{y}) \in Q$, where

$$\mathbf{c}_j = \begin{bmatrix} \mathbf{x}_0 \\ \mathbf{y}_0 \end{bmatrix} + t_j \begin{bmatrix} \mathbf{x} - \mathbf{x}_0 \\ \mathbf{y} - \mathbf{y}_0 \end{bmatrix}, \quad 0 < t_j < 1, \quad j = 1, 2, 3. \tag{15.2}$$

Let

$$A(t_1,t_2,t_3) = \begin{bmatrix} \frac{\partial g_1}{\partial x_1}(\mathbf{c}_1) & \frac{\partial g_1}{\partial x_2}(\mathbf{c}_1) \\ \frac{\partial g_2}{\partial x_1}(\mathbf{c}_2) & \frac{\partial g_2}{\partial x_2}(\mathbf{c}_2) \\ \frac{\partial g_3}{\partial x_1}(\mathbf{c}_3) & \frac{\partial g_3}{\partial x_2}(\mathbf{c}_3) \end{bmatrix}$$

and

$$B(t_1,t_2,t_3) = \begin{bmatrix} \frac{\partial g_1}{\partial y_1}(\mathbf{c}_1) & \frac{\partial g_1}{\partial y_2}(\mathbf{c}_1) & \frac{\partial g_1}{\partial y_3}(\mathbf{c}_1) \\ \frac{\partial g_2}{\partial y_1}(\mathbf{c}_2) & \frac{\partial g_2}{\partial y_2}(\mathbf{c}_2) & \frac{\partial g_2}{\partial y_3}(\mathbf{c}_2) \\ \frac{\partial g_3}{\partial y_1}(\mathbf{c}_3) & \frac{\partial g_3}{\partial y_2}(\mathbf{c}_3) & \frac{\partial g_3}{\partial y_3}(\mathbf{c}_3) \end{bmatrix}$$

so that

$$\mathbf{g}(\mathbf{x},\mathbf{y}) - \mathbf{g}(\mathbf{x}_0,\mathbf{y}_0) = A(t_1,t_2,t_3)(\mathbf{x}-\mathbf{x}_0) + B(t_1,t_2,t_3)(\mathbf{y}-\mathbf{y}_0).$$

Thus, if

$$\mathbf{g}(\mathbf{x}_0,\mathbf{y}_0) = \mathbf{0}$$

and $B(t_1,t_2,t_3)$ is invertible, then in order to find values of $\mathbf{y}$ that will make $\mathbf{g}(\mathbf{x},\mathbf{y}) = \mathbf{0}$, it is tempting to set

$$\mathbf{y} = \mathbf{y}_0 - B(t_1,t_2,t_3)^{-1}A(t_1,t_2,t_3)(\mathbf{x}-\mathbf{x}_0). \tag{15.3}$$

Indeed the invertibility of the matrix $B(t_1,t_2,t_3)$ is guaranteed by choosing $\|\mathbf{x}-\mathbf{x}_0\| < \gamma$ and $\|\mathbf{y}-\mathbf{y}_0\| < \delta$ with γ and δ small enough, since $B_0 = B(0,0,0)$ is invertible and the partial derivatives $\frac{\partial g_i}{\partial y_j}$ are continuous in the vicinity of the point $(\mathbf{x}_0,\mathbf{y}_0)$. However, formula (15.3) is deceptive, because the vectors $\mathbf{c}_1,\mathbf{c}_2,\mathbf{c}_3$ depend upon $\mathbf{y}$. That is to say, in formula (15.2) the right-hand side also depends upon $\mathbf{y}$ except in the special case that $\mathbf{g}(\mathbf{x},\mathbf{y})$ is of the form

$$\mathbf{g}(\mathbf{x},\mathbf{y}) = A_0(\mathbf{x}-\mathbf{x}_0) + B_0(\mathbf{y}-\mathbf{y}_0)$$

for some pair of matrix valued functions $A_0 = A_0(\mathbf{x})$ and $B_0 = B_0(\mathbf{x})$ that are independent of $\mathbf{y}$. Thus, we are forced to do something a little more clever.

15.2. The main theorem

Theorem 15.1. *Let*

$$\mathbf{g}(\mathbf{x},\mathbf{y}) = \begin{bmatrix} g_1(x_1,\ldots,x_q,y_1,\ldots,y_p) \\ \vdots \\ g_p(x_1,\ldots,x_q,y_1,\ldots,y_p) \end{bmatrix}$$

be a vector valued function that maps the set

$$Q = \{(\mathbf{x},\mathbf{y}) \in \mathbb{R}^q \times \mathbb{R}^p : \|\mathbf{x}-\mathbf{x}_0\| < \alpha \quad \text{and} \quad \|\mathbf{y}-\mathbf{y}_0\| < \beta\}$$

into $\mathbb{R}^p$ such that:

(1) $\mathbf{g} \in \mathcal{C}^1(Q)$.

(2) $\mathbf{g}(\mathbf{x}_0, \mathbf{y}_0) = \mathbf{0}$.

(3) *The* $p \times p$ *matrix*

$$G_2'(\mathbf{x}, \mathbf{y}) = \begin{bmatrix} \frac{\partial g_1}{\partial y_1}(\mathbf{x}, \mathbf{y}) & \cdots & \frac{\partial g_1}{\partial y_p}(\mathbf{x}, \mathbf{y}) \\ \vdots & & \vdots \\ \frac{\partial g_p}{\partial y_1}(\mathbf{x}, \mathbf{y}) & \cdots & \frac{\partial g_p}{\partial y_p}(\mathbf{x}, \mathbf{y}) \end{bmatrix}$$

is invertible at the point $(\mathbf{x}_0, \mathbf{y}_0)$.

Then there exists a pair of positive numbers γ *and* δ *such that for every* $\mathbf{x}$ *in the ball*

$$B_\gamma(\mathbf{x}_0) = \{\mathbf{x} \in \mathbb{R}^q : \|\mathbf{x} - \mathbf{x}_0\| < \gamma\}$$

there exists exactly one point $\mathbf{y} = \varphi(\mathbf{x})$ *in the ball*

$$B_\delta(\mathbf{y}_0) = \{\mathbf{y} \in \mathbb{R}^p : \|\mathbf{y} - \mathbf{y}_0\| < \delta\}$$

such that $\mathbf{g}(\mathbf{x}, \varphi(\mathbf{x})) = \mathbf{0}$. *Moreover,*

$$\text{if } \mathbf{g} \in \mathcal{C}^k(Q), \quad \text{then } \varphi \in \mathcal{C}^k(B_\gamma(\mathbf{x}_0)). \tag{15.4}$$

Proof. Let

$$B = G_2'(\mathbf{x}_0, \mathbf{y}_0)$$

and, in order to invoke the fixed point theorem, let

$$\mathbf{f}(\mathbf{x}, \mathbf{y}) = \mathbf{y} - B^{-1}\mathbf{g}(\mathbf{x}, \mathbf{y}).$$

Then

$$\mathbf{g}(\mathbf{x}, \mathbf{y}) = \mathbf{0} \quad \text{if and only if} \quad \mathbf{f}(\mathbf{x}, \mathbf{y}) = \mathbf{y}.$$

Moreover, the array of partial derivatives

$$F_2'(\mathbf{x}, \mathbf{y}) = \begin{bmatrix} \frac{\partial f_1}{\partial y_1} & \cdots & \frac{\partial f_1}{\partial y_p} \\ \vdots & & \vdots \\ \frac{\partial f_p}{\partial y_1} & \cdots & \frac{\partial f_p}{\partial y_p} \end{bmatrix},$$

is simply related to $G_2'(\mathbf{x}, \mathbf{y})$:

$$F_2'(\mathbf{x}, \mathbf{y}) = I_p - B^{-1}G_2'(\mathbf{x}, \mathbf{y}).$$

Therefore, since $F_2'(\mathbf{x}_0, \mathbf{y}_0) = 0$, assumption (1) guarantees the existence of $\gamma > 0$ and $\delta > 0$ such that

(1) $\|F_2'(\mathbf{x}, \mathbf{y})\| < K < 1$ and

(2) $\|B^{-1}g(\mathbf{x}, \mathbf{y}_0)\| \leq (1 - K)\frac{\delta}{2}$

for $\|\mathbf{x}-\mathbf{x}_0\| < \gamma$ and $\|\mathbf{y}-\mathbf{y}_0\| < \delta$. Now fix any point $\mathbf{x}$ in the ball $B_\gamma(\mathbf{x}_0)$ and let $\mathbf{h}(\mathbf{y}) = \mathbf{f}(\mathbf{x},\mathbf{y})$ and

$$H'(\mathbf{y}) = \begin{bmatrix} \frac{\partial h_1}{\partial y_1}(\mathbf{y}) & \cdots & \frac{\partial h_1}{\partial y_p}(\mathbf{y}) \\ \vdots & & \vdots \\ \frac{\partial h_p}{\partial y_1}(\mathbf{y}) & \cdots & \frac{\partial h_p}{\partial y_p}(\mathbf{y}) \end{bmatrix}.$$

Then, by Theorem 14.7, the inequality

$$\|\mathbf{h}(\mathbf{b}) - \mathbf{h}(\mathbf{a})\| \le \|H'(\mathbf{c})\|\|\mathbf{b}-\mathbf{a}\|$$

holds for any two points $\mathbf{a}$ and $\mathbf{b}$ in the ball $B_\delta(\mathbf{y}_0)$, where $\mathbf{c} = \mathbf{a} + t_0(\mathbf{b}-\mathbf{a})$ for some $0 < t_0 < 1$ is a point on the line segment joining the points $\mathbf{a}$ and $\mathbf{b}$ and consequently also belongs to this ball:

$$\begin{aligned} \|\mathbf{c}-\mathbf{y}_0\| &= \|(1-t_0)(\mathbf{a}-\mathbf{y}_0) + t_0(\mathbf{b}-\mathbf{y}_0)\| \\ &\le (1-t_0)\|(\mathbf{a}-\mathbf{y}_0)\| + t_0\|(\mathbf{b}-\mathbf{y}_0)\| \\ &< (1-t_0)\delta + t_0\delta = \delta\,. \end{aligned}$$

Thus, as

$$H'(\mathbf{c}) = F_2'(\mathbf{x},\mathbf{c}) \quad \text{and} \quad \|F_2'(\mathbf{x},\mathbf{c})\| < K\,,$$

it follows that

$$\|\mathbf{h}(\mathbf{b}) - \mathbf{h}(\mathbf{a})\| \le K\|\mathbf{b}-\mathbf{a}\| \tag{15.5}$$

for every pair of points $\mathbf{a}$ and $\mathbf{b}$ in the open ball $B_\delta(\mathbf{y}_0)$. Now let

$$E = \overline{B_{\delta_1}(\mathbf{y}_0)} = \{\mathbf{y} \in \mathbb{R}^p : \|\mathbf{y}-\mathbf{y}_0\| \le \delta_1\}$$

for any choice of δ_1 that meets the inequality $\frac{\delta}{2} < \delta_1 < \delta$ and break the proof into five parts in order to clarify the logic.

(a) $\mathbf{h}$ maps the closed set E into itself.

If $\mathbf{y} \in E$, then

$$\begin{aligned} \|\mathbf{h}(\mathbf{y}) - \mathbf{y}_0\| &\le \|\mathbf{h}(\mathbf{y}) - \mathbf{h}(\mathbf{y_0})\| + \|\mathbf{h}(\mathbf{y_0}) - \mathbf{y}_0\| \\ &\le K\|\mathbf{y}-\mathbf{y}_0\| + \|\mathbf{h}(\mathbf{y_0}) - \mathbf{y}_0\| \\ &\le K\delta_1 + \|\mathbf{h}(\mathbf{y_0}) - \mathbf{y}_0\| \end{aligned}$$

and hence, as

$$\|\mathbf{h}(\mathbf{y_0}) - \mathbf{y}_0\| = \|\mathbf{f}(\mathbf{x},\mathbf{y}_0) - \mathbf{f}(\mathbf{x}_0,\mathbf{y}_0)\| = \|B^{-1}\mathbf{g}(\mathbf{x},\mathbf{y}_0)\| \le (1-K)\frac{\delta}{2}$$

and $\delta/2 < \delta_1$,

$$\|\mathbf{h}(\mathbf{y}) - \mathbf{y}_0\| \le K\delta_1 + (1-K)\frac{\delta}{2} \le K\delta_1 + (1-K)\delta_1 \le \delta_1\,.$$

(b) The inequality (15.5) guarantees that $\|\mathbf{h}(\mathbf{b}) - \mathbf{h}(\mathbf{a})\| \le K\|\mathbf{b}-\mathbf{a}\|$ holds for every pair of points $\mathbf{a},\mathbf{b} \in E$.

(c) Invoke the fixed point theorem to $\mathbf{h}(\mathbf{y})$ to conclude that for each $\mathbf{x}$ in the ball $B_\gamma(\mathbf{x}_0)$, there exists a unique $\mathbf{y} = \varphi(\mathbf{x})$ in the closed ball E such that $\mathbf{h}(\mathbf{y}) = \mathbf{y}$.

(d) Strengthen conclusion (c) to conclude that for each vector $\mathbf{x}$ in the ball $B_\gamma(\mathbf{x}_0)$, there exists exactly one vector $\mathbf{y} \in \mathbb{R}^p = \varphi(\mathbf{x})$ in the ball $B_\delta(\mathbf{y}_0)$ such that

$$\mathbf{g}(\mathbf{x}, \varphi(\mathbf{x})) = 0.$$

Since $E \subset B_\delta(\mathbf{y}_0)$, Item (c) guarantees that for each $\mathbf{x}$ in the ball $B_\gamma(\mathbf{x}_0)$, there exists at least one $\mathbf{y} = \varphi(\mathbf{x})$ in $B_\delta(\mathbf{y}_0)$ such that

$$\mathbf{f}(\mathbf{x}, \mathbf{y}) = \mathbf{y}\,.$$

However, if there were two such fixed points $\mathbf{y}_*$ and $\mathbf{w}_*$ in $B_\delta(\mathbf{y}_0)$, then the inequality (15.5) implies that

$$\|\mathbf{y}_* - \mathbf{w}_*\| = \|\mathbf{h}(\mathbf{y}_*) - \mathbf{h}(\mathbf{w}_*)\| \le K\|\mathbf{y}_* - \mathbf{w}_*\|\,, \quad \text{with} \quad K < 1\,,$$

which is viable only if $\mathbf{y}_* = \mathbf{w}_*$. Therefore, uniqueness prevails in $B_\delta(\mathbf{y}_0)$ too. But this is equivalent to the stated conclusion.

(e) Verify the implication (15.4).

To obtain the differentiability of φ, observe that

$$\mathbf{0} = \mathbf{g}(\mathbf{b}, \varphi(\mathbf{b})) - \mathbf{g}(\mathbf{a}, \varphi(\mathbf{a})) = \begin{bmatrix} \nabla g_1(\mathbf{c}_1) \\ \vdots \\ \nabla g_p(\mathbf{c}_p) \end{bmatrix} \begin{bmatrix} (\mathbf{b} - \mathbf{a}) \\ \varphi(\mathbf{b}) - \varphi(\mathbf{a}) \end{bmatrix},$$

where

$$\mathbf{c}_i = \begin{bmatrix} \mathbf{a} \\ \varphi(\mathbf{a}) \end{bmatrix} + t_i \begin{bmatrix} \mathbf{b} - \mathbf{a} \\ \varphi(\mathbf{b}) - \varphi(\mathbf{a}) \end{bmatrix} \quad \text{for} \quad t_i \in (0,1), \quad i = 1, \dots, p,$$

and hence, upon setting

$$L = -\begin{bmatrix} \frac{\partial g_1}{\partial x_1}(\mathbf{c}_1) & \cdots & \frac{\partial g_1}{\partial x_q}(\mathbf{c}_1) \\ \vdots & & \vdots \\ \frac{\partial g_p}{\partial x_1}(\mathbf{c}_p) & \cdots & \frac{\partial g_p}{\partial x_q}(\mathbf{c}_p) \end{bmatrix} \quad \text{and} \quad M = \begin{bmatrix} \frac{\partial g_1}{\partial y_1}(\mathbf{c}_1) & \cdots & \frac{\partial g_1}{\partial y_p}(\mathbf{c}_1) \\ \vdots & & \vdots \\ \frac{\partial g_p}{\partial y_1}(\mathbf{c}_p) & \cdots & \frac{\partial g_p}{\partial y_p}(\mathbf{c}_p) \end{bmatrix},$$

that

$$L\,[\mathbf{b} - \mathbf{a}] = M\,[\varphi(\mathbf{b}) - \varphi(\mathbf{a})]\,. \tag{15.6}$$

The matrices

$$L = L(\mathbf{c}_1, \dots, \mathbf{c}_p) \quad \text{and} \quad M = M(\mathbf{c}_1, \dots, \mathbf{c}_p)$$

are matrices of sizes $p \times q$ and $p \times p$, respectively, that depend upon $\mathbf{a}$ and $\mathbf{b}$. If γ is small enough, then M is invertible and $\|M^{-1}L\| \le K_1$ for all points

a and **b** in $B_\gamma(\mathbf{x}_0)$. Thus,

$$\|\varphi(\mathbf{b}) - \varphi(\mathbf{a})\| \leq K_1\|\mathbf{b} - \mathbf{a}\|,$$

which establishes the continuity of $\varphi(\mathbf{x})$ in $B_\gamma(\mathbf{x}_0)$. In particular, this guarantees that

$$\mathbf{c}_i \longrightarrow \begin{bmatrix} \mathbf{a} \\ \varphi(\mathbf{a}) \end{bmatrix} \quad \text{as} \quad \mathbf{b} \longrightarrow \mathbf{a}.$$

The calculation of the partial derivatives of $\varphi(\mathbf{x})$ is completed by choosing

$$\mathbf{b} = \mathbf{a} + \varepsilon\mathbf{e}_k, \;\; k = 1, \ldots, q,$$

in formula (15.6), where $\mathbf{e}_k$ is the k'th column of the identity matrix I_q and ε is a small real number. Then, for example, if $k = 1$, we obtain the formula

$$-\begin{bmatrix} \frac{\partial g_1}{\partial x_1}(\mathbf{c}_1) & \cdots & \frac{\partial g_1}{\partial x_q}(\mathbf{c}_1) \\ \vdots & & \vdots \\ \frac{\partial g_p}{\partial x_1}(\mathbf{c}_p) & \cdots & \frac{\partial g_p}{\partial x_q}(\mathbf{c}_p) \end{bmatrix} \begin{bmatrix} 1 \\ 0 \\ \vdots \\ 0 \end{bmatrix}$$

$$= \begin{bmatrix} \frac{\partial g_1}{\partial y_1}(\mathbf{c}_1) & \cdots & \frac{\partial g_1}{\partial y_p}(\mathbf{c}_1) \\ \vdots & & \vdots \\ \frac{\partial g_p}{\partial y_1}(\mathbf{c}_p) & \cdots & \frac{\partial g_p}{\partial y_p}(\mathbf{c}_p) \end{bmatrix} \begin{bmatrix} \frac{\varphi_1(a_1+\varepsilon, a_2, \ldots, a_q) - \varphi_1(a_1, a_2, \ldots, a_q)}{\varepsilon} \\ \vdots \\ \frac{\varphi_p(a_1+\varepsilon, a_2, \ldots, a_q) - \varphi_p(a_1, a_2, \ldots, a_q)}{\varepsilon} \end{bmatrix}.$$

Letting $\varepsilon \downarrow 0$ we obtain formulas for $\frac{\partial \varphi_1}{\partial x_1}, \ldots, \frac{\partial \varphi_p}{\partial x_1}$. Formulas for the other partial derivatives are obtained from the other choices of $\mathbf{e}_k$. □

Exercise 15.1. Let a, b and c be real numbers and let

$$g(x, y) = (x - a)^2 + (y - b)^2 - c^2.$$

Show that if

$$g(x_0, y_0) = 0 \quad \text{and} \quad \frac{\partial g}{\partial y}(x_0, y_0) \neq 0,$$

then to each point x that is sufficiently close to x_0 (i.e., $|x - x_0| < \delta$ for small enough δ) there is exactly one point y such that $g(x, y) = 0$. Since this single point y is uniquely determined by the x, this is the same as to say that $y = \varphi(x)$ for such x.

Exercise 15.2. Show that in the setting of Exercise 15.1, one can solve for x as a function of y in the neighborhood of any point (x_0, y_0) such that

$$g(x_0, y_0) = 0 \quad \text{and} \quad \frac{\partial g}{\partial x}(x_0, y_0) \neq 0.$$

Exercise 15.3. Let $g_1(x, y_1, y_2) = x^2(y_1^2 + y_2^2) - 5$ and $g_2(x, y_1, y_2) = (x - y_2)^2 + y_1^2 - 2$. Show that in a neighborhood of the point $x = 1$, $y_1 = -1$, $y_2 = 2$, the curve of intersection of the two surfaces $g_1(x, y_1, y_2) = 0$ and $g_2(x, y_1, y_2) = 0$ can be described by a pair of functions $y_1 = \varphi_1(x)$ and $y_2 = \varphi_2(x)$.

Exercise 15.4. Let

$$S = \left\{ \begin{bmatrix} x_1 \\ x_2 \\ x_3 \\ x_4 \end{bmatrix} \in \mathbb{R}^4 : \begin{array}{c} x_1 - 2x_2 + 2x_3^2 - x_4 = 0 \\ \text{and} \\ x_1 - 2x_2 + 2x_3 - x_4 = 0 \end{array} \right\}, \mathbf{u} = \begin{bmatrix} 1 \\ 1 \\ 1 \\ 1 \end{bmatrix} \text{ and } \mathbf{v} = \begin{bmatrix} 1 \\ 1 \\ 0 \\ -1 \end{bmatrix}.$$

Use the implicit function theorem to show that it is possible to solve for x_3 and x_4 as functions of x_1 and x_2 for points in S that are close to $\mathbf{u}$ and to points in S that are close to $\mathbf{v}$ and write down formulas for these functions for each of these two cases.

15.3. A generalization of the implicit function theorem

Theorem 15.2. *Let Q be an open subset of $\mathbb{R}^n$ that contains the point $\mathbf{u}^\circ$, let $g_i \in \mathcal{C}^1(Q)$ for $i = 1, \ldots, k$ and suppose that*

$$\operatorname{rank} \begin{bmatrix} \frac{\partial g_1}{\partial u_1}(\mathbf{u}) & \cdots & \frac{\partial g_1}{\partial u_n}(\mathbf{u}) \\ \vdots & & \vdots \\ \frac{\partial g_k}{\partial u_1}(\mathbf{u}) & \cdots & \frac{\partial g_k}{\partial u_n}(\mathbf{u}) \end{bmatrix} = p \quad \textit{for} \quad \mathbf{u} \in B_r(\mathbf{u})^\circ \subset Q$$

for some $r > 0$ and let $q = n - p$. Then there exists a permutation σ of the indices $\{1, \ldots, n\}$ of the components of $\mathbf{u}$ and a pair of numbers $\gamma > 0$ and $\delta > 0$ such that if

(15.7)
$$x_i = u_{\sigma(i)} \quad \textit{for} \quad i = 1, \ldots, q \quad \textit{and} \quad y_i = u_{\sigma(q+i)} \quad \textit{for} \quad i = 1, \ldots, p,$$

then:

(1) $$\operatorname{rank} \begin{bmatrix} \frac{\partial g_1}{\partial y_1}(\mathbf{u}^\circ) & \cdots & \frac{\partial g_1}{\partial y_p}(\mathbf{u}^\circ) \\ \vdots & & \vdots \\ \frac{\partial g_k}{\partial y_1}(\mathbf{u}^\circ) & \cdots & \frac{\partial g_k}{\partial y_p}(\mathbf{u}^\circ) \end{bmatrix} = p.$$

(2) *For each point in the ball $B_\gamma(\mathbf{x}_0) = \{\mathbf{x} \in \mathbb{R}^q : \|\mathbf{x} - \mathbf{x}^\circ\| < \gamma\}$, there exists exactly one point $\mathbf{y} = \varphi(\mathbf{x})$ in the ball $\{\mathbf{y} \in \mathbb{R}^p : \|\mathbf{y} - \mathbf{y}^\circ\| < \delta\}$ such that*

$$g_i(\mathbf{u}) = 0 \quad \textit{for} \quad i = 1, \ldots, k \quad \textit{when} \quad \begin{bmatrix} u_{\sigma(1)} \\ \vdots \\ u_{\sigma(n)} \end{bmatrix} = \begin{bmatrix} \mathbf{x} \\ \varphi(\mathbf{x}) \end{bmatrix}.$$

(3) $\varphi \in \mathcal{C}^1(B_\gamma(\mathbf{x}_0))$.

Proof. The basic idea is to first reorder the functions $g_1, \ldots, g_k$ so that

$$\operatorname{rank} \begin{bmatrix} \frac{\partial g_1}{\partial u_1}(\mathbf{u}) & \cdots & \frac{\partial g_1}{\partial u_n}(\mathbf{u}) \\ \vdots & & \vdots \\ \frac{\partial g_p}{\partial u_1}(\mathbf{u}) & \cdots & \frac{\partial g_p}{\partial u_n}(\mathbf{u}) \end{bmatrix} = p$$

and then to relabel the independent variables in accordance with (15.7) so that

$$\operatorname{rank} \begin{bmatrix} \frac{\partial g_1}{\partial y_1}(\mathbf{u}^\circ) & \cdots & \frac{\partial g_1}{\partial y_p}(\mathbf{u}^\circ) \\ \vdots & & \vdots \\ \frac{\partial g_p}{\partial y_1}(\mathbf{u}^\circ) & \cdots & \frac{\partial g_p}{\partial y_p}(\mathbf{u}^\circ) \end{bmatrix} = p\,.$$

The existence of a function $\varphi(\mathbf{x})$ such that (2) and (3) hold for $i = 1, \ldots, p$ then follows from the implicit function theorem.

To complete the proof when $k > p$, it remains to check that (2) holds for $i = p+1, \ldots, k$. To this end, fix i in this range; let

$$\mathbf{u}(t) = \begin{bmatrix} u_1(t) \\ \vdots \\ u_n(t) \end{bmatrix} \quad \text{with} \quad \begin{bmatrix} u_{\sigma(1)}(t) \\ \vdots \\ u_{\sigma(n)}(t) \end{bmatrix} = \begin{bmatrix} \mathbf{x}(t) \\ \varphi(\mathbf{x}(t)) \end{bmatrix} \quad \text{for} \quad 0 \le t \le 1\,,$$

where $\mathbf{x}(0) = \mathbf{x}^\circ$ and $\mathbf{x}(t)$ is a smooth curve inside the ball $B_\gamma(\mathbf{x}^\circ)$; and set $h(t) = g_i(\mathbf{u}(t))$. Then, since

$$(\nabla g_i)(\mathbf{u}(t)) = \sum_{j=1}^{p} a_j(\mathbf{u}(t))(\nabla g_j)(\mathbf{u}(t))$$

for $0 \le t \le 1$ and $h(0) = 0$,

$$\begin{aligned} h(t) &= \int_0^t \frac{d}{ds} h(s) ds \\ &= \int_0^t (\nabla g_i)(\mathbf{u}(s))\mathbf{u}'(s) ds \\ &= \int_0^t \left\{ \sum_{j=1}^{p} a_j(\mathbf{u}(s))(\nabla g_j)(\mathbf{u}(s)) \right\} \mathbf{u}'(s) ds \\ &= \int_0^t \sum_{j=1}^{p} a_j(\mathbf{u}(s)) \left\{ (\nabla g_j)(\mathbf{u}(s))\mathbf{u}'(s) \right\} ds \\ &= 0\,, \end{aligned}$$

because

$$(\nabla g_j)(\mathbf{u}(s))\mathbf{u}'(s) = \frac{d}{ds} g_j(\mathbf{u}(s)) = 0 \quad \text{for} \quad j = 1, \ldots, p \quad \text{and} \quad 0 < s < 1 .$$

□

15.4. Continuous dependence of solutions

The implicit function theorem is often a useful tool to check the continuous dependence of the solution of an equation on the coefficients appearing in the equation. Suppose, for example, that $X \in \mathbb{R}^{2\times 2}$ is a solution of the matrix equation

$$A^T X + XA = B$$

for some fixed choice of the matrices $A, B \in \mathbb{R}^{2\times 2}$. Then we shall invoke the implicit function theorem to show that if A changes only a little, then X will also change only a little. To this end, let

$$F(A, X) = A^T X + XA - B$$

so that

$$f_{ij}(A, X) = \mathbf{e}_i^T (A^T X + XA - B)\mathbf{e}_j , \quad i, j = 1, 2 ,$$

where $\mathbf{e}_1, \mathbf{e}_2$ denote the standard basis vectors in $\mathbb{R}^2$. Then, upon writing

$$X = \begin{bmatrix} x_{11} & x_{12} \\ x_{21} & x_{22} \end{bmatrix} ,$$

one can readily check that

$$\begin{aligned} \frac{\partial f_{ij}}{\partial x_{st}} &= \mathbf{e}_i^T (A^T \mathbf{e}_s \mathbf{e}_t^T + \mathbf{e}_s \mathbf{e}_t^T A)\mathbf{e}_j \\ &= a_{si} \mathbf{e}_t^T \mathbf{e}_j + a_{tj} \mathbf{e}_i^T \mathbf{e}_s . \end{aligned}$$

Thus,

$$\begin{aligned} \frac{\partial f_{ij}}{\partial x_{11}} &= a_{1i} \mathbf{e}_1^T \mathbf{e}_j + a_{1j} \mathbf{e}_i^T \mathbf{e}_1 \\ \frac{\partial f_{ij}}{\partial x_{12}} &= a_{1i} \mathbf{e}_2^T \mathbf{e}_j + a_{2j} \mathbf{e}_i^T \mathbf{e}_1 \\ \frac{\partial f_{ij}}{\partial x_{21}} &= a_{2i} \mathbf{e}_1^T \mathbf{e}_j + a_{1j} \mathbf{e}_i^T \mathbf{e}_2 \\ \frac{\partial f_{ij}}{\partial x_{22}} &= a_{2i} \mathbf{e}_2^T \mathbf{e}_j + a_{2j} \mathbf{e}_i^T \mathbf{e}_2 . \end{aligned}$$

Correspondingly,

$$\begin{bmatrix} \frac{\partial f_{11}}{\partial x_{11}} & \frac{\partial f_{11}}{\partial x_{12}} & \frac{\partial f_{11}}{\partial x_{21}} & \frac{\partial f_{11}}{\partial x_{22}} \\ \frac{\partial f_{12}}{\partial x_{11}} & \frac{\partial f_{12}}{\partial x_{12}} & \frac{\partial f_{12}}{\partial x_{21}} & \frac{\partial f_{12}}{\partial x_{22}} \\ \frac{\partial f_{21}}{\partial x_{11}} & \frac{\partial f_{21}}{\partial x_{12}} & \frac{\partial f_{21}}{\partial x_{21}} & \frac{\partial f_{21}}{\partial x_{22}} \\ \frac{\partial f_{22}}{\partial x_{11}} & \frac{\partial f_{22}}{\partial x_{12}} & \frac{\partial f_{22}}{\partial x_{21}} & \frac{\partial f_{22}}{\partial x_{22}} \end{bmatrix} = \begin{bmatrix} 2a_{11} & a_{21} & a_{21} & 0 \\ a_{12} & a_{11}+a_{22} & 0 & a_{21} \\ a_{12} & 0 & a_{11}+a_{22} & a_{21} \\ 0 & a_{12} & a_{12} & 2a_{22} \end{bmatrix}.$$

Now suppose that $F(A_0, X_0) = 0$ and that the matrix on the right in the last identity is invertible when the terms a_{ij} are taken from A_0. Then the implicit function theorem guarantees the existence of a pair of numbers $\gamma > 0$ and $\delta > 0$ such that for every matrix $A \in \mathbb{R}^{2\times 2}$ in the ball $\|A - A_0\| < \gamma$ there exists a unique $X = \Phi(A)$ in the ball $\|X - X_0\| < \delta$ such that $F(A, X) = 0$ and hence that $\Phi(A)$ is a continuous function of X in the ball $\|A - A_0\| < \gamma$.

15.5. The inverse function theorem

Theorem 15.3. *Suppose that the $p \times 1$ real vector valued function*

$$\mathbf{f}(\mathbf{x}) = \begin{bmatrix} f_1(x_1, \dots, x_p) \\ \vdots \\ f_p(x_1, \dots, x_p) \end{bmatrix}$$

is in $\mathcal{C}^1(B_\alpha(\mathbf{x}_0))$ for some $\alpha > 0$ and that the Jacobian matrix

$$J_{\mathbf{f}}(\mathbf{x}) = \begin{bmatrix} \frac{\partial f_1}{\partial x_1}(\mathbf{x}) & \cdots & \frac{\partial f_1}{\partial x_p}(\mathbf{x}) \\ \vdots & & \vdots \\ \frac{\partial f_p}{\partial x_1}(\mathbf{x}) & \cdots & \frac{\partial f_p}{\partial x_p}(\mathbf{x}) \end{bmatrix}$$

is invertible at the point $\mathbf{x}_0$. Let $\mathbf{y}_0 = \mathbf{f}(\mathbf{x}_0)$. Then there exist a pair of numbers $\gamma > 0$ and $\delta > 0$ such that for each point $\mathbf{y} \in \mathbb{R}^p$ in the ball $B_\delta(\mathbf{y}_0)$ there exists exactly one point $\mathbf{x}$ in the ball $B_\gamma(\mathbf{x}_0)$ such that $\mathbf{y} = \mathbf{f}(\mathbf{x})$. Moreover, the function $\mathbf{x} = \vartheta(\mathbf{y})$ is in $\mathcal{C}^1(B_\delta(\mathbf{y}_0))$.

Proof. Let $\mathbf{g}(\mathbf{x}, \mathbf{y}) = \mathbf{f}(\mathbf{x}) - \mathbf{y}$ and let

$$G_1'(\mathbf{x}, \mathbf{y}) = \begin{bmatrix} \frac{\partial g_1}{\partial x_1}(\mathbf{x}, \mathbf{y}) & \cdots & \frac{\partial g_1}{\partial x_p}(\mathbf{x}, \mathbf{y}) \\ \vdots & & \vdots \\ \frac{\partial g_p}{\partial x_1}(\mathbf{x}, \mathbf{y}) & \cdots & \frac{\partial g_p}{\partial x_p}(\mathbf{x}, \mathbf{y}) \end{bmatrix}.$$

Then, since $\mathbf{g}(\mathbf{x}_0, \mathbf{y}_0) = 0$ and the matrix $G_1'(\mathbf{x}_0, \mathbf{y}_0) = J(\mathbf{x}_0)$ is invertible, the implicit function theorem guarantees the existence of a pair of positive numbers γ and δ such that for each vector $\mathbf{y} \in B_\delta(\mathbf{y}_0)$, there exists exactly one point $\mathbf{x} = \vartheta(\mathbf{y})$ such that $\mathbf{g}(\vartheta(\mathbf{y}), \mathbf{y}) = 0$ and that moreover, $\vartheta(\mathbf{y})$ will have continous first order partial derivatives in $B_\delta(\mathbf{y}_0)$. But this is equivalent to the asserted statement. □

Exercise 15.5. Let $\mathbf{g}(\mathbf{x})$ be a continous mapping of $\mathbb{R}^p$ into $\mathbb{R}^p$ such that all the partial derivatives $\frac{\partial g_i}{\partial x_j}$ exist and are continuous on $\mathbb{R}^p$. Write

$$\mathbf{g}(\mathbf{x}) = \begin{bmatrix} g_1(x_1, \dots, x_p) \\ \vdots \\ g_p(x_1, \dots, x_p) \end{bmatrix}, \quad J_{\mathbf{g}}(\mathbf{x}) = \begin{bmatrix} \frac{\partial g_1}{\partial x_1}(\mathbf{x}) & , \dots & \frac{\partial g_1}{\partial x_p}(\mathbf{x}) \\ \vdots & & \vdots \\ \frac{\partial g_1}{\partial x_1}(\mathbf{x}) & , \dots & \frac{\partial g_1}{\partial x_p}(\mathbf{x}) \end{bmatrix},$$

$\mathbf{y}^\circ = \mathbf{g}(\mathbf{x}^\circ)$ and suppose that the matrix $B = J_{\mathbf{g}}(\mathbf{x}^\circ)$ is invertible and that $\|I_p - B^{-1}J_{\mathbf{g}}(\mathbf{x})\| < 1/2$ for every point $\mathbf{x}$ in the closed ball $\overline{B_\delta(\mathbf{x}^\circ)}$. Show that if $\rho = \delta/(2\|B^{-1}\|)$, then for each fixed $\mathbf{y}$ in the closed ball $\overline{B_\rho(\mathbf{y}^\circ)}$, there exists exactly one point $\mathbf{x} \in \overline{B_\delta(\mathbf{x}^\circ)}$ such that $\mathbf{g}(\mathbf{x}) = \mathbf{y}$. [HINT: Show that for each point $\mathbf{y} \in \overline{B_\rho(\mathbf{y}^\circ)}$, the function $h(\mathbf{x}) = \mathbf{x} - B^{-1}(g(\mathbf{x}) - \mathbf{y})$ has a fixed point in $\overline{B_\delta(\mathbf{x}^\circ)}$.]

Exercise 15.6. Let

$$\mathbf{g}(\mathbf{x}) = \begin{bmatrix} x_1^2 - x_2 \\ x_2 + x_3 \\ x_3^2 - 2x_3 + 1 \end{bmatrix}, \quad \mathbf{x}^\circ = \begin{bmatrix} 1 \\ -1 \\ -1 \end{bmatrix} \quad \text{and} \quad \mathbf{y}^\circ = \mathbf{g}(\mathbf{x}^\circ).$$

(a) Calculate $J_{\mathbf{g}}(\mathbf{x})$, $B = J_{\mathbf{g}}(\mathbf{x}^\circ)$ and B^{-1}.

(b) Show that $\|B^{-1}\|^2 < 5/3$.

(c) Show that if $\mathbf{y} \in \mathbb{R}^3$ is fixed and $\mathbf{h}(\mathbf{x}) = \mathbf{x} - B^{-1}(\mathbf{g}(\mathbf{x}) - \mathbf{y})$, then $J_{\mathbf{h}}(\mathbf{x}) = B^{-1}(J_{\mathbf{g}}(\mathbf{x}^\circ) - J_{\mathbf{g}}(\mathbf{x}))$.

(d) Show that $\|J_{\mathbf{h}}(\mathbf{x})\| \le 2\|B^{-1}\|\|\mathbf{x} - \mathbf{x}^\circ\|$.

(e) Show that if $2\|B^{-1}\|\delta < 1/2$, then for each fixed point $\mathbf{y}$ in the closed ball $\overline{B_\rho(\mathbf{y}^\circ)}$, with $\rho = \delta/(2\|B^{-1}\|)$, there exists exactly one point $\mathbf{x}$ in the closed ball $\overline{B_\delta(\mathbf{x}^\circ)}$ such that $\mathbf{g}(\mathbf{x}) = \mathbf{y}$.

Exercise 15.7. Let $\mathbf{u}^\circ \in \mathbb{R}^2$ and let $\mathbf{f} \in \mathcal{C}^2(B_r(\mathbf{u}^\circ)$ and suppose that $\begin{bmatrix} (\nabla f_1)(\mathbf{u}) \\ (\nabla f_2)(\mathbf{v}) \end{bmatrix}$ is invertible for every pair of vectors $\mathbf{u}, \mathbf{v} \in B_r(\mathbf{u}^\circ)$. Show that if $\mathbf{a}, \mathbf{b} \in B_r(\mathbf{u}^\circ)$, then $\mathbf{f}(\mathbf{a}) = \mathbf{f}(\mathbf{b}) \iff \mathbf{a} = \mathbf{b}$.

Exercise 15.8. Show that the condition in Exercise 15.7 cannot be weakened to $\begin{bmatrix} (\nabla f_1)(\mathbf{u}) \\ (\nabla f_2)(\mathbf{u}) \end{bmatrix}$ is invertible for every vector $\mathbf{u} \in B_r(\mathbf{u}^\circ)$. [HINT: Consider the function $\mathbf{f}(\mathbf{x})$ with components $f_1(\mathbf{x}) = x_1 \cos x_2$ and $f_2(\mathbf{x}) = x_1 \sin x_2$ in a ball of radius 2π centered at the point $(3\pi, 2\pi)$.

Exercise 15.9. Calculate the Jacobian matrix $J_{\mathbf{f}}(\mathbf{x})$ of the function $\mathbf{f}(\mathbf{x})$ with components $f_i(x_1, x_2, x_3) = x_i/(1 + x_1 + x_2 + x_3)$ for $i = 1, 2, 3$ that are defined at all points $\mathbf{x} \in \mathbb{R}^3$ with $x_1 + x_2 + x_3 \neq -1$.

Exercise 15.10. Show that the vector valued function that is defined in Exercise 15.8 defines a one to one map from its domain of definition in $\mathbb{R}^3$ and find the inverse mapping.

15.6. Roots of polynomials

Theorem 15.4. *The roots of the polynomial*

$$f(\lambda) = \lambda^n + a_1\lambda^{n-1} + \cdots + a_n$$

vary continuously with the coefficients $a_1, \ldots, a_n$.

This theorem is of great importance in applications. It guarantees that a small change in the coefficients $a_1, \ldots, a_n$ of the polynomial causes only a small change in the roots of the polynomial. It is usually proved by Rouché's theorem from the theory of complex variables; see e.g. pp. 153-154 of [**7**] and Appendix B. Below, we shall treat a special case of this theorem in which the polynomial has distinct roots via the implicit function theorem. The full result will be established later by invoking a different circle of ideas in Chapter 17. Another approach is considered in Exercise 17.15.

15.7. An instructive example

To warm up, consider first the polynomial

$$p(\lambda) = \lambda^3 + \lambda^2 - 4\lambda + 6.$$

It has three roots:

$$\lambda_1 = 1 + i, \ \lambda_2 = 1 - i, \ \text{and} \ \lambda_3 = -3.$$

This means that the equation

$$(\mu + i\nu)^3 + a(\mu + i\nu)^2 + b(\mu + i\nu) + c = 0 \tag{15.8}$$

in terms of the 5 real variables μ, ν, a, b, c is satisfied by the choices:

$$\begin{aligned}
\mu &= 1, \quad \nu = 1, \quad a = 1, \ b = -4, \ c = 6 \\
\mu &= 1, \quad \nu = -1, \ a = 1, \ b = -4, \ c = 6 \\
\mu &= -3, \ \nu = 0, \quad a = 1, \ b = -4, \ c = 6 .
\end{aligned}$$

To put this into the setting of the implicit function theorem, let us express

$$\begin{aligned}
f(a, b, c, \mu + i\nu) &= (\mu + i\nu)^3 + a(\mu + i\nu)^2 + b(\mu + i\nu) + c \\
&= \mu^3 + 3\mu^2 i\nu + 3\mu(i\nu)^2 + (i\nu)^3 \\
&\quad + a(\mu^2 + 2\mu i\nu + (i\nu)^2) + b(\mu + i\nu) + c
\end{aligned}$$

in terms of its real and imaginary parts as

$$f(a,b,c,\mu+i\nu) = f_1(a,b,c,\mu,\nu) + if_2(a,b,c,\mu,\nu),$$

where

$$f_1(a,b,c,\mu,\nu) = \mu^3 - 3\mu\nu^2 + a\mu^2 - a\nu^2 + b\mu + c$$

and

$$f_2(a,b,c,\mu,\nu) = 3\mu^2\nu - \nu^3 + 2a\mu\nu + b\nu.$$

Thus, we have converted the study of the solutions of the roots of the equation

$$\lambda^3 + a\lambda^2 + b\lambda + c = 0$$

with real coefficients a, b, c to the study of the solutions of the system

$$\begin{aligned} f_1(a,b,c,\mu,\nu) &= 0 \\ f_2(a,b,c,\mu,\nu) &= 0\,. \end{aligned}$$

The implicit function theorem guarantees the continuous dependence of the pair (μ, ν) on (a, b, c) in the vicinity of a solution provided that the matrix

$$\triangle(a,b,c,\mu,\nu) = \begin{bmatrix} \frac{\partial f_1}{\partial \mu} & \frac{\partial f_1}{\partial \nu} \\ \frac{\partial f_2}{\partial \mu} & \frac{\partial f_2}{\partial \nu} \end{bmatrix}$$

is invertible. Let us explore this at the point $a = 1, b = -4, c = 6, \mu = 1, \nu = 1$. To begin with

$$\frac{\partial f_1}{\partial \mu} = 3\mu^2 - 3\nu^2 + 2a\mu + b\,, \tag{15.9}$$

$$\frac{\partial f_1}{\partial \nu} = -6\mu\nu - 2a\nu\,, \tag{15.10}$$

$$\frac{\partial f_2}{\partial \mu} = 6\mu\nu + 2a\nu\,, \tag{15.11}$$

$$\frac{\partial f_2}{\partial \nu} = 3\mu^2 - 3\nu^2 + 2a\mu + b\,. \tag{15.12}$$

Therefore,

$$\frac{\partial f_1}{\partial \mu}(1,-4,6,1,1) = -2\,,$$
$$\frac{\partial f_1}{\partial \nu}(1,-4,6,1,1) = -8\,,$$
$$\frac{\partial f_2}{\partial \mu}(1,-4,6,1,1) = 8\,,$$
$$\frac{\partial f_2}{\partial \nu}(1,-4,6,1,1) = -2$$

and

$$\triangle(1,-4,6,1,1) = \det\begin{bmatrix} -2 & -8 \\ 8 & -2 \end{bmatrix} = 2^2 + 8^2 = 68\,. \tag{15.13}$$

Thus we can conclude that if the coefficients a, b, c of the polynomial

$$\lambda^3 + a\lambda^2 + b\lambda + c$$

change a little bit from $1, -4, 6$, then the root in the vicinity of $1 + i$ will only change a little bit. Similar considerations apply to the other two roots in this example.

Exercise 15.11. Show that there exists a pair of numbers $\gamma > 0$ and $\delta > 0$ such that the polynomial $\lambda^3 + a\lambda^2 + b\lambda + c$ with real coefficients has exactly one root $\lambda = \mu + i\nu$ in the ball $\mu^2 + (\nu - 2)^2 < \delta$ if $(a-1)^2 + (b-4)^2 + (c-4)^2 < \gamma$. [HINT: The polynomial $\lambda^3 + \lambda^2 + 4\lambda + 4$ has three distinct roots: $2i$, $-2i$ and -1.]

15.8. A more sophisticated approach

The next step is to see if we can redo this example in a more transparent way that will enable us to generalize the procedure. The answer is yes, and the key rests in looking carefully at the formulas (15.9)–(15.12) and noting that

$$\frac{\partial f_1}{\partial \mu} = \frac{\partial f_2}{\partial v} \quad \text{and} \quad \frac{\partial f_1}{\partial \nu} = -\frac{\partial f_2}{\partial \mu}$$

and hence that the determinant of interest is equal to

$$\begin{aligned} \frac{\partial f_1}{\partial \mu}\frac{\partial f_2}{\partial \nu} - \frac{\partial f_1}{\partial \nu}\frac{\partial f_2}{\partial \mu} &= \left(\frac{\partial f_1}{\partial \mu}\right)^2 + \left(\frac{\partial f_2}{\partial \mu}\right)^2 \\ &= \left|\frac{\partial f_1}{\partial \mu} + i\frac{\partial f_2}{\partial \mu}\right|^2 = \left|\frac{\partial f}{\partial \mu}\right|^2 \\ &= \left|\frac{\partial f}{\partial \lambda}\right|^2. \end{aligned}$$

Moreover, in the case at hand,

$$\begin{aligned} f(1,-4,6,\lambda) &= \lambda^3+\lambda^2-4\lambda+6 \\ &= (\lambda-\lambda_1)(\lambda-\lambda_2)(\lambda-\lambda_3) \end{aligned}$$

and

$$\frac{\partial f}{\partial \lambda}(1,-4,6,\lambda_1) = (\lambda_1-\lambda_2)(\lambda_1-\lambda_3) \neq 0$$

because the roots are distinct.

Lemma 15.5. *Let*

$$\begin{aligned} f(\lambda) &= \lambda^n + a_1\lambda^{n-1} + \cdots + a_n \\ &= (\mu+i\nu)^n + a_1(\mu+i\nu)^{n-1} + \cdots + a_n. \end{aligned}$$

Then

$$\frac{\partial f}{\partial \mu} = \frac{\partial f}{\partial \lambda} \quad \textit{and} \quad \frac{\partial f}{\partial \nu} = i\frac{\partial f}{\partial \lambda}.$$

Proof. By the chain rule,

$$\frac{\partial f}{\partial \mu} = \frac{\partial f}{\partial \lambda}\frac{\partial \lambda}{\partial \mu} = \frac{\partial f}{\partial \lambda} \quad \text{and} \quad \frac{\partial f}{\partial \nu}\frac{\partial f}{\partial \lambda}\frac{\partial \lambda}{\partial \nu} = i\frac{\partial f}{\partial \lambda}.$$

Thus, if we write

$$f(\lambda) = f_1(\mu,\nu) + i f_2(\mu,\nu)$$

where $f_1(\mu,\nu)$ and $f_2(\mu,\nu)$ are now both real functions, we see that

$$\frac{\partial f_1}{\partial \mu} + i\frac{\partial f_2}{\partial \mu} = \frac{1}{i}\left(\frac{\partial f_1}{\partial \nu} + i\frac{\partial f_2}{\partial \nu}\right).$$

Matching real and imaginary parts, we obtain

$$\frac{\partial f_1}{\partial \mu} = \frac{\partial f_2}{\partial \nu} \quad \text{and} \quad \frac{\partial f_2}{\partial \mu} = -\frac{\partial f_1}{\partial \nu}$$

in this case also. These are the well-known **Cauchy-Riemann equations**, which will resurface in Chapter 17. In particular, this analysis leads to the conclusion that, for the vector function $\mathbf{f}$ with components f_1 and f_2,

$$|\det J_{\mathbf{f}}(\mu,\nu)|^2 = \left|\frac{\partial f}{\partial \lambda}\right|^2,$$

which is nonzero at simple roots of the polynomial $f(\lambda)$. Thus, the implicit function theorem guarantees that the roots of a polynomial $f(\lambda)$ depend continuously on the coefficients of the polynomial if $f(\lambda)$ has distinct roots.

15.9. Dynamical systems

Dynamical systems are equations of the form

$$\mathbf{x}'(t) = \mathbf{f}(\mathbf{x}(t)) \quad \text{for} \quad t \geq 0 \tag{15.14}$$

or, in the discrete case,

$$\mathbf{x}_{k+1} = \mathbf{f}(\mathbf{x}_k) \quad \text{for} \quad k = 0, 1, \dots , \tag{15.15}$$

where $\mathbf{f}$ maps an open set $\Omega \subset \mathbb{R}^n$ into $\mathbb{R}^n$ and is constrained to be smooth enough to guarantee the existence and uniqueness of a solution to the stated equation for each given initial condition $\mathbf{x}(0) \in \Omega$.

Example 15.6. Let $\mathbf{f}$ be a continuous mapping of $\mathbb{R}^n$ into itself such that

$$\|\mathbf{f}(\mathbf{x}) - f(\mathbf{y})\| \leq \gamma \|\mathbf{x} - \mathbf{y}\|$$

for all vectors $\mathbf{x}, \mathbf{y} \in \mathbb{R}^n$ and let $0 < b < \infty$. Then there exists exactly one continuous vector valued function $\mathbf{x}(t)$ such that

$$\mathbf{x}(t) = \mathbf{v} + \int_0^t \mathbf{f}(\mathbf{x}(s))ds \quad \text{for } 0 \leq t \leq b. \tag{15.16}$$

Moreover, $\mathbf{x} \in \mathcal{C}((0, \infty))$ and

$$\mathbf{x}'(t) = \mathbf{f}(\mathbf{x}(t)) \quad \text{for} \quad 0 < t < b. \tag{15.17}$$

Proof. Let $\mathbf{x}_0(t) = \mathbf{v}$ for $0 \leq t \leq b$ and let

$$\mathbf{x}_{k+1}(t) = \mathbf{v} + \int_0^t f(\mathbf{x}_k(s))ds \quad \text{for} \quad 0 \leq t \leq b \quad \text{and} \quad k = 0, 1, \dots .$$

Then the vector valued functions $\mathbf{x}_k(t)$, $k = 0, 1, \dots$, are continuous on the interval $0 \leq t \leq b$ and

$$\mathbf{x}_1(t) - \mathbf{x}_0(t) = \int_0^t \mathbf{f}(\mathbf{v})ds = \mathbf{f}(\mathbf{v})t \, .$$

Therefore, upon setting $\beta = \|\mathbf{f}(\mathbf{v})\|$, one can readily see that

$$\begin{aligned} \|\mathbf{x}_2(t) - \mathbf{x}_1(t)\| &\leq \int_0^t \|\mathbf{f}(\mathbf{x}_1(s)) - \mathbf{f}(\mathbf{x}_0(s))\|ds \\ &\leq \beta\gamma \int_0^t s ds = \beta\gamma t^2/2 \end{aligned}$$

and, upon iterating this procedure, that

$$\|\mathbf{x}_{k+1}(t) - \mathbf{x}_k(t)\| \leq \frac{\beta}{\gamma} \frac{(\gamma t)^{k+1}}{(k+1)!} \, .$$

Consequently,

$$\begin{aligned}\|\mathbf{x}_{k+\ell}(t)-\mathbf{x}_k(t)\| &\le \frac{\beta}{\gamma}\sum_{j=k+1}^{k+\ell}\frac{(\gamma t)^j}{j!}\\ &\le \frac{\beta}{\gamma}\frac{(\gamma t)^{k+1}}{(k+1)!}\,e^{\gamma t}\\ &\le \frac{\beta}{\gamma}\frac{(\gamma b)^{k+1}}{(k+1)!}\,e^{\gamma b},\end{aligned}$$

which can be made arbitrarily small by choosing k large enough. This suffices to guarantee that the continuous functions $\mathbf{x}_k(t)$ converge uniformly to a continuous limit $\mathbf{x}(t)$ on the interval $0 \le t \le b$. Moreover,

$$\begin{aligned}\lim_{h\to 0}\frac{\mathbf{x}(t+h)-\mathbf{x}(t)}{h} &= \lim_{h\to 0}\frac{1}{h}\int_t^{t+h}\mathbf{f}(\mathbf{x}(s))ds\\ &= \mathbf{f}(\mathbf{x}(t))\end{aligned}$$

for each point $t \in (0, b)$; i.e.,

$$\mathbf{x}'(t) = \mathbf{f}(\mathbf{x}(t)) \quad \text{for } 0 < t < b.$$

To obtain uniqueness, suppose that $\mathbf{x}(t)$ and $\mathbf{y}(t)$ satisfy (15.16) and let

$$\delta = \max\{\|\mathbf{x}(s)-\mathbf{y}(s)\| : \ 0 \le s \le b\}.$$

Then

$$\begin{aligned}\|\mathbf{x}(t)-\mathbf{y}(t)\| &= \left\|\int_0^t\{\mathbf{f}(\mathbf{x}(s))-\mathbf{f}(\mathbf{y}(s))\}ds\right\|\\ &\le \int_0^t\gamma\|\mathbf{x}(s)-\mathbf{y}(s)\|ds\\ &\le \gamma\delta t.\end{aligned}$$

Therefore, upon invoking this bound in the basic inequality

$$\|\mathbf{x}(t)-\mathbf{y}(t)\| \le \gamma\int_0^t \|\mathbf{x}(s)-\mathbf{y}(s)\|ds\,,$$

we obtain the inequality

$$\|\mathbf{x}(t)-\mathbf{y}(t)\| \le \delta\gamma^2\int_0^t sds = \delta\frac{(\gamma t)^2}{2!}$$

and, upon iterating this procedure,

$$\begin{aligned}\|\mathbf{x}(t)-\mathbf{y}(t)\| &\le \delta\frac{(\gamma t)^k}{k!}\\ &\le \delta\frac{(\gamma b)^k}{k!} \quad \text{for } 0 \le t \le b,\end{aligned}$$

which tends to zero as $k \uparrow \infty$. Therefore, $\mathbf{x}(t) = \mathbf{y}(t)$.

15.10. Lyapunov functions

Let $\mathbf{f}$ be a continuous map of an open set $\Omega \subset \mathbb{R}^n$ into $\mathbb{R}^n$ and suppose that $\mathbf{f}(\mathbf{w}_0) = \mathbf{0}$ at a point $\mathbf{w}_0 \in \Omega$. A real-valued function $\varphi \in \mathcal{C}^1(\Omega)$ is said to be a **Lyapunov function** for the dynamical system

$$\mathbf{x}'(t) = \mathbf{f}(\mathbf{x}(t)), \quad 0 \leqslant t < \infty, \quad \text{at } \mathbf{w}_0 \text{ if}$$

(1) $\varphi(\mathbf{w}_0) = 0$ and $\varphi(\mathbf{x}) > 0$ for $\mathbf{x} \in \Omega \setminus \{\mathbf{w}_0\}$.

(2) $\langle \nabla\varphi, \mathbf{f}(\mathbf{x})\rangle \leqslant 0$ for $\mathbf{x} \in \Omega \setminus \{\mathbf{w}_0\}$.

A Lyapunov function $\varphi(\mathbf{x})$ on Ω is said to be a **strict Lyapunov function** if the inequality in (2) is strict, i.e., if

(3) $\langle (\nabla\varphi)(\mathbf{x}), \mathbf{f}(\mathbf{x})\rangle < 0$ for $\mathbf{x} \in \Omega \setminus \{\mathbf{w}_0\}$.

Theorem 15.7. *Let $\mathbf{x}(t)$, $0 \le t < \infty$, be a solution of the dynamical system*

$$\mathbf{x}'(t) = \mathbf{f}(\mathbf{x}(t)) \quad \textit{for} \quad t \ge 0 \quad \textit{with} \quad \mathbf{x}(0) = \mathbf{x}_0$$

and let $\varphi(\mathbf{x})$ be a Lyapunov function for this system at the point $\mathbf{w}_0$. Then:

(1) *Given any $\varepsilon > 0$, there exists a δ such that*

$$\|\mathbf{x}(0) - \mathbf{w}_0\| < \delta \Longrightarrow \|\mathbf{x}(t) - \mathbf{w}_0\| < \varepsilon \quad \textit{for all } t \geqslant 0.$$

(2) *If $\varphi(\mathbf{x})$ is a strict Lyapunov function for the system, then there exists a $\delta > 0$ such*

$$\|\mathbf{x}(0) - \mathbf{w}_0\| < \delta \Longrightarrow \mathbf{x}(t) \to \mathbf{w}_0 \quad \textit{as } t \uparrow \infty.$$

Proof. Let $\varepsilon > 0$ be given and let $0 < \varepsilon_1 \leqslant \varepsilon$ be such that $B_{\varepsilon_1}(\mathbf{w}_0) \subset \Omega$. Let $\alpha = \min\{\varphi(\mathbf{x}) : \|\mathbf{x} - \mathbf{w}_0\| = \varepsilon_1\}$ and choose $0 < \delta < \varepsilon_1$ such that

$$\max\{\varphi(\mathbf{x}) : \|\mathbf{x} - \mathbf{w}_0\| \le \delta\} = \alpha_1 < \alpha.$$

Let $\mathbf{x}(t)$ denote the trajectory of the given dynamical system with initial value $\mathbf{x}(0) \in B_\delta(\mathbf{w}_0)$. Then, since

$$\frac{d}{dt}\varphi(\mathbf{x}(t)) = \langle (\nabla\varphi)(\mathbf{x}), f(\mathbf{x})\rangle,$$

it follows that

$$\begin{aligned} \varphi(\mathbf{x}(t_2)) &= \varphi(\mathbf{x}(t_1)) + \int_{t_1}^{t_2} \frac{d}{dt}\varphi(\mathbf{x}(s))ds \\ &= \varphi(\mathbf{x}(t_1)) + \int_{t_1}^{t_2} \langle (\nabla\varphi)(\mathbf{x}(s)), f(\mathbf{x}(s))\rangle ds \\ &\leqslant \varphi(\mathbf{x}(t_1)) \quad \text{for } t_1 < t_2, \end{aligned}$$

and hence that

$$\varphi(\mathbf{x}(t)) \leqslant \varphi(\mathbf{x}(0)) \le \alpha_1 \quad \text{for all } t \geqslant 0.$$

Therefore, $\|\mathbf{x}(t) - \mathbf{w}_0\| < \varepsilon_1$ for all $t \geqslant 0$. This completes the proof of (1).

Suppose next that $\varphi(\mathbf{x})$ is a strict Lyapunov function on Ω and that $\|\mathbf{x}(t)-\mathbf{w}_0\| \leq \varepsilon$ for all $t \geq 0$. Then, in order to establish (2), it follows from (1) that it suffices to show that there does not exist an $\varepsilon_0 > 0$ such that

$$\|\mathbf{x}(t)-\mathbf{w}_0\| \geqslant \varepsilon_0 \quad \text{for all sufficiently large } t.$$

This is because the function

$$h(\mathbf{x}) = \langle (\nabla\varphi)(\mathbf{x}), \mathbf{f}(\mathbf{x})\rangle$$

is subject to the lower bound

$$h(\mathbf{x}) \geq \rho > 0 \quad \text{for} \quad \varepsilon_0 \leqslant \|\mathbf{x}-\mathbf{w}_0\| \leq \varepsilon$$

and hence

$$|\varphi(\mathbf{x}(t_2)) - \varphi(\mathbf{x}(t_1))| = \left|\int_{t_1}^{t_2} h(\mathbf{x}(s))ds\right| \geq (t_2-t_1)\rho, \quad t_2 > t_1,$$

which is not consistent with the fact that $\varphi(\mathbf{x}(t))$ tends to a limit as $t \uparrow \infty$. $\square$

Exercise 15.12. Show that if $A \in \mathbb{R}^{n\times n}$ and $A = A^T$, then

$$\langle A\mathbf{x},\mathbf{x}\rangle \geqslant 0 \quad \text{for all } \mathbf{x}\in\mathbb{R}^n \iff \langle A\mathbf{x},\mathbf{x}\rangle \geqslant 0 \quad \text{for all } \mathbf{x}\in\mathbb{C}^n.$$

Exercise 15.13. Show that if $A \in \mathbb{R}^{n\times n}$ and $A^T = A$, then

$$\langle A(\mathbf{x}+i\mathbf{y}), \mathbf{x}+i\mathbf{y}\rangle = \langle A\mathbf{x},\mathbf{x}\rangle + \langle A\mathbf{y},\mathbf{y}\rangle \text{ for all vectors } \mathbf{x},\mathbf{y}\in\mathbb{R}^n.$$

Exercise 15.14. Let A, $B \in \mathbb{R}^{n\times n}$, $B = B^T$. Show that $\varphi(\mathbf{x}) = \langle B\mathbf{x},\mathbf{x}\rangle$ is a Lyapunov function for the dynamical system $\mathbf{x}'(t) = A\mathbf{x}(t)$, $0 \leq t < \infty$, at the point $\mathbf{0}$ if and only if $B \succ 0$ and $BA + A^TB \preceq O$.

Exercise 15.15. Show that in the setting of Exercise 15.14, $\varphi(\mathbf{x})$ is a strict Lyapunov function for the considered system at the point $\mathbf{0}$ if and only if

$$B \succ O \text{ and } BA + A^TB \prec O.$$

Exercise 15.16. A dynamical system $\mathbf{x}'(t) = \mathbf{f}(\mathbf{x}(t))$, $t \geq 0$, is said to be a **gradient system** if $\mathbf{f}(\mathbf{x}) = -\nabla g(\mathbf{x})$ for some real valued function $g \in \mathcal{C}^1(\mathbb{R}^n)$. Show that for such a system

$$\frac{d}{dt}g(\mathbf{x}(t)) \leqslant 0 \quad \text{for every} \quad t > 0.$$

Exercise 15.17. Let $g \in \mathcal{C}^1(\mathbb{R}^n)$ and let $\mathbf{x}_0$ be an isolated minimum of $g(\mathbf{x})$. Show that $g(\mathbf{x}) - g(\mathbf{x}_0)$ is a strict Lyapunov function for the system $\mathbf{x}'(t) = -\nabla g(\mathbf{x}(t))$ in a neighborhood of the point $\mathbf{x}_0$.

15.11. Bibliographical notes

The presented proof of Theorem 15.1 was adapted from Saaty and Bram [**62**]. The treatment of Lyapunov functions was adapted from La Salle and Lefschetz [**46**].

Chapter 16

Extremal problems

So I wrote this tune–took me three months. I wanted to keep it simple, elegant. Complex things are easy to do. Simplicity's the real challenge. I worked on it every day until I began to get it right. Then I worked on it some more.... Finally, one night I played it.

R. J. Waller [**70**], p. 168.

This chapter is devoted primarily to classical extremal problems and extremal problems with constraints, which are resolved by the method of Lagranges multipliers. Applications to conjugate gradients and dual extremal problems are also considered.

16.1. Classical extremal problems

Let $f(\mathbf{x}) = f(x_1, \ldots, x_n)$ be a real-valued function of the variables $x_1, \ldots, x_n$ that is defined in some open set $\Omega \subset \mathbb{R}^n$ and suppose that $f \in \mathcal{C}^1(\Omega)$ and $\mathbf{a} \in \Omega$. Then Theorem 14.4 guarantees that the **directional derivative**:

$$(D_{\mathbf{u}}f)(\mathbf{a}) = \lim_{\varepsilon \downarrow 0} \frac{f(\mathbf{a}+\varepsilon\mathbf{u}) - f(\mathbf{a})}{\varepsilon} \tag{16.1}$$

exists for every choice of $\mathbf{u} \in \mathbb{R}^n$ with$\|\mathbf{u}\| = 1$ and supplies the formula

$$(D_{\mathbf{u}}f)(\mathbf{a}) = (\nabla f)(\mathbf{a})\mathbf{u}\,. \tag{16.2}$$

If $\mathbf{a}$ is a local maximum, then

$$\mathbf{f}(\mathbf{a}) \geq f(\mathbf{a}+\varepsilon\mathbf{u})$$

for all unit vectors $\mathbf{u}$ and all sufficiently small positive numbers ε. Thus,

$$\varepsilon > 0 \Longrightarrow \frac{f(\mathbf{a}+\varepsilon\mathbf{u}) - f(\mathbf{a})}{\varepsilon} \leq 0 \Longrightarrow (D_{\mathbf{u}}f)(\mathbf{a}) = (\nabla f)(\mathbf{a})\mathbf{u} \leq 0$$

for all unit vectors $\mathbf{u} \in \mathbb{R}^n$. However, since the same inequality holds when $\mathbf{u}$ is replaced by $-\mathbf{u}$, it follows that the last inequality must in fact be an equality: If $\mathbf{a}$ is a local maximum, then

$$(\nabla f)(\mathbf{a})\mathbf{u} = (D_{\mathbf{u}} f)(\mathbf{a}) = 0$$

for all directions $\mathbf{u}$. Therefore, as similar arguments lead to the same conclusion when $\mathbf{a}$ is a local minimum point for $f(\mathbf{x})$, we obtain the following result:

Theorem 16.1. *Let Q be an open subset of $\mathbb{R}^n$ and let $f \in \mathcal{C}^1(Q)$. If a vector $\mathbf{a} \in Q$ is a local maximum or a local minimum for $f(\mathbf{x})$, then*

$$(\nabla f)(\mathbf{a}) = 0_{1\times n}\,. \tag{16.3}$$

WARNING: The condition (16.3) is necessary but not sufficient for $\mathbf{a}$ to be a local extreme point (i.e., a local maximum or a local minimum). Thus, for example, the point $(0,0)$ is not a local extreme point for the function $f(x_1,x_2) = x_1^2 - x_2^2$, even though $(\nabla f)(0,0) = \begin{bmatrix} 0 & 0 \end{bmatrix}$.

More can be said if $f \in \mathcal{C}^2(Q)$, because then, if $B_r(\mathbf{a}) \subset Q$ and $\mathbf{b} \in B_r(\mathbf{a})$, Taylor's formula with remainder applied to the function

$$h(t) = f(\mathbf{a} + t(\mathbf{b} - \mathbf{a}))$$

implies that

$$h(1) = h(0) + h'(0)\cdot 1 + h''(t_0)\cdot \frac{1^2}{2!}$$

for some point $t_0 \in (0,1)$. But this is the same as to say that

$$f(\mathbf{b}) = f(\mathbf{a}) + \sum_{j=1}^{n} \frac{\partial f}{\partial x_j}(\mathbf{a})(b_j - a_j) + \frac{1}{2}\sum_{i,j=1}^{n} (b_i - a_i)\frac{\partial^2 f}{\partial x_i \partial x_j}(\mathbf{c})(b_j - a_j),$$

where

$$\mathbf{c} = \mathbf{a} + t_0(\mathbf{b} - \mathbf{a})$$

is a point on the open line segment between $\mathbf{a}$ and $\mathbf{b}$. Thus, upon writing the gradient $(\nabla f)(\mathbf{a})$ as a $1 \times n$ row vector:

$$\nabla f(\mathbf{a}) = \begin{bmatrix} \frac{\partial f}{\partial x_1}(\mathbf{a}) & \cdots & \frac{\partial f}{\partial x_n}(\mathbf{a}) \end{bmatrix}, \tag{16.4}$$

and introducing the **Hessian**

$$H_f(\mathbf{c}) = \begin{bmatrix} \frac{\partial^2 f}{\partial x_1 \partial x_1}(\mathbf{c}) & \cdots & \frac{\partial^2 f}{\partial x_1 \partial x_n}(\mathbf{c}) \\ \vdots & & \vdots \\ \frac{\partial^2 f}{\partial x_n \partial x_1}(\mathbf{c}) & \cdots & \frac{\partial^2 f}{\partial x_n \partial x_n}(\mathbf{c}) \end{bmatrix}, \tag{16.5}$$

we can rewrite the formula for $f(\mathbf{b})$ as

$$f(\mathbf{b}) = f(\mathbf{a}) + (\nabla f)(\mathbf{a})(\mathbf{b}-\mathbf{a}) + \frac{1}{2}\langle H_{\mathbf{f}}(\mathbf{c})(\mathbf{b}-\mathbf{a}), (\mathbf{b}-\mathbf{a})\rangle. \tag{16.6}$$

Let us now choose

$$\mathbf{b} = \mathbf{a} + \varepsilon\mathbf{u},$$

where $\mathbf{u}$ is a unit vector and ε is a **positive** number that will eventually tend to zero. Then the last formula implies that

$$\frac{f(\mathbf{a}+\varepsilon\mathbf{u}) - f(\mathbf{a})}{\varepsilon} = (\nabla f)(\mathbf{a})\mathbf{u} + \frac{1}{2}\varepsilon\langle H_f(\mathbf{c})\mathbf{u}, \mathbf{u}\rangle.$$

Exercise 16.1. Show that if $A \in \mathbb{R}^{n\times n}$, then the following two conditions are equivalent:

(1) $A = A^T$ and $\langle A\mathbf{u}, \mathbf{u}\rangle > 0$ for every nonzero vector $\mathbf{u} \in \mathbb{R}^n$.

(2) $\langle A\mathbf{u}, \mathbf{u}\rangle > 0$ for every nonzero vector $\mathbf{u} \in \mathbb{C}^n$.

In view of Exercise 16.1, the notation $A \succ 0$ may be used for real symmetric matrices that are positive definite over $\mathbb{R}^n$ as well as for matrices $A \in \mathbb{C}^{n\times n}$ that are positive definite over $\mathbb{C}^n$. Correspondingly we define

$$\mathbb{R}^{n\times n}_{\succ} = \{A \in \mathbb{R}^{n\times n} : A \succ O\}. \tag{16.7}$$

Exercise 16.2. Let $A, B \in \mathbb{R}^{n\times n}$ and suppose that $A \succ O$, $B = B^T$ and $\|A - B\| < \lambda_{min}$, where λ_{min} denotes the smallest eigenvalue of A. Show that $B \succ O$. [HINT: $\langle B\mathbf{u}, \mathbf{u}\rangle = \langle A\mathbf{u}, \mathbf{u}\rangle + \langle (B - A)\mathbf{u}, \mathbf{u}\rangle$.]

Exercise 16.3. Let $A, B \in \mathbb{R}^{n\times n}$ and suppose that $A \succ O$ and $B = B^T$. Show that $A + \varepsilon B \succ O$ if $|\varepsilon|$ is sufficiently small.

If $f \in \mathcal{C}^2(B_r(\mathbf{a}))$ and $r > \varepsilon$, then, in view of formula (16.6),

$$(\nabla f)(\mathbf{a}) = 0_{1\times n} \Longrightarrow f(\mathbf{a}+\varepsilon\mathbf{u}) - f(\mathbf{a}) = \frac{1}{2}\varepsilon^2\langle H_f(\mathbf{c})\mathbf{u}, \mathbf{u}\rangle \tag{16.8}$$

for some point $\mathbf{c}$ on the open line segment between $\mathbf{a} + \varepsilon\mathbf{u}$ and $\mathbf{a}$ and leads easily to the following conclusions:

Theorem 16.2. *Let $f(\mathbf{x}) = f(x_1, \ldots, x_n)$ belong to $\mathcal{C}^2(Q)$, where $Q \subseteq \mathbb{R}^n$ is an open set that contains the point $\mathbf{a}$. Then:*

(16.9)

$$(\nabla f)(\mathbf{a}) = 0_{1\times n} \text{ and } H_f(\mathbf{a}) \succ 0 \Longrightarrow \mathbf{a} \text{ is a local minimum for } f(\mathbf{x}).$$

(16.10)

$$(\nabla f)(\mathbf{a}) = 0_{1\times n} \text{ and } H_f(\mathbf{a}) \prec 0 \Longrightarrow \mathbf{a} \text{ is a local maximum for } f(\mathbf{x}).$$

Proof. The proof of (16.9) follows from (16.8) and Exercise 16.2. The latter is applicable because $H_f(\mathbf{a}) \succ O$ and $H_f(\mathbf{c})$ is a real symmetric matrix that tends to $H_f(\mathbf{a})$ when $\varepsilon \to 0$. The verification of (16.10) is similar. □

Theorem 16.2 implies that the behavior of a smooth function $f(\mathbf{x})$ in the vicinity of a point $\mathbf{a}$ at which $(\nabla f)(\mathbf{a}) = 0_{1\times n}$ depends critically on the eigenvalues of the real symmetric matrix $H_f(\mathbf{a})$.

Example 16.3. Let $f(u,v) = \alpha(u-1)^2 + \beta(v-2)^3$ with nonzero coefficients $\alpha \in \mathbb{R}$ and $\beta \in \mathbb{R}$. Then

$$\frac{\partial f}{\partial u}(u,v) = 2\alpha(u-1) \quad \text{and} \quad \frac{\partial f}{\partial v}(u,v) = 3\beta(v-2)^2 .$$

Hence,

$$(\nabla f)(u,v) = [0 \quad 0] \text{ if } u = 1 \text{ and } v = 2 .$$

However, the point $(1,2)$ is not a local maximum point or a local minimum point for the function $f(u,v)$. The Hessian

$$H_f(u,v) = \begin{bmatrix} \dfrac{\partial^2 f}{\partial u^2} & \dfrac{\partial^2 f}{\partial u \partial v} \\ \dfrac{\partial^2 f}{\partial v \partial u} & \dfrac{\partial^2 f}{\partial v^2} \end{bmatrix} = \begin{bmatrix} 2\alpha & 0 \\ 0 & 6\beta(v-2) \end{bmatrix},$$

and

$$H_f(1,2) = \begin{bmatrix} 2\alpha & 0 \\ 0 & 0 \end{bmatrix}$$

is neither positive definite nor negative definite.

Exercise 16.4. Show that the Hessian $H_g(1,2)$ of the function

$$g(u,v) = \alpha(u-1)^2 + \beta(v-2)^4$$

is the same as the Hessian $H_f(1,2)$ of the function considered in the preceding example.

WARNING: A local minimum or maximum point is not necessarily an absolute minimum or maximum point: In the figure, $f(x)$ has a local minimum at $x = c$ and a local maximum at $x = b$. However, the absolute maximum value of $f(x)$ in the closed interval $a \le x \le d$ is attained at the point d and the absolute minimum value of $f(x)$ in this interval is attained at the point $x = a$.

Exercise 16.5. Show that if $\alpha = 1$ and $\beta = 1$, then the point $(1,2)$ is a local minimum for the function

$$g(u,v) = (u-1)^2 + (v-2)^4,$$

but it is not a local minimum point for the function

$$f(u,v) = (u-1)^2 + (v-2)^3.$$

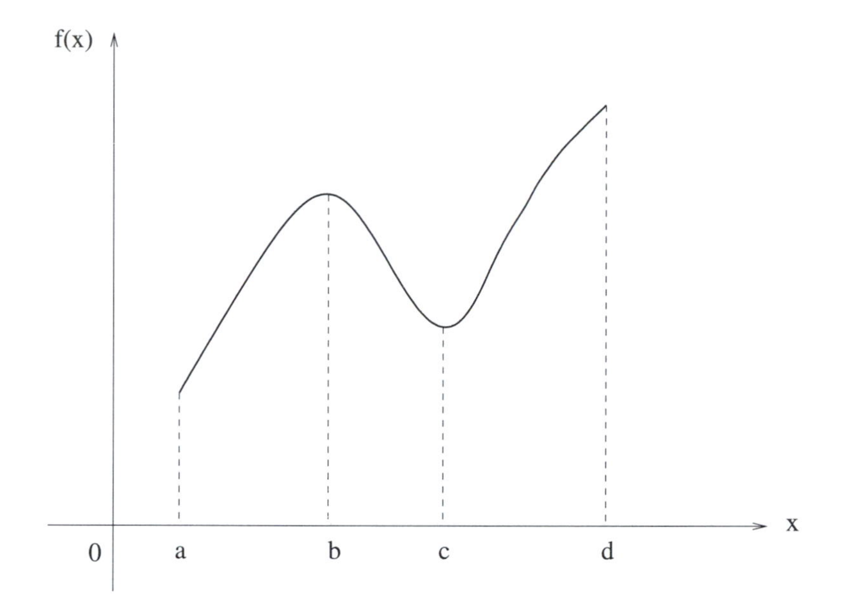

Exercise 16.6. Let $f \in \mathcal{C}^2(\mathbb{R}^2)$ and suppose that $(\nabla f)(a,b) = \begin{bmatrix} 0 & 0 \end{bmatrix}$, and let λ_1 and λ_2 denote the eigenvalues of $H_f(a,b)$. Show that the point (a,b) is: (i) a local minimum for f if $\lambda_1 > 0$ and $\lambda_2 > 0$, (ii) a local maximum for f if $\lambda_1 < 0$ and $\lambda_2 < 0$, (iii) neither a local maximum nor a local minimum if $|\lambda_1 \lambda_2| > 0$, but $\lambda_1 \lambda_2 < 0$.

Exercise 16.7. In many textbooks on calculus the conclusions formulated in Exercise 16.6 are given in terms of the second-order partial derivatives $\alpha = (\partial^2 f/\partial x^2)(a,b)$, $\beta = (\partial^2 f/\partial y^2)(a,b)$ and $\gamma = (\partial^2 f/\partial x \partial y)(a,b)$ by the conditions (i) $\alpha > 0$, $\beta > 0$ and $\alpha\beta - \gamma^2 > 0$; (ii) $\alpha < 0$, $\beta < 0$ and $\alpha\beta - \gamma^2 > 0$; (iii) $\alpha\beta - \gamma^2 < 0$, respectively. Show that the two formulations are equivalent.

Exercise 16.8. Let Q be an open convex subset of $\mathbb{R}^n$ and suppose that $f \in \mathcal{C}^2(Q)$ and $H_f(\mathbf{x}) \succ O$ for every point $\mathbf{x} \in Q$. Show that if $\mathbf{a} \in Q$, then

$$(\nabla f)(\mathbf{a}) = \mathbf{0} \Longrightarrow f(\mathbf{b}) > f(\mathbf{a}) \quad \text{for every point} \quad \mathbf{b} \in Q\,. \tag{16.11}$$

16.2. Extremal problems with constraints

In this section we shall consider extremal problems with constraints, using the method of **Lagrange multipliers**. Let $\mathbf{a}$ be an extreme point of the function $f(x_1, \ldots, x_n)$ when the variables $(x_1, \ldots, x_n)$ are subject to the constraint

$$g(x_1, \ldots, x_n) = 0.$$

Geometrically, this amounts to evaluating to $f(x_1, \ldots, x_n)$ at the points of the surface determined by the constraint $g(x_1, \ldots, x_n) = 0$. Thus, for example, if $g(x_1, \dots, x_n) = x_1^2 + \cdots + x_n^2 - 1$, the surface is a sphere of radius 1.

If $\mathbf{x}(t)$, $-1 \le t \le 1$, is any smooth curve on this surface passing through the point $\mathbf{a}$ with $\mathbf{x}(0) = \mathbf{a}$, then, in view of the formulas

$$g(\mathbf{x}(t)) = 0 \quad \text{and} \quad \frac{d}{dt}g(\mathbf{x}(t)) = (\nabla g)(\mathbf{x}(t))\mathbf{x}'(t) \quad \text{for all} \quad t \in (-1,1)\,,$$

it follows that

$$\frac{d}{dt}g(\mathbf{x}(t)) = (\nabla g)(\mathbf{x}(t))\mathbf{x}'(t) = 0$$

for all $t \in (-1,1)$ and hence, in particular, that

$$(\nabla g)(\mathbf{a})\mathbf{x}'(0) = 0\,.$$

At the same time, since $\mathbf{a}$ is a local extreme point for $f(\mathbf{x})$ we also have

$$0 = \frac{d}{dt}f(\mathbf{x}(0)) = (\nabla f)(\mathbf{a})\mathbf{x}'(0).$$

Thus, if the set of possible vectors $\mathbf{x}'(0)$ fill out an $n-1$ dimensional space, then

$$(\nabla f)(\mathbf{a}) = \lambda(\nabla g)(\mathbf{a})$$

for some constant λ. Our next objective is to present a precise version of this argument, with the help of the implicit function theorem.

Theorem 16.4. *Let $f(\mathbf{u}) = f(u_1, \dots, u_n)$, $g_1(\mathbf{u}) = g_1(u_1, \dots, u_n), \dots,$ $g_k(\mathbf{u}) = g_k(u_1, \dots, u_n)$ be real-valued functions in $\mathcal{C}^1(Q)$ for some open set Q in $\mathbb{R}^n$, where $k < n$. Let*

$$S = \{(u_1, \dots, u_n) \in Q : g_j(u_1, \dots, u_n) = 0 \textit{ for } j = 1, \dots, k\}$$

and assume that:

(1) *There exists a point $\mathbf{a} \in S$ and a number $\alpha > 0$ such that the open ball $B_\alpha(\mathbf{a})$ is a subset of Q and either $f(\mathbf{u}) \ge f(\mathbf{a})$ for all $\mathbf{u} \in S \cap B_\alpha(\mathbf{a})$ or $f(\mathbf{u}) \le f(\mathbf{a})$ for all $\mathbf{u} \in S \cap B_\alpha(\mathbf{a})$.*

(2) $\operatorname{rank} \begin{bmatrix} (\nabla g_1)(\mathbf{u}) \\ \vdots \\ (\nabla g_k)(\mathbf{u}) \end{bmatrix} = p$ *for all points $\mathbf{u}$ in the ball $B_\alpha(\mathbf{a})$.*

Then there exists a set of k constants $\lambda_1, \dots, \lambda_k$ such that

$$(\nabla f)(\mathbf{a}) = \lambda_1(\nabla g_1)(\mathbf{a}) + \cdots + \lambda_k(\nabla g_k)(\mathbf{a})\,.$$

Proof. The general implicit function theorem guarantees the existence of a pair of constants $\gamma > 0$ and $\delta > 0$ and a permutation matrix $P \in \mathbb{R}^{n \times n}$ such that if

(16.12)

$$P\mathbf{u} = \begin{bmatrix} \mathbf{x} \\ \mathbf{y} \end{bmatrix} \quad \text{with} \quad \mathbf{x} \in \mathbb{R}^q, \quad \mathbf{y} \in \mathbb{R}^p, \quad p + q = n \quad \text{and} \quad P\mathbf{a} = \begin{bmatrix} \mathbf{x}^\circ \\ \mathbf{y}^\circ \end{bmatrix},$$

then for each point $\mathbf{x}$ in the ball $B_\gamma(\mathbf{x}_0)$ there exists exactly one point $\mathbf{y} = \varphi(\mathbf{x})$ in the ball $B_\delta(\mathbf{y}^\circ)$ such that

$$g_i(\mathbf{u}) = 0 \quad \text{for} \quad i = 1, \ldots, k \quad \text{when} \quad P\mathbf{u} = \begin{bmatrix} \mathbf{x} \\ \varphi(\mathbf{x}) \end{bmatrix}.$$

Moreover, $\varphi(\mathbf{x}) \in \mathcal{C}^1(B_\gamma(\mathbf{x}_0))$.

Let $\mathbf{x}(t)$, $-1 \leq t \leq 1$, be a curve in $B_\gamma(\mathbf{x}_0)$ with $\mathbf{x}(0) = \mathbf{x}^\circ$ and let

$$\mathbf{u}(t) = P^T \begin{bmatrix} \mathbf{x}(t) \\ \varphi(\mathbf{x}(t)) \end{bmatrix}.$$

Then,

$$\frac{d}{dt} f(\mathbf{u}(t))|_{t=0} = (\nabla f)(\mathbf{a})\mathbf{u}'(0) = 0 \tag{16.13}$$

and

$$g_i(\mathbf{u}(t)) = 0 \quad \text{for} \quad -1 < t < 1 \quad \text{and} \quad i = 1, \ldots, k.$$

Therefore,

$$\frac{d}{dt} g_i(\mathbf{u}(t)) = (\nabla g_i)(\mathbf{u}(t))\mathbf{u}'(t) = 0 \quad \text{for} \quad -1 < t < 1 \quad \text{and} \quad i = 1, \ldots, k.$$

In particular,

$$(\nabla g_i)(\mathbf{a})\mathbf{u}'(0) = (\nabla g_i)(\mathbf{u}(0))\mathbf{u}'(0) = 0$$

for $i = 1, \ldots, k$. Therefore, the vector $\mathbf{u}'(0)$ belongs to the null space $\mathcal{N}_A$ of the $k \times n$ matrix

$$A = \begin{bmatrix} (\nabla g_1)(\mathbf{a}) \\ \vdots \\ (\nabla g_k)(\mathbf{a}) \end{bmatrix}.$$

Since rank $A = p$, by assumption, dim $\mathcal{N}_A = q$.

Next, consider the curve

$$\mathbf{x}(t) = \mathbf{x}^\circ + t\mathbf{w}, \quad -1 \leq t \leq 1,$$

which belongs to the ball $\{\mathbf{x} \in \mathbb{R}^q : \|\mathbf{x} - \mathbf{x}^\circ\| < \gamma\}$ for every vector $\mathbf{w} \in \mathbb{R}^q$ with $\|\mathbf{w}\| < \gamma$. Therefore, since $P^T = P^{-1}$, it follows that

$$\mathbf{u}'(0) = P^T \begin{bmatrix} \mathbf{x}'(0) \\ \mathbf{y}'(0) \end{bmatrix} = P^T \begin{bmatrix} \mathbf{w} \\ B\mathbf{w} \end{bmatrix},$$

where

$$B = \begin{bmatrix} \frac{\partial \varphi_1}{\partial \mathbf{x}_1}(\mathbf{x}^\circ) & \cdots & \frac{\partial \varphi_1}{\partial \mathbf{x}_q}(\mathbf{x}^\circ) \\ \vdots & & \vdots \\ \frac{\partial \varphi_p}{\partial \mathbf{x}_1}(\mathbf{x}^\circ) & \cdots & \frac{\partial \varphi_p}{\partial \mathbf{x}_q}(\mathbf{x}^\circ) \end{bmatrix}.$$

Thus, the considered set of vectors $\mathbf{u}'(0)$ span $\mathcal{N}_A$. Consequently, if $\mathbf{v} \in \mathbb{R}^n$ and

$$\mathbf{v}^T \mathbf{u}'(0) = 0$$

for all vectors $\mathbf{u}'(0)$ of the indicated form, then $\mathbf{v} \in \mathcal{R}_{A^T}$. Therefore, $(\nabla f)(\mathbf{a}) \in \mathcal{R}_{A^T}$, which is equivalent to the asserted conclusion. □

16.3. Examples

Example 16.5. Let $A \in \mathbb{R}^{p \times q}$ and let

$$f(\mathbf{x}) = \langle A\mathbf{x}, A\mathbf{x}\rangle \text{ and } g(\mathbf{x}) = \langle \mathbf{x}, \mathbf{x}\rangle - 1.$$

The problem is to find

$$\max f(\mathbf{x})$$

subject to the constraint

$$g(\mathbf{x}) = 0.$$

Discussion. The first order of business is to verify the following formulas for the gradients, **written as column vectors**:

$$(\nabla f)(\mathbf{x}) = 2A^T A\mathbf{x} \tag{16.14}$$

$$(\nabla g)(\mathbf{x}) = 2\mathbf{x}. \tag{16.15}$$

Next, Theorem 16.4 guarantees that if $\mathbf{a}$ is an extreme point for this problem, then there exists a constant α such that

$$(\nabla f)(\mathbf{a}) = \alpha(\nabla g)(\mathbf{a}).$$

But this in turn implies that

$$2A^T A\mathbf{a} = 2\alpha\mathbf{a},$$

and hence, since $\mathbf{a} \neq \mathbf{0}$, that α is an eigenvalue of $A^T A$. The maximum value is obtained by choosing $\alpha = s_1^2$, where s_1 is the largest singular value of A.

Example 16.6. Let A be a real $p \times q$ matrix of rank $r > 1$ and let

$$f(\mathbf{x}) = \langle A\mathbf{x}, A\mathbf{x}\rangle, \ g_0(\mathbf{x}) = \langle \mathbf{x}, \mathbf{x}\rangle - 1 \text{ and } g_1(\mathbf{x}) = \langle \mathbf{x}, \mathbf{u}_1\rangle,$$

where

$$A^T A\mathbf{u}_1 = s_1^2 \mathbf{u}_1 \text{ and } \|\mathbf{u}_1\| = 1.$$

The problem is to find

$$\max f(\mathbf{x})$$

subject to the constraints

$$g_0(\mathbf{x}) = 0 \quad \text{and} \quad g_1(\mathbf{x}) = 0.$$

Discussion. By Theorem 16.4, there exists a pair of real constants α and β such that

$$(\nabla f)(\mathbf{a}) = \alpha(\nabla g_0)(\mathbf{a}) + \beta(\nabla g_1)(\mathbf{a}) \tag{16.16}$$

at each extreme point $\mathbf{a}$ of the given problem. Therefore,

$$2A^T A\mathbf{a} = 2\alpha\mathbf{a} + \beta\mathbf{u}_1, \tag{16.17}$$

where the constraints supply the supplementary information that

$$\langle \mathbf{a}, \mathbf{a} \rangle = 1 \quad \text{and} \quad \langle \mathbf{a}, \mathbf{u}_1 \rangle = 0.$$

Since $A^T A$ is a real symmetric (and hence a Hermitian) matrix, it has a set of q orthonormal eigenvectors $\mathbf{u}_1, \ldots, \mathbf{u}_q$ such that the corresponding eigenvalues are the squares of the singular values of A:

$$A^T A\mathbf{u}_j = s_j^2 \mathbf{u}_j, \; j = 1, \ldots, q,$$

where $s_1 \geq s_2 \geq \cdots \geq s_q \geq 0$. Thus, as $\mathbf{u}_1, \ldots, \mathbf{u}_q$ is an orthonormal basis for $\mathbb{C}^q$,

$$\mathbf{a} = \sum_{j=1}^{q} c_j \mathbf{u}_j \,, \quad \text{where} \quad c_j = \langle \mathbf{a}, \mathbf{u}_j \rangle \quad \text{for} \quad j = 1, \ldots, q \,.$$

This last formula exhibits once again the advantage of working with an orthonormal basis: it is easy to calculate the coefficients. In particular, the constraint $g_1(\mathbf{a}) = 0$ forces $c_1 = 0$ and hence we are left with

$$\mathbf{a} = \sum_{j=2}^{q} c_j \mathbf{u}_j .$$

Substituting this into formula (16.17), we obtain

$$2A^T A \sum_{j=2}^{q} c_j \mathbf{u}_j = 2\alpha \sum_{j=2}^{q} c_j \mathbf{u}_j + \beta \mathbf{u}_1,$$

which reduces to

$$2\sum_{j=2}^{q} c_j s_j^2 \mathbf{u}_j = 2\alpha \sum_{j=2}^{q} c_j \mathbf{u}_j + \beta \mathbf{u}_1 .$$

Therefore, we must have $\beta = 0$ and $c_j s_j^2 = \alpha c_j$ for $j = 2, \ldots, q$. Moreover, since the constraint

$$g_0(\mathbf{a}) = 0 \Longrightarrow \sum_{j=2}^{q} c_j^2 = 1,$$

we see that

$$\langle A\mathbf{a}, A\mathbf{a} \rangle = \sum_{j=2}^{q} c_j^2 s_j^2 \leq \left(\sum_{j=2}^{q} c_j^2 \right) s_2^2 = s_2^2$$

and hence that the maximum value of $\langle A\mathbf{a}, A\mathbf{a}\rangle$ subject to the two given constraints is obtained by choosing $\alpha = s_2^2, c_2 = 1$ and $c_j = 0$ for $j = 2, \dots, q$. This gives a geometric interpretation to s_2. Analogous interpretations hold for the other singular values of A.

Exercise 16.9. Let A be the $(n+1)\times(n+1)$ matrix with entries $a_{ij} = 1/(i+j+1)$ for $i, j = 0, \dots, n$. Show that $A \succ O$ and that the largest singular value s_1 of A is subject to the bound

(16.18)

$$s_1 \le \sqrt{k+1/2}\sum_{j=0}^{n} \frac{1}{(k+j+1)\sqrt{j+1/2}} \quad \text{for some integer } k\,, \quad 0 \le k \le n\,.$$

[HINT: Let u_j, $j = 0, \dots, n$ denote the entries in an eigenvector corresponding to s_1 and choose k so that $\sqrt{k+1/2}|u_k| = \max\{\sqrt{j+1/2}|u_j|\}$.]

Remark 16.7. The inequality (16.18) combined with the sequence of inequalities

$$\begin{aligned}\sum_{j=0}^{n} \frac{1}{(k+j+1)\sqrt{j+1/2}} &< \int_{-1/2}^{n+1/2} \frac{dx}{(x+k+1/2)\sqrt{x+1/2}} \\ &= \int_{0}^{\sqrt{(n+1/2}} \frac{2dy}{y^2+k+1/2} < \frac{\pi}{\sqrt{k+1/2}}\end{aligned}$$

provides another proof that the norm of the Hilbert matrix considered in Exercises 8.38–8.40 is less than π.

Exercise 16.10. Let $A \in \mathbb{R}^{p\times p}$ and $B = A + A^T$. Show that

$$(16.19) \qquad \langle A\mathbf{x}, \mathbf{x}\rangle = \frac{1}{2}\langle B\mathbf{x}, \mathbf{x}\rangle \quad \text{for every} \quad \mathbf{x} \in \mathbb{R}^p\,.$$

Exercise 16.11. Let $A \in \mathbb{R}^{4\times 4}$, let $\lambda_1 > \lambda_2 > \lambda_3 > \lambda_4$ denote the eigenvalues of the matrix $B = A + A^T$ and let $\mathbf{u}_1, \mathbf{u}_2, \mathbf{u}_3, \mathbf{u}_4$ denote a corresponding set of normalized eigenvectors; i.e., $\langle \mathbf{u}_j, \mathbf{u}_j\rangle = 1$ and $B\mathbf{u}_j = \lambda_j\mathbf{u}_j$ for $j = 1, \dots, 4$. Find the maximum value of the function $f(\mathbf{x}) = \langle A\mathbf{x}, \mathbf{x}\rangle$ over all vectors $\mathbf{x} \in \mathbb{R}^4$ that are subject to the two constraints $\langle \mathbf{x}, \mathbf{x}\rangle = 1$ and $\langle \mathbf{x}, \mathbf{v}\rangle = 0$, where $\mathbf{v} = (3/5)\mathbf{u}_1 + (4/5)\mathbf{u}_2$.

Exercise 16.12. In the setting of Exercise 16.11, find the minimum value of $f(\mathbf{x})$ over all vectors $\mathbf{x} \in \mathbb{R}^4$ that are subject to the same two constraints that are considered there.

Example 16.8. The problem of maximizing $f(X) = (\det X)^2$ over the set of all matrices $X \in \mathbb{R}^{p\times p}$ that meet the constraint $\operatorname{trace} XX^T = 1$ fits naturally into the setting of Lagrange multipliers upon setting $g(X) = \operatorname{trace} XX^T - 1$.

Discussion. This is a problem with p^2 variables, x_{ij}, $i, j = 1, \ldots, p$. It turns out to be convenient to organize the gradient of $f(X)$ as a $p \times p$ matrix with ij entry equal to $\frac{\partial f}{\partial x_{ij}}$. To warm up to the problem at hand, let $h(X) = \det X$ and note, for example, that the formula

$$h(X) = x_{11}M_{11} - x_{12}M_{12} + x_{13}M_{13} + \cdots + (-1)^{1+p}x_{1p}M_{1p}$$

for the determinant in terms of the minors M_{ij} clearly implies that

$$\frac{\partial h}{\partial x_{12}} = -M_{12}$$

and that, in general,

$$\frac{\partial h}{\partial x_{ij}} = (-1)^{i+j}M_{ij}\,. \tag{16.20}$$

Since the maximum value of $f(X)$ under the given constraint is nonzero, there is no loss in generality in assuming that X is invertible. Then, in view of Theorem 5.9,

$$(\nabla h)(X) = (\det X)\,(X^T)^{-1} = h(X)(X^T)^{-1}$$

and hence

$$(\nabla f)(X) = 2(\det X)\,(\nabla h)(X) = 2f(X)(X^T)^{-1}\,.$$

Thus, as $(\nabla f)(X) = \alpha(\nabla g)(X)$ at a critical point X_0 of the constrained problem and $(\nabla g)(X) = 2X$, it follows that

$$2f(X_0)(X_0^T)^{-1} = \alpha 2X_0\,.$$

Therefore,

$$f(X_0)I_p = \alpha X_0X_0^T\,,$$

which, upon computing the trace of both sides, implies that $pf(X_0) = \alpha$ and hence that

$$f(X_0)I_p = pf(X_0)X_0X_0^T\,.$$

Consequently, $X_0X_0^T = p^{-1}I_p$ and

$$(\det X_0)^2 = \det X_0 \det X_0^T = p^{-p}\,.$$

It is readily confirmed that p^{-p} is indeed the maximum value of the function under consideration by using singular value decompositions; see Exercise ...

Exercise 16.13. Find the maximum value of the function $f(X) = \ln(\det X)^2$ over all matrices $X \in \mathbb{R}^{p\times p}$ with trace $XX^T = 1$.

Exercise 16.14. Find the maximum value of the function $f(X) = \text{trace}\, XX^T$ over all matrices $X \in \mathbb{R}^{p\times p}$ with $\det X = 1$.

Exercise 16.15. Let $C \in \mathbb{R}^{n\times n}$, $Q = \{X \in \mathbb{R}^{n\times n} : \det X > 0\}$ and let

$$f(X) = \langle X, C\rangle - \ln\det X \quad \text{for} \quad X \in Q\,,$$

where $\langle X, C\rangle = \text{trace}\,\{C^T X\}$. Show that if $(\nabla f)(X)$ is written as an $n \times n$ matrix with ij entry equal to $\partial f/\partial x_{ij}$, then $(\nabla f)(X) = C - (X^T)^{-1}$.

Exercise 16.16. Find $\max\{\text{trace}\, A : A \in \mathbb{R}^{n\times n} \quad \text{and} \quad \text{trace}\,(A^T A) = 1\}$.

Exercise 16.17. Let $B \in \mathbb{R}_{\succ}^{n\times n}$, $Q = \{A \in \mathbb{R}_{\succ}^{n\times n} : a_{ij} = b_{ij} \quad \text{for} \quad |i-j| \le m\}$ and let $f(X) = \ln\det X$. Use the method of Lagranges multipliers to show that if $\widetilde{A} \in Q$ and $f(\widetilde{A}) \ge f(A)$ for every matrix $A \in Q$, then $\mathbf{e}_i^T(\widetilde{A})^{-1}\mathbf{e}_j = 0$ for $|i-j| > m$.

Exercise 16.18. Let $\alpha_1 < \cdots < \alpha_k$ and let $p(x) = c_0 + c_1x + \cdots + c_nx^n$ be a polynomial of degree $n \ge k$ with coefficients $c_0, \ldots, c_n \in \mathbb{R}$. Show that if $p(\alpha_j) = \beta_j$ for $j = 1, \ldots, k$, then

$$\sum_{j=0}^n c_j^2 \ge \mathbf{b}^T(U^TU)^{-1}\mathbf{b}, \quad \text{where} \quad U = \begin{bmatrix} 1 & \cdots & 1 \\ \alpha_1 & & \alpha_k \\ \vdots & & \vdots \\ \alpha_1^n & & \alpha_k^n \end{bmatrix} \quad \text{and} \quad \mathbf{b} = \begin{bmatrix} \beta_1 \\ \vdots \\ \beta_k \end{bmatrix},$$

and find a polynomial with coefficients that achieve the exhibited minimum.

Exercise 16.19. Let $\mathbf{e}_i$ denote the i'th column of I_n for $i = 1, \ldots, n$ and let

$$Q = \{\mathbf{x} \in \mathbb{R}^n : \mathbf{e}_i^T\mathbf{x} > 0 \quad \text{for} \quad i = 1, \ldots, n \quad \text{and} \quad \sum_{i=1}^n \mathbf{e}_i^T\mathbf{x} = 1\}\,.$$

Show that if $\mathbf{a} \in Q$, $\mathbf{b} \in Q$ and $\mathbf{u} \in \mathbb{R}^n$ are vectors with components $a_1, \ldots, a_n$, $b_1 \ldots, b_n$ and $u_1 \ldots, u_n$, respectively, then

$$\max_{\mathbf{u}\in\mathbb{R}^n} \left\{ \langle \mathbf{u}, \mathbf{a}\rangle - \ln\left(\sum_{i=1}^n e^{u_i}b_i\right)\right\} = \sum_{i=1}^n a_i \ln\frac{a_i}{b_i}\,.$$

[HINT: First rewrite the function that is to be maximized as $\sum_{i=1}^n a_i \ln\frac{c_i}{b_i}$, where $c_i = e^{u_i}b_i/\{\sum_{s=1}^n e^{u_s}b_s\}$ and note that the vector $\mathbf{c}$ with components c_i, $i = 1, \ldots, n$, belongs to Q. Therefore, the quantity of interest can be identified with

$$\max \sum_{i=1}^n a_i \ln\frac{x_i}{b_i}$$

over $\mathbf{x} \in \mathbb{R}^n$ with $x_i > 0$ and subject to $g(\mathbf{x}) = x_1 + \cdots + x_n - 1 = 0$.

Exercise 16.20. Use Lagrange's method to find the point on the line of intersection of the two planes $a_1x_1 + a_2x_2 + a_3x_3 + a_0 = 0$, $b_1x_1 + b_2x_2 + b_3x_3 + b_0 = 0$ which is nearest to the origin. You may assume that the two planes really intersect, but you should explain where this enters into the calculation.

Exercise 16.21. Use Lagrange's method to find the shortest distance from the point $(0, b)$ on the y axis to the parabola $x^2 - 4y = 0$.

Exercise 16.22. Use Lagrange's method to find the maximum value of $\langle A\mathbf{x}, \mathbf{x}\rangle$ subject to the conditions $\langle \mathbf{x}, \mathbf{x}\rangle = 1$ and $\langle \mathbf{u}_1, \mathbf{x}\rangle = 0$, where $\mathbf{u}_1$ is a nonzero vector in $\mathcal{N}_{(s_1^2 I_n - A^T A)}$, s_1 is the largest singular value of A and $A = A^T \in \mathbb{R}^{n\times n}$.

16.4. Krylov subspaces

The k'th Krylov subspace of $\mathbb{R}^n$ generated by a nonzero vector $\mathbf{u} \in \mathbb{R}^n$ and a matrix $A \in \mathbb{R}^{n\times n}$ is defined by the formula

$$\mathcal{H}_k = \text{span}\{\mathbf{u}, A\mathbf{u}, \dots, A^{k-1}\mathbf{u}\} \quad \text{for} \quad k = 1, 2, \dots .$$

Clearly, $\dim \mathcal{H}_1 = 1$ and $\dim \mathcal{H}_k \leq k$ for $k = 2, 3, \dots$.

Lemma 16.9. *If $\mathcal{H}_{k+1} = \mathcal{H}_k$ for some positive integer k, then $\mathcal{H}_j = \mathcal{H}_k$ for every integer $j \geq k$.*

Proof. If $\mathcal{H}_{k+1} = \mathcal{H}_k$ for some positive integer k, then $A^k\mathbf{u} = c_{k-1}A^{k-1}\mathbf{u} + \cdots + c_0\mathbf{u}$ for some choice of coefficients $c_0, \dots, c_{k-1} \in \mathbb{R}$. Therefore, $A^{k+1}\mathbf{u} = c_{k-1}A^k + \cdots + c_0 A\mathbf{u}$, which implies that $\mathcal{H}_{k+2} = \mathcal{H}_{k+1}$ and hence that $\mathcal{H}_{k+2} = \mathcal{H}_k$. The same argument implies that $\mathcal{H}_{k+3} = \mathcal{H}_{k+2} = \mathcal{H}_{k+1}$ and so on down the line, to complete the proof. □

Exercise 16.23. Let $A \in \mathbb{R}^{n\times n}$, let $\mathbf{u} \in \mathbb{R}^n$ and let $k \geq 1$ be a positive integer. Show that if $A \succ 0$, then the matrix

$$\begin{bmatrix} \langle A\mathbf{u}, \mathbf{u}\rangle & \langle A^2\mathbf{u}, \mathbf{u}\rangle & \cdots & \langle A^k\mathbf{u}, \mathbf{u}\rangle \\ \langle A\mathbf{u}, A\mathbf{u}\rangle & \langle A^2\mathbf{u}, A\mathbf{u}\rangle & \cdots & \langle A^k\mathbf{u}, A\mathbf{u}\rangle \\ \vdots & & \cdots & \vdots \\ \langle A\mathbf{u}, A^{k-1}\mathbf{u}\rangle & \langle A^2\mathbf{u}, A^{k-1}\mathbf{u}\rangle & \cdots & \langle A^k\mathbf{u}, A^{k-1}\mathbf{u}\rangle \end{bmatrix}$$

is invertible if and only if the vectors $\mathbf{u}, A\mathbf{u}, \dots, A^{k-1}\mathbf{u}$ are linearly independent in $\mathbb{R}^n$.

16.5. The conjugate gradient method

The method of conjugate gradients is an iterative approach to solving equations of the form $A\mathbf{x} = \mathbf{b}$ for matrices $A \in \mathbb{R}_{\succ}^{n\times n}$ when $\mathbf{b} \in \mathbb{R}^n$. It is presented in this chapter because the justification of this method is based on solving an appropriately chosen sequence of minimization problems.

Let $A \in \mathbb{R}^{n\times n}$, $\mathbf{b} \in \mathbb{R}^n$ and

$$\varphi(\mathbf{x}) = \frac{1}{2}\langle A\mathbf{x}, \mathbf{x}\rangle - \langle \mathbf{b}, \mathbf{x}\rangle \tag{16.21}$$

for every $\mathbf{x} \in \mathbb{R}^n$. Then the gradient $(\nabla\varphi)(\mathbf{x})$, written as a column vector, is readily seen to be equal to

$$(\nabla\varphi)(\mathbf{x}) = \begin{bmatrix} \frac{\partial\varphi}{\partial x_1} \\ \vdots \\ \frac{\partial\varphi}{\partial x_n} \end{bmatrix} = \frac{1}{2}(A + A^T)\mathbf{x} - \mathbf{b}\,,$$

and the Hessian $H_\varphi(\mathbf{x})$ is equal to

$$H_\varphi(\mathbf{x}) = \frac{1}{2}(A + A^T)\,.$$

Therefore, upon invoking the general formula (16.6), one can readily check that

$$\varphi(\mathbf{y}) - \varphi(\mathbf{x}) = \langle(\nabla\varphi)(\mathbf{x}), \mathbf{y} - \mathbf{x}\rangle + \frac{1}{4}\langle(A + A^T)(\mathbf{y} - \mathbf{x}), (\mathbf{y} - \mathbf{x})\rangle\,.$$

Thus, if the given matrix $A \in \mathbb{R}_{\succ}^{n\times n}$, which will be the case for the rest of this section, then the function $\varphi(\mathbf{x})$ attains its minimum value at exactly one point $\mathbf{a} \in \mathbb{R}^n$, namely when $A\mathbf{a} = \mathbf{b}$. This is the one and only point in $\mathbb{R}^n$ at which the gradient $\nabla\varphi(\mathbf{x}) = A\mathbf{x} - \mathbf{b}$ vanishes. Consequently, the unique solution $\mathbf{x}$ of the equation $A\mathbf{x} = \mathbf{b}$ is the unique point at which $\varphi(\mathbf{x})$ attains its minimum value. The **conjugate gradient** method exploits this fact in order to solve for the solution of the equation by solving for this minimum recursively, i.e., via a sequence of minimization problems.

Lemma 16.10. *Let $A \in \mathbb{R}_{\succ}^{n\times n}$ and let Q be a closed nonempty convex subset of $\mathbb{R}^n$. Then there exists exactly one point $\mathbf{q} \in Q$ at which the function $\varphi(\mathbf{x})$ defined by formula* (16.21) *attains its minimum value, i.e., at which*

$$\varphi(\mathbf{q}) \le \varphi(\mathbf{x}) \quad \textit{for every} \quad \mathbf{x} \in Q\,.$$

Proof. Let $\lambda_1 \le \cdots \le \lambda_n$ denote the eigenvalues of A. Then, $\lambda_1 > 0$ and

$$\begin{aligned} \varphi(\mathbf{x}) &= \frac{1}{2}\langle A\mathbf{x}, \mathbf{x}\rangle - \langle \mathbf{b}, \mathbf{x}\rangle \\ &\ge \frac{1}{2}\lambda_1\|\mathbf{x}\|_2^2 - \|\mathbf{b}\|_2\|\mathbf{x}\|_2 \\ &= \|\mathbf{x}\|_2(\frac{1}{2}\lambda_1\|\mathbf{x}\|_2 - \|\mathbf{b}\|_2)\,, \end{aligned}$$

which clearly tends to ∞ as $\|\mathbf{x}\|_2$ tends to ∞. Thus, if $\mathbf{y} \in Q$, there exists a number $R > 0$ such that $\varphi(\mathbf{x}) > \varphi(\mathbf{y})$ if $\|\mathbf{x}\|_2 \ge R$. Consequently, $\varphi(\mathbf{x})$ will achieve its lowest values in the set $Q \cap \{\mathbf{x} : \|\mathbf{x}\|_2 \le R\}$. Thus, as this set is closed and bounded and $\varphi(\mathbf{x})$ is a continuous function of $\mathbf{x}$, $\varphi(\mathbf{x})$ will attain its minimum value on this set.

The next step is to verify that $\varphi(\mathbf{x})$ attains its minimum at exactly one point in the set Q. The proof is based on the fact that $\varphi(\mathbf{x})$ is **strictly convex**; i.e.,

$$\varphi(t\mathbf{u} + (1-t)\mathbf{v}) < t\varphi(\mathbf{u}) + (1-t)\varphi(\mathbf{v}) \tag{16.22}$$

for every pair of distinct vectors $\mathbf{u}$ and $\mathbf{v}$ in $\mathbb{R}^n$ and every number $t \in (0,1)$. Granting this statement, which is left to the reader as an exercise, one can readily see that if $\varphi(\mathbf{u}) = \varphi(\mathbf{v}) = \gamma$ for two distinct vectors $\mathbf{u}$ and $\mathbf{v}$ in Q and if $t \in (0,1)$, then $t\mathbf{u} + (1-t)\mathbf{v} \in Q$ and

$$\begin{aligned} \gamma &\leq \varphi(t\mathbf{u} + (1-t)\mathbf{v}) \\ &< t\varphi(\mathbf{u}) + (1-t)\varphi(\mathbf{v}) \\ &= t\gamma + (1-t)\gamma = \gamma, \end{aligned}$$

which is clearly impossible. Therefore $\mathbf{u} = \mathbf{v}$. □

Exercise 16.24. Verify that the function $\varphi(\mathbf{x})$ defined by formula (16.21) is strictly convex. [HINT: Check that the identity

(16.23)

$$\varphi(t\mathbf{u} + (1-t)\mathbf{v}) = t\varphi(\mathbf{u}) + (1-t)\varphi(\mathbf{v}) - \frac{1}{2}t(1-t)\langle A(\mathbf{u}-\mathbf{v}), \mathbf{u}-\mathbf{v}\rangle$$

is valid for every pair of vectors $\mathbf{u}$ and $\mathbf{v}$ in $\mathbb{R}^n$ and every number $t \in \mathbb{R}$.]

Lemma 16.11. *Let $A \in \mathbb{R}_{\succ}^{n\times n}$, $\mathbf{b} \in \mathbb{R}^n$, and for any vector $\mathbf{x}_0 \in \mathbb{R}^n$ such that $\mathbf{u} = A\mathbf{x}_0 - \mathbf{b} \neq \mathbf{0}$, $\mathbf{x}_0 \in \mathbb{R}^n$, let $\mathcal{H}_j$, $j = 1, 2, \ldots$, denote the Krylov subspaces based on the matrix A and the vector $\mathbf{u}$. Let ℓ denotes the smallest positive integer such that $\mathcal{H}_{\ell+1} = \mathcal{H}_\ell$. Then the vector $\mathbf{a} = A^{-1}\mathbf{b}$ meets the following constraints:*

(1) $\mathbf{a} \in \mathbf{x}_0 + \mathcal{H}_\ell$.

(2) $\mathbf{a} \notin \mathbf{x}_0 + \mathcal{H}_j$ *for* $j = 1, \ldots, \ell - 1$.

Proof. By assumption, $A^\ell\mathbf{u} \in \mathcal{H}_\ell$, and hence

$$A^\ell\mathbf{u} = c_{\ell-1}A^{\ell-1}\mathbf{u} + \cdots + c_0\mathbf{u}$$

for some choice of coefficients $c_{\ell-1}, \ldots, c_0 \in \mathbb{R}$. Moreover, $c_0 \neq 0$, because otherwise $A^\ell\mathbf{u} = c_{\ell-1}A^{\ell-1}\mathbf{u} + \cdots + c_1A\mathbf{u}$, and hence, as A is invertible, it follows that $A^{\ell-1}\mathbf{u} = c_{\ell-2}A^{\ell-1}\mathbf{u} + \cdots + c_1\mathbf{u}$. But this implies that $\mathcal{H}_\ell = \mathcal{H}_{\ell-1}$, which contradicts the definition of ℓ. Thus, the polynomial

$$p(x) = x^\ell - \{c_{\ell-1}x^{\ell-1} + \cdots + c_0\}$$

meets the conditions $p(0) \neq 0$ and $p(A)\mathbf{u} = \mathbf{0}$. Therefore,

$$r(x) = \frac{p(x) - p(0)}{x}$$

is a polynomial of degree $\ell - 1$ and

$$p(0)\mathbf{u} + Ar(A)\mathbf{u} = p(A)\mathbf{u} = \mathbf{0}\,.$$

But this is the same as to say that

$$p(0)A(\mathbf{x}_0 - \mathbf{a}) = -Ar(A)\mathbf{u}$$

and hence, as A is invertible, that

$$\mathbf{a} - \mathbf{x}_0 = p(0)^{-1}r(A)\mathbf{u}\,;$$

that is to say, $\mathbf{a} \in \mathbf{x}_0 + \mathcal{H}_\ell$, as claimed in (1).

To verify (2), suppose that $\mathbf{a} \in \mathbf{x}_0 + \mathcal{H}_j$ for some integer $j \geq 1$; i.e.,

$$\mathbf{a} - \mathbf{x}_0 = c_{j-1}A^{j-1}\mathbf{u} + \cdots + c_0\mathbf{u}$$

for some set of coefficients $c_0, \ldots, c_{j-1} \in \mathbb{R}$ and hence that

$$\mathbf{u} = A(\mathbf{x}_0 - \mathbf{a}) = -c_{j-1}A^j\mathbf{u} - \cdots - c_0A\mathbf{u}\,.$$

But this implies that the vectors $\mathbf{u}, \ldots, A^j\mathbf{u}$ are linearly dependent. Therefore, $j \geq \ell$. □

Lemma 16.10 guarantees that for each integer j, $j = 1, \ldots, \ell$, there exists a unique point $\mathbf{x}_j$ in $\mathbf{x}_0 + \mathcal{H}_j$ at which $\varphi(\mathbf{x})$ attains its minimum value, i.e.,

$$\varphi(\mathbf{x}_j) = \min\{\varphi(\mathbf{x}) : \mathbf{x} \in \mathbf{x}_0 + \mathcal{H}_j\}\,.$$

Lemma 16.12. *If $k \leq \ell$, then $\dim \mathcal{H}_k = k$ and:*

(1) $\mathbf{x}_k - \mathbf{x}_0 \in \mathcal{H}_k$.

(2) $\langle A\mathbf{x}_k - \mathbf{b}, \mathbf{y}\rangle = 0$ *for every vector* $\mathbf{y} \in \mathcal{H}_k$.

(3) $A\mathbf{x}_k - \mathbf{b} = \mathbf{u} + A(\mathbf{x}_k - \mathbf{x}_0)$.

(4) $A\mathbf{x}_k - \mathbf{b} \in \mathcal{H}_{k+1}$.

(5) $\langle A(\mathbf{x}_{k+1} - \mathbf{x}_k), \mathbf{y}\rangle = 0$ *for every vector* $\mathbf{y} \in \mathcal{H}_k$.

(6) *The vectors*

$$\mathbf{r}_j = A\mathbf{x}_j - \mathbf{b}\,, \quad j = 0, \ldots, k-1\,, \tag{16.24}$$

form an orthogonal basis for the space $\mathcal{H}_k$ with respect to the standard inner product in $\mathbb{R}^n$.

Proof. Item (1) is by definition. Item (2) follows from the formula

$$\langle \nabla\varphi(\mathbf{x})_k, \mathbf{v}\rangle = \lim_{\varepsilon \to 0} \frac{\varphi(\mathbf{x}_k + \varepsilon\mathbf{v}) - \varphi(\mathbf{x}_k)}{\varepsilon}\,,$$

which is valid for every vector $\mathbf{v} \in \mathcal{H}_k$, and the definition of the point $\mathbf{x}_k$. The remaining assertions are easy if tackled in the presented order and are left to the reader. □

Exercise 16.25. Verify assertions (3)–(6) of Lemma 16.12.

Exercise 16.26. Show that properties (1) and (2) of Lemma 16.12 serve to uniquely determine $\mathbf{x}_k$.

Let $\mathbf{p}_0 = \mathbf{u}$ and let $\mathbf{p}_j$, $j = 0, \ldots, k-1$, be a basis for $\mathcal{H}_k$ that is orthogonal with respect to the inner product $\langle A\mathbf{x}, \mathbf{y}\rangle$, i.e.,

$$\langle A\mathbf{p}_i, \mathbf{p}_j\rangle = 0 \quad \text{for} \quad i, j = 0, \ldots, \ell-1 \quad \text{and} \quad i \neq j\,,$$

and suppose further that these vectors are normalized so that $\mathbf{r}_j - \mathbf{p}_j \in \mathcal{H}_j$.

Lemma 16.13. *If $k \leq \ell - 1$, then $\mathbf{x}_{k+1} - \mathbf{x}_k = c_k\mathbf{p}_k$.*

Proof. If $k \leq \ell - 1$, then $\mathbf{x}_{k+1} - \mathbf{x}_k \in \mathcal{H}_{k+1}$. Therefore,

$$\mathbf{x}_{k+1} - \mathbf{x}_k = \sum_{j=0}^{k} \gamma_j \mathbf{p}_j$$

for some choice of real constants $\gamma_0, \ldots, \gamma_k$, since $\mathbf{p}_0, \ldots, \mathbf{p}_k$ is a basis for $\mathcal{H}_{k+1}$. Moreover,

$$\langle A(\mathbf{x}_{k+1} - \mathbf{x}_k), \mathbf{p}_i\rangle = \gamma_i \langle A\mathbf{p}_i, \mathbf{p}_i\rangle$$

for $i = 0, \ldots, k$. Therefore, by (5) of Lemma 16.12, $\gamma_i = 0$ for $i < k$, and hence the result follows with $c_k = \gamma_k$. □

Exercise 16.27. Show that if $k \leq \ell - 1$, then

$$\mathbf{x}_{k+1} - \mathbf{x}_k = c_k\mathbf{p}_k \quad \text{with} \quad c_k = -\frac{\langle \mathbf{r}_k, \mathbf{p}_k\rangle}{\langle A\mathbf{p}_k, \mathbf{p}_k\rangle}\,. \tag{16.25}$$

Lemma 16.14. *If $k \leq \ell - 1$, then $\mathbf{p}_k - \mathbf{r}_k = d_{k-1}\mathbf{p}_{k-1}$.*

Proof. By construction, $\mathbf{p}_k - \mathbf{r}_k \in \mathcal{H}_k$. Therefore,

$$\mathbf{p}_k - \mathbf{r}_k = \sum_{i=0}^{k-1} \delta_i \mathbf{p}_i$$

for some choice of coefficients $\delta_i \in \mathbb{R}$. However, if $i \leq k-2$, then

$$\begin{aligned}
\langle A(\mathbf{p}_k - \mathbf{r}_k), \mathbf{p}_i\rangle &= -\langle A\mathbf{r}_k, \mathbf{p}_i\rangle \\
&= -\langle \mathbf{r}_k, A\mathbf{p}_i\rangle \\
&= 0\,,
\end{aligned}$$

since $\mathbf{r}_k$ is orthogonal to $\mathcal{H}_{k-1}$ with respect to the standard inner product. Therefore, $\mathbf{p}_k - \mathbf{r}_k = \delta_{k-1}\mathbf{p}_{k-1}$, which is of the claimed form. □

Exercise 16.28. Show that if $k \leq \ell - 1$, then

$$\mathbf{p}_k - \mathbf{r}_k = d_{k-1}\mathbf{p}_{k-1} \quad \text{with} \quad d_{k-1} = -\frac{\langle A\mathbf{r}_k, \mathbf{p}_{k-1}\rangle}{\langle A\mathbf{p}_{k-1}, \mathbf{p}_{k-1}\rangle}. \tag{16.26}$$

Exercise 16.29. Show directly that $\varphi(\mathbf{x}_j) \leq \varphi(\mathbf{x}_{j-1})$ for $j = 1, \dots, \ell$.

Moral: The preceding analysis provides a recursive scheme for calculating $\mathbf{a}$ via formulas (16.24)–(16.26): Choose any vector $\mathbf{x}_0 \in \mathbb{R}^n$ and set $\mathbf{r}_0 = \mathbf{p}_0 = A\mathbf{x}_0 - \mathbf{b}$. Then run the recursions for ℓ steps. The desired solution $\mathbf{a} = \mathbf{x}_\ell$. Thus, it is not necessary to solve the extremal problems that were considered above.

Exercise 16.30. Use formula (16.24) and the recursions (16.25) and (16.26) to solve the equation $A\mathbf{x} = \mathbf{b}$ when $A = \begin{bmatrix} 2 & 1 \\ 1 & 1 \end{bmatrix}$ and $\mathbf{b} = -\begin{bmatrix} 1 \\ 1 \end{bmatrix}$.

16.6. Dual extremal problems

Recall that if $\mathcal{X}$ is a finite dimensional normed linear space over $\mathbb{F}$, then every linear functional f on $\mathcal{X}$ is bounded. The set of linear functionals on such a finite dimensional space $\mathcal{X}$ is denoted by $\mathcal{X}'$.

Theorem 16.15. *Let $\mathcal{X}$ be a finite dimensional normed linear space over $\mathbb{C}$, let $\mathcal{U}$ be a subspace of $\mathcal{X}$ and let $\mathcal{U}^\circ = \{f \in \mathcal{X}' : f(\mathbf{u}) = 0$ for every $\mathbf{u} \in \mathcal{U}\}$ Then:*

(1) *For each vector $\mathbf{x} \in \mathcal{X}$,*

$$\min_{\mathbf{u} \in \mathcal{U}} \|\mathbf{x} - \mathbf{u}\| = \max_{\substack{f \in \mathcal{U}^\circ \\ \|f\| \leq 1}} |f(\mathbf{x})|.$$

(2) *For each bounded linear functional $f \in \mathcal{X}'$,*

$$\min_{g \in \mathcal{U}^\circ} \|f - g\| = \max_{\substack{\mathbf{u} \in \mathcal{U} \\ \|\mathbf{u}\| \leq 1}} |f(\mathbf{u})|.$$

Proof. To establish (1), let $\mathbf{x} \in \mathcal{X}$ and suppose further that $\mathbf{x} \notin \mathcal{U}$, because otherwise (1) is self-evident, since both sides of the asserted equality are equal to zero. Then for any $f \in \mathcal{U}^\circ$ with $\|f\| \leq 1$ and any $\mathbf{u} \in \mathcal{U}$,

$$\begin{aligned} |f(\mathbf{x})| = |f(\mathbf{x} - \mathbf{u})| &\leq \|f\| \, \|\mathbf{x} - \mathbf{u}\| \\ &\leq \|\mathbf{x} - \mathbf{u}\|. \end{aligned}$$

Therefore,

$$|f(\mathbf{x})| \leq d, \tag{16.27}$$

where

$$d = \inf\{\|\mathbf{x} - \mathbf{u}\| : \mathbf{u} \in \mathcal{U}\} = \min\{\|\mathbf{x} - \mathbf{u}\| : \mathbf{u} \in \mathcal{U}\},$$

since $\mathcal{X}$ is finite dimensional. Now, since $\mathbf{x} \notin \mathcal{U}$, we may define the linear functional f_0 on the subspace

$$\mathcal{U}_1 = \{\alpha\mathbf{x} + \beta\mathbf{u} : \alpha, \beta \in \mathbb{C} \quad \text{and} \quad \mathbf{u} \in \mathcal{U}\}$$

by the formula

$$f_0(\alpha\mathbf{x} + \beta\mathbf{u}) = \alpha d.$$

Then

$$\begin{aligned} |f_0(\alpha\mathbf{x} + \beta\mathbf{u})| &= |\alpha| d \\ &\leq |\alpha| \, \|\mathbf{x} - \mathbf{y}\| = \|\alpha\mathbf{x} - \alpha\mathbf{y}\| \end{aligned}$$

for every choice of $\mathbf{y} \in \mathcal{U}$. If $\alpha \neq 0$, the particular choice

$$\mathbf{y} = -\frac{\beta}{\alpha}\mathbf{u}$$

yields the inequality

$$|f_0(\alpha\mathbf{x} + \beta\mathbf{u})| \leq \|(\alpha\mathbf{x} + \beta\mathbf{u})\|,$$

which is valid for all choices of $\alpha, \beta \in \mathbb{C}$ and $\mathbf{u} \in \mathcal{U}$. Theorem 7.20 (the junior version of the Hahn-Banach Theorem) guarantees the existence of a bounded linear functional f_1 on the full space $\mathcal{X}$ such that

$$f_1(\alpha\mathbf{x} + \beta\mathbf{u}) = f_0(\alpha\mathbf{x} + \beta\mathbf{u})$$

for all choices of $\alpha, \beta \in \mathbb{C}$ and $\mathbf{u} \in \mathcal{U}$ and

$$\|f_1\| = \sup\{|f_0(\alpha\mathbf{x} + \beta\mathbf{u})| : \|\alpha\mathbf{x} + \beta\mathbf{u}\| \leq 1\}.$$

Thus, $f_1 \in \mathcal{U}^\circ$, $\|f_1\| \leq 1$ and $f_1(\mathbf{x}) = f_0(\mathbf{x}) = d$. Therefore, the upper bound on the left-hand side of equation (16.27) is attained by $f_1(\mathbf{x})$.

Next, to verify (2), fix $f \in \mathcal{X}'$ and suppose $f \notin \mathcal{U}^\circ$; otherwise both sides of (2) are equal to zero. Then

$$|f(\mathbf{u})| = |f(\mathbf{u}) - g(\mathbf{u})| \leq \|f - g\|\|\mathbf{u}\|$$

for every choice of $g \in \mathcal{U}^\circ$ and $\mathbf{u} \in \mathcal{U}$. Therefore,

$$\sup\{|f(\mathbf{u})| : \mathbf{u} \in \mathcal{U} \quad \text{and} \quad \|\mathbf{u}\| \leq 1\} \leq \inf\{\|f - g\| : g \in \mathcal{U}^\circ\}. \tag{16.28}$$

Now let $f_0 = f|_{\mathcal{U}}$, the restriction of f to the subspace $\mathcal{U}$. By Theorem 7.20, there exists a bounded linear functional $f_1 \in \mathcal{X}'$ such that

$$f_1(\mathbf{u}) = f_0(\mathbf{u}) = f(\mathbf{u})$$

for every vector $\mathbf{u} \in \mathcal{U}$ and

$$\sup\{|f_1(\mathbf{x})| : \mathbf{x} \in \mathcal{X} \text{ and } \|\mathbf{x}\| \leq 1\} = \sup\{|f_0(\mathbf{u})| : \mathbf{u} \in \mathcal{U} \text{ and } \|\mathbf{u}\| \leq 1\},$$

i.e.,

$$\|f_1\| = \sup\{|f(\mathbf{u})| : \mathbf{u} \in \mathcal{U} \quad \text{and} \quad \|\mathbf{u}\| \leq 1\}.$$

Moreover, since $f - f_1 \in \mathcal{U}^\circ$, it follows that

$$\begin{aligned}\inf\{\|f-g\| : g \in \mathcal{U}^\circ\} &\le \|f-(f-f_1)\| = \|f_1\| \\ &= \sup\{|f(\mathbf{u})| : \mathbf{u} \in \mathcal{U} \quad \text{and} \quad \|\mathbf{u}\| \le 1\}.\end{aligned}$$

Therefore, equality prevails in (16.28) and the infimum is a minimum since it is attained by $g = f - f_1$. Finally, since $\mathcal{X}$ is finite dimensional, the supremum (16.28) is attained also and hence is a maximum. □

Exercise 16.31. Show that the conclusions of Theorem 16.15 are also valid when $\mathcal{X}$ is a normed linear space over $\mathbb{R}$. [HINT: $f \in \mathcal{U}^\circ \iff -f \in \mathcal{U}^\circ$ and $\mathbf{u} \in \mathcal{U} \iff -\mathbf{u} \in \mathcal{U}$.]

Exercise 16.32. Show that if $A \in \mathbb{C}^{p\times q}$ and $\mathbf{b} \in \mathbb{C}^p$, then

$$\min\{\|A\mathbf{x}-\mathbf{b}\|_1 : \mathbf{x} \in \mathbb{C}^q\} = \max\{|\mathbf{y}^H\mathbf{b}| : \mathbf{y} \in \mathcal{N}_{A^H} \text{ and } \|\mathbf{y}\|_\infty \le 1\}.$$

Exercise 16.33. Show that if $A \in \mathbb{C}^{p\times q}$, $\mathbf{b} \in \mathbb{C}^p$, $s > 1$ and $1/s + 1/t = 1$, then

$$\min\{\|A\mathbf{x}-\mathbf{b}\|_s : \mathbf{x} \in \mathbb{C}^q\} = \max\{|\mathbf{y}^H\mathbf{b}| : \mathbf{y} \in \mathcal{N}_{A^H} \text{ and } \|\mathbf{y}\|_t \le 1\}.$$

Exercise 16.34. Show that if $A \in \mathbb{C}^{p\times q}$, $\mathbf{c} \in \mathbb{C}^q$, $s > 1$ and $1/s + 1/t = 1$, then

$$\min\{\|\mathbf{c}-\mathbf{y}\|_s : \mathbf{y} \in \mathcal{N}_A\} = \max\{|\mathbf{c}^H A^H \mathbf{x}| : \mathbf{x} \in \mathbb{C}^p \text{ and } \|\mathbf{x}\|_t \le 1\}.$$

Remark 16.16. Theorem 16.15 is also valid if $\mathcal{X}$ is a Banach space over $\mathbb{C}$ or $\mathbb{R}$, but with inf in place of min in (1) and sup in place of max in (2): If $\mathbf{x} \in \mathcal{X}$, then

$$\inf\{\|\mathbf{x}-\mathbf{u}\| : \mathbf{u} \in \mathcal{U} \text{ and } \|f\| \le 1\} = \max\{|f(\mathbf{x})| : f \in \mathcal{U}^\circ \text{ and } \|f\| \le 1\},$$

whereas if $f \in \mathcal{X}'$, then

$$\min\{\|f-g\| : g \in \mathcal{U}^\circ \quad \text{and} \quad \mathbf{u} \in \mathcal{U}\} = \sup\{|f(\mathbf{u})| : \mathbf{u} \in \mathcal{U} \quad \text{and} \quad \|\mathbf{u}\| \le 1\}.$$

The proof is much the same, except that the Hahn-Banach theorem is invoked in place of the finite dimensional version considered here.

16.7. Bibliographical notes

The formulation of Exercise 16.9 is one step down the line, so to speak. It exploits the information supplied by Example 16.5. The inequalities in Exercise 16.9 and Remark 16.7 are taken from [**37**]. They credit this approach to a paper of J.W.S. Cassels. Exercise 16.17 is connected with the maximum entropy extensions considered in Section 12.7. The first papers on this problem by J. P. Burg were based on variational methods. Section 16.5 was adapted from the discussion in [**48**] and [**68**]. Exercise 16.19 is taken from Ellis [**29**].

Chapter 17

Matrix valued holomorphic functions

... like most normally constituted writers Martin had no use for any candid opinion that was not wholly favorable.

Patrick O'Brian [**54**], p. 162

The main objective of this chapter is to introduce matrix valued contour integrals of the form

$$\int_{\Gamma} \varphi(\lambda)(\lambda I_n - A)^{-1} d\lambda\,,$$

where Γ is a simple closed smooth curve that does not pass through any of the eigenvalues of A and $\varphi(\lambda)$ is analytic in an open set that contains the curve Γ and its interior. Integral formulas of this type are extremely useful, both in the present context of matrices as well as for more general classes of operators. In this chapter they will be used to establish the continuous dependence of the eigenvalues of a matrix on the entries in the matrix and then, subsequently, to give an elegant proof of the formula for the spectral radius and some formulas for the fractional powers of matrices. We begin, however, with a brief introduction to the theory of scalar analytic functions of one complex variable. A number of useful supplementary facts are surveyed in Appendix 2.

17.1. Differentiation

A complex valued function $f(\lambda)$ of the complex variable λ that is defined in an open set Ω of the complex plane $\mathbb{C}$ is said to be **holomorphic** (or analytic) in Ω if the limit

$$f'(\lambda) = \lim_{\xi \to 0} \frac{f(\lambda + \xi) - f(\lambda)}{\xi} \tag{17.1}$$

exists for every point $\lambda \in \Omega$. This is a very strong constraint because the variable ξ in this difference ratio is complex and the definition requires the limit to be the same regardless of how ξ tends to zero. Thus, for example, if $f(\lambda)$ is analytic in Ω, then the directional derivative

$$(D_\theta f)(\lambda) = \lim_{r \downarrow 0} \frac{f(\lambda + re^{i\theta}) - f(\lambda)}{r} \tag{17.2}$$

exists and

$$f'(\lambda) = e^{-i\theta}(D_\theta f)(\lambda) \tag{17.3}$$

for every angle θ. In particular, because of (17.1),

$$(D_0 f)(\lambda) = e^{-i\pi/2}(D_{\pi/2} f)(\lambda); \tag{17.4}$$

i.e., in more conventional notation

$$\frac{\partial f}{\partial x}(\lambda) = -i\frac{\partial f}{\partial y}(\lambda) . \tag{17.5}$$

This last formula can also be obtained by more familiar manipulations by writing $\lambda = x + iy$, $\xi = \alpha + i\beta$, $f(\lambda) = g(x,y) + ih(x,y)$, where $x, y, \alpha, \beta, g(x,y)$ and $h(x,y)$ are all real, and noting that

(17.6)

$$\frac{f(\lambda + \xi) - f(\lambda)}{\xi} = \frac{g(x+\alpha, y+\beta) - g(x,y)}{\alpha + i\beta} + \frac{ih(x+\alpha, y+\beta) - h(x,y)}{\alpha + i\beta} .$$

Then, since the limit when $\beta = 0$ and $\alpha \to 0$ must agree with the limit when $\alpha = 0$ and $\beta \to 0$, it follows that

$$\frac{\partial g}{\partial x}(x,y) + i\frac{\partial h}{\partial x}(x,y) = \frac{1}{i}\left\{\frac{\partial g}{\partial y}(x,y) + i\frac{\partial h}{\partial y}(x,y)\right\} , \tag{17.7}$$

which is just another way of writing formula (17.6). Clearly formula (17.7) (and hence also formula (17.6)) is equivalent to the pair of equations

$$\frac{\partial g}{\partial x}(x,y) = \frac{\partial h}{\partial y}(x,y) \quad \text{and} \quad \frac{\partial h}{\partial x}(x,y) = -\frac{\partial g}{\partial y}(x,y) . \tag{17.8}$$

These are the Cauchy-Riemann equations. But it's easiest to remember them in the form (17.5). There is a converse statement:

If (17.8) *holds in some open set* Ω *and if the indicated partial derivatives are continuous in* Ω, *then* $f(\lambda) = g(x,y) + ih(x,y)$ *is holomorphic in* Ω.

For sharper statements, see e.g. [**60**].

Theorem 17.1. *Let* Ω *be an open nonempty subset of* $\mathbb{C}$ *and let* $\mathfrak{h}_\Omega$ *denote the set of functions that are holomorphic (i.e., analytic) in* Ω. *Then:*

(1) $f \in \mathfrak{h}_\Omega \Longrightarrow f(\lambda)$ *is continuous in* Ω.

(2) $f \in \mathfrak{h}_\Omega, g \in \mathfrak{h}_\Omega \Longrightarrow \alpha f + \beta g \in \mathfrak{h}_\Omega$ *for every choice of* $\alpha, \beta \in \mathbb{C}$.

(3) $f \in \mathfrak{h}_\Omega, g \in \mathfrak{h}_\Omega \Longrightarrow fg \in \mathfrak{h}_\Omega$ *and* $(fg)'(\lambda) = f'(\lambda)g(\lambda) + f(\lambda)g'(\lambda)$.

(4) $f \in \mathfrak{h}_\Omega$, $f(\lambda) \neq 0$ *for any point* $\lambda \in \Omega \Longrightarrow (1/f) \in \mathfrak{h}_\Omega$.

(5) $g \in \mathfrak{h}_\Omega, g(\Omega) \subset \Omega_1$, Ω_1 *open*, $f \in \mathfrak{h}_{\Omega_1} \Longrightarrow f \circ g \in \mathfrak{h}_\Omega$.

(6) $f \in \mathfrak{h}_\Omega, \alpha \in \Omega \Longrightarrow R_\alpha f \in \mathfrak{h}_\Omega$, *where*

$$(R_\alpha f)(\lambda) = \begin{cases} \dfrac{f(\lambda) - f(\alpha)}{\lambda - \alpha} & \text{if } \lambda \neq \alpha \\ f'(\alpha) & \text{if } \lambda = \alpha \end{cases} . \tag{17.9}$$

(7) $f \in \mathfrak{h}_\Omega \Longrightarrow f' \in \mathfrak{h}_\Omega$.

Proof. Let $\alpha \in \Omega$. Then $\alpha + \xi \in \Omega$ if $|\xi|$ is small enough and

$$\begin{aligned} |f(\alpha+\xi) - f(\alpha)| &= \left| \frac{f(\alpha+\xi) - f(\alpha)}{\xi} \right| |\xi| \\ &\leq \left| \frac{f(\alpha+\xi) - f(\alpha)}{\xi} - f'(\alpha) \right| |\xi| + |f'(\alpha)||\xi| , \end{aligned}$$

which clearly tends to zero as $|\xi| \longrightarrow 0$. Therefore, (1) holds. The next four items in this list are easy to verify directly from the definition. Items (6) and (7) require more extensive treatment. They will be discussed after we introduce contour integration. □

We turn now to some examples.

Example 17.2. The polynomial $p(\lambda) = a_0 + a_1\lambda + \cdots + a_n\lambda^n$ is holomorphic in the whole complex plane $\mathbb{C}$.

Discussion. In view of Items 2 and 3 of Theorem 17.1, it suffices to verify this for polynomials $p_1(\lambda) = a_0 + a_1\lambda$ of degree one. But this is easy:

$$\frac{p_1(\lambda+\xi) - p_1(\lambda)}{\xi} = \frac{a_0 + a_1(\lambda+\xi) - a_0 - a_1\lambda}{\xi} = a_1 .$$

Therefore, the powers

$$\lambda^k = \lambda \cdot \lambda \cdots \lambda , \quad k = 1, 2, \ldots , n ,$$

are holomorphic in $\mathbb{C}$ as is $p(\lambda)$.

Exercise 17.1. Verify directly that

$$\lim_{\xi \to 0} \frac{(\lambda+\xi)^n - \lambda^n}{\xi} = n\lambda^{n-1} \quad \text{for every positive integer} \quad n \geq 2 .$$

Example 17.3. $f(\lambda) = \dfrac{1}{\lambda - \alpha}$ is analytic in $\mathbb{C} \setminus \{\alpha\}$.

Discussion. Clearly

$$\begin{aligned}\frac{f(\lambda+\xi)-f(\lambda)}{\xi} &= \frac{1}{\xi}\left\{\frac{1}{\lambda+\xi-\alpha}-\frac{1}{\lambda-\alpha}\right\}\\ &= \frac{\lambda-\alpha-(\lambda+\xi-\alpha)}{\xi(\lambda+\xi-\alpha)(\lambda-\alpha)}\\ &= \frac{-1}{(\lambda+\xi-\alpha)(\lambda-\alpha)} \rightarrow \frac{-1}{(\lambda-\alpha)^2},\end{aligned}$$

as $\xi \rightarrow 0$.

Example 17.4. $f(\lambda) = \dfrac{(\lambda-\alpha_1)\cdots(\lambda-\alpha_k)}{(\lambda-\beta_1)\cdots(\lambda-\beta_n)}$ is analytic in $\mathbb{C}\setminus\{\beta_1,\ldots,\beta_n\}$.

Discussion. This is immediate from Item 3 of Theorem 17.1 and the preceding two examples.

Example 17.5. $f(\lambda) = e^{\alpha\lambda}$ is analytic in $\mathbb{C}$ for every fixed $\alpha \in \mathbb{C}$ and $f'(\lambda) = \alpha f(\lambda)$.

Discussion. In view of Item 5 of Theorem 17.1, it suffices to verify that e^λ is analytic. To this end, write $\lambda = \mu + i\nu$ so that

$$f(\lambda) = e^\lambda = e^\mu e^{i\nu} = e^\mu(\cos\nu + i\sin\nu)$$

and, consequently,

$$\frac{\partial f}{\partial \mu}(\lambda) = f(\lambda) \quad \text{and} \quad \frac{\partial f}{\partial \nu}(\lambda) = e^\mu(-\sin\nu + i\cos\nu) = if(\lambda)\,.$$

Thus, as the Cauchy-Riemann equations are satisfied, $f(\lambda)$ is analytic in $\mathbb{C}$. A perhaps more satisfying proof can be based on the exhibited formula for $f(\lambda)$ and Taylor's formula with remainder applied to the real valued functions $e^\mu, \cos\nu, \sin\nu$, to check the existence of the limit in formula (17.1). Yet another approach is to first write

$$\frac{e^{\alpha(\lambda+\xi)}-e^{\alpha\lambda}}{\xi} = e^{\alpha\lambda}\left\{\frac{e^{\alpha\xi}-1}{\xi}\right\},$$

and then to verify the term inside the curly brackets tends to α as ξ tends to 0, using the power series expansion for the exponential. It all depends upon what you are willing to assume to begin with.

17.2. Contour integration

A **directed curve** (or contour) Γ in the complex plane $\mathbb{C}$ is the set of points $\{\gamma(t) : a \le t \le b\}$ traced out by a complex valued function $\gamma \in \mathcal{C}([a,b])$ as t runs from a to b. The curve is said to be **closed** if $\gamma(a) = \gamma(b)$, it is said to be **smooth** if $\gamma \in \mathcal{C}^1([a,b])$, it is said to be **simple** if γ is one to one on the open interval $a < t < b$; i.e., if $a < t_1, t_2 < b$ and $\gamma(t_1) = \gamma(t_2)$, then $t_1 = t_2$. The simplest contours are line segments and arcs of circles. Thus, for example, if Γ is the horizontal line segment directed from $\alpha_1 + i\beta$ to $\alpha_2 + i\beta$ and $\alpha_1 < \alpha_2$, then we may choose

$$\gamma(t) = t + i\beta\,, \ \alpha_1 \le t \le \alpha_2 \quad \text{or} \quad \gamma(t) = \alpha_1 + t(\alpha_2 - \alpha_1) + i\beta\,, \ 0 \le t \le 1\,.$$

The second parametrization is valid even if $\alpha_1 > \alpha_2$. If Γ is the vertical line segment directed from $\alpha + i\beta_1$ to $\alpha + i\beta_2$ and $\beta_1 < \beta_2$, then we may choose $\gamma(t) = \alpha + it$, $\beta_1 \le t \le \beta_2$. If Γ is a circular arc of radius R directed from $Re^{i\alpha}$ to $Re^{i\beta}$ and $\alpha < \beta$, then we may choose $\gamma(t) = Re^{it}$, $\alpha \le t \le \beta$.

A curve Γ is said to be **piecewise smooth** if it is a finite union of smooth curves, such as a polygon.

The contour integral $\int_\Gamma f(\lambda)d\lambda$ of a continuous complex valued function that is defined on a smooth curve Γ that is parametrized by $\gamma \in \mathcal{C}^1([a,b])$ is defined by the formula

$$\int_\Gamma f(\lambda)d\lambda = \int_a^b f(\gamma(t))\gamma'(t)dt\,. \tag{17.10}$$

The numerical value of the integral depends upon the curve Γ, but not upon the particular choice of the (one to one) function $\gamma(t)$ that is used to describe the curve, as the following exercise should help to clarify.

Exercise 17.2. Use the rules of contour integration to calculate the integral (17.10) when $f(\lambda) = \lambda$ and (a) $\gamma(t) = t$ for $1 \le t \le 2$; (b) $\gamma(t) = t^2$ for $1 \le t \le \sqrt{2}$; (c) $\gamma(t) = e^t$ for $0 \le t \le \ln 2$ and (d) $\gamma(t) = 1 + \sin t$ for $0 \le t \le \pi/2$.

Exercise 17.3. Use the rules of contour integration to calculate the integral (17.10) when $f(\lambda) = \lambda$ and Γ is the rectangle directed counterclockwise with vertices $-a - ib, a - ib, a + ib, -a + ib$, where $a > 0$ and $b > 0$.

Exercise 17.4. Repeat the preceding exercise for $f(\lambda) = \lambda^n$, n an integer (positive, zero or negative) and the same curve Γ.

Theorem 17.6. *Let $f(\lambda)$ be analytic in some open nonempty set Ω. Let Γ be a simple smooth closed curve in Ω such that all the points enclosed by Γ also belong to Ω. Then*

$$\int_\Gamma f(\lambda)d\lambda = 0\,.$$

Discussion. Consider first the special case when Γ is a rectangle with vertices $a_1+ib_1, a_2+ib_1, a_2+ib_2, a_1+ib_2$, with $a_1 < a_2$, $b_1 < b_2$ and suppose that the curve is directed counterclockwise. Then the integral over the two horizontal segments of Γ is equal to

$$\int_{a_1}^{a_2} f(x+ib_1)dx - \int_{a_1}^{a_2} f(x+ib_2)dx = \int_{a_1}^{a_2} \left\{ \int_{b_1}^{b_2} -\frac{\partial f}{\partial y}(x+iy)dy \right\} dx \ ,$$

whereas the integral over the vertical segments of Γ is equal to

$$\int_{b_1}^{b_2} f(a_2+iy)idy - \int_{b_1}^{b_2} f(a_1+iy)idy = \int_{b_1}^{b_2} i \left\{ \int_{a_1}^{a_2} \frac{\partial f}{\partial x}(x+iy)dx \right\} dy \ .$$

Therefore, assuming that it is legitimate to interchange the order of integration on the right-hand side of the first of these two formulas, we see that

$$\begin{aligned} \int_\Gamma f(\lambda)d\lambda &= \int_{b_1}^{b_2} \int_{a_1}^{a_2} \left\{ -\frac{\partial f}{\partial y}(x+iy) + i\frac{\partial f}{\partial x}(x+iy) \right\} dxdy \\ &= 0 \ , \end{aligned}$$

by the Cauchy-Riemann equations (17.5). The conclusion for general Γ is obtained by approximating it as in Figure 1. The point is that the sum of the contour integrals over every little box is equal to zero, since the integral around the edges of each box is equal to zero. However, in the sum, each interior edge is integrated over twice, once in each direction. Therefore, the integrals over the interior edges cancel out and you are left with an integral over the outside edges, which is approximately equal to the integral over the curve Γ and in fact tends to the integral over this curve as the boxes shrink to zero.

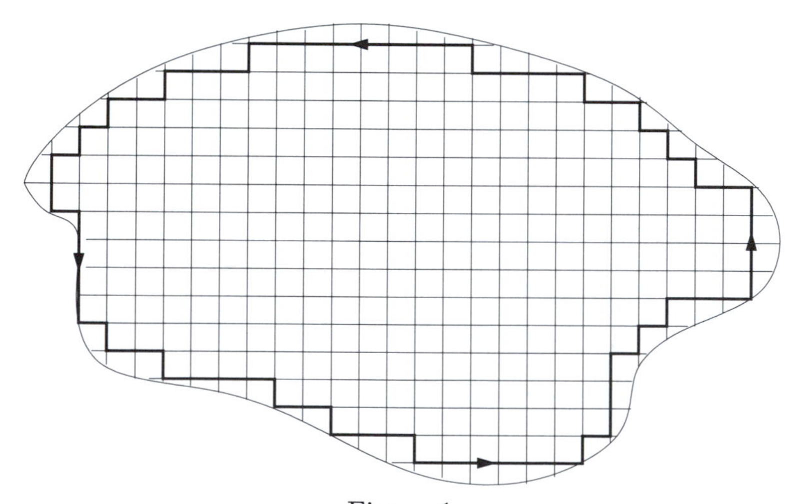

Figure 1

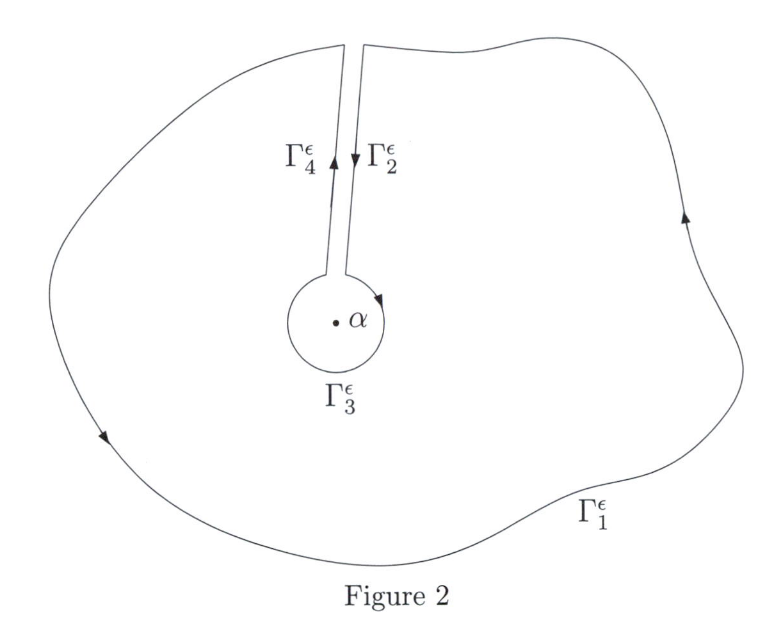

Figure 2

Theorem 17.7. *Let $f(\lambda)$ be holomorphic in some open nonempty set Ω. Let Γ be a simple closed curve in Ω that is directed counterclockwise such that all the points enclosed by Γ also belong to Ω. Then*

$$\frac{1}{2\pi i}\int_\Gamma \frac{f(\lambda)}{\lambda-\alpha}d\lambda = \begin{cases} f(\alpha) & \text{if } \alpha \text{ is inside } \Gamma \\ 0 & \text{if } \alpha \text{ is outside } \Gamma \ . \end{cases}$$

Proof. Let α be inside Γ and introduce a new curve $\Gamma_1^\epsilon + \Gamma_2^\epsilon + \Gamma_3^\epsilon + \Gamma_4^\epsilon$, as in Figure 2, such that α is outside this new curve, Γ_3^ϵ is "most" of a circle of radius r centered at the point α and

$$\frac{1}{2\pi i}\int_\Gamma \frac{f(\lambda)}{(\lambda-\alpha)}d\lambda = \lim_{\epsilon\to 0}\frac{1}{2\pi i}\int_{\Gamma_1^\epsilon} \frac{f(\lambda)}{(\lambda-\alpha)}d\lambda \ .$$

Then, by the construction,

$$\int_{\Gamma_1^\epsilon} \frac{f(\lambda)}{(\lambda-\alpha)}d\lambda = -\int_{\Gamma_2^\epsilon} \frac{f(\lambda)}{(\lambda-\alpha)}d\lambda - \int_{\Gamma_3^\epsilon} \frac{f(\lambda)}{(\lambda-\alpha)}d\lambda - \int_{\Gamma_4^\epsilon} \frac{f(\lambda)}{(\lambda-\alpha)}d\lambda \, .$$

Now, as ϵ tends to zero, the first and third integrals on the right cancel each other out and the second integral

$$\int_{\Gamma_3^\epsilon} \frac{f(\lambda)}{(\lambda-\alpha)}d\lambda \longrightarrow -\int_0^{2\pi} \frac{f(\alpha+re^{i\theta})}{(\alpha+re^{i\theta}-\alpha)}ire^{i\theta}d\theta$$

as ϵ tends to 0. The final formula follows from the fact that

$$\int_0^{2\pi} \frac{f(\alpha+re^{i\theta})}{(\alpha+re^{i\theta}-\alpha)}ire^{i\theta}d\theta \longrightarrow 2\pi i f(\alpha)$$

as r tends to 0.

It remains to consider the case that the point α is outside Γ. But then the statement is immediate from Theorem 17.6. □

Theorem 17.8. *Let Ω be a nonempty open subset of $\mathbb{C}$ and let $f \in \mathfrak{h}_\Omega$. Then $f' \in \mathfrak{h}_\Omega$.*

Proof. Let $\omega \in \Omega$ and let $\Gamma = \omega + re^{i\theta}$, $0 \le \theta < 2\pi$, where r is chosen small enough so that $\lambda - \omega \in \Omega$ when $|\lambda - \omega| \le r$. Then

$$f(\omega) = \frac{1}{2\pi i}\int_\Gamma \frac{f(\lambda)}{\lambda - \omega} d\lambda$$

and

$$\begin{aligned}\frac{f(\omega+\xi) - f(\omega)}{\xi} &= \frac{1}{2\pi i}\int \frac{f(\lambda)}{\xi}\left\{\frac{1}{\lambda-\omega-\xi} - \frac{1}{\lambda-\omega}\right\}d\lambda \\ &= \frac{1}{2\pi i}\int_\Gamma \frac{f(\lambda)}{(\lambda-\omega-\xi)(\lambda-\omega)}d\lambda \\ &\to \frac{1}{2\pi i}\int_\Gamma \frac{f(\lambda)}{(\lambda-\omega)^2}d\lambda\end{aligned}$$

as $\xi \to 0$. Therefore,

$$f'(\omega) = \frac{1}{2\pi i}\int_\Gamma \frac{f(\lambda)}{(\lambda-\omega)^2}d\lambda$$

and hence

$$\frac{f'(\omega+\xi) - f'(\omega)}{\xi} = \frac{1}{2\pi i}\int_\gamma \frac{f(\lambda)}{\xi}\left\{\frac{1}{(\lambda-\omega-\xi)^2} - \frac{1}{(\lambda-\omega)^2}\right\}d\lambda,$$

which tends to the limit

$$\frac{2}{2\pi i}\int_\Gamma \frac{f(\lambda)}{(\lambda-\omega)^3}d\lambda$$

as ξ tends to 0. □

Corollary 17.9. In the setting of Theorem 17.8,

$$\frac{f^{(k)}(\omega)}{k!} = \frac{1}{2\pi i}\int_\Gamma \frac{f(\lambda)}{(\lambda-\omega)^{k+1}}d\lambda \tag{17.11}$$

for $k = 0, 1, \ldots$.

Exercise 17.5. Verify Corollary 17.9.

Exercise 17.6. Verify assertion (6) in Theorem 17.1.

Theorem 17.10. *Let Ω be a nonempty open subset of $\mathbb{C}$ and let $f(\lambda) = g(\lambda)/h(\lambda)$, where $g \in \mathfrak{h}_\Omega$ and*

$$h(\lambda) = (\lambda - \alpha_1)^{k_1}\cdots(\lambda-\alpha_n)^{k_n}$$

is a polynomial with n distinct roots $\alpha_1, \ldots, \alpha_n$. Let Γ be a simple closed smooth curve directed counterclockwise such that Γ and all the points inside Γ belong to Ω. Suppose further that $\alpha_1, \ldots, \alpha_\ell$ are inside Γ and $\alpha_{\ell+1}, \ldots, \alpha_n$ lie outside Γ. Then

$$\frac{1}{2\pi i}\int_\Gamma f(\lambda)d\lambda = \sum_{j=1}^{\ell} \operatorname{Res}(f, \alpha_j),$$

where

$$\operatorname{Res}(f, \alpha_j) = \lim_{\lambda\to\alpha_j} \frac{\{(\lambda-\alpha_j)^{k_j} f(\lambda)\}^{(k_j-1)}}{(k_j-1)!},$$

and the superscript $k_j - 1$ in the formula indicates the order of differentiation.

Discussion. The number $\operatorname{Res}(f, \alpha_j)$ is called the **residue** of f at the point α_j. The basic strategy is much the same as the proof of Theorem 17.10 except that now ℓ little discs have to be extracted, one for each of the distinct zeros of $h(\lambda)$ inside the curve Γ. This leads to the formula

$$\frac{1}{2\pi i}\int_\Gamma f(\lambda)d\lambda = \sum_{j=1}^{\ell} \frac{1}{2\pi i}\int_{\Gamma_j} f(\lambda)d\lambda,$$

where Γ_j is a small circle of radius r_j centered at α_j that is directed counterclockwise, and it is assumed that $r_j < (1/2)\min\{|\alpha_i - \alpha_k| : i, k = 1, \ldots, \ell\}$ and that $\{\lambda \in \mathbb{C} : |\lambda - \alpha_i| \le r_i\}$ lies inside Γ for $i = 1, \ldots, \ell$. Now let

$$f_j(\lambda) = (\lambda - \alpha_j)^{k_j} f(\lambda).$$

Then, since $f_j(\lambda)$ is holomorphic in an open set that contains Γ_j and all its interior points, formula (17.11) yields the evaluation

$$\frac{1}{2\pi i}\int_{\Gamma_j} f(\lambda)d\lambda = \frac{1}{2\pi i}\int_{\Gamma_j} \frac{f_j(\lambda)}{(\lambda-\alpha_j)^{k_j}}d\lambda = \frac{f_j^{(k_j-1)}(\alpha_j)}{(k_j-1)!},$$

for $j = 1, \ldots, \ell$, which coincides with the advertised formula.

17.3. Evaluating integrals by contour integration

Having come so far, it is worth expending a little extra energy to review some evaluations of integrals that emerge as a very neat application of contour integration and also serve as a good introduction to some of the basic formulas of Fourier analysis, which will be the subject of the next section.

Example 17.11. $\displaystyle\int_{-\infty}^{\infty} \frac{1}{x^2+1}dx = \pi.$

Discussion. Let

$$f(\lambda) = \frac{1}{\lambda^2+1}, \quad g(\lambda) = (\lambda - i)f(\lambda) = \frac{1}{\lambda + i}$$

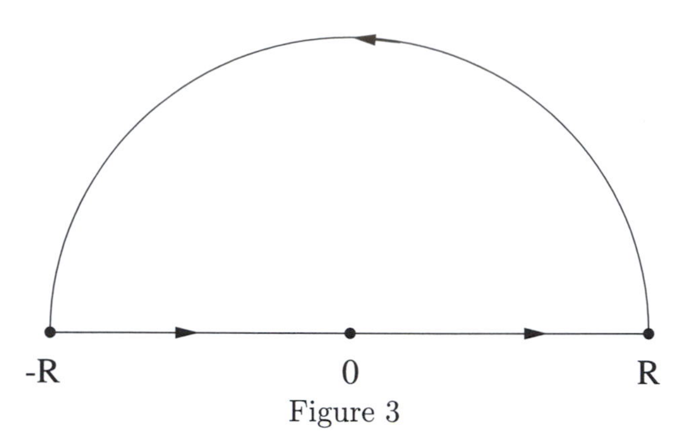

Figure 3

and let Γ_R denote the semicircle of radius R in the upper half plane, including the base $(-R, R)$, directed counterclockwise as depicted in Figure 3. Then

$$\int_{-R}^{R} \frac{1}{x^2+1} dx = I_R + II_R,$$

where

$$\begin{aligned} I_R &= \int_{\Gamma_R} f(\lambda) d\lambda = \int_{\Gamma_R} \frac{g(\lambda)}{\lambda - i} d\lambda \\ &= 2\pi i g(i) = \pi \quad \text{if} \quad R > 1\,, \end{aligned}$$

since g is holomorphic in $\mathbb{C} \setminus \{-i\}$, and

$$II_R = -\int_{C_R} f(\lambda) d\lambda,$$

the integral over the circular arc $C_R = Re^{i\theta}$, $0 \leq \theta \leq \pi$, tends to zero as $R \uparrow \infty$, since

$$\begin{aligned} \left| \int_{C_R} f(\lambda) d\lambda \right| &= \left| \int_0^{\pi} f(Re^{i\theta}) i Re^{i\theta} d\theta \right| \\ &\leq \int_0^{\pi} \frac{1}{R^2 - 1} R d\theta \quad \text{if} \quad R > 1. \end{aligned}$$

Example 17.12. $\displaystyle\int_{-\infty}^{\infty} \frac{e^{itx}}{x^2+1} dx = \pi e^{-|t|}$ if $t \in \mathbb{R}$.

Discussion. Let

$$f(\lambda) = \frac{e^{it\lambda}}{\lambda^2 + 1} \quad \text{and} \quad g(\lambda) = (\lambda - i) f(\lambda) = \frac{e^{it\lambda}}{\lambda + i}$$

and let Γ_R denote the contour depicted in Figure 3. Then, since g is holomorphic in $\mathbb{C} \setminus \{-i\}$, the strategy introduced in Example 17.11 yields the

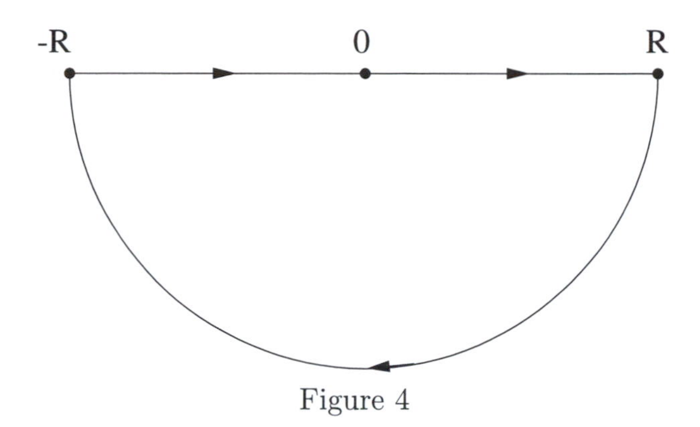

Figure 4

evaluations

$$\int_{\Gamma_R} f(\lambda)d\lambda = \int_{\Gamma_R} \frac{g(\lambda)}{\lambda - i}d\lambda = 2\pi i g(i) \quad \text{if} \quad R > 1$$

and

$$\begin{aligned} \left|\int_{C_R} f(\lambda)d\lambda\right| &= \left|\int_0^\pi f(Re^{i\theta})iRe^{i\theta}d\theta\right| \\ &\leq \int_0^\pi \frac{e^{-tR\sin\theta}}{R^2-1}Rd\theta. \end{aligned}$$

Thus, if $t > 0$, then

$$\left|\int_{C_R} f(\lambda)d\lambda\right| \leq \int_0^\pi \frac{R}{R^2-1}d\theta,$$

which tends to zero as $R \uparrow \infty$. If $t < 0$, then this bound is no longer valid; however, the given integral may be evaluated by completing the line segment $[-R \quad R]$ with a semicircle in the lower half plane.

Exercise 17.7. Show that

$$\int_{-\infty}^{\infty} \frac{e^{itx}}{x^2+1}dx = \pi e^t \quad \text{if} \quad t < 0$$

by integrating along the curve Γ_R shown in Figure 4.

Example 17.13. $\displaystyle\int_{-\infty}^{\infty} \frac{1-\cos tx}{x^2}dx = \pi|t|$ if $t \in \mathbb{R}$.

Discussion. Let

$$f(\lambda) = \frac{1-\cos t\lambda}{\lambda^2} = \frac{2 - e^{it\lambda} - e^{-it\lambda}}{2\lambda^2}.$$

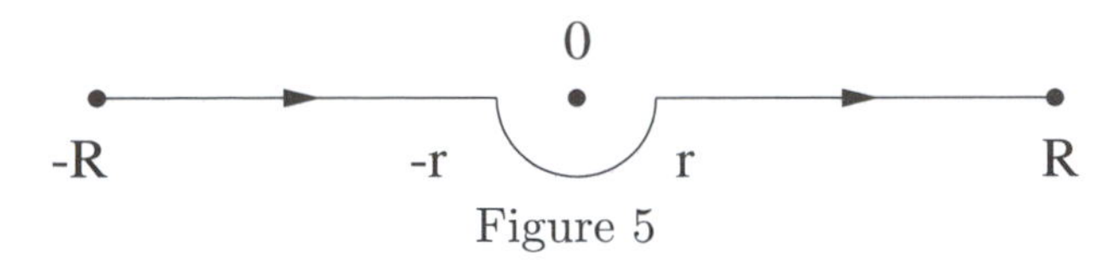

Figure 5

Then f is holomorphic in $\mathbb{C}$ and, following the strategy of Example 1,

$$\int_{-R}^{R} f(x)dx = I_R + II_R,$$

where

$$I_R = \int_{\Gamma_R} f(\lambda)d\lambda$$

and

$$II_R = -\int_{C_R} f(\lambda)d\lambda = -\int_0^{\pi} f(Re^{i\theta})iRe^{i\theta}d\theta.$$

However, this does not lead to any useful conclusions because II_R does not tend to zero as $R \uparrow \infty$ (due to the presence of both $e^{it\lambda}$ and $e^{-it\lambda}$ inside the integral). This is in fact good news because $I_R = 0$. It is tempting to split $f(\lambda)$ into the two pieces as

$$f(\lambda) = f_1(\lambda) + f_2(\lambda)$$

with

$$f_1(\lambda) = \frac{1 - e^{it\lambda}}{2\lambda^2} \quad \text{and} \quad f_2(\lambda) = \frac{1 - e^{-it\lambda}}{2\lambda^2}$$

and then, if say $t > 0$, to integrate $f_1(\lambda)$ around a contour in the upper half plane and $f_2(\lambda)$ around a contour in the lower half plane. However, this does not work because the integrals

$$\int_{-R}^{R} f_1(x)dx \quad \text{and} \int_{-R}^{R} f_2(x)dx$$

are not well defined (because of the presence of a **pole** at zero). This new difficulty is resolved by first noting that

$$\int_{-R}^{R} f(x)dx = \int_{L_R} f(\lambda)d\lambda\,,$$

where L_R is the directed path in Figure 5. Since this path detours the troublesome point $\lambda = 0$,

$$\int_{-R}^{R} f(x)dx = \int_{L_R} \{f_1(\lambda) + f_2(\lambda)\}\, d\lambda = \int_{L_R} f_1(\lambda)d\lambda + \int_{L_R} f_2(\lambda)d\lambda\,,$$

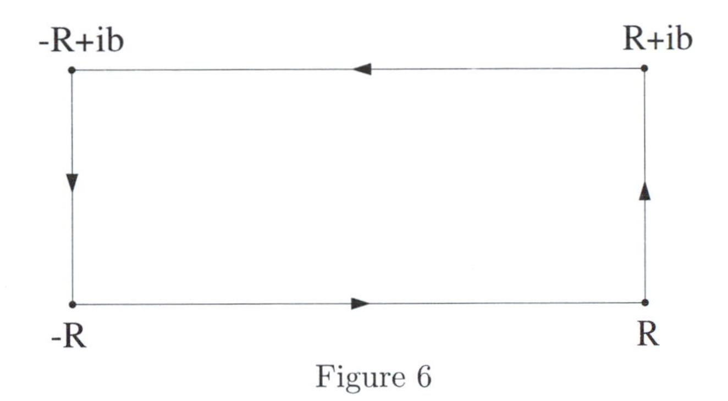

Figure 6

and hence if $t > 0$, $R > 0$ and C_R depicts the circular arc in Figure 3,

$$\begin{aligned}\int_{L_R} f_1(\lambda)d\lambda &= \int_{L_R} f_1(\lambda)d\lambda + \int_{C_R} f_1(\lambda)d\lambda - \int_{C_R} f_1(\lambda)d\lambda \\ &= \pi t - \int_0^{\pi} f_1(Re^{i\theta})iRe^{i\theta}d\theta \\ &\longrightarrow \pi t \text{ as } R \uparrow \infty ;\end{aligned}$$

whereas, if D_R denotes the circular arc depicted in Figure 4,

$$\begin{aligned}\int_{L_R} f_2(\lambda)d\lambda &= \int_{L_R} f_2(\lambda) + \int_{D_R} f_2(\lambda)d\lambda - \int_{\pi}^{2\pi} f_2(Re^{i\theta})iRe^{i\theta}d\theta \\ &= 0 + \int_{\pi}^{2\pi} f_2(Re^{i\theta})iRe^{i\theta}d\theta \\ &\longrightarrow 0 \text{ as } R \uparrow \infty .\end{aligned}$$

This completes the evaluation if $t > 0$. The result for $t < 0$ may be obtained by exploiting the fact that $1 - \cos tx$ is an even function of t.

Exercise 17.8. Verify the evaluation of the integral given in the preceding example by exploiting the fact that

$$\int_{-\infty}^{\infty} \frac{1 - \cos tx}{x^2} dx = \lim_{\varepsilon \downarrow 0} \int_{-\infty}^{\infty} \frac{1 - \cos tx}{x^2 + \varepsilon^2} dx .$$

Example 17.14. $\displaystyle\int_{-\infty}^{\infty} e^{-a(x-ib)^2} dx = \int_{-\infty}^{\infty} e^{-ax^2} dx$ if $a > 0$ and $b \in \mathbb{R}$.

Discussion. Let Γ_R denote the counterclockwise rectangular path indicated in Figure 6 and let $f(\lambda) = e^{-a\lambda^2}$. Then, since $f(\lambda)$ is holomorphic in the whole complex plane $\mathbb{C}$,

$$\begin{aligned}0 = \int_{\Gamma_R} f(\lambda)d\lambda &= \int_{-R}^{R} f(x)dx + \int_0^b f(R + iy)idy \\ &- \int_{-R}^{R} f(x + ib)dx - \int_0^b f(-R + iy)idy .\end{aligned}$$

Next, invoking the bound

$$|\exp\{-a(R+ib)^2\}| = |\exp\{-a(R^2+2iRb-b^2)\}| = \exp\{-a(R^2-b^2)\}\,,$$

one can readily see that the integrals over the vertical segments of the rectangle tend to zero as $R\uparrow\infty$ and hence that

$$\int_{-\infty}^{\infty} e^{-ax^2}dx = \int_{-\infty}^{\infty} f(x)dx = \int_{-\infty}^{\infty} f(x+ib)dx \int_{-\infty}^{\infty} e^{-a(x+ib)^2}dx\,,$$

for every choice of $b>0$. Since the same argument works for $b<0$, the verification of the asserted formula is complete.

17.4. A short detour on Fourier analysis

Let

$$\widehat{f}(\mu) = \int_{-\infty}^{\infty} e^{i\mu x} f(x)dx \tag{17.12}$$

denote the **Fourier transform** of f (whenever the integral is meaningful) and let

$$f_{ab}(x) = \begin{cases} 1 & \text{for } a\le x\le b \\ 0 & \text{elsewhere}\end{cases} \qquad \text{and} \quad t=b-a\,.$$

Then

$$\widehat{f_{ab}}(\mu) = \int_{-\infty}^{\infty} e^{i\mu x} f_{ab}(x)dx = \int_a^b e^{i\mu x}dx = \frac{e^{i\mu b}-e^{i\mu a}}{i\mu}$$

and

$$\begin{aligned}\int_{-\infty}^{\infty}\left|\widehat{f_{ab}}(\mu)\right|^2 d\mu &= \int_{-\infty}^{\infty}\left|\frac{e^{i\mu b}-e^{i\mu a}}{i\mu}\right|^2 d\mu \\ &= \int_{-\infty}^{\infty}\left|\frac{e^{i\mu t}-1}{i\mu}\right|^2 d\mu \\ &= 2\int_{-\infty}^{\infty}\frac{1-\cos\mu t}{\mu^2}d\mu \\ &= 2\pi t \quad \text{(by the formula in Example 17.13)} \\ &= 2\pi\int_{-\infty}^{\infty}|f_{ab}(x)|^2\,dx\,.\end{aligned}$$

Exercise 17.9. Show that if $a<b\le c<d$, then

$$\int_{-\infty}^{\infty}\widehat{f_{cd}}(\mu)\overline{\widehat{f_{ab}}(\mu)}d\mu = 0\,.$$

[HINT: Let

$$g(\mu) = \widehat{f_{cd}}(\mu)\overline{\widehat{f_{ab}}(\mu)} = \left\{\frac{e^{i\mu d}-e^{i\mu c}}{i\mu}\right\}\overline{\left\{\frac{e^{i\mu b}-e^{i\mu a}}{i\mu}\right\}}$$

$$= \frac{e^{i\mu(d-b)} - e^{i\mu(c-b)} - e^{i\mu(d-a)} + e^{i\mu(c-a)}}{\mu^2}$$

and exploit the fact that $g(\lambda)$ is holomorphic in $\mathbb{C}$ and the coefficients of $i\mu$ in the exponential terms in $g(\mu)$ are all nonnegative.]

Exercise 17.10. Show that

$$\lim_{R\uparrow\infty} \frac{1}{2\pi} \int_{-R}^{R} e^{-i\mu x} \widehat{f}_{ab}(\mu) d\mu = f_{ab}(x)$$

for all points $x \in \mathbb{R}$ other than a and b.

In view of the formulas in Exercises 17.9 and 17.10, it is now easy to check that

$$\lim_{R\uparrow\infty} \frac{1}{2\pi} \int_{-R}^{R} e^{-i\mu x} \widehat{f}(\mu) d\mu = f(x) \tag{17.13}$$

and

$$\frac{1}{2\pi} \int_{-\infty}^{\infty} |\widehat{f}(\mu)|^2 d\mu = \int_{-\infty}^{\infty} |f(x)|^2 dx \tag{17.14}$$

for functions f of the form

$$f(x) = \sum_{j=1}^{n} c_j f_{a_j b_j}(x),$$

where $a_1 < b_1 \leq a_2 < b_2 \leq \cdots \leq a_n < b_n$ and $c_1, \ldots, c_n$ is any set of complex numbers. The first formula (17.13) exhibits a way of recovering $f(x)$ from its Fourier transform $\widehat{f}(\mu)$. Accordingly, the auxiliary transform

$$g^{\vee}(\mu) = \frac{1}{2\pi} \int_{-\infty}^{\infty} e^{-i\mu x} g(\mu) d\mu \tag{17.15}$$

(appropriately interpreted) is termed the **inverse Fourier transform**. The second formula (17.14) is commonly referred to as the **Parseval/Plancherel or Pareseval-Plancherel** formula. It exhibits the fact that

$$\|(2\pi)^{-\frac{1}{2}} \widehat{f}\|_2 = \|f\|_2$$

where

$$\|f\|_2 = \left\{ \int_{-\infty}^{\infty} |f(x)|^2 dx \right\}^{\frac{1}{2}},$$

for the class of functions under consideration. However, the conclusion is valid for the class of f which belong to the space L^2 of f such that $|f|^2$ is integrable in the sense of Lebesgue on the line $\mathbb{R}$.

Exercise 17.11. Show that (17.14) holds if and only if

$$\frac{1}{2\pi}\int_{-\infty}^{\infty}\widehat{f}(\mu)\overline{\widehat{g}(\mu)}d\mu = \int_{-\infty}^{\infty} f(x)\overline{g(x)}dx \tag{17.16}$$

holds for every pair of piecewise constant functions $f(x)$ and $g(x)$. [HINT: This is just (8.5).]

The space L^2 has the pleasant feature that $f \in L^2 \iff \widehat{f} \in L^2$. An even pleasanter class for Fourier analysis is the **Schwartz class** $\mathcal{S}$ of infinitely differentiable functions $f(x)$ on $\mathbb{R}$ such that

$$\lim_{x\uparrow+\infty} |x^j f^{(k)}(x)| = \lim_{x\downarrow-\infty} |x^j f^{(k)}(x)| = 0$$

for every pair of nonnegative integers j and k.

Exercise 17.12. Show that if $f \in \mathcal{S}$, then its Fourier transform $\widehat{f}(\lambda)$ enjoys the following properties:

(a) $(-i\lambda)^j\widehat{f}(\lambda) = \int_{-\infty}^{\infty} e^{i\lambda x}f^{(j)}(x)dx$ for $j = 1, 2, \ldots$.

(b) $(-iD_\lambda)^k\widehat{f} = \int_{-\infty}^{\infty} e^{i\lambda x}x^k f(x)dx$ for $k = 1, 2, \ldots$.

(c) $\widehat{f} \in \mathcal{S}$.

You may take it as known that if $f \in \mathcal{S}$, then the derivative

$$D_\lambda\widehat{f} = \lim_{\xi\to 0}\frac{\widehat{f}(\lambda+\xi) - \widehat{f}(\lambda)}{\xi}$$

can be brought inside the integral that defines the transform.

Exercise 17.13. Show that if $f(x)$ and $g(x)$ belong to the Schwartz class $\mathcal{S}$, then the **convolution**

$$(f \circ g)(x) = \int_{-\infty}^{\infty} f(x-y)g(y)dy \tag{17.17}$$

belongs to the class $\mathcal{S}$ and that

$$\widehat{(f \circ g)}(\lambda) = \widehat{f}(\lambda)\widehat{g}(\lambda)\,. \tag{17.18}$$

Exercise 17.14. Show that if $f(x) = e^{-x^2/2}$, then $\widehat{f}(\mu) = e^{-\mu^2/2}\widehat{f}(0)$. [HINT: Exploit the formula that was established in Example 17.14.]

17.5. Contour integrals of matrix valued functions

The contour integral

$$\int_\Gamma F(\lambda)d\lambda$$

of a $p \times q$ matrix valued function

$$F(\lambda) = \begin{bmatrix} f_{11}(\lambda) & \cdots & f_{1q}(\lambda) \\ \vdots & & \vdots \\ f_{p1}(\lambda) & \cdots & f_{qq}(\lambda) \end{bmatrix}$$

is defined by the formula

$$\int_\Gamma F(\lambda)d\lambda = \begin{bmatrix} a_{11} & \cdots & a_{1q} \\ \vdots & & \vdots \\ a_{p1} & \cdots & a_{pq} \end{bmatrix},$$

where

$$a_{ij} = \int_\Gamma f_{ij}(\lambda)d\lambda\ ,\ i = 1,\ldots,p,\ j = 1,\ldots,q;$$

i.e., each entry is integrated separately. It is readily checked that

$$\int_\Gamma \{F(\lambda) + G(\lambda)\}d\lambda = \int_\Gamma F(\lambda)d\lambda + \int_\Gamma G(\lambda)d\lambda$$

and that if B and C are appropriately sized constant matrices, then

$$\int_\Gamma BF(\lambda)Cd\lambda = B\left(\int_\Gamma F(\lambda)d\lambda\right)C\,.$$

Moreover, if $\varphi(\lambda)$ is a scalar valued function and $C \in \mathbb{C}^{p\times q}$, then

$$\int_\Gamma \varphi(\lambda)Cd\lambda = \left(\int_\Gamma \varphi(\lambda)d\lambda\right)C\,.$$

Lemma 17.15. *If $\Gamma = \{\gamma(t) : a \le t \le b\}$ is a simple smooth curve that is parametrized by a function $\gamma(t) \in \mathcal{C}^1([a,b])$ and $F(\lambda)$ is continous $p \times q$ matrix valued function on Γ, then*

$$\left\|\int_\Gamma F(\lambda)d\lambda\right\| \le \int_a^b \|F(\gamma(t))\||\gamma'(t)|dt\,.$$

Proof. This is a straightforward consequence of the triangle inequality applied to the Riemann sums that are used to approximate the integral. □

Lemma 17.16. *Let*

$$J = C_\mu^{(k)} = \mu I_k + N$$

be a Jordan cell and let Γ be a simple smooth counterclockwise directed curve in the complex plane $\mathbb{C}$ that does not intersect the point μ. Then

$$\frac{1}{2\pi i}\int_\Gamma (\lambda I_k - J)^{-1}d\lambda = \begin{cases} I_k & \text{if } \mu \text{ is inside } \Gamma \\ O_{k\times k} & \text{if } \mu \text{ is outside } \Gamma\,. \end{cases} \tag{17.19}$$

Proof. Clearly,

$$\lambda I_k - J = (\lambda - \mu)I_k - N$$

and, since $N^k = O_{k\times k}$,

$$\begin{aligned}(\lambda I_k - J)^{-1} &= (\lambda-\mu)^{-1}\left(I_k - \frac{N}{\lambda-\mu}\right)^{-1} \\ &= \frac{I_k}{\lambda-\mu} + \frac{N}{(\lambda-\mu)^2} + \cdots + \frac{N^{k-1}}{(\lambda-\mu)^k}\,.\end{aligned}$$

Therefore,

$$\frac{1}{2\pi i}\int_\Gamma (\lambda I_k - J)^{-1} d\lambda = \sum_{j=1}^{k}\left\{\frac{1}{2\pi i}\int_\Gamma \frac{1}{(\lambda-\mu)^j}d\lambda\right\} N^{j-1}\,.$$

But this yields the asserted formula, since

$$\frac{1}{2\pi i}\int_\Gamma \frac{1}{(\lambda-\mu)^j}d\lambda = 0 \quad \text{if} \quad j > 1$$

and

$$\frac{1}{2\pi i}\int_\Gamma \frac{1}{\lambda-\mu}d\lambda = \begin{cases} 1 & \text{if } \mu \text{ is inside } \Gamma \\ 0 & \text{if } \mu \text{ is outside } \Gamma\,. \end{cases}$$

□

Let $A \in \mathbb{C}^{n\times n}$ admit a Jordan decomposition of the form

$$A = UJU^{-1} = [U_1 \cdots U_\ell]\begin{bmatrix} J_1 & & \\ & \ddots & \\ & & J_\ell \end{bmatrix}\begin{bmatrix} V_1 \\ \vdots \\ V_\ell \end{bmatrix}, \tag{17.20}$$

where $J_1, \ldots, J_\ell$ denote the Jordan cells of J, $U_1, \ldots, U_\ell$ denote the corresponding block columns of U and $V_1, \ldots, V_\ell$ denote the corresponding block rows of U^{-1}. Consequently,

$$A = \sum_{i=1}^{\ell} U_i J_i V_i \tag{17.21}$$

and, if J_i is a Jordan cell of size $n_i \times n_i$, then

$$(\lambda I_n - A)^{-1} = U(\lambda I_n - J)^{-1}U^{-1} = \sum_{i=1}^{\ell} U_i(\lambda I_{n_i} - J_i)^{-1}V_i\,. \tag{17.22}$$

Note that if A has k distinct eigenvalues with geometric multiplicities $\gamma_1, \ldots,$ γ_k, then $\ell = \gamma_1 + \cdots + \gamma_k$ in formula (17.22).

Lemma 17.17. *Let $A \in \mathbb{C}^{n\times n}$ admit a Jordan decomposition of the form (17.20), where the $n_i \times n_i$ Jordan cell $J_i = \beta_i I_{n_i} + N_{n_i}$, and let Γ be a simple* *smooth counterclockwise directed curve in* $\mathbb{C}$ *that does not intersect any of the eigenvalues of* A. *Then*

$$\frac{1}{2\pi i}\int_\Gamma (\lambda I_n - A)^{-1} d\lambda = \sum_{i=1}^{\ell} U_i X_i V_i \,, \tag{17.23}$$

where

$$X_i = \begin{cases} I_{n_i} & \text{if } \beta_i \text{ is inside } \Gamma \\ 0_{n_i\times n_i} & \text{if } \beta_i \text{ is outside } \Gamma \end{cases} .$$

Proof. This is an easy consequence of formula (17.22) and Lemma 17.16. □

It is readily checked that the sum on the right-hand side of formula (17.23) is a projection:

$$\left(\sum_{i=1}^{\ell} U_i X_i V_i\right)^2 = \sum_{i=1}^{\ell} U_i X_i V_i \,.$$

Therefore, the integral on the left-hand side of (17.23),

$$P_\Gamma^A = \frac{1}{2\pi i}\int_\Gamma (\lambda I_n - A)^{-1} d\lambda \,, \tag{17.24}$$

is also a projection. It is termed the **Riesz projection**.

Lemma 17.18. *Let* $A \in \mathbb{C}^{n\times n}$, *let* $\det(\lambda I_n - A) = (\lambda - \lambda_1)^{\alpha_1} \cdots (\lambda - \lambda_k)^{\alpha_k}$, *where the points* $\lambda_1 \ldots \lambda_k$ *are distinct, and let* Γ *be a simple smooth counterclockwise directed curve that does not intersect any of the eigenvalues of* A. *Then*

$$\operatorname{rank} P_\Gamma^A = \sum_{i\in G} \alpha_i \quad \text{where} \quad G = \{i : \lambda_i \quad \text{is inside} \quad \Gamma\} \,. \tag{17.25}$$

Proof. The conclusion rests on the observation that

$$\begin{aligned} \operatorname{rank} \sum_{i=1}^{\ell} U_i X_i V_i &= \operatorname{rank}\,(U\{\operatorname{diag}\{X_1, \ldots, X_\ell\}\}\,V) \\ &= \operatorname{rank}\,\{\operatorname{diag}\{X_1, \ldots, X_\ell\}\} \,, \end{aligned}$$

since U and V are invertible. Therefore, the rank of the indicated sum is equal to the sum of the sizes of the nonzero X_i that intervene in the formula for P_Γ^A, which agrees with formula (17.25). □

17.6. Continuous dependence of the eigenvalues

In this section we shall use the projection formulas P_Γ^A to establish the continuous dependence of the eigenvalues of a matrix $A \in \mathbb{C}^{n\times n}$ on A. The strategy is to show that if $B \in \mathbb{C}^{n\times n}$ is sufficiently close to A, then

$\|P_\Gamma^A - P_\Gamma^B\| < 1$, and hence, by Lemma 9.16, rank P_Γ^A = rank P_Γ^B. The precise formulation is given in the next theorem, in which

$$D_\epsilon(\mu) = \{\lambda \in \mathbb{C} : |\lambda - \mu| < \epsilon\}.$$

Theorem 17.19. *Let $A, B \in \mathbb{C}^{n\times n}$ and let*

$$\det(\lambda I_n - A) = (\lambda - \lambda_1)^{\alpha_1} \cdots (\lambda - \lambda_k)^{\alpha_k},$$

where the k points $\lambda_1 \ldots, \lambda_k$ are distinct. Then for every $\epsilon > 0$, there exists a $\delta > 0$ such that if $\|A - B\| < \delta$, then

$$\sigma(B) \subset \bigcup_{i=1}^{k} D_\epsilon(\lambda_i).$$

Moreover, if

$$\varepsilon_0 = \frac{1}{2}\min\{|\lambda_i - \lambda_j| : i \neq j\}$$

and $\varepsilon < \varepsilon_0$, then each disk $D_\epsilon(\lambda_i)$ will contain exactly α_i points of spectrum of B, counting multiplicities.

Proof. Let $r < \varepsilon_0$ and let $\Gamma_j = \Gamma_j(r)$ denote a circle of radius r centered at λ_j and directed counterclockwise. This constraint on r insures that the $\Gamma_j \cap \Gamma_k = \emptyset$ if $j \neq k$ and that λ_j is the only eigenvalue of A inside the closed disc $\overline{D_r(\lambda_j)} = \{\lambda \in \mathbb{C} : |\lambda - \lambda_j| \leq r\}$. The remainder of the proof rests on the following four estimates that are taken from Lemma 7.18:

(1) If $\lambda \in \Gamma_j$, then $\lambda I_n - A$ is invertible and there exists a constant γ_j such that $\|(\lambda I_n - A)^{-1}\| \leq \gamma_j < \infty$ for $j = 1, \ldots, k$.

(2) If $\gamma = \max\{\gamma_j : j = 1, \ldots, k\}$ and $\|A - B\| < \frac{1}{\gamma}$, then $\lambda I_n - B$ is invertible and

$$\|(\lambda I_n - B)^{-1}\| \leq \frac{\gamma}{1 - \gamma\|B - A\|}$$

for every point $\lambda \in \cup_{j=1}^k \Gamma_j$.

(3) The bound

$$\|(\lambda I_n - A)^{-1} - (\lambda I_n - B)^{-1}\| \leq \frac{\gamma^2\|B - A\|}{1 - \gamma\|B - A\|}$$

is in force for every point $\lambda \in \cup_{j=1}^k \Gamma_j$.

(4) In view of (3) and Lemma 17.15,

$$\begin{aligned} \|P_{\Gamma_j}^A - P_{\Gamma_j}^B\| &= \left\| \frac{1}{2\pi i} \int_{\Gamma_j} (\lambda I_n - A)^{-1} - (\lambda I_n - B)^{-1} d\lambda \right\| \\ &\leq \frac{\gamma^2\|B - A\| r}{1 - \gamma\|B - A\|}. \end{aligned}$$

At first glance, the last upper bound might look strange, unless you keep in mind that the number γ **also depends upon** r and can be expected to increase as r decreases. Nevertheless, this bound can be made less than one by choosing

$$\|B - A\| < \frac{1}{\gamma + \gamma^2 r}\,.$$

Thus, given any $\varepsilon > 0$, let $r < \min\{\varepsilon, \varepsilon_0\}$ and $\delta < (\gamma + \gamma^2 r)^{-1}$; then, by Lemma 9.16,

$$\operatorname{rank} P^B_{\Gamma_j} = \operatorname{rank} P^A_{\Gamma_j} = \alpha_j$$

for $j = 1, \ldots, k$, and hence by Lemma 17.18,

$$\begin{aligned}\operatorname{rank} P^B_{\Gamma_j} &= \text{the sum of the algebraic multiplicities of the eigenvalues} \\ &\quad \text{of B inside } D_\epsilon(\lambda_j)\,,\end{aligned}$$

which yields the desired result. □

Exercise 17.15. Show that the roots $\lambda_1, \ldots, \lambda_n$ of the polynomial $p(\lambda) = a_0 + a_1\lambda + \cdots + a_n\lambda^n$ with $a_n \neq 0$ depend continuously on the coefficients $a_0, \ldots, a_n$ of the polynomial. [HINT: Exploit companion matrices.]

17.7. More on small perturbations

In this section we shall continue to investigate the effect of small changes in the entries of a matrix.

Lemma 17.20. *If $A \in \mathbb{C}^{n\times n}$ has n distinct eigenvalues, then there exists a number $\delta > 0$ such that every matrix $B \in \mathbb{C}^{n\times n}$ for which $\|A - B\| < \delta$ also has n distinct eigenvalues.*

Proof. This is an easy consequence of Theorem 17.19. □

The next result shows that any matrix $A \in \mathbb{C}^{n\times n}$ can be approximated arbitrarily well by a matrix $B \in \mathbb{C}^{n\times n}$ with n distinct eigenvalues; i.e., the class of $n \times n$ matrices with n distinct eigenvalues is **dense** in $\mathbb{C}^{n\times n}$.

Lemma 17.21. *If $A \in \mathbb{C}^{n\times n}$ and $\delta > 0$ are specified, then there exists a matrix $B \in \mathbb{C}^{n\times n}$ with n distinct eigenvalues such that $\|A - B\| < \delta$.*

Proof. Let $A = PJP^{-1}$ be the Jordan decomposition of the matrix A and let D be a diagonal matrix such that the diagonal entries of the matrix $D + J$ are all distinct and

$$|d_{ii}| \le (\|P\|\|P^{-1}\|)^{-1}\delta$$

for $i = 1, \ldots, n$. Then the matrix $B = P(D + J)P^{-1}$ has n distinct eigenvalues and

$$\|A - B\| = \|PDP^{-1}\| \le \|P\|\|D\|\|P^{-1}\| \le \delta\,.$$

□

Note that this lemma does not say that every matrix B that meets the inequality $\|A - B\| < \delta$ has n distinct eigenvalues.

Since the class of $n \times n$ matrices with n distinct eigenvalues is a subclass of the set of diagonalizable matrices, it is reasonable to ask whether or not a diagonalizable matrix remains diagonalizable if some of its entries are changed just a little. The answer is not always!

Exercise 17.16. Let $A = \begin{bmatrix} 1 & 0 \\ 0 & 1 \end{bmatrix}$ and $B = \begin{bmatrix} 1 & \beta \\ 0 & 1 \end{bmatrix}$, where $\beta \neq 0$. Show that $\|A - B\| = |\beta|$, but that A is diagonalizable, whereas B is not.

Exercise 17.17. Let $A \in \mathbb{C}^{p\times q}$ and suppose that $\operatorname{rank} A = k$ and $k < \min\{p, q\}$. Show that for every $\epsilon > 0$ there exists a matrix $B \in \mathbb{C}^{p\times q}$ such that $\|A - B\| < \epsilon$ and $\operatorname{rank} B = k + 1$. [HINT: Use the singular value decomposition of A.]

Some conclusions

A subset $\mathcal{X}$ of the set of $p \times q$ matrices $\mathbb{C}^{p\times q}$ is said to be **open** if for every matrix $A \in \mathcal{X}$ there exists a number $\delta > 0$ such that the **open ball**

$$\mathcal{B}_\delta(A) = \{B \in \mathbb{C}^{p\times q} : \|A - B\| < \delta\}$$

is also a subset of $\mathcal{X}$. The meaning of this condition is that if the entries in a matrix $A \in \mathcal{X}$ are changed only slightly, then the new perturbed matrix will also belong to the class $\mathcal{X}$. This is a significant property in applications and computations because the entries in any matrix that is obtained from data or from a numerical algorithm are only known approximately. The preceding analysis implies that:

(1) $\{A \in \mathbb{C}^{p\times q} : \operatorname{rank} A = \min\{p, q\}\}$ is an open subset of $\mathbb{C}^{p\times q}$.

(2) $\{A \in \mathbb{C}^{p\times q} : \operatorname{rank} A < \min\{p, q\}\}$ is not an open subset of $\mathbb{C}^{p\times q}$.

(3) $\{A \in \mathbb{C}^{n\times n} :$ with n distinct eigenvalues$\}$ is an open subset of $\mathbb{C}^{n\times n}$.

(4) $\{A \in \mathbb{C}^{n\times n} : A$ is diagonalizable$\}$ is not an open subset of $\mathbb{C}^{n\times n}$.

The set $\{A \in \mathbb{C}^{n\times n} :$ with n distinct eigenvalues$\}$ is particularly useful because it is both open and **dense** in $\mathbb{C}^{n\times n}$, thanks to Lemma 17.21. Open dense sets are said to be **generic**.

Exercise 17.18. Show that $\{A \in \mathbb{C}^{n\times n} : A$ is invertible$\}$ is a generic set.

17.8. Spectral radius redux

In this section we shall use the methods of complex analysis to obtain a simple proof of the formula

$$r_\sigma(A) = \lim_{k\uparrow\infty} \|A^k\|^{1/k}$$

for the spectral radius

$$r_\sigma(A) = \max\{|\lambda| : \lambda \in \sigma(A)\}$$

of a matrix $A \in \mathbb{C}^{n\times n}$. Since the inequality

$$r_\sigma(A) \le \|A^k\|^{1/k} \text{ for } k = 1, 2, \ldots$$

is easily verified, it suffices to show that

$$\lim_{k\uparrow\infty} \|A^k\|^{1/k} \le r_\sigma(A)\,. \tag{17.26}$$

Lemma 17.22. *If $A \in \mathbb{C}^{n\times n}$ and $\sigma(A)$ belongs to the set of points enclosed by a simple smooth counterclockwise directed closed curve Γ, then*

$$A^k = \frac{1}{2\pi i}\int_\Gamma \lambda^k(\lambda I_n - A)^{-1} d\lambda\,. \tag{17.27}$$

Proof. Let $g(\lambda) = \lambda^k$; let Γ_r denote a circle of radius r centered at zero and directed counterclockwise, and suppose that $r > \|A\|$. Then

$$\begin{aligned}
\frac{1}{2\pi i}\int_{\Gamma_r} \lambda^k(\lambda I_n - A)^{-1} d\lambda &= \frac{1}{2\pi i}\int_{\Gamma_r} g(\lambda)\sum_{j=0}^{\infty}\frac{A^j}{\lambda^{j+1}}d\lambda \\
&= \sum_{j=0}^{\infty}\left\{\frac{1}{2\pi i}\int_{\Gamma_r}\frac{g(\lambda)}{\lambda^{j+1}}d\lambda\right\}A^j \\
&= \sum_{j=0}^{\infty}\frac{g^{(j)}(0)}{j!}A^j = A^k\,.
\end{aligned}$$

The assumption $r > \|A\|$ is used to guarantee the uniform convergence of the sum $\sum_{j=0}^{\infty}\lambda^{-j}A^j$ on Γ_r and subsequently to justify the interchange in the order of summation and integration. Thus, to this point, we have established formula (17.27) for $\Gamma = \Gamma_r$ and $r > \|A\|$. However, since $\lambda^k(\lambda I_n - A)^{-1}$ is holomorphic in an open set that contains the points between and on the curves Γ and Γ_r, it follows that

$$\frac{1}{2\pi i}\int_{\Gamma} \lambda^k(\lambda I_n - A)^{-1} d\lambda = \frac{1}{2\pi i}\int_{\Gamma_r} \lambda^k(\lambda I_n - A)^{-1} d\lambda\,.$$

□

Corollary 17.23. If $A \in \mathbb{C}^{n\times n}$ and $\sigma(A)$ belongs to the set of points enclosed by a simple smooth counterclockwise directed closed curve Γ, then

$$p(A) = \frac{1}{2\pi i}\int_\Gamma p(\lambda)(\lambda I_n - A)^{-1} d\lambda \tag{17.28}$$

for every polynomial $p(\lambda)$.

Theorem 17.24. *If $A \in \mathbb{C}^{n\times n}$, then*

$$\lim_{k\uparrow\infty} \|A^k\|^{1/k} = r_\sigma(A)\,; \tag{17.29}$$

i.e., the limit exists and is equal to the modulus of the maximum eigenvalue of A.

Proof. Fix $\epsilon > 0$, let $r = r_\sigma(A) + \epsilon$, and let

$$\gamma_r = \max\{\|(\lambda I_n - A)^{-1}\| : |\lambda| = r\}\,.$$

Then, by formula (17.27),

$$\begin{aligned} \|A^k\| &= \left\| \frac{1}{2\pi i} \int_0^{2\pi} (re^{i\theta})^k (re^{i\theta} I_n - A)^{-1} ire^{i\theta} d\theta \right\| \\ &\le \frac{1}{2\pi} \int_0^{2\pi} r^k \|(re^{i\theta} I_n - A)^{-1}\| r d\theta \\ &\le r^{k+1}\gamma_r\,. \end{aligned}$$

Thus,

$$\|A^k\|^{1/k} \le r(r\gamma_r)^{1/k}$$

and, as

$$(r\gamma_r)^{1/k} \to 1 \text{ as } k\uparrow\infty\,,$$

it follows that

$$\limsup_{k\uparrow\infty} \|A^k\|^{1/k} \le r = r_\sigma(A) + \epsilon\,.$$

The inequality

$$\limsup_{k\uparrow\infty} \|A^k\|^{1/k} \le r_\sigma(A)$$

is then obtained by letting $\epsilon \downarrow 0$. Therefore, since $r_\sigma(A) \le \|A^k\|^{1/k}$ for $k = 1, 2, \ldots$, it follows that

$$r_\sigma(A) \le \liminf_{k\uparrow\infty} \|A^k\|^{1/k} \le \limsup_{k\uparrow\infty} \|A^k\|^{1/k} \le r_\sigma(A)\,,$$

which serves to establish formula (17.29). □

Exercise 17.19. Show that if $A \in \mathbb{C}^{n\times n}$ and $\sigma(A)$ belongs to the set of points enclosed by a simple smooth counterclockwise directed closed curve Γ, then

$$\frac{1}{2\pi i}\int_\Gamma e^\lambda (\lambda I_n - A)^{-1} d\lambda = \sum_{j=0}^{\infty} \frac{A^j}{j!}\,. \tag{17.30}$$

Let $A \in \mathbb{C}^{n\times n}$ and let $f(\lambda)$ be holomorphic in an open set Ω that contains $\sigma(A)$. Then, in view of formulas (17.28) and (17.30) it is reasonable to **define**

$$f(A) = \frac{1}{2\pi i}\int_\Gamma f(\lambda)(\lambda I_n - A)^{-1}d\lambda\,, \tag{17.31}$$

where Γ is any simple smooth counterclockwise directed closed curve in Ω that encloses $\sigma(A)$ such that every point inside Γ also belongs to Ω. This definition is independent of the choice of Γ and is consistent with the definitions of $f(A)$ considered earlier.

Exercise 17.20. Show that if, in terms of the notation introduced in (17.21), $A = U_1C_\alpha^{(p)}V_1 + U_2C_\beta^{(q)}V_2$, then

$$f(A) = U_1\sum_{j=0}^{p-1}\frac{f^{(j)}(\alpha)}{j!}(C_0^{(p)})^jV_1 + U_2\sum_{j=0}^{q-1}\frac{f^{(j)}(\alpha)}{j!}(C_0^{(q)})^jV_2 \tag{17.32}$$

for every function $f(\lambda)$ that is holomorphic in an open set that contains the points α and β.

Exercise 17.21. Show that in the setting of Exercise 17.20

$$\det(\lambda I_n - f(A)) = (\lambda - f(\alpha))^p(\lambda - f(\beta))^q\,. \tag{17.33}$$

Exercise 17.22. Show that if $A \in \mathbb{C}^{n\times n}$ and $f(\lambda)$ is holomorphic in an open set that contains $\sigma(A)$, then

(17.34)

$$\begin{aligned}\det(\lambda I_n - A) &= (\lambda-\lambda_1)^{\alpha_1}\cdots(\lambda-\lambda_k)^{\alpha_k}\\ &\Longrightarrow \det(\lambda I_n - f(A)) = (\lambda - f(\lambda_1))^{\alpha_1}\cdots(\lambda - f(\lambda_k))^{\alpha_k}\,.\end{aligned}$$

[HINT: The main ideas are contained in Exercises 17.20 and 17.21. The rest is just more elaborate bookkeeping.]

Theorem 17.25. (The spectral mapping theorem) *Let $A \in \mathbb{C}^{n\times n}$ and let $f(\lambda)$ be holomorphic in an open set that contains $\sigma(A)$. Then*

$$\mu \in \sigma(f(A)) \Longleftrightarrow f(\mu) \in \sigma(A)\,. \tag{17.35}$$

Proof. This is immediate from formula (17.34). □

17.9. Fractional powers

Exercise 17.23. Let $A \in \mathbb{C}^{n\times n}$. Show that if $A \succ O$, then

$$A^{1/2} = \frac{1}{2\pi i}\int_\Gamma \sqrt{\lambda}(\lambda I_n - A)^{-1}d\lambda$$

for any simple closed smooth curve Γ in the open right half plane that includes the eigenvalues of A in its interior.

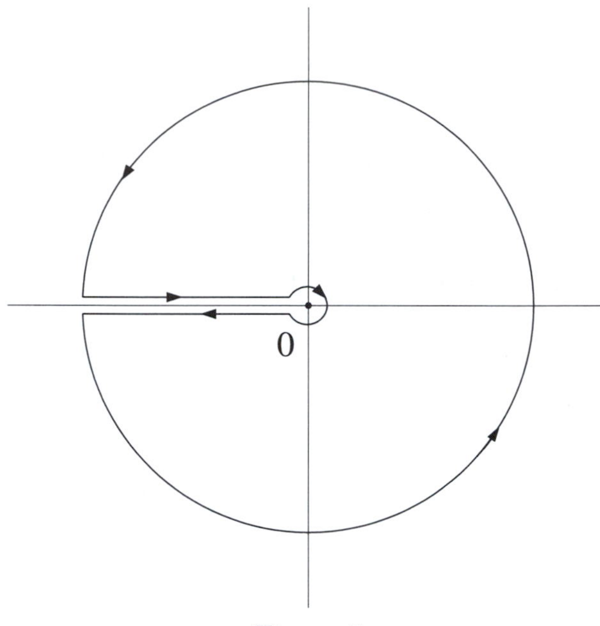

Figure 7

Exercise 17.24. Let A, $B \in \mathbb{C}^{n\times n}$ and suppose that $A \succ B \succ O$. Show that if $0 < t < 1$, then

$$A^t - B^t = \frac{1}{2\pi i}\int_\Gamma \lambda^t \left\{(\lambda I_n - A)^{-1} - (\lambda I_n - B)^{-1}\right\} d\lambda\,, \tag{17.36}$$

where Γ indicates the curve in Figure 7, and then, by passing to appropriate limits, obtain the formula

$$A^t - B^t = \frac{\sin \pi t}{\pi}\int_0^\infty x^t (xI_n + A)^{-1}(A - B)(xI_n + B)^{-1} dt\,. \tag{17.37}$$

Exercise 17.25. Use formula (17.37) to show that if A, $B \in \mathbb{C}^{n\times n}$, then

$$A \succ B \succ O \Longrightarrow A^t \succ B^t \quad \text{for} \quad 0 < t < 1\,. \tag{17.38}$$

Chapter 18

Matrix equations

... confusion between creativity and originality. Being original entails saying something that nobody has said before. Originality ... must be exhibited, or feigned, for academic advancement. Creativity, by contrast, reflects the inner experience of the individual overcoming a challenge. Creativity is not diminished when one achieves ... what has already been discovered...

Shalom Carmy [**15**], p. 26

In this chapter we shall analyze the existence and uniqueness of solutions to a number of matrix equations that occur frequently in applications. The notation

$$\Pi_+ = \{\lambda \in \mathbb{C} : \lambda + \overline{\lambda} > 0\} \quad \text{and} \quad \Pi_- = \{\lambda \in \mathbb{C} : \lambda + \overline{\lambda} < 0\}$$

for the open right and open left half plane, respectively, will be useful.

18.1. The equation $X - AXB = C$

Theorem 18.1. *Let $A \in \mathbb{C}^{p\times p}$, $B \in \mathbb{C}^{q\times q}$ and $C \in \mathbb{C}^{p\times q}$; let $\alpha_1, \ldots, \alpha_p$ and $\beta_1, \ldots, \beta_q$ denote the eigenvalues of the matrices A and B (repeated according to their algebraic multiplicity), respectively; and let T denote the linear transformation from $\mathbb{C}^{p\times q}$ into $\mathbb{C}^{p\times q}$ that is defined by the rule*

$$T : X \in \mathbb{C}^{p\times q} \longrightarrow X - AXB \in \mathbb{C}^{p\times q}\,. \tag{18.1}$$

Then

(18.2)
$$\mathcal{N}_T = \{O_{p\times q}\} \Longleftrightarrow \alpha_i\beta_j \neq 1 \quad \textit{for} \quad i = 1, \ldots, p \quad \textit{and} \quad j = 1, \ldots, q\,.$$

Proof. Let $A\mathbf{u}_i = \alpha_i \mathbf{u}_i$ and $B^T\mathbf{v}_j = \beta_j \mathbf{v}_j$ for some pair of nonzero vectors $\mathbf{u}_i \in \mathbb{C}^p$ and $\mathbf{v}_j \in \mathbb{C}^q$ and let $X = \mathbf{u}_i\mathbf{v}_j^T$. Then the formula

$$TX = \mathbf{u}_i\mathbf{v}_j^T - A\mathbf{u}_i\mathbf{v}_j^T B = (1-\alpha_i\beta_j)\mathbf{u}_i\mathbf{v}_j^T$$

clearly implies that the condition stated in (18.2) is necessary for $\mathcal{N}_T \neq \{O_{p\times q}\}$.

To prove the sufficiency of this condition, invoke Jordan decompositions $A = UJU^{-1}$ and $B = V\widetilde{J}V^{-1}$ of these matrices. Then, since

$$\begin{aligned} X - AXB = O \iff& X - UJU^{-1}XV\widetilde{J}V^{-1} = O \\ \iff& U^{-1}XV - J(U^{-1}XV)\widetilde{J} = O\,, \end{aligned}$$

the proof is now completed by setting $Y = U^{-1}XV$ and writing

(18.3) $$J = \begin{bmatrix} J_1 & O & \cdots & O \\ O & J_2 & \cdots & O \\ \vdots & & & \vdots \\ O & O & \cdots & J_k \end{bmatrix} \quad \text{and} \quad \widetilde{J} = \begin{bmatrix} \widetilde{J}_1 & O & \cdots & O \\ O & \widetilde{J}_2 & \cdots & O \\ \vdots & & & \vdots \\ O & O & \cdots & \widetilde{J}_\ell \end{bmatrix}$$

in block diagonal form in terms of their Jordan cells $J_1, \ldots, J_k$ and $\widetilde{J}_1, \ldots, \widetilde{J}_\ell$, respectively. Then, upon expressing Y in compatible block form with blocks Y_{ij}, it is readily seen that

(18.4)
$$Y - JY\widetilde{J} = O \iff Y_{ij} - J_iY_{ij}\widetilde{J}_j = O \quad \text{for} \quad i = 1, \ldots, k \text{ and } j = 1, \ldots \ell\,.$$

Thus, if the Jordan cells $J_i = C_{\alpha_i}^{(p_i)} = \alpha_i I_{p_i} + N$ and $\widetilde{J}_j = C_{\beta_j}^{(q_j)} = \beta_j I_{q_j} + \widetilde{N}$, then

$$Y_{ij} - J_iY_{ij}\widetilde{J}_j = Y_{ij} - (\alpha_i I_{p_i} + N)Y_{ij}\widetilde{J}_j = Y_{ij}(I_{q_j} - \alpha_i\widetilde{J}_j) - NY_{ij}\widetilde{J}_j\,.$$

However, if $1 - \alpha_i\beta_j \neq 0$, then the matrix $I_{q_j} - \alpha_i\widetilde{J}_j = (1-\alpha_i\beta_j)I_{q_j} - \alpha_i\widetilde{N}$ is invertible, and hence, upon setting

$$M = \widetilde{J}_j(I_{q_j} - \alpha_i\widetilde{J}_j)^{-1}$$

the equation $Y_{ij} - J_iY_{ij}\widetilde{J}_j = O$ reduces to

$$Y_{ij} = NY_{ij}M\,, \quad \text{which iterates to} \quad Y_{ij} = N^kY_{ij}M^k$$

and hence implies that $Y_{ij} = O$, since $N^k = O$ for large enough k. Therefore, $Y = O$ and $X = UYV^{-1} = O$. This completes the proof of the sufficiency of the condition $\alpha_i\beta_j \neq 1$ to insure that $\mathcal{N}_T = \{O_{p\times q}\}$, i.e., that $X = O$ is the only solution of the equation $X - AXB = O$. □

Theorem 18.2. *Let $A \in \mathbb{C}^{p\times p}$, $B \in \mathbb{C}^{q\times q}$ and $C \in \mathbb{C}^{p\times q}$ and let $\alpha_1, \ldots, \alpha_p$ and $\beta_1, \ldots, \beta_q$ denote the eigenvalues of the matrices A and B (repeated according to their algebraic multiplicity), respectively. Then the equation*

$$X - AXB = C \tag{18.5}$$

has a unique solution $X \in \mathbb{C}^{p\times q}$ if and only if $\alpha_i\beta_j \neq 1$ for every choice of i and j.

Proof. This is immediate from Theorem 18.1 and the principle of conservation of dimension: If T is the linear transformation that is defined by the rule (18.1), then

$$pq = \dim \mathcal{N}_T + \dim \mathcal{R}_T\,.$$

Therefore, T maps onto $\mathbb{C}^{p\times q}$ if and only if $\mathcal{N}_T = O$, i.e., if and only if $\alpha_i\beta_j \neq 1$ for every choice of i and j. □

Corollary 18.3. Let $A \in \mathbb{C}^{p\times p}$, $C \in \mathbb{C}^{p\times p}$ and let $\alpha_1, \ldots, \alpha_p$ denote the eigenvalues of the matrix A. Then the **Stein equation**

$$X - A^H XA = C \tag{18.6}$$

has a unique solution $X \in \mathbb{C}^{p\times p}$ if and only if $1 - \overline{\alpha_i}\alpha_j \neq 0$ for every choice of i and j.

Exercise 18.1. Verify the corollary.

Exercise 18.2. Let $A = C_\alpha^{(2)}$ and $B = C_\beta^{(2)}$ and suppose that $\alpha\beta = 1$. Show that the equation $X - AXB = C$ has no solutions if either $c_{21} \neq 0$ or $\alpha c_{11} \neq \beta c_{22}$.

Exercise 18.3. Let $A = C_\alpha^{(2)}$ and $B = C_\beta^{(2)}$ and suppose that $\alpha\beta = 1$. Show that if $c_{21} = 0$ and $\alpha c_{11} = \beta c_{22}$, then the equation $X - AXB = C$ has infinitely many solutions.

Exercise 18.4. Find the unique solution $X \in \mathbb{C}^{p\times p}$ of equation (18.6) when $A = C_0^{(p)}$, $C = \mathbf{e}_1\mathbf{u}^H + \mathbf{u}\mathbf{e}_1^H + \mathbf{e}_p\mathbf{e}_p^H$ and $\mathbf{u}^H = \begin{bmatrix} 0 & \overline{s_1} & \cdots & \overline{s_{p-1}} \end{bmatrix}$.

18.2. The Sylvester equation $AX - XB = C$

The strategy for studying the equation $AX - XB = C$ is much the same as for the equation $X - AXB = C$.

Theorem 18.4. *Let $A \in \mathbb{C}^{p\times p}$, $B \in \mathbb{C}^{q\times q}$ and let $\alpha_1, \ldots, \alpha_p$ and $\beta_1, \ldots, \beta_q$ denote the eigenvalues of the matrices A and B (repeated according to their algebraic multiplicity), respectively, and let T denote the linear transformation from $\mathbb{C}^{p\times q}$ into $\mathbb{C}^{p\times q}$ that is defined by the rule*

$$T : X \in \mathbb{C}^{p\times q} \longrightarrow AX - XB \in \mathbb{C}^{p\times q}\,. \tag{18.7}$$

Then

(18.8) $\mathcal{N}_T = \{O_{p\times q}\} \Longleftrightarrow \alpha_i - \beta_j \neq 0 \quad for \quad i = 1,\dots,p, \quad j = 1,\dots,q.$

Proof. Let $A\mathbf{u}_i = \alpha_i\mathbf{u}_i$ and $B^T\mathbf{v}_j = \beta_j\mathbf{v}_j$ for some pair of nonzero vectors $\mathbf{u}_i \in \mathbb{C}^p$ and $\mathbf{v}_j \in \mathbb{C}^q$ and let $X = \mathbf{u}_i\mathbf{v}_j^T$. Then the formula

$$TX = A\mathbf{u}_i\mathbf{v}_j^T - \mathbf{u}_i\mathbf{v}_j^T B = (\alpha_i - \beta_j)\mathbf{u}_i\mathbf{v}_j^T$$

clearly implies that the condition stated in (18.8) is necessary for $\mathcal{N}_T = \{O_{p\times q}\}$.

To prove the sufficiency of this condition, invoke Jordan decompositions $A = UJU^{-1}$ and $B = V\widetilde{J}V^{-1}$ of these matrices. Then, since

$$\begin{aligned} AX - XB = O &\iff UJU^{-1}X - XV\widetilde{J}V^{-1} = O \\ &\iff J\left(U^{-1}XV\right) - \left(U^{-1}XV\right)\widetilde{J} = O, \end{aligned}$$

the proof is now completed by setting $Y = U^{-1}XV$ and writing J and $\widetilde{J}$ in block diagonal form in terms of their Jordan cells $J_1,\dots,J_k$ and $\widetilde{J}_1,\dots,\widetilde{J}_\ell$, respectively, just as in (18.3). Then, upon expressing Y in compatible block form with blocks Y_{ij}, it is readily seen that

(18.9)
$$JY - Y\widetilde{J} = O \Longleftrightarrow J_iY_{ij} - Y_{ij}\widetilde{J}_j = O \quad \text{for} \quad i = 1,\dots,k \text{ and } j = 1,\dots\ell.$$

Thus, if $J_i = \alpha_i I_{p_i} + N$ and $\widetilde{J}_j = \beta_j I_{q_j} + \widetilde{N}$, then

$$J_iY_{ij} - Y_{ij}\widetilde{J}_j = (\alpha_i I_{p_i} + N)Y_{ij} - Y_{ij}\widetilde{J}_j = Y_{ij}(\alpha_i I_{q_j} - \widetilde{J}_j) + NY_{ij}.$$

However, if $\alpha_i - \beta_j \neq 0$, then the matrix $\alpha_i I_{q_i} - \widetilde{J}_j = (\alpha_i - \beta_j)I_{q_j} - \widetilde{N}$ is invertible, and hence, upon setting

$$M = -(\alpha_i I_{q_j} - \widetilde{J}_j)^{-1},$$

the equation reduces to

$$Y_{ij} = NY_{ij}M, \quad \text{which iterates to} \quad Y_{ij} = N^kY_{ij}M^k \quad \text{for} \quad k = 2,3,\dots$$

and hence implies that $Y_{ij} = O$, since $N^k = O$ for large enough k. This completes the proof of the sufficiency of the condition $\alpha_i - \beta_j \neq 0$ to insure that $\mathcal{N}_T = \{O_{p\times q}\}$. □

Theorem 18.5. *Let $A \in \mathbb{C}^{p\times p}$, $B \in \mathbb{C}^{q\times q}$ and $C \in \mathbb{C}^{p\times q}$ and let $\alpha_1,\dots,\alpha_p$ and $\beta_1,\dots,\beta_q$ denote the eigenvalues of the matrices A and B (repeated according to their algebraic multiplicity), respectively. Then the equation*

$$AX - XB = C$$

has a unique solution $X \in \mathbb{C}^{p\times q}$ if and only if $\alpha_i - \beta_j \neq 0$ for any choice of i and j.

Proof. This is an immediate corollary of Theorem 18.4 and the principle of conservation of dimension. The details are left to the reader. □

Exercise 18.5. Complete the proof of Theorem 18.5.

Exercise 18.6. Let $A \in \mathbb{C}^{n\times n}$. Show that the **Lyapunov equation**

$$A^H X + XA = Q \tag{18.10}$$

has a unique solution for each choice of $Q \in \mathbb{C}^{n\times n}$ if and only if $\sigma(A) \cap \sigma(-A^H) = \emptyset$.

Lemma 18.6. *If $A, Q \in \mathbb{C}^{n\times n}$, and if $\sigma(A) \subset \Pi_-$ and $-Q \succeq O$, then the Lyapunov equation* (18.10) *has a unique solution $X \in \mathbb{C}^{n\times n}$. Moreover this solution is positive semidefinite with respect to $\mathbb{C}^n$.*

Proof. Since $\sigma(A) \subset \Pi_-$, the matrix

$$Z = -\int_0^\infty e^{tA^H} Q e^{tA} dt$$

is well defined and is positive semidefinite with respect to $\mathbb{C}^n$. Moreover,

$$\begin{aligned}
A^H Z &= -\int_0^\infty A^H e^{tA^H} Q e^{tA} dt \\
&= -\int_0^\infty \left(\frac{d}{dt} e^{tA^H}\right) Q e^{tA} dt \\
&= -\left\{ e^{tA^H} Q e^{tA} \Big|_{t=0}^\infty - \int_0^\infty e^{tA^H} \frac{d}{dt}(Q e^{tA}) dt \right\} \\
&= Q + \int_0^\infty e^{tA^H} Q e^{tA} dt A \\
&= Q - ZA\,.
\end{aligned}$$

Thus, the matrix Z is a solution of the Lyapunov equation (18.10) and hence, as the assumption $\sigma(A) \subset \Pi_-$ implies that $\sigma(A) \cap \sigma(A^H) = \phi$, there is only one, by Exercise 18.6. Therefore, $X = Z$ is positive semidefinite. □

A number of refinements of this lemma may be found in [**45**].

Exercise 18.7. Let $A \in \mathbb{C}^{n\times n}$. Show that if $\sigma(A) \subset \Pi_+$, the open right half plane, then the equation $A^H X + XA = Q$ has a unique solution for every choice of $Q \in \mathbb{C}^{n\times n}$ and that this solution can be expressed as

$$X = \int_0^\infty e^{-tA^H} Q e^{-tA} dt$$

for every choice of $Q \in \mathbb{C}^{n\times n}$. [HINT: Integrate the formula

$$A^H \int_0^\infty e^{-tA^H} Q e^{-tA} dt = -\int_0^\infty \frac{d}{dt}\left(e^{-tA^H} Q\right) e^{-tA} dt$$

by parts.]

Exercise 18.8. Show that in the setting of Exercise 18.7, the solution X can also be expressed as

$$X = -\frac{1}{2\pi}\int_{-\infty}^{\infty}(i\mu I_n + A^H)^{-1}Q(i\mu I_n - A)^{-1}d\mu\,.$$

Exercise 18.9. Let $A = \mathrm{diag}\{A_{11}, A_{22}\}$ be a block diagonal matrix in $\mathbb{C}^{n\times n}$ with $\sigma(A_{11}) \subset \Pi_+$ and $\sigma(A_{22}) \subset \Pi_-$, let $Q \in \mathbb{C}^{n\times n}$ and let $Y \in \mathbb{C}^{n\times n}$ and $Z \in \mathbb{C}^{n\times n}$ be solutions of the Lyapunov equation $A^HX + XA = Q$. Show that if Y and Z are written in block form consistent with the block decomposition of A, then $Y_{11} = Z_{11}$ and $Y_{22} = Z_{22}$.

Exercise 18.10. Let A, $Q \in \mathbb{C}^{n\times n}$. Show that if $\sigma(A) \cap i\mathbb{R} = \emptyset$ and if Y and Z are both solutions of the same Lyapunov equation $A^HX + XA = Q$ such that $Y - Z \succeq O$, then $Y = Z$. [HINT: To warm up, suppose first that $A = \mathrm{diag}\{A_{11}, A_{22}\}$, where $\sigma(A_{11}) \subset \Pi_+$ and $\sigma(A_{22}) \subset \Pi_-$ and consider Exercise 18.9.]

Exercise 18.11. Let $A = \sum_{j=1}^{3} \mathbf{e}_j\mathbf{e}_{j+1}^T = C_0^{(4)}$ and let T denote the linear transformation from $\mathbb{C}^{4\times 4}$ into itself that is defined by the formula $TX = A^HX - XA$.

(a) Calculate $\dim \mathcal{N}_T$.

(b) Show that a matrix $X \in \mathbb{C}^{4\times 4}$ with entries x_{ij} is a solution of the matrix equation $A^HX - XA = \begin{bmatrix} 0 & -a & -b & -c \\ a & 0 & 0 & 0 \\ b & 0 & 0 & 0 \\ c & 0 & 0 & 0 \end{bmatrix}$ if and only if X is a Hankel matrix with $x_{11} = a$, $x_{12} = b$ and $x_{13} = c$.

Exercise 18.12. Let $A = C_0^{(4)}$ and let T denote the linear transformation from $\mathbb{C}^{4\times 4}$ into itself that is defined by the formula $TX = X - A^HXA$.

(a) Calculate $\dim \mathcal{N}_T$.

(b) Show that a matrix $X \in \mathbb{C}^{4\times 4}$ is a solution of the matrix equation $X - A^HXA = \begin{bmatrix} a & b & c & d \\ e & 0 & 0 & 0 \\ f & 0 & 0 & 0 \\ g & 0 & 0 & 0 \end{bmatrix}$ if and only if X is a Toeplitz matrix.

18.3. Special classes of solutions

Let $A \in \mathbb{C}^{n\times n}$ and let:

- $\mathcal{E}_+(A)$ = the number of zeros of $\det(\lambda I_n - A)$ in Π_+.
- $\mathcal{E}_-(A)$ = the number of zeros of $\det(\lambda I_n - A)$ in Π_-.

- $\mathcal{E}_0(A)$ = the number of zeros of det $(\lambda I_n - A)$ in $i\mathbb{R}$.

The triple $(\mathcal{E}_+(A), \mathcal{E}_-(A), \mathcal{E}_0(A))$ is called the **inertia** of A; since multiplicities are counted

$$\mathcal{E}_+(A) + \mathcal{E}_-(A) + \mathcal{E}_0(A) = n\,.$$

Theorem 18.7. *Let $A \in \mathbb{C}^{n\times n}$ and suppose that $\sigma(A) \cap i\mathbb{R} = \phi$. Then there exists a Hermitian matrix $G \in \mathbb{C}^{n\times n}$ such that*

(1) $A^H G + GA \succ O$.

(2) $\mathcal{E}_+(G) = \mathcal{E}_+(A)$, $\mathcal{E}_-(G) = \mathcal{E}_-(A)$ *and* $\mathcal{E}_0(G) = \mathcal{E}_0(A) = 0$.

Proof. Suppose first that $\mathcal{E}_+(A) = p \geq 1$ and $\mathcal{E}_-(A) = q \geq 1$. Then the assumption $\sigma(A) \cap i\mathbb{R} = \phi$ guarantees that $p + q = n$ and that A admits a Jordan decomposition UJU^{-1} of the form

$$A = U \begin{bmatrix} J_1 & O \\ O & J_2 \end{bmatrix} U^{-1}$$

with $J_1 \in \mathbb{C}^{p\times p}$, $\sigma(J_1) \subset \Pi_+$, $J_2 \in \mathbb{C}^{q\times q}$ and $\sigma(J_2) \subset \Pi_-$.

Let $P_{11} \in \mathbb{C}^{p\times p}$ be positive definite over $\mathbb{C}^p$ and $P_{22} \in \mathbb{C}^{q\times q}$ be positive definite over $\mathbb{C}^q$. Then

$$X_{11} = \int_0^\infty e^{-tJ_1^H} P_{11} e^{-tJ_1} dt$$

is a positive definite solution of the equation

$$J_1^H X_{11} + X_{11} J_1 = P_{11}$$

and

$$X_{22} = -\int_0^\infty e^{tJ_2^H} P_{22} e^{tJ_2} dt$$

is a negative definite solution of the equation

$$J_2^H X_{22} + X_{22} J_2 = P_{22}\,.$$

Let

$$X = \text{diag}\,\{X_{11}, X_{22}\} \text{ and } P = \text{diag}\,\{P_{11}, P_{22}\}\,.$$

Then

$$J^H X + XJ = P$$

and hence

$$(U^H)^{-1} J^H U^H (U^H)^{-1} X U^{-1} + (U^H)^{-1} X U^{-1} U J U^{-1} = (U^H)^{-1} P U^{-1}\,.$$

Thus, the matrix $G = (U^H)^{-1} X U^{-1}$ is a solution of the equation

$$A^H G + GA = (U^H)^{-1} P U^{-1}$$

and, with the help of Sylvester's inertia theorem (which is discussed in Chapter 20), is readily seen to fulfill all the requirements of the theorem. The cases $p = 0,\ q = n$ and $p = n,\ q = 0$ are left to the reader. □

Exercise 18.13. Complete the proof of Theorem 18.7 by verifying the cases $p = 0$ and $p = n$.

18.4. Riccati equations

In this section we shall investigate the existence and uniqueness of solutions $X \in \mathbb{C}^{n\times n}$ to the Riccati equation

$$A^H X + XA + XRX + Q = O\,, \tag{18.11}$$

in which $A, R, Q \in \mathbb{C}^{n\times n}$, $R = R^H$ and $Q = Q^H$. This class of equations has important applications. Moreover, the exploration of their properties has the added advantage of serving both as a useful review and a nonartificial application of a number of concepts considered earlier.

The study of the Riccati equation (18.11) is intimately connected with the invariant subspaces of the matrix

$$G = \begin{bmatrix} A & R \\ -Q & -A^H \end{bmatrix}, \tag{18.12}$$

which is often referred to as the **Hamiltonian matrix** in the control theory literature. The first order of business is to verify that the eigenvalues of G are symmetrically distributed with respect to the imaginary axis $i\mathbb{R}$, or, to put it more precisely:

Lemma 18.8. *The roots of the polynomial* $p(\lambda) = \det(\lambda I_{2n} - G)$ *are symmetrically distributed with respect to* $i\mathbb{R}$.

Proof. This is a simple consequence of the identity

$$SGS^{-1} = -G^H\,,$$

where

$$S = \begin{bmatrix} O & -I_n \\ I_n & O \end{bmatrix}. \tag{18.13}$$

□

Exercise 18.14. Verify the identity $SGS^{-1} = -G^H$ and the assertion of Lemma 18.8.

If $\sigma(G) \cap i\mathbb{R} = \emptyset$, then Lemma 18.8 guarantees that G admits a Jordan decomposition of the form

$$G = U \begin{bmatrix} J_1 & O \\ O & J_2 \end{bmatrix} U^{-1}\,, \tag{18.14}$$

where $J_1, J_2 \in \mathbb{C}^{n\times n}$, $\sigma(J_1) \subset \Pi_-$ and $\sigma(J_2) \subset \Pi_+$.

It turns out that the upper left-hand $n \times n$ corner X_1 of the matrix U will play a central role in the subsequent analysis; i.e., upon writing

$$U \begin{bmatrix} I_n \\ O \end{bmatrix} = \begin{bmatrix} X_1 \\ X_2 \end{bmatrix} \quad \text{and} \quad \Lambda = J_1$$

so that

$$G \begin{bmatrix} X_1 \\ X_2 \end{bmatrix} = \begin{bmatrix} X_1 \\ X_2 \end{bmatrix} \Lambda \quad \text{and} \quad \sigma(\Lambda) \subset \Pi_-, \tag{18.15}$$

the case in which X_1 is invertible will be particularly significant.

Lemma 18.9. *If $\sigma(G) \cap i\mathbb{R} = \emptyset$ and formula (18.15) is in force for some matrix $\Lambda \in \mathbb{C}^{n \times n}$ (that is not necessarily in Jordan form), then*

$$X_1^H X_2 = X_2^H X_1 \,. \tag{18.16}$$

Proof. Let

$$Z = X_2^H X_1 - X_1^H X_2 \,.$$

Then

$$\begin{aligned} Z\Lambda &= [X_1^H \quad X_2^H] S \begin{bmatrix} X_1 \\ X_2 \end{bmatrix} \Lambda \\ &= [X_1^H \quad X_2^H] SG \begin{bmatrix} X_1 \\ X_2 \end{bmatrix} \\ &= -[X_1^H \quad X_2^H] G^H S \begin{bmatrix} X_1 \\ X_2 \end{bmatrix} \\ &= -\Lambda^H [X_1^H \quad X_2^H] S \begin{bmatrix} X_1 \\ X_2 \end{bmatrix} \\ &= -\Lambda^H Z \,. \end{aligned}$$

Consequently, the matrix Z is a solution of the equation

$$Z\Lambda + \Lambda^H Z = O \,.$$

However, since $\sigma(\Lambda) \subset \Pi_-$ and hence $\sigma(\Lambda^H) \subset \Pi_-$, it follows from Theorem 18.5 that $Z = O$ is the only solution of the last equation. □

Theorem 18.10. *If $\sigma(G) \cap i\mathbb{R} = \emptyset$ and the matrix X_1 in formula (18.15) is invertible, then:*

(1) *The matrix $X = X_2 X_1^{-1}$ is a solution of the Riccati equation (18.11).*

(2) $X = X^H$.

(3) $\sigma(A + RX) \subset \Pi_-$.

Proof. If X_1 is invertible and $X = X_2X_1^{-1}$, then formula (18.15) implies that

$$G\begin{bmatrix} I_n \\ X \end{bmatrix} = \begin{bmatrix} I_n \\ X \end{bmatrix} X_1\Lambda X_1^{-1}$$

and hence, upon filling in the block entries in G and writing this out in detail, that

$$\begin{aligned} A + RX &= X_1\Lambda X_1^{-1} \\ -Q - A^H X &= X(X_1\Lambda X_1^{-1}) \,. \end{aligned}$$

Therefore,

$$-Q - A^H X = X(A + RX) \,,$$

which serves to verify (1).

Assertion (2) is immediate from Lemma 18.9, whereas (3) follows from the formula $A + RX = X_1\Lambda X_1^{-1}$ and the fact that $\sigma(\Lambda) \subset \Pi_-$. □

Theorem 18.10 established conditions that guarantee the existence of a solution X to the Riccati equation (18.11) such that $\sigma(A + RX) \subset \Pi_-$. There is a converse:

Theorem 18.11. *If $X = X^H$ is a solution of the Riccati equation* (18.11) *such that $\sigma(A + RX) \subset \Pi_-$, then $\sigma(G) \cap i\mathbb{R} = \emptyset$ and the matrix X_1 in formula* (18.15) *is invertible.*

Proof. If X is a solution of the Riccati equation with the stated properties, then

$$\begin{aligned} G\begin{bmatrix} I_n \\ X \end{bmatrix} &= \begin{bmatrix} A & R \\ -Q & -A^H \end{bmatrix}\begin{bmatrix} I_n \\ X \end{bmatrix} = \begin{bmatrix} A + RX \\ -Q - A^H X \end{bmatrix} \\ &= \begin{bmatrix} I_n \\ X \end{bmatrix}(A + RX) \,. \end{aligned}$$

Moreover, upon invoking the Jordan decomposition

$$A + RX = PJ_1P^{-1} \,,$$

we see that

$$G\begin{bmatrix} I_n \\ X \end{bmatrix} P = \begin{bmatrix} I_n \\ X \end{bmatrix} PJ_1$$

which serves to identify the columns of the matrix

$$\begin{bmatrix} I_n \\ X \end{bmatrix} P = \begin{bmatrix} P \\ XP \end{bmatrix}$$

as a full set of eigenvectors and generalized eigenvectors of the matrix G corresponding to the eigenvalues of G in Π_-. Thus, $X_1 = P$ is invertible and, in view of Lemma 18.8, $\sigma(G) \cap i\mathbb{R} = \emptyset$. □

Theorem 18.12. *The Riccati equation* (18.11) *has at most one solution* $X \in \mathbb{C}^{n\times n}$ *such that* $X = X^H$ *and* $\sigma(A + RX) \subset \Pi_-$.

Proof. Let X and Y be a pair of Hermitian solutions of the Riccati equation (18.11) such that $\sigma(A + RX) \subset \Pi_-$ and $\sigma(A + RY) \subset \Pi_-$. Then, since

$$A^H X + XA + XRX + Q = O$$

and

$$A^H Y + YA + YRY + Q = O\ ,$$

it is clear that

$$A^H(X - Y) + (X - Y)A + XRX - YRY = O\ .$$

However, as $Y = Y^H$, this last equation can also be reexpressed as

$$(A + RY)^H(X - Y) + (X - Y)(A + RX) = O\ ,$$

which exhibits $X - Y$ as the solution of an equation of the form

$$BZ + ZC = O$$

with $\sigma(B) \subset \Pi_-$ and $\sigma(C) \subset \Pi_-$. Theorem 18.5 insures that this equation has at most one solution. Thus, as $Z = O_{n\times n}$ is a solution, it is in fact the only solution. Therefore $X = Y$, as claimed. □

The preceding analysis leaves open the question as to when the conditions imposed on the Hamiltonian matrix G are satisfied. The next theorem provides an answer to this question when $R = -BB^H$ and $Q = C^H C$.

Theorem 18.13. *Let* $A \in \mathbb{C}^{n\times n}$, $B \in \mathbb{C}^{n\times k}$, $C \in \mathbb{C}^{r\times n}$ *and suppose that*

(a) $\operatorname{rank}\begin{bmatrix} A - \lambda I_n \\ C \end{bmatrix} = n$ *for every point* $\lambda \in i\mathbb{R}$ *and*

(b) $\operatorname{rank}[A - \lambda I_n \quad B] = n$ *for every point* $\lambda \in \overline{\Pi_+}$, *the closed right half plane.*

Then there exists exactly one Hermitian solution X *of the Riccati equation*

$$A^H X + XA - XBB^H X + C^H C = O \tag{18.17}$$

such that $\sigma(A - BB^H X) \subset \Pi_-$. *Moreover, this solution* X *is positive semidefinite over* $\mathbb{C}^n$, *and if* A, B *and* C *are real matrices, then* $X \in \mathbb{R}^{n\times n}$.

Proof. Let

$$G = \begin{bmatrix} A & -BB^H \\ -C^H C & -A^H \end{bmatrix}$$

and suppose first that (a) and (b) are in force and that

$$\begin{bmatrix} A & -BB^H \\ -C^H C & -A^H \end{bmatrix}\begin{bmatrix} \mathbf{x} \\ \mathbf{y} \end{bmatrix} = \lambda \begin{bmatrix} \mathbf{x} \\ \mathbf{y} \end{bmatrix}$$

for some choice of $\mathbf{x} \in \mathbb{C}^n$, $\mathbf{y} \in \mathbb{C}^n$ and $\lambda \in \mathbb{C}$. Then

$$(A - \lambda I_n)\mathbf{x} = BB^H\mathbf{y}$$

and

$$(A^H + \lambda I_n)\mathbf{y} = -C^H C\mathbf{x} \ .$$

Therefore,

$$\langle (A - \lambda I_n)\mathbf{x}, \mathbf{y}\rangle = \langle BB^H\mathbf{y}, \mathbf{y}\rangle = \|B^H\mathbf{y}\|_2^2$$

and

$$\langle (A + \overline{\lambda} I_n)\mathbf{x}, \mathbf{y}\rangle = \langle \mathbf{x}, (A^H + \lambda I_n)\mathbf{y}\rangle = -\langle \mathbf{x}, C^H C\mathbf{x}\rangle = -\|C\mathbf{x}\|_2^2 \ .$$

Thus,

$$\begin{aligned} -(\lambda + \overline{\lambda})\langle \mathbf{x}, \mathbf{y}\rangle &= \langle (A - \lambda I_n)\mathbf{x}, \mathbf{y}\rangle - \langle (A + \overline{\lambda} I_n)\mathbf{x}, \mathbf{y}\rangle \\ &= \|B^H\mathbf{y}\|_2^2 + \|C\mathbf{x}\|_2^2 \end{aligned}$$

and hence

$$\lambda + \overline{\lambda} = 0 \Longrightarrow B^H\mathbf{y} = \mathbf{0} \text{ and } C\mathbf{x} = \mathbf{0} \ ,$$

which in turn implies that

$$\begin{bmatrix} A - \lambda I_n \\ C \end{bmatrix} \mathbf{x} = \mathbf{0} \quad \text{and} \quad \mathbf{y}^H [A + \overline{\lambda} I_n \quad B] = \mathbf{0}$$

when $\lambda \in i\mathbb{R}$. However, in view of (a) and (b), this is viable only if $\mathbf{x} = \mathbf{0}$ and $\mathbf{y} = \mathbf{0}$. Consequently, $\sigma(G) \cap i\mathbb{R} = \emptyset$.

The next step is to show that if (a) and (b) are in force and if

$$\begin{bmatrix} A & -BB^H \\ -C^H C & -A^H \end{bmatrix} \begin{bmatrix} X_1 \\ X_2 \end{bmatrix} = \begin{bmatrix} X_1 \\ X_2 \end{bmatrix} \Lambda \quad \text{and} \quad \text{rank} \begin{bmatrix} X_1 \\ X_2 \end{bmatrix} = n \, ,$$

where $X_1, X_2, \Lambda \in \mathbb{C}^{n \times n}$ and $\sigma(\Lambda) \subset \Pi_-$, then X_1 is invertible.

Suppose that $\mathbf{u} \in \mathcal{N}_{X_1}$. Then

$$-BB^H X_2\mathbf{u} = X_1 \Lambda \mathbf{u} \ ,$$

and hence, as $X_2^H X_1 = X_1^H X_2$ by Lemma 18.9,

$$\begin{aligned} -\|B^H X_2 \mathbf{u}\|^2 &= -\mathbf{u}^H X_2^H BB^H X_2\mathbf{u} = \mathbf{u}^H X_2^H X_1 \Lambda \mathbf{u} \\ &= \mathbf{u}^H X_1^H X_2 \Lambda \mathbf{u} = \mathbf{0} \ . \end{aligned}$$

Thus, $B^H X_2 \mathbf{u} = \mathbf{0}$ and

$$X_1 \Lambda \mathbf{u} = -BB^H X_2 \mathbf{u} = \mathbf{0} \, ,$$

which means that $\mathcal{N}_{X_1}$ is invariant under Λ and hence that either $\mathcal{N}_{X_1} = \{\mathbf{0}\}$ or that $\Lambda\mathbf{v} = \lambda\mathbf{v}$ for some point $\lambda \in \Pi_-$ and some nonzero vector $\mathbf{v} \in \mathcal{N}_{X_1}$. In the latter case,

$$-BB^H X_2 \mathbf{v} = X_1 \Lambda \mathbf{v} = \lambda X_1 \mathbf{v} = \mathbf{0}$$

and

$$-A^H X_2\mathbf{v} = X_2\Lambda\mathbf{v} = \lambda X_2\mathbf{v} \ ,$$

which is the same as to say

$$\mathbf{v}^H X_2^H [A + \overline{\lambda} I_n \quad B] = \mathbf{0}^H$$

for some point $\lambda \in \Pi_-$. Therefore, since $-\overline{\lambda} \in \Pi_+$, assumption (b) implies that $X_2\mathbf{v} = \mathbf{0}$. Therefore,

$$\begin{bmatrix} X_1 \\ X_2 \end{bmatrix} \mathbf{v} = \mathbf{0} \Longrightarrow \mathbf{v} = \mathbf{0} \Longrightarrow \mathcal{N}_{X_1} = \{\mathbf{0}\} \Longrightarrow X_1 \quad \text{is invertible} .$$

Thus, in view of Theorems 18.10 and 18.12, there exists exactly one Hermitian solution X of the Riccati equation (18.17) such that $\sigma(A - BB^H X) \subset \Pi_-$. If the matrices A, B and C are real, then the matrix $\overline{X}$ is also a Hermitian solution of the Riccati equation (18.17) such that $\sigma(A - BB^H\overline{X}) \subset \Pi_-$. Therefore, in this case, $X \in \mathbb{R}^{n\times n}$.

It remains only to verify that this solution X is positive semidefinite with respect to $\mathbb{C}^n$. To this end, it is convenient to reexpress the Riccati equation

$$A^H X + XA - XBB^H X + C^H C = O$$

as

$$(A - BB^H X)^H X + X(A - BB^H X) = -C^H C - XBB^H X \ ,$$

which is of the form

$$A_1^H X + XA_1 = Q \ ,$$

where

$$\sigma(A_1) \subset \Pi_- \quad \text{and} \quad -Q \succeq O \, .$$

The desired result then follows by invoking Lemma 18.6. □

Exercise 18.15. Let $A \in \mathbb{C}^{n\times n}$, $B \in \mathbb{C}^{n\times k}$. Show that if $\sigma(A) \cap i\mathbb{R} = \emptyset$ and $\operatorname{rank}\left[A - \lambda I_n \quad B\right] = n$ for every point $\lambda \in \overline{\Pi_+}$, then there exists exactly one Hermitian solution X of the Riccati equation $A^H X + XA - XBB^H X = O$ such that $\sigma(A - BB^H X) \subset \Pi_-$.

For future applications, it will be convenient to have another variant of Theorem 18.13.

Theorem 18.14. *Let $A \in \mathbb{C}^{n\times n}$, $B \in \mathbb{C}^{n\times k}$, $Q \in \mathbb{C}^{n\times n}$, $R \in \mathbb{C}^{k\times k}$; and suppose that $Q \succeq O$, $R \succ O$,*

(a) $\operatorname{rank} \begin{bmatrix} A - \lambda I_n \\ Q \end{bmatrix} = n$ *for every point* $\lambda \in i\mathbb{R}$

and

(b) $\operatorname{rank}\,[A - \lambda I_n \quad B] = n$ *for every point* $\lambda \in \overline{\Pi_+}$.

Then there exists exactly one Hermitian solution X of the Riccati equation

$$A^H X + XA - XBR^{-1}B^H X + Q = O$$

such that $\sigma(A - BR^{-1}B^H X) \subset \Pi_-$. Moreover, this solution X is positive semidefinite over $\mathbb{C}^n$, and if A, B and C are real matrices, then $X \in \mathbb{R}^{n\times n}$.

Proof. Since $Q \succeq O$ there exists a matrix $C \in \mathbb{C}^{r\times n}$ such that $C^H C = Q$ and rank $C =$ rank $Q = r$. Thus, upon setting $B_1 = BR^{-1/2}$, we see that

$$\begin{bmatrix} A & -BR^{-1}B^H \\ -Q & -A^H \end{bmatrix} = \begin{bmatrix} A & -B_1 B_1^H \\ -C^H C & -A^H \end{bmatrix}$$

is of the form considered in Theorem 18.13. Moreover, since

$$\begin{bmatrix} A - \lambda I_n \\ C \end{bmatrix} \mathbf{u} = \mathbf{0} \Longleftrightarrow \begin{bmatrix} A - \lambda I_n \\ Q \end{bmatrix} \mathbf{u} = \mathbf{0}\,,$$

condition (a) implies that

$$\text{rank} \begin{bmatrix} A - \lambda I_n \\ C \end{bmatrix} = n \quad \text{for every point } \lambda \in i\mathbb{R}.$$

Furthermore, as

$$\text{rank}\ [A - \lambda I_n \quad B] = \text{rank}\ [A - \lambda I_n \quad BR^{-1/2}],$$

assumption (b) guarantees that

$$\text{rank}\ [A - \lambda I_n \quad B_1] = n \text{ for every point } \lambda \in \overline{\Pi_+}.$$

The asserted conclusion is now an immediate consequence of Theorem 18.13. □

Exercise 18.16. Show that if $N, Y, M \in \mathbb{C}^{n\times n}$, then the $n \times n$ matrix valued function

$$X(t) = e^{tN} Y e^{tM}$$

is a solution of the differential equation

$$X'(t) = NX(t) + X(t)M$$

that meets the initial condition $X(0) = Y$.

18.5. Two lemmas

The two lemmas in this section are prepared for use in the next section.

Lemma 18.15. *Let $A, Q \in \mathbb{C}^{n\times n}$, $B, L \in \mathbb{C}^{n\times k}$, $R \in \mathbb{C}^{k\times k}$,*

$$E = \begin{bmatrix} Q & L \\ L^H & R \end{bmatrix}$$

and suppose that $E \succeq O$, $R \succ O$ and that

$$\text{rank}\, E = \text{rank}\, Q + \text{rank}\, R. \tag{18.18}$$

Then the formulas

$$\text{rank}\begin{bmatrix} \widetilde{A}-\lambda I_n \\ \widetilde{Q} \end{bmatrix} = \text{rank}\begin{bmatrix} A-\lambda I_n \\ Q \end{bmatrix} \tag{18.19}$$

and

$$\text{rank}\,[\widetilde{A}-\lambda I_n \quad B] = \text{rank}\,[A-\lambda I_n \quad B] \tag{18.20}$$

are valid for the matrices

$$\widetilde{A} = A - BR^{-1}L^H \quad \textit{and} \quad \widetilde{Q} = Q - LR^{-1}L^H \tag{18.21}$$

and every point $\lambda \in \mathbb{C}$.

Proof. The formula

$$\begin{bmatrix} Q & L \\ L^H & R \end{bmatrix} = \begin{bmatrix} I_n & LR^{-1} \\ O & I_k \end{bmatrix}\begin{bmatrix} \widetilde{Q} & O \\ O & R \end{bmatrix}\begin{bmatrix} I_n & O \\ R^{-1}L^H & I_n \end{bmatrix}$$

implies that

$$\text{rank}\,E = \text{rank}\,\widetilde{Q} + \text{rank}\,R \quad \text{and} \quad \widetilde{Q} \succeq O\,.$$

Thus, in view of assumption (18.18),

$$\text{rank}\,Q = \text{rank}\,\widetilde{Q}$$

and, since $Q = \widetilde{Q} + LR^{-1}L^H$ is the sum of two positive semidefinite matrices,

$$\mathcal{N}_Q = \mathcal{N}_{\widetilde{Q}} \cap \mathcal{N}_{L^H} \subseteq \mathcal{N}_{\widetilde{Q}}\,.$$

However, since

$$\text{rank}\,Q = \text{rank}\,\widetilde{Q} \Longrightarrow \dim \mathcal{N}_Q = \dim \mathcal{N}_{\widetilde{Q}}\,,$$

the last inclusion is in fact an equality:

$$\mathcal{N}_Q = \mathcal{N}_{\widetilde{Q}} \quad \text{and} \quad \mathcal{N}_{\widetilde{Q}} \subseteq \mathcal{N}_{L^H}$$

and hence,

$$\begin{bmatrix} \widetilde{A}-\lambda I_n \\ \widetilde{Q} \end{bmatrix}\mathbf{u} = \mathbf{0} \Longleftrightarrow \begin{bmatrix} A-\lambda I_n \\ Q \end{bmatrix}\mathbf{u} = \mathbf{0}.$$

The conclusion (18.19) now follows easily from the principle of conservation of dimension.

The second conclusion (18.20) is immediate from the identity

$$[\widetilde{A}-\lambda I_n \quad B] = [A-\lambda I_n \quad B]\begin{bmatrix} I_n & O \\ -R^{-1}L^H & I_k \end{bmatrix}.$$

□

Lemma 18.16. *Assume that the matrices A, $\widetilde{A}$, Q, $\widetilde{Q}$, B, L, R and E are as in Lemma 18.15 and that*

$$\text{rank } E = \text{rank } Q + \text{rank} R, \tag{18.22}$$

$$\text{rank} \begin{bmatrix} A - \lambda I_n \\ Q \end{bmatrix} = n \textit{ for every point } \lambda \in i\mathbb{R} \tag{18.23}$$

and

$$\text{rank } [A - \lambda I_n \quad B] = n \textit{ for every point } \lambda \in \overline{\Pi_+}\,. \tag{18.24}$$

Then there exists exactly one Hermitian solution $X \in \mathbb{C}^{n\times n}$ of the Riccati equation

$$\widetilde{A}^H X + X\widetilde{A} - XBR^{-1}B^H X + \widetilde{Q} = O \tag{18.25}$$

such that $\sigma(\widetilde{A} - BR^{-1}B^H X) \subset \Pi_-$. Moreover, this solution X is positive semidefinite over $\mathbb{C}^n$, and if the matrices A, B, Q, L and R are real, then $X \in \mathbb{R}^{n\times n}$.

Proof. Under the given assumptions, Lemma 18.15 guarantees that

$$\text{rank} \begin{bmatrix} \widetilde{A} - \lambda I_n \\ \widetilde{Q} \end{bmatrix} = n \text{ for every point } \lambda \in i\mathbb{R}$$

and

$$\text{rank } [\widetilde{A} - \lambda I_n \quad B] = n \text{ for every point } \lambda \in \overline{\Pi_+}.$$

Therefore, Theorem 18.14 is applicable with $\widetilde{A}$ in place of A and $\widetilde{Q}$ in place of Q. □

18.6. An LQR problem

Let $A \in \mathbb{R}^{n\times n}$ and $B \in \mathbb{R}^{n\times k}$ and let

$$\mathbf{x}(t) = e^{tA}\mathbf{x}(0) + \int_0^t e^{(t-s)A}B\mathbf{u}(s)ds\,, \quad 0 \le t < \infty\,,$$

be the solution of the first-order vector system of equations

$$\mathbf{x}'(t) = A\mathbf{x}(t) + B\mathbf{u}(t),\ t \ge 0\,,$$

in which the vector $\mathbf{x}(0) \in \mathbb{R}^n$ and the vector valued function $\mathbf{u}(t) \in \mathbb{R}^k$, $t \ge 0$, are specified. The LQR (linear quadratic regulator) problem in control engineering is to choose $\mathbf{u}$ to minimize the value of the integral

$$Z(t) = \int_0^t [\mathbf{x}^T(s)\ \mathbf{u}^T(s)] \begin{bmatrix} Q & L \\ L^T & R \end{bmatrix} \begin{bmatrix} \mathbf{x}(s) \\ \mathbf{u}(s) \end{bmatrix} ds \tag{18.26}$$

when

$$\begin{bmatrix} Q & L \\ L^T & R \end{bmatrix} \succeq O,$$

$Q = Q^T \in \mathbb{R}^{n\times n}$, $L \in \mathbb{R}^{n\times k}$, $R = R^T \in \mathbb{R}^{k\times k}$ and R is assumed to be invertible.

The first step in the analysis of this problem is to reexpress it in simpler form by invoking the Schur complement formula:

$$\begin{bmatrix} Q & L \\ L^T & R \end{bmatrix} = \begin{bmatrix} I_n & LR^{-1} \\ O & I_k \end{bmatrix} \begin{bmatrix} Q - LR^{-1}L^T & O \\ O & R \end{bmatrix} \begin{bmatrix} I_n & O \\ R^{-1}L^T & I_k \end{bmatrix}.$$

Then, upon setting

$$\widetilde{A} = A - BR^{-1}L^T, \quad \widetilde{Q} = Q - LR^{-1}L^T$$

and

$$\mathbf{v}(s) = R^{-1}L^T\mathbf{x}(s) + \mathbf{u}(s),$$

the integral (18.26) can be reexpressed more conveniently as

$$(18.27)\qquad Z(t) = \int_0^t \begin{bmatrix} \mathbf{x}^T(s) & \mathbf{v}^T(s) \end{bmatrix} \begin{bmatrix} \widetilde{Q} & O \\ O & R \end{bmatrix} \begin{bmatrix} \mathbf{x}(s) \\ \mathbf{v}(s) \end{bmatrix} ds,$$

where the vectors $\mathbf{x}(s)$ and $\mathbf{v}(s)$ are linked by the equation

$$(18.28)\qquad \mathbf{x}'(s) = \widetilde{A}\mathbf{x}(s) + B\mathbf{v}(s),$$

i.e.,

$$(18.29)\qquad \mathbf{x}(t) = e^{t\widetilde{A}}\mathbf{x}(0) + \int_0^t e^{(t-s)\widetilde{A}}B\mathbf{v}(s)ds.$$

Lemma 18.17. *Let X be the unique Hermitian solution of the Riccati equation* (18.25) *such that $\sigma(\widetilde{A} - BR^{-1}B^TX) \subset \Pi_-$. Then $X \in \mathbb{R}^{n\times n}$ and*

(18.30)

$$Z(t) = \mathbf{x}(0)^TX\mathbf{x}(0) - \mathbf{x}(t)^TX\mathbf{x}(t) + \int_0^t \|R^{-1/2}(B^TX\mathbf{x}(s) + R\mathbf{v}(s))\|_2^2\, ds.$$

Proof. Let $\varphi(t) = \frac{d}{ds}\{\mathbf{x}(s)^TX\mathbf{x}(s)\}$. Then

$$\begin{aligned}
\varphi'(t) &= \mathbf{x}'(s)^TX\mathbf{x}(s) + \mathbf{x}(s)^TX\mathbf{x}'(s) \\
&= (\widetilde{A}\mathbf{x}(s) + B\mathbf{v}(s))^TX\mathbf{x}(s) + \mathbf{x}(s)^TX(\widetilde{A}\mathbf{x}(s) + B\mathbf{v}(s)) \\
&= \mathbf{x}(s)^T(\widetilde{A}^TX + X\widetilde{A})\mathbf{x}(s) + \mathbf{v}(s)^TB^TX\mathbf{x}(s) + \mathbf{x}(s)^TXB\mathbf{v}(s) \\
&= \mathbf{x}(s)^T(XBR^{-1}B^TX - \widetilde{Q})\mathbf{x}(s) + \mathbf{v}(s)^TB^TX\mathbf{x}(s) + \mathbf{x}(s)^TXB\mathbf{v}(s) \\
&= (\mathbf{x}(s)^TXB + \mathbf{v}(s)^TR)R^{-1}(B^TX\mathbf{x}(s) + R\mathbf{v}(s)) \\
&\quad -\mathbf{x}(s)^T\widetilde{Q}\mathbf{x}(s) - \mathbf{v}(s)^TR\mathbf{v}(s).
\end{aligned}$$

Therefore,

$$
\begin{aligned}
Z(t) &= \int_0^t \{\mathbf{x}(s)^T \widetilde{Q} \mathbf{x}(s) + \mathbf{v}(s)^T R \mathbf{v}(s)\} ds \\
&= -\int_0^t \frac{d}{ds} \{\mathbf{x}(s)^T X \mathbf{x}(s)\} ds \\
&\quad + \int_0^t (B^T X \mathbf{x}(s) + R\mathbf{v}(s))^T R^{-1} (B^T X \mathbf{x}(s) + R\mathbf{v}(s)) ds,
\end{aligned}
$$

which is the same as the advertised formula. □

Theorem 18.18. *If the hypotheses of Lemma 18.16 are met and if* $X \in \mathbb{R}^{n \times n}$ *is the unique Hermitian solution of the Riccati equation* (18.25) *with* $\sigma(\widetilde{A} - BR^{-1}B^T X) \subset \Pi_-$, *then:*

(1) $Z(t) \geqslant \mathbf{x}(0)^T X \mathbf{x}(0) - \mathbf{x}(t)^T X \mathbf{x}(t)$ *with equality if* $\mathbf{v}(s) = -R^{-1} B^T X \mathbf{x}(s)$ *for* $0 \le s \le t$.

(2) *If* $\mathbf{v}(s) = -R^{-1} B^T X \mathbf{x}(s)$ *for* $0 \le s < \infty$, *then* $Z(\infty) = \mathbf{x}(0)^T X \mathbf{x}(0)$.

Proof. The first assertion is immediate from formula (18.30). Moreover, if $\mathbf{v}(s)$ is chosen as specified in assertion (2), then $\mathbf{x}(t)$ is a solution of the vector differential equation

$$\mathbf{x}'(t) = (\widetilde{A} - BR^{-1}B^T X)\mathbf{x}(t),$$

and hence, as $\sigma(\widetilde{A} - BR^{-1}B^T X) \subset \Pi_-$, it is readily checked that $\mathbf{x}(t) \to 0$ as $t \uparrow \infty$. □

18.7. Bibliographical notes

The discussion of Riccati equations and the LQR problem was partially adapted from the monograph [**74**], which is an excellent source of supplementary information on both of these topics. The monograph [**44**] is recommended for more advanced studies. Exercise 18.4 is adapted from [**1**].

Chapter 19

Realization theory

... was so upset when her mother married one that she took to mathematics and Hebrew directly ... though she was the prettiest girl for miles around...

Patrick O'Brian [**53**], p. 122

A function $f(\lambda)$ is said to be **rational** if it can be expressed as the ratio of two polynomials:

$$f(\lambda) = \frac{\alpha_0 + \alpha_1\lambda + \cdots + \alpha_k\lambda^k}{\beta_0 + \beta_1\lambda + \cdots + \beta_n\lambda^n}. \tag{19.1}$$

A rational function $f(\lambda)$ is said to be **proper** if $f(\lambda)$ tends to a finite limit as $\lambda \to \infty$; it is said to be **strictly proper** if $f(\lambda) \longrightarrow 0$ as $\lambda \to \infty$. If $\alpha_k \neq 0$, $\beta_n \neq 0$ and the numerator and denominator have no common factors, then the degree of $f(\lambda)$ is defined by the rule

$$\deg f(\lambda) = \max\{k, n\}. \tag{19.2}$$

In this case $f(\lambda)$ is proper if $n \geq k$. A $p \times q$ mvf (matrix valued function)

$$F(\lambda) = \begin{bmatrix} f_{11}(\lambda) & \cdots & f_{1q}(\lambda) \\ \vdots & & \vdots \\ f_{p1}(\lambda) & \cdots & f_{pq}(\lambda) \end{bmatrix}$$

is said to be **rational** if each of its entries is rational. It is said to be **proper** if each of its entries is proper or, equivalently, if $F(\lambda)$ tends to a finite limit as $\lambda \to \infty$ and **strictly proper** if $F(\lambda) \longrightarrow O$ as $\lambda \to \infty$.

Theorem 19.1. *Let $F(\lambda)$ be a $p \times q$ rational mvf that is proper. Then there exists an integer $n > 0$ and matrices $D \in \mathbb{C}^{p\times q}$, $C \in \mathbb{C}^{p\times n}$, $A \in \mathbb{C}^{n\times n}$ and*

$B \in \mathbb{C}^{n\times q}$ *such that*

$$F(\lambda) = D + C(\lambda I_n - A)^{-1}B \,. \tag{19.3}$$

Proof. Let us suppose first that

$$F(\lambda) = \frac{X_k}{(\lambda-\omega)^k} + \cdots + \frac{X_1}{\lambda-\omega} + X_0 \,, \tag{19.4}$$

where the X_j are constant $p \times q$ matrices. Let A denote the $kp \times kp$ block Jordan cell of the form

$$A = \begin{bmatrix} \omega I_p & I_p & O & O & \cdots & O \\ O & \omega I_p & I_p & O & \cdots & O \\ \vdots & \vdots & \vdots & & & I_p \\ O & O & O & O & \cdots & \omega I_p \end{bmatrix} .$$

Let $n = kp$ and let

$$N = A - \omega I_n \,.$$

Then

$$\begin{aligned} (\lambda I_n - A)^{-1} &= ((\lambda-\omega)I_n - N)^{-1} \\ &= (\lambda-\omega)^{-1}\left\{ I_n + \frac{N}{\lambda-\omega} + \cdots + \frac{N^{k-1}}{(\lambda-\omega)^{k-1}} \right\} , \end{aligned}$$

since $N^k = O$. Therefore, the top block row of $(\lambda I_n - A)^{-1}$ is equal to

$$[I_p \quad O \quad \cdots \quad O](\lambda I_n - A)^{-1} = \begin{bmatrix} \frac{I_p}{\lambda-\omega} & \frac{I_p}{(\lambda-\omega)^2} & \cdots & \frac{I_p}{(\lambda-\omega)^k} \end{bmatrix} .$$

Thus, upon setting

$$D = X_0 \,, \; C = [I_p \quad O \quad \cdots \quad O] \quad \text{and} \quad B = \begin{bmatrix} X_1 \\ \vdots \\ X_k \end{bmatrix} ,$$

it is readily seen that the mvf $F(\lambda)$ specified in (19.4) can be expressed in the form (19.3) for the indicated choices of A, B, C and D.

A proper rational $p \times q$ mvf $F(\lambda)$ will have poles at a finite number of distinct points $\omega_1, \ldots, \omega_\ell \in \mathbb{C}$ and there will exist mvf's

$$F_1(\lambda) = \frac{X_{1k_1}}{(\lambda-\omega_1)^{k_1}} + \cdots + \frac{X_{11}}{\lambda-\omega_1}, \ldots, F_\ell(\lambda) = \frac{X_{\ell k_\ell}}{(\lambda-\omega_\ell)^{k_\ell}} + \cdots + \frac{X_{\ell 1}}{\lambda-\omega_\ell}$$

with matrix coefficients $X_{ij} \in \mathbb{C}^{p\times q}$ such that

$$F(\lambda) - \sum_{j=1}^{\ell} F_j(\lambda)$$

is holomorphic and bounded in the whole complex plane. Therefore, by Liouville's theorem (applied to each entry of $F(\lambda)-\sum_{j=1}^{\ell} F_j(\lambda)$ separately),

$$F(\lambda) - \sum_{j=1}^{\ell} F_j(\lambda) = D,$$

a constant $p \times q$ matrix. Moreover,

$$D = \lim_{\lambda\to\infty} F(\lambda),$$

and, by the preceding analysis, there exist matrices $C_j \in \mathbb{C}^{p\times n_j}$, $A_j \in \mathbb{C}^{n_j\times n_j}$, $B_j \in \mathbb{C}^{n_j\times q}$ and $n_j = pk_j$ for $j = 1, \ldots, \ell$, such that

$$F_j(\lambda) = C_j(\lambda I - A_j)^{-1}B_j \quad \text{and} \quad n = n_1 + \cdots + n_\ell.$$

Formula (19.3) for $F(\lambda)$ emerges upon setting

$$C = [C_1 \ \cdots \ C_\ell], \ A = \operatorname{diag}\{A_1, \ldots, A_\ell\} \quad \text{and} \quad B = \begin{bmatrix} B_1 \\ \vdots \\ B_\ell \end{bmatrix}.$$

□

Formula (19.3) is called a **realization** of $F(\lambda)$. It is far from unique.

Exercise 19.1. Check that the mvf $F(\lambda)$ defined by formula (19.3) does not change if C is replaced by CS, A by $S^{-1}AS$ and B by $S^{-1}B$ for some invertible matrix $S \in \mathbb{C}^{n\times n}$.

Exercise 19.2. Show that if $F_1(\lambda) = D_1 + C_1(\lambda I_{n_1} - A_1)^{-1}B_1$ is a $p \times q$ mvf and $F_2(\lambda) = D_2 + C_2(\lambda I_{n_2} - A_2)^{-1}B_2$ is a $q \times r$ mvf, then

$$F_1(\lambda)\,F_2(\lambda) = D_3 + C_3(\lambda I_n - A_3)^{-1}B_3,$$

where

$$D_3 = D_1D_2, \quad C_3 = [C_1 \quad D_1C_2], \quad A_3 = \begin{bmatrix} A_1 & -B_1C_2 \\ O & A_2 \end{bmatrix}, \quad B_3 = \begin{bmatrix} B_1D_2 \\ B_2 \end{bmatrix}$$

and $n = n_1 + n_2$.

Exercise 19.3. Show that if $D \in \mathbb{C}^{p\times p}$ is invertible, then

(19.5)

$$\left\{D + C(\lambda I_n - A)^{-1}B\right\}^{-1} = D^{-1} - D^{-1}C(\lambda I_n - [A - BD^{-1}C])^{-1}BD^{-1}.$$

Let $C \in \mathbb{C}^{p\times n}$, $A \in \mathbb{C}^{n\times n}$ and $B \in \mathbb{C}^{n\times q}$. Then the pair (A, B) is said to be **controllable** if the **controllability matrix**

$$\mathfrak{C} = [B \quad AB \quad \cdots \quad A^{n-1}B]$$

is right invertible, i.e., if

$$\operatorname{rank} \mathfrak{C} = n.$$

The pair (C, A) is said to be **observable** if the **observability matrix**

$$\mathfrak{O} = \begin{bmatrix} C \\ CA \\ \vdots \\ CA^{n-1} \end{bmatrix}$$

is left invertible, i.e., if its null space

$$\mathcal{N}_{\mathfrak{O}} = \{\mathbf{0}\} .$$

Exercise 19.4. Show that (C, A) is an observable pair if and only if the pair (A^H, C^H) is controllable.

Lemma 19.2. *The following are equivalent:*

(1) (A, B) *is controllable.*

(2) $\operatorname{rank}[A - \lambda I_n \quad B] = n$ *for every point* $\lambda \in \mathbb{C}$.

(3) *The rows of the mvf* $(\lambda I_n - A)^{-1}B$ *are linearly independent in the sense that if* $\mathbf{u}^H(\lambda I_n - A)^{-1}B = \mathbf{0}^H$ *for* $\mathbf{u} \in \mathbb{C}^n$ *and all* λ *in some open nonempty subset* Ω *of* $\mathbb{C}$ *that does not contain any eigenvalues of* A, *then* $\mathbf{u} = \mathbf{0}$.

(4) $\int_0^t e^{sA}BB^He^{sA^H}ds \succ O$ *for every* $t > 0$.

(5) *For each vector* $\mathbf{v} \in \mathbb{C}^n$ *and each* $t > 0$, *there exists an* $m \times 1$ *vector valued function* $\mathbf{u}(s)$ *on the interval* $0 \leq s \leq t$ *such that*

$$\int_0^t e^{(t-s)A}B\mathbf{u}(s)ds = \mathbf{v} .$$

(6) *The matrix* $\mathfrak{C}\mathfrak{C}^H$ *is invertible.*

Proof. (1)$\Longrightarrow$(2). Let $\mathbf{u} \in \mathbb{C}^n$ be orthogonal to the columns of the mvf $[A - \lambda I_n \quad B]$ for some point $\lambda \in \mathbb{C}$. Then

$$\mathbf{u}^H A = \lambda \mathbf{u}^H \text{ and } \mathbf{u}^H B = \mathbf{0}^H .$$

Therefore,

$$\mathbf{u}^H A^k B = \lambda^k \mathbf{u}^H B = \mathbf{0}^H$$

for $k = 0, \ldots, n-1$, i.e., $\mathbf{u}^H\mathfrak{C} = \mathbf{0}^H$. Thus, $\mathbf{u} = \mathbf{0}$.

(2)$\Longrightarrow$(1). Suppose that $\mathbf{u}^H A^k B = \mathbf{0}^H$ for $k = 0, \ldots, n-1$ for some nonzero vector $\mathbf{u} \in \mathbb{C}^n$. Then $\mathcal{N}_{\mathfrak{C}^H}$ is nonempty and hence, since $\mathcal{N}_{\mathfrak{C}^H}$ is invariant under A^H, A^H has an eigenvector in this nullspace. Thus, there exists a nonzero vector $\mathbf{v} \in \mathbb{C}^n$ and a point $\alpha \in \mathbb{C}$ such that

$$A^H\mathbf{v} = \alpha\mathbf{v} \text{ and } \mathfrak{C}^H\mathbf{v} = \mathbf{0} .$$

But this implies that

$$\mathbf{v}^H[A - \alpha I \quad B] = \mathbf{0}^H ,$$

which is incompatible with (2). Thus, $\mathcal{N}_{\mathfrak{C}^H} = \{\mathbf{0}\}$; i.e., (A, B) is controllable.

(1)$\Longleftrightarrow$(3). This follows from the observation that

$$\begin{aligned}
\mathbf{u}^H(\lambda I_n - A)^{-1}B &= \mathbf{0}^H \quad \text{for} \quad \lambda \in \Omega \\
&\Longleftrightarrow \mathbf{u}^H(\lambda I_n - A)^{-1}B = \mathbf{0}^H \quad \text{for} \quad \lambda \in \mathbb{C} \setminus \sigma(A) \\
&\Longleftrightarrow \mathbf{u}^H \sum_{j=0}^{\infty} \frac{A^j}{\lambda^j} B = \mathbf{0}^H \quad \text{for} \quad |\lambda| > \|A\| \\
&\Longleftrightarrow \mathbf{u}^H A^k B = \mathbf{0}^H \quad \text{for} \quad k = 0, \ldots, n-1 \\
&\Longleftrightarrow \mathbf{u}^H \mathfrak{C} = \mathbf{0}^H .
\end{aligned}$$

The verification of (4), (5) and (6) is left to the reader. □

Exercise 19.5. Show that (A, B) is controllable if and only if condition (4) in Lemma 19.2 is met.

Exercise 19.6. Show that (A, B) is controllable if and only if condition (5) in Lemma 19.2 is met.

Exercise 19.7. Show that (A, B) is controllable if and only if $\mathfrak{C}\mathfrak{C}^H$ is invertible.

Lemma 19.3. *The following are equivalent:*

(1) (C, A) *is observable.*

(2) $\operatorname{rank} \begin{bmatrix} A - \lambda I_n \\ C \end{bmatrix} = n$ *for every point* $\lambda \in \mathbb{C}$.

(3) *The columns of the mvf* $C(\lambda I_n - A)^{-1}$ *are linearly independent in the sense that if* $\mathbf{u} \in \mathbb{C}^n$ *and* $C(\lambda I_n - A)^{-1}\mathbf{u} = \mathbf{0}$ *for all points* λ *in some open nonempty subset* Ω *of* $\mathbb{C}$ *that does not contain any eigenvalues of* A, *then* $\mathbf{u} = \mathbf{0}$.

(4) $\int_0^t Ce^{sA}e^{sA^H}C^H ds \succ O$ *for every* $t > 0$.

(5) *For each vector* $\mathbf{v} \in \mathbb{C}^n$ *and each* $t > 0$, *there exists a* $p \times 1$ *vector valued function* $\mathbf{u}(s)$ *on the interval* $0 \le s \le t$ *such that*

$$\int_0^t e^{(t-s)A^H} C^H \mathbf{u}(s) ds = \mathbf{v} .$$

(6) *The matrix* $\mathfrak{O}^H \mathfrak{O}$ *is invertible.*

Proof. (1)$\Longrightarrow$(2). Let

$$\begin{bmatrix} A-\lambda I_n \\ C \end{bmatrix} \mathbf{u} = \mathbf{0}$$

for some vector $\mathbf{u} \in \mathbb{C}^n$ and some point $\lambda \in \mathbb{C}$. Then $A\mathbf{u} = \lambda\mathbf{u}$ and $C\mathbf{u} = 0$. Therefore,

$$CA^k\mathbf{u} = \lambda^k C\mathbf{u} = \mathbf{0} \quad \text{for} \quad k = 1, 2, \ldots$$

also and hence $\mathbf{u} = \mathbf{0}$ by (1).

(2)$\Longrightarrow$(1). Clearly $\mathcal{N}_{\mathfrak{O}}$ is invariant under A. Therefore, if $\mathcal{N}_{\mathfrak{O}} \neq \{\mathbf{0}\}$, then it contains an eigenvector of A. But this means that there is a nonzero vector $\mathbf{v} \in \mathcal{N}_{\mathfrak{O}}$ such that $A\mathbf{v} = \alpha\mathbf{v}$, and hence that

$$\begin{bmatrix} A-\alpha I \\ C \end{bmatrix} \mathbf{v} = \mathbf{0}\,.$$

But this is incompatible with (2). Therefore, (C, A) is observable.

(3)$\Longleftrightarrow$(1). This follows from the observation that

(19.6)

$$C(\lambda I_n - A)^{-1}\mathbf{u} = \mathbf{0} \quad \text{for} \quad \lambda \in \Omega \Longleftrightarrow CA^k\mathbf{u} = \mathbf{0} \quad \text{for} \quad k = 0, \ldots, n-1\,.$$

The details and the verification that (4), (5) and (6) are each equivalent to observability are left to the reader. □

Exercise 19.8. Verify the equivalence (19.6) and then complete the proof that (3) is equivalent to (1) in Lemma 19.3.

Exercise 19.9. Show that in Lemma 19.3, (4) is equivalent to (1).

Exercise 19.10. Show that in Lemma 19.3, (5) is equivalent to (1).

Exercise 19.11. Show that the pair (C, A) is observable if and only if the matrix $\mathcal{O}^H\mathcal{O}$ is invertible.

Exercise 19.12. Let $F(\lambda) = I_p + C(\lambda I_n - A)^{-1}B$ and $G(\lambda) = I_p - C_1(\lambda I_n - A_1)^{-1}B_1$. Show that if $C_1 = C$ and (C, A) is an observable pair, then $F(\lambda)G(\lambda) = I_p$ if and only if $B_1 = B$ and $A_1 = A - BC$.

A realization $F(\lambda) = D + C(\lambda I_n - A)^{-1}B$ of a $p \times q$ rational nonconstant mvf $F(\lambda)$ is said to be an **observable realization** if the pair (C, A) is observable; it is said to be a **controllable realization** if the pair (A, B) is observable.

Theorem 19.4. *Let $F(\lambda)$ be a nonconstant proper rational $p \times q$ mvf such that*

$$F(\lambda) = D_1 + C_1(\lambda I_{n_1} - A_1)^{-1}B_1 = D_2 + C_2(\lambda I_{n_2} - A_2)^{-1}B_2$$

and suppose that both of these realizations are controllable and observable. Then:

(1) $D_1 = D_2$ *and* $n_1 = n_2$.

(2) *There exists exactly one invertible* $n_1 \times n_1$ *matrix* Y *such that*

$$C_1 = C_2 Y \ , \ A_1 = Y^{-1} A_2 Y \text{ and } B_1 = Y^{-1} B_2 \ .$$

Proof. It is readily checked that

$$F(\infty) = D_1 = D_2$$

and that

$$C_1 A_1^j B_1 = C_2 A_2^j B_2 \ , \ j = 0, 1, \dots \ .$$

Let

$$\Sigma_2 = [B_2 \quad A_2 B_2 \quad \cdots \quad A_2^{n_1-1} B_2], \quad \Omega_2 = \begin{bmatrix} C_2 \\ C_2 A_2 \\ \vdots \\ C_2 A_2^{n_1-1} \end{bmatrix}$$

and bear in mind that $\Sigma_2 \neq \mathfrak{C}_2$ and $\Omega_2 \neq \mathfrak{O}_2$ unless $n_2 = n_1$. Nevertheless, the identity

$$\mathfrak{O}_1 \mathfrak{C}_1 = \Omega_2 \Sigma_2$$

holds. Moreover, under the given assumptions, the observability matrix $\mathfrak{O}_1$ is left invertible, whereas the controllability matrix $\mathfrak{C}_1$ is right invertible. Thus, the inclusions

$$\mathcal{R}_{\mathfrak{O}_1} = \mathcal{R}_{\mathfrak{O}_1 \mathfrak{C}_1 \mathfrak{C}_1^H} \subseteq \mathcal{R}_{\mathfrak{O}_1 \mathfrak{C}_1} \subseteq \mathcal{R}_{\mathfrak{O}_1}$$

imply that

$$\text{rank}(\mathfrak{O}_1) = \text{rank}(\mathfrak{O}_1 \mathfrak{C}_1 \mathfrak{C}_1^H) \leq \text{rank}(\mathfrak{O}_1 \mathfrak{C}_1) \leq \text{rank}(\mathfrak{O}_1) = n_1 \ ,$$

which implies in turn that

$$n_1 = \text{rank}(\mathfrak{O}_1 \mathfrak{C}_1) = \text{rank}(\Omega_2 \Sigma_2) \leq n_2 \ .$$

However, since the roles played by the two realizations in the preceding analysis can be reversed, we must also have

$$n_2 \leq n_1 \ .$$

Therefore, equality prevails and so $\mathfrak{O}_2 = \Omega_2$ and $\mathfrak{C}_2 = \Sigma_2$.

Next, observe that the identity

$$\mathfrak{O}_1 B_1 = \mathfrak{O}_2 B_2 \, ,$$

implies that

$$B_1 = X B_2 \, , \text{ where } X = (\mathfrak{O}_1^H \mathfrak{O}_1)^{-1} \mathfrak{O}_1^H \mathfrak{O}_2 \ .$$

Similarly, the identity

$$C_1 \mathfrak{C}_1 = C_2 \mathfrak{C}_2$$

implies that

$$C_1 = C_2 Y \, , \text{ where } Y = \mathfrak{C}_2 \mathfrak{C}_1^H (\mathfrak{C}_1 \mathfrak{C}_1^H)^{-1} \ .$$

Moreover,

$$\begin{aligned} XY &= (\mathfrak{O}_1^H\mathfrak{O}_1)^{-1}\mathfrak{O}_1^H\mathfrak{O}_2\mathfrak{C}_2\mathfrak{C}_1^H(\mathfrak{C}_1\mathfrak{C}_1^H)^{-1} \\ &= (\mathfrak{O}_1^H\mathfrak{O}_1)^{-1}\mathfrak{O}_1^H\mathfrak{O}_1\mathfrak{C}_1\mathfrak{C}_1^H(\mathfrak{C}_1\mathfrak{C}_1^H)^{-1} \\ &= I_{n_1} \, . \end{aligned}$$

Thus, $X = Y^{-1}$.

Finally, from the formula

$$\mathfrak{O}_1 A_1 \mathfrak{C}_1 = \mathfrak{O}_2 A_2 \mathfrak{C}_2$$

we obtain

$$\begin{aligned} A_1 &= (\mathfrak{O}_1^H\mathfrak{O}_1)^{-1}\mathfrak{O}_1^H\mathfrak{O}_2 A_2\mathfrak{C}_2\mathfrak{C}_1^H(\mathfrak{C}_1\mathfrak{C}_1^H)^{-1} \\ &= XA_2Y = Y^{-1}A_2Y \, . \end{aligned}$$

□

Exercise 19.13. Verify the asserted uniqueness of the invertible matrix Y that is constructed in the proof of (2) of Theorem 19.4.

19.1. Minimal realizations

Let

$$F(\lambda) = D + C(\lambda I_n - A)^{-1}B \tag{19.7}$$

for some choice of the matrices $C \in \mathbb{C}^{p\times n}$, $A \in \mathbb{C}^{n\times n}$, $B \in \mathbb{C}^{n\times q}$ and $D \in \mathbb{C}^{p\times q}$. Then this realization is said to be **minimal** if the integer n is as small as possible, and then the number n is termed the **McMillan degree** of $F(\lambda)$.

Theorem 19.5. *A realization* (19.7) *for a proper rational nonconstant function $F(\lambda)$ is minimal if and only if the pair (C, A) is observable and the pair (A, B) is controllable.*

For ease of future reference, it is convenient to first prove two preliminary lemmas that are of independent interest.

Lemma 19.6. *The controllability matrix $\mathfrak{C}$ has rank $k < n$ if and only if there exists an invertible matrix T such that*

(1) $T^{-1}AT = \begin{bmatrix} A_{11} & A_{12} \\ O & A_{22} \end{bmatrix}$, $T^{-1}B = \begin{bmatrix} B_1 \\ O \end{bmatrix}$, *where* $A_{11} \in \mathbb{C}^{k\times k}$, $B_1 \in \mathbb{C}^{k\times q}$ *and*

(2) *the pair (A_{11}, B_1) is controllable.*

Proof. Suppose first that $\mathfrak{C}$ has rank $k < n$ and let X be an $n \times k$ matrix whose columns are selected from the columns of $\mathfrak{C}$ in such a way that

$$\text{rank } X = \text{rank } \mathfrak{C} = k\,.$$

Next, let $\ell = n - k$ and let Y be an $n \times \ell$ matrix such that the $n \times n$ matrix

$$T = [X \quad Y]$$

is invertible and express T^{-1} in block row form as

$$T^{-1} = \begin{bmatrix} U \\ V \end{bmatrix},$$

where $U \in \mathbb{C}^{k\times n}$ and $V \in \mathbb{C}^{\ell\times n}$. Then, the formula

$$I_n = T^{-1}T = \begin{bmatrix} U \\ V \end{bmatrix} [X \quad Y] = \begin{bmatrix} UX & UY \\ VX & VY \end{bmatrix}$$

implies that

$$UX = I_k\,, \quad UY = O_{k\times\ell}\,, \quad VX = O_{\ell\times k} \quad \text{and} \quad VY = I_\ell\,.$$

Moreover, since

$$A\mathfrak{C} = \mathfrak{C}E\,, \quad \mathfrak{C} = XF\,, \quad X = \mathfrak{C}G \quad \text{and} \quad B = \mathfrak{C}L \tag{19.8}$$

for appropriate choices of the matrices E, F, G and L, it follows easily that

$$AX = A\mathfrak{C}G = \mathfrak{C}EG = XFEG \quad \text{and} \quad B = \mathfrak{C}L\,.$$

Therefore,

$$VAX = VXFEG = O \quad \text{and} \quad VB = VXFL = O\,.$$

Thus, in a self-evident notation,

$$\begin{aligned} T^{-1}AT &= \begin{bmatrix} U \\ V \end{bmatrix} A[X \quad Y] \\ &= \begin{bmatrix} UAX & UAY \\ VAX & VAY \end{bmatrix} = \begin{bmatrix} A_{11} & A_{12} \\ O & A_{22} \end{bmatrix} \end{aligned}$$

and

$$T^{-1}B = \begin{bmatrix} U \\ V \end{bmatrix} B = \begin{bmatrix} UB \\ VB \end{bmatrix} = \begin{bmatrix} B_1 \\ O \end{bmatrix},$$

where $A_{11} \in \mathbb{C}^{k\times k}$ and $B_1 \in \mathbb{C}^{k\times q}$. Furthermore, since

$$T^{-1}A^jB = (T^{-1}AT)^jT^{-1}B \quad \text{for} \quad j = 0, 1, \ldots, n-1\,,$$

it is readily checked that

$$T^{-1}\mathfrak{C} = \begin{bmatrix} B_1 & A_{11}B_1 & \cdots & A_{11}^{n-1}B_1 \\ O & O & \cdots & O \end{bmatrix}$$

and hence that

$$k = \text{rank}\,\mathfrak{C} = \text{rank}\,T^{-1}\mathfrak{C} = \text{rank}\,\begin{bmatrix} B_1 & A_{11}B_1 & \cdots & A_{11}^{k-1}B_1 \end{bmatrix}.$$

Thus, (A_{11}, B_1) is controllable. This completes the proof in one direction. The converse is easy. $\square$

Exercise 19.14. Verify the assertions in formula (19.8).

Lemma 19.7. *The observability matrix $\mathfrak{O}$ has rank $k < n$ if and only if there exists an invertible matrix T such that*

(1) $CT = [C_1 \quad O]$, $T^{-1}AT = \begin{bmatrix} A_{11} & O \\ A_{21} & A_{22} \end{bmatrix}$, *where* $C_1 \in \mathbb{C}^{p\times k}$, $A_{11} \in \mathbb{C}^{k\times k}$ *and*

(2) *the pair (C_1, A_{11}) is observable.*

Proof. Suppose first that the observability matrix $\mathfrak{O}$ has rank $k < n$ and let U be a $k \times n$ matrix whose rows are selected from the rows of $\mathfrak{O}$ in such a way that

$$\text{rank } U = \text{rank } \mathfrak{O} = k \; .$$

Then, let $\ell = n - k$ and let V be any $\ell \times n$ matrix such that the $n \times n$ matrix $\begin{bmatrix} U \\ V \end{bmatrix}$ is invertible, and set

$$T = [X \quad Y] = \begin{bmatrix} U \\ V \end{bmatrix}^{-1} \; ,$$

where $X \in \mathbb{C}^{n\times k}$ and $Y \in \mathbb{C}^{n\times \ell}$. The next step is to check that

$$CY = O \quad \text{and} \quad UAY = O \; .$$

The details may be worked out by adapting the arguments used to prove Lemma 19.6 and are left to the reader, as is the rest of the proof. $\square$

Exercise 19.15. Complete the proof of Lemma 19.6.

Proof of Theorem 19.5. If the realization (19.7) is not controllable, then, by Lemma 19.6, there exists an invertible matrix $T \in \mathbb{C}^{n\times n}$ such that

$$T^{-1}AT = \begin{bmatrix} A_{11} & A_{12} \\ O & A_{22} \end{bmatrix} \quad \text{and} \quad T^{-1}B = \begin{bmatrix} B_1 \\ O \end{bmatrix}$$

and the pair $(A_{11}, B_1) \in \mathbb{C}^{k\times k} \times \mathbb{C}^{k\times q}$ is controllable. Let

$$CT = [C_1 \quad C_2] \; .$$

Then, as

$$(\lambda I_n - T^{-1}AT)^{-1} = \begin{bmatrix} (\lambda I_k - A_{11})^{-1} & (\lambda I_k - A_{11})^{-1}A_{12}(\lambda I_\ell - A_{22})^{-1} \\ O & (\lambda I_\ell - A_{22})^{-1} \end{bmatrix} ,$$

it is easily seen that

$$\begin{aligned} C(\lambda I_n - A)^{-1}B &= \begin{bmatrix} C_1 & C_2 \end{bmatrix} \begin{bmatrix} \lambda I_k - A_{11} & -A_{12} \\ O & \lambda I_\ell - A_{22} \end{bmatrix}^{-1} \begin{bmatrix} B_1 \\ O \end{bmatrix} \\ &= C_1(\lambda I_k - A_{11})^{-1}B_1 \,, \end{aligned}$$

which contradicts the presumed minimality of the original realization. Therefore, a minimal realization must be controllable.

The proof that a minimal realization must be observable follows from Lemma 19.7 in much the same way.

Conversely, a realization that is both controllable and observable must be minimal, thanks to Theorem 19.4 and the fact that a minimal realization must be controllable and observable, as was just proved above. □

Exercise 19.16. Let $(C, A) \in \mathbb{C}^{p\times n} \times \mathbb{C}^{n\times n}$ be an observable pair and let $\mathbf{u} \in \mathbb{C}^n$. Show that $C(\lambda I_n - A)^{-1}\mathbf{u}$ has a pole at α if and only if $(\lambda I_n - A)^{-1}\mathbf{u}$ has a pole at α. [HINT: First show that it suffices to focus on the case that $A = C_\alpha^{(n)}$ is a single Jordan cell.]

Exercise 19.17. Let $(C, A) \in \mathbb{C}^{p\times n} \times \mathbb{C}^{n\times n}$ be an observable pair and let $\mathbf{u}(\lambda) = \mathbf{u}_0 + \lambda\mathbf{u}_1 + \cdots + \lambda^k\mathbf{u}_k$ be a vector polynomial with coefficients in $\mathbb{C}^n$. Show that $C(\lambda I_n - A)^{-1}\mathbf{u}(\lambda)$ has a pole at α if and only if $(\lambda I_n - A)^{-1}\mathbf{u}(\lambda)$ has a pole at α. [HINT: Try Exercise 19.16 first to warm up.]

Theorem 19.8. *Let the realization*

$$F(\lambda) = D + C(\lambda I_n - A)^{-1}B$$

be minimal. Then the poles of $F(\lambda)$ coincide with the eigenvalues of A.

Discussion. Since $F(\lambda)$ is holomorphic in $\mathbb{C} \setminus \sigma(A)$, every pole of $F(\lambda)$ must be an eigenvalue of A. To establish the converse statement, there is no loss of generality in assuming that A is in Jordan form. Suppose, for the sake of definiteness, that

$$A = \operatorname{diag}\{C_{\omega_1}^{(3)}, C_{\omega_2}^{(2)}\} \quad \text{with} \quad \omega_1 \neq \omega_2 \,.$$

Then, upon writing

$$C = \begin{bmatrix} \mathbf{c}_1, \dots, \mathbf{c}_5 \end{bmatrix} \quad \text{and} \quad B^T = \begin{bmatrix} \mathbf{b}_1, \dots, \mathbf{b}_5 \end{bmatrix} \,,$$

one can readily see that

$$\begin{aligned} C(\lambda I_5 - A)^{-1}B &= \frac{\mathbf{c}_1\mathbf{b}_3^T}{(\lambda-\omega_1)^3} + \frac{\mathbf{c}_1\mathbf{b}_2^T + \mathbf{c}_2\mathbf{b}_3^T}{(\lambda-\omega_1)^2} + \frac{\mathbf{c}_1\mathbf{b}_1^T + \mathbf{c}_2\mathbf{b}_2^T + \mathbf{c}_3\mathbf{b}_3^T}{(\lambda-\omega_1)} \\ &+ \frac{\mathbf{c}_4\mathbf{b}_5^T}{(\lambda-\omega_2)^2} + \frac{\mathbf{c}_4\mathbf{b}_4^T + \mathbf{c}_5\mathbf{b}_5^T}{(\lambda-\omega_2)} \,. \end{aligned}$$

Moreover, item (2) in Lemma 19.3 implies that

$$\operatorname{rank}\begin{bmatrix} \lambda-\omega_1 & -1 & 0 & 0 & 0 \\ 0 & \lambda-\omega_1 & -1 & 0 & 0 \\ 0 & 0 & \lambda-\omega_1 & 0 & 0 \\ 0 & 0 & 0 & \lambda-\omega_2 & -1 \\ 0 & 0 & 0 & 0 & \lambda-\omega_2 \\ \mathbf{c}_1 & \mathbf{c}_2 & \mathbf{c}_3 & \mathbf{c}_4 & \mathbf{c}_5 \end{bmatrix} = n$$

for every point $\lambda \in \mathbb{C}$ and hence, in particular that the vectors $\mathbf{c}_1$ and $\mathbf{c}_4$ are both different from zero. Similarly, item (2) in Lemma 19.2 implies that the vectors $\mathbf{b}_3^T$ and $\mathbf{b}_5^T$ are both nonzero. Therefore, the matrices $\mathbf{c}_1\mathbf{b}_3^T$ and $\mathbf{c}_4\mathbf{b}_5^T$ are nonzero, and thus both ω_1 and ω_2 are poles of $F(\lambda)$. □

Theorem 19.8 does not address multiplicities of the poles versus the multiplicities of the eigenvalues. This issue is taken up in Theorem 19.9.

Theorem 19.9. *Let $F(\lambda) = D + C(\lambda I_n - A)^{-1}B$ be an $m \times l$ rational mvf and suppose that $A \in \mathbb{C}^{n\times n}$ has k distinct eigenvalues $\lambda_1, \dots, \lambda_k$ with algebraic multiplicities $\alpha_1, \dots, \alpha_k$, respectively. Let*

$$\Omega_j = \begin{bmatrix} F_{\alpha_j}^{(j)} & F_{\alpha_j-1}^{(j)} & \cdots & F_1^{(j)} \\ O & F_{\alpha_j}^{(j)} & \cdots & F_2^{(j)} \\ \vdots & \ddots & \ddots & \vdots \\ O & \cdots & O & F_{\alpha_j}^{(j)} \end{bmatrix}$$

be the block Toeplitz matrix based on the matrix coefficients $F_i^{(j)}$ of $(\lambda - \lambda_j)^{-i}, i = 1, \dots, \alpha_j$, in the Laurent expansion of $F(\lambda)$ in the vicinity of the point λ_j. Then the indicated realization of $F(\lambda)$ is minimal if and only if

$$\operatorname{rank} \Omega_j = \alpha_j \quad \textit{for } j = 1, \cdots, k.$$

Proof. Let

$$\begin{aligned} A &= UJU^{-1} \\ &= [U_1 \cdots U_k] \begin{bmatrix} J_1 & & \\ & \ddots & \\ & & J_k \end{bmatrix} \begin{bmatrix} V_1 \\ \vdots \\ V_k \end{bmatrix} \\ &= \textstyle\sum_{j=1}^k U_j J_j V_j \,, \end{aligned}$$

where $U_j \in \mathbb{C}^{n\times\alpha_j}$, $V_j \in \mathbb{C}^{\alpha_j\times n}$ and $J_j \in \mathbb{C}^{\alpha_j\times\alpha_j}$ are the components in a Jordan decomposition of A that are connected with λ_j for $j = 1, \dots, k$. Then, just as in Chapter 17,

$$V_i U_j = \begin{cases} I_{\alpha_j} & \text{if } i = j \\ O_{\alpha_i\times\alpha_j} & \text{if } i \neq j \end{cases}$$

and the $n \times n$ matrices

$$P_j = U_j V_j, \; j = 1, \cdots, k,$$

are projections, i.e., $P_j^2 = P_j$, even though here J_j may be include several Jordan cells.

The formula

$$\begin{aligned} C(\lambda I_n - A)^{-1}B &= CU(\lambda I_n - J)^{-1}U^{-1}B \\ &= C\left\{\sum_{i=1}^{k} U_i(\lambda I_{\alpha_i} - J_i)^{-1}V_i\right\}B \\ &= C\left\{\sum_{i=1}^{k} U_i((\lambda - \lambda_i)I_{\alpha_i} - \widetilde{N}_i)^{-1}V_i\right\}B \end{aligned}$$

implies that the poles in the Laurent expansion of $F(\lambda)$ in the vicinity of the point λ_j are given by

$$\begin{aligned} &CU_j((\lambda - \lambda_j)I_{\alpha_j} - \widetilde{N}_j)^{-1}V_jB \\ &\quad = CU_j\left\{\frac{I_{\alpha_j}}{\lambda - \lambda_j} + \frac{\widetilde{N}_j}{(\lambda - \lambda_j)^2} + \cdots + \frac{\widetilde{N}_j^{\alpha_j - 1}}{(\lambda - \lambda_j)^{\alpha_j}}\right\}V_jB\,. \end{aligned}$$

Consequently,

$$\Omega_j = \begin{bmatrix} X\widetilde{N}^{\kappa}Y & X\widetilde{N}^{\kappa-1}Y & \cdots & XY \\ O & X\widetilde{N}^{\kappa}Y & \cdots & X\widetilde{N}Y \\ \vdots & \vdots & \ddots & \vdots \\ O & O & \cdots & X\widetilde{N}^{\kappa}Y \end{bmatrix}$$

with $X = CU_j$, $Y = V_jB$, $\widetilde{N} = \widetilde{N}_j$ and $\kappa = \alpha_j - 1$. Therefore, since $\widetilde{N}^j = O$ for $j > \kappa$,

$$\Omega_j = \begin{bmatrix} X \\ X\widetilde{N} \\ \vdots \\ X\widetilde{N}^{\kappa} \end{bmatrix} [\widetilde{N}^{\kappa}Y \quad \widetilde{N}^{\kappa-1}Y \quad \cdots \quad Y]. \tag{19.9}$$

Next, let

(19.10)

$$\mathfrak{O}_j = \begin{bmatrix} CU_j \\ CU_jJ_j \\ \vdots \\ CU_jJ_j^{\alpha_j-1} \end{bmatrix} \quad \text{and} \quad \mathfrak{C}_j = [V_jB \quad J_jV_jB \quad \cdots \quad J_j^{\alpha_j-1}V_jB]\,,$$

for $j = 1, \cdots, k$. It is readily checked that

$$(19.11) \qquad (C, A) \text{ is observable} \iff \text{rank } \mathfrak{O}_j = \alpha_j \text{ for } j = 1, \cdots, k$$

and

$$(19.12) \qquad (A, B) \text{ is controllable} \iff \text{rank } \mathfrak{C}_j = \alpha_j \text{ for } j = 1, \cdots, k\,.$$

Moreover, in view of Exercises 19.20 and 19.21 (below), the ranks of the two factors in formula (19.9) for Ω_j are equal to rank $\mathfrak{O}_j$ and rank $\mathfrak{C}_j$, respectively. Therefore,

$$\text{rank } (\Omega_j) \le \min\{\text{rank } \mathfrak{O}_j, \ \text{rank } \mathfrak{C}_j\} \le \alpha_j\,,$$

and hence as

$$\text{rank } \mathfrak{O}_j \le \alpha_j \text{ and rank } \mathfrak{C}_j \le \alpha_j,$$

it is readily seen that

$$\text{rank } \Omega_j = \alpha_j \iff \text{rank } \mathfrak{O}_j = \alpha_j \text{ and rank } \mathfrak{C}_j = \alpha_j.$$

Therefore, the indicated realization is minimal if and only if rank $\Omega_j = \alpha_j$ for $j = 1, \cdots, k$. ☐

Exercise 19.18. Show that the coefficients $F_\ell^{(j)}$ of $(\lambda - \lambda_j)^{-\ell}$ in the Laurent expansion of the matrix valued function $F(\lambda)$ considered in Theorem 19.9 are given by the formula

$$F_\ell^{(j)} = CP_j(A - \lambda_j I_n)^{\ell-1}B \text{ for } \ell = 1, \cdots, \alpha_j,$$

where $P_j = U_j V_j$ is the projector defined in the proof of the theorem.

Exercise 19.19. Show that the projector P_j defined in the proof of Theorem 19.9 is the Riesz projector that is defined by the formula

$$P_j = \frac{1}{2\pi i}\int_{\Gamma_j} (\lambda I_n - A)^{-1} d\lambda\,,$$

if Γ_j is a small enough circle centered at λ_j and directed counterclockwise.

Exercise 19.20. Show that if $C \in \mathbb{C}^{p\times n}$ and $A \in \mathbb{C}^{n\times n}$, then

$$\text{rank} \begin{bmatrix} C \\ CA \\ \vdots \\ CA^\ell \end{bmatrix} = \text{rank} \begin{bmatrix} C \\ C(\alpha I_n + A) \\ \\ C(\alpha I_n + A)^\ell \end{bmatrix}$$

for every positive integer ℓ and every point $\alpha \in \mathbb{C}$.

Exercise 19.21. Show that if $A \in \mathbb{C}^{n\times n}$ and $B \in \mathbb{C}^{n\times q}$, then

$$\text{rank } [B \ AB \ \cdots \ A^\ell B] = \text{rank } [B \ (\alpha I_n + A)B \ \cdots \ (\alpha I_n + A)^\ell B]$$

for every positive integer ℓ and every point $\alpha \in \mathbb{C}$.

19.2. Stabilizable and detectable realizations

A pair of matrices $(A, B) \in \mathbb{C}^{n\times n} \times \mathbb{C}^{n\times r}$ is said to be **stabilizable** if there exists a matrix $K \in \mathbb{C}^{r\times n}$ such that $\sigma(A + BK) \subset \Pi_-$.

Lemma 19.10. *Let $(A, B) \in \mathbb{C}^{n\times n} \times \mathbb{C}^{n\times r}$. Then the following two conditions are equivalent:*

(1) *(A, B) is stabilizable.*

(2) $\operatorname{rank}[A - \lambda I_n \quad B] = n$ *for every point $\lambda \in \overline{\Pi_+}$.*

Proof. Suppose first that (A, B) is stabilizable and that $\sigma(A+BK) \subset \Pi_-$ for some $K \in \mathbb{C}^{r\times n}$. Then the formula

$$[A + BK - \lambda I_n \qquad B] = [A - \lambda I_n \qquad B] \begin{bmatrix} I_n & O \\ K & I_r \end{bmatrix}$$

clearly implies that

$$\operatorname{rank}[A - \lambda I_n \qquad B] = \operatorname{rank}[A + BK - \lambda I_n \qquad B]$$

and hence that (1) $\Longrightarrow$ (2), since $A + BK - \lambda I_n$ is invertible for $\lambda \in \overline{\Pi_+}$.

Suppose next that (2) is in force. Then, by Theorem 18.13 there exists a solution X of the Riccati equation (18.17) with $C = I_n$ such that $\sigma(A - BB^H X) \subset \Pi_-$. Therefore, the pair (A, B) is stabilizable. □

A pair of matrices $(C, A) \in \mathbb{C}^{m\times n} \times \mathbb{C}^{n\times n}$ is said to be **detectable** if there exists a matrix $L \in \mathbb{C}^{n\times m}$ such that $\sigma(A + LC) \subset \Pi_-$.

Lemma 19.11. *Let $(C, A) \in \mathbb{C}^{m\times n} \times \mathbb{C}^{n\times n}$. Then the following two conditions are equivalent:*

(1) *(C, A) is detectable.*

(2) $\operatorname{rank}\begin{bmatrix} A - \lambda I_n \\ C \end{bmatrix} = n$ *for every point $\lambda \in \overline{\Pi_+}$.*

Proof. Suppose first that $\sigma(A + LC) \subset \Pi_-$ for some $L \in \mathbb{C}^{n\times m}$. Then the formula

$$\begin{bmatrix} I_n & L \\ O & I_m \end{bmatrix} \begin{bmatrix} A - \lambda I_n \\ C \end{bmatrix} = \begin{bmatrix} A + LC - \lambda I_n \\ C \end{bmatrix}$$

clearly implies that

$$\operatorname{rank}\begin{bmatrix} A - \lambda I_n \\ C \end{bmatrix} = \operatorname{rank}\begin{bmatrix} A + LC - \lambda I_n \\ C \end{bmatrix}$$

for every point $\lambda \in \mathbb{C}$ and hence that (1) $\Longrightarrow$ (2).

Suppose next that (2) is in force. Then

$$\operatorname{rank}[A^H - \lambda I_n \qquad C^H] = n$$

for every point $\lambda \in \overline{\Pi_+}$, and hence, by Theorem 18.13 with

$$G = \begin{bmatrix} A^H & -C^H C \\ -I_n & -A \end{bmatrix},$$

there exists a Hermitian solution X of the Riccati equation

$$AX + XA^H - XC^H CX + I_n = O$$

such that

$$\sigma(A^H - C^H CX) \subset \Pi_-.$$

Therefore

$$\sigma(A - XC^H C) \subset \Pi_-.$$

□

Exercise 19.22. Show that (A, B) is stabilizable if and only if (B^H, A^H) is detectable.

Exercise 19.23. Show that if (C, A) is detectable and (A, B) is stabilizable, then there exist matrices K and L such that

$$\sigma\left(\begin{bmatrix} A & -BK \\ LC & A - BK - LC \end{bmatrix}\right) \subset \Pi_- .$$

[HINT:

$$\begin{bmatrix} I_n & O \\ I_n & -I_n \end{bmatrix}^{-1} \begin{bmatrix} A & -BK \\ LC & A - BK - LC \end{bmatrix} \begin{bmatrix} I_n & O \\ I_n & -I_n \end{bmatrix} = \begin{bmatrix} A - BK & BK \\ O & A - LC \end{bmatrix}.]$$

19.3. Reproducing kernel Hilbert spaces

A Hilbert space $\mathcal{H}$ of complex $m \times 1$ vector valued functions that are defined on a nonempty subset Ω of $\mathbb{C}$ is said to be a **reproducing kernel Hilbert space** if there exists an $m \times m$ mvf $K_\omega(\lambda)$ that is defined on $\Omega \times \Omega$ such that for every choice of $\omega \in \Omega$ and $\mathbf{u} \in \mathbb{C}^m$ the following two conditions are fulfilled:

(1) $K_\omega \mathbf{u} \in \mathcal{H}$.

(2) $\langle f, K_\omega \mathbf{u} \rangle_{\mathcal{H}} = \mathbf{u}^H f(\omega)$ for every $f \in \mathcal{H}$.

An $m \times m$ mvf $K_\omega(\lambda)$ that meets these two conditions is called a **reproducing kernel** for $\mathcal{H}$.

Lemma 19.12. *Let $\mathcal{H}$ be a reproducing kernel Hilbert space of $\mathbb{C}^m$ valued functions that are defined on a nonempty subset Ω of $\mathbb{C}$ with reproducing kernel $K_\omega(\lambda)$. Then*

(1) *The reproducing kernel is unique; i.e., if $L_\omega(\lambda)$ is also a reproducing kernel for $\mathcal{H}$, then $K_\omega(\lambda) = L_\omega(\lambda)$ for all points $\lambda, \omega \in \Omega$.*

(2) *$K_\omega(\lambda) = K_\lambda(\omega)^H$ for all points $\lambda, \omega \in \Omega$.*

(3) $\sum_{i,j=1}^{n} \mathbf{u}_i^H K_{\omega_j}(\omega_i)\mathbf{u}_j \geq 0$ *for every choice of the points* $\omega_1, \ldots, \omega_n \in \Omega$ *and the vectors* $\mathbf{u}_1 \ldots, \mathbf{u}_n \in \mathbb{C}^m$.

Proof. If $L_\omega(\lambda)$ and $K_\omega(\lambda)$ are reproducing kernels for the reproducing kernel Hilbert space $\mathcal{H}$, then

$$\begin{aligned} \mathbf{v}^H L_\alpha(\beta)\mathbf{u} &= \langle L_\alpha \mathbf{u}, K_\beta \mathbf{v}\rangle_{\mathcal{H}} = \overline{\langle K_\beta \mathbf{v}, L_\alpha \mathbf{u}\rangle_{\mathcal{H}}} \\ &= \overline{\mathbf{u}^H K_\beta(\alpha)\mathbf{v}} = (\mathbf{u}^H K_\beta(\alpha)\mathbf{v})^H \\ &= \mathbf{v}^H K_\beta(\alpha)^H \mathbf{u} \end{aligned}$$

for every choice of $\mathbf{u}$ and $\mathbf{v}$ in $\mathbb{C}^m$. Therefore,

$$L_\alpha(\beta) = K_\beta(\alpha)^H$$

for every choice of α and β in Ω. In particular, this implies that

$$K_\alpha(\beta) = K_\beta(\alpha)^H$$

and hence, upon invoking both of the last two identities, that

$$K_\alpha(\beta) = K_\beta(\alpha)^H = L_\alpha(\beta)$$

for every choice of α and β in Ω. This completes the proof of (1) and (2). The third assertion is left to the reader. □

Exercise 19.24. Justify assertion (3) of Lemma 19.12.

In this section we shall focus on a class of finite dimensional reproducing kernel Hilbert spaces.

Theorem 19.13. *Every finite dimensional Hilbert space* $\mathcal{H}$ *of strictly proper rational* $m \times 1$ *vector valued functions can be identified as a space*

$$\mathcal{M}(X) = \mathcal{M}_\Gamma(X) = \{F(\lambda)X\mathbf{u} : \mathbf{u} \in \mathbb{C}^n\}, \tag{19.13}$$

endowed with the inner product

$$\langle FX\mathbf{u}, FX\mathbf{v}\rangle_{\mathcal{M}(X)} = \mathbf{v}^H X\mathbf{u}, \tag{19.14}$$

where

$$\Gamma = (C, A) \in \mathbb{C}^{m\times n} \times \mathbb{C}^{n\times n} \quad \textit{is an observable pair,} \tag{19.15}$$

$$F(\lambda) = C(\lambda I_n - A)^{-1} \tag{19.16}$$

and

$$X \succeq O \quad \textit{is an } n\times n \textit{ matrix with} \quad \operatorname{rank} X = \dim \mathcal{H}. \tag{19.17}$$

Proof. Let $\{\mathbf{f}_1(\lambda), \ldots, \mathbf{f}_r(\lambda)\}$ be a basis for an r dimensional inner product space $\mathcal{M}$ of strictly proper $m \times 1$ rational vector valued functions. Then, in view of Theorem 19.1, there exists a set of matrices $C \in \mathbb{C}^{m\times n}$, $A \in \mathbb{C}^{n\times n}$ and $B \in \mathbb{C}^{n\times r}$ such that the $m \times r$ matrix valued function with columns $\mathbf{f}_1(\lambda), \ldots, \mathbf{f}_r(\lambda)$ admits a minimal realization of the form

$$\begin{bmatrix}\mathbf{f}_1(\lambda) & \cdots & \mathbf{f}_r(\lambda)\end{bmatrix} = C(\lambda I_n - A)^{-1}B\,.$$

Moreover, if G denotes the Gram matrix with entries

$$g_{ij} = \langle \mathbf{f}_j, \mathbf{f}_i\rangle_{\mathcal{M}}\,, \quad \text{for} \quad i, j = 1, \ldots, r\,,$$

and $F(\lambda) = C(\lambda I_n - A)^{-1}$, then

$$\mathcal{H} = \{F(\lambda)B\mathbf{u} : \mathbf{u} \in \mathbb{C}^r\}$$

and

$$\langle FB\mathbf{x}, FB\mathbf{y}\rangle_{\mathcal{H}} = \mathbf{y}^H G\mathbf{x} \quad \text{for every choice of} \quad \mathbf{x}, \mathbf{y} \in \mathbb{C}^r\,. \tag{19.18}$$

Let $X = BG^{-1}B^H$. Then it is readily checked that $\mathcal{N}_{B^H} = \mathcal{N}_X$ and hence that $\mathcal{R}_B = \mathcal{R}_X$. Thus $X \succeq O$, $\operatorname{rank} X = r$ and formulas (19.13) and (19.14) drop out easily upon taking $\mathbf{x} = G^{-1}B^H\mathbf{u}$ and $\mathbf{y} = G^{-1}B^H\mathbf{v}$ in formula (19.18). □

Exercise 19.25. Verify formulas (19.13) and (19.14) and check that the inner product in the latter is well defined; i.e., if $FX\mathbf{u}_1 = FX\mathbf{u}_2$ and $FX\mathbf{v}_1 = FX\mathbf{v}_2$, then $\mathbf{v}_1^H X\mathbf{u}_1 = \mathbf{v}_2^H X\mathbf{u}_2$.

Theorem 19.14. *The space $\mathcal{M}(X)$ defined by formula* (19.13) *endowed with the inner product* (19.14) *is a reproducing kernel Hilbert space with reproducing kernel*

$$K_\omega^{\mathcal{M}}(\lambda) = F(\lambda)XF(\omega)^H\,. \tag{19.19}$$

Proof. Clearly $K_\omega^{\mathcal{M}}(\lambda)\mathbf{u} \in \mathcal{M}$ (as a function of λ) for every choice of $\mathbf{u} \in \mathbb{C}^m$ and $\omega \in \Omega$. Let $f \in \mathcal{M}$. Then $f(\lambda) = F(\lambda)X\mathbf{v}$ for some vector $\mathbf{v} \in \mathbb{C}^n$. Then, in view of formula (19.18),

$$\begin{aligned}\langle f, K_\omega^{\mathcal{M}}\mathbf{u}\rangle_{\mathcal{M}} &= \langle FX\mathbf{v}, FXF(\omega)^H\mathbf{u}\rangle_{\mathcal{M}} \\ &= \mathbf{u}^H F(\omega)X\mathbf{v} = \mathbf{u}^H f(\omega)\,,\end{aligned}$$

for every vector $\mathbf{u} \in \mathbb{C}^m$ and every point $\omega \in \Omega$, as needed. □

19.4. de Branges spaces

A matrix $J \in \mathbb{C}^{m\times m}$ is said to be a **signature matrix** if

$$J = J^H \text{ and } J^H J = I_m\,. \tag{19.20}$$

Exercise 19.26. Show that if $J \in \mathbb{C}^{m\times m}$ is a signature matrix, then either $J = \pm I_m$ or $J = U\mathrm{diag}\{I_p, -I_q\}U^H$, with U unitary, $p \geq 1$, $q \geq 1$ and $p+q=m$.

The finite dimensional reproducing kernel Hilbert space $\mathcal{M}(X)$ will be called a **de Branges space** $\mathcal{H}(\Theta)$ if there exists a proper rational $m \times m$ mvf $\Theta(\lambda)$ and an $m \times m$ signature matrix J such that

$$F(\lambda)XF(\omega)^H = \frac{J - \Theta(\lambda)J\Theta(\omega)^H}{\lambda + \overline{\omega}}. \tag{19.21}$$

Theorem 19.15. *The finite dimensional reproducing kernel Hilbert space $\mathcal{M}(X)$ is a de Branges space $\mathcal{H}(\Theta)$ if and only if the Hermitian matrix X is a solution of the Riccati equation*

$$XA^H + AX + XC^HJCX = O. \tag{19.22}$$

Moreover, if X is a solution of the Riccati equation (19.22)*, then $\Theta(\lambda)$ is uniquely specified by the formula*

$$\Theta(\lambda) = I_m - F(\lambda)XC^HJ, \tag{19.23}$$

up to a constant factor $K \in \mathbb{C}^{m\times m}$ on the right, which is subject to the constraint

$$KJK^H = J.$$

Proof. Suppose first that there exists a proper rational $m \times m$ matrix valued function $\Theta(\lambda)$ that satisfies the identity (19.21). Then

$$(\lambda + \overline{\omega})F(\lambda)X(\overline{\omega}I_n - A^H)^{-1}C^H = J - \Theta(\lambda)J\Theta(\omega)^H, \tag{19.24}$$

and hence, upon letting ω tend to infinity, it follows that

$$F(\lambda)XC^H = J - \Theta(\lambda)J\Theta(\infty)^H.$$

The identity (19.24) also implies that

$$J - \Theta(\lambda)J\Theta(-\overline{\lambda})^H = O$$

and consequently, upon letting $\lambda = i\nu$ with $\nu \in \mathbb{R}$, that

$$\Theta(i\nu)J\Theta(i\nu)^H = J$$

and hence that

$$\Theta(\infty)J\Theta(\infty)^H = J.$$

Thus, $\Theta(\infty)$ is invertible and the last formula can also be written as

$$\Theta(\infty)^HJ\Theta(\infty) = J.$$

But this in turn implies that

$$\begin{aligned} F(\lambda)XC^HJ\Theta(\infty) &= (J - \Theta(\lambda)J\Theta(\infty)^H)J\Theta(\infty) \\ &= \Theta(\infty) - \Theta(\lambda). \end{aligned}$$

Therefore, $\Theta(\lambda)$ is uniquely specified by the formula

$$\Theta(\lambda) = (I_m - F(\lambda)XC^H J)\Theta(\infty),$$

up to a multiplicative constant factor $K = \Theta(\infty)$ on the right, which meets the constraint $KJK^H = J$. Moreover, since $\Theta(\infty)J\Theta(\infty)^H = J$,

$$\begin{aligned}\Theta(\lambda)J\Theta(\omega)^H &= (I_m - F(\lambda)XC^H J)\Theta(\infty)J\Theta(\infty)^H(I_m - JCXF(\omega)^H)\\ &= J - F(\lambda)XC^H - CXF(\omega)^H + F(\lambda)XC^H JCXF(\omega)^H\\ &= J - F(\lambda)\{\cdots\}F(\omega)^H,\end{aligned}$$

where

$$\begin{aligned}\{\cdots\} &= X(\overline{\omega}I_n - A^H) + (\lambda I_n - A)X - XC^H JCX\\ &= (\lambda + \overline{\omega})X - (XA^H + AX + XC^H JCX).\end{aligned}$$

Therefore,

(19.25)
$$\begin{aligned}\frac{J - \Theta(\lambda)J\Theta(\omega)^H}{\lambda + \overline{\omega}} &= F(\lambda)XF(\omega)^H\\ &\quad - \frac{F(\lambda)(XA^H + AX + XC^H JCX)F(\omega)^H}{\lambda + \overline{\omega}}.\end{aligned}$$

Thus, upon comparing the last formula with the identity (19.21), it is readily seen that

$$F(\lambda)(XA^H + AX + XC^H JCX)F(\omega)^H = O$$

and hence, as (C, A) is observable, that X must be a solution of the Riccati equation (19.22).

Conversely, if X is a solution of the Riccati equation (19.22) and if $\Theta(\lambda)$ is then defined by formula (19.23), the calculations leading to the formula (19.25) serve to justify formula (19.21). □

19.5. R_α invariance

Lemma 19.16. *Let $\mathcal{M}(X)$ denote the space defined in terms of an observable pair $\Gamma = (C, A) \in \mathbb{C}^{m\times n} \times \mathbb{C}^{n\times n}$ and an $n \times n$ matrix $X \succeq O$ as in Theorem 19.13. Then the following conditions are equivalent:*

(1) *The space $\mathcal{M}(X)$ is invariant under the action of the backward shift operator*

(19.26)
$$(R_\alpha f)(\lambda) = \frac{f(\lambda) - f(\alpha)}{\lambda - \alpha}$$

for every point $\alpha \in \mathbb{C} \setminus \sigma(A)$.

(2) *The space $\mathcal{M}(X)$ is invariant under the action of the operator R_α for at least one point $\alpha \in \mathbb{C} \setminus \sigma(A)$.*

(3) $AX = XA_0$ *for some matrix* $A_0 \in \mathbb{C}^{n\times n}$.

(4) $AX = XA_0$ *for some matrix* $A_0 \in \mathbb{C}^{n\times n}$ *with* $\sigma(A_0) \subseteq \sigma(A)$.

Proof. Let $h_j(\lambda) = F(\lambda)X\mathbf{e}_j$ for $j = 1, \ldots, n$, where $\mathbf{e}_j$ denotes the j'th column of the identity matrix I_n and suppose that (2) is in force for some point $\alpha \in \mathbb{C} \setminus \sigma(A)$. Then, since

$$\begin{aligned}(R_\alpha h_j)(\lambda) &= \frac{F(\lambda) - F(\alpha)}{\lambda - \alpha} X\mathbf{e}_j \\ &= F(\lambda)\left\{\frac{(\alpha I_n - A) - (\lambda I_n - A)}{\lambda - \alpha}\right\}(\alpha I_n - A)^{-1}X\mathbf{e}_j \\ &= -F(\lambda)(\alpha I_n - A)^{-1}X\mathbf{e}_j\,,\end{aligned}$$

the invariance assumption guarantees the existence of a set of vectors $\mathbf{v}_j \in \mathbb{C}^n$ such that

$$-F(\lambda)(\alpha I_n - A)^{-1}X\mathbf{e}_j = F(\lambda)X\mathbf{v}_j \quad \text{for} \quad j = 1, \ldots, n\,.$$

Consequently,

$$-(\alpha I_n - A)^{-1}X\mathbf{e}_j = X\mathbf{v}_j \quad \text{for} \quad j = 1, \ldots, n\,,$$

and hence

$$-(\alpha I_n - A)^{-1}X\begin{bmatrix}\mathbf{e}_1 & \cdots & \mathbf{e}_n\end{bmatrix} = X\begin{bmatrix}\mathbf{v}_1 & \cdots & \mathbf{v}_n\end{bmatrix}\,,$$

which is the same as to say that

$$-(\alpha I_n - A)^{-1}X = XQ_\alpha$$

for some matrix $Q_\alpha \in \mathbb{C}^{n\times n}$. In view of Lemma 20.14, there is no loss of generality in assuming that Q_α is invertible and consequently that

$$AX = X(\alpha I_n + Q_\alpha^{-1}),$$

which serves to justify the implication (2) $\Longrightarrow$ (3). The equivalence (3) $\Longleftrightarrow$ (4) is covered by another application of Lemma 20.14. The remaining implications (4) $\Longrightarrow$ (1) $\Longrightarrow$ (2) are easy and are left to the reader. $\square$

Exercise 19.27. Complete the proof of Lemma 19.16 by justifying the implications (4) $\Longrightarrow$ (1) $\Longrightarrow$ (2).

19.6. Factorization of $\Theta(\lambda)$

We shall assume from now on that $X \in \mathbb{C}^{n\times n}$ is a positive semidefinite solution of the Riccati equation

$$XA^H + AX + XC^HJCX = O, \tag{19.27}$$

and shall obtain a factorization of the matrix valued $\Theta(\lambda)$ based on a decomposition of the space $\mathcal{M}(X)$. In particular, formula (19.27) implies that

$$AX = X(-A^H - C^HJCX)$$

and hence, in view of Lemma 19.16, that $\mathcal{M}(X)$ is invariant under the action of R_α for every choice of $\alpha \in \mathbb{C} \setminus \sigma(A)$. Thus, if $\alpha \in \mathbb{C} \setminus \sigma(A)$, Lemma 19.16 guarantees the existence of a nonzero vector valued function $\mathbf{g}_1 \in \mathcal{M}$ and a scalar $\mu_1 \in \mathbb{C}$ such that

$$(R_\alpha \mathbf{g}_1)(\lambda) = \mu_1 \mathbf{g}_1(\lambda).$$

Therefore, since

$$\mathbf{g}_1(\lambda) = F(\lambda) X \mathbf{u}_1$$

for some vector $\mathbf{u}_1 \in \mathbb{C}^n$ such that $X\mathbf{u}_1 \neq \mathbf{0}$, it follows that

$$(R_\alpha \mathbf{g}_1)(\lambda) = -F(\lambda)(\alpha I_n - A)^{-1} X \mathbf{u}_1 = \mu_1 F(\lambda) X \mathbf{u}_1$$

and hence that

$$-(\alpha I_n - A)^{-1} X \mathbf{u}_1 = \mu_1 X \mathbf{u}_1.$$

But this in turn implies that $\mu_1 \neq 0$. Thus, the previous formula can be rewritten as

$$AX\mathbf{u}_1 = \omega_1 X \mathbf{u}_1 \text{ with } \omega_1 = \alpha + \frac{1}{\mu_1} \tag{19.28}$$

and, consequently,

$$\mathbf{g}_1(\lambda) = F(\lambda) X \mathbf{u}_1 = \frac{CX\mathbf{u}_1}{\lambda - \omega_1} \tag{19.29}$$

and

$$\|\mathbf{g}_1\|_{\mathcal{M}}^2 = \langle FX\mathbf{u}_1,\ FX\mathbf{u}_1 \rangle_{\mathcal{M}} = \mathbf{u}_1^H X \mathbf{u}_1\,. \tag{19.30}$$

Let

$$\mathcal{M}_1 = \{\beta \mathbf{g}_1 : \beta \in \mathbb{C}\}$$

denote the one dimensional subspace of $\mathcal{M}$ spanned by $\mathbf{g}_1(\lambda)$ and let Π_1 denote the orthogonal projection of $\mathcal{M}$ onto $\mathcal{M}_1$. Then

$$\begin{aligned} \Pi_1 FX\mathbf{v} &= \frac{\langle FX\mathbf{v}, \mathbf{g}_1 \rangle_{\mathcal{M}}}{\langle \mathbf{g}_1, \mathbf{g}_1 \rangle_{\mathcal{M}}} \mathbf{g}_1 = \frac{\mathbf{u}_1^H X \mathbf{v}}{\mathbf{u}_1^H X \mathbf{u}_1} FX\mathbf{u}_1 \\ &= FX\mathbf{u}_1 (\mathbf{u}_1^H X \mathbf{u}_1)^{-1} \mathbf{u}_1^H X \mathbf{v} = FX_1 \mathbf{v} \end{aligned}$$

with

$$X_1 = X\mathbf{u}_1 (\mathbf{u}_1^H X \mathbf{u}_1)^{-1} \mathbf{u}_1^H X. \tag{19.31}$$

Let

$$Q_1 = \mathbf{u}_1 (\mathbf{u}_1^H X \mathbf{u}_1)^{-1} \mathbf{u}_1^H X\,.$$

Then, since Q_1 is a projection, i.e., $Q_1^2 = Q_1$, it is readily checked that

$$X_1 = XQ_1 = XQ_1^2 = X_1 Q_1 = Q_1^H X_1$$

and

$$Q_1^H X Q_1 = X_1^H Q_1 = X_1 Q_1 = X_1\,.$$

Thus, with the aid of the formula

$$AX_1 = \omega_1 X_1,$$

it follows that

$$\begin{aligned} Q_1^H AXQ_1 &= Q_1^H AX_1 = \omega_1 Q_1^H X_1 \\ &= \omega_1 X_1 = AX_1 \end{aligned}$$

and

$$Q_1^H XA^H Q_1 = X_1 A^H.$$

Consequently, upon multiplying the Riccati equation (19.27) on the left by Q_1^H and on the right by Q_1, it follows that

$$\begin{aligned} O &= Q_1^H XA^H Q_1 + Q_1^H AXQ_1 + Q_1^H XC^H JCXQ_1 \\ &= X_1 A^H + AX_1 + X_1 C^H JCX_1 ; \end{aligned}$$

i.e., X_1 is a rank one solution of the Riccati equation (19.27). Therefore,

$$\mathcal{M}_1 = \mathcal{H}(\vartheta_1)$$

is a de Branges space based upon the matrix valued function

$$\vartheta_1(\lambda) = I_m - F(\lambda)X_1 C^H J = I_m - \frac{CX_1 C^H J}{\lambda - \omega_1} .$$

Let $\mathcal{M}_1^\perp$ denote the orthogonal complement of $\mathcal{M}_1$ in $\mathcal{M}(X)$. Then

$$\mathcal{M}_1^\perp = \{(I - \Pi_1)FX\mathbf{u} : \mathbf{u} \in \mathbb{C}^n\} = \{FX_2\mathbf{u} : \mathbf{u} \in \mathbb{C}^n\} , \tag{19.32}$$

where

$$X_2 = X - X_1 = X - Q_1^H XQ_1 = (I_n - Q_1^H)X(I_n - Q_1) \succeq O .$$

Let

$$\mathcal{M}_2 = \vartheta_1(\lambda)^{-1}\mathcal{M}^\perp$$

and

$$\Theta_2(\lambda) = \vartheta_1(\lambda)^{-1}\Theta(\lambda) .$$

By formula (19.5),

$$\vartheta_1(\lambda)^{-1} = I_m + C(\lambda I_n - A_1)^{-1}X_1 C^H J ,$$

where

$$A_1 = A + X_1 C^H JC .$$

Moreover, by straightforward calculations that are left to the reader,

$$\mathcal{M}_2 = \{C(\lambda I_n - A_1)^{-1}X_2\mathbf{u} : \mathbf{u} \in \mathbb{C}^{n \times n}\} , \tag{19.33}$$

$$\Theta_2(\lambda) = I_m - C(\lambda I_n - A_1)^{-1}X_2 C^H J \tag{19.34}$$

and, since both X and X_1 are solutions of (19.27), $X_2 \succeq O$ is a solution of the Riccati equation

$$X_2 A_1^H + A_1 X_2 + X_2 C^H J C X_2 = O \tag{19.35}$$

with $\operatorname{rank} X_2 = r - 1$. Thus, as the pair

$$(C, A_1) = (C, A + X_1 C^H J C)$$

is observable, we can define

$$F_1(\lambda) = C(\lambda I_n - A_1)^{-1}$$

and the space

$$\mathcal{M}_2 = \{F_1(\lambda) X_2 \mathbf{u} : \mathbf{u} \in \mathbb{C}^n\}$$

endowed with the inner product

$$\langle F_1 X_2 \mathbf{u},\ F_1 X_2 \mathbf{v} \rangle_{\mathcal{M}_2} = \mathbf{v}^H X_2 \mathbf{u}$$

is also R_α invariant for each point $\alpha \in \mathbb{C} \setminus \sigma(A_1)$. Therefore,

$$\mathcal{M}_2 = \mathcal{H}(\Theta_2),$$

and the factorization procedure can be iterated to obtain a factorization

$$\Theta(\lambda) = \vartheta_1(\lambda) \cdots \vartheta_k(\lambda)$$

of $\Theta(\lambda)$ as a product of k elementary factors of McMillan degree one with $k = \operatorname{rank} X$.

Exercise 19.28. Verify the statements in (19.32).

Exercise 19.29. Verify formula (19.34). [HINT: The trick in this calculation (and others of this kind) is to note that in the product, the two terms

$$C(\lambda I_n - A)^{-1} X C^H J + C(\lambda I_n - A_1)^{-1} X_1 C^H J C(\lambda I_n - A)^{-1} X C^H J$$

can be reexpressed as

$$C(\lambda I_n - A_1)^{-1} \{\lambda I_n - A_1 + X_1 C^H J C\} (\lambda I_n - A)^{-1} X C^H J\,,$$

which simplifies beautifully.]

Exercise 19.30. Show that if $f \in \mathcal{M}_2$, then

$$\|\vartheta_1 f\|_{\mathcal{M}} = \|f\|_{\mathcal{M}_2}.$$

[HINT: First check that

$$\vartheta_1(\lambda) C(\lambda I_n - A_1)^{-1} X_2 = C(\lambda I_n - A)^{-1} X_2$$

and then exploit the fact that

$$X_2 = X(I_n - Q_1) = (I_n - Q_1^H) X (I_n - Q_1).]$$

Exercise 19.31. Show that $\operatorname{rank} X_2 = \operatorname{rank} X - \operatorname{rank} X_1$.

Exercise 19.32. Let $A \in \mathbb{C}^{n\times n}$, $C \in \mathbb{C}^{m\times n}$, $J \in \mathbb{C}^{m\times m}$, let $\lambda_1, \ldots, \lambda_k$ denote the distinct eigenvalues of A; and let P be a solution of the Stein equation $P - A^H PA = C^H JC$. Show that if $1 - \lambda_i\overline{\lambda_j} \neq 0$ for $i, j = 1, \ldots, k$, then $\mathcal{N}_{\mathfrak{O}} \subseteq \mathcal{N}_P$. [HINT: $\mathcal{N}_{\mathfrak{O}}$ is invariant under A.]

Exercise 19.33. Let $A \in \mathbb{C}^{n\times n}$, $C \in \mathbb{C}^{m\times n}$, $J \in \mathbb{C}^{m\times m}$; and let P be a solution of the Lyapunov equation $A^H P + PA = C^H JC$. Show that if $\sigma(A) \cap \sigma(-A^H) = \emptyset$, then $\mathcal{N}_{\mathfrak{O}} \subseteq \mathcal{N}_P$. [HINT: $\mathcal{N}_{\mathfrak{O}}$ is invariant under A.]

19.7. Bibliographical notes

The monographs [**74**] and [**72**] are good sources of supplementary information on realization theory and applications to control theory. Condition (2) in Lemmas 19.2, 19.3, 19.10, 19.11 and variations thereof are usually referred to as Hautus tests or Popov-Belevich-Hautus tests. Theorem 19.9 is adapted from [**6**]. Exercise 19.23 is adapted from Theorem 4.3 in Chapter 3 of [**61**]. The connection between finite dimensional de Branges spaces and Riccati equations is adapted from [**22**]. This connection lends itself to a rather clean framework for handling a number of bitangential interpolation problems; see e.g., [**23**]; the treatment of factorization in the last section is adapted from the article [**20**], which includes extensions of the factorization discussed here to nonsquare matrix valued functions.

Chapter 20

Eigenvalue location problems

When I'm finished [shooting] that bridge ... I'll have made it into something of my own, by lens choice, or camera angle, or general composition, and most likely by some combination of all those.

I don't just take things as given, I try to make them into something that reflects my personal consciousness, my spirit. I try to find the poetry in the image.

Waller [**70**], p. 50

If $A \in \mathbb{C}^{n\times n}$ and $A = A^H$, then $\sigma(A) \subset \mathbb{R}$ and hence:

- $\mathcal{E}_+(A)$ = the number of positive eigenvalues of A, counting multiplicities;
- $\mathcal{E}_-(A)$ = the number of negative eigenvalues of A, counting multiplicities;
- $\mathcal{E}_0(A)$ = the number of zero eigenvalues of A, counting multiplicities.

Thus,

$$\mathcal{E}_+(A) + \mathcal{E}_-(A) = \operatorname{rank} A \quad \text{and} \quad \mathcal{E}_0(A) = \dim \mathcal{N}_A .$$

20.1. Interlacing

Theorem 20.1. *Let B be the upper left $k \times k$ corner of a $(k+1)\times(k+1)$ Hermitian matrix A and let $\lambda_1(A) \le \cdots \le \lambda_{k+1}(A)$ and $\lambda_1(B) \le \cdots \le$*

$\lambda_k(B)$ denote the eigenvalues of A and B, respectively. Then

$$\lambda_j(A) \le \lambda_j(B) \le \lambda_{j+1}(A) \quad \textit{for} \quad j = 1, \ldots, k\,. \tag{20.1}$$

Proof. Let

$$\mathfrak{a}(\mathcal{X}) = \max\{\langle A\mathbf{x}, \mathbf{x}\rangle : \mathbf{x} \in \mathcal{X} \quad \text{and} \quad \|\mathbf{x}\| = 1\}$$

for each subspace $\mathcal{X}$ of $\mathbb{C}^{k+1}$ and let

$$\mathfrak{b}(\mathcal{Y}) = \max\{\langle B\mathbf{y}, \mathbf{y}\rangle : \mathbf{y} \in \mathcal{Y} \quad \text{and} \quad \|\mathbf{y}\| = 1\}\,,$$

for each subspace $\mathcal{Y}$ of $\mathbb{C}^k$. Let $\mathcal{S}_j$ denote the set of all j-dimensional subspaces of $\mathbb{C}^{k+1}$ for $j = 1, \ldots, k+1$; let $\mathcal{T}_j$ denote the set of all j-dimensional subspaces of $\mathbb{C}^k$ for $j = 1, \ldots, k$; and let $\mathcal{S}_j^\circ$ denote the set of all j-dimensional subspaces of $\mathbb{C}^{k+1}$ for $j = 1, \ldots, k$ that are orthogonal to $\mathbf{e}_{k+1}$, the $k+1$'st column of I_{k+1}. Then, by the Courant-Fischer theorem,

$$\begin{aligned} \lambda_j(A) &= \min_{\mathcal{X}\in\mathcal{S}_j} \mathfrak{a}(\mathcal{X}) \\ &\le \min_{\mathcal{X}\in\mathcal{S}_j^\circ} \mathfrak{a}(\mathcal{X}) = \min_{\mathcal{Y}\in\mathcal{T}_j} \mathfrak{b}(\mathcal{Y}) \\ &= \lambda_j(B) \quad \text{for} \quad j = 1, \ldots, k\,. \end{aligned}$$

The second inequality in (20.1) depends upon the observation that for each $j+1$-dimensional subspace $\mathcal{X}$ of $\mathbb{C}^{k+1}$, there exists at least one j-dimensional subspace $\mathcal{Y}$ of $\mathbb{C}^k$ such that

$$\left\{\begin{bmatrix}\mathbf{y}\\0\end{bmatrix} : \mathbf{y} \in \mathcal{Y}\right\} \subseteq \mathcal{X}\,. \tag{20.2}$$

Thus, as

$$\langle B\mathbf{y}, \mathbf{y}\rangle = \left\langle A\begin{bmatrix}\mathbf{y}\\0\end{bmatrix}, \begin{bmatrix}\mathbf{y}\\0\end{bmatrix}\right\rangle \Longrightarrow \mathfrak{b}(\mathcal{Y}) \le \mathfrak{a}(\mathcal{X})$$

for $\mathbf{y} \in \mathcal{Y}$ and such a pair of spaces $\mathcal{Y}$ and $\mathcal{X}$, it follows that

$$\lambda_j(B) = \min_{\mathcal{Y}\in\mathcal{T}_j} \mathfrak{b}(\mathcal{Y}) \le \mathfrak{a}(\mathcal{X})\,.$$

Therefore, as this lower bound is valid for each subspace $\mathcal{X} \in \mathcal{S}_{j+1}$, it is also valid for the minimum over all $\mathcal{X} \in \mathcal{S}_{j+1}$, i.e.,

$$\lambda_j(B) \le \min_{\mathcal{X}\in\mathcal{S}_{j+1}} \mathfrak{a}(\mathcal{X}) = \lambda_{j+1}(A)\,.$$

□

Exercise 20.1. Find a 2-dimensional subspace $\mathcal{Y}$ of $\mathbb{C}^3$ such that (20.2) holds for each of the following two choices of $\mathcal{X}$:

$$\mathcal{X} = \text{span}\left\{\begin{bmatrix}1\\1\\1\\1\end{bmatrix}, \begin{bmatrix}1\\1\\0\\1\end{bmatrix}, \begin{bmatrix}1\\0\\1\\1\end{bmatrix}\right\} \quad \text{and} \quad \mathcal{X} = \text{span}\left\{\begin{bmatrix}1\\1\\1\\0\end{bmatrix}, \begin{bmatrix}1\\1\\0\\0\end{bmatrix}, \begin{bmatrix}1\\0\\0\\0\end{bmatrix}\right\}.$$

Exercise 20.2. Show that if $\mathcal{X}$ is a $j+1$-dimensional subspace of $\mathbb{C}^{k+1}$ with basis $\mathbf{u}_1, \ldots, \mathbf{u}_{k+1}$, then there exists a j-dimensional subspace $\mathcal{Y}$ of $\mathbb{C}^k$ such that (20.2) holds. [HINT: Build a basis $\mathbf{v}_1, \ldots, \mathbf{v}_k$ for $\mathcal{Y}$, with Exercise 20.1 as a guide.]

Exercise 20.3. Let $A = A^H \in \mathbb{C}^{n\times n}$ and $B = B^H \in \mathbb{C}^{n\times n}$. Show that if $\lambda_1(A) \leq \cdots \leq \lambda_n(A)$ and $\lambda_1(B) \leq \cdots \leq \lambda_n(B)$, then

$$\lambda_j(A) + \lambda_1(B) \leq \lambda_j(A+B) \leq \lambda_j(A) + \lambda_n(B) \quad \text{for} \quad j = 1, \ldots, n\,.$$

[HINT: Invoke the Courant-Fischer theorem.]

Theorem 20.2. *Let A_1 and A_2 be $n \times n$ Hermitian matrices with eigenvalues $\lambda_1^{(1)} \leq \cdots \leq \lambda_n^{(1)}$ and $\lambda_1^{(2)} \leq \cdots \leq \lambda_n^{(2)}$, respectively, and suppose further that $A_1 - A_2 \succeq 0$. Then:*

(1) $\lambda_j^{(1)} \geq \lambda_j^{(2)}$ *for* $j = 1, \ldots, n$.

(2) *If also* $\operatorname{rank}(A_1 - A_2) = r$, *then* $\lambda_j^{(1)} \leq \lambda_{j+r}^{(2)}$ *for* $j = 1, \ldots, n-r$.

Proof. The first assertion is a straightforward consequence of the Courant-Fischer theorem and is left to the reader as an exercise. To verify (2), let $B = A_1 - A_2$. Then, since $\operatorname{rank} B = r$ by assumption, $\dim \mathcal{N}_B = n - r$, and hence, for any k dimensional subspace $\mathcal{Y}$ of $\mathbb{C}^n$,

$$\begin{aligned} \dim(\mathcal{Y} \cap \mathcal{N}_B) &= \dim \mathcal{Y} + \dim \mathcal{N}_B - \dim(Y + \mathcal{N}_B) \\ &\geq k + n - r - n \\ &= k - r\,. \end{aligned}$$

Thus, if $k > r$ and $\mathcal{S}_j$ denotes the set of all j-dimensional subspaces of $\mathbb{C}^n$, then

$$\begin{aligned} \lambda_{k-r}^{(1)} &= \min_{\mathcal{U} \in \mathcal{S}_{k-r}} \max\{\langle A_1\mathbf{y}, \mathbf{y}\rangle : \mathbf{y} \in \mathcal{U} \text{ and } \|\mathbf{y}\| = 1\} \\ &\leq \max\{\langle A_1\mathbf{y}, \mathbf{y}\rangle : \mathbf{y} \in \mathcal{Y} \cap \mathcal{N}_B \text{ and } \|\mathbf{y}\| = 1\} \\ &= \max\{\langle A_2\mathbf{y}, \mathbf{y}\rangle + \langle B\mathbf{y}, \mathbf{y}\rangle : \mathbf{y} \in \mathcal{Y} \cap \mathcal{N}_B \text{ and } \|\mathbf{y}\| = 1\} \\ &= \max\{\langle A_2\mathbf{y}, \mathbf{y}\rangle : \mathbf{y} \in \mathcal{Y} \cap \mathcal{N}_B \text{ and } \|\mathbf{y}\| = 1\} \\ &\leq \max\{\langle A_2\mathbf{y}, \mathbf{y}\rangle : \mathbf{y} \in \mathcal{Y} \text{ and } \|\mathbf{y}\| = 1\}\,. \end{aligned}$$

Therefore, since this inequality is valid for every choice of $\mathcal{Y} \in \mathcal{S}_k$, it follows that

$$\lambda_{k-r}^{(1)} \leq \lambda_k^{(2)} \quad \text{for} \quad k = r+1, \ldots, n\,.$$

But this is equivalent to the asserted upper bound in (2). □

Exercise 20.4. Verify the first assertion of Theorem 20.2.

Exercise 20.5. Let $A = B + \gamma\mathbf{u}\mathbf{u}^H$, where $B \in \mathbb{C}^{n\times n}$ is Hermitian, $\mathbf{u} \in \mathbb{C}^n$, $\gamma \in \mathbb{R}$; and let $\lambda_1(A) \leq \cdots \leq \lambda_n(A)$, $\lambda_1(B) \leq \cdots \leq \lambda_n(B)$ denote the eigenvalues of A and B, respectively. Show that

(a) If $\gamma \geq 0$, then $\lambda_j(B) \leq \lambda_j(A) \leq \lambda_{j+1}(B)$ for $j = 1, \dots, n-1$ and $\lambda_n(B) \leq \lambda_n(A)$.

(b) If $\gamma \leq 0$, then $\lambda_{j-1}(B) \leq \lambda_j(A) \leq \lambda_j(B)$ for $j = 2, \dots, n$ and $\lambda_1(A) \leq \lambda_1(B)$.

Exercise 20.6. Show that in the setting of Exercise 20.5,

$$\lambda_j(A) = \lambda_j(B) + c_j\gamma, \quad \text{where} \quad c_j \geq 0 \quad \text{and} \quad \sum_{j=1}^{n} c_j = \mathbf{u}^H\mathbf{u}.$$

[HINT: $\sum_{j=1}^n \lambda_j(A) = \sum_{j=1}^n \langle A\mathbf{u}_j, \mathbf{u}_j\rangle = \sum_{j=1}^n \langle A\mathbf{v}_j, \mathbf{v}_j\rangle$ for any two orthonormal sets of vectors $\{\mathbf{u}_1, \dots, \mathbf{u}_n\}$ and $\{\mathbf{v}_1, \dots, \mathbf{v}_n\}$ in $\mathbb{C}^n$.]

Exercise 20.7. Let A° be a pseudoinverse of a matrix $A = A^H \in \mathbb{C}^{n\times n}$ such that A° is also Hermitian. Show that $\mathcal{E}_\pm(A) = \mathcal{E}_\pm(A^\circ)$ and $\mathcal{E}_0(A) = \mathcal{E}_0(A^\circ)$.

A **tridiagonal** Hermitian matrix $A_n \in \mathbb{R}^{n\times n}$ of the form

$$A_n = \sum_{j=1}^{n} a_j\mathbf{e}_j\mathbf{e}_j^T + \sum_{j=1}^{n-1} b_j(\mathbf{e}_j\mathbf{e}_{j+1}^T + \mathbf{e}_{j+1}\mathbf{e}_j^T) = \begin{bmatrix} a_1 & b_1 & 0 & \cdots & 0 & 0 \\ b_1 & a_2 & b_2 & \cdots & 0 & 0 \\ 0 & b_2 & a_3 & \cdots & 0 & 0 \\ \vdots & & & & & \vdots \\ 0 & 0 & 0 & \cdots & b_{n-1} & a_n \end{bmatrix}$$

with $b_j > 0$ and $a_j \in \mathbb{R}$ is termed a **Jacobi matrix**.

Exercise 20.8. Show that a Jacobi matrix A_{n+1} has $n+1$ distinct eigenvalues $\lambda_1 < \cdots < \lambda_{n+1}$ and that if $\mu_1 < \cdots < \mu_n$ denote the eigenvalues of the Jacobi matrix A_n, then $\lambda_j < \mu_j < \lambda_{j+1}$ for $j = 1, \dots, n$.

20.2. Sylvester's law of inertia

Theorem 20.3. *Let A and $B = C^HAC$ be Hermitian matrices of sizes $n \times n$ and $m \times m$, respectively. Then*

$$\mathcal{E}_+(A) \geq \mathcal{E}_+(B) \quad \textit{and} \quad \mathcal{E}_-(A) \geq \mathcal{E}_-(B),$$

with equality in both if $\operatorname{rank} A = \operatorname{rank} B$.

Proof. Since A and B are Hermitian matrices, there exists a pair of invertible matrices $U \in \mathbb{C}^{n\times n}$ and $V \in \mathbb{C}^{m\times m}$ such that

$$A = U^H \begin{bmatrix} I_{s_1} & O & O \\ O & -I_{t_1} & O \\ O & O & O \end{bmatrix} U \quad \text{and} \quad B = V^H \begin{bmatrix} I_{s_2} & O & O \\ O & -I_{t_2} & O \\ O & O & O \end{bmatrix} V,$$

where $s_1 = \mathcal{E}_+(A)$, $t_1 = \mathcal{E}_-(A)$, $s_2 = \mathcal{E}_+(B)$ and $t_2 = \mathcal{E}_-(B)$. Therefore,

$$\begin{bmatrix} I_{s_2} & O & O \\ O & -I_{t_2} & O \\ O & O & O \end{bmatrix} = Q^H \begin{bmatrix} I_{s_1} & O & O \\ O & -I_{t_1} & O \\ O & O & O \end{bmatrix} Q\,,$$

where $Q = UCV^{-1}$. Thus, upon expressing the $n \times m$ matrix in block form as

$$Q = \begin{bmatrix} Q_{11} & Q_{12} & Q_{13} \\ Q_{21} & Q_{22} & Q_{23} \\ Q_{31} & Q_{32} & Q_{33} \end{bmatrix},$$

where the heights of the block rows are s_1, t_1 and $n - s_1 - t_1$ and the widths of the block columns are s_2, t_2 and $m - s_2 - t_2$, respectively, it is readily seen that

$$I_{s_2} = Q_{11}^H Q_{11} - Q_{21}^H Q_{21} \quad \text{and} \quad I_{t_2} = -Q_{12}^H Q_{12} + Q_{22}^H Q_{22}\,.$$

Therefore,

$$Q_{11}^H Q_{11} = I_{s_2} + Q_{21}^H Q_{21} \quad \text{and} \quad Q_{22}^H Q_{22} = I_{t_2} + Q_{12}^H Q_{21}\,.$$

The first of these formulas implies that $\mathcal{N}_{Q_{11}} = \{\mathbf{0}\}$, and hence the principle of conservation of dimension applied to the $s_1 \times s_2$ matrix Q_{11} implies that $s_2 = \operatorname{rank} Q_{11}$. But this in turn implies that Q_{11} must have s_2 linearly independent rows and hence that $s_1 \geq s_2$. By similar reasoning, $t_1 \geq t_2$. Finally, if $\operatorname{rank} A = \operatorname{rank} B = r$, then $t_1 = r - s_1$ and $t_2 = r - s_2$ and hence

$$t_1 \geq t_2 \Longleftrightarrow r - s_1 \geq r - s_2 \Longleftrightarrow s_2 \geq s_1\,.$$

Thus, under this extra condition, equality prevails. □

Corollary 20.4. (**Sylvester's law of inertia**) Let A and $B = C^H AC$ be two $n \times n$ Hermitian matrices and suppose that C is invertible. Then

$$\mathcal{E}_+(A) = \mathcal{E}_+(B)\,, \quad \mathcal{E}_-(A) = \mathcal{E}_-(B) \quad \text{and} \quad \mathcal{E}_0(A) = \mathcal{E}_0(B)\,.$$

20.3. Congruence

A pair of matrices $A \in \mathbb{C}^{n\times n}$ and $B \in \mathbb{C}^{n\times n}$ is said to be **congruent** if there exists an invertible matrix $C \in \mathbb{C}^{n\times n}$ such that $A = C^H BC$. This connection will be denoted by the symbol $A \sim B$.

Remark 20.5. In terms of congruence, Sylvester's law of inertia is:

If $A, B \in \mathbb{C}^{n\times n}$, $B = B^H$ and $A \sim B$, then $A = A^H$, $\mathcal{E}_\pm(A) = \mathcal{E}_\pm(B)$ and $\mathcal{E}_0(A) = \mathcal{E}_0(B)$.

Exercise 20.9. Let $U \in \mathbb{C}^{n\times n}$ be a unitary matrix and let $A \in \mathbb{C}^{n\times n}$. Show that if $A \succ O$ and $AU \succ O$, then $U = I_n$.

Lemma 20.6. *If $B \in \mathbb{C}^{p\times q}$, then*

$$E = \begin{bmatrix} O & B \\ B^H & O \end{bmatrix} \sim \begin{bmatrix} BB^H & O \\ O & -B^H B \end{bmatrix} \quad \textit{and} \quad \mathcal{E}_\pm(E) = \operatorname{rank} B\,.$$

Proof. The first step is to note that, in terms of the Moore-Penrose inverse $B^\dagger$ of B,

$$\begin{bmatrix} I_p & (B^\dagger)^H \\ -B^H & I_q \end{bmatrix} \begin{bmatrix} O & B \\ B^H & O \end{bmatrix} \begin{bmatrix} I_p & -B \\ B^\dagger & I_q \end{bmatrix} = 2 \begin{bmatrix} BB^\dagger & O \\ O & -B^H B \end{bmatrix}$$

and, by Theorem 5.5,

$$\det \begin{bmatrix} I_p & -B \\ B^\dagger & I_q \end{bmatrix} = \det(I_q + B^\dagger B) > 0\,.$$

The conclusion then follows upon multiplying both sides of the preceding identity by the matrix

$$\frac{1}{\sqrt{2}} \begin{bmatrix} Y & O \\ O & I_q \end{bmatrix}, \quad \text{where} \quad Y = (BB^H)^{1/2} + I_p - BB^\dagger = Y^H\,.$$

This does the trick, since Y is invertible and

$$YBB^\dagger Y = (BB^H)^{1/2} BB^\dagger (BB^H)^{1/2} = BB^H\,.$$

□

Exercise 20.10. Show that the matrix Y that is defined in the proof of Lemma 20.6 is invertible.

Exercise 20.11. Furnish a second proof of Lemma 20.6 by showing that if $p+q=n$, then $\det(\lambda I_n - E) = \lambda^{(p-q)} \det(\lambda^2 I_q - B^H B)$. [HINT: Show that if $\operatorname{rank} B = k$, then $\mathcal{E}_\pm(E) = k$ and $\mathcal{E}_0(E) = n - 2k$.]

Lemma 20.7. *If $B \in \mathbb{C}^{p\times q}$ and $C = C^H \in \mathbb{C}^{n\times n}$, then*

$$E = \begin{bmatrix} O & O & B \\ O & C & O \\ B^H & O & O \end{bmatrix} \sim \begin{bmatrix} BB^H & O & O \\ O & C & O \\ O & O & -B^H B \end{bmatrix}$$

and hence

$$\mathcal{E}_\pm(E) = \mathcal{E}_\pm(C) + \operatorname{rank} B\,.$$

Proof. This is an easy variant of Lemma 20.6: Just multiply the given matrix on the right by the invertible constant matrix

$$K = \begin{bmatrix} I_p & O & -B \\ O & I_n & O \\ B^\dagger & O & I_q \end{bmatrix}$$

and on the left by K^H to start things off. This yields the formula

$$K^H E K = \begin{bmatrix} 2BB^\dagger & O & O \\ O & C & O \\ O & O & -2B^H B \end{bmatrix},$$

which leads easily to the desired conclusion upon multiplying $K^H EK$ on the left and the right by the invertible constant Hermitian matrix

$$\frac{1}{\sqrt{2}} \mathrm{diag}\,\{Y, \sqrt{2} I_n, I_q\},$$

where Y is defined in the proof of Lemma 20.6. □

Exercise 20.12. Show that congruence is an equivalence relation, i.e., (i) $A \sim A$; (ii) $A \sim B \Longrightarrow B \sim A$; and (iii) $A \sim B$, $B \sim C \Longrightarrow A \sim C$.

Exercise 20.13. Let $A = A^H$, $C = C^H$ and $E = \begin{bmatrix} A & B \\ B^H & C \end{bmatrix}$. Show that $E = E^H$ and that $\mathcal{E}_\pm(E) \geq \mathcal{E}_\pm(A)$ and $\mathcal{E}_\pm(E) \geq \mathcal{E}_\pm(C)$. [HINT: Exploit Theorem 20.3.]

Exercise 20.14. Let $\mathbf{e}_j$ denote the j'th column of I_n. Show that the eigenvalues of $Z_n = \sum_{j=1}^n \mathbf{e}_j \mathbf{e}_{n+1-j}^T$ must be equal to either 1 or -1 without calculating them. [HINT: Show that $Z_n^H = Z_n$ and $Z_n^H Z_n = I_n$.]

Exercise 20.15. Let Z_n be the matrix defined in Exercise 20.14.

(a) Show that if $n = 2k$ is even, then $\mathcal{E}_+(Z_n) = \mathcal{E}_-(Z_n) = k$.

(b) Show that if $n = 2k+1$ is odd, then $\mathcal{E}_+(Z_n) = k+1$ and $\mathcal{E}_-(Z_n) = k$.

[HINT: Verify and exploit the identity $\begin{bmatrix} I_k & Z_k \\ -Z_k & I_k \end{bmatrix} \begin{bmatrix} O & Z_k \\ Z_k & O \end{bmatrix} \begin{bmatrix} I_k & -Z_k \\ Z_k & I_k \end{bmatrix} = 2\,\mathrm{diag}\,\{I_k, -I_k\}$.]

Exercise 20.16. Confirm the conclusions in Exercise 20.14 by calculating $\det(\lambda I_n - Z_n)$. [HINT: If $n = 2k$, then $\lambda I_n - Z_n = \begin{bmatrix} \lambda I_k & -Z_k \\ -Z_k & \lambda I_k \end{bmatrix}$.]

20.4. Counting positive and negative eigenvalues

Lemma 20.8. *Let $A = A^H \in \mathbb{C}^{k\times k}$, $B \in \mathbb{C}^{k\times \ell}$; and let*

$$E = \begin{bmatrix} A & B \\ B^H & O \end{bmatrix}. \tag{20.3}$$

Then

$$\mathcal{E}_\pm(E) \geq \mathrm{rank}\,(B). \tag{20.4}$$

Proof. Let $m = \operatorname{rank} B$ and suppose that $m \geq 1$, because otherwise there is nothing to prove. Then, the singular value decomposition of B yields the factorization $B = VSU^H$, where V and U are unitary matrices of sizes $k \times k$ and $\ell \times \ell$, respectively,

$$S = \begin{bmatrix} D & O_{m\times \ell'} \\ O_{k'\times m} & O_{k'\times \ell'} \end{bmatrix}, \quad k' = k - m, \ \ell' = \ell - m,$$

D is an $m \times m$ positive definite diagonal matrix, and for the sake of definiteness, it is assumed that $k' \geq 1$ and $\ell' \geq 1$. Thus,

$$\begin{aligned} E &= \begin{bmatrix} V & O \\ O & U \end{bmatrix} \begin{bmatrix} V^HAV & S \\ S^H & O \end{bmatrix} \begin{bmatrix} V^H & O \\ O & U^H \end{bmatrix} \\ &\sim \begin{bmatrix} V^HAV & S \\ S^H & O \end{bmatrix} \\ &= \begin{bmatrix} \widetilde{A}_{11} & \widetilde{A}_{12} & D & O \\ \widetilde{A}_{21} & \widetilde{A}_{22} & O & O \\ D & O & O & O \\ O & O & O & O \end{bmatrix}, \end{aligned}$$

where the block decomposition of $\widetilde{A} = V^HAV$ is chosen to be compatible with that of S, i.e., $\widetilde{A}_{11} \in \mathbb{C}^{m\times m}$, $\widetilde{A}_{12} \in \mathbb{C}^{m\times \ell'}$, $\widetilde{A}_{22} \in \mathbb{C}^{\ell\times\ell}$. Moreover, $\widetilde{A}_{11} = \widetilde{A}_{11}^H$, $\widetilde{A}_{12} = \widetilde{A}_{21}^H$, $\widetilde{A}_{22} = \widetilde{A}_{22}^H$ and, since D is invertible, the last matrix on the right is congruent to

$$\begin{bmatrix} O & O & D & O \\ O & \widetilde{A}_{22} & O & O \\ D & O & O & O \\ O & O & O & O \end{bmatrix}, \quad \text{which is congruent to} \quad \begin{bmatrix} \widetilde{A}_{22} & O & O & O \\ O & O & D & O \\ O & D & O & O \\ O & O & O & O \end{bmatrix}$$

as may be easily verified by symmetric block Gaussian elimination for the first congruence and appropriately chosen permutations for the second; the details are left to the reader as an exercise. Therefore, by Lemma 20.6,

$$\begin{aligned} \mathcal{E}_\pm(E) &= \mathcal{E}_\pm(\widetilde{A}_{22}) + \mathcal{E}_\pm\left(\begin{bmatrix} O & D \\ D & O \end{bmatrix}\right) \\ &= \mathcal{E}_\pm(\widetilde{A}_{22}) + \operatorname{rank}(D) \geq \operatorname{rank}(D). \end{aligned}$$

This completes the proof, since $\operatorname{rank}(D) = \operatorname{rank}(B)$. □

Exercise 20.17. Show that if $B, C \in \mathbb{C}^{n\times k}$ and $A = BC^H + CB^H$, then $\mathcal{E}_\pm(A) \leq k$. [HINT: Write $A = [B\ C]J[B\ C]^H$ for an appropriately chosen signature matrix J and invoke Theorem 20.3.]

Exercise 20.18. Show that if $B \in \mathbb{C}^{p\times q}$ and $B^\dagger$ denotes the Moore-Penrose inverse of B, then $\begin{bmatrix} O & B \\ B^H & O \end{bmatrix} \sim \begin{bmatrix} BB^\dagger & O \\ O & -B^\dagger B \end{bmatrix}$.

Exercise 20.19. Evaluate $\mathcal{E}_\pm(E_j)$ and $\mathcal{E}_0(E_j)$ for the matrices

$$E_1 = \begin{bmatrix} I_k & I_k \\ I_k & O \end{bmatrix} \quad \text{and} \quad E_2 = \begin{bmatrix} I_\ell & O & O & I_\ell \\ O & I_m & O & O \\ O & O & O & O \\ I_\ell & O & O & O \end{bmatrix}.$$

Lemma 20.9. *Let $A = A^H$, $D = D^H$ and*

$$E = \begin{bmatrix} A & B & C \\ B^H & D & O \\ C^H & O & O \end{bmatrix}.$$

Then

$$\mathcal{E}_\pm(E) \geq \mathcal{E}_\pm(D) + \operatorname{rank} C\,.$$

Proof. Let $C = VSU^H$ be the singular value decomposition of C. Then

$$E \sim \begin{bmatrix} \widetilde{A} & \widetilde{B} & S \\ \widetilde{B}^H & D & O \\ S^H & O & O \end{bmatrix},$$

where $\widetilde{A} = V^H AV$ and $\widetilde{B} = V^H B$. Thus, if

$$S = \begin{bmatrix} F & O \\ O & O \end{bmatrix},$$

where F is a positive definite diagonal matrix and $\widetilde{A}$ and $\widetilde{B}$ are written in compatible block form as

$$\widetilde{A} = \begin{bmatrix} \widetilde{A}_{11} & \widetilde{A}_{12} \\ \widetilde{A}_{21} & \widetilde{A}_{22} \end{bmatrix} \quad \text{and} \quad \begin{bmatrix} \widetilde{B}_1 \\ \widetilde{B}_2 \end{bmatrix},$$

respectively, then $\widetilde{A}_{11} = \widetilde{A}_{11}^H$, $\widetilde{A}_{12} = \widetilde{A}_{21}^H$, $\widetilde{A}_{22} = \widetilde{A}_{22}^H$ and

$$(20.5) \qquad E \sim \begin{bmatrix} \widetilde{A}_{11} & \widetilde{A}_{12} & \widetilde{B}_1 & F & O \\ \widetilde{A}_{21} & \widetilde{A}_{22} & \widetilde{B}_2 & O & O \\ \widetilde{B}_1^H & \widetilde{B}_2^H & D & O & O \\ F & O & O & O & O \\ O & O & O & O & O \end{bmatrix} \sim \begin{bmatrix} O & O & O & F & O \\ O & \widetilde{A}_{22} & \widetilde{B}_2 & O & O \\ O & \widetilde{B}_2^H & D & O & O \\ F & O & O & O & O \\ O & O & O & O & O \end{bmatrix}.$$

Therefore, upon applying Theorem 20.3 to the identity

$$\begin{bmatrix} O & O & F \\ O & D & O \\ F & O & O \end{bmatrix} = \begin{bmatrix} I & O & O \\ O & O & O \\ O & I & O \\ O & O & I \\ O & O & O \end{bmatrix}^T \begin{bmatrix} O & O & O & F & O \\ O & \widetilde{A}_{22} & \widetilde{B}_2 & O & O \\ O & \widetilde{B}_2^H & D & O & O \\ F & O & O & O & O \\ O & O & O & O & O \end{bmatrix} \begin{bmatrix} I & O & O \\ O & O & O \\ O & I & O \\ O & O & I \\ O & O & O \end{bmatrix},$$

and then invoking Lemma 20.7, it is readily seen that

$$\mathcal{E}_\pm(E) \geq \operatorname{rank} F + \mathcal{E}_\pm(D)\,.$$

This completes the proof, since $\operatorname{rank} F = \operatorname{rank} C$. □

Lemma 20.10. *Let $E = E^H$ be given by formula (20.3) with* $\operatorname{rank} A \geq 1$ *and let $A^\dagger$ denote the Moore-Penrose inverse of A. Then*

$$\mathcal{E}_\pm(E) \geq \mathcal{E}_\pm(A) + \mathcal{E}_\mp(BA^\dagger B^H). \tag{20.6}$$

Proof. Since $A = A^H$, there exists a unitary matrix U and an invertible diagonal matrix D such that

$$A = U \begin{bmatrix} D & O \\ O & O \end{bmatrix} U^H.$$

Thus, upon writing

$$U = [U_1 \quad U_2]$$

and

$$U^H B = \begin{bmatrix} U_1^H B \\ U_2^H B \end{bmatrix} = \begin{bmatrix} B_1 \\ B_2 \end{bmatrix}$$

in compatible block form, it is readily seen that

$$E = \begin{bmatrix} U & O \\ O & I \end{bmatrix} \begin{bmatrix} D & O & B_1 \\ O & O & B_2 \\ B_1^H & B_2^H & O \end{bmatrix} \begin{bmatrix} U^H & O \\ O & I \end{bmatrix},$$

i.e.,

$$E \sim \begin{bmatrix} D & O & B_1 \\ O & O & B_2 \\ B_1^H & B_2^H & O \end{bmatrix}.$$

Moreover, since D is invertible,

$$\begin{bmatrix} D & O & B_1 \\ O & O & B_2 \\ B_1^H & B_2^H & O \end{bmatrix} \sim \begin{bmatrix} D & O \\ O & E_1 \end{bmatrix},$$

where

$$E_1 = \begin{bmatrix} O & B_2 \\ B_2^H & O \end{bmatrix} - \begin{bmatrix} O \\ B_1^H \end{bmatrix} D^{-1} [O \quad B_1] = \begin{bmatrix} O & B_2 \\ B_2^H & -B_1^H D^{-1} B_1 \end{bmatrix}.$$

Thus,

$$\mathcal{E}_\pm(E) = \mathcal{E}_\pm(D) + \mathcal{E}_\pm(E_1)\,,$$

which justifies both the equality

$$\mathcal{E}_\pm(E) = \mathcal{E}_\pm(A) + \mathcal{E}_\pm(E_1) \tag{20.7}$$

and the inequality

$$\mathcal{E}_\pm(E) \geq \mathcal{E}_\pm(A). \tag{20.8}$$

Much the same sort of analysis implies that

$$\mathcal{E}_\pm(E_1) \geq \mathcal{E}_\pm(-B_1^H D^{-1} B_1) = \mathcal{E}_\mp(B_1^H D^{-1} B_1)$$

and hence, as

$$\begin{aligned} B_1^H D^{-1} B_1 &= [B_1^H \; B_2^H] \begin{bmatrix} D^{-1} & O \\ O & O \end{bmatrix} \begin{bmatrix} B_1 \\ B_2 \end{bmatrix} \\ &= B^H U \begin{bmatrix} D^{-1} & O \\ O & O \end{bmatrix} U^H B \\ &= B^H A^\dagger B, \end{aligned}$$

implies that

$$\mathcal{E}_\pm(E_1) \geq \mathcal{E}_\mp(B^H A^\dagger B).$$

The asserted inequality (20.6) now drops out easily upon inserting the last inequality into the identity (20.7). □

Exercise 20.20. Let $B \in \mathbb{C}^{p\times q}$, $C = C^H \in \mathbb{C}^{q\times q}$ and let $E = \begin{bmatrix} O & B \\ B^H & C \end{bmatrix}$. Show that $\mathcal{E}_\pm(E) \geq \max\{\mathcal{E}_\pm(C), \operatorname{rank}(B)\}$.

20.5. Exploiting continuity

In this section we shall illustrate by example how to exploit the fact that the eigenvalues of a matrix $A \in \mathbb{C}^{n\times n}$ are continuous functions of the entries in the matrix to obtain information on the location of its eigenvalues. But see also Exercise 20.22. The facts in Appendix A may be helpful.

Theorem 20.11. *Let*

$$E = \begin{bmatrix} A & B & I_p \\ B^H & I_p & O \\ I_p & O & O \end{bmatrix},$$

in which $A, B \in \mathbb{C}^{p\times p}$ and $A = A^H$. Then $\mathcal{E}_+(E) = 2p$ and $\mathcal{E}_-(E) = p$.

Proof. Let

$$E(t) = \begin{bmatrix} tA & tB & I_p \\ tB^H & I_p & O \\ I_p & O & O \end{bmatrix}.$$

Then it is readily checked that:

(1) $E(t)$ is invertible for every choice of $t \in \mathbb{R}$.

(2) $\mathcal{E}_+(E(0)) = 2p$ and $\mathcal{E}_-(E(0)) = p$.

(3) The set $\Omega_1 = \{t \in \mathbb{R} : \mathcal{E}_+(E(t)) = 2p\}$ is an open subset of $\mathbb{R}$.

(4) The set $\Omega_2 = \{t \in \mathbb{R} : \mathcal{E}_+(E(t)) < 2p\}$ is an open subset of $\mathbb{R}$.

(5) $\mathbb{R} = \Omega_1 \cup \Omega_2$.

(6) $\Omega_1 \neq \emptyset$.

Thus, as the connected set $\mathbb{R} = \Omega_1 \cup \Omega_2$ is the union of two open sets, Item (6) implies that $\Omega_2 = \emptyset$. Therefore, $\Omega_1 = \mathbb{R}$. $\square$

Exercise 20.21. Verify the six items that are listed in the proof of Theorem 20.11.

Exercise 20.22. Show that the matrix $E(t)$ that is introduced in the proof of Theorem 20.11 is congruent to $E(0)$ for every choice of $t \in \mathbb{R}$.

20.6. Geršgorin disks

Let $A \in \mathbb{C}^{n\times n}$ with entries a_{ij} and let

$$\rho_i(A) = \sum_{j=1}^{n} |a_{ij}| - |a_{ii}| \quad \text{for} \quad i = 1, \ldots, n\,.$$

Then the set

$$\Gamma_i(A) = \{\lambda \in \mathbb{C} : |\lambda - a_{ii}| \leq \rho_i(A)\}$$

is called the i'th **Geršgorin disk** of A.

Theorem 20.12. *If $A \in \mathbb{C}^{n\times n}$, then:*

(1) *$\sigma(A) \subseteq \cup_{i=1}^{n} \Gamma_i(A)$.*

(2) *A union Ω_1 of k Geršgorin disks that has no points in common with the union Ω_2 of the remaining $n-k$ Geršgorin disks contains exactly k eigenvalues of A.*

Proof. If $\lambda \in \sigma(A)$, then there exists a nonzero vector $\mathbf{u} \in \mathbb{C}^n$ with components $u_1, \ldots, u_n$ such that $(\lambda I_n - A)\mathbf{u} = \mathbf{0}$. Suppose that

$$|u_k| = \max\{|u_j| : j = 1, \ldots, n\}\,.$$

Then the identity

$$(\lambda - a_{kk})u_k = \sum_{j=1}^{n} a_{kj}u_j - a_{kk}u_k$$

implies that

$$\begin{aligned} |(\lambda - a_{kk})u_k| &\leq \sum_{j=1}^{n} |a_{kj}||u_j| - |a_{kk}||u_k| \\ &\leq \rho_k(A)|u_k|\,. \end{aligned}$$

Therefore,

$$\lambda \in \Gamma_k(A) \subset \bigcup_{i=1}^{n} \Gamma_i(A)\,.$$

This completes the proof of (1).

Next, to verify (2), let $D = \operatorname{diag}\{a_{11}, \ldots, a_{nn}\}$ and let

$$B(t) = D + t(A - D) \quad \text{for} \quad 0 \le t \le 1 .$$

Then

$$\sigma(B(t)) \subseteq \Omega_1(t) \cup \Omega_2(t) \quad \text{for} \quad 0 \le t \le 1 ,$$

where $\Omega_j(t)$ denotes the union of the disks with centers in $\Omega_j(1)$ but with radii $\rho_i(B(t)) = t\rho_i(B(1))$ for $0 \le t \le 1$. Clearly $\Omega_1(0)$ contains exactly k eigenvalues of $D = B(0)$, and $\Omega_2(0)$ contains exactly $n-k$ eigenvalues of $D = B(0)$. Moreover, since the eigenvalues of $B(t)$ must belong to $\Omega_1(t) \cup \Omega_2(t)$ and vary continuously with t, the assertion follows from the inclusions

$$\Omega_1(t) \subseteq \Omega_1(1) \quad \text{and} \quad \Omega_2(t) \subseteq \Omega_2(1) \quad \text{for} \quad 0 \le t \le 1$$

and the fact that $\Omega_1(1) \cap \Omega_2(1) = \emptyset$. □

Exercise 20.23. Show that the spectral radius $r_\sigma(A)$ of a matrix $A \in \mathbb{C}^{n\times n}$ is subject to the bound

$$r_\sigma(A) \le \max \left\{ \sum_{j=1}^{n} |a_{ij}| : i = 1, \ldots, n \right\} .$$

Exercise 20.24. Show that the spectral radius $r_\sigma(A)$ of a matrix $A \in \mathbb{C}^{n\times n}$ is subject to the bound

$$r_\sigma(A) \le \max \{ \sum_{i=1}^{n} |a_{ij}| : j = 1, \ldots, n \} .$$

Exercise 20.25. Let $A \in \mathbb{C}^{n\times n}$. Show that if $a_{ii} > \rho_i(A)$ for $i = 1, \ldots, n$, then A is invertible.

Exercise 20.26. Let $A \in \mathbb{C}^{n\times n}$. Show that A is a diagonal matrix if and only if $\sigma(A) = \cup_{i=1}^{n} \Gamma_i(A)$.

20.7. The spectral mapping principle

In this section we shall give an elementary proof of the spectral mapping principle for polynomials because it fits the theme of this chapter; see Theorem 17.25 for a stronger result.

Theorem 20.13. *Let $A \in \mathbb{C}^{n\times n}$, let $\lambda_1, \ldots, \lambda_k$ denote the distict eigenvalues of A and let $p(\lambda)$ be a polynomial. Then*

(20.9)

$$\begin{aligned} \det(\lambda I_n - A) \quad &= \quad (\lambda - \lambda_1)^{\alpha_1} \cdots (\lambda - \lambda_k)^{\alpha_k} \\ &\Longrightarrow \det(\lambda I_n - p(A)) = (\lambda - p(\lambda_1))^{\alpha_1} \cdots (\lambda - p(\lambda_k))^{\alpha_k} ; \end{aligned}$$

i.e.,

$$\sigma(p(A)) = p(\sigma(A)) . \tag{20.10}$$

Proof. Let V be an invertible matrix such that $V^{-1}AV = J$ is in Jordan form. Then it is readily checked that $J^k = D^k + T_k$ for $k = 1, 2. \ldots,$ where $D = \operatorname{diag}\{d_{11}, \ldots, d_{nn}\}$ is a diagonal matrix and T_k is strictly upper triangular. Thus, if the polynomial $p(\lambda) = a_0 + a_1\lambda + \cdots + a_\ell\lambda^\ell$, then

$$p(J) = p(D) + \sum_{j=1}^{\ell} a_j T_j \,.$$

Consequently,

$$\begin{aligned} \det(\lambda I_n - p(A)) &= \det(\lambda I_n - p(J)) = \det(\lambda I_n - p(D)) \\ &= (\lambda - p(d_{11})) \cdots (\lambda - p(d_{nn}))\,, \end{aligned}$$

which yields (20.9) and (20.10). □

20.8. $AX = XB$

Lemma 20.14. *Let $A, X, B \in \mathbb{C}^{n\times n}$ and suppose that $AX = XB$. Then there exists a matrix $C \in \mathbb{C}^{n\times n}$ such that $AX = X(B+C)$ and $\sigma(B+C) \subseteq \sigma(A)$.*

Proof. If X is invertible, then $\sigma(A) = \sigma(B)$; i.e., the matrix $C = O$ does the trick. Suppose therefore that X is not invertible and that $C_\beta^{(k)}$ is a $k \times k$ Jordan cell in the Jordan decomposition of $B = UJU^{-1}$ such that $\beta \notin \sigma(A)$. Then there exists a set of k linearly independent vectors $\mathbf{u}_1, \ldots, \mathbf{u}_k$ in $\mathbb{C}^n$ such that

$$B[\mathbf{u}_1 \cdots \mathbf{u}_k] = [\mathbf{u}_1 \cdots \mathbf{u}_k]C_\beta^{(k)}$$

and

$$\begin{aligned} AX[\mathbf{u}_1 \cdots \mathbf{u}_k] &= XB[\mathbf{u}_1 \cdots \mathbf{u}_k] \\ &= X[\mathbf{u}_1 \cdots \mathbf{u}_k]C_\beta^{(k)}\,; \end{aligned}$$

i.e.,

$$AX\mathbf{u}_1 = \beta X\mathbf{u}_1$$

and

$$AX\mathbf{u}_j = \beta X\mathbf{u}_j + \mathbf{u}_{j-1} \text{ for } j = 2, \ldots, k\,.$$

Since $\beta \notin \sigma(A)$, it is readily checked that $X\mathbf{u}_j = \mathbf{0}$ for $j = 1, \ldots, k$. Thus, if

$$B_1 = B + [\mathbf{u}_1 \cdots \mathbf{u}_k](C_\alpha^{(k)} - C_\beta^{(k)}) \begin{bmatrix} \vec{\mathbf{v}}_1 \\ \vdots \\ \vec{\mathbf{v}}_k \end{bmatrix},$$

where $\vec{\mathbf{v}}_1, \ldots, \vec{\mathbf{v}}_k$ are the rows in $V = U^{-1}$ corresponding to the columns $\mathbf{u}_1, \ldots, \mathbf{u}_k$, and $\alpha \in \sigma(A)$, then

$$XB_1 = XB = AX,$$

and the diagonal entry of the block under consideration in the Jordan decomposition of B_1 now belongs to $\sigma(A)$ and not to $\sigma(B)$. Moreover, none of the other Jordan blocks in the Jordan decomposition of B are affected by this change. The same procedure can now be applied to change the diagonal entry of any Jordan cell in the Jordan decomposition of B_1 from a point that is not in $\sigma(A)$ to a point that is in $\sigma(A)$. The proof is completed by iterating this procedure. □

Exercise 20.27. Let $A, X, B \in \mathbb{C}^{n\times n}$. Show that if $AX = XB$ and the columns of $V \in \mathbb{C}^{n\times k}$ form a basis for $\mathcal{N}_X$, then there exists a matrix $L \in \mathbb{C}^{k\times n}$ such that $\sigma(B + VL) \subseteq \sigma(A)$.

20.9. Inertia theorems

Theorem 20.15. *Let $A \in \mathbb{C}^{n\times n}$ and suppose that $G \in \mathbb{C}^{n\times n}$ is a Hermitian matrix such that $A^HG + GA \succ O$. Then:*

(1) *G is invertible.*

(2) *$\sigma(A) \cap i\mathbb{R} = \emptyset$.*

(3) *$\mathcal{E}_+(G) = \mathcal{E}_+(A)$, $\mathcal{E}_-(G) = \mathcal{E}_-(A)$ and (in view of (1) and (2)) $\mathcal{E}_0(G) = \mathcal{E}_0(A) = 0$.*

Proof. Let

$$Q = A^HG + GA.$$

Then

$$G\mathbf{u} = \mathbf{0} \Longrightarrow \mathbf{u}^HQ\mathbf{u} = 0 \Longrightarrow \mathbf{u} = \mathbf{0}$$

since $Q \succ O$. Therefore (1) holds. Similarly

$$A\mathbf{u} = \lambda\mathbf{u}, \mathbf{u} \neq \mathbf{0} \Longrightarrow (\lambda + \overline{\lambda})\mathbf{u}^HG\mathbf{u} = \mathbf{u}^HQ\mathbf{u} > 0\,.$$

Therefore, $\lambda + \overline{\lambda} \neq 0$; i.e., (2) holds.

Suppose now for the sake of definiteness that A has p eigenvalues in the open right half plane Π_+ and q eigenvalues in the open left half plane with $p \geq 1$ and $q \geq 1$. Then $p + q = n$ and $A = UJU^{-1}$ with a Jordan matrix

$$J = \begin{bmatrix} J_1 & O \\ O & J_2 \end{bmatrix},$$

where J_1 is a $p \times p$ Jordan matrix with $\sigma(J_1) \subset \Pi_+$ and J_2 is a $q \times q$ Jordan matrix with $\sigma(J_2) \subset \Pi_-$. Thus, the formula $A^HG + GA = Q$ can be rewritten as

$$J^H(U^HGU) + (U^HGU)J = U^HQU$$

and hence, upon writing

$$U^HGU = \begin{bmatrix} K_{11} & K_{12} \\ K_{21} & K_{22} \end{bmatrix} \text{ and } U^HQU = \begin{bmatrix} P_{11} & P_{12} \\ P_{21} & P_{22} \end{bmatrix}$$

in block form that is compatible with the decomposition of J,

$$\begin{bmatrix} J_1^H & O \\ O & J_2^H \end{bmatrix} \begin{bmatrix} K_{11} & K_{12} \\ K_{21} & K_{22} \end{bmatrix} + \begin{bmatrix} K_{11} & K_{12} \\ K_{21} & K_{22} \end{bmatrix} \begin{bmatrix} J_1 & O \\ O & J_2 \end{bmatrix} = \begin{bmatrix} P_{11} & P_{12} \\ P_{21} & P_{22} \end{bmatrix}.$$

Therefore,

$$J_1^H K_{11} + K_{11} J_1 = P_{11}$$

and

$$J_2^H K_{22} + K_{22} J_2 = P_{22}.$$

In view of Exercise 18.6, both of these equations have unique solutions. Moreover, it is readily checked that $K_{11} \succ O$, $K_{22} \prec O$ and

$$\begin{bmatrix} K_{11} & K_{12} \\ K_{21} & K_{22} \end{bmatrix} = C^H \begin{bmatrix} K_{11} & O \\ O & K_{22} - K_{21} K_{11}^{-1} K_{12} \end{bmatrix} C,$$

with

$$C = \begin{bmatrix} I_p & K_{11}^{-1} K_{12} \\ O & I_q \end{bmatrix}.$$

Consequently, the Sylvester inertia theorem implies that

$$\mathcal{E}_+(G) = \mathcal{E}_+(U^H G U) = \mathcal{E}_+(K_{11}) = p = \mathcal{E}_+(A)$$

and

$$\mathcal{E}_-(G) = \mathcal{E}_-(U^H G U) = \mathcal{E}_-(K_{22} - K_{21} K_{11}^{-1} K_{12}) = q = \mathcal{E}_-(A).$$

The same conclusions prevail when either $p = 0$ or $q = 0$. The details are left to the reader. □

Exercise 20.28. Verify that $K_{11} \succ O$ and $K_{22} \prec O$, as was asserted in the proof of Theorem 20.15. [HINT: $K_{11} = \int_0^\infty e^{-sJ_1^H} P_{11} e^{-sJ_1} ds$.]

Exercise 20.29. Complete the details of the proof of Theorem 20.15 when $p = 0$.

A more elaborate argument yields the following supplementary conclusion:

Theorem 20.16. *Let $A \in \mathbb{C}^{n \times n}$ and let $G \in \mathbb{C}^{n \times n}$ be a Hermitian matrix such that:*

(1) $A^H G + GA \succeq O$.

(2) $\sigma(A) \cap i\mathbb{R} = \emptyset$.

(3) *G is invertible.*

Then $\mathcal{E}_+(A) = \mathcal{E}_+(G)$, $\mathcal{E}_-(A) = \mathcal{E}_-(G)$ (and $\mathcal{E}_0(A) = \mathcal{E}_0(G) = 0$ by assumption).

Proof. The asserted result is an immediate consequence of the following lemma. □

Lemma 20.17. *Let $A \in \mathbb{C}^{n\times n}$ and let $G \in \mathbb{C}^{n\times n}$ be a Hermitian matrix such that $A^HG + GA \succeq O$. Then the following implications hold:*

(1) $\sigma(A) \cap i\mathbb{R} = \emptyset \Longrightarrow \mathcal{E}_+(G) \le \mathcal{E}_+(A)$ *and* $\mathcal{E}_-(G) \le \mathcal{E}_-(A)$.

(2) *G is invertible* $\Longrightarrow \mathcal{E}_+(G) \ge \mathcal{E}_+(A)$ *and* $\mathcal{E}_-(G) \ge \mathcal{E}_-(A)$.

Proof. Suppose first that $\sigma(A) \cap i\mathbb{R} = \emptyset$. Then, by Theorem 18.7, there exists a Hermitian matrix G_0 such that $A^HG_0 + G_0A \succ O$. Therefore,

$$A^H(G + \varepsilon G_0) + (G + \varepsilon G_0)A = A^HG + GA + \varepsilon(A^HG_0 + G_0A) \succ O$$

for every $\varepsilon > 0$. Thus, by Theorem 20.15,

$$\mathcal{E}_\pm(G + \varepsilon G_0) = \mathcal{E}_\pm(A)$$

for every $\varepsilon > 0$. Moreover, since the eigenvalues of $G + \varepsilon G_0$ are continuous functions of ε and $G+\varepsilon G_0$ is invertible for every $\varepsilon > 0$, the desired conclusion follows by letting $\varepsilon \downarrow 0$.

Next, if G is invertible and $A_\varepsilon = A + \varepsilon G^{-1}$, then

$$A_\varepsilon^HG + GA_\varepsilon = A^HG + GA + 2\varepsilon I_n \succ O$$

for every choice of $\varepsilon > 0$. Therefore, by Theorem 20.15, $\mathcal{E}_+(G) = \mathcal{E}_+(A_\varepsilon)$ and $\mathcal{E}_-(G) = \mathcal{E}_-(A_\varepsilon)$ for every choice of $\varepsilon > 0$. Then, since the eigenvalues of A_ε are continuous functions of ε, the inequalities in (2) follow by letting $\varepsilon \downarrow 0$. □

20.10. An eigenvalue assignment problem

A basic problem in control theory amounts to shifting the eigenvalues of a given matrix A to preassigned values, or a preassigned region, by an appropriately chosen additive perturbation of the matrix, which in practice is implemented by feedback. Since the eigenvalues of A are the roots of its characteristic polynomial, this corresponds to shifting $\det(\lambda I_n - A)$ to a polynomial $c_0 + \cdots + c_{n-1}\lambda^{n-1} + \lambda^n$ with suitable roots.

Theorem 20.18. *Let*

$$\mathfrak{C} = [\mathbf{b}\ A\mathbf{b} \cdots A^{n-1}\mathbf{b}]$$

be the controllability matrix of a pair of matrices $(A, \mathbf{b}) \in \mathbb{C}^{n\times n} \times \mathbb{C}^n$. Then

$$A\mathfrak{C} = \mathfrak{C}\, S_f^T \quad \text{and} \quad S_f^TH_f = H_fS_f\,, \tag{20.11}$$

where S_f denotes the companion matrix based on the polynomial

$$f(\lambda) = \det(\lambda I_n - A) = a_0 + a_1\lambda + \cdots + a_{n-1}\lambda^{n-1} + \lambda^n$$

and H_f denotes the Hankel matrix

$$(20.12) \qquad H_f = \begin{bmatrix} a_1 & a_2 & \cdots & a_{n-1} & 1 \\ a_2 & a_3 & \cdots & 1 & 0 \\ \vdots & & & & \vdots \\ 1 & 0 & \cdots & 0 & 0 \end{bmatrix}$$

based on the coefficients of $f(\lambda)$. If $(A, \mathbf{b})$ is controllable, then $\mathfrak{C}$ is invertible.

Proof. Let $N = C_0^{(n)} = \sum_{j=1}^{n-1} \mathbf{e}_j \mathbf{e}_{j+1}^T$ be the $n \times n$ matrix with ones on the first superdiagonal and zeros elsewhere. Then clearly,

$$\begin{aligned} A\mathfrak{C} &= A[\mathbf{b}\ A\mathbf{b} \cdots A^{n-1}\mathbf{b}] = [A\mathbf{b}\ A^2\mathbf{b} \cdots A^n\mathbf{b}] \\ &= [A\mathbf{b} \cdots A^{n-1}\mathbf{b}\ \mathbf{0}] + [\mathbf{0} \cdots \mathbf{0}\ A^n\mathbf{b}] \\ &= \mathfrak{C}\, N^T + [\mathbf{0} \cdots \mathbf{0}\ A^n\mathbf{b}] = \mathfrak{C} S_f^T\,, \end{aligned}$$

since

$$A^n\mathbf{b} = -(a_0 I_n + a_1 A + \cdots + a_{n-1}A^{n-1})\mathbf{b}$$

by the Cayley-Hamilton theorem, and, consequently,

$$[\mathbf{0} \cdots \mathbf{0}\ A^n\mathbf{b}] = \mathfrak{C} \begin{bmatrix} 0 & \cdots & 0 & -a_0 \\ \vdots & & \vdots & \vdots \\ 0 & \cdots & 0 & -a_{n-1} \end{bmatrix}.$$

Moreover, if $\mathfrak{C}$ is controllable, then it is invertible, since $\mathfrak{C}$ is a square matrix.

The verification of the second identity in (20.11) is left to the reader. $\square$

Exercise 20.30. Verify the second identity in (20.11).

Exercise 20.31. Let $A = \begin{bmatrix} 1 & 1 & 0 \\ 0 & 1 & 1 \\ 0 & 0 & 1 \end{bmatrix}$ and let $\mathbf{b} = \begin{bmatrix} 0 \\ 0 \\ 1 \end{bmatrix}$. Find a vector $\mathbf{u} \in \mathbb{C}^3$ such that $\sigma(A + \mathbf{b}\mathbf{u}^H) = \{2, 3, 4\}$.

Theorem 20.19. *If $(A, \mathbf{b}) \in \mathbb{C}^{n \times n} \times \mathbb{C}^n$ is controllable, then for each choice of $c_0, c_1, \cdots, c_{n-1} \in \mathbb{C}$, there exists a vector $\mathbf{u} \in \mathbb{C}^n$ such that*

$$\det(\lambda I_n - A - \mathbf{b}\mathbf{u}^H) = c_0 + c_1\lambda + \cdots + c_{n-1}\lambda^{n-1} + \lambda^n.$$

Proof. Let S_g denote the companion matrix based on the polynomial

$$g(\lambda) = c_0 + c_1\lambda + \cdots + c_{n-1}\lambda^{n-1} + \lambda^n.$$

Then it suffices to show that there exists a vector $\mathbf{u} \in \mathbb{C}^n$ such that $A + \mathbf{b}\mathbf{u}^H$ is similar to S_g. Since $A\mathfrak{C}H_f = \mathfrak{C}H_f S_f$ by Theorem 20.18, and $\mathfrak{C}H_f$ is invertible, it is enough to check that

$$(A + \mathbf{b}\mathbf{u}^H)\mathfrak{C}H_f = \mathfrak{C}H_f S_g$$

or, equivalently, that

$$\mathbf{b}\mathbf{u}^H \mathfrak{C} H_f = \mathfrak{C} H_f (S_g - S_f)\,.$$

In order for this equality to hold, it is necessary that

$$\mathbf{b}\mathbf{u}^H = \mathfrak{C} H_f (S_g - S_f) H_f^{-1} \mathfrak{C}^{-1}\,;$$

i.e.,

$$\mathbf{u}^H = (\mathbf{b}^H\mathbf{b})^{-1}\mathbf{b}^H \mathfrak{C} H_f (S_g - S_f) H_f^{-1} \mathfrak{C}^{-1}\,.$$

It remains to check that this choice of $\mathbf{u}$ really works, i.e.,

$$\left(A + \mathbf{b}(\mathbf{b}^H\mathbf{b})^{-1}\mathbf{b}^H \mathfrak{C} H_f (S_g - S_f) H_f^{-1} \mathfrak{C}^{-1}\right)\mathfrak{C} H_f = \mathfrak{C} H_f S_g\,.$$

The main steps in the verification are to check that

(1) $S_g - S_f = \mathbf{e}_n \mathbf{w}^T$, where $\mathbf{w}^T = \begin{bmatrix} a_0 - c_0 & \cdots & a_{n-1} - c_{n-1} \end{bmatrix}$.

(2) $H_f(S_g - S_f) = \mathbf{e}_1 \mathbf{w}^T$.

(3) $\mathfrak{C} H_f (S_g - S_f) = \mathbf{b}\mathbf{w}^T$.

(4) $\mathbf{b}(\mathbf{b}^H\mathbf{b})^{-1}\mathbf{b}^H \mathfrak{C} H_f (S_g - S_f) = \mathbf{b}\mathbf{w}^T = \mathfrak{C} H_f (S_g - S_f)$.

The details are left to the reader as an exercise. □

Exercise 20.32. Complete the proof of Theorem 20.19 by verifying the four items in the list furnished towards the end of the proof of that theorem.

Theorem 20.20. *If $(A, B) \in \mathbb{C}^{n\times n} \times \mathbb{C}^{n\times k}$ is a controllable pair with $k > 1$, then there exist a matrix $C \in \mathbb{C}^{k\times n}$ and a vector $\mathbf{b} \in \mathcal{R}_B$ such that $(A + BC, \mathbf{b})$ is a controllable pair.*

Discussion. Suppose for the sake of definiteness that $A \in \mathbb{C}^{10\times 10}$, $B = [\mathbf{b}_1 \cdots \mathbf{b}_5]$, and then permute the columns in the controllability matrix to obtain the matrix

$$\begin{bmatrix} \mathbf{b}_1 & A\mathbf{b}_1 & \cdots & A^9\mathbf{b}_1 & \cdots & \mathbf{b}_5 & A\mathbf{b}_5 & \cdots & A^9\mathbf{b}_5 \end{bmatrix}\,.$$

Then, moving from left to right, discard vectors that may be expressed as linear combinations of vectors that sit to their left. Suppose further that

$$\begin{aligned} A^3\mathbf{b}_1 &\in \operatorname{span}\{\mathbf{b}_1, A\mathbf{b}_1, A^2\mathbf{b}_1\} \\ A^5\mathbf{b}_2 &\in \operatorname{span}\{\mathbf{b}_1, A\mathbf{b}_1, A^2\mathbf{b}_1, \mathbf{b}_2, \ldots, A^4\mathbf{b}_2\} \\ A^2\mathbf{b}_3 &\in \operatorname{span}\{\mathbf{b}_1, A\mathbf{b}_1, A^2\mathbf{b}_1, \mathbf{b}_2, \ldots, A^4\mathbf{b}_2, \mathbf{b}_3, A\mathbf{b}_3\} \end{aligned}$$

and that the matrix

$$Q = [\mathbf{b}_1\ A\mathbf{b}_1\ A^2\mathbf{b}_1\ \mathbf{b}_2\ A\mathbf{b}_2\ A^2\mathbf{b}_2\ A^3\mathbf{b}_2\ A^4\mathbf{b}_2\ \mathbf{b}_3\ A\mathbf{b}_3]$$

is invertible. Let $\mathbf{e}_j$ denote the j'th column of I_5 and let $\mathbf{f}_k$ denote the k'th column of I_{10} and set

$$G = [\mathbf{0}\ \mathbf{0}\ \mathbf{e}_2\ \mathbf{0}\ \mathbf{0}\ \mathbf{0}\ \mathbf{0}\ \mathbf{e}_3\ \mathbf{0}\ \mathbf{0}]$$

and $C = GQ^{-1}$. Then

$$\begin{aligned}(A+BC)\mathbf{b}_1 &= A\mathbf{b}_1 + BGQ^{-1}Q\mathbf{f}_1 = A\mathbf{b}_1\\ (A+BC)^2\mathbf{b}_1 &= (A+BC)A\mathbf{b}_1 = A^2\mathbf{b}_1 + BGQ^{-1}Q\mathbf{f}_2\\ &= A^2\mathbf{b}_1\\ (A+BC)^3\mathbf{b}_1 &= (A+BC)A^2\mathbf{b}_1 = A^3\mathbf{b}_1 + BGQ^{-1}Q\mathbf{f}_3\\ &= A^3\mathbf{b}_1 + B\mathbf{e}_2 = A^3\mathbf{b}_1 + \mathbf{b}_2.\end{aligned}$$

Similar considerations lead to the conclusion that

$$[\mathbf{b}_1 \ (A+BC)\mathbf{b}_1 \ \cdots \ (A+BC)^9\mathbf{b}_1] = QU\,,$$

where U is an upper triangular matrix with ones on the diagonal. Therefore, since QU is invertible, $(A+BC, \mathbf{b}_1)$ is a controllable pair.

Theorem 20.21. *Let $(A, B) \in \mathbb{C}^{n\times n} \times \mathbb{C}^{n\times k}$ be a controllable pair and let $\{\mu_1, \dots, \mu_n\}$ be any set of points in $\mathbb{C}^n$ (not necessarily distinct). Then there exists a matrix $K \in \mathbb{C}^{k\times n}$ such that*

$$\det(\lambda I_n - A - BK) = (\lambda - \mu_1)\cdots(\lambda - \mu_n).$$

Proof. If $k = 1$, then the conclusion is given in Theorem 20.19. If $k > 1$, then, by Theorem 20.20, there exists a matrix $C \in \mathbb{C}^{k\times n}$ and a vector $\mathbf{b} \in \mathcal{R}_B$ such that $(A+BC,\ \mathbf{b})$ is a controllable pair. Therefore, by Theorem 20.19, there exists a vector $\mathbf{u} \in \mathbb{C}^n$ such that

$$\det(\lambda I_n - A - BC - \mathbf{b}\mathbf{u}^H) = (\lambda - \mu_1)\cdots(\lambda - \mu_n).$$

However, $BC + \mathbf{b}\mathbf{u}^H = B(C + \mathbf{v}\mathbf{u}^H)$ for some $\mathbf{v} \in \mathbb{C}^k$, since $\mathbf{b} \in \mathcal{R}_B$. Thus, the proof may be completed by choosing $K = C + \mathbf{v}\mathbf{u}^H$. □

20.11. Bibliographical notes

Theorem 20.15 is due to Ostrowski and Schneider, and independently, to M. G. Krein; see p. 445 of [**45**] for a discussion of the history and a converse statement. Theorem 20.16 is due to Carlson and Schneider. The proofs presented here are adapted from the presentation in [**45**]. The discussion of Theorem 20.20 is partially adapted from Heymann [**38**]. A number of the congruence theorems were taken from the paper [**27**].

Chapter 21

Zero location problems

Just because it hurts, it doesn't mean that it's good for you.

Michael Dym

In this chapter we shall present two recipes for counting the number of common roots of two polynomials and applications thereof. In particular, we shall discuss stable polynomials and a criterion of Kharitonov for checking the stability of a family of polynomials with coefficients that are only specified within certain bounds.

21.1. Bezoutians

Let

$$f(\lambda) = f_0 + f_1\lambda + \cdots + f_n\lambda^n \,, \quad f_n \neq 0 \,,$$

be a polynomial of degree n with coefficients $f_0, \ldots, f_n \in \mathbb{C}$ and let

$$g(\lambda) = g_0 + g_1\lambda + \cdots + g_n\lambda^n$$

be a polynomial of degree less than or equal to n with coefficients $g_0, \ldots, g_n \in \mathbb{C}$, at least one of which is nonzero. Then the matrix $B \in \mathbb{C}^{n\times n}$ with entries b_{ij}, $i, j = 0, \ldots, n-1$, that is uniquely defined by the formula

$$(21.1)\quad \frac{f(\lambda)g(\mu) - g(\lambda)f(\mu)}{\lambda - \mu} = \sum_{i,j=0}^{n-1} \lambda^i b_{ij} \mu^j = [1 \quad \lambda \cdots \lambda^{n-1}]B \begin{bmatrix} 1 \\ \mu \\ \vdots \\ \mu^{n-1} \end{bmatrix}$$

is called the **Bezoutian** of the polynomials $f(\lambda)$ and $g(\lambda)$ and will be denoted by the symbol $B(f,g)$.

The first main objective of this chapter is to verify the formula

$$\dim \mathcal{N}_B = \nu(f,g)\,,$$

where

$$\nu(f,g) \;=\; \text{the number of common roots of the polynomials } f(\lambda) \text{ and } g(\lambda), \text{ counting multiplicities.}$$

In particular, it is readily seen that if $f(\alpha) = 0$ and $g(\alpha) = 0$, then

$$[1 \quad \lambda \cdots \lambda^{n-1}]B \begin{bmatrix} 1 \\ \alpha \\ \vdots \\ \alpha^{n-1} \end{bmatrix} = 0$$

for every point $\lambda \in \mathbb{C}$ and hence that

$$B \begin{bmatrix} 1 \\ \alpha \\ \vdots \\ \alpha^{n-1} \end{bmatrix} = 0\,.$$

Moreover, if $f(\alpha) = f'(\alpha) = 0$ and $g(\alpha) = g'(\alpha) = 0$, then the identity

$$\frac{f(\lambda)g'(\mu) - g(\lambda)f'(\mu)}{\lambda - \mu} + \frac{f(\lambda)g(\mu) - g(\lambda)f(\mu)}{(\lambda - \mu)^2}$$

$$= [1 \quad \lambda \cdots \lambda^{n-1}]B \begin{bmatrix} 0 \\ 1 \\ 2\mu \\ \vdots \\ (n-1)\mu^{n-2} \end{bmatrix},$$

which is obtained by differentiating both sides of formula (21.1) with respect to μ, implies that

$$[1 \quad \lambda \cdots \lambda^{n-1}]B \begin{bmatrix} 0 \\ 1 \\ 2\alpha \\ \vdots \\ (n-1)\alpha^{n-2} \end{bmatrix} = 0 \tag{21.2}$$

for every point $\lambda \in \mathbb{C}$. Therefore, $\dim \mathcal{N}_B \geq 2$, since the vector

$$(21.3) \quad \mathbf{v}(\alpha) = \begin{bmatrix} 1 \\ \alpha \\ \vdots \\ \alpha^{n-1} \end{bmatrix} \quad \text{and its derivative} \quad \mathbf{v}^{(1)}(\alpha) = \begin{bmatrix} 0 \\ 1 \\ 2\alpha \\ \vdots \\ (n-1)\alpha^{n-2} \end{bmatrix}$$

with respect to α both belong to $\mathcal{N}_B$ and are linearly independent.

Much the same sort of reasoning leads rapidly to the conclusion that if $f(\alpha) = f^{(1)}(\alpha) = \cdots f^{(k-1)}(\alpha) = 0$ and $g(\alpha) = g^{(1)}(\alpha) = \cdots g^{(k-1)}(\alpha) = 0$, then the vectors $\mathbf{v}(\alpha), \ldots, \mathbf{v}(\alpha)^{(k-1)}$ all belong to $\mathcal{N}_B$. Thus, as these vectors are linearly independent if $k \leq n$, $\dim \mathcal{N}_B \geq k$. Moreover, if $f(\beta) = f^{(1)}(\beta) = \cdots f^{(j-1)}(\beta) = 0$ and $g(\beta) = g^{(1)}(\beta) = \cdots g^{(j-1)}(\beta) = 0$, then the vectors $\mathbf{v}(\beta), \ldots, \mathbf{v}^{(j-1)}(\beta)$ all belong to $\mathcal{N}_B$. Therefore, since this set of vectors is linearly independent of the set $\mathbf{v}(\alpha), \ldots, \mathbf{v}^{(k-1)}(\alpha)$, as will be shown in the next section, $\dim \mathcal{N}_B \geq k+j$. Proceeding this way, it is rapidly deduced that

$$\dim \mathcal{N}_B \geq \nu(f,g) \, .$$

The verification of equality is more subtle and will require a number of steps. The first step is to obtain a formula for $B(f, \varphi_k)$ for the monomial $\varphi_k(\lambda) = \lambda^k$, $k = 0, \ldots, n$, in terms of the **Hankel** matrices

(21.4)

$$H_f^{[k,n]} = \begin{bmatrix} f_k & f_{k+1} & \cdot & \cdot & f_n \\ f_{k+1} & & & \cdot & 0 \\ \cdot & & & \cdot & \cdot \\ \cdot & & \cdot & & \cdot \\ f_n & 0 & \cdot & \cdot & 0 \end{bmatrix}, \quad \widetilde{H_f}^{[0,k-1]} = \begin{bmatrix} 0 & \cdot & \cdot & \cdot & f_0 \\ \cdot & & & \cdot & f_1 \\ \cdot & & \cdot & & \cdot \\ \cdot & \cdot & & & \cdot \\ f_0 & f_1 & \cdot & \cdot & f_{k-1} \end{bmatrix}$$

(for $k = 1, \ldots, n$), the projectors

$$(21.5) \qquad E^{(j,k)} = \sum_{i=j}^{k} \mathbf{e}_i \mathbf{e}_i^T \quad \text{for} \quad 1 \leq j \leq k \leq n$$

based on the columns $\mathbf{e}_i$ of I_n and the $n \times n$ matrices

$$(21.6) \qquad N = C_0^{(n)} = \begin{bmatrix} 0 & 1 & 0 & \cdots & 0 \\ 0 & 0 & 1 & \cdots & 0 \\ \vdots & \vdots & & \ddots & \\ 0 & 0 & & & 1 \\ 0 & 0 & & \cdots & 0 \end{bmatrix} = \sum_{i=1}^{n-1} \mathbf{e}_i \mathbf{e}_{i+1}^T \, .$$

Theorem 21.1. *If $f(\lambda) = f_0 + \cdots + f_n\lambda^n$ is a polynomial of degree n, i.e., $f_n \neq 0$, and $n \geq 2$, and if $\varphi_k(\lambda) = \lambda^k$, then*

$$B(f, \varphi_k) = \begin{cases} H_f^{[1,n]} & \text{if} \quad k = 0\,, \\ \begin{bmatrix} \widetilde{H_f}^{[0,k-1]} & O \\ O & H_f^{[k+1,n]} \end{bmatrix} & \text{if} \quad 1 \leq k \leq n-1\,, \\ -\widetilde{H_f}^{[0,n-1]} & \text{if} \quad k = n\,. \end{cases} \tag{21.7}$$

Proof. Suppose first that $1 \leq k \leq n-1$. Then

$$\frac{f(\lambda)\mu^k - \lambda^k f(\mu)}{\lambda - \mu} = \sum_{j=0}^{n} f_j \left\{ \frac{\lambda^j \mu^k - \lambda^k \mu^j}{\lambda - \mu} \right\} = ① + ②,$$

where

$$① = \sum_{j=0}^{k-1} f_j \left\{ \frac{\lambda^j \mu^k - \lambda^k \mu^j}{\lambda - \mu} \right\}$$

and

$$② = \sum_{j=k+1}^{n} f_j \left\{ \frac{\lambda^j \mu^k - \lambda^k \mu^j}{\lambda - \mu} \right\}.$$

If $j < k$, as in ①, then

$$\begin{aligned} \frac{\lambda^j \mu^k - \lambda^k \mu^j}{\lambda - \mu} &= \lambda^j \mu^j \left\{ \frac{\mu^{k-j} - \lambda^{k-j}}{\lambda - \mu} \right\} \\ &= -\lambda^j \mu^j \{ \lambda^{k-j-1} + \lambda^{k-j-2}\mu + \cdots + \mu^{k-j-1} \} \\ &= -\{ \lambda^{k-1}\mu^j + \lambda^{k-2}\mu^{j+1} + \cdots + \lambda^j \mu^{k-1} \}\,. \end{aligned}$$

On the other hand, if $j > k$, then

$$\begin{aligned} \frac{\lambda^j \mu^k - \lambda^k \mu^j}{\lambda - \mu} &= \lambda^k \mu^k \left\{ \frac{\lambda^{j-k} - \mu^{j-k}}{\lambda - \mu} \right\} \\ &= \lambda^k \mu^k \{ \lambda^{j-k-1} + \lambda^{j-k-2}\mu + \cdots + \mu^{j-k-1} \} \\ &= \lambda^{j-1}\mu^k + \lambda^{j-2}\mu^{k+1} + \cdots + \lambda^k \mu^{j-1}\,. \end{aligned}$$

Thus,

$$\begin{aligned} \frac{f(\lambda)\mu^k - \lambda^k f(\mu)}{\lambda - \mu} &= \sum_{s,t=0}^{n-1} \lambda^s b_{s,t} \mu^t \\ &= -\sum_{j=0}^{k-1} f_j \{\lambda^{k-1}\mu^j + \lambda^{k-2}\mu^{j+1} + \cdots + \lambda^j \mu^{k-1}\} \\ &\quad + \sum_{j=k+1}^{n} f_j \{\lambda^{j-1}\mu^k + \lambda^{j-2}\mu^{k+1} + \cdots + \lambda^k \mu^{j-1}\}, \end{aligned}$$

and hence the nonzero entries in the Bezoutian matrix B are specified by the formulas

$$b_{k-1,j} = b_{k-2,j+1} = \cdots = b_{j,k-1} = -f_j \text{ for } j = 0, \ldots, k-1 \tag{21.8}$$

and

$$b_{j-1,k} = b_{j-2,k+1} = \cdots = b_{k,j-1} = f_j \text{ for } j = k+1, \ldots, n\,. \tag{21.9}$$

But this is exactly the same as the matrix that is depicted in the statement of the theorem for the case $1 \leq k \leq n-1$.

It remains to consider the cases $k = 0$ and $k = n$:

$$k = 0 \Longrightarrow ① = 0 \quad \text{and} \quad ② = \sum_{j=1}^{n} f_j \left\{ \frac{\lambda^j - \mu^j}{\lambda - \mu} \right\}$$

and hence that formula (21.9) is in force for $k = 0$. Similarly,

$$k = n \Longrightarrow ① = \sum_{j=0}^{n-1} f_j \left\{ \frac{\lambda^j \mu^n - \lambda^n \mu^j}{\lambda - \mu} \right\} \text{ and } ② = 0$$

and hence that formula (21.8) is in force for $k = n$. □

Corollary 21.2. $B(f, g)$ is a symmetric Hankel matrix.

Proof. Formula (21.7) clearly implies that $B(f, \varphi_k)$ is a symmetric Hankel matrix. Therefore, the same conclusion holds for

$$B(f, g) = \sum_{k=0}^{n} g_j B(f, \varphi_j)\,,$$

since a linear combination of $n \times n$ symmetric Hankel matrices is a symmetric Hankel matrix. □

Remark 21.3. It is convenient to set

$$H_f = H_f^{[1,n]} \quad \text{and} \quad \widetilde{H}_f = \widetilde{H}_f^{[0,n-1]} \tag{21.10}$$

for short. Then the formula for $k = 1, \dots, n-1$ can be expressed more succinctly as

(21.11)
$$B(f, \varphi_k) = E^{(k+1,n)} H_f N^k - N^{n-k} \widetilde{H}_f E^{(1,k)} \quad \text{if} \quad k = 1, \dots, n-1 .$$

Exercise 21.1. Verify the formulas
$$(N^T)^k N^k = \begin{bmatrix} 0 & 0 \\ 0 & I_{n-k} \end{bmatrix} = E^{(k+1,n)} \quad \text{for} \quad k = 1, \dots, n-1 .$$

Exercise 21.2. Verify the formulas
$$N^k (N^T)^k = \begin{bmatrix} I_{n-k} & 0 \\ 0 & 0 \end{bmatrix} = E^{(1,n-k)} \quad \text{for} \quad k = 1, \dots, n-1 .$$

Exercise 21.3. Show that formula (21.11) can be expressed as

(21.12)
$$B(f, \varphi_k) = (N^T)^k N^k H_f N^k - N^{n-k} \widetilde{H}_f N^{n-k} (N^T)^{n-k} \quad \text{for} \quad k = 0, \dots, n .$$

21.2. A derivation of the formula for H_f based on realization

If $N \in \mathbb{R}^{n \times n}$ is the matrix defined by formula (21.6), then
$$(I_n - \lambda N^T)^{-1} = \begin{bmatrix} 1 & 0 & \cdots & 0 \\ \lambda & 1 & & \vdots \\ \vdots & & \ddots & 0 \\ \lambda^{n-1} & \lambda^{n-2} & \cdots & 1 \end{bmatrix} .$$

Thus a polynomial $f(\lambda) = f_0 + f_1 \lambda \cdots + f_n \lambda^n$ admits the **realization**

(21.13)
$$f_0 + f_1 \lambda \cdots + f_n \lambda^n = f_0 + \lambda [f_1 \cdots f_n] (I_n - \lambda N^T)^{-1} \mathbf{e}_1 .$$

In this section we shall use (21.13) to obtain a new proof the first formula in (21.7).

Exercise 21.4. Verify formula (21.13).

Lemma 21.4. *If $f(\lambda)$ is a polynomial of degree $n \geq 1$, then*
$$B(f, 1) = H_f .$$

Proof. In view of formula (21.13),
$$\begin{aligned} f(\lambda) - f(\mu) &= [f_1 \ \cdots \ f_n] \{ \lambda (I_n - \lambda N^T)^{-1} - \mu (I - \mu N^T)^{-1} \} \mathbf{e}_1 \\ &= [f_1 \ \cdots \ f_n] \{ (\lambda - \mu)(I_n - \lambda N^T)^{-1} (I_n - \mu N^T)^{-1} \} \mathbf{e}_1 \end{aligned}$$

and, consequently,

(21.14)
$$\frac{f(\lambda) - f(\mu)}{\lambda - \mu} = [f_1 \ \cdots \ f_n] (I - \lambda N^T)^{-1} (I - \mu N^T)^{-1} \mathbf{e}_1 .$$

The next step is to verify the identity

$$[f_1 \ \cdots \ f_n](I_n - \lambda N^T)^{-1} = [1 \quad \lambda \ \cdots \ \lambda^{n-1}]H_f \,. \tag{21.15}$$

This follows easily from the observation that

$$[f_1 \ \cdots \ f_n](N^T)^k = \mathbf{e}_{k+1}^T H_f \quad \text{for} \quad k = 0, \ldots, n-1 \ : \tag{21.16}$$

$$\begin{aligned} [f_1 \ \cdots \ f_n](I_n - \lambda N^T)^{-1} &= [f_1 \ \cdots \ f_n]\sum_{j=0}^{n-1}(\lambda N^T)^j \\ &= \sum_{j=0}^{n-1} \lambda^j \mathbf{e}_{j+1} H_f = [1\, \lambda \cdots \lambda^{n-1}]H_f \,. \end{aligned}$$

Thus, substituting formula (21.15) into formula (21.14),

$$\begin{aligned} \frac{f(\lambda) - f(\mu)}{\lambda - \mu} &= [1\ \lambda\ \cdots\ \lambda^{n-1}]H_f(I_n - \mu N^T)^{-1}\mathbf{e}_1 \\ &= [1\ \lambda\ \cdots\ \lambda^{n-1}]H_f \begin{bmatrix} 1 \\ \mu \\ \vdots \\ \mu^{n-1} \end{bmatrix} \end{aligned}$$

i.e., $B(f, \varphi_0) = H_f$ as claimed. □

21.3. The Barnett identity

The next order of business is to establish the identity

$$B(f, \varphi_k) = H_f S_f^k \quad \text{for} \quad k = 0, \ldots, n \,. \tag{21.17}$$

The case $k = 0$ has already been established twice: once in Theorem 21.1 and then again in Lemma 21.4.

Lemma 21.5. *The identity*

$$E^{(1,k)} H_f S_f^k = -N^{n-k} \widetilde{H}_f E^{(1,k)}$$

holds for $k = 1, \ldots, n$.

Proof. It is readily checked by direct calculation that the asserted identity is valid when $k = 1$, since

$$\mathbf{e}_1^T H_f S_f = -[f_0 \quad O_{1\times(n-1)}] = -\mathbf{e}_n^T \widetilde{H}_f \mathbf{e}_1 \mathbf{e}_1^T \,.$$

Thus, proceeding by induction, assume that the formula is valid for $k - 1$ and let

$$E^{(1,k)} H_f S_f^k = ① + ②$$

where

$$① = E^{(1,k-1)} H_f S_f^k \quad \text{and} \quad ② = \mathbf{e}_k \mathbf{e}_k^T H_f S_f^k \,.$$

By the induction hypothesis,

$$E^{(1,k-1)}H_fS_f^k = -N^{(n-k+1)}\widetilde{H}_fE^{(1,k-1)}S_f\,.$$

Therefore, since

$$E^{(1,k-1)}S_f = E^{(1,k-1)}N = NE^{(1,k)} \quad \text{and} \quad \widetilde{H}_fN = N^T\widetilde{H}_f\,,$$

the term ① can be reexpressed as

$$\begin{aligned} ① &= -N^{(n-k)}N\widetilde{H}_fNE^{(1,k)} \\ &= -N^{(n-k)}NN^T\widetilde{H}_fE^{(1,k)} \\ &= -N^{(n-k)}E^{(1,n-1)}\widetilde{H}_fE^{(1,k)}\,. \end{aligned}$$

Next, the key to evaluating ② is the observation that

$$\begin{aligned} \mathbf{e}_k^TH_fS_f^k &= [f_k\cdots f_n \;\; 0_{1\times(k-1)}]S_f^k \\ &= [f_k\cdots f_n \;\; 0_{1\times(k-1)}]N^{k-1}S_f \\ &= [0_{1\times(k-1)} \;\; f_k\cdots f_n]S_f \\ &= -[f_0\cdots f_{k-1} \;\; 0_{1\times(n-1)}] \\ &= -\mathbf{e}_n^T\widetilde{H}_fE^{(1,k)} \end{aligned}$$

and

$$\mathbf{e}_k = N^{n-k}\mathbf{e}_n\,.$$

Thus,

$$① + ② = -N^{n-k}\left\{E^{(1,n-1)} + \mathbf{e}_n\mathbf{e}_n^T\right\}\widetilde{H}_fE^{(1,k)} = -N^{n-k}\widetilde{H}_fE^{(1,k)}\,,$$

as needed. □

Lemma 21.6. *The identity*

$$E^{(k+1,n)}H_fS_f^k = E^{(k+1,n)}H_fN^k$$

holds for $k = 0,\cdots,n-1$.

Proof. The special triangular structure of H_f yields the identity

$$E^{(k+1,n)}H_f = E^{(k+1,n)}H_fE^{(1,n-k)}\,.$$

The asserted conclusion then follows from the fact that

$$E^{(1,n-k)}S_f^k = E^{(1,n-k)}N^k\,.$$

□

Theorem 21.7. *The identity*

$$H_fS_f^k = E^{(k+1,n)}H_fN^k - N^{n-k}\widetilde{H}_fE^{(1,k)}$$

holds for $k = 0,1,\ldots,n$, *with the understanding that* $E^{(n+1,n)} = E^{(1,0)} = 0$.

Proof. This is an immediate consequence of the preceding two lemmas. □

Theorem 21.8. (The Barnett identity) *If $f(\lambda) = f_0 + f_1\lambda + \cdots + f_n\lambda^n$ is a polynomial of degree n (i.e., $f_n \neq 0$) and $g(\lambda)$ is a polynomial of degree $\leq n$ with at least one nonzero coefficient, then*

$$B(f,g) = H_f\, g(S_f)\,. \tag{21.18}$$

Proof. By formula (21.11), the identity

$$B(f,\varphi_k) = E^{(k+1,n)} H_f N^k - N^{n-k}\widetilde{H_f} E^{(1,k)}$$

holds for $k = 0, 1, \ldots, n$, with self-evident conventions at $k = 0$ and $k = n$ (see (21.12). Thus, in view of Theorem 21.7,

$$B(f,\varphi_k) = H_f S_f^k \quad \text{for} \quad k = 0, \ldots, n.$$

Therefore,

$$B(f,g) = \sum_{k=0}^{n} g_k B(f,\varphi_k) = \sum_{k=0}^{n} g_k H_f S_f^k,$$

as needed. □

21.4. The main theorem on Bezoutians

This section is devoted to the statement and proof of the main theorem on Bezoutians. It is convenient to first establish a lemma for Jordan cells.

Lemma 21.9. *If $g(\lambda)$ is a polynomial and $N = C_0^{(p)}$, then*

$$g(C_\lambda^{(p)}) = \sum_{j=0}^{p-1} \frac{g^{(j)}(\lambda)}{j!} N^j = \begin{bmatrix} g(\lambda) & \frac{g^{(1)}(\lambda)}{1!} & \cdots & \frac{g^{(p-1)}(\lambda)}{(p-1)!} \\ 0 & g(\lambda) & \cdots & \frac{g^{(p-2)}(\lambda)}{(p-2)!} \\ \vdots & & \ddots & \vdots \\ 0 & 0 & \cdots & g(\lambda) \end{bmatrix} \tag{21.19}$$

and

$$\operatorname{rank} g(C_\lambda^{(p)}) = \begin{cases} p & \text{if} \quad g(\lambda) \neq 0 \\ p-s & \text{if} \quad g(\lambda) = \cdots = g^{(s-1)}(\lambda) = 0 \quad \text{but} \quad g^{(s)}(\lambda) \neq 0 \end{cases},$$

where, in the last line, s is an integer such that $1 \leq s \leq p$.

Proof. Let Γ denote a circle of radius $R > |\lambda|$ that is centered at the origin and is directed counterclockwise. Then

$$\begin{aligned} g(\lambda I_p + N) &= \frac{1}{2\pi i}\int_\Gamma g(\zeta)(\zeta I_p - \lambda I_p - N)^{-1} d\zeta \\ &= \sum_{j=0}^{p-1} \frac{1}{2\pi i}\int_\Gamma \frac{g(\zeta)}{(\zeta-\lambda)^{j+1}} N^j d\zeta \\ &= \sum_{j=0}^{n} \frac{g^{(j)}(\lambda)}{j!} N^j \,. \end{aligned}$$

The formula for the rank of $g(C_\lambda^{(p)})$ is clear from the fact that the matrix under consideration is an upper triangular Toeplitz matrix. Thus, for example, if $p = 3$, then

$$g(\lambda I_3 + N) = \begin{bmatrix} g(\lambda) & g^{(1)}(\lambda) & g^{(2)}(\lambda)/2! \\ 0 & g(\lambda) & g^{(1)}(\lambda) \\ 0 & 0 & g(\lambda) \end{bmatrix}.$$

But this clearly exhibits the fact that

$$\text{rank } g(\lambda I_3 + N) = \begin{cases} 3 & \text{if} \quad g(\lambda) \neq 0 \\ 2 & \text{if} \quad g(\lambda) = 0 \text{ and } g^{(1)}(\lambda) \neq 0 \\ 1 & \text{if} \quad g(\lambda) = g^{(1)}(\lambda) = 0 \text{ and } g^{(2)}(\lambda) \neq 0 \end{cases}.$$

□

Exercise 21.5. Confirm formula (21.19) for the polynomial

$$g(\lambda) = \sum_{k=0}^{n} g_k \lambda^k \quad \text{by writing} \quad g(\lambda I_p + N) = \sum_{k=0}^{n} g_k(\lambda I_p + N)^k$$

and invoking the binomial formula. [REMARK: This is a good exercise in manipulating formulas, but it's a lot more work.]

Theorem 21.10. *If $f(\lambda)$ is a polynomial of degree n (i.e., $f_n \neq 0$) and $g(\lambda)$ is a polynomial of degree $\leq n$ with at least one nonzero coefficient, then* $\dim \mathcal{N}_{B(f,g)}$ *is equal to the number of common roots of $f(\lambda)$ and $g(\lambda)$ counting multiplicities.*

Proof. The proof rests on formula (21.18). There are three main ingredients in understanding how to exploit this formula:

(1) The special structure of Jordan forms J_f of companion matrix S_f: If
$$f(\lambda) = f_n(\lambda - \lambda_1)^{m_1} \cdots (\lambda - \lambda_k)^{m_k}$$
with k distinct roots $\lambda_1, \cdots, \lambda_k$, where $f_n \neq 0$ and $m_1 + \cdots + m_k = n$, then (up to permutation of the blocks) the Jordan form
$$J_f = \text{diag}\{C_{\lambda_1}^{(m_1)}, C_{\lambda_2}^{(m_2)}, \cdots, C_{\lambda_k}^{(m_k)}\}\,.$$

(2) The special structure of $g(S_f)$: If $S_f = UJ_fU^{-1}$, then $g(S_f) = Ug(J_f)U^{-1}$. Therefore,

$$g(J_f) = \operatorname{diag}\{g(C_{\lambda_1}^{(m_1)}), \cdots, g(C_{\lambda_k}^{(m_k)})\}$$

and correspondingly,

$$\dim \mathcal{N}_{B(f,g)} = \dim \mathcal{N}_{g(S_f)} = \sum_{j=1}^{k} \dim \mathcal{N}_{g(C_{\lambda_j}^{(m_j)})} .$$

(3) The formulas in Lemma 21.9, which clarifies the connection between $\dim \mathcal{N}_{g(C_{\lambda_j}^{(m_j)})}$ and the order of λ_j as a zero of $g(\lambda)$.

Let

$$\nu_j = \dim \mathcal{N}_{g(\lambda_j I_{m_j} + N)} \text{ for } j = 1, \ldots, k.$$

Then, clearly,

$$\nu_j = m_j - \operatorname{rank} g(C_{\lambda_j}^{(m_j)}) \quad \text{and} \quad \nu_j > 0 \Longleftrightarrow g(\lambda_j) = 0 .$$

Moreover, if $\nu_j > 0$, then

$$g(\lambda) = (\lambda - \lambda_j)^{\nu_j} h_j(\lambda), \text{ where } h_j(\lambda_j) \neq 0;$$

i.e., $g(\lambda)$ has a zero of order ν_j at the point λ_j. □

Exercise 21.6. Show that if $A \in \mathbb{C}^{n \times n}$, $B \in \mathbb{C}^{n \times n}$ and $AB = BA$, then

$$g(B) = \sum_{k=0}^{n} \frac{g^{(k)}(A)}{k!}(B - A)^k$$

for every polynomial $g(\lambda)$ of degree $\leq n$.

Exercise 21.7. Use Theorem 21.10 and formula (21.18) to calculate the number of common roots of the polynomials $f(x) = 2 - 3x + x^3$ and $g(x) = -2 + x + x^2$.

21.5. Resultants

The $2n \times 2n$ matrix

$$R(f,g) = \left[\begin{array}{cccc:cccc} f_0 & f_1 & \cdots & f_{n-1} & f_n & 0 & \cdots & 0 \\ 0 & f_0 & \cdots & f_{n-2} & f_{n-1} & f_n & \cdots & 0 \\ \vdots & \vdots & & \vdots & \vdots & \vdots & & \vdots \\ 0 & 0 & \cdots & f_0 & f_1 & f_2 & \cdots & f_n \\ \hdashline g_0 & g_1 & \cdots & g_{n-1} & g_n & 0 & \cdots & 0 \\ 0 & g_0 & \cdots & g_{n-2} & g_{n-1} & g_n & \cdots & 0 \\ \vdots & \vdots & & \vdots & \vdots & \vdots & & \vdots \\ 0 & 0 & \cdots & g_0 & g_1 & g_2 & \cdots & g_n \end{array}\right]$$

based on the coefficients of the polynomials $f(\lambda) = f_0 + f_1\lambda + \cdots + f_n\lambda^n$ and $g(\lambda) = g_0 + g_1\lambda + \cdots + g_n\lambda^n$ is termed the **resultant** of $f(\lambda)$ and $g(\lambda)$.

Theorem 21.11. *If $f(\lambda)$ is a polynomial of degree n (i.e., $f_n \neq 0$) and $g(\lambda)$ is a polynomial of degree $\leq n$ with at least one nonzero coefficient, then $\dim \mathcal{N}_{R(f,g)}$ = the number of common roots of $f(\lambda)$ and $g(\lambda)$ counting multiplicities.*

The proof rests on a number of matrix identities that are used to show that

$$\dim \mathcal{N}_{B(f,g)} = \dim \mathcal{N}_{R(f,g)} \,.$$

It is convenient to express the relevant matrices in terms of the $n \times n$ matrices:

$$(21.20)\quad Z = \begin{bmatrix} 0 & 0 & \cdots & 0 & 1 \\ 0 & 0 & \cdots & 1 & 0 \\ \vdots & & & & \vdots \\ 0 & 1 & \cdots & 0 & 0 \\ 1 & 0 & \cdots & 0 & 0 \end{bmatrix} \quad \text{and} \quad N = \begin{bmatrix} 0 & 1 & 0 & \cdots & 0 \\ 0 & 0 & 1 & \cdots & 0 \\ \vdots & \vdots & \ddots & \ddots & \\ 0 & 0 & \cdots & 0 & 1 \\ 0 & 0 & \cdots & 0 & 0 \end{bmatrix},$$

$$(21.21)\quad H_f = \begin{bmatrix} f_1 & f_2 & \cdot & \cdot & f_n \\ f_2 & & & \cdot & 0 \\ \cdot & & & \cdot & \cdot \\ \cdot & \cdot & & & \cdot \\ f_n & 0 & \cdot & \cdot & 0 \end{bmatrix} = \sum_{j=0}^{n-1} f_{n-j}\, Z(N^T)^j,$$

$$(21.22)\quad \widetilde{H_f} = \begin{bmatrix} 0 & \cdot & \cdot & \cdot & f_0 \\ \cdot & & & \cdot & f_1 \\ \cdot & & \cdot & & \cdot \\ \cdot & \cdot & & & \cdot \\ f_0 & f_1 & \cdot & \cdot & f_{n-1} \end{bmatrix} = \sum_{j=0}^{n-1} f_j\, ZN^j,$$

$$(21.23)\quad T_f = \begin{bmatrix} f_0 & f_1 & \cdots & f_{n-1} \\ 0 & f_0 & \cdots & f_{n-2} \\ \vdots & \vdots & & \vdots \\ 0 & \cdot & \cdots & f_0 \end{bmatrix} = \sum_{j=0}^{n-1} f_j\, N^j$$

and

$$(21.24)\quad G_f = ZH_f = \begin{bmatrix} f_n & 0 & \cdots & 0 \\ f_{n-1} & f_n & & 0 \\ \vdots & \vdots & \ddots & \vdots \\ f_1 & f_2 & \cdots & f_n \end{bmatrix} = \sum_{j=0}^{n-1} f_{n-j}\, (N^T)^j.$$

Lemma 21.12. *The matrices Z and N satisfy the relations*

$$ZN = N^T Z \text{ and } ZN^T = NZ. \tag{21.25}$$

Proof. In terms of the standard basis $\mathbf{e}_j, j = 1, \dots, n$, of $\mathbb{C}^n$,

$$Z = \sum_{j=1}^{n} \mathbf{e}_j\, \mathbf{e}_{n-j+1}^T \text{ and } N = \sum_{i=1}^{n-1} \mathbf{e}_i\, \mathbf{e}_{i+1}^T.$$

Therefore,

$$\begin{aligned} ZN &= \sum_{j=1}^{n} \mathbf{e}_j\, \mathbf{e}_{n-j+1}^T \sum_{i=1}^{n-1} \mathbf{e}_i\, \mathbf{e}_{i+1}^T \\ &= \sum_{j=2}^{n} \mathbf{e}_j (\mathbf{e}_{n-j+1}^T\, \mathbf{e}_{n-j+1}) \mathbf{e}_{n-j+2}^T \\ &= \sum_{j=2}^{n} \mathbf{e}_j\, \mathbf{e}_{n-j+2}^T, \end{aligned}$$

which is a real symmetric matrix:

$$(ZN)^T = \sum_{j=2}^{n} \mathbf{e}_{n-j+2}\, \mathbf{e}_j^T = \sum_{i=2}^{n} \mathbf{e}_i\, \mathbf{e}_{n-i+2}^T = ZN\,.$$

Therefore,

$$ZN = N^T Z^T = N^T Z,$$

since $Z = Z^T$. This justifies the first formula in (21.25). The proof of the second formula is similar. □

Lemma 21.13. *If $f(\lambda) = f_0 + f_1\lambda + \cdots + f_n(\lambda)^n$ and $g(\lambda) = g_0 + g_1\lambda + \cdots + g_n(\lambda)^n$, then the Bezoutian*

$$B(f,g) = H_f T_g - H_g T_f \tag{21.26}$$

and

$$O = H_f G_g - H_g G_f\,. \tag{21.27}$$

Proof. Clearly,

$$\begin{aligned} \frac{f(\lambda)g(\mu) - g(\lambda)f(\mu)}{\lambda - \mu} &= \frac{f(\lambda) - f(\mu)}{\lambda - \mu} g(\mu) - \frac{g(\lambda) - g(\mu)}{\lambda - \mu} f(\mu) \\ &= \mathbf{v}(\lambda)^T \{H_f \mathbf{v}(\mu) g(\mu) - H_g \mathbf{v}(\mu) f(\mu)\}, \end{aligned}$$

in view of formulas (21.3) and (21.7). Moreover,

$$\mathbf{v}(\mu)g(\mu) = \begin{bmatrix} 1 \\ \mu \\ \vdots \\ \mu^{n-1} \end{bmatrix} g(\mu) = \begin{bmatrix} g_0 & g_1 & \cdots & g_{n-1} \\ 0 & g_0 & \cdots & g_{n-2} \\ \vdots & \ddots & \ddots & \vdots \\ 0 & \cdots & 0 & g_0 \end{bmatrix} \begin{bmatrix} 1 \\ \mu \\ \vdots \\ \mu^{n-1} \end{bmatrix} + \begin{bmatrix} g_n & 0 & \cdots & 0 \\ g_{n-1} & g_n & \ddots & \vdots \\ \vdots & \vdots & \ddots & 0 \\ g_1 & g_2 & \cdots & g_n \end{bmatrix} \begin{bmatrix} \mu^n \\ \mu^{n+1} \\ \vdots \\ \mu^{2n-1} \end{bmatrix},$$

i.e.,

$$\mathbf{v}(\mu)g(\mu) = T_g\mathbf{v}(\mu) + \mu^n G_g\mathbf{v}(\mu). \tag{21.28}$$

Consequently,

$$\begin{aligned} \frac{f(\lambda)g(\mu) - g(\lambda)f(\mu)}{\lambda - \mu} &= \mathbf{v}(\lambda)^T\{H_fT_g - H_gT_f\}\,\mathbf{v}(\mu) \\ &+ \mu^n\mathbf{v}(\lambda)^T\{H_fG_g - H_gG_f\}\,\mathbf{v}(\mu). \end{aligned}$$

Thus, in order to complete the proof, it suffices to verify (21.27). However, by formulas (21.21) and (21.24),

$$\begin{aligned} H_fG_g - H_gG_f &= \left(\sum_{j=0}^{n-1} f_{n-j}Z(N^T)^j\right)\left(\sum_{k=0}^{n-1} g_{n-k}(N^T)^k\right) \\ &- \left(\sum_{k=0}^{n-1} g_{n-k}Z(N^T)^k\right)\left(\sum_{j=0}^{n-1} f_{n-j}(N^T)^j\right) \\ &= \sum_{j=0}^{n-1}\sum_{k=0}^{n-1} f_{n-j}\, g_{n-k}\, Z((N^T)^{j+k} - (N^T)^{k+j}) \\ &= O, \end{aligned}$$

as claimed. □

Proof of Theorem 21.11. Lemma 21.13 clearly implies that

$$\begin{bmatrix} H_f & -H_g \\ O & I_n \end{bmatrix}\begin{bmatrix} T_g & G_g \\ T_f & G_f \end{bmatrix} = \begin{bmatrix} B_{(f,g)} & O \\ T_f & G_f \end{bmatrix}$$

or, equivalently, that

$$\begin{bmatrix} -H_g & H_f \\ I_n & O \end{bmatrix}\begin{bmatrix} T_f & G_f \\ T_g & G_g \end{bmatrix} = \begin{bmatrix} B_{(f,g)} & O \\ T_f & G_f \end{bmatrix}.$$

Therefore, since H_f and G_f are invertible when $f_n \neq 0$ and

$$R(f,g) = \begin{bmatrix} T_f & G_f \\ T_g & G_g \end{bmatrix},$$

it follows that

$$\dim \mathcal{N}_{R(f,g)} = \dim \mathcal{N}_{B(f,g)},$$

as claimed. □

Exercise 21.8. Show that the polynomial $p_n(\lambda) = \sum_{j=0}^{n} \frac{\lambda^j}{j!}$ has simple roots. [HINT: It is enough to show that $R(p_n, p_n')$ is invertible.]

Exercise 21.9. Use Theorem 21.11 to calculate the number of common roots of the polynomials $f(x) = 2 - 3x + x^3$ and $g(x) = -2 + x + x^2$.

21.6. Other directions

Theorem 21.14. *Let $f(\lambda) = f_0 + \cdots + f_n\lambda^n$ be a polynomial of degree n (i.e., $f_n \neq 0$), let $g(\lambda) = g_0 + \cdots + g_n\lambda^n$ be a polynomial of degree less than or equal to n that is not identically zero and let $B = B(f,g)$. Then*

$$S_f^T B - BS_f = O \tag{21.29}$$

and B is a solution of the matrix equation

$$S_f^H B - BS_f = f_n(\overline{\mathbf{v}} - \mathbf{v})\{[g_0 \ \cdots \ g_{n-1}] + g_n\mathbf{v}^T\}, \tag{21.30}$$

where

$$\mathbf{v}^T = \mathbf{e}_n^T S_f = -\frac{1}{f_n}[f_0 \ \cdots \ f_{n-1}].$$

Proof. In view of Corollary 21.2, $B = B^T$. Therefore, since $B(f,\varphi_k) = H_f S_f^k$, by Theorem 21.8, it follows in particular that

$$H_f = H_f^T \text{ and } H_f S_f = (H_f S_f)^T$$

and hence, upon reexpressing the last identity, that

$$H_f S_f = (H_f S_f)^T = S_f^T H_f^T = S_f^T H_f, \tag{21.31}$$

which serves to justify the first assertion.

The next step is to observe that

$$S_f^T = N^T + \mathbf{v}\mathbf{e}_n^T$$

and

$$S_f^H = \overline{S_f^T} = N^T + \overline{\mathbf{v}}\mathbf{e}_n^T,$$

and hence that

$$S_f^H = S_f^T + (\overline{\mathbf{v}} - \mathbf{v})\,\mathbf{e}_n^T.$$

Therefore,

$$S_f^H B = S_f^T B + (\overline{\mathbf{v}} - \mathbf{v})\,\mathbf{e}_n^T B = BS_f + (\overline{\mathbf{v}} - \mathbf{v})\,\mathbf{e}_n^T B.$$

Moreover, by Theorem 21.8,

$$\mathbf{e}_n^T B = \mathbf{e}_n^T H_f g(S_f) = f_n \, \mathbf{e}_1^T g(S_f) = f_n \sum_{k=0}^{n} g_k \mathbf{e}_1^T S_f^k .$$

The asserted formula (21.30) now follows easily from the fact that

$$\mathbf{e}_1^T \; S_f^k = \begin{cases} \mathbf{e}_{1+k}^T & \text{if} \quad k = 0, \ldots, n-1 \\ \\ \mathbf{v}^T & \text{if} \quad k = n. \end{cases}$$

□

Theorem 21.15. *Let $f(\lambda) = f_0 + \cdots + f_n\lambda^n$ be a polynomial of degree n (i.e., $f_n \neq 0$), let*

$$f^{\#}(\lambda) = \overline{f(\overline{\lambda})}$$

and let $B = B(f, f^{\#})$. Then:

(1) *$-iB$ is a Hermitian matrix.*

(2) *The Bezoutian $B = B(f, f^{\#})$ is a solution of the matrix equation*

$$S_f^H X - X S_f = |f_n|^2 (\overline{\mathbf{v}} - \mathbf{v})(\mathbf{v} - \overline{\mathbf{v}})^T \tag{21.32}$$

for $X \in \mathbb{C}^{n \times n}$.

(3) *The Hermitian matrix $\widetilde{B} = -iB$ is a solution of the matrix equation*

$$\widetilde{S}^H X + X\widetilde{S} = |f_n|^2 (\overline{\mathbf{v}} - \mathbf{v})(\overline{\mathbf{v}} - \mathbf{v})^H$$

for $X \in \mathbb{C}^{n \times n}$, where $\widetilde{S} = -iS_f$.

(4) *If $B = B(f, f^{\#})$ is invertible, then*

$$\mathcal{E}_+(\widetilde{B}) \quad = \quad \text{the number of roots of } f(\lambda) \text{ in } \mathbb{C}_+$$

and

$$\mathcal{E}_-(\widetilde{B}) \quad = \quad \text{the number of roots of } f(\lambda) \text{ in } \mathbb{C}_- .$$

Proof. Let $g(\lambda) = \sum_{j=0}^{n} g_k \lambda^k$. Then $g(\lambda) = f^{\#}(\lambda)$ if and only if $g_j = \overline{f_j}$ for $j = 0, \ldots, n$. In this case, the right-hand side of equation (21.30) is equal to

$$\begin{aligned} f_n(\overline{\mathbf{v}} - \mathbf{v})(\overline{f}_n \mathbf{v}^T + [\overline{f}_0 \cdots \overline{f}_{n-1}]) &= |f_n|^2 (\overline{\mathbf{v}} - \mathbf{v})(\mathbf{v} - \overline{\mathbf{v}})^T \\ &= |f_n|^2 (\overline{\mathbf{v}} - \mathbf{v})(\overline{\mathbf{v}} - \mathbf{v})^H . \end{aligned}$$

Thus (2) holds and (3) follows easily from (2):

$$\widetilde{S}^H \widetilde{B} + \widetilde{B}\widetilde{S} = S_f^H B - B S_f = |f_n|^2 (\overline{\mathbf{v}} - \mathbf{v})(\overline{\mathbf{v}} - \mathbf{v})^H \succeq O .$$

Moreover, if B is invertible, then, by Theorem 21.10, $f(\lambda)$ and $f^{\#}(\lambda)$ have no common roots; i.e., if $\lambda_1, \ldots, \lambda_k$ denote the distinct roots of $f(\lambda)$, then

$$\{\lambda_1, \ldots, \lambda_k\} \cap \{\overline{\lambda_1}, \ldots, \overline{\lambda_k}\} = \emptyset .$$

In particular, $f(\lambda)$ has no real roots and therefore $\sigma(S_f) \cap \mathbb{R} = \emptyset$ and hence $\sigma(\widetilde{S}) \cap i\mathbb{R} = \emptyset$. Therefore, by Theorem 20.15,

$$\mathcal{E}_+(\widetilde{S}) = \mathcal{E}_+(\widetilde{B}) \quad \text{and} \quad \mathcal{E}_-(\widetilde{S}) = \mathcal{E}_-(\widetilde{B}).$$

But this is easily seen to be equivalent to assertion (4), since

$$-i\lambda \in \Pi_+ \Longleftrightarrow \lambda \in \mathbb{C}_+ \quad \text{and} \quad f(\lambda) = 0 \Longleftrightarrow \lambda \in \sigma(S_f) \Longleftrightarrow -i\lambda \in \sigma(\widetilde{C}).$$

Item (1) is left to the reader. □

21.7. Bezoutians for real polynomials

Theorem 21.16. *Let*

$$a(\lambda) = \sum_{j=0}^{n} a_j \lambda^j \text{ and } b(\lambda) = \sum_{j=0}^{n} b_j \lambda^j$$

be a pair of polynomials with real coefficients and suppose that $a_n \neq 0$ and $|b_0| + \cdots + |b_n| > 0$. Then

(1) $B(a, b) = B(a, b)^H$.

(2) $B(a, b) \prec O$ *if and only if the polynomial $a(\lambda)$ has n real roots $\lambda_1 < \cdots < \lambda_n$ and $a'(\lambda_j)b(\lambda_j) < 0$ for $j = 1, \ldots, n$.*

Proof. Let $B = B(a, b)$ and recall the formula in (21.3) for $\mathbf{v}(\lambda)$. The formulas

$$\begin{aligned}
\mathbf{v}(\lambda)^T B \mathbf{v}(\mu) &= \frac{a(\lambda)b(\mu) - b(\lambda)a(\mu)}{\lambda - \mu} \\
&= \left(\frac{a(\overline{\lambda})b(\overline{\mu}) - b(\overline{\lambda})a(\overline{\mu})}{\overline{\lambda} - \overline{\mu}} \right)^H \\
&= \left(\frac{a(\overline{\mu})b(\overline{\lambda}) - b(\overline{\mu})a(\overline{\lambda})}{\overline{\mu} - \overline{\lambda}} \right)^H \\
&= \left(\mathbf{v}(\overline{\mu})^T B \mathbf{v}(\overline{\lambda}) \right)^H \\
&= \mathbf{v}(\lambda)^T B^H \mathbf{v}(\mu)
\end{aligned}$$

imply that

$$\mathbf{v}(\lambda)^T B \mathbf{v}(\mu) = \mathbf{v}(\lambda)^T B^H \mathbf{v}(\mu)$$

for every choice of λ and μ in $\mathbb{C}$ and hence that (1) holds.

To verify (2), suppose first that $B \prec O$. Then the formulas

$$\frac{a(\lambda)b(\overline{\lambda}) - b(\lambda)a(\overline{\lambda})}{\lambda - \overline{\lambda}} = \mathbf{v}(\overline{\lambda})^H B \mathbf{v}(\overline{\lambda}) < 0 \quad \text{if} \quad \lambda \neq \overline{\lambda}$$

imply that $a(\lambda) \neq 0$ if $\lambda \notin \mathbb{R}$, because

$$a(\lambda) = 0 \Longleftrightarrow \overline{a(\lambda)} = 0 \Longleftrightarrow a(\overline{\lambda}) = 0$$

(since $a(\lambda)$ has real coefficients) and this would contradict the last inequality. Therefore, the roots of $a(\lambda)$ are real. Moreover, if $a(\mu) = 0$ for some point $\mu \in \mathbb{R}$, then

$$\begin{aligned} a'(\mu)b(\mu) &= \lim_{\lambda \to \mu} \frac{(a(\lambda) - a(\mu))b(\mu)}{\lambda - \mu} \\ &= \lim_{\lambda \to \mu} \frac{a(\lambda)b(\mu) - b(\lambda)a(\mu)}{\lambda - \mu} \\ &= \mathbf{v}(\mu)^H B \mathbf{v}(\mu) < 0 \,. \end{aligned}$$

(Thus, the roots of $a(\lambda)$ are simple.)

Next, suppose conversely that the roots of $a(\lambda)$ are real and $a'(\mu)b(\mu) < 0$ at every root μ. Then the roots of $a(\lambda)$ are simple and hence, ordering them as $\lambda_1 < \lambda_2 < \cdots < \lambda_n$, the formulas

$$\mathbf{v}(\lambda_j)^H B \mathbf{v}(\lambda_k) = \frac{a(\lambda_j)b(\lambda_k) - b(\lambda_j)a(\lambda_k)}{\lambda_j - \lambda_k} = 0 \quad \text{if} \quad j \neq k$$

and

$$\mathbf{v}(\lambda_j)^H B \mathbf{v}(\lambda_j) = a'(\lambda_j)b(\lambda_j) < 0$$

imply that the matrix $V^H B V$, in which

$$V = [\mathbf{v}(\lambda_1) \cdots \mathbf{v}(\lambda_n)]$$

is the $n \times n$ Vandermonde matrix with columns $\mathbf{v}(\lambda_j)$, is a diagonal matrix with negative entries on the diagonal. Therefore, $B \prec O$, since V is invertible. □

Exercise 21.10. Show that if, in the setting of Theorem 21.16, $B(a, b) \prec O$ and $b_n = 0$, then $b_{n-1} \neq 0$ and the polynomial $b(\lambda)$ has $n - 1$ real roots $\mu_1, \ldots, \mu_{n-1}$ which interlace the roots of $a(\lambda)$, i.e.,

$$\lambda_1 < \mu_1 < \lambda_2 < \cdots < \lambda_{n-1} < \mu_{n-1} < \lambda_n.$$

Exercise 21.11. Show that if, in the setting of Theorem 21.16, $b_n \neq 0$, then $B(a, b) \prec O$ if and only if $b(\lambda)$ has n real roots $\mu_1 < \cdots < \mu_n$ and $b'(\mu_j)a(\mu_j) > 0$ for $j = 1, \ldots, n$.

21.8. Stable polynomials

A polynomial is said to be **stable** if all of its roots lie in the open left half plane Π_-.

Theorem 21.17. *Let $p(\lambda) = p_0 + p_1\lambda + \cdots + p_n\lambda^n$ be a polynomial of degree n with coefficients $p_j \in \mathbb{R}$ and $p_n \neq 0$ and let*

$$a(\lambda) = \frac{p(i\lambda) + p(-i\lambda)}{2} \quad \text{and} \quad b(\lambda) = \frac{p(i\lambda) - p(-i\lambda)}{2i} \,. \tag{21.33}$$

Then $p(\lambda)$ is stable if and only if the Bezoutian $B(a, b) \prec O$.

Remark 21.18. If $n = 2m$, then

$$a(\lambda) = p_0 - p_2\lambda^2 + \cdots + (-1)^m p_{2m}\lambda^{2m}$$

and

$$b(\lambda) = p_1\lambda - p_3\lambda^3 + \cdots + (-1)^{m-1} p_{2m-1}\lambda^{2m-1}.$$

If $n = 2m+1$, then

$$a(\lambda) = p_0 - p_2\lambda^2 + \cdots + (-1)^m p_{2m}\lambda^{2m}$$

and

$$b(\lambda) = p_1\lambda - p_3\lambda^3 + \cdots + (-1)^m p_{2m+1}\lambda^{2m+1}.$$

Proof. Let $f(\lambda) = p(i\lambda)$ and $f^\#(\lambda) = \overline{f(\overline{\lambda})}$. Then it is readily checked that:

(1) $a(\lambda) = \dfrac{f(\lambda) + f^\#(\lambda)}{2}$.

(2) $b(\lambda) = \dfrac{f(\lambda) - f^\#(\lambda)}{2i}$.

(3) $B(a,b) = -\dfrac{B(f,f^\#)}{2i}$.

(4) $p(\lambda)$ is stable if and only if the roots of $f(\lambda)$ belong to the open upper half plane $\mathbb{C}_+$.

Now, to begin the real work, suppose that the Bezoutian $B(a,b) \prec O$. Then $-iB(f,f^\#) \succ O$ and hence, in view of Theorem 21.15, the roots of the polynomial $f(\lambda)$ are all in $\mathbb{C}_+$. Therefore, the roots of $p(\lambda)$ are all in Π_-; i.e., $p(\lambda)$ is stable.

The main step in the proof of the converse is to show that if $p(\lambda)$ is stable, then $B(a,b)$ is invertible because if $B(a,b)$ is invertible, then $B(f,f^\#)$ is invertible, and hence, by Theorem 21.15,

$$\begin{aligned} \mathcal{E}_+(-iB(f,f^\#)) &= \text{the number of roots of } f(\lambda) \text{ in } \mathbb{C}_+ \\ &= \text{the number of roots of } p(\lambda) \text{ in } \Pi_- \\ &= n; \end{aligned}$$

i.e., $-iB(f,f^\#) \succ O$ and therefore $B(a,b) \prec O$.

To complete the proof, it suffices to show that $a(\lambda)$ and $b(\lambda)$ have no common roots. Suppose to the contrary that $a(\alpha) = b(\alpha) = 0$ for some point $\alpha \in \mathbb{C}$. Then

$$p(i\alpha) = a(\alpha) + ib(\alpha) = 0\,.$$

Therefore, $i\alpha \in \Pi_-$. However, since $a(\lambda)$ and $b(\lambda)$ have real coefficients, it follows that

$$p(i\overline{\alpha}) = a(\overline{\alpha}) + ib(\overline{\alpha}) = \overline{a(\alpha)} + i\overline{b(\alpha)} = 0$$

also. Therefore, if α is a common root of $a(\lambda)$ and $b(\lambda)$, then $i\alpha$ and $i\overline{\alpha}$ both belong to Π_-; i.e., the real part of $i\alpha$ is negative and the real part of $i\overline{\alpha}$ is negative. But this is impossible. □

Exercise 21.12. Verify the four assertions (1)–(4) that are listed at the beginning of the proof of Theorem 21.17. [HINT: To check (4) note that if $p(\lambda) = (\lambda-\lambda_1)^{m_1}\cdots(\lambda-\lambda_k)^{m_k}$, then $p(i\lambda) = (i\lambda+i^2\lambda_1)^{m_1}\cdots(i\lambda+i^2\lambda_k)^{m_k}$.]

Exercise 21.13. Show that if $p(\lambda)$ and $q(\lambda)$ are two stable polynomials with either the same even part or the same odd part, then $tp(\lambda)+(1-t)q(\lambda)$ is stable when $0 \le t \le 1$. (In other words, the set of stable polynomials with real coefficients that have the same even (respectively odd) part is a convex set.)

21.9. Kharitonov's theorem

A problem of great practical interest is to determine when a given polynomial

$$p(\lambda) = p_0 + p_1\lambda + \cdots + p_n\lambda^n$$

is stable. Moreover, in realistic problems the coefficients may only be known approximately, i.e., within the bounds

$$\underline{p}_j \le p_j \le \overline{p}_j \quad \text{for} \quad j = 0, \dots, n\,. \tag{21.34}$$

Thus, in principle it is necessary to show that every polynomial that meets the stated constraints on its coefficients has all its roots in Π_-. A remarkable theorem of Kharitonov states that it is enough to check four extremal cases.

Theorem 21.19. **(Kharitonov)** *Let* $\underline{p}_j \le \overline{p}_j$, $j = 0, \dots, n$, *be given, and let*

$$\varphi_1(\lambda) = \underline{p}_0 + \overline{p}_2\lambda^2 + \underline{p}_4\lambda^4 + \cdots$$

$$\varphi_2(\lambda) = \overline{p}_0 + \underline{p}_2\lambda^2 + \overline{p}_4\lambda^4 + \cdots$$

$$\psi_1(\lambda) = \underline{p}_1\lambda + \overline{p}_3\lambda^3 + \underline{p}_5\lambda^5 + \cdots$$

$$\psi_2(\lambda) = \overline{p}_1\lambda + \underline{p}_3\lambda^3 + \overline{p}_5\lambda^5 + \cdots\,.$$

Then every polynomial

$$p(\lambda) = p_0 + p_1\lambda + \cdots + p_n\lambda^n$$

with coefficients p_j that are subject to the constraints (21.34) *is stable if and only if the four polynomials*

$$p_{jk}(\lambda) = \varphi_j(\lambda) + \psi_k(\lambda)\,, \quad j,k = 1,2,$$

are stable.

Proof. One direction of the asserted equivalence is self-evident. The strategy for proving the other rests on Theorem 21.17, which characterizes the stability of a polynomial $p(\lambda)$ in terms of the Bezoutian $B(a,b)$ of the pair of associated polynomials

$$a(\lambda) = \frac{p(i\lambda)+p(-i\lambda)}{2} \quad \text{and} \quad b(\lambda) = \frac{p(i\lambda)-p(-i\lambda)}{2i}.$$

However, since $\varphi_j(\lambda)$ is even and $\psi_k(\lambda)$ is odd,

$$\begin{aligned} \frac{p_{jk}(i\lambda)+p_{jk}(-i\lambda)}{2} &= \frac{\varphi_j(i\lambda)+\varphi_j(-i\lambda)}{2} = \varphi_j(i\lambda) \\ \frac{p_{jk}(i\lambda)-p_{jk}(-i\lambda)}{2i} &= \frac{\psi_k(i\lambda)-\psi_k(-i\lambda)}{2i} = -i\psi_k(i\lambda). \end{aligned}$$

Thus, upon setting

$$\mathfrak{a}_j(\lambda) = \varphi_j(i\lambda) \quad \text{and} \quad \mathfrak{b}_k(\lambda) = -i\psi_k(i\lambda),$$

it remains to show that

$$B(\mathfrak{a}_j, \mathfrak{b}_k) \prec 0 \quad \text{for} \quad j,k = 1,2 \Longrightarrow B(a,b) \prec 0.$$

Theorem 21.16 is useful for this.

Suppose for the sake of definiteness that $n = 2m$ and $\underline{p}_n > 0$. Then clearly $\mathfrak{a}_1(\lambda) \leq a(\lambda) \leq \mathfrak{a}_2(\lambda)$ for every point $\lambda \in \mathbb{R}$. Moreover, since the Bezoutians $B(\mathfrak{a}_1, \mathfrak{b}_1) \prec O$ and $B(\mathfrak{a}_2, \mathfrak{b}_1) \prec O$, $\mathfrak{b}_1(\lambda)$ has $n-1$ distinct real roots and the polynomials $\mathfrak{a}_1(\lambda)$ and $\mathfrak{a}_2(\lambda)$ each have n distinct real roots that interlace the roots of $\mathfrak{b}_1(\lambda)$. Thus, by Exercise 21.10, the polynomial $a(\lambda)$ also has n real roots $\lambda_1, \ldots, \lambda_n$ that interlace the roots of $\mathfrak{b}_1(\lambda)$ and meet the constraint $a'(\lambda_j)\mathfrak{b}_1(\lambda_j) < 0$. Therefore, by Theorem 21.16, $B(a, \mathfrak{b}_1) \prec O$. Similar considerations imply that $a'(\lambda_j)\mathfrak{b}_2(\lambda_j) < 0$ and $B(a, \mathfrak{b}_2) \prec O$. Moreover, since $\mathfrak{b}_1(\lambda) \leq b(\lambda) \leq \mathfrak{b}_2(\lambda)$ for $\lambda > 0$, $\mathfrak{b}_2(\lambda) \leq b(\lambda) \leq \mathfrak{b}_1(\lambda)$ for $\lambda < 0$ and $\mathfrak{b}_1(0) = b(0) = \mathfrak{b}_2(0) = 0$, the roots of $a(\lambda)$ interlace the roots of $b(\lambda)$ and $a'(\lambda_j)b(\lambda_j) < 0$ for $j = 1, \ldots, n$. Thus, Theorem 21.16 guarantees that $B(a,b) \prec O$. This completes the proof for the case under consideration. The other cases may be handled in much the same way. The details are left to the reader. □

21.10. Bibliographical notes

My main source for early versions of this chapter was the book [**45**]. However, a number of the proofs changed considerably as the chapter evolved. Formula (12.10) and its block matrix analogue (12.37) exhibit the inverse of a Toeplitz matrix as a Bezoutian that is tailored to the unit circle rather than to $\mathbb{R}$ or $i\mathbb{R}$. The application of Bezoutians to verify Kharitonov's theorem is due to A. Olshevsky and V. Olshevsky [**55**]. There are also versions of Kharitonov's theorem for polynomials with complex coefficients. Exercise 21.13 is based on an observation in the paper [**16**].

Chapter 22

Convexity

All extremism, fanaticism and obscurantism come from a lack of security. A person who is secure cannot be an extremist. He uses his heart and his mind in a normal fashion.

I resent very much that certain roshei yeshiva and certain teachers want to impose their will upon the boys.. ... It is up to the individual to make the choice.

Rabbi Joseph B. Soloveitchik, cited in [**59**], p. 237 and p. 240

• **WARNING** The warnings in preceding chapters should be kept in mind.

22.1. Preliminaries

Recall that a subset Q of a vector space $\mathcal{U}$ over $\mathbb{F}$ is said to be **convex** if

$$\mathbf{u}, \mathbf{v} \in Q \Longrightarrow \lambda\mathbf{u} + (1-\lambda)\mathbf{v} \in Q \text{ for every } \lambda \text{ such that } 0 \leq \lambda \leq 1\,;$$

and if $\mathcal{U}$ is a normed linear space over $\mathbb{F}$ with norm $\|\ \|$ and $\mathbf{v} \in \mathcal{U}$, then $\overline{B_r(\mathbf{v})} = \{\mathbf{u} \in \mathcal{U} : \|\mathbf{u} - \mathbf{v}\| \leq r\}$ is a closed convex subset of $\mathcal{U}$, whereas $B_r(\mathbf{v}) = \{\mathbf{u} \in \mathcal{U} : \|\mathbf{u} - \mathbf{v}\| < r\}$ is an open convex subset of $\mathcal{U}$ for every choice of $r > 0$

Exercise 22.1. Verify that the two sets indicated just above are both convex. [HINT: $\lambda\mathbf{u}_1 + (1-\lambda)\mathbf{u}_2 - \mathbf{v} = \lambda(\mathbf{u}_1 - \mathbf{v}) + (1-\lambda)(\mathbf{u}_2 - \mathbf{v})$.]

Exercise 22.2. Show that $Q = \{A \in \mathbb{C}^{n\times n} : A \succeq O\}$is a convex set.

Exercise 22.3. Show that

$$Q = \left\{(A, B, C) \in \mathbb{C}^{n\times n} \times \mathbb{C}^{n\times n} \times \mathbb{C}^{n\times n} : \|A\| \leq 1, \|B\| \leq 1 \text{ and } \|C\| \leq 1\right\}$$

is a convex set.

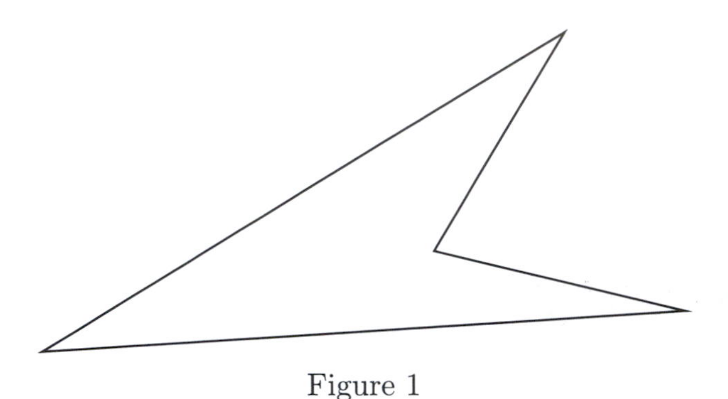

Figure 1

Exercise 22.4. Show that the four-sided figure in Figure 1 is not convex.

A **convex combination of n vectors** is a sum of the form

$$\sum_{i=1}^{n} \lambda_i \mathbf{v}_i \quad \text{with} \quad \lambda_i \geq 0 \quad \text{and} \quad \sum_{i=1}^{n} \lambda_i = 1\,. \tag{22.1}$$

Lemma 22.1. *Let Q be a nonempty subset of a vector space $\mathcal{U}$ over $\mathbb{F}$. Then Q is convex if and only if it is closed under convex combinations, i.e., if and only if for every integer $n \geq 1$,*

$$\mathbf{v}_1, \ldots, \mathbf{v}_n \in Q \Longrightarrow \sum_{i=1}^{n} \lambda_i \mathbf{v}_i \in Q$$

for every choice of nonnegative numbers $\lambda_1, \ldots, \lambda_n$ such that $\lambda_1 + \cdots + \lambda_n = 1$.

Proof. Suppose first that Q is convex, and that $\mathbf{v}_1, \ldots, \mathbf{v}_n \in Q$. Then, if $n > 2$, $\lambda_1 < 1$ and $\mu_j = \lambda_j/(1 - \lambda_j)$ for $j = 2, \ldots, n$, the formula

$$\lambda_1 \mathbf{v}_1 + \lambda_2 \mathbf{v}_2 + \cdots + \lambda_n \mathbf{v}_n = \lambda_1 \mathbf{v}_1 + (1 - \lambda_1) \left\{ \frac{\lambda_2 \mathbf{v}_2 + \cdots + \lambda_n \mathbf{v}_n}{1 - \lambda_1} \right\}$$

implies that

$$\lambda_1 \mathbf{v}_1 + \cdots + \lambda_n \mathbf{v}_n \in Q \Longleftrightarrow \mu_2 \mathbf{v}_2 + \cdots + \mu_n \mathbf{v}_n \in Q\,.$$

Thus, Q is closed under convex combinations of n vectors if and only if it is closed under convex combinations of $n - 1$ vectors. Therefore, Q is closed under convex combinations of n vectors if and only if it is closed under convex combinations of 2 vectors, i.e., if and only if Q is convex. $\square$

Lemma 22.2. *Let $\mathbf{x} \in \mathbb{C}^n$, let Q be a nonempty subset of $\mathbb{C}^n$ and let*

$$d = \inf\{\|\mathbf{x} - \mathbf{u}\| : \mathbf{u} \in Q\}\,.$$

Then the following conclusions hold:

(a) *If Q is closed, then there exists at least one vector $\mathbf{u}_0 \in Q$ such that*

$$d = \|\mathbf{x} - \mathbf{u}_0\| \; .$$

(b) *If Q is closed and convex, then there exists exactly one vector $\mathbf{u}_0 \in Q$ such that*

$$d = \|\mathbf{x} - \mathbf{u}_0\| \; .$$

Proof. Choose a sequence of vectors $\mathbf{u}_1, \mathbf{u}_2, \ldots \in Q$ such that

$$\|\mathbf{x} - \mathbf{u}_k\| \leq d + \frac{1}{k}$$

for $k = 1, 2, \ldots$. Then the bound

$$\|\mathbf{u}_k\| = \|\mathbf{u}_k - \mathbf{x} + \mathbf{x}\| \leq \|\mathbf{u}_k - \mathbf{x}\| + \|\mathbf{x}\| \leq d + \frac{1}{k} + \|\mathbf{x}\|$$

guarantees that the vectors $\mathbf{u}_k$ are bounded and hence that a subsequence $\mathbf{u}_{k_1}, \mathbf{u}_{k_2}, \ldots$ converges to a limit $\mathbf{u}_0$, which must belong to Q if Q is closed. The bounds

$$d \leq \|\mathbf{x} - \mathbf{u}_0\| \leq \|\mathbf{x} - \mathbf{u}_{k_j}\| + \|\mathbf{u}_{k_j} - \mathbf{u}_0\| \leq d + \frac{1}{k_j} + \|\mathbf{u}_{k_j} - \mathbf{u}_0\|$$

serve to complete the proof of (a).

Suppose next that Q is both closed and convex and that $d = \|\mathbf{x} - \mathbf{u}_0\| = \|\mathbf{x} - \mathbf{v}_0\|$. Then, by the parallelogram law,

$$\begin{aligned} 4d^2 &= 2\|\mathbf{x} - \mathbf{u}_0\|^2 + 2\|\mathbf{x} - \mathbf{v}_0\|^2 = \|\mathbf{x} - \mathbf{u}_0 + \mathbf{x} - \mathbf{v}_0\|^2 + \|\mathbf{v}_0 - \mathbf{u}_0\|^2 \\ &= 4\left\|\mathbf{x} - \frac{\mathbf{u}_0 + \mathbf{v}_0}{2}\right\|^2 + \|\mathbf{v}_0 - \mathbf{u}_0\|^2 \geq 4d^2 + \|\mathbf{v}_0 - \mathbf{u}_0\|^2 . \end{aligned}$$

Therefore, $0 \geq \|\mathbf{v}_0 - \mathbf{u}_0\|$, which proves uniqueness. □

22.2. Convex functions

A real-valued function $f(\mathbf{x})$ that is defined on a convex subset Q of a vector space $\mathcal{U}$ over $\mathbb{R}$ is said to be **convex** if

$$f(t\mathbf{x} + (1-t)\mathbf{y}) \leq tf(\mathbf{x}) + (1-t)f(\mathbf{y}) \tag{22.2}$$

for $\mathbf{x}, \mathbf{y} \in Q$ and $0 \leq t \leq 1$, or, equivalently, if **Jensen's inequality** holds:

$$f(\sum_{i=1}^{n} \lambda_i \mathbf{x}_i) \leq \sum_{i=1}^{n} \lambda_i f(\mathbf{x}_i) \tag{22.3}$$

for every convex combination $\sum_{i=1}^{n} \lambda_i \mathbf{x}_i$ of vectors in Q.

Lemma 22.3. *Let $f(x)$ be a convex function on an open subinterval Q of $\mathbb{R}$ and let $a < c < b$ be three points in Q. Then*

$$\frac{f(c) - f(a)}{c - a} \leq \frac{f(b) - f(a)}{b - a} \leq \frac{f(b) - f(c)}{b - c} . \tag{22.4}$$

Proof. Let $0 < t < 1$ and $c = ta + (1-t)b$. Then the inequality

$$f(c) \leq tf(a) + (1-t)f(b)$$

implies that

$$f(b) - f(c) \geq t(f(b) - f(a)),$$

or, equivalently, that

$$\frac{f(b) - f(c)}{t(b-a)} \geq \frac{f(b) - f(a)}{b-a},$$

which serves to prove the second inequality, since $t(b-a) = b-c$. The proof of the first inequality is established in much the same way. The first step is to observe that

$$f(c) - f(a) \leq (1-t)(f(b) - f(a)).$$

The rest is left to the reader as an exercise. □

Exercise 22.5. Complete the proof of Lemma 22.3.

Lemma 22.4. *Let $Q = (\alpha, \beta)$ be an open subinterval of $\mathbb{R}$ and let $f \in \mathcal{C}^2(Q)$. Then $f(x)$ is convex on Q if and only if $f''(x) \geq 0$ at every point $x \in Q$.*

Proof. Suppose first that $f(x)$ is convex on Q and let $a < c < b$ be three points in Q. Then upon letting $c \downarrow a$ in the inequality (22.4), one can readily see that

$$f'(a) \leq \frac{f(b) - f(a)}{b-a}.$$

Next, upon letting $c \uparrow b$ in the same set of inequalities, it follows that

$$\frac{f(b) - f(a)}{b-a} \leq f'(b).$$

Thus, $f'(a) \leq f'(b)$ when $a \leq b$ and $f(x)$ is convex. Therefore, $f''(x) \geq 0$ at every point $x \in Q$.

Conversely, if $f''(x) \geq 0$ at every point $x \in Q$ and if $a < c < b$ are three points in Q, then, by the mean value theorem,

$$\frac{f(c) - f(a)}{c-a} = f'(\xi) \quad \text{for some point} \quad \xi \in (a, c)$$

and

$$\frac{f(b) - f(c)}{b-c} = f'(\eta) \quad \text{for some point} \quad \eta \in (c, b).$$

Therefore, since $f'(\xi) \leq f'(\eta)$, it follows that

$$(f(c) - f(a))(b-c) \leq (f(b) - f(c))(c-a).$$

But this in turn implies that

$$f(c)(b-a) \leq (b-c)f(a) + (c-a)f(b),$$

which, upon setting $c = ta + (1-t)b$ for any choice of $t \in (0,1)$, is easily seen to be equivalent to the requisite condition for convexity. □

Exercise 22.6. Let $f(x) = x^r$ on the set $Q = (0, \infty)$. Show that $f(x)$ is convex on Q if and only if $r \geq 1$ or $r \leq 0$ and that $-f(x)$ is convex on Q if and only if $0 \leq r \leq 1$.

Lemma 22.5. *Let Q be an open nonempty convex subset of $\mathbb{R}^n$, and let $f \in \mathcal{C}^2(Q)$ be a real-valued function on Q with Hessian $H_f(\mathbf{x})$. Then*

$$(22.5) \qquad f \quad \text{is convex on } Q \text{ if and only if} \quad H_f(\mathbf{x}) \succeq 0 \quad \text{on} \quad Q$$

for every point $\mathbf{x} \in Q$.

Proof. Let $I_{\mathbf{x},\mathbf{y}} = \{t : \mathbf{x} + t\mathbf{y} \in Q\}$ for any pair of distinct vectors $\mathbf{x}, \mathbf{y} \in Q$. Then it is readily checked that $I_{\mathbf{x},\mathbf{y}}$ is an open nonempty convex subset of $\mathbb{R}$. Moreover, since the function

$$g(t) = f(\mathbf{x} + t\mathbf{y})$$

satisfies the identities

$$g(\mu t_1 + (1-\mu)t_2) = f(\mu(\mathbf{x} + t_1\mathbf{y}) + (1-\mu)(\mathbf{x} + t_2\mathbf{y}))$$

and

$$\mu g(t_1) + (1-\mu)g(t_2) = \mu f(\mathbf{x} + t_1\mathbf{y}) + (1-\mu)f(\mathbf{x} + t_2\mathbf{y}),$$

it is also readily seen that f is convex on Q if and only if g is convex on $I_{\mathbf{x},\mathbf{y}}$ for every choice of $\mathbf{x}, \mathbf{y} \in Q$. Thus, in view of Lemma 22.4 it follows that f is convex on Q if and only if

$$\left(\frac{\partial^2 g}{\partial t^2}\right)(0) \geq 0 \quad \text{for every choice of} \quad \mathbf{x}, \mathbf{y} \in Q.$$

But this serves to complete the proof, since

$$\left(\frac{\partial^2 g}{\partial t^2}\right)(0) = \sum_{i,j=1}^{n} y_i \left(\frac{\partial^2 f}{\partial x_i \partial x_j}\right)(\mathbf{x}) y_j.$$

□

Corollary 22.6. Let Q be a closed nonempty convex subset of $\mathbb{R}^n$, and let $f \in \mathcal{C}^2(Q)$ be a convex real-valued function on Q. Then f attains its minimum value in Q.

22.3. Convex sets in $\mathbb{R}^n$

Lemma 22.7. *Let Q be a closed nonempty convex subset of $\mathbb{R}^n$, let $\mathbf{x} \in \mathbb{R}^n$ and let $\mathbf{u_x}$ be the unique element in Q that is closest to $\mathbf{x}$. Then*

$$(22.6) \qquad \langle \mathbf{x} - \mathbf{u_x}, \mathbf{u} - \mathbf{u_x} \rangle \leq 0 \text{ for every } \mathbf{u} \in Q.$$

Proof. Let $\mathbf{u} \in Q$. Then clearly $(1-\lambda)\mathbf{u_x} + \lambda\mathbf{u} \in Q$ for every number λ in the interval $0 \le \lambda \le 1$. Therefore,

$$\begin{aligned} \|\mathbf{x} - \mathbf{u_x}\|_2^2 &\le \|\mathbf{x} - (1-\lambda)\mathbf{u_x} - \lambda\mathbf{u}\|_2^2 \\ &= \|\mathbf{x} - \mathbf{u_x} - \lambda(\mathbf{u} - \mathbf{u_x})\|_2^2 \\ &= \|\mathbf{x} - \mathbf{u_x}\|_2^2 - 2\lambda\langle \mathbf{x} - \mathbf{u_x}, \mathbf{u} - \mathbf{u_x}\rangle + \lambda^2\|\mathbf{u} - \mathbf{u_x}\|_2^2 \,, \end{aligned}$$

since $\lambda \in \mathbb{R}$ and all the vectors belong to $\mathbb{R}^n$. But this in turn implies that

$$2\lambda\langle \mathbf{x} - \mathbf{u_x}, \mathbf{u} - \mathbf{u_x}\rangle \le \lambda^2\|\mathbf{u} - \mathbf{u_x}\|_2^2$$

and hence that

$$2\langle \mathbf{x} - \mathbf{u_x}, \mathbf{u} - \mathbf{u_x}\rangle \le \lambda\|\mathbf{u} - \mathbf{u_x}\|_2^2$$

for every λ in the interval $0 < \lambda \le 1$. (The restriction $\lambda > 0$ is imposed in order to permit division by λ in the line preceding the last one.) The desired inequality (22.6) now drops out easily upon letting $\lambda \downarrow 0$. □

Exercise 22.7. Let B be a closed nonempty convex subset of $\mathbb{R}^n$, let $\mathbf{a}_0 \in \mathbb{R}^n$ and let $h(\mathbf{x}) = \langle \mathbf{a}_0 - \mathbf{b}_0, \mathbf{x}\rangle$, where $\mathbf{b}_0$ is the unique vector in B that is closest to $\mathbf{a}_0$. Show that if $\mathbf{a}_0 \notin B$, then there exists a number $\delta > 0$ such that $h(\mathbf{a}_0) \ge \delta + h(\mathbf{b})$ for every vector $\mathbf{b} \in B$.

Lemma 22.8. *Let $\mathcal{U}$ be a nonempty subspace of $\mathbb{R}^n$ and let $\mathbf{x} \in \mathbb{R}^n$. Then there exists a unique vector $\mathbf{u_x} \in \mathcal{U}$ that is closest to $\mathbf{x}$. Moreover,*

$$\langle \mathbf{x} - \mathbf{u_x}, \mathbf{u}\rangle = 0 \text{ for every vector } \mathbf{u} \in \mathcal{U} \,. \tag{22.7}$$

Proof. The existence and uniqueness of $\mathbf{u_x}$ follows from Lemma 22.7, since a nonempty subspace of $\mathbb{R}^n$ is a closed nonempty convex set. Lemma 22.7 implies that

$$\langle \mathbf{x} - \mathbf{u_x}, \mathbf{u}\rangle = \langle \mathbf{x} - \mathbf{u_x}, \mathbf{u} + \mathbf{u_x} - \mathbf{u_x}\rangle \le 0$$

for every vector $\mathbf{u} \in \mathcal{U}$. Therefore, since $\mathcal{U}$ is a subspace, the supplementary inequality

$$\langle \mathbf{x} - \mathbf{u_x}, -\mathbf{u}\rangle \le 0$$

is also in force for every vector $\mathbf{u} \in \mathcal{U}$. □

Lemma 22.9. *Let Q be a closed nonempty convex subset of $\mathbb{R}^n$, let $\mathbf{x}, \mathbf{y} \in \mathbb{R}^n$ and let $\mathbf{u_x}$ and $\mathbf{u_y}$ denote the unique elements in Q that are closest to $\mathbf{x}$ and $\mathbf{y}$, respectively. Then*

$$\|\mathbf{u_x} - \mathbf{u_y}\| \le \|\mathbf{x} - \mathbf{y}\| \,. \tag{22.8}$$

Proof. Let

$$\alpha = \langle \mathbf{x} - \mathbf{u_x}, \mathbf{u_y} - \mathbf{u_x}\rangle \text{ and } \beta = \langle \mathbf{y} - \mathbf{u_y}, \mathbf{u_x} - \mathbf{u_y}\rangle \,.$$

In view of Lemma 22.7, $\alpha \leq 0$ and $\beta \leq 0$. Therefore

$$\begin{aligned} \|\mathbf{x}-\mathbf{y}\|^2 &= \|(\mathbf{x}-\mathbf{u_x})-(\mathbf{y}-\mathbf{u_y})+(\mathbf{u_x}-\mathbf{u_y})\|_2^2 \\ &= \|(\mathbf{x}-\mathbf{u_x})-(\mathbf{y}-\mathbf{u_y})\|_2^2-\alpha-\beta+\|\mathbf{u_x}-\mathbf{u_y}\|_2^2 \\ &\geq \|\mathbf{u_x}-\mathbf{u_y}\|_2^2 , \end{aligned}$$

as claimed. □

The inequality (22.8) implies that if Q is a closed nonempty convex subset of $\mathbb{R}^n$, then the mapping from $\mathbf{x} \in \mathbb{R}^n \to \mathbf{u_x} \in Q$ is continuous. This fact will be used to advantage in the next section, which deals with separation theorems.

22.4. Separation theorems in $\mathbb{R}^n$

The next theorem extends Exercise 22.7.

Theorem 22.10. *Let A and B be disjoint nonempty closed convex sets in $\mathbb{R}^n$ such that B is also compact. Then there exists a point $\mathbf{a}_0 \in A$ and a point $\mathbf{b}_0 \in B$ such that*

$$\begin{aligned} \langle \mathbf{a}_0-\mathbf{b}_0, \mathbf{a}\rangle &\geq \frac{1}{2}\{\|\mathbf{a}_0\|_2^2-\|\mathbf{b}_0\|_2^2+\|\mathbf{a}_0-\mathbf{b}_0\|_2^2\} \\ &> \frac{1}{2}\{\|\mathbf{a}_0\|_2^2-\|\mathbf{b}_0\|_2^2-\|\mathbf{a}_0-\mathbf{b}_0\|_2^2\} \geq \langle \mathbf{a}_0-\mathbf{b}_0, \mathbf{b}\rangle \end{aligned}$$

for every choice of $\mathbf{a} \in A$ and $\mathbf{b} \in B$.

Proof. Let $f_A(\mathbf{x})$ denote the unique point in A that is closest to $\mathbf{x} \in \mathbb{R}^n$. Then, by Lemma 22.7,

$$\langle \mathbf{x}-f_A(\mathbf{x}), \mathbf{a}-f_A(\mathbf{x})\rangle \leq 0 \quad \text{for every} \quad \mathbf{a} \in A .$$

Moreover, since $f_A(\mathbf{x})$ is a continuous function of $\mathbf{x}$ by Lemma 22.9,

$$g(\mathbf{x}) = \|\mathbf{x}-f_A(\mathbf{x})\|$$

is a continuous scalar valued function of $\mathbf{x} \in \mathbb{R}^n$. In particular, g is continuous on the compact set B, and hence there exists a vector $\mathbf{b}_0 \in B$ such that

$$\|\mathbf{b}_0-f_A(\mathbf{b}_0)\| \leq \|\mathbf{b}-f_A(\mathbf{b})\|$$

for every $\mathbf{b} \in B$. Let $\mathbf{a}_0 = f_A(\mathbf{b}_0)$. Then

$$\|\mathbf{b}_0-\mathbf{a}_0\| \leq \|\mathbf{b}-f_A(\mathbf{b})\| \leq \|\mathbf{b}-\mathbf{a}_0\|$$

for every choice of $\mathbf{b} \in B$, and hence, as B is convex,

$$\begin{aligned} \|\mathbf{b}_0-\mathbf{a}_0\|^2 &\leq \|(1-\lambda)\mathbf{b}_0+\lambda\mathbf{b}-\mathbf{a}_0\|_2^2 \\ &= \|\lambda(\mathbf{b}-\mathbf{b}_0)-(\mathbf{a}_0-\mathbf{b}_0)\|_2^2 \\ &= \lambda^2\|\mathbf{b}-\mathbf{b}_0\|_2^2-2\lambda\langle \mathbf{b}-\mathbf{b}_0, \mathbf{a}_0-\mathbf{b}_0\rangle+\|\mathbf{a}_0-\mathbf{b}_0\|_2^2 \end{aligned}$$

for $0 \le \lambda \le 1$. But this reduces to the inequality

$$2\lambda\langle \mathbf{a}_0 - \mathbf{b}_0, \mathbf{b} - \mathbf{b}_0\rangle \le \lambda^2 \|\mathbf{b} - \mathbf{b}_0\|_2^2$$

and hence implies that

$$2\langle \mathbf{a}_0 - \mathbf{b}_0, \mathbf{b} - \mathbf{b}_0\rangle \le \lambda \|\mathbf{b} - \mathbf{b}_0\|_2^2$$

for every λ in the interval $0 < \lambda \le 1$. Thus, upon letting $\lambda \downarrow 0$, we obtain the auxiliary inequality

$$\langle \mathbf{a}_0 - \mathbf{b}_0, \mathbf{b} - \mathbf{b}_0\rangle \le 0 \text{ for every } \mathbf{b} \in B\,,$$

which, in turn, yields the inequality

$$\langle \mathbf{a}_0 - \mathbf{b}_0, \mathbf{b}\rangle \le \langle \mathbf{a}_0 - \mathbf{b}_0, \mathbf{b}_0\rangle \text{ for every } \mathbf{b} \in B\,. \tag{22.9}$$

Moreover, Lemma 22.7 implies that

$$\langle \mathbf{b}_0 - \mathbf{a}_0, \mathbf{a} - \mathbf{a}_0\rangle \le 0 \text{ for every } \mathbf{a} \in A$$

and hence that

$$\langle \mathbf{a}_0 - \mathbf{b}_0, \mathbf{a}\rangle \ge \langle \mathbf{a}_0 - \mathbf{b}_0, \mathbf{a}_0\rangle \text{ for every } \mathbf{a} \in A\,. \tag{22.10}$$

Next, since

$$2\langle \mathbf{a}_0, \mathbf{b}_0\rangle = \|\mathbf{a}_0\|_2^2 + \|\mathbf{b}_0\|_2^2 - \|\mathbf{a}_0 - \mathbf{b}_0\|_2^2$$

and $\mathbf{a}_0 \ne \mathbf{b}_0$, it is readily checked that

$$\begin{aligned}
\langle \mathbf{a}_0 - \mathbf{b}_0, \mathbf{a}_0\rangle &= \|\mathbf{a}_0\|_2^2 - \frac{1}{2}\{\|\mathbf{a}_0\|_2^2 + \|\mathbf{b}_0\|_2^2 - \|\mathbf{a}_0 - \mathbf{b}_0\|_2^2\} \\
&= \frac{1}{2}\{\|\mathbf{a}_0\|_2^2 - \|\mathbf{b}_0\|_2^2 + \|\mathbf{a}_0 - \mathbf{b}_0\|_2^2\} \\
&> \frac{1}{2}\{\|\mathbf{a}_0\|_2^2 - \|\mathbf{b}_0\|_2^2 - \|\mathbf{a}_0 - \mathbf{b}_0\|_2^2\} \\
&= \langle \mathbf{a}_0 - \mathbf{b}_0, \mathbf{b}_0\rangle\ .
\end{aligned}$$

The asserted conclusion now drops out easily upon combining the last chain of inequalities with the inequalities (22.9) and (22.10). □

Theorem 22.11. *Let A and B be disjoint nonempty closed convex sets in $\mathbb{R}^n$ such that B is compact. Then there exists a linear functional $f(\mathbf{x})$ on $\mathbb{R}^n$ and a pair of numbers $c_1, c_2 \in \mathbb{R}$ such that*

$$f(\mathbf{a}) \ge c_1 > c_2 \ge f(\mathbf{b}) \tag{22.11}$$

for every choice of $\mathbf{a} \in A$ and $\mathbf{b} \in B$.

Proof. By Theorem 22.10, there exists a pair of points $\mathbf{a}_0 \in A$ and $\mathbf{b}_0 \in B$ such that

$$\langle \mathbf{a}_0 - \mathbf{b}_0, \mathbf{a}\rangle \ge c_1 > c_2 \ge \langle \mathbf{a}_0 - \mathbf{b}_0, \mathbf{b}\rangle$$

for every $\mathbf{a} \in A$ and $\mathbf{b} \in B$, where

$$2c_1 = \|\mathbf{a}_0\|_2^2 - \|\mathbf{b}_0\|_2^2 + \|\mathbf{a}_0 - \mathbf{b}_0\|_2^2$$

and

$$2c_2 = \|\mathbf{a}_0\|_2^2 - \|\mathbf{b}_0\|_2^2 - \|\mathbf{a}_0 - \mathbf{b}_0\|_2^2 \ .$$

The inequality $c_1 > c_2$ holds because $A \cap B = \emptyset$. The proof is now easily completed by defining

$$f(\mathbf{x}) = \langle \mathbf{x}, \mathbf{a}_0 - \mathbf{b}_0 \rangle \ .$$

□

22.5. Hyperplanes

The conclusions of Theorem 22.11 are often stated in terms of **hyperplanes**: A subset Q of an n dimensional vector space $\mathcal{V}$ over $\mathbb{F}$ is said to be a hyperplane in $\mathcal{V}$ if there exists an element $\mathbf{v} \in \mathcal{V}$ such that the set

$$Q - \mathbf{v} = \{\mathbf{u} - \mathbf{v} : \mathbf{u} \in Q\}$$

is an $n-1$ dimensional subspace of $\mathcal{V}$. Or, to put it another way, Q is a hyperplane in $\mathcal{V}$ if and only if there exists a vector $\mathbf{v} \in \mathcal{V}$ and an $n-1$ dimensional subspace $\mathcal{W}$ of $\mathcal{V}$ such that

$$Q = \mathbf{v} + \mathcal{W} \ .$$

Lemma 22.12. *Let $\mathcal{X}$ and $\mathcal{Y}$ be two subspaces of a vector space $\mathcal{V}$ over $\mathbb{F}$, let $\mathbf{u}, \mathbf{v} \in \mathcal{V}$ and suppose that*

$$\mathbf{u} + \mathcal{X} = \mathbf{v} + \mathcal{Y} \ .$$

Then $\mathcal{X} = \mathcal{Y}$ and $\mathbf{u} - \mathbf{v} \in \mathcal{X}$.

Proof. Let $\mathbf{x} \in \mathcal{X}$. Then, under the given assumptions, there exists a pair of vectors $\mathbf{y}_1, \mathbf{y}_2 \in \mathcal{Y}$ such that

$$\mathbf{u} = \mathbf{v} + \mathbf{y}_1 \text{ and } \mathbf{u} + \mathbf{x} = \mathbf{v} + \mathbf{y}_2 \ .$$

Therefore,

$$\mathbf{x} = \mathbf{y}_2 - \mathbf{y}_1$$

and hence, since $\mathcal{Y}$ is a vector space, $\mathcal{X} \subseteq \mathcal{Y}$. But, by much the same argument, $\mathcal{Y} \subseteq \mathcal{X}$ and consequently $\mathcal{X} = \mathcal{Y}$ and thus $\mathbf{u} - \mathbf{v} \in \mathcal{X}$. □

Lemma 22.13. *Let $\mathcal{V}$ be an n dimensional vector space over $\mathbb{F}$, let $\alpha \in \mathbb{F}$ and let f be a linear functional on $\mathcal{V}$ that is not identically zero. Then the set*

$$Q_f(\alpha) = \{\mathbf{v} \in \mathcal{V} : f(\mathbf{v}) = \alpha\}$$

is a hyperplane. Conversely if Q is a hyperplane in $\mathcal{V}$, then there exists a point $\alpha \in \mathbb{F}$ and a linear functional f on $\mathcal{V}$ such that $Q = Q_f(\alpha)$.

Proof. To show that $Q_f(\alpha)$ is a hyperplane, let $\mathbf{u}, \mathbf{w} \in Q_f(\alpha)$. Then $\mathbf{w} - \mathbf{u} \in Q_f(0)$, since

$$f(\mathbf{w} - \mathbf{u}) = f(\mathbf{w}) - f(\mathbf{u}) = \alpha - \alpha = 0\,.$$

Therefore,

$$Q_f(\alpha) \subseteq \mathbf{u} + Q_f(0)$$

and hence, as the opposite inclusion $\mathbf{u} + Q_f(0) \subseteq Q_f(\alpha)$ is easily verified, it follows that

$$Q_f(\alpha) = \mathbf{u} + Q_f(0)\,.$$

Thus, as $Q_f(0)$ is an $n-1$ dimensional subspace of $\mathcal{V}$, $Q_f(\alpha)$ is a hyperplane.

Conversely, if Q is a hyperplane, then there exists a vector $\mathbf{u} \in \mathcal{V}$ and an $n-1$ dimensional subspace $\mathcal{W}$ of $\mathcal{V}$ such that

$$Q = \mathbf{u} + \mathcal{W}\,.$$

There are two cases to consider: (a) $\mathbf{u} \notin \mathcal{W}$ and (b) $\mathbf{u} \in \mathcal{W}$. In case (a),

$$\mathcal{V} = \{\alpha\mathbf{u} + \mathbf{w} : \alpha \in \mathbb{F} \text{ and } \mathbf{w} \in \mathcal{W}\}\,.$$

Moreover, if

$$\alpha_1\mathbf{u} + \mathbf{w}_1 = \alpha_2\mathbf{u} + \mathbf{w}_2\,,$$

then

$$(\alpha_1 - \alpha_2)\mathbf{u} = \mathbf{w}_2 - \mathbf{w}_1$$

and hence as $\mathbf{w}_2 - \mathbf{w}_1 \in \mathcal{W}$ and $\mathbf{u} \notin \mathcal{W}$, $\alpha_1 = \alpha_2$; i.e., the coefficient of $\mathbf{u}$ in the representation of a vector $\mathbf{v} \in \mathcal{V}$ as a linear combination of $\mathbf{u}$ and a vector in $\mathcal{W}$ is unique. Consequently, the functional $f_{\mathbf{u}}$ that is defined by the rule

$$f_{\mathbf{u}}(\mathbf{v}) = \alpha \quad \text{for vectors} \quad \mathbf{v} = \alpha\mathbf{u} + \mathbf{w} \quad \text{with} \quad \mathbf{w} \in \mathcal{W}$$

is a well-defined linear functional on $\mathcal{V}$ and

$$Q = \{\mathbf{v} \in \mathcal{V} : f_{\mathbf{u}}(\mathbf{v}) = 1\}\,.$$

This completes case (a).

In case (b) choose a vector $\mathbf{y} \notin \mathcal{W}$ and note that every vector $\mathbf{v} \in \mathcal{V}$ admits a unique representation of the form

$$\mathbf{v} = \alpha\mathbf{y} + \mathbf{w} \quad \text{with} \quad \mathbf{w} \in \mathcal{W}\,.$$

Correspondingly, the formula $f_{\mathbf{y}}(\alpha\mathbf{y} + \mathbf{w}) = \alpha$ defines a linear functional on $\mathcal{V}$ and

$$Q = \{\mathbf{v} : f_{\mathbf{y}}(\mathbf{v}) = 0\}\,.$$

□

Theorem 22.11 states that if $c_2 < \alpha < c_1$, then the hyperplane $Q_f(\alpha)$ that is defined in terms of the linear functional f that meets the constraints in (22.11) **separates** the two sets A and B.

22.6. Support hyperplanes

A hyperplane H in $\mathbb{R}^n$ is said to be a **support hyperplane** of a nonempty proper convex subset Q of $\mathbb{R}^n$ if every vector in Q sits inside one of the closed halfspaces determined by H. In other words, if $H = \{\mathbf{x} \in \mathbb{R}^n : \langle \mathbf{x}, \mathbf{u}\rangle = c\}$, then either $\langle \mathbf{x}, \mathbf{u}\rangle \leq c$ for every vector in Q, or $\langle \mathbf{x}, \mathbf{u}\rangle \geq c$ for every vector in Q.

Theorem 22.14. *Let $\mathbf{a} \in \mathbb{R}^n$ belong to the boundary of a nonempty convex subset Q of $\mathbb{R}^n$. Then:*

(1) *There exists a point $\mathbf{b} \in \mathbb{R}^n$ such that $\|\mathbf{b} - \mathbf{x}\| \geq \|\mathbf{b} - \mathbf{a}\|$ for every point $\mathbf{x} \in \overline{Q}$, the closure of Q.*

(2) *The hyperplane*

$$H = \{\mathbf{a} + \mathbf{x} : \mathbf{x} \in \mathbb{R}^n \quad \text{and} \quad \langle \mathbf{x}, \mathbf{b} - \mathbf{a}\rangle = 0\}$$

is a support hyperplane for Q through the point $\mathbf{a}$.

Proof. Let $0 < \varepsilon < 1$ and let $\mathbf{c} \in B_\varepsilon(\mathbf{a})$ be such that $c \notin \overline{Q}$. By Theorem 22.11, there exists a vector $\mathbf{u} \in \mathbb{R}^n$ such that the hyperplane $\widetilde{H} = \{\mathbf{x} \in \mathbb{R}^n : \langle \mathbf{x} - \mathbf{c}, \mathbf{u}\rangle = 0\}$ does not intersect $\overline{Q}$. It is left to the reader to check that

$$\min\{\|\mathbf{x} - \mathbf{a}\| : \mathbf{x} \in \widetilde{H}\} = \frac{\langle \mathbf{c} - \mathbf{a}, \mathbf{u}\rangle}{\|\mathbf{u}\|} \leq \varepsilon\,. \tag{22.12}$$

Thus, if $\mathbf{d} \in \mathbb{R}^n$ is the point in the intersection of $B_1(\mathbf{a})$ and the line through $\mathbf{a}$ and the point in $\widetilde{H}$ that achieves the minimum distance given in (22.12), then $\|\mathbf{d} - \mathbf{x}\| \geq 1 - \varepsilon$ for every $\mathbf{x} \in \overline{Q}$. Thus, as ϵ can be an arbitrarily small positive number, and the function

$$f_{\overline{Q}}(\mathbf{x}) = \min\{\|\mathbf{x} - \mathbf{q}\| : \mathbf{q} \in \overline{Q}\}$$

is continuous, it follows that

$$\max\{f_{\overline{Q}}(\mathbf{x}) : \|\mathbf{x} - \mathbf{a}\| = 1\} = 1$$

and is attained by some point $\mathbf{b} \in \mathbb{R}^n$ with $\|\mathbf{b} - \mathbf{a}\| = 1$. Thus,

$$\|\mathbf{b} - \mathbf{x}\| \geq f_{\overline{Q}}(\mathbf{b}) \geq \|\mathbf{b} - \mathbf{a}\| = 1 \quad \text{for} \quad \mathbf{x} \in \overline{Q}\,.$$

This completes the proof of (1).

Next, Lemma 22.7 implies that $\langle \mathbf{b} - \mathbf{a}, \mathbf{x} - \mathbf{a}\rangle \leq 0$ for every point $\mathbf{x} \in \overline{Q}$, which serves to complete the proof, since the given hyperplane can also be written as

$$H = \{\mathbf{x} \in \mathbb{R}^n : \langle \mathbf{x} - \mathbf{a}, \mathbf{b} - \mathbf{a}\rangle = 0\}\,.$$

□

Exercise 22.8. Verify formula (22.12).

22.7. Convex hulls

Let Q be a subset of a vector space $\mathcal{V}$ over $\mathbb{F}$. The **convex hull of** Q is the smallest convex set in $\mathcal{V}$ that contains Q. Since the intersection of two convex sets is convex, the convex hull of Q can also be defined as the intersection of all convex sets in $\mathcal{V}$ that contain Q. The symbol conv Q will be used to denote the convex hull of a set Q.

Lemma 22.15. *Let Q be a subset of $\mathbb{F}^n$. Then the convex hull of Q is equal to the set of all convex combinations of elements in Q, i.e.,*

$$\text{(22.13)} \qquad \text{conv } Q = \left\{ \sum_{i=1}^{n} t_i \mathbf{x}_i : n \geq 1,\ t_i \geq 0,\ \sum_{1}^{n} t_i = 1 \ \textit{and}\ \mathbf{x}_i \in Q \right\}.$$

Proof. It is readily seen that the set on the right-hand side of (22.13) is a convex set: if

$$\mathbf{u} = \sum_{i=1}^{n} t_i \mathbf{x}_i \text{ is a convex combination of } \mathbf{x}_1, \ldots, \mathbf{x}_n \in Q$$

and

$$\mathbf{y} = \sum_{j=1}^{k} s_j \mathbf{y}_j \text{ is a convex combination of } \mathbf{y}_1, \ldots, \mathbf{y}_k \in Q\ ,$$

then

$$\lambda \mathbf{u} + (1-\lambda)\mathbf{v} = \sum_{i=1}^{n} \lambda t_i \mathbf{x}_i + \sum_{j=1}^{k} (1-\lambda) s_j \mathbf{y}_j$$

is again a convex combination of elements of Q for every choice of λ in the interval $0 \leq \lambda \leq 1$, since for such λ, $\lambda t_i \geq 0$, $(1-\lambda)s_j \geq 0$ and

$$\lambda t_1 + \cdots + \lambda t_n + (1-\lambda)s_1 + \cdots + (1-\lambda)s_k = 1\ .$$

Thus, the right-hand side of (22.13) is a convex set that contains Q. Moreover, since every convex set that contains Q must contain the convex combinations of elements in Q, the right hand side of (22.13) is the smallest convex set that contains Q. □

Theorem 22.16. (**Carathéodory**) *Let Q be a nonempty subset of $\mathbb{R}^n$. Then every vector $\mathbf{x} \in \operatorname{conv} Q$ is a convex combination of at most $n+1$ vectors in Q.*

Proof. Let

$$Q_1 = \text{conv} \left\{ \begin{bmatrix} 1 \\ \mathbf{x} \end{bmatrix} : \mathbf{x} \in Q \right\}.$$

Then

$$\begin{aligned}\mathbf{x} \in \operatorname{conv} Q &\implies \begin{bmatrix} 1 \\ \mathbf{x} \end{bmatrix} \in \operatorname{conv} Q_1 \\ &\implies \begin{bmatrix} 1 \\ \mathbf{x} \end{bmatrix} = \sum_{j=1}^{k} \alpha_j \begin{bmatrix} 1 \\ \mathbf{x}_j \end{bmatrix} \text{ with } \alpha_j > 0,\ \mathbf{x}_j \in Q \text{ and } \sum_{j=1}^{k} \alpha_j = 1\,.\end{aligned}$$

If the vectors

$$\mathbf{y}_j = \begin{bmatrix} 1 \\ \mathbf{x}_j \end{bmatrix}, \quad j = 1, \ldots, k\,,$$

are linearly independent, then $k \le n+1$, as claimed. If not, then there exists a set of coefficients $\beta_1, \ldots, \beta_k \in \mathbb{R}$ such that

$$\sum_{j=1}^{k} \beta_j \mathbf{y}_j = \mathbf{0} \quad \text{and} \quad \mathcal{P} = \{j : \beta_j > 0\} \neq \emptyset\,.$$

Let

$$\gamma = \min\left\{ \frac{\alpha_j}{\beta_j} : j \in \mathcal{P} \right\}.$$

Then $\alpha_j - \gamma\beta_j \ge 0$ for $j = 1, \ldots, k$, and, since at least one of these numbers is equal to zero, the formula

$$\begin{bmatrix} 1 \\ \mathbf{x} \end{bmatrix} = \sum_{j=1}^{k} (\alpha_j - \gamma\beta_j) \begin{bmatrix} 1 \\ \mathbf{x}_j \end{bmatrix} \tag{22.14}$$

displays the vector on the left as a convex combination of at most $k-1$ vectors. If these vectors are linearly independent, then there are at most $n+1$ of them. If not, the same argument can be repeated to eliminate additional vectors from the representation until a representation of the form (22.14), but with $k \le n+1$, is obtained. The resulting identities

$$1 = \sum_{j=1}^{k} (\alpha_j - \gamma\beta_j) \quad \text{and} \quad \mathbf{x} = \sum_{j=1}^{k} (\alpha_j - \gamma\beta_j)\mathbf{x}_j$$

serve to complete the proof. □

Exercise 22.9. Let Q be a nonempty subset of $\mathbb{R}^n$. Show that every nonzero vector $\mathbf{x} \in \operatorname{conv} Q$ can be expressed as a linear combination $\sum_{j=1}^{\ell} \alpha_j \mathbf{x}_j$ of ℓ linearly independent vectors in Q with positive coefficients. [HINT: The justification is a variant of the proof of Theorem 22.16.]

The conclusions in Exercise 22.9 can be extended to cones: If Q is a nonempty subset of $\mathbb{R}^n$, then the **cone** generated by Q is the set

$$K_Q = \left\{ \sum_{j=1}^{\ell} \alpha_j \mathbf{x}_j : \mathbf{x}_j \in Q \quad \text{and} \quad \alpha_j > 0 \right\}$$

of all finite linear combinations of vectors in Q with positive coefficients.

Exercise 22.10. Let Q be a nonempty subset of $\mathbb{R}^n$. Show that every nonzero vector $\mathbf{x} \in K_Q$ can be expressed as a linear combination $\sum_{j=1}^{\ell} \alpha_j \mathbf{x}_j$ of ℓ linearly independent vectors in Q with positive coefficients.

Lemma 22.17. *Let Q be a nonempty subset of $\mathbb{R}^n$. Then*

(1) *Q open $\Longrightarrow$ conv Q is open.*

(2) *Q compact $\Longrightarrow$ conv Q is compact.*

Proof. Suppose first that Q is open and that $\mathbf{x} = \sum_{j=1}^{\ell} c_j \mathbf{x}_j$ is a convex combination of vectors $\mathbf{x}_1, \ldots, \mathbf{x}_\ell \in Q$. Then there exists an $\varepsilon > 0$ such that $\mathbf{x}_j + \mathbf{u} \in Q$ for $j = 1, \ldots, \ell$ and every vector $\mathbf{u} \in \mathbb{R}^n$ with $\|\mathbf{u}\| < \varepsilon$. Therefore,

$$\mathbf{x} + \mathbf{u} = \sum_{j=1}^{\ell} c_j(\mathbf{x}_j + \mathbf{u}) \in Q$$

for every vector $\mathbf{u} \in \mathbb{R}^n$ with $\|\mathbf{u}\| < \varepsilon$; i.e., $\operatorname{conv} Q$ is open.

Suppose next that Q is compact and that $\{\mathbf{x}_j\}$, $j = 1, 2, \ldots$, is an infinite sequence of vectors in $\operatorname{conv} Q$ that converges to a vector $\mathbf{x}_0 \in \mathbb{R}^n$. Then, by Theorem 22.16, there exists a sequence of matrices $A_j \in \mathbb{R}^{n\times(n+1)}$ with columns in Q and a sequence of vectors $\mathbf{c}_j \in \mathbb{R}^{n+1}$ with nonnegative coefficients and $\|\mathbf{c}_j\|_1 = 1$ such that $\mathbf{x}_j = A_j \mathbf{c}_j$ for $j = 1, 2, \ldots$. By the presumed compactness of Q and the compactness of the $\{\mathbf{c} \in \mathbb{R}^{n+1} : \|\mathbf{c}\|_1 = 1\}$, there exists a subsequence $n_1 < n_2, \cdots$ such that $A_{n_j} \to A$, $\mathbf{c}_{n_j} \to \mathbf{c}$ and

$$\mathbf{x}_0 = \lim_{j \longrightarrow \infty} \mathbf{x}_{n_j} = \lim_{j \longrightarrow \infty} A_{n_j} \mathbf{c}_{n_j} = A\mathbf{c}\,,$$

where $A \in \mathbb{R}^{n\times(n+1)}$ with columns in Q and $\mathbf{c} \in \mathbb{R}^{n+1}$ with nonnegative coefficients and $\|\mathbf{c}\|_1 = 1$. Thus, $\mathbf{x}_0 = A\mathbf{c} \in \operatorname{conv} Q$. This proves that $\operatorname{conv} Q$ is closed. Since $\operatorname{conv} Q$ is also clearly a bounded subset of $\mathbb{R}^{n+1}$, it must be compact. □

22.8. Extreme points

Let Q be a convex subset of $\mathbb{R}^n$. A vector $\mathbf{u} \in Q$ is said to be an **extreme point** of Q if

$$0 < \alpha < 1,\ \mathbf{x},\ \mathbf{y} \in Q \quad \text{and} \quad \mathbf{u} = \alpha\mathbf{x} + (1-\alpha)\mathbf{y} \Longrightarrow \mathbf{x} = \mathbf{y} = \mathbf{u}\,.$$

Lemma 22.18. *Every nonempty compact convex subset of $\mathbb{R}^n$ contains at least one extreme point.*

Proof. Let Q be a nonempty compact convex subset of $\mathbb{R}^n$ and let $f(\mathbf{x}) = \|\mathbf{x}\|$ for every vector $\mathbf{x} \in Q$. The inequality

$$|f(\mathbf{x}) - f(\mathbf{y})| = |\|\mathbf{x}\| - \|\mathbf{y}\|| \le \|\mathbf{x} - \mathbf{y}\|$$

implies that f is continuous on Q. Therefore f attains its maximum value on Q; i.e., there exists a vector $\mathbf{u} \in Q$ such that $\|\mathbf{u}\| \geq \|\mathbf{x}\|$ for every $\mathbf{x} \in Q$.

The next step is to show that $\mathbf{u}$ is an extreme point of Q. It suffices to restrict attention to the case $\mathbf{u} \neq \mathbf{0}$. But if $\mathbf{u} = \alpha\mathbf{x} + (1-\alpha)\mathbf{y}$ for some pair of vectors $\mathbf{x}$ and $\mathbf{y}$ in Q and some α with $0 < \alpha < 1$, then the inequalities

$$\|\mathbf{u}\| = \|\alpha\mathbf{x} + (1-\alpha)\mathbf{y}\| \leq \alpha\|\mathbf{x}\| + (1-\alpha)\|\mathbf{y}\| \leq \|\mathbf{u}\|$$

clearly imply that $\|\mathbf{x}\| = \|\mathbf{y}\| = \|\mathbf{u}\|$. Thus,

$$\begin{aligned} \langle \mathbf{u}, \mathbf{u} \rangle &= \alpha\langle \mathbf{x}, \mathbf{u} \rangle + (1-\alpha)\langle \mathbf{y}, \mathbf{u} \rangle \leq \alpha\|\mathbf{u}\|\|\mathbf{x}\| + (1-\alpha)\|\mathbf{u}\|\|\mathbf{y}\| \\ &= \alpha\|\mathbf{u}\|^2 + (1-\alpha)\|\mathbf{u}\|^2 = \|\mathbf{u}\|^2 . \end{aligned}$$

Therefore, equality is attained in the Cauchy-Schwarz inequality, and hence $\mathbf{x} = a\mathbf{u}$ and $\mathbf{y} = b\mathbf{u}$ for some choice of a and b in $\mathbb{R}$. However, since $\|\mathbf{x}\| = \|\mathbf{y}\| = \|\mathbf{u}\|$, it is readily seen that the only viable possibilities are $a = b = 1$. Therefore, $\mathbf{u}$ is an extreme point of Q. □

Lemma 22.19. *The set of extreme points of a nonempty compact convex subset of* $\mathbb{R}^2$ *is closed.*

Discussion. The stated conclusion is self-evident if A is either a single point or a subset of a line in $\mathbb{R}^2$. Suppose, therefore, that neither of these cases prevails and let $\mathbf{a} = \lim_{k\uparrow\infty} \mathbf{a}_k$ be the limit of a sequence $\mathbf{a}_1, \mathbf{a}_2, \ldots$ of extreme points of A. Then $\mathbf{a}$ must belong to the boundary of A; i.e., for every $r > 0$ the open unit ball $B_r(\mathbf{a})$ of radius $r > 0$ centered at $\mathbf{a}$ contains points in A and in $\mathbb{R}^2 \setminus A$. Therefore, there exists a line L through the point $\mathbf{a}$ such that all the points in A sit on one side of L. Without loss of generality, we may assume that $L = \{\mathbf{x} \in \mathbb{R}^2 : x_1 = \gamma\}$ for some $\gamma \in \mathbb{R}$ and that $A \subset \{\mathbf{x} \in \mathbb{R}^2 : x_1 \leq \gamma\}$. Thus, if

$$\mathbf{a} = \alpha\mathbf{b} + (1-\alpha)\mathbf{c} \quad \text{with} \quad \mathbf{b}, \mathbf{c} \in A \quad \text{and} \quad 0 < \alpha < 1 ,$$

then, since the first coordinates are subject to the constraints $a_1 = \gamma$, $b_1 \leq \gamma$ and $c_1 \leq \gamma$, it is readily seen that $a_1 = b_1 = c_1$, i.e., $\mathbf{b}, \mathbf{c} \in A \cap L$. Moreover, if $\mathbf{a}$ is not an extreme point of A, then there exists a choice of points $\mathbf{b}, \mathbf{c} \in A \cap L$ with $b_2 > a_2 > c_2$ and a point $\mathbf{d} \in A$ such that $B_r(\mathbf{d}) \subset A$ for some $r > 0$. Consequently,

$$\mathbf{a}_k \in \text{conv}\,\{\mathbf{b}, \mathbf{c}, \mathbf{d}\} \cup \{\mathbf{x} \in \mathbb{R}^2 : c_2 < x_2 < b_2\}$$

for all sufficiently large k. But this is not possible, since the $\mathbf{a}_k$ were presumed to be extreme points of A.

The conclusions of Lemma 22.19 do not propagate to higher dimensions:

Exercise 22.11. Let $Q_1 = \{\mathbf{x} \in \mathbb{R}^3 : x_1^2 + x_2^2 \leq 1 \quad \text{and} \quad x_3 = 0\}$, $Q_2 = \{\mathbf{x} \in \mathbb{R}^3 : x_1 = 1, x_2 = 0 \quad \text{and} \quad -1 \leq x_3 \leq 1\}$. Show that the set of extreme points of the set $\text{conv}\,(Q_1 \cup Q_2)$ is not a closed subset of $\mathbb{R}^3$.

We turn next to a finite dimensional version of the **Krein-Milman theorem**.

Theorem 22.20. *Let Q be a nonempty compact convex set in $\mathbb{R}^n$. Then Q is equal to the convex hull of its extreme points.*

Discussion. By Lemma 22.18, the set E of extreme points of Q is nonempty. Let $\overline{E}$ denote the closure of E and let $F = \operatorname{conv} \overline{E}$ denote the convex hull of $\overline{E}$ and suppose that there exists a vector $\mathbf{q}_0 \in Q$ such that $\mathbf{q}_0 \notin F$. Then, since F is closed, Theorem 22.11 guarantees that there exists a real linear functional h on $\mathbb{R}^n$ and a number $\delta \in \mathbb{R}$ such that

$$h(\mathbf{q}_0) > \delta \geq h(\mathbf{x}) \quad \text{for every} \quad \mathbf{x} \in F\,. \tag{22.15}$$

Let

$$\gamma = \sup\{h(\mathbf{x}) : \mathbf{x} \in Q\} \quad \text{and} \quad E_h = \{\mathbf{x} \in Q : h(\mathbf{x}) = \gamma\}\,.$$

The inequality (22.15) implies that $\gamma > \delta$ and hence that $E \cap E_h = \emptyset$.

On the other hand, it is readily checked that E_h is a compact convex set. Therefore, by Lemma 22.18, it contains extreme points. The next step is to check that if $\mathbf{x}_0 \in E_h$ is an extreme point of E_h, then it is also an extreme point of Q: If

$$\mathbf{x}_0 = \alpha\mathbf{u} + (1-\alpha)\mathbf{v}$$

for some pair of vectors $\mathbf{u}, \mathbf{v} \in Q$ and $0 < \alpha < 1$, then the identity

$$h(\mathbf{x}_0) = \alpha h(\mathbf{u}) + (1-\alpha)h(\mathbf{v})$$

implies that $h(\mathbf{x}_0) = h(\mathbf{u}) = h(\mathbf{v}) = \gamma$ and hence that $\mathbf{u}, \mathbf{v} \in E_h$. Therefore, since $\mathbf{x}_0$ is an extreme point for E_h, $\mathbf{x}_0 = \mathbf{u} = \mathbf{v}$. Thus, $\mathbf{x}_0 \in E$, which proves that $E \cap E_h \neq \emptyset$; i.e.,

$$\text{if} \quad F \neq Q\,, \quad \text{then} \quad E \cap E_h = \emptyset \quad \text{and} \quad E \cap E_h \neq \emptyset\,,$$

which is clearly impossible. Therefore, $F = Q$; i.e., $\operatorname{conv} \overline{E} = Q$.

It remains to show that $Q = \operatorname{conv} E$. In view of Lemma 22.19, this is the case if $Q \subset \mathbb{R}^2$, because then $\overline{E} = E$. Proceeding inductively, suppose that in fact $Q = \operatorname{conv} E$ if Q is a subset of $\mathbb{R}^k$ for $k < p$, let Q be a subset of $\mathbb{R}^p$ and let $\mathbf{q} \in \overline{E}$. Then $\mathbf{q}$ belongs to the boundary of Q and, by translating and rotating Q appropriately, we can assume that $\mathbf{q}$ belongs to the hyperplane $H = \{\mathbf{x} \in \mathbb{R}^p : x_1 = 0\}$ and that Q is a subset of the halfspace $H_- = \{\mathbf{x} \in \mathbb{R}^p : x_1 \leq 0\}$. Thus, $Q \cap H$ can be identified with a compact convex subset of $\mathbb{R}^{p-1}$. Let E' denote the extreme points of $Q \cap H$. By the induction hypothesis $Q \cap H = \operatorname{conv} E'$. Therefore, since $\mathbf{q} \in Q \cap H$, and $\operatorname{conv} E' \subseteq \operatorname{conv} E$, it follows that

$$\overline{E} \subseteq \operatorname{conv} E\,, \tag{22.16}$$

and hence that $Q = \operatorname{conv} E$, as claimed. □

Theorem 22.21. *Let Q be a nonempty compact convex set in $\mathbb{R}^n$. Then every vector in Q is a convex combination of at most $n+1$ extreme points of Q.*

Proof. This is an immediate consequence of Carathéodory's theorem and the Krein-Milman theorem. □

22.9. Brouwer's theorem for compact convex sets

A simple argument serves to extend the Brouwer fixed point theorem to compact convex subsets of $\mathbb{R}^n$.

Theorem 22.22. *Let Q be a nonempty compact convex subset of $\mathbb{R}^n$ and let $\mathbf{f}$ be a continuous mapping of Q into Q. Then there exists at least one point $\mathbf{q} \in Q$ such that $\mathbf{f}(\mathbf{q}) = \mathbf{q}$.*

Proof. Since Q is compact, there exists an $r > 0$ such that the closed ball $\overline{B_r(\mathbf{0})} = \{\mathbf{x} \in \mathbb{R}^n : \|\mathbf{x}\| \leq r\}$ contains Q. Then, by Lemma 22.2, for each point $\mathbf{x} \in \overline{B_r(\mathbf{0})}$ there exists a unique vector $\mathbf{q_x} \in Q$ that is closest to $\mathbf{x}$. Moreover, by Lemma 22.9, the function $\mathbf{g}$ from $\overline{B_r(\mathbf{0})}$ into Q that is defined by the rule $\mathbf{g}(\mathbf{x}) = \mathbf{q_x}$ is continuous. Therefore the composite function $\mathbf{h}(\mathbf{x}) = \mathbf{f}(\mathbf{g}(\mathbf{x}))$ is a continuous map of $\overline{B_r(\mathbf{0})}$ into itself and therefore has a fixed point in $\overline{B_r(\mathbf{0})}$. But this serves to complete the proof:

$$\mathbf{f}(\mathbf{g}(\mathbf{x})) = \mathbf{x} \Longrightarrow \mathbf{f}(\mathbf{q_x}) = \mathbf{x} \Longrightarrow \mathbf{x} \in Q \Longrightarrow \mathbf{x} = \mathbf{q_x} \Longrightarrow \mathbf{f}(\mathbf{x}) = \mathbf{x}\,.$$

□

Exercise 22.12. Show that the function

$$\mathbf{f}(\mathbf{x}) = \begin{bmatrix} (x_1 + x_2)/2 \\ \\ \sqrt{x_1 x_2} \end{bmatrix}$$

maps the set $Q = \{\mathbf{x} \in \mathbb{R}^2 : 1 \leq x_1 \leq 2 \quad \text{and} \quad 1 \leq x_2 \leq 2\}$ into itself and then invoke Theorem 22.22 to establish the existence of fixed points in this set and find them.

Exercise 22.13. Show that the function $\mathbf{f}$ defined in Exercise 22.12 does not satisfy the constraint $\|\mathbf{f}(\mathbf{x}) - \mathbf{f}(\mathbf{y})\| < \gamma\|\mathbf{x} - \mathbf{y}\|$ for all vectors $\mathbf{x}$, $\mathbf{y}$ in the set Q that is considered there if $\gamma < 1$. [HINT: Consider the number of fixed points.]

22.10. The Minkowski functional

Let $\mathcal{X}$ be a normed linear space over $\mathbb{F}$ and let $Q \subseteq \mathcal{X}$. Then the functional

$$p_Q(\mathbf{x}) = \inf\{t > 0 : \frac{\mathbf{x}}{t} \in Q\}$$

is called the **Minkowski functional**. If the indicated set of t is empty, then $p_Q(\mathbf{x}) = \infty$.

Lemma 22.23. *Let Q be a convex subset of a normed linear space $\mathcal{X}$ over $\mathbb{F}$ such that*

$$Q \supseteq B_r(\mathbf{0}) \quad \textit{for some} \quad r > 0$$

and let $\operatorname{int} Q$ *and* $\overline{Q}$, *denote the interior and the closure of* Q, *respectively. Then:*

(1) $p_Q(\mathbf{x}+\mathbf{y}) \le p_Q(\mathbf{x}) + p_Q(\mathbf{y})$ *for* $\mathbf{x}, \mathbf{y} \in \mathcal{X}$.

(2) $p_Q(\alpha\mathbf{x}) = \alpha p_Q(\mathbf{x})$ *for* $\alpha \ge 0$ *and* $\mathbf{x} \in \mathcal{X}$.

(3) $p_Q(\mathbf{x})$ *is continuous.*

(4) $\{\mathbf{x} \in \mathcal{X} : p_Q(\mathbf{x}) < 1\} = \operatorname{int} Q$.

(5) $\{\mathbf{x} \in \mathcal{X} : p_Q(\mathbf{x}) \le 1\} = \overline{Q}$.

(6) *If Q is also bounded, then* $p_Q(\mathbf{x}) = 0 \Rightarrow \mathbf{x} = \mathbf{0}$.

Proof. Let $\mathbf{x}, \mathbf{y} \in \mathcal{X}$ and suppose that $\alpha^{-1}\mathbf{x} \in Q$ and $\beta^{-1}\mathbf{y} \in Q$ for some choice of $\alpha > 0$ and $\beta > 0$. Then, since Q is convex,

$$\frac{\mathbf{x}+\mathbf{y}}{\alpha+\beta} = \frac{\alpha}{\alpha+\beta}(\alpha^{-1}\mathbf{x}) + \frac{\beta}{\alpha+\beta}(\beta^{-1}\mathbf{y})$$

belongs to Q and hence,

$$p_Q(\mathbf{x}+\mathbf{y}) \le \alpha + \beta\ .$$

Consequently, upon letting α run through a sequence of values $\alpha_1 \ge \alpha_2 \ge \cdots$ that tend to $p_Q(\mathbf{x})$ and letting β run through a sequence of values $\beta_1 \ge \beta_2 \ge \cdots$ that tend to $p_Q(\mathbf{y})$, one can readily see that

$$p_Q(\mathbf{x}+\mathbf{y}) \le p_Q(\mathbf{x}) + p_Q(\mathbf{y})\ .$$

Suppose next that $\alpha > 0$ and $p_Q(\mathbf{x}) = a$. Then there exists a sequence of numbers $t_1, t_2, \ldots$ such that $t_j > 0$,

$$\frac{\mathbf{x}}{t_j} \in Q \text{ and } \lim_{j\uparrow\infty} t_j = a\ .$$

Therefore, since

$$\frac{\alpha\mathbf{x}}{\alpha t_j} \in Q \quad \text{and} \quad \lim_{j\uparrow\infty} \alpha t_j = \alpha a\ ,$$

$$f_Q(\alpha\mathbf{x}) \le \alpha f_Q(\mathbf{x})\ .$$

However, the same argument yields the opposite inequality:

$$\alpha f_Q(\mathbf{x}) = \alpha f_Q(\alpha^{-1}\alpha\mathbf{x}) \le \alpha\alpha^{-1} f_Q(\alpha\mathbf{x}) = f_Q(\alpha\mathbf{x})\ .$$

Therefore, equality prevails. This completes the proof of (2) when $\alpha > 0$. However, (2) holds when $\alpha = 0$, because $p_Q(\mathbf{0}) = 0$.

If $\mathbf{x} \neq \mathbf{0}$, then

$$p_Q(\mathbf{x}) \leq 2r\|\mathbf{x}\|, \quad \text{since} \quad \frac{\mathbf{x}}{2r\|\mathbf{x}\|} \in B_r(\mathbf{0}).$$

Therefore, since the last inequality is clearly also valid if $\mathbf{x} = \mathbf{0}$, and, as follows from (1),

$$|p_Q(\mathbf{x}) - p_Q(\mathbf{y})| \leq p_Q(\mathbf{x} - \mathbf{y}),$$

it is easily seen that $p_Q(\mathbf{x})$ is a continuous function of $\mathbf{x}$ on $\mathcal{X}$.

Items (4) and (5) are left to the reader.

Finally, to verify (6), suppose that $p_Q(\mathbf{x}) = 0$. Then there exists a sequence of points $\alpha_1 \geq \alpha_2 \geq \cdots$ decreasing to 0 such that $\alpha_j^{-1}\mathbf{x} \in Q$. Therefore, since Q is bounded, say $Q \subseteq \{\mathbf{x} : \|\mathbf{x}\| \leq C\}$, the inequality $\|\alpha_j^{-1}\mathbf{x}\| \leq C$ implies that $\|\mathbf{x}\| \leq \alpha_j C$ for $j = 1, 2 \ldots$ and hence that $\mathbf{x} = \mathbf{0}$. □

Exercise 22.14. Complete the proof of Lemma 22.23 by verifying items (4) and (5).

Exercise 22.15. Show that in the setting of Lemma 22.23, $p_Q(\mathbf{x}) < 1 \Longrightarrow \mathbf{x} \in Q$ and $\mathbf{x} \in Q \Longrightarrow p_Q(\mathbf{x}) \leq 1$.

The proof of the next theorem that is presented below serves to illustrate the use of the Minkowski functional. The existence of a support hyperplane is already covered by Theorem 22.14.

Theorem 22.24. *Let Q be a convex subset of $\mathbb{R}^n$ such that $B_r(\mathbf{q}) \subset Q$ for some $\mathbf{q} \in \mathbb{R}^n$ and some $r > 0$. Let $\mathbf{v} \in \mathbb{R}^n$ and $\mathcal{U}$ be a k-dimensional subspace of $\mathbb{R}^n$ such that $0 \leq k < n$ and the set $V = \mathbf{v} + \mathcal{U}$ has no points in common with $\operatorname{int} Q$. Then there exist a vector $\mathbf{y} \in \mathbb{R}^n$ and a constant $c \in \mathbb{R}$ such that $\langle \mathbf{x}, \mathbf{y} \rangle = c$ if $\mathbf{x} \in V$ and $\langle \mathbf{x}, \mathbf{y} \rangle < c$ if $\mathbf{x} \in \operatorname{int} Q$.*

Proof. By an appropriate translation of the problem we may assume that $\mathbf{q} = \mathbf{0}$, and hence that $\mathbf{0} \notin V$. Thus, there exists a linear functional f on the vector space $\mathcal{W} = \{\alpha\mathbf{v} + \mathcal{U} : \alpha \in \mathbb{R}\}$ such that

$$V = \{\mathbf{w} \in \mathcal{W} : f(\mathbf{w}) = 1\}.$$

The next step is to check that

$$f(\mathbf{x}) \leq p_Q(\mathbf{x}) \quad \text{for every vector} \quad \mathbf{x} \in \mathcal{W}. \tag{22.17}$$

Since $p_Q(\mathbf{x}) \geq 0$, the inequality (22.17) is clearly valid if $f(\mathbf{x}) \leq 0$. On the other hand, if $\mathbf{x} \in \mathcal{W}$ and $f(\mathbf{x}) = \alpha > 0$, then $\alpha^{-1}\mathbf{x} \in V$ and thus, as $V \cap \operatorname{int} Q = \emptyset$,

$$\alpha > 0 \Longrightarrow \frac{1}{\alpha} p_Q(\mathbf{x}) = p_Q\left(\frac{\mathbf{x}}{\alpha}\right) \geq 1 = f\left(\frac{\mathbf{x}}{\alpha}\right) = \frac{1}{\alpha} f(\mathbf{x});$$

i.e.,

$$f(\mathbf{x}) > 0 \Longrightarrow f(\mathbf{x}) \le p_Q(\mathbf{x}) .$$

Thus, (22.17) is verified. Therefore, by the variant of the Hahn-Banach theorem discussed in Exercise 7.29, there exists a linear functional F on $\mathbb{R}^n$ such that $F(\mathbf{x}) = f(\mathbf{x})$ for $\mathbf{x} \in \mathcal{W}$ and $F(\mathbf{x}) \le p_Q(\mathbf{x})$ for every $\mathbf{x} \in \mathbb{R}^n$. Let $H = \{\mathbf{x} \in \mathbb{R}^n : F(\mathbf{x}) = 1\}$. Then $F(\mathbf{x}) \le p_Q(\mathbf{x}) < 1$ when $\mathbf{x} \in \operatorname{int} Q$, whereas $F(\mathbf{x}) = 1$ when $\mathbf{x} \in V$, since $V \subseteq H$. Thus, as $F(\mathbf{x}) = \langle \mathbf{x}, \mathbf{y} \rangle$ for some vector $\mathbf{y} \in \mathbb{R}^n$, it follows that $\langle \mathbf{x}, \mathbf{y} \rangle < 1$ when $\mathbf{x} \in \operatorname{int} Q$ and $\langle \mathbf{x}, \mathbf{y} \rangle = 1$ when $\mathbf{x} \in H$. □

22.11. The Gauss-Lucas theorem

Theorem 22.25. *Let $f(\lambda) = a_0 + a_1\lambda + \cdots + a_n\lambda^n$ be a polynomial of degree $n \ge 1$ with coefficients $a_i \in \mathbb{C}$ for $i = 1, \cdots, n$ and $a_n \ne 0$. Then the roots of the derivative $f'(\lambda)$ lie in the convex hull of the roots of $f(\lambda)$.*

Proof. Let $\lambda_1, \cdots, \lambda_n$ denote the roots of $f(\lambda)$, allowing repetitions as needed. Then

$$f(\lambda) = a_n(\lambda - \lambda_1)\cdots(\lambda - \lambda_n)$$

and

$$\begin{aligned} \frac{f'(\lambda)}{f(\lambda)} &= \frac{1}{\lambda - \lambda_1} + \cdots + \frac{1}{\lambda - \lambda_n} \\ &= \frac{\overline{\lambda} - \overline{\lambda_1}}{|\lambda - \lambda_1|^2} + \cdots + \frac{\overline{\lambda} - \overline{\lambda_n}}{|\lambda - \lambda_n|^2} . \end{aligned}$$

Thus, if $f'(\mu) = 0$ and $f(\mu) \ne 0$, then

$$\begin{aligned} &\mu\left\{\frac{1}{|\mu - \lambda_1|^2} + \cdots + \frac{1}{|\mu - \lambda_n|^2}\right\} \\ &\quad = \frac{\lambda_1}{|\mu - \lambda_1|^2} + \cdots + \frac{\lambda_n}{|\mu - \lambda_n|^2}, \end{aligned}$$

which, upon setting

$$t_j = \frac{1}{|\mu - \lambda_j|^2}\left\{\frac{1}{|\mu - \lambda_1|^2} + \cdots + \frac{1}{|\mu - \lambda_n|^2}\right\}^{-1}$$

for $j = 1, \cdots, n$ exhibits μ as the convex sum

$$\mu = t_1\lambda_1 + \cdots + t_n\lambda_n .$$

This completes the proof since the conclusion for the case $f'(\mu) = f(\mu) = 0$ is self-evident. □

22.12. The numerical range

Let $A \in \mathbb{C}^{n\times n}$. The set

$$W(A) = \{\langle A\mathbf{x}, \mathbf{x}\rangle : \mathbf{x} \in \mathbb{C}^n \text{ and } \|\mathbf{x}\| = 1\}$$

is called the numerical range of A. The objective of this section is to show that $W(A)$ is a convex subset of $\mathbb{C}$. We begin with a special case.

Lemma 22.26. *Let $B \in \mathbb{C}^{n\times n}$ and let $\mathbf{x}, \mathbf{y} \in \mathbb{C}^n$ be nonzero vectors such that $\langle B\mathbf{x}, \mathbf{x}\rangle = 1$ and $\langle B\mathbf{y}, \mathbf{y}\rangle = 0$. Then for each number λ in the interval $0 < \lambda < 1$, there exists a vector $\mathbf{v}_\lambda \in \mathbb{C}^n$ with $\|\mathbf{v}_\lambda\| = 1$ such that $\langle B\mathbf{v}_\lambda, \mathbf{v}_\lambda\rangle = \lambda$.*

Proof. Let

$$\mathbf{u}_t = t\gamma\mathbf{x} + (1-t)\mathbf{y}$$

where $|\gamma| = 1$ and $0 \le t \le 1$. Then

$$\begin{aligned}
\langle B\mathbf{u}_t, \mathbf{u}_t\rangle &= t^2\langle B\mathbf{x}, \mathbf{x}\rangle + t(1-t)\{\gamma\langle B\mathbf{x}, \mathbf{y}\rangle + \overline{\gamma}\langle B\mathbf{y}, \mathbf{x}\rangle\} + (1-t)^2\langle B\mathbf{y}, \mathbf{y}\rangle \\
&= t^2 + t(1-t)\{\gamma\langle B\mathbf{x}, \mathbf{y}\rangle + \overline{\gamma}\langle B\mathbf{y}, \mathbf{x}\rangle\}\,.
\end{aligned}$$

The next step is to show that there exists a choice of γ such that

$$\gamma\langle B\mathbf{x}, \mathbf{y}\rangle + \overline{\gamma}\langle B\mathbf{y}, \mathbf{x}\rangle$$

is a real number. To this end it is convenient to write

$$B = C + iD$$

in terms of its **real** and **imaginary** parts

$$C = \frac{B + B^H}{2} \quad \text{and} \quad D = \frac{B - B^H}{2i}\,.$$

Then, since C and D are both Hermitian matrices,

$$\begin{aligned}
\gamma\langle B\mathbf{x}, \mathbf{y}\rangle + \overline{\gamma}\langle B\mathbf{y}, \mathbf{x}\rangle &= \gamma\langle C\mathbf{x}, \mathbf{y}\rangle + i\gamma\langle D\mathbf{x}, \mathbf{y}\rangle + \overline{\gamma}\langle C\mathbf{y}, \mathbf{x}\rangle + i\overline{\gamma}\langle D\mathbf{y}, \mathbf{x}\rangle \\
&= \gamma c + \overline{\gamma c} + i\{\gamma d + \overline{\gamma d}\}\,,
\end{aligned}$$

where

$$c = \langle C\mathbf{x}, \mathbf{y}\rangle \text{ and } d = \langle D\mathbf{x}, \mathbf{y}\rangle$$

are both independent of t. Now, in order to eliminate the imaginary component, set

$$\gamma = \begin{cases} 1 & \text{if } d = 0 \\ i|d|^{-1}\overline{d} & \text{if } d \ne 0\,. \end{cases}$$

Then, for this choice of γ,

$$\langle B\mathbf{u}_t, \mathbf{u}_t\rangle = t^2 + t(1-t)(\gamma c + \overline{\gamma c})\,.$$

Moreover, since $\langle B\mathbf{x}, \mathbf{x}\rangle = 1$ and $\langle B\mathbf{y}, \mathbf{y}\rangle = 0$, the vectors $\mathbf{x}$ and $\mathbf{y}$ are linearly independent. Thus,

$$\|\mathbf{u}_t\|_2^2 = t^2 + (1-t)t\{\langle \gamma\mathbf{x}, \mathbf{y}\rangle + \langle \mathbf{y}, \gamma\mathbf{x}\rangle\} + (1-t)^2 \neq 0$$

for every choice of t in the interval $0 \le t \le 1$. Therefore, the vector

$$\mathbf{v}_t = \frac{\mathbf{u}_t}{\|\mathbf{u}_t\|}$$

is a well-defined unit vector and

$$\langle B\mathbf{v}_t, \mathbf{v}_t\rangle = \frac{t^2 + t(1-t)\{\gamma c + \overline{\gamma} c\}}{t^2 + t(1-t)\{\langle \gamma\mathbf{x}, \mathbf{y}\rangle + \langle \mathbf{y}, \gamma\mathbf{x}\rangle\} + (1-t)^2}$$

is a continuous-real valued function of t on the interval $0 \le t \le 1$ such that

$$\langle B\mathbf{v}_0, \mathbf{v}_0\rangle = 0 \text{ and } \langle B\mathbf{v}_1, \mathbf{v}_1\rangle = 1\ .$$

Therefore, the equation

$$\langle B\mathbf{v}_t, \mathbf{v}_t\rangle = \lambda$$

has at least one solution t in the interval $0 \le t \le 1$ for every choice of λ in the interval $0 \le \lambda \le 1$. □

Theorem 22.27. **(Toeplitz-Hausdorff)** *The numerical range $W(A)$ of a matrix $A \in \mathbb{C}^{n\times n}$ is a convex subset of $\mathbb{C}$.*

Proof. The objective is to show that if $\|\mathbf{x}\| = \|\mathbf{y}\| = 1$ and if

$$\langle A\mathbf{x}, \mathbf{x}\rangle = a \quad \text{and} \quad \langle A\mathbf{y}, \mathbf{y}\rangle = b\,,$$

then for each choice of the number t in the interval $0 \le t \le 1$, there exists a vector $\mathbf{u}_t$ such that

$$\|\mathbf{u}_t\| = 1 \text{ and } \langle B\mathbf{u}_t, \mathbf{u}_t\rangle = t\mathbf{a} + (1-t)b\,.$$

If $a = b$, then $ta + (1-t)b = a = b$, and hence we can choose $\mathbf{u}_t = \mathbf{x}$ or $\mathbf{u}_t = \mathbf{y}$. Suppose therefore that $a \neq b$ and let

$$B = \alpha A + \beta I_n$$

where α, β are solutions of the system of equations

$$\begin{aligned} a\alpha + \beta &= 1 \\ b\alpha + \beta &= 0\,. \end{aligned}$$

Then

$$\begin{aligned} \langle B\mathbf{x}, \mathbf{x}\rangle &= \langle \alpha A\mathbf{x}, \mathbf{x}\rangle + \beta\langle \mathbf{x}, \mathbf{x}\rangle \\ &= \alpha a + \beta = 1 \end{aligned}$$

and

$$\begin{aligned} \langle B\mathbf{y}, \mathbf{y}\rangle &= \langle \alpha A\mathbf{y}, \mathbf{y}\rangle + \beta\langle \mathbf{y}, \mathbf{y}\rangle \\ &= \alpha b + \beta = 0\ . \end{aligned}$$

Therefore, by Lemma 22.26, for each choice of t in the interval $0 \le t \le 1$, there exists a vector $\mathbf{w}_t$ such that

$$\|\mathbf{w}_t\| = 1 \text{ and } \langle B\mathbf{w}_t, \mathbf{w}_t\rangle = t\,.$$

But this in turn is the same as to say

$$\begin{aligned}\alpha\langle A\mathbf{w}_t, \mathbf{w}_t\rangle + \beta\langle \mathbf{w}_t, \mathbf{w}_t\rangle &= t + (1-t)0 \\ &= t(\alpha a + \beta) + (1-t)(b\alpha + \beta) \\ &= \alpha\{ta + (1-t)b\} + \beta\,.\end{aligned}$$

Thus, as $\langle \mathbf{w}_t, \mathbf{w}_t\rangle = 1$ and $\alpha \neq 0$,

$$\langle A\mathbf{w}_t, \mathbf{w}_t\rangle = ta + (1-t)b\,,$$

as claimed. □

22.13. Eigenvalues versus numerical range

The eigenvalues of a matrix $A \in \mathbb{C}^{n\times n}$ clearly belong to the numerical range $W(A)$ of A, i.e.,

$$\sigma(A) \subseteq W(A)\,.$$

Therefore,

$$\operatorname{conv}\sigma(A) \subseteq W(A) \quad \text{for every} \quad A \in \mathbb{C}^{n\times n}\,, \tag{22.18}$$

since $W(A)$ is convex. In general, however, these two sets can be quite different. If

$$A = \begin{bmatrix} 0 & 0 \\ 1 & 0 \end{bmatrix},$$

for example, then

$$\sigma(A) = 0 \text{ and } W(A) = \{a\bar{b} : a, b \in \mathbb{C} \text{ and } |a|^2 + |b|^2 = 1\}\,.$$

The situation for normal matrices is markedly different:

Theorem 22.28. *Let $A \in \mathbb{C}^{n\times n}$ be a normal matrix, i.e., $AA^H = A^HA$. Then the convex hull of $\sigma(A)$ is equal to the numerical range of A, i.e.,*

$$\operatorname{conv}\sigma(A) = W(A)\,.$$

Proof. Since A is normal, it is unitarily equivalent to a diagonal matrix; i.e., there exists a unitary matrix $U \in \mathbb{C}^{n\times n}$ such that

$$U^HAU = \operatorname{diag}\{\lambda_1, \dots, \lambda_n\}\,.$$

The columns $\mathbf{u}_1, \dots, \mathbf{u}_n$ of U form an orthonormal basis for $\mathbb{C}^n$. Thus, if $\mathbf{x} \in \mathbb{C}^n$ and $\|\mathbf{x}\| = 1$, then

$$\mathbf{x} = \sum_{i=1}^{n} c_i\mathbf{u}_i\,,$$

is a linear combination of $\mathbf{u}_1, \ldots, \mathbf{u}_n$,

$$\begin{aligned} \langle A\mathbf{x}, \mathbf{x}\rangle &= \langle A\sum_{i=1}^{n} c_i\mathbf{u}_i, \sum_{j=1}^{n} c_j\mathbf{u}_j\rangle = \sum_{i=1}^{n} \lambda_i c_i \overline{c}_j \langle \mathbf{u}_i, \mathbf{u}_j\rangle \\ &= \sum_{i=1}^{n} \lambda_i |c_i|^2 \end{aligned}$$

and

$$\sum_{i=1}^{n} |c_i|^2 = \|\mathbf{x}\|^2 = 1 \,.$$

Therefore, $W(A) \subseteq \operatorname{conv}(\sigma(A))$ and hence, as the opposite inclusion (22.18) is already known to be in force, the proof is complete. □

Exercise 22.16. Verify the inclusion $W(A) \subseteq \operatorname{conv}(\sigma(A))$ for normal matrices $A \in \mathbb{C}^{n\times n}$ by checking directly that every convex combination $\sum_{i=1}^{n} t_i\lambda_i$ of the eigenvalues $\lambda_1, \ldots, \lambda_n$ of A belongs to $W(A)$. [HINT:

$$\sum_{i=1}^{n} t_i\lambda_i = \sum_{i=1}^{n} t_i\langle A\mathbf{u}_i, \mathbf{u}_i\rangle = \langle A\sum_{i=1}^{n} \sqrt{t_i}\mathbf{u}_i, \sum_{j=1}^{n} \sqrt{t_j}\mathbf{u}_j\rangle \,.]$$

Exercise 22.17. Find the numerical range of the matrix $\begin{bmatrix} 0 & 0 & i \\ 1 & 0 & 0 \\ 0 & 1 & 0 \end{bmatrix}$.

22.14. The Heinz inequality

Lemma 22.29. *Let $A = A^H \in \mathbb{C}^{p\times p}$, $B = B^H \in \mathbb{C}^{q\times q}$ and $Q \in \mathbb{C}^{p\times q}$. Then the following inequalities hold under the extra conditions indicated in each item.*

(1) *If also $p = q$, $Q = Q^H$ and A is invertible, then*

$$2\|Q\| \le \|AQA^{-1} + A^{-1}QA\| \,. \tag{22.19}$$

(2) *If also $p = q$ and A is invertible, then* (22.19) *is still in force.*

(3) *If A and B are invertible, then*

$$2\|Q\| \le \|AQB^{-1} + A^{-1}QB\| \,. \tag{22.20}$$

(4) *If $A \succeq O$ and $B \succeq O$, then*

$$2\|AQB\| \le \|A^2Q + QB^2\| \,. \tag{22.21}$$

Proof. Let $\lambda \in \sigma(Q)$. Then $\lambda \in \sigma(A^{-1}QA)$, and hence there exists a unit vector $\mathbf{x} \in \mathbb{C}^p$ such that

$$\lambda = \langle AQA^{-1}\mathbf{x}, \mathbf{x}\rangle$$

and

$$\overline{\lambda} = \langle \mathbf{x}, AQA^{-1}\mathbf{x}\rangle = \langle A^{-1}QA\mathbf{x}, \mathbf{x}\rangle\,.$$

Therefore, since $\lambda = \overline{\lambda}$,

$$|2\lambda| = |\langle (AQA^{-1} + A^{-1}QA)\mathbf{x}, \mathbf{x}\rangle| \le \|AQA^{-1} + A^{-1}QA\|\,,$$

which leads easily to the inequality (22.19).

To extend (1) to matrices $Q \in \mathbb{C}^{p\times p}$ that are not necessarily Hermitian, apply (1) to the matrices

$$\mathcal{Q} = \begin{bmatrix} O & Q \\ Q^H & O \end{bmatrix} \quad \text{and} \quad \mathcal{A} = \begin{bmatrix} A & O \\ O & A \end{bmatrix}.$$

This leads easily to (2), since

$$\|\mathcal{Q}\| = \|Q\| \quad \text{and} \quad \|\mathcal{A}\mathcal{Q}\mathcal{A}^{-1} + \mathcal{A}^{-1}\mathcal{Q}\mathcal{A}\| = \|AQA^{-1} + A^{-1}QA\|\,.$$

Next, (3) follows from (2) by setting

$$\mathcal{Q} = \begin{bmatrix} O & Q \\ O & O \end{bmatrix} \quad \text{and} \quad \mathcal{A} = \begin{bmatrix} A & O \\ O & B \end{bmatrix}.$$

Finally to obtain (4), let $A_\varepsilon = A + \varepsilon I_p$ and $B_\varepsilon = B + \varepsilon I_q$ with $\varepsilon > 0$. Then, since A_ε and B_ε are invertible Hermitian matrices, we can invoke (22.20) to obtain the inequality

$$2\|A_\varepsilon Q B_\varepsilon\| \le \|A_\varepsilon^2 Q B_\varepsilon B_\varepsilon^{-1} + A_\varepsilon^{-1} A_\varepsilon Q B_\varepsilon^2\| = \|A_\varepsilon^2 Q + Q B_\varepsilon^2\|\,,$$

which tends to the asserted inequality as $\varepsilon \downarrow 0$. □

Theorem 22.30. **(Heinz)** *Let $A \in \mathbb{C}^{p\times p}$, $Q \in \mathbb{C}^{p\times q}$, $B \in \mathbb{C}^{q\times q}$ and suppose that $A \succeq O$ and $B \succeq O$. Then*

$$\|A^tQB^{1-t} + A^{1-t}QB^t\| \le \|AQ + QB\| \quad \textit{for} \quad 0 \le t \le 1\,. \tag{22.22}$$

Proof. Let $f(t) = \|A^tQB^{1-t} + A^{1-t}QB^t\|$, let $0 \le a < b \le 1$ and set $c = (a+b)/2$ and $d = b - c$. Then, as $c = a + d$ and $1 - c = 1 - b + d$,

$$\begin{aligned} f(c) &= \|A^cQB^{1-c} + A^{1-c}QB^c\| = \|A^d\left(A^aQB^{1-b} + A^{1-b}QB^a\right)B^d\| \\ &\le \frac{1}{2}\|A^{2d}\left(A^aQB^{1-b} + A^{1-b}QB^a\right) + \left(A^aQB^{1-b} + A^{1-b}QB^a\right)B^{2d}\| \\ &= \frac{1}{2}\|A^bQB^{1-b} + A^{1-a}QB^a + A^aQB^{1-a} + A^{1-b}QA^b\| \\ &\le \frac{f(a) + f(b)}{2}\,; \end{aligned}$$

i.e., $f(t)$ is a convex function on the interval $0 \le t \le 1$. Thus, as the upper bound in formula (22.22) is equal to $f(0) = f(1)$ and $f(t)$ is continuous on the interval $0 \le t \le 1$, it is readily seen that $f(t) \le f(0)$ for every point t in the interval $0 \le t \le 1$. □

Theorem 22.31. *Let $A \in \mathbb{C}^{p\times p}$, $B \in \mathbb{C}^{p\times p}$ and suppose that $A \succ O$ and $B \succ O$. Then*

$$\|A^sB^s\| \le \|AB\|^s \quad \textit{for} \quad 0 \le s \le 1\,. \tag{22.23}$$

Proof. Let $Q = \{u : 0 \le u \le 1 \text{ for which (22.23) is in force}\}$ and let s and t be a pair of points in Q. Then, with the help of the auxiliary inequality

$$\begin{aligned}\|A^{(s+t)/2}B^{(s+t)/2}\|^2 &= \|B^{(s+t)/2}A^{s+t}B^{(s+t)/2}\| = r_\sigma(B^{(s+t)/2}A^{s+t}B^{(s+t)/2}) \\ &= r_\sigma(B^sA^{s+t}B^t) \le \|B^sA^{s+t}B^t\|\,,\end{aligned}$$

it is readily checked that Q is convex. The proof is easily completed, since $0 \in Q$ and $1 \in Q$. □

Exercise 22.18. Show that if A and B are as in Theorem 22.31, then $\varphi(s) = \|A^sB^s\|^{1/s}$ is an increasing function of s for $s > 0$.

22.15. Bibliographical notes

A number of the results stated in this chapter can be strengthened. The monograph by Webster [**71**] is an eminently readable source of supplementary information on convexity in $\mathbb{R}^n$. Exercise 22.11 was taken from [**71**]. Applications of convexity to optimization may be found in [**10**] and the references cited therein. The proof of Theorem 22.14 is adapted from the expository paper [**11**]; the proof of Theorem 22.24 is adapted from [**48**].

The presented proof of the Krein-Milman theorem, which works in more general settings (with convex hull replaced by the closure of the convex hull), is adapted from [**73**]. The presented proof of the convexity of numerical range is based on an argument that is sketched briefly in [**36**]. Halmos credits it to C. W. R. de Boor. The presented proof works also for bounded operators in Hilbert space; see also McIntosh [**50**] for another very attractive approach. The proof of the Heinz inequality is taken from a beautiful short paper [**32**] that establishes the equivalence of the inequalities in (1)–(4) of Lemma 22.29 with the Heinz inequality (22.22) for bounded operators in Hilbert space and sketches the history. The elegant passage from (22.21) to (22.22) is credited to an unpublished paper of A. McIntosh. The proof of Theorem 22.31 is adapted from [**33**].

The notion of convexity can be extended to matrix valued functions: a function f that maps a convex set Q of symmetric $p \times p$ matrices into a set of symmetric $q \times q$ matrices is said to be convex if

$$f(tX + (1-t)Y) \le tf(X) + (1-t)f(Y)$$

for every $t \in (0,1)$ when X and Y belong to Q. Thus, for example, the function $f(X) = X^r$ is convex on the set $Q = \{X \in \mathbb{R}^{n\times n} : X \succeq O\}$ is convex if $1 \le r \le 2$ or $-1 \le r \le 0$ and $-f$ is convex if $0 \le r \le 1$, see [**3**].

Chapter 23

Matrices with nonnegative entries

Be wary of writing many books, there is no end, and much study is wearisome to the flesh. Ecclesiastes 12:12

Matrices with nonnegative entries play an important role in numerous applications. This chapter is devoted to the study of some of their special properties.

A rectangular matrix $A \in \mathbb{R}^{n\times m}$ with entries a_{ij}, $i = 1,\dots,n$, $j = 1,\dots,m$, is said to be **nonnegative** if $a_{ij} \geq 0$ for $i = 1,\dots,n$ and $j = 1,\dots,m$; A is said to be **positive** if $a_{ij} > 0$ for $i = 1,\dots,n$ and $j = 1,\dots,m$. The notation $A \geq O$ and $A > O$ are used to designate nonnegative matrices A and positive matrices A, respectively. Note the distinction with the notation $A \succ O$ for positive definite and $A \succeq O$ for positive semidefinite matrices that was introduced earlier. The symbols $A \geq B$ and $A > B$ will be used to indicate that $A - B \geq O$ and $A - B > O$, respectively.

A nonnegative square matrix $A \in \mathbb{R}^{n\times n}$ is said to be **irreducible** if for every pair of indices $i, j \in \{1,\dots,n\}$ there exists an integer $k \geq 1$ such that the ij entry of A^k is positive; i.e., in terms of the standard basis $\mathbf{e}_i$, $i = 1,\dots,n$, of $\mathbb{R}^n$, if

$$\langle A^k \mathbf{e}_j, \mathbf{e}_i\rangle > 0 \quad \text{for some positive integer } k \text{ that may depend upon } ij\,.$$

This is less restrictive than assuming that there exists an integer $k \geq 1$ such that $A^k > O$.

Exercise 23.1. Show that the matrix

$$A = \begin{bmatrix} 0 & 1 \\ 1 & 0 \end{bmatrix}$$

is a nonnegative irreducible matrix, but that A^k is never a positive matrix.

Exercise 23.2. Show that the matrix

$A = \begin{bmatrix} 0 & 1 \\ 1 & 1 \end{bmatrix}$ is irreducible, but the matrix $B = \begin{bmatrix} 1 & 1 \\ 0 & 1 \end{bmatrix}$ is not irreducible

and, more generally, that every triangular nonnegative matrix is not irreducible.

Lemma 23.1. *Let $A \in \mathbb{R}^{n\times n}$ be a nonnegative irreducible matrix. Then $(I_n + A)^{n-1}$ is positive.*

Proof. Suppose to the contrary that the ij entry of $(I_n + A)^{n-1}$ is equal to zero for some choice of i and j. Then, in view of the formula

$$(I_n + A)^{n-1} = \sum_{k=0}^{n-1} \binom{n-1}{k} A^{n-1-k} = \sum_{k=0}^{n-1} \binom{n-1}{k} A^k ,$$

it is readily seen that

$$\langle (I_n + A)^{n-1}\mathbf{e}_j, \mathbf{e}_i\rangle = \sum_{k=0}^{n-1} \binom{n-1}{k} \langle A^k \mathbf{e}_j, \mathbf{e}_i\rangle = 0$$

and hence that

$$\langle A^k \mathbf{e}_i, \mathbf{e}_j\rangle = 0 \quad \text{for} \quad k = 0, \ldots, n-1 .$$

Therefore, by the Cayley-Hamilton theorem,

$$\langle A^k \mathbf{e}_i, \mathbf{e}_j\rangle = 0 \quad \text{for} \quad k = 0, 1, \ldots ,$$

which contradicts the assumed irreducibility. □

Exercise 23.3. Let $A \in \mathbb{R}^{n\times n}$ have nonnegative entries and let $D \in \mathbb{R}^{n\times n}$ be a diagonal matrix with strictly positive diagonal entries. Show that

$$A \text{ is irreducible} \iff AD \text{ is irreducible} \iff DA \text{ is irreducible}.$$

23.1. Perron-Frobenius theory

The main objective of this section is to show that if A is a square nonnegative irreducible matrix, then $r_\sigma(A)$, the spectral radius of A, is an eigenvalue of A with algebraic multiplicity equal to one and that it is possible to choose a corresponding eigenvector $\mathbf{u}$ to have strictly positive entries.

Theorem 23.2. (**Perron-Frobenius**) *Let $A \in \mathbb{R}^{n\times n}$ be a nonnegative irreducible matrix. Then $A \neq O_{n\times n}$ and:*

(1) $r_\sigma(A) \in \sigma(A)$.

(2) *There exists a positive vector* $\mathbf{u} \in \mathbb{R}^n$ *such that* $A\mathbf{u} = r_\sigma(A)\mathbf{u}$.

(3) *There exists a positive vector* $\mathbf{v} \in \mathbb{R}^n$ *such that* $A^T\mathbf{v} = r_\sigma(A)\mathbf{v}$.

(4) *The algebraic multiplicity of* $r_\sigma(A)$ *as an eigenvalue of* A *is equal to one.*

The proof of this theorem is divided into a number of lemmas. Let

$$\mathcal{B} = \{\mathbf{x} \in \mathbb{R}^n : \|\mathbf{x}\|_2 = 1\}, \quad \mathcal{C} = \{\mathbf{x} \in \mathbb{R}^n : \mathbf{x} \geq 0\}$$

and, for $\mathbf{x} \in \mathcal{C}$, let

$$\delta_A(\mathbf{x}) = \min_i \left\{ \frac{\langle A\mathbf{x}, \mathbf{e}_i \rangle}{\langle \mathbf{x}, \mathbf{e}_i \rangle} : \langle \mathbf{x}, \mathbf{e}_i \rangle > 0 \right\} .$$

Lemma 23.3. *Let* $A \in \mathbb{R}^{n\times n}$ *be a nonnegative irreducible matrix and let* $\mathbf{x} \in \mathcal{C} \cap \mathcal{B}$ *be such that* $A\mathbf{x} - \delta_A(\mathbf{x})\mathbf{x} \neq \mathbf{0}$. *Then*

$$\mathbf{y} = \frac{(I_n + A)^{n-1}\mathbf{x}}{\|(I_n + A)^{n-1}\mathbf{x}\|_2}$$

is a positive vector that belongs to $\mathcal{C} \cap \mathcal{B}$ *and* $\delta_A(\mathbf{y}) > \delta_A(\mathbf{x})$.

Proof. Clearly,

$$A\mathbf{x} - \delta_A(\mathbf{x})\mathbf{x} \geq \mathbf{0} \text{ for every } \mathbf{x} \in \mathcal{C},$$

and so too for every $\mathbf{x} \in \mathcal{C} \cap \mathcal{B}$. Moreover, if $A\mathbf{x} - \delta_A(\mathbf{x})\mathbf{x} \neq \mathbf{0}$, then $(I_n + A)^{n-1}(A\mathbf{x} - \delta_A(\mathbf{x})\mathbf{x}) > \mathbf{0}$. Consequently, the vector $\mathbf{y}$ defined above belongs to $\mathcal{C} \cap \mathcal{B}$ and, since $\mathbf{y} > 0$ and $A\mathbf{y} - \delta_A(\mathbf{x})\mathbf{y} > \mathbf{0}$, the inequality

$$\frac{\sum_{j=1}^n a_{ij} y_j}{y_i} > \delta_A(\mathbf{x})$$

is in force for $i = 1, \ldots, n$. Therefore $\delta_A(\mathbf{y}) > \delta_A(\mathbf{x})$, as asserted. □

Exercise 23.4. Show that if $A \in \mathbb{R}^{n\times n}$ is a nonnegative irreducible matrix, then $\delta_A(\mathbf{x}) \leq \|A\|$ for every $\mathbf{x} \in \mathcal{C} \cap \mathcal{B}$.

Exercise 23.5. Show that if $A \in \mathbb{R}^{n\times n}$ is a nonnegative irreducible matrix, then $\delta_A(\mathbf{x}) \leq \max_i\{\sum_{j=1}^n a_{ij}\}$.

Exercise 23.6. Evaluate $\delta_A(\mathbf{x})$ for the matrix $A = \begin{bmatrix} 2 & 1 \\ 0 & 1 \end{bmatrix}$ and the vectors $\mathbf{x} = \begin{bmatrix} 1 \\ \varepsilon \end{bmatrix}$ for $\varepsilon \geq 0$.

Let

$$\mathcal{C}_A = \{(I_n + A)^{n-1}\mathbf{x} : \mathbf{x} \in \mathcal{C}\} .$$

Exercise 23.7. Let $A \in \mathbb{R}^{n\times n}$ be a nonnegative irreducible matrix. Show that δ_A is continuous on $\mathcal{C}_A \cap \mathcal{B}$ but is not necessarily continuous on the set $\mathcal{C} \cap \mathcal{B}$. [HINT: Exercise 23.6 serves to illustrate the second assertion.]

Lemma 23.4. *Let $A \in \mathbb{R}^{n\times n}$ be a nonnegative irreducible matrix. Then there exists a vector $\mathbf{y} \in \mathcal{C}_A \cap \mathcal{B}$ such that*

$$\delta_A(\mathbf{y}) \geq \delta_A(\mathbf{x}) \quad \textit{for every} \quad \mathbf{x} \in \mathcal{C} \cap \mathcal{B}\,.$$

Proof. Let

$$\rho_A = \sup\{\delta_A(\mathbf{x}) : \mathbf{x} \in \mathcal{C} \cap \mathcal{B}\}\,.$$

Then, by the definition of supremum, there exists a sequence $\mathbf{x}_1, \mathbf{x}_2, \ldots$ of vectors in $\mathcal{C} \cap \mathcal{B}$ such that $\delta_A(\mathbf{x}_j) \to \rho_A$ as $j \uparrow \infty$. Moreover, by passing to a subsequence if need be, there is no loss of generality in assuming that $\delta_A(\mathbf{x}_1) \leq \delta_A(\mathbf{x}_2) \leq \cdots$ and that $\mathbf{x}_j \to \mathbf{x}$, as $j \uparrow \infty$. Let

$$\mathbf{y}_j = \frac{(I_n + A)^{n-1}\mathbf{x}_j}{\|(I_n + A)^{n-1}\mathbf{x}_j\|_2} \quad \text{for} \quad j = 1, 2, \ldots \quad \text{and} \quad \mathbf{y} = \frac{(I_n + A)^{n-1}\mathbf{x}}{\|(I_n + A)^{n-1}\mathbf{x}\|_2}\,.$$

Then it is left to the reader to check that:

(a) The vectors $\mathbf{y}$ and $\mathbf{y}_1, \mathbf{y}_2, \ldots$ all belong to $\mathcal{C}_A \cap \mathcal{B}$.

(b) $\mathbf{y}_j \to \mathbf{y}$, as $j \uparrow \infty$.

(c) $\delta_A(\mathbf{y}_j) \to \delta_A(\mathbf{y})$, as $j \uparrow \infty$.

Therefore, the vector $\mathbf{y}$ exhibits all the advertised properties. □

Exercise 23.8. Complete the proof of Lemma 23.4 by justifying assertions (a), (b) and (c). [HINT: Exploit Exercise 23.7 for part (c).]

Lemma 23.5. *Let $A \in \mathbb{R}^{n\times n}$ be a nonnegative irreducible matrix; then ρ_A is an eigenvalue of A. Moreover, if $\mathbf{u} \in \mathcal{C} \cap \mathcal{B}$ is such that $\delta_A(\mathbf{u}) \geq \delta_A(\mathbf{x})$ for every $\mathbf{x} \in \mathcal{C} \cap \mathcal{B}$, then $A\mathbf{u} = \delta_A(\mathbf{u})\mathbf{u}$ and $\mathbf{u} \in \mathcal{C}_A \cap \mathcal{B}$.*

Proof. By the definition of $\delta_A(\mathbf{u})$,

$$A\mathbf{u} - \delta_A(\mathbf{u})\mathbf{u} \geq \mathbf{0}\,.$$

Moreover, if $A\mathbf{u} - \delta_A(\mathbf{u})\mathbf{u} \neq \mathbf{0}$ and

$$\mathbf{y} = \frac{(I_n + A)^{n-1}\mathbf{u}}{\|(I_n + A)^{n-1}\mathbf{u}\|_2}\,, \quad \text{then} \quad \delta_A(\mathbf{y}) > \delta_A(\mathbf{u})\,,$$

by Lemma 23.3. But this contradicts the presumed maximality of $\delta_A(\mathbf{u})$. Therefore $A\mathbf{u} = \delta_A(\mathbf{u})\mathbf{u}$. The last equality implies further that

$$(I_n + A)^{n-1}\mathbf{u} = (1 + \delta_A(\mathbf{u}))^{n-1}\mathbf{u}$$

and hence, as $1 + \delta_A(\mathbf{u}) > 0$, that $\mathbf{u} \in \mathcal{C}_A \cap \mathcal{B}$. □

Lemma 23.6. *Let $A \in \mathbb{R}^{n\times n}$ be a nonnegative irreducible matrix. Then $\rho_A = r_\sigma(A)$, the spectral radius of A.*

Proof. Let $\lambda \in \sigma(A)$. Then there exists a nonzero vector $\mathbf{x} \in \mathbb{C}^n$ such that $\lambda\mathbf{x} = A\mathbf{x}$; i.e.,

$$\lambda x_i = \sum_{j=1}^{n} a_{ij}x_j \quad \text{for} \quad i = 1, \ldots, n\,.$$

Therefore, since $a_{ij} \geq 0$,

$$|\lambda||x_i| = \left|\sum_{j=1}^{n} a_{ij}x_j\right| \leq \sum_{j=1}^{n} a_{ij}|x_j| \quad \text{for} \quad i = 1, \ldots, n\,,$$

which in turn implies that

$$|\lambda| \leq \frac{1}{|x_i|}\sum_{j=1}^{n} a_{ij}|x_j| \quad \text{if} \quad |x_i| \neq 0\,.$$

Thus, the vector $\mathbf{v}$ with components $v_i = |x_i|$ belongs to $\mathcal{C}$ and meets the inequality

$$|\lambda| \leq \delta_A(\mathbf{v}) \leq \rho_A\,.$$

Since the inequality $|\lambda| \leq \rho_A$ is valid for all eigenvalues of A, it follows that $r_\sigma(A) \leq \rho_A$. However, since ρ_A has already been shown to be an eigenvalue of A, equality must prevail: $r_\sigma(A) = \rho_A$. □

Lemma 23.7. *Let $A \in \mathbb{R}^{n\times n}$ be a nonnegative irreducible matrix and let $\mathbf{x}$ be a nonzero vector in $\mathbb{C}^n$ such that $A\mathbf{x} = r_\sigma(A)\mathbf{x}$. Then there exists a number $c \in \mathbb{C}$ such that $cx_i > 0$ for every entry x_i in the vector $\mathbf{x}$ and $c\mathbf{x} \in \mathcal{C}_A$.*

Proof. Let $\mathbf{x}$ be a nonzero vector in $\mathbb{C}^n$ such that $A\mathbf{x} = r_\sigma(A)\mathbf{x}$ and let $\mathbf{v} \in \mathbb{R}^n$ be the vector with components $v_i = |x_i|$, for $i = 1, \ldots, n$. Then $\mathbf{v} \in \mathcal{C}$ and

$$r_\sigma(A)v_i = r_\sigma(A)|x_i| = \left|\sum_{j=1}^{n} a_{ij}x_j\right| \leq \sum_{j=1}^{n} a_{ij}v_j\;,\; i = 1, \ldots, n\;,$$

i.e., the vector $A\mathbf{v} - r_\sigma(A)\mathbf{v}$ is nonnegative. But this in turn implies that either $A\mathbf{v} - r_\sigma(A)\mathbf{v} = \mathbf{0}$ or $(I_n + A)^{n-1}(A\mathbf{v} - r_\sigma(A)\mathbf{v}) > \mathbf{0}$. The second possibility leads to a contradiction of the fact that $r_\sigma(A) = \rho_A$. Therefore, equality must prevail. Moreover, in this case, the subsequent equality

$$(I_n + A)^{n-1}\mathbf{v} = \gamma\mathbf{v} \quad \text{with} \quad \gamma = (1 + r_\sigma(A))^{n-1}$$

implies that $\mathbf{v} \in \mathcal{C}_A$ and hence that $v_i > 0$ for $i = 1, \ldots, n$. Furthermore, the two formulas $A\mathbf{x} = r_\sigma(A)\mathbf{x}$ and $A\mathbf{v} = r_\sigma(A)\mathbf{v}$ imply that $A^k\mathbf{x} = r_\sigma(A)^k\mathbf{x}$ and $A^k\mathbf{v} = r_\sigma(A)^k\mathbf{v}$ for $k = 0, 1, \ldots$. Therefore, $\mathbf{x}$ and $\mathbf{v}$ are eigenvectors of the matrix $B = (I_n + A)^{n-1}$, corresponding to the eigenvalue γ; i.e.,

$B\mathbf{x} = \gamma\mathbf{x}$ and $B\mathbf{v} = \gamma\mathbf{v}$. Then, since $v_i = |x_i|$, the last two formulas imply that

$$\sum_{j=1}^{n} b_{ij}|x_j| = \gamma|x_i| = \left|\sum_{j=1}^{n} b_{ij}x_j\right| \quad \text{for} \quad i = 1, \dots, n\,.$$

Moreover, since $b_{ij} > 0$ for all entries in the matrix B, and $|x_j| > 0$ for $j = 1, \dots, n$, since $\mathbf{v} \in \mathcal{C}_A$, the numbers $c_j = b_{ij}x_j$, $j = 1, \dots, n$, are all nonzero and satisfy the constraint

$$|c_1 + \cdots + c_n| = |c_1| + |c_2| + \cdots + |c_n|\,.$$

But this is only possible if $c_j = |c_j|e^{i\theta}$ for $j = 1, \dots, n$ for some fixed θ, i.e., if and only if $x_j = |x_j|e^{i\theta}$ for $j = 1, \dots, n$. Therefore, the number $c = e^{-i\theta}$ fulfills the assertion of the lemma: $c\mathbf{x} \in \mathcal{C}_A$. □

Exercise 23.9. Show that if the complex numbers $c_1, \dots, c_n \in \mathbb{C}\backslash\{0\}$, then $|c_1 + \cdots + c_n| = |c_1| + |c_2| + \cdots + |c_n|$ if and only if there exists a number $\theta \in [0, 2\pi)$ such that $c_j = e^{i\theta}|c_j|$ for $j = 1, \dots, n$.

Lemma 23.8. *Let $A \in \mathbb{R}^{n\times n}$ be a nonnegative irreducible matrix. Then the geometric multiplicity of $r_\sigma(A)$ as an eigenvalue of A is equal to one, i.e.,* $\dim \mathcal{N}_{(r_\sigma(A)I_n - A)} = 1$.

Proof. Let $\mathbf{u}$ and $\mathbf{v}$ be any two nonzero vectors in $\mathbb{C}^n$ such that $A\mathbf{u} = r_\sigma(A)\mathbf{u}$ and $A\mathbf{v} = r_\sigma(A)\mathbf{v}$. Then the entries $u_1, \dots, u_n$ of $\mathbf{u}$ and the entries $v_1, \dots, v_n$ of $\mathbf{v}$ are all nonzero and $v_1\mathbf{u} - u_1\mathbf{v}$ is also in the null space of the matrix $A - r_\sigma(A)I_n$. Thus, by the preceding lemma, either $v_1\mathbf{u} - u_1\mathbf{v} = \mathbf{0}$ or there exists a number $c \in \mathbb{C}$ such that $c(v_1u_j - u_1v_j) > 0$ for $j = 1, \dots, n$. However, the second situation is clearly impossible since the first entry in the vector $c(v_1\mathbf{u} - u_1\mathbf{v})$ is equal to zero. Thus, $\mathbf{u}$ and $\mathbf{v}$ are linearly dependent. □

Lemma 23.9. *Let $A \in \mathbb{R}^{n\times n}$ and $B \in \mathbb{R}^{n\times n}$ be nonnegative matrices such that A is irreducible and $A - B$ is nonnegative (i.e., $A \geq B \geq O$). Then:*

(1) $r_\sigma(A) \geq r_\sigma(B)$.

(2) $r_\sigma(A) = r_\sigma(B) \Longleftrightarrow A = B$.

Proof. Let $\beta \in \sigma(B)$ with $|\beta| = r_\sigma(B)$, let $B\mathbf{y} = \beta\mathbf{y}$ for some nonzero vector $\mathbf{y} \in \mathbb{C}^n$ and let $\mathbf{v} \in \mathbb{R}^n$ be the vector with $v_i = |y_i|$ for $i = 1, \dots, n$. Then, in the usual notation,

$$\beta y_i = \sum_{j=1}^{n} b_{ij}y_j \quad \text{for} \quad i = 1, \dots, n$$

and so
(23.1)

$$r_\sigma(B)v_i = |\beta||y_i| = \left|\sum_{j=1}^n b_{ij}y_j\right| \le \sum_{j=1}^n b_{ij}|y_j| \le \sum_{j=1}^n a_{ij}|y_j| = \sum_{j=1}^n a_{ij}v_j\,.$$

Therefore, since $\mathbf{v} \in \mathcal{C}$, this implies that

$$r_\sigma(B) \le \delta_A(\mathbf{v})$$

and hence that $r_\sigma(B) \le r_\sigma(A)$.

Suppose next that $r_\sigma(A) = r_\sigma(B)$. Then the inequality (23.1) implies that

$$A\mathbf{v} - r_\sigma(A)\mathbf{v} \ge \mathbf{0}\,.$$

But this forces $A\mathbf{v} - r_\sigma(A)\mathbf{v} = \mathbf{0}$ because otherwise Lemma 23.6 yields a contradiction to the already established inequality $r_\sigma(A) \ge \delta_A(\mathbf{u})$ for every $\mathbf{u} \in \mathcal{C}$. But this in turn implies that

$$\sum_{i=1}^n (a_{ij} - b_{ij})v_j = 0 \quad \text{for} \quad i = 1, \dots, n$$

and thus, as $a_{ij} - b_{ij} \ge 0$ and $v_j > 0$, we must have $a_{ij} = b_{ij}$ for every choice of $i, j \in \{1, \dots, n\}$, i.e., $r_\sigma(A) = r_\sigma(B) \Longrightarrow A = B$. The other direction is self-evident. □

Lemma 23.10. *Let $A \in \mathbb{R}^{n\times n}$ be a nonnegative irreducible matrix. Then the algebraic multiplicity of $r_\sigma(A)$ as an eigenvalue of A is equal to one.*

Proof. It suffices to show that $r_\sigma(A)$ is a simple root of the characteristic polynomial $\varphi(\lambda) = \det(\lambda I_n - A)$ of the matrix A. Let

$$C(\lambda) = \begin{bmatrix} c_{11}(\lambda) & \cdots & c_{1n}(\lambda) \\ \vdots & & \vdots \\ c_{n1}(\lambda) & \cdots & c_{nn}(\lambda) \end{bmatrix},$$

where

$$c_{ij}(\lambda) = (-1)^{i+j}(\lambda I_n - A)_{ji}$$

and $(\lambda I_n - A)_{ji}$ denotes the determinant of the $(n-1)\times(n-1)$ matrix that is obtained from $\lambda I_n - A$ by deleting the j'th row and the i'th column of $\lambda I_n - A$. Then, as

$$(\lambda I_n - A)C(\lambda) = C(\lambda)(\lambda I_n - A) = \varphi(\lambda)I_n\,,$$

it follows that

$$(r_\sigma(A)I_n - A)C(r_\sigma(A)) = O_{n\times n}$$

and hence that each nonzero column of the matrix $C(r_\sigma(A))$ is an eigenvector of A corresponding to the eigenvalue $r_\sigma(A)$. Therefore, in view of Lemma 23.7, each column of $C(r_\sigma(A))$ is a constant multiple of the unique vector $\mathbf{u} \in \mathcal{C}_A \cap \mathcal{B}$ such that $A\mathbf{u} = r_\sigma(A)\mathbf{u}$. Next, upon differentiating the formula $C(\lambda)(\lambda I_n - A) = \varphi(\lambda)I_n$ with respect to λ, we obtain

$$C'(\lambda)(\lambda I_n - A) + C(\lambda) = \varphi'(\lambda)I_n$$

and, consequently,

$$C(r_\sigma(A))\mathbf{u} = \varphi'(r_\sigma(A))\mathbf{u} \ .$$

Thus, in order to prove that $\varphi'(r_\sigma(A)) \neq 0$, it suffices to show that at least one entry in the vector $C(r_\sigma(A))\mathbf{u}$ is not equal to zero. However, since A^T is also a nonnegative irreducible matrix and $r_\sigma(A) = r_\sigma(A^T)$, much the same sort of analysis leads to the auxiliary conclusion that there exists a unique vector $\mathbf{v} \in \mathcal{C}_{A^T} \cap \mathcal{B}$ such that $A^T\mathbf{v} = r_\sigma(A)\mathbf{v}$ and consequently that each column of $C(r_\sigma(A))^T$ is a constant multiple of $\mathbf{v}$. But this is the same as to say that each row of $C(r_\sigma(A))$ is a constant multiple of $\mathbf{v}^T$. Thus, in order to show that the bottom entry in $C(r_\sigma(A))\mathbf{u}$ is not equal to zero, it suffices to show that $c_{nn}(r_\sigma(A)) \neq 0$.

By definition,

$$c_{nn}(r_\sigma(A)) = \det(r_\sigma(A)I_{n-1} - \widetilde{A}) \ ,$$

where $\widetilde{A}$ is the $(n-1) \times (n-1)$ matrix that is obtained from A by deleting its n'th row and its n'th column. Let

$$B = \begin{bmatrix} \widetilde{A} & O_{(n-1)\times 1} \\ O_{1\times(n-1)} & 0 \end{bmatrix} .$$

Then clearly $B \geq O$ and $A - B \geq O$. Moreover, $A - B \neq O$, since A is irreducible and B is not. Thus, by Lemma 23.9, $r_\sigma(A) > r_\sigma(B)$ and consequently, $r_\sigma(A)I_n - B$ is invertible. But this in turn implies that

$$r_\sigma(A)c_{nn}(r_\sigma(A)) = \det(r_\sigma(A)I_n - B) \neq 0 \ ,$$

and hence that $c_{nn}(r_\sigma(A)) \neq 0$, as needed to complete the proof. □

Exercise 23.10. Show that if $A \in \mathbb{R}^{n\times n}$ is a nonnegative irreducible matrix, then, in terms of the notation used in the proof of Lemma 23.10,

$$C(r_\sigma(A)) = \gamma \mathbf{u}\mathbf{v}^T$$

for some $\gamma > 0$.

The proof of Theorem 23.2 is an easy consequence of the preceding lemmas. Under additional assumptions one can show more:

Theorem 23.11. *Let $A \in \mathbb{R}^{n\times n}$ be a nonnegative irreducible matrix such that A^k is positive for some integer $k \geq 1$, and let $\lambda \in \sigma(A)$. Then*

$$\lambda \neq r_\sigma(A) \Rightarrow |\lambda| < r_\sigma(A) \ .$$

Exercise 23.11. Prove Theorem 23.11.

Exercise 23.12. Let $A \in \mathbb{R}^{n\times n}$ be a nonnegative irreducible matrix with spectral radius $r_\sigma(A) = 1$. Let $B = A - \mathbf{x}\mathbf{y}^T$, where $\mathbf{x} = A\mathbf{x}$ and $\mathbf{y} = A^T\mathbf{y}$ are positive eigenvectors of the matrices A and A^T, respectively, corresponding to the eigenvalue 1 such that $\mathbf{y}^T\mathbf{x} = 1$. Show that:

(a) $\sigma(B) \subset \sigma(A) \cup \{0\}$, but that $1 \notin \sigma(B)$.

(b) $\lim_{N\to\infty} \frac{1}{N}\sum_{k=1}^N B^k = 0$.

(c) $B^k = A^k - \mathbf{x}\mathbf{y}^T$ for $k = 1, 2, \ldots$.

(d) $\lim_{N\to\infty} \frac{1}{N}\sum_{k=1}^N A^k = \mathbf{x}\mathbf{y}^T$.

[HINT: If $r_\sigma(B) < 1$, then it is readily checked that $B^k \to O$ as $k \to \infty$. However, if $r_\sigma(B) = 1$, then B may have complex eigenvalues of the form $e^{i\theta}$ and a more careful analysis is required that exploits the fact that $\lim_{N\to\infty} \frac{1}{N}\sum_{k=1}^N e^{ik\theta} = 0$ if $e^{i\theta} \neq 1$.]

23.2. Stochastic matrices

A nonnegative matrix $P \in \mathbb{R}^{n\times n}$ with entries p_{ij}, $i, j = 1, \ldots, n$, is said to be a **stochastic matrix** if

$$\sum_{j=1}^{n} p_{ij} = 1 \quad \text{for every} \quad i \in \{1, \ldots, n\}\,. \tag{23.2}$$

Stochastic matrices play a prominent role in the theory of Markov chains with a finite number of states.

Exercise 23.13. Let $P \in \mathbb{R}^{n\times n}$ be a stochastic matrix and let $\mathbf{e}_i$ denote the i'th column of I_n for $i = 1, \ldots, n$. Show that $\mathbf{e}_i^T P^k \mathbf{e}_j \le 1$ for $k = 1, 2, \ldots$. [HINT: Justify and exploit the formula $\mathbf{e}_i^T P^{k+1}\mathbf{e}_j = \sum_{s=1}^n \mathbf{e}_i^T P \mathbf{e}_s \mathbf{e}_s^T P^k \mathbf{e}_j$.]

Exercise 23.14. Show that the spectral radius $r_\sigma(P)$ of a stochastic matrix P is equal to one. [HINT: Invoke the bounds established in Exercise 23.13 to justify the inequality $r_\sigma(P) \le 1$.]

Exercise 23.15. Show that if $P \in \mathbb{R}^{n\times n}$ is an irreducible stochastic matrix with entries p_{ij} for $i, j = 1, \ldots, n$, then there exists a positive vector $\mathbf{u} \in \mathbb{R}^n$ with entries u_i for $i, \ldots, n$ such that $u_j = \sum_{i=1}^n u_i p_{ij}$ for $j, \ldots, n$. [HINT: Exploit Theorem 23.2 and Exercises 23.14 and 23.15.]

Exercise 23.16. Show that the matrix $P = \begin{bmatrix} 1/2 & 0 & 1/2 \\ 1/4 & 1/2 & 1/4 \\ 1/8 & 3/8 & 1/2 \end{bmatrix}$ is an irreducible stochastic matrix and find a positive vector $\mathbf{u} \in \mathbb{R}^3$ that meets the conditions discussed in Exercise 23.15.

23.3. Doubly stochastic matrices

A nonnegative matrix $P \in \mathbb{R}^{n\times n}$ is said to be a **doubly stochastic matrix** if both P and P^T are stochastic matrices, i.e., if (23.2) and

$$\sum_{i=1}^{n} p_{ij} = 1 \quad \text{for every} \quad j \in \{1, \dots, n\} \tag{23.3}$$

are both in force. The main objective of this section is to establish a theorem of Birkhoff and von Neumann that states that every doubly stochastic matrix is a convex combination of permutation matrices. It turns out that the notion of permanents is a convenient tool for obtaining this result.

If $A \in \mathbb{C}^{n\times n}$, the **permanent** of A, abbreviated $\text{per}(A)$ or $\text{per}\, A$, is defined by the rule

$$\text{per}\,(A) = \sum_{\sigma\in\Sigma_n} a_{1\sigma(1)} \cdots a_{n\sigma(n)}\,, \tag{23.4}$$

where the summation is taken over the set Σ_n of all $n!$ permutations σ of the integers $\{1, \dots, n\}$. This differs from the formula (5.2) for $\det A$ because the term $d(P_\sigma)$ is replaced by the number one.

There is also a formula for computing $\text{per}\, A$ that is analogous to the formula for computing determinants by expanding by minors:

$$\text{per}\, A = \sum_{j=1}^{n} a_{ij} \text{per}\, A_{\langle ij\rangle} \quad \text{for each choice of } i\,, \tag{23.5}$$

where $A_{\langle ij\rangle}$ denotes the $(n-1)\times(n-1)$ matrix that is obtained from A by deleting the i'th row and the j'th column.

Exercise 23.17. Show that if $A \in \mathbb{F}^{n\times n}$, then (i) $\text{per}\, A$ is a multilinear functional of the rows of A; (ii) $\text{per}\, PA = \text{per}\, A = \text{per}\, AP$ for every $n\times n$ permutation matrix P; and (iii) $\text{per}\, I_n = 1$.

Exercise 23.18. Let $A = \begin{bmatrix} B & 0 \\ C & D \end{bmatrix}$, where B and D are square matrices. Show that

$$\text{per}\, A = \text{per}\, B \cdot \text{per}\, D\,.$$

Exercise 23.19. Let $A \in \mathbb{R}^{2\times 2}$ be a nonnegative matrix. Show that $\text{per}\, A = 0$ if and only if A contains a 1×2 submatrix of zeros or a 2×1 submatrix of zeros.

Exercise 23.20. Let $A \in \mathbb{R}^{3\times 3}$ be a nonnegative matrix. Show that $\text{per}\, A = 0$ if and only if A contains an $r\times s$ submatrix of zeros where $r + s = 3 + 1$.

Lemma 23.12. *Let $A \in \mathbb{R}^{n\times n}$ be a nonnegative matrix. Then $\text{per}\, A = 0$ if and only if there exists an $r\times s$ zero submatrix of A with $r + s = n + 1$.*

Proof. Suppose first that $\operatorname{per} A = 0$. Then, by Exercise 23.19, the claim is true for 2×2 matrices A. Suppose that in fact the assertion is true for $k \times k$ matrices when $k < n$ and let $A \in \mathbb{R}^{n \times n}$. Then formula (23.5) implies that $a_{ij} \operatorname{per} A_{\langle ij \rangle} = 0$ for every choice of $i, j = 1, \ldots, n$. If $a_{ij} = 0$, for $j = 1, \ldots, n$ and some i, then A has an $n \times 1$ submatrix of zeros. If $a_{ij} \neq 0$ for some j, then $\operatorname{per} A_{\langle ij \rangle} = 0$ and so by the induction assumption $A_{\langle ij \rangle}$ has an $r \times s$ submatrix of zeros with $r + s = n$. By permuting rows and columns we can assume $O_{r \times s}$ is the upper right-hand block of A, i.e.,

$$P_1 A P_2 = \begin{bmatrix} B & O \\ C & D \end{bmatrix}$$

for some pair of permutation matrices P_1 and P_2. Thus, as

$$\operatorname{per} B \operatorname{per} D = \operatorname{per}(P_1 A P_2) = \operatorname{per} A = 0\,,$$

it follows that either $\operatorname{per} B = 0$ or $\operatorname{per} D = 0$.

Suppose, for the sake of definiteness, that $\operatorname{per} B = 0$. Then, since $B \in \mathbb{R}^{r \times r}$ and $r < n$, the induction assumption guarantees the existence of an $i \times j$ submatrix of zeros in B with $i + j = r + 1$. Therefore, by permuting columns we obtain a zero submatrix of A of size $i \times (j + s)$. This fits the assertion, since

$$i + j + s = r + 1 + s = n + 1\,.$$

The other cases are handled similarly.

To establish the converse, suppose now that A has an $r \times s$ submatrix of zeros with $r + s = n + 1$. Then, by permuting rows and columns, we can without loss of generality assume that

$$A = \begin{bmatrix} B & O \\ C & D \end{bmatrix}, \quad r + s = n + 1\,.$$

Thus, as $B \in \mathbb{R}^{r \times (n-s)}$ and $r \times (n - s) = r \times (r - 1)$, any product of the form

$$a_{1\sigma(1)} a_{2\sigma(2)} \cdots a_{r\sigma(r)} \cdots a_{n\sigma(n)}$$

is equal to zero, since at least one of the first r terms in the product sits in the zero block. □

Lemma 23.13. *Let $A \in \mathbb{R}^{n \times n}$ be a doubly stochastic matrix. Then $\operatorname{per} A > 0$.*

Proof. If $\operatorname{per} A = 0$, then, by the last lemma, we can assume that

$$A = \begin{bmatrix} B & O \\ C & D \end{bmatrix},$$

where $B \in \mathbb{R}^{r\times(n-s)}$, $C \in \mathbb{R}^{(n-r)\times(n-s)}$ and $r+s=n+1$. Let Σ_G denote the sum of the entries in the matrix G. Then, since A is doubly stochastic,

$$r = \Sigma_B \leq \Sigma_B + \Sigma_C = n - s\,;$$

i.e., $r+s=n$, which is not compatible with the assumption $\operatorname{per} A = 0$. Therefore, $\operatorname{per} A > 0$, as claimed. □

Theorem 23.14. (Birkhoff-von Neumann) *Let $P \in \mathbb{R}^{n\times n}$ be a doubly stochastic matrix. Then P is a convex combination of finitely many permutation matrices.*

Proof. If P is a permutation matrix, then the assertion is self-evident. If P is not a permutation matrix, then, in view of Lemma 23.13 and the fact that P is doubly stochastic, there exists a permutation σ of the integers $\{1,\ldots,n\}$ such that $1 > p_{1\sigma(1)}p_{2\sigma(2)}\cdots p_{n\sigma(n)} > 0$. Let

$$\lambda_1 = \min\{p_{1\sigma(1)},\ldots,p_{n\sigma(n)}\}$$

and let Π_1 be the permutation matrix with 1's in the $i\sigma(i)$ position for $i=1,\ldots,n$. Then it is readily checked that

$$P_1 = \frac{P-\lambda_1\Pi_1}{1-\lambda_1}$$

is a doubly stochastic matrix with at least one more zero entry than P and that

$$P = \lambda_1\Pi_1 + (1-\lambda_1)P_1\ .$$

If P_1 is not a permutation matrix, then the preceding argument can be repeated; i.e., there exists a number λ_2, $0<\lambda_2<1$, and a permutation matrix Π_2 such that

$$P_2 = \frac{P_1-\lambda_2\Pi_2}{1-\lambda_2}$$

is a doubly stochastic matrix with at least one more zero entry than P_1. Then

$$P = \lambda_1\Pi_1 + (1-\lambda_1)\{\lambda_2\Pi_2 + (1-\lambda_2)P_2\}\ .$$

Clearly this procedure must terminate after a finite number of steps. □

Exercise 23.21. Let Q denote the set of doubly stochastic $n\times n$ matrices.

(a) Show that Q is a convex set and that every $n\times n$ permutation matrix is an extreme point of Q.

(b) Show that if $P\in Q$ and P is not a permutation matrix, then P is not an extreme point of Q.

(c) Give a second proof of Theorem 23.14 based on the Krein-Milman theorem.

23.4. An inequality of Ky Fan

Let $A, B \in \mathbb{R}^{n\times n}$ be a pair of symmetric matrices with eigenvalues $\mu_1 \geq \mu_2 \geq \cdots \geq \mu_n$ and $\nu_1 \geq \nu_2 \geq \cdots \geq \nu_n$, respectively. Then the Cauchy-Schwarz inequality applied to the inner product space $\mathbb{C}^{n\times n}$ with inner product

$$\langle A, B\rangle = \text{trace}\{B^H A\}$$

leads easily to the inequality

$$\text{trace}\{AB\} \leq \left\{\sum_{j=1}^{n} \mu_j^2\right\}^{1/2} \left\{\sum_{j=1}^{n} \nu_j^2\right\}^{1/2} .$$

In this section we shall use the Birkhoff-von Neumann theorem and the Hardy-Littlewood-Polya rearrangement lemma to obtain a sharper result for real symmetric matrices.

The Hardy-Littlewood-Polya rearrangement lemma, which extends the observation that

$$a_1 \geq a_2 \text{ and } b_1 \geq b_2 \Longrightarrow a_1b_1 + a_2b_2 \geq a_1b_2 + a_2b_1 \tag{23.6}$$

to longer ordered sequences of numbers, can be formulated as follows:

Lemma 23.15. *Let* $\mathbf{a}$ *and* $\mathbf{b}$ *be vectors in* $\mathbb{R}^n$ *with entries* $a_1 \geq a_2 \geq \cdots \geq a_n$ *and* $b_1 \geq b_2 \geq \cdots \geq b_n$, *respectively. Then*

$$\mathbf{a}^T P\mathbf{b} \leq \mathbf{a}^T\mathbf{b} \tag{23.7}$$

for every $n \times n$ *permutation matrix* P.

Proof. Let $P = \sum_{j=1}^n \mathbf{e}_j\mathbf{e}_{\sigma(j)}^T$ for some one to one mapping σ of the integers $\{1, \ldots, n\}$ onto themselves, and suppose that $P \neq I_n$. Then there exists a smallest positive integer k such that $\sigma(k) \neq k$. If $k > 1$, this means that $\sigma(1) = 1, \ldots, \sigma(k-1) = k-1$ and $k = \sigma(\ell)$, for some integer $\ell > k$. Therefore, $b_k = b_{\sigma(\ell)} \geq b_{\sigma(k)}$ and hence the inequality (23.6) implies that

$$\begin{aligned} a_k b_{\sigma(k)} + a_\ell b_{\sigma(\ell)} &\leq a_k b_{\sigma(\ell)} + a_\ell b_{\sigma(k)} \\ &= a_k b_k + a_\ell b_{\sigma(k)} . \end{aligned}$$

In the same way, one can rearrange the remaining terms to obtain the inequality (23.7). □

Lemma 23.16. *Let* $A, B \in \mathbb{R}^{n\times n}$ *be symmetric matrices with eigenvalues*

$$\mu_1 \geq \mu_2 \geq \cdots \geq \mu_n \text{ and } \nu_1 \geq \nu_2 \geq \cdots \geq \nu_n ,$$

respectively. Then

$$\text{trace}\{AB\} \leq \mu_1\nu_1 + \cdots + \mu_n\nu_n , \tag{23.8}$$

with equality if and only if there exists an $n \times n$ *orthogonal matrix* U *that diagonalizes both matrices and preserves the order of the eigenvalues in each.*

Proof. Under the given assumptions there exists a pair of $n \times n$ orthogonal matrices U and V such that

$$A = UD_AU^T \text{ and } B = VD_BV^T ,$$

where

$$D_A = \text{diag}\{\mu_1, \ldots, \mu_n\} \text{ and } D_B = \text{diag}\{\nu_1, \ldots, \nu_n\} .$$

Thus

$$\begin{aligned} \text{trace}\{AB\} &= \text{trace}\{UD_AU^TVD_BV^T\} \\ &= \text{trace}\{D_AWD_BW^T\} \\ &= \sum_{i,j=1}^{n} \mu_i w_{ij}^2 \nu_j , \end{aligned}$$

where w_{ij} denotes the ij entry of the matrix $W = U^TV$. Moreover, since W is an orthogonal matrix, the matrix $Z \in \mathbb{R}^{n\times n}$ with entries $z_{ij} = w_{ij}^2$, $i, j = 1, \ldots, n$, is a doubly stochastic matrix and consequently, by the Birkhoff-von Neumann theorem,

$$Z = \sum_{s=1}^{\ell} \lambda_s P_s$$

is a convex combination of permutation matrices. Thus, upon setting $\mathbf{x}^T = [\mu_1, \ldots, \mu_n]$ and $\mathbf{y}^T = [\nu_1, \ldots, \nu_n]$ and invoking Lemma 23.15, it is readily seen that

$$\begin{aligned} \text{trace}\{AB\} &= \sum_{s=1}^{\ell} \lambda_s \mathbf{x}^T P_s \mathbf{y} \\ &\leq \sum_{s=1}^{\ell} \lambda_s \mathbf{x}^T \mathbf{y} = \mathbf{x}^T \mathbf{y} , \end{aligned}$$

as claimed.

The case of equality is left to the reader. □

Remark 23.17. A byproduct of the proof is the observation that the Schur product $U \circ U$ of an orthogonal matrix U with itself is a doubly stochastic matrix. Doubly stochastic matrices of this special form are often referred to as **orthostochastic** matrices. Not every doubly stochastic matrix is an orthostochastic matrix; see Exercise 23.23 The subclass of orthostochastic matrices play a special role in the next section.

Exercise 23.22. Show that in the setting of Lemma 23.16,

$$\text{trace}\{AB\} \geq \mu_1\nu_n + \cdots + \mu_n\nu_1 .$$

Exercise 23.23. Show that the doubly stochastic matrix $A = \frac{1}{6}\begin{bmatrix} 0 & 3 & 3 \\ 3 & 1 & 2 \\ 3 & 2 & 1 \end{bmatrix}$ is not an orthostochastic matrix.

23.5. The Schur-Horn convexity theorem

Let $A = [a_{ij}], i, j = 1, \ldots, n$ be a real symmetric matrix with eigenvalues $\mu_1, \ldots, \mu_n$. Then, by Theorem 9.7, there exists an orthogonal matrix $Q \in \mathbb{R}^{n\times n}$ such that $A = QDQ^T$, where $D = \operatorname{diag}\{\mu_1, \ldots, \mu_n\}$. Thus,

$$a_{ii} = \sum_{j=1}^{n} q_{ij}^2 \mu_j \quad \text{for} \quad i = 1, \ldots, n$$

and the vector $\mathbf{d}_A$ with components a_{ii} for $= 1, \ldots, n$ is given by the formula

$$\mathbf{d}_A = \begin{bmatrix} a_{11} \\ \vdots \\ a_{nn} \end{bmatrix} = B \begin{bmatrix} \mu_1 \\ \vdots \\ \mu_n \end{bmatrix},$$

where B denotes the orthostochastic matrix with entries

$$b_{ij} = q_{ij}^2 \quad \text{for} \quad i, j = 1, \ldots, n.$$

This observation is due to Schur [**63**]. By Theorem 23.14,

$$B = \sum_{\sigma\in\Sigma_n} c_\sigma P_\sigma$$

is a convex combination of permutation matrices P_σ. Thus, upon writing P_σ in terms of the standard basis $\mathbf{e}_i$, $i = 1, \ldots, n$ for $\mathbb{R}^n$ as

$$P_\sigma = \sum_{i=1}^{n} \mathbf{e}_i \mathbf{e}_{\sigma(i)}^T,$$

it is readily checked that the vector

$$\mathbf{d}_A = \sum_{\sigma\in\Sigma_n} c_\sigma \begin{bmatrix} \mu_{\sigma(1)} \\ \vdots \\ \mu_{\sigma(n)} \end{bmatrix}$$

is a convex combination of the vectors corresponding to the eigenvalues of the matrix A and all their permutations. In other words,

$$\mathbf{d}_A \in \operatorname{conv}\left\{ \begin{bmatrix} \mu_{\sigma(1)} \\ \vdots \\ \mu_{\sigma(n)} \end{bmatrix} : \sigma \in \Sigma_n \right\}. \tag{23.9}$$

There is a converse statement due to Horn [**39**], but in order to state it, we must first introduce the notion of majorization.

Given a sequence $\{x_1, \ldots, x_n\}$ of real numbers, let $\{\widetilde{x}_1, \ldots, \widetilde{x}_n\}$ denote the rearrangement of the sequence in "decreasing" order: $\widetilde{x}_1 \geq \cdots \geq \widetilde{x}_n$. Thus, for example, if $n = 4$ and $\{x_1, \ldots, x_4\} = \{5, 3, 6, 1\}$, then $\{\widetilde{x}_1, \ldots, \widetilde{x}_4\} = \{6, 5, 3, 1\}$. A sequence $\{x_1, \ldots, x_n\}$ of real numbers is said to **majorize** a sequence $\{y_1, \ldots, y_n\}$ of real numbers if

$$\widetilde{x}_1 = \cdots \widetilde{x}_k \quad \geq \quad \widetilde{y}_1 + \cdots + \widetilde{y}_k \quad \text{for} \quad k = 1, \ldots, n-1$$

and

$$\widetilde{x}_1 = \cdots \widetilde{x}_n \quad = \quad \widetilde{y}_1 + \cdots + \widetilde{y}_n \, .$$

Exercise 23.24. Show that if $A \in \mathbb{R}^{n \times n}$ is a doubly stochastic matrix and if $\mathbf{y} = A\mathbf{x}$ for some $\mathbf{x} \in \mathbb{R}^n$, then the set of entries $\{x_1, \ldots, x_n\}$ in the vector $\mathbf{x}$ majorizes the set of entries $\{y_1, \ldots, y_n\}$ in the vector $\mathbf{y}$. [HINT: If $x_1 \geq \cdots x_n$, $y_1 \geq \cdots y_n$ and $1 \leq i \leq k \leq n$, then

$$y_i = \sum_{j=1}^{n} a_{ij} x_j \leq \sum_{j=1}^{k-1} a_{ij} x_j + x_k \sum_{j=k}^{n} a_{ij} = \sum_{j=1}^{k-1} a_{ij}(x_i - x_k) + x_k \, .$$

Now exploit the fact that $\sum_{i=1}^{k} a_{ij} \leq 1$ to bound $y_1 + \cdots + y_k$.]

Lemma 23.18. *Let $\{x_1, \ldots, x_n\}$ and $\{y_1, \ldots, y_n\}$ be two sequences of real numbers such that $\{x_1, \ldots, x_n\}$ majorizes $\{y_1, \ldots, y_n\}$. Then there exists a set of $n-1$ orthonormal vectors $\mathbf{u}_1, \ldots, \mathbf{u}_{n-1}$ and a permutation $\sigma \in \Sigma_n$ such that*

$$y_{\sigma(i)} = \sum_{j=1}^{n} (\mathbf{u}_i)_j^2 x_j \quad \textit{for} \quad i = 1, \ldots, n-1 \, . \tag{23.10}$$

Discussion. Without loss of generality, we may assume that $x_1 \geq \cdots \geq x_n$ and $y_1 \geq \cdots \geq y_n$. To ease the presentation, we shall focus on the case $n = 4$. Then the given assumptions imply that

$$x_1 \geq y_1 \, , \quad x_1 + x_2 \geq y_1 + y_2 \, , \quad x_1 + x_2 + x_3 \geq y_1 + y_2 + y_3$$

and, because of the equality $x_1 + \cdots + x_4 = y_1 + \cdots + y_4$,

$$x_4 \leq y_4 \, , \quad x_3 + x_4 \leq y_3 + y_4 \, , \quad x_2 + x_3 + x_4 \leq y_2 + y_3 + y_4 \, .$$

The rest of the argument depends upon the location of the points y_1, y_2, y_3 with respect to the points $x_1, \ldots, x_4$. We shall suppose that $x_1 > x_2 > x_3 > x_4$ and shall consider three cases:

Case 1: $y_3 \leq x_3$ and $y_2 \leq x_2$. Clearly, there exists a choice of c_3, $c_4 \in \mathbb{R}$ such that

$$y_3 = c_3^2 x_3 + c_4^2 x_4 \quad \text{with} \quad c_3^2 + c_4^2 = 1 \, .$$

Moreover, if $w = u^2 x_2 + v^2(\gamma^2 x_3 + \delta^2 x_4)$, where

$$\gamma c_3 + \delta c_4 = 0 \quad \text{and} \quad \gamma^2 + \delta^2 = u^2 + v^2 = 1 \, ,$$

then the vectors $\mathbf{u}_1$ and $\mathbf{u}_2$ defined by the formulas

$$\mathbf{u}_1^T = \begin{bmatrix} 0 & 0 & c_3 & c_4 \end{bmatrix} \quad \text{and} \quad \mathbf{u}_2^T = \begin{bmatrix} 0 & u & v\gamma & v\delta \end{bmatrix}$$

are orthonormal and, since

$$\gamma^2 x_3 + \delta^2 x_4 = (1 - c_3^2)x_3 + (1 - c_4^2)x_4 = x_3 + x_4 - y_3 \leq y_4 \leq y_2$$

and $y_2 \leq x_2$, there exists a choice of u, v with $u^2 + v^2 = 1$ such that $y_2 = w$, i.e.,

$$y_2 = b_2^2 x_2 + b_3^3 x_3 + b_4^2 x_4 \, .$$

Now, let

$$w = u^2 x_1 + v^2(\beta^2 x_2 + \gamma^2 x_3 + \delta^2 x_4)\, ,$$

where the numbers β, γ, δ, u and v are redefined by the constraints

$$\gamma^2 c_3 + \delta^2 c_4 = \beta^2 b_2 + \gamma^2 b_3 + \delta^2 b_4 = 0 \quad \text{and} \quad \beta^2 + \gamma^2 + \delta^2 = u^2 + v^2 = 1\, .$$

Then the vectors

$$\mathbf{u}_1 = \begin{bmatrix} 0 \\ b_2 \\ b_3 \\ b_4 \end{bmatrix}, \quad \mathbf{u}_2 = \begin{bmatrix} 0 \\ 0 \\ c_3 \\ c_4 \end{bmatrix}, \quad \text{and} \quad \mathbf{u}_3 = \begin{bmatrix} u \\ v\beta \\ v\gamma \\ v\delta \end{bmatrix}$$

are orthonormal and, since

$$\begin{aligned} \beta^2 x_2 + \gamma^2 x_3 + \delta^2 x_4 &= (1 - b_2^2)x_2 + (1 - b_3^2 - c_3^2)x_3 + (1 - b_4^2 - c_4^2)x_4 \\ &= x_2 + x_3 + x_4 - y_2 - y_3 \leq y_4 \leq y_1 \end{aligned}$$

and $y_1 \leq x_1$, there exists a choice of real numbers u and v with $u^2 + v^2 = 1$ such that $w = \mathbf{u}_3^T \mathbf{x} = y_1$.

Case 2: $y_1 \geq x_2$ and $y_2 \geq x_3$. This is a reflected version of Case 1, and much the same sort of analysis as in that case leads rapidly to the existence of representations

$$\begin{aligned} y_1 &= a_1^2 x_1 + a_2^2 x_2 \quad \text{with} \quad a_1^2 + a_2^2 = 1\, , \\ y_2 &= b_1^2 x_1 + b_2^2 x_2 + b_3^2 x_3 \quad \text{with} \quad b_1^2 + b_2^2 + b_3^2 = 1 \end{aligned}$$

and $a_1 b_1 + a_2 b_2 = 0$. Therefore, if

$$w = u^2(\alpha^2 x_1 + \beta^2 x_2 + \gamma^2 x_3) + v^2 x_4$$

with

$$\alpha a_1 + \beta a_2 = \alpha b_1 + \beta b_2 + \gamma b_3 = 0 \quad \text{and} \quad \alpha^2 + \beta^2 + \gamma^2 = u^2 + v^2 = 1\, ,$$

then the vectors

$$\mathbf{u}_1 = \begin{bmatrix} a_1 \\ a_2 \\ 0 \\ 0 \end{bmatrix}, \quad \mathbf{u}_2 = \begin{bmatrix} b_1 \\ b_2 \\ b_3 \\ 0 \end{bmatrix}, \quad \text{and} \quad \mathbf{u}_3 = \begin{bmatrix} u\alpha \\ u\beta \\ u\gamma \\ v \end{bmatrix}$$

are orthonormal and, since

$$\begin{aligned}\alpha^2 x_1 + \beta^2 x_2 + \gamma^2 x_3 &= (1 - a_1^2 - b_1^2)x_1 + (1 - a_2^2 - b_2^2)x_2 + (1 - b_3^2)x_3 \\ &= x_1 + x_2 + x_3 - y_1 - y_2 \geq y_3\end{aligned}$$

and $y_3 \geq x_4$, there exists a choice of u, v with $u^2 + v^2 = 1$ such that $w = \mathbf{u}_3^T \mathbf{x} = y_3$.

Case 3: $x_3 \leq y_3 \leq y_2 \leq y_1 \leq x_2$. If $x_2 + x_3 \leq y_1 + y_2$, then we may write

$$\begin{aligned}y_2 &= b_2^2 x_2 + b_3^2 x_3 \quad \text{with} \quad b_2^2 + b_3^2 = 1 \\ y_1 &= a_1^2 x_1 + a_2^2 x_2 + a_3^2 x_3 \quad \text{with} \quad a_1^2 + a_2^2 + a_3^2 = 1\end{aligned}$$

and $a_2 b_2 + a_3 b_3 = 0$. Thus, if

$$w = u^2(\alpha^2 x_1 + \beta^2 x_2 + \gamma^2 x_3) + v^2 x_4 \,,$$

where

$$\alpha a_1 + \beta a_2 + \gamma a_3 = \beta b_2 + \gamma b_3 = 0 \quad \text{and} \quad \alpha^2 + \beta^2 + \gamma^2 = u^2 + v^2 = 1 \,,$$

then the vectors

$$\mathbf{u}_1 = \begin{bmatrix} a_1 \\ a_2 \\ a_3 \\ 0 \end{bmatrix}, \quad \mathbf{u}_2 = \begin{bmatrix} 0 \\ b_2 \\ b_3 \\ 0 \end{bmatrix}, \quad \text{and} \quad \mathbf{u}_3 = \begin{bmatrix} u\alpha \\ u\beta \\ u\gamma \\ v \end{bmatrix}$$

are orthonormal and, since

$$\begin{aligned}\alpha^2 x_1 + \beta^2 x_2 + \gamma^2 x_3 &= (1 - a_1^2)x_1 + (1 - a_2^2 - b_2^2)x_2 + (1 - a_3^3 - b_3^2)x_3 \\ &= x_1 + x_2 + x_3 - y_1 - y_2 \geq y_3\end{aligned}$$

and $x_4 \leq y_3$, there exists a choice of real numbers u and v with $u^2 + v^2 = 1$ such that $w = \mathbf{u}_3^T \mathbf{x} = y_3$.

On the other hand, if $x_2 + x_3 \geq y_1 + y_2$, then

$$\begin{aligned}y_2 &= b_2^2 x_2 + b_3^2 x_3 \quad \text{with} \quad b_2^2 + b_3^2 = 1 \\ y_3 &= c_2^2 x_2 + b_c^2 x_3 + c_4^2 x_4 \quad \text{with} \quad c_2^2 + c_3^2 + c_4^2 = 1\end{aligned}$$

and $b_2 c_2 + b_3 c_3 + b_4 c_4 = 0$ and the construction is completed by choosing

$$y_1 = u^2 x_1 + v^2(\beta^2 x_2 + \gamma^2 x_3 + \delta^2 x_4)$$

with

$$\beta b_2 + \gamma b_3 + \delta b_4 = \beta c_2 + \gamma c_3 + \delta c_4 = 0 \,, \quad \beta^2 + \gamma^2 + \delta^2 = 1$$

and an appropriate choice of u, v with $u^2 + v^2 = 1$.

Theorem 23.19. **(Schur-Horn)** *Let $\{\mu_1, \ldots, \mu_n\}$ be any set of real numbers (not necessarily distinct) and let $\mathbf{a} \in \mathbb{R}^n$. Then*

$$\mathbf{a} \in \operatorname{conv}\left\{ \begin{bmatrix} \mu_{\sigma(1)} \\ \vdots \\ \mu_{\sigma(n)} \end{bmatrix} : \sigma \in \Sigma_n \right\}$$

if and only if there exists a symmetric matrix $A \in \mathbb{R}^{n \times n}$ with eigenvalues $\mu_1, \ldots, \mu_n$ such that $\mathbf{d}_A = \mathbf{a}$.

Proof. Suppose first that $\mathbf{a}$ belongs to the indicated convex hull. Then

$$\mathbf{a} = \sum_{\sigma \in \Sigma_n} c_\sigma P_\sigma \begin{bmatrix} \mu_1 \\ \vdots \\ \mu_n \end{bmatrix} = P \begin{bmatrix} \mu_1 \\ \vdots \\ \mu_n \end{bmatrix},$$

where $P = \sum_{\sigma \in \Sigma_n} c_\sigma P_\sigma$ is a convex combination of permutation matrices P_σ and hence, by Exercise 23.24, $\{\mu_1, \ldots, \mu_n\}$ majorizes $\{a_1, \ldots, a_n\}$. Therefore, by Lemma 23.18, there exists a set $\{\mathbf{u}_1, \ldots, \mathbf{u}_{n-1}\}$ of $n-1$ orthonormal vectors in $\mathbb{R}^n$ and a permutation $\sigma \in \Sigma_n$ such that

$$a_{\sigma(i)} = \sum_{j=1}^{n} (\mathbf{u}_i)_j^2 \mu_j \quad \text{for} \quad i = 1, \ldots, n-1.$$

Let $\mathbf{u}_n \in \mathbb{R}^n$ be a vector of norm one that is orthogonal to $\mathbf{u}_1, \ldots, \mathbf{u}_{n-1}$. Then

$$\begin{aligned} \sum_{j=1}^{n} (\mathbf{u}_n)_j^2 \mu_j &= \sum_{j=1}^{n} \left(1 - \sum_{i=1}^{n-1} (\mathbf{u}_i)_j^2\right) \mu_j \\ &= \mu_1 + \cdots + \mu_n - (a_{\sigma(1)} + \cdots + a_{\sigma(n-1)}) = a_{\sigma(n)}. \end{aligned}$$

This completes the proof in one direction. The other direction is covered by the first few lines of this section. □

Exercise 23.25. The components $\{x_1, \ldots, x_n\}$ of $\mathbf{x} \in \mathbb{R}^n$ majorize the components $\{y_1, \ldots, y_n\}$ of $\mathbf{y} \in \mathbb{R}^n$ if and only if

$$\mathbf{y} \in \operatorname{conv}\{P_\sigma \mathbf{x} : \sigma \in \Sigma_n\}.$$

Exercise 23.26. Verify Lemma 23.18 when $\{x_1, \ldots, x_4\}$ majorizes $\{y_1, \ldots, y_4\}$ and $x_1 \geq x_2 \geq y_1 \geq y_2 \geq y_3 \geq x_3 \geq x_4$. [HINT: Express y_1 as a convex combination of x_1 and x_4 and y_3 as a convex combination of x_2 and x_3.]

23.6. Bibliographical notes

Applications of Perron-Frobenius theory to the control of a group of autonomous wheeled vehicles are found in the paper [**47**]. The presented proof of Fan's inequality is adapted from an exercise with hints in [**10**]. Exercise

23.25 is a theorem of Rado. Exercise 23.24 is taken from [**37**]. Exercise 23.23 is taken from [**63**]. Additional discussion on the history of the Schur-Horn convexity theorem and references to generalizations may be found in [**28**]. Related applications are discussed in [**13**].

The definitive account of permanents up till about 1978 was undoubtedly the book *Permanents* by Henryk Minc [**51**]. However, in 1980/81 two proofs of van der Waerden's conjecture, which states that

The permanent of a doubly stochastic $n \times n$ matrix is bounded below by $n!/n^n$, with equality if and only if each entry in the matrix is equal to $1/n$,

were published. The later book *Nonnegative Matrices* [**52**] includes a proof of this conjecture.

Appendix A

Some facts from analysis

... a liberal arts school that administers a light education to students lured by fine architecture and low admission requirements. ... Among financially gifted parents of academically challenged students along the Eastern Seaboard, the college is known as ... a place where ..., barring a felony conviction, ... [your child] ... will get to wear a black gown and attend graduation....

Garrison Keillor [**42**], p. 7

A.1. Convergence of sequences of points

A **sequence of points** $x_1, x_2, \ldots \in \mathbb{R}$ is said to

- be **bounded** if there exists a finite number $M > 0$ such that $|x_j| \leq M$ for $j = 1, 2, \ldots,$
- be **monotonic** if either $x_1 \leq x_2 \leq \cdots$ or $x_1 \geq x_2 \geq \cdots,$
- **converge** to a limit x if for every $\varepsilon > 0$ there exists an integer N such that
$$|x_j - x| < \varepsilon \text{ if } \quad j \geq N\,.$$
- be a **Cauchy sequence** (or a **fundamental sequence**) if for every $\varepsilon > 0$ there exists an integer N such that
$$|x_{j+k} - x_j| < \varepsilon \text{ if } \quad j \geq N \quad \text{and} \quad k \geq 1\,.$$

It is easy to see that every convergent sequence is a Cauchy sequence. The converse is true also. The principle facts regarding convergence are:

- A sequence of points $x_1, x_2, \ldots \in \mathbb{R}$ converges to a limit $x \in \mathbb{R}$ if and only if it is a Cauchy sequence.
- Every bounded sequence of points in $\mathbb{R}$ has a convergent subsequence.
- If every convergent subsequence of a bounded sequence of points converges to the same limit, then the sequence converges to the same limit.
- Every bounded monotonic sequence converges to a finite limit.

A.2. Convergence of sequences of functions

A **sequence of functions** $f_1(x), f_2(x), \cdots$ that is defined on a set $Q \subset \mathbb{R}$ is said to **converge** to a limit $f(x)$ on Q if $f_n(x)$ converges to $f(x)$ at each point $x \in Q$, i.e., if for each point $x \in Q$ and every $\varepsilon > 0$, there exists an integer N such that

$$|f_j(x) - f(x)| < \varepsilon \quad \text{if} \quad j \geq N.$$

In general, the number N depends upon x.

The sequence $f_1(x), f_2(x), \ldots$ is said to **converge uniformly** to $f(x)$ on Q if for every $\varepsilon > 0$ there exists an integer N that is independent of the choice of $x \in Q$ such that

$$|f_j(x) - f(x)| < \varepsilon \quad \text{for} \quad j \geq N \quad \text{and every} \quad x \in Q\,.$$

A.3. Convergence of sums

A sum $\sum_{j=1}^{\infty} a_j$ of points $a_j \in \mathbb{R}$ is said to converge to a limit $a \in \mathbb{R}$ if the partial sums

$$S_n = \sum_{j=1}^{n} a_j$$

tend to a as $n \uparrow \infty$, i.e., if for every $\varepsilon > 0$, there exists an integer N such that

$$|S_n - a| < \varepsilon \quad \text{for} \quad n > N.$$

The Cauchy criterion for convergence then translates to: for every $\varepsilon > 0$ there exists an integer N such that

$$|S_{n+k} - S_n| < \varepsilon \quad \text{for } n > N \quad \text{and} \quad k \geq 1$$

or, equivalently,

$$\left| \sum_{j=n+1}^{n+k} a_j \right| < \varepsilon \quad \text{for} \quad n > N \text{ and} \quad k \geq 1\,.$$

In particular, a sufficient (but not necessary) condition for convergence is that for every $\varepsilon > 0$ there exists an integer N such that

$$\sum_{j=n+1}^{n+k} |a_j| < \varepsilon \quad \text{for} \quad n > N \quad \text{and} \quad k \geq 1 \quad \text{or, equivalently,} \quad \sum_{j=1}^{\infty} |a_j| < \infty .$$

A.4. Sups and infs

If Q is a bounded set of points in $\mathbb{R}$, then:

> m is said to be a **lower bound** for Q if $m \leq x$ for every $x \in Q$.
>
> M is said to be an **upper bound** for Q if $x \leq M$ for every $x \in Q$.

Moreover, there exists a unique **greatest lower bound** $\widetilde{m}$ for Q and a unique **least upper bound** $\widetilde{M}$ for Q. These numbers are referred to as the **infimum** and **supremum** of Q, respectively. They are denoted by the symbols

$$\widetilde{m} = \inf\{x : x \in Q\} \quad \text{and} \quad \widetilde{M} = \sup\{x : x \in Q\} ,$$

respectively, and may be characterized by the conditions

infimum: $\widetilde{m} = \inf\{x : x \in Q\}$ if and only if $\widetilde{m} \leq x$ for every $x \in Q$, but for every $\varepsilon > 0$ there exists at least one point $x \in Q$ such that $x < \widetilde{m} + \varepsilon$.

supremum: $\widetilde{M} = \sup\{x : x \in Q\}$ if and only if $x \leq \widetilde{M}$ for every $x \in Q$, but for every $\varepsilon > 0$ there exists at least one point $x \in Q$ such that $x > \widetilde{M} - \varepsilon$.

Let $x_1, x_2, \cdots$ be a sequence of points in $\mathbb{R}$ such that $|x_j| \leq M < \infty$ and let

$$M_k = \sup\{x_k, x_{k+1}, \cdots\} \quad \text{for} \quad k = 1, 2, \ldots .$$

Then

$$M_1 \geq M_2 \geq M_3 \geq \cdots \geq -M ;$$

i.e., $M_1, M_2, \ldots$ is a bounded monotone sequence. Therefore $\lim_{j\uparrow\infty} M_j$ exists, even though the original sequence $x_1, x_2, \ldots$ may not have a limit. (Think of the sequence $0, 1, 0, 1, \ldots$.) This number is called the **limit superior** and is written

$$\limsup_{j\uparrow\infty} x_j = \lim_{j\uparrow\infty} \sup\{x_j, x_{j+1} \cdots\}.$$

The **limit inferior** is defined similarly:

$$\liminf_{j\uparrow\infty} x_j = \lim_{j\uparrow\infty} \inf\{x_j, x_{j+1} \cdots\}.$$

A.5. Topology

Let

$$B_r(y) = \{x \in \mathbb{R} : |x - y| < r\} \quad \text{and} \quad \overline{B_r(y)} = \{x \in \mathbb{R} : |x - y| \leq r\}.$$

A point $y \in \mathbb{R}$ is said to be a **limit point** of a set $Q \subset \mathbb{R}$ if for every choice of $r > 0$, $B_r(y)$ contains at least one point of Q other than y. A subset Q of $\mathbb{R}$ is said to be:

open: if for every point $y \in Q$ there exists an $r > 0$ (which usually depends upon y) such that $B_r(y) \subset Q$.

closed: if Q contains all its limit points.

bounded: if $Q \subset B_R(0)$ for some $R > 0$.

If $A_1, A_2 \ldots$ are open sets, then finite intersections $\cap_{i=1}^n A_i$ and (even) infinite unions $\cup_{i=1}^\infty A_i$ are open. If $B_1, B_2, \ldots$ are closed sets, then finite unions $\cup_{i=1}^n B_i$ and (even) infinite intersections $\cap_{i=1}^\infty B_i$ are closed.

An open set $\Omega \subset \mathbb{C}$ is said to be **connected** if there does not exist a pair of disjoint nonempty open sets A and B such that $\Omega = A \cup B$. In other words, if $\Omega = A \cup B$, A, B are open and $A \cap B = \emptyset$, then either $A = \emptyset$ or $B = \emptyset$.

A.6. Compact sets

A collection of open sets $\{B_\alpha : \alpha \in A\}$ is said to be an **open covering** of Q if $Q \subset \bigcup_{\alpha \in A} B_\alpha$. Q is said to be **compact** if every open covering contains a finite collection of open sets $B_{\alpha_1}, \ldots, B_{\alpha_n}$ such that $Q \subset \bigcup_{j=1}^n B_{\alpha_j}$.

- Let $Q \subset \mathbb{R}$. Then Q is compact if and only if Q is closed and bounded.
- A continuous real-valued function $f(x)$ that is defined on a compact set K attains its maximum value at some point in K and its minimum value at some point in K.

A.7. Normed linear spaces

The definitions introduced above for the vector space $\mathbb{R}$ have natural analogues in the vector space $\mathbb{C}$ and in normed linear spaces over $\mathbb{R}$ or $\mathbb{C}$. Thus, for example, if $\mathcal{X}$ is a normed linear space over $\mathbb{C}$ and $\mathbf{y} \in \mathcal{X}$, it is natural to let

$$B_r(\mathbf{y}) = \{\mathbf{x} \in \mathcal{X} : \|\mathbf{x} - \mathbf{y}\| < r\} \quad \text{and} \quad \overline{B_r(\mathbf{y})} = \{\mathbf{x} \in \mathcal{X} : \|\mathbf{x} - \mathbf{y}\| \leq r\}.$$

Then a point $\mathbf{y} \in \mathcal{X}$ is said to be a **limit point** of a set $Q \subset \mathcal{X}$ if for every choice of $r > 0$, $B_r(\mathbf{y})$ contains at least one point of Q other than $\mathbf{y}$. A subset Q of $\mathbb{R}$ is said to be:

open: if for every point $\mathbf{y} \in Q$ there exists an $r > 0$ (which usually depends upon $\mathbf{y}$) such that $B_r(\mathbf{y}) \subset Q$.

closed: if Q contains all its limit points.

bounded: if $Q \subset B_R(\mathbf{0})$ for some $R > 0$.

compact: if every open covering of Q contains a finite open covering of Q.

A sequence of vectors $\mathbf{x}_1, \mathbf{x}_2, \ldots$ in $\mathcal{X}$ is said to be a **Cauchy sequence** if for every $\varepsilon > 0$ there exists an integer N such that

$$\|\mathbf{x}_{n+k} - \mathbf{x}_n\| < \varepsilon \quad \text{for every} \quad k \geq 1 \quad \text{when} \quad n \geq N\,.$$

A normed linear space $\mathcal{X}$ over $\mathbb{C}$ or $\mathbb{R}$ is said to be **complete** if every Cauchy sequence converges to a point $\mathbf{x} \in \mathcal{X}$.

Finite dimensional normed linear spaces over $\mathbb{C}$ or $\mathbb{R}$ are complete. Therefore, their properties are much the same as those recorded for $\mathbb{R}$ earlier.

Thus, for example, if $\mathcal{X}$ is a finite dimensional normed linear space over $\mathbb{C}$ or $\mathbb{R}$, then every bounded sequence has a convergent subsequence. Moreover, if every convergent subsequence of a bounded sequence tends to the same limit, then the full original sequence converges to that limit.

A point $\mathbf{a} \in \mathcal{X}$ is said to be a **boundary point** of a subset A of $\mathcal{X}$ if

$$B_r(\mathbf{a}) \cap A \neq \emptyset \quad \text{and} \quad B_r(\mathbf{a}) \cap (\mathcal{X} \setminus A) \neq \emptyset \quad \text{for every} \quad r > 0\,,$$

where $\mathcal{X} \setminus A$ denotes the set of points that are in $\mathcal{X}$ but are not in A.

A point $\mathbf{a} \in \mathcal{X}$ is said to be an **interior point** of A if $B_r(\mathbf{a}) \subset A$ for some $r > 0$. The symbol $\operatorname{int} A$ will be used to denote the set of all interior points of A. This set is called the **interior** of A.

Appendix B

More complex variables

The game was not as close as the score indicated.

Rud Rennie (after observing a 19 to 1 rout), cited in [**40**], p. 48

This appendix is devoted to some supplementary facts on complex variable theory to supplement the brief introduction given in Chapter 17.

B.1. Power series

An infinite series of the form $\sum_{n=0}^{\infty} a_n(\lambda-\omega)^n$ is called a **power series** and the number R that is defined by the formula

$$\frac{1}{R} = \limsup_{k\uparrow\infty}\{|a_k|^{1/k}\},$$

with the understanding that

$$R=\infty \quad \text{if} \quad \limsup_{k\uparrow\infty}|a_k|^{1/k}=0 \quad \text{and} \quad R=0 \quad \text{if} \quad \limsup_{k\uparrow\infty}|a_k|^{1/k}=\infty,$$

is termed the **radius of convergence** of the power series. The name stems from the fact that the series converges if $|\lambda-\omega|<R$ and diverges if $\lambda-\omega|>R$. Thus, for example, if $0<R<\infty$ and $0<r<R$, then the partial sums

$$f_n(\lambda)=\sum_{k=0}^{n} a_k(\lambda-\omega)^k, \; n=0,1,\dots,$$

form a Cauchy sequence for each point λ in the closure

$$\overline{B_r(\omega)}=\{\lambda\in\mathbb{C}:\;|\lambda-\omega|\le r\} \quad \text{of} \quad B_r(\omega)=\{\lambda\in\mathbb{C}:\;|\lambda-\omega|<r\}$$

if $r < R$: if $r < r_1 < R$, then $|a_j| \le r_1^{-j}$ for all $j \ge n$ if n is chosen large enough and hence

$$|f_{n+k}(\lambda) - f_n(\lambda)| = \left| \sum_{j=n+1}^{n+k} a_j(\lambda-\omega)^j \right| \le \sum_{j=n+1}^{n+k} \left(\frac{r}{r_1}\right)^j ,$$

which tends to zero as $n \uparrow \infty$.

On the other hand, if

$$|\lambda - \omega| = r > R \quad \text{and} \quad \frac{1}{r} < \frac{1}{r_1} < \frac{1}{R},$$

then

$$|a_k|^{1/k} \ge \frac{1}{r_1} \text{ infinitely often } \Longrightarrow |a_k||\lambda-\omega|^k \ge \left(\frac{r}{r_1}\right)^k > 1 \text{ infinitely often.}$$

Therefore, the power series diverges if $|\lambda - \omega| > R$.

The cases $R = 0$ and $R = \infty$ are left to the reader.

The next order of business is to check that every holomorphic function generates a power series. More precisely:

Lemma B.1. *Let f be holomorphic in an open set $\Omega \subset \mathbb{C}$, let $\omega \in \Omega$ and suppose that $\overline{B_R(\omega)} \subset \Omega$ for some $R > 0$. Then*

$$f(\lambda) = \sum_{n=0}^{\infty} \frac{f^{(n)}(\omega)}{n!}(\lambda-\omega)^n \quad \text{for} \quad |\lambda - \omega| < R.$$

Proof. Let Γ_R denote a circle centered at ω of radius R directed counterclockwise and let $|\lambda - \omega| < R$. Then

$$\begin{aligned} f(\lambda) &= \frac{1}{2\pi i}\int_{\Gamma_R} \frac{f(\zeta)}{\zeta-\lambda} d\zeta \\ &= \frac{1}{2\pi i}\int_{\Gamma_R} \frac{f(\zeta)}{\zeta-\omega-(\lambda-\omega)} d\zeta \\ &= \frac{1}{2\pi i}\int_{\Gamma_R} \frac{f(\zeta)}{(\zeta-\omega)(1-\frac{\lambda-\omega}{\zeta-\omega})} d\zeta \\ &= \frac{1}{2\pi i}\int_{\Gamma_R} \frac{f(\zeta)}{\zeta-\omega} \sum_{n=0}^{\infty}\left(\frac{\lambda-\omega}{\zeta-\omega}\right)^n d\zeta \\ &= \sum_{n=0}^{\infty}\left\{\frac{1}{2\pi i}\int_{\Gamma_R} \frac{f(\zeta)}{(\zeta-\omega)^{n+1}} d\zeta\right\}(\lambda-\omega)^n , \end{aligned}$$

which coincides with the asserted formula. □

Conversely, every convergent power series $\sum_{k=0}^{\infty} a_k(\lambda-\omega)^k$ defines a holomorphic function:

Lemma B.2. *Let $\sum_{k=0}^{\infty} a_k(\lambda-\omega)^k$ be a power series with radius of convergence R. Then the partial sums*

$$f_n(\lambda) = \sum_{k=0}^{n} a_k(\lambda - \omega)^k, \; n = 0, 1, \ldots,$$

converge to a holomorphic function $f(\lambda)$ at every point $\lambda \in B_R(\omega)$ and the partial sums

$$f_n'(\lambda) = \sum_{k=1}^{n} k a_k(\lambda - \omega)^{k-1}, \; n = 1, 2, \ldots,$$

converge to its derivative $f'(\lambda)$ at every point $\lambda \in B_R(\omega)$. Moreover, the convergence of both of these sequences of polynomials is uniform in $\overline{B_r(\omega)}$ for every $r < R$.

Proof. Since $\lim_{n\uparrow\infty} n^{1/n} = 1$, it follows that

$$\limsup_{n\uparrow\infty} (n|a_n|)^{1/n} = \limsup_{n\uparrow\infty} |a_n|^{1/n}$$

and hence that the two power series

$$\sum_{k=0}^{\infty} a_k(\lambda - \omega)^k \text{ and } \sum_{k=1}^{\infty} k a_k(\lambda - \omega)^{k-1}$$

have the same radius of convergence R. Moreover, the limits

$$f(\lambda) = \lim_{n\uparrow\infty} f_n(\lambda) \quad \text{and} \quad g(\lambda) = \lim_{n\uparrow\infty} f_n'(\lambda)$$

are both holomorphic in $B_R(\omega)$, since the convergence is uniform in $\overline{B_r(\omega)}$ if $0 < r < R$. Furthermore, if $\lambda \in B_R(\omega)$ and Γ_r denotes a circle of radius r centered at ω and directed counterclockwise, then

$$\begin{aligned} f'(\lambda) &= \frac{1}{2\pi i} \int_{\Gamma_r} \frac{f(\zeta)}{(\zeta - \lambda)^2} d\zeta \\ &= \lim_{n\uparrow\infty} \frac{1}{2\pi i} \int_{\Gamma_r} \frac{f_n(\zeta)}{(\zeta - \lambda)^2} d\zeta \\ &= \lim_{n\uparrow\infty} f_n'(\lambda) = g(\lambda)\,. \end{aligned}$$

□

B.2. Isolated zeros

Lemma B.3. *Let f be holomorphic in an open connected set Ω and suppose that $f(\lambda) = 0$ for every point λ in an infinite set $\Omega_0 \subset \Omega$ that contains a limit point. Then $f(\lambda) = 0$ for every point $\lambda \in \Omega$.*

Proof. Let

$$A = \{\omega \in \Omega : \; f^{(j)}(\omega) = 0 \text{ for } j = 0, 1, \dots\}$$

and let

$$B = \{\omega \in \Omega : \; f^{(k)}(\omega) \neq 0 \text{ for at least one integer } k \geq 0\}.$$

Clearly $A \cup B = \Omega$, $A \cap B = \emptyset$ and B is open. Moreover, since $\omega \in A$ if and only if there exists a radius $r_\omega > 0$ such that $B_{r_\omega}(\omega) \subset \Omega$ and $f(\lambda) = 0$ for every point $\lambda \in B_{r_\omega}(\omega)$, it follows that A is also open. The proof is completed by showing that if $\alpha_1, \alpha_2, \cdots$ is a sequence of points in Ω_0 that tend to a limit $\alpha \in \Omega_0$, then $\alpha \notin B$. Therefore $\alpha \in A$, i.e., $A = \Omega$ and $B = \emptyset$. The verification that $\alpha \notin B$ follows by showing that the zeros of $f(\lambda)$ in B are isolated. The details are carried out in a separate lemma. $\square$

Lemma B.4. *Let $f(\lambda)$ admit a power series expansion of the form*

$$f(\lambda) = \frac{f^{(k)}(\omega)}{k!}(\lambda - \omega)^k + \frac{f^{(k+1)}(\omega)}{(k+1)!}(\lambda - \omega)^{k+1} + \dots$$

in the ball $B_r(\omega)$ and suppose that $f^{(k)}(\omega) \neq 0$ for some nonnegative integer k. Then there exists a number $\rho_\omega > 0$ such that $f(\lambda) \neq 0$ for all points λ in the annulus $0 < |\lambda - \omega| < \rho_\omega$.

Proof. If $0 < r_1 < r$, then $\overline{B_{r_1}(\omega)} \subset B_r(\omega)$ and the coefficients in the exhibited power series are subject to the bound

$$\left| \frac{f^{(j)}(\omega)}{j!} \right| \leq \frac{M}{r_1^j},$$

where

$$M = \max\{|f(\zeta)| : \zeta \in \mathbb{C} \text{ and } |\omega - \zeta| = r_1\}.$$

Therefore, if $|\lambda - \omega| < r_1$, then

$$\begin{aligned} \left| \sum_{j=k+1}^{\infty} \frac{f^{(j)}(\omega)}{j!}(\lambda - \omega)^j \right| &\leq \sum_{j=k+1}^{\infty} \frac{M}{r_1^j} |\lambda - \omega|^j \\ &\leq M \left(\frac{|\lambda - \omega|}{r_1} \right)^{k+1} \frac{r_1}{r_1 - |\lambda - \omega|}. \end{aligned}$$

Consequently,

$$\begin{aligned} &\left| \frac{f^{(k)}(\omega)(\lambda - \omega)^k}{k!} + \frac{f^{(k+1)}(\omega)(\lambda - \omega)^{k+1}}{(k+1)!} + \cdots \right| \\ &\geq |\lambda - \omega|^k \left\{ \left| \frac{f^{(k)}(\omega)}{k!} \right| - \frac{M|\lambda - \omega|}{r_1^{k+1}} \frac{r_1}{r_1 - |\lambda - \omega|} \right\} \end{aligned}$$

which is not equal to zero for $0<|\lambda-\omega|<r_2$ when r_2 is chosen small enough. □

Exercise B.1. Let f be holomorphic in $B_r(\omega)$. Show that if $|f(\lambda)|=M$ for every point $\lambda\in B_r(\omega)$, then $f(\lambda)$ is constant in $B_r(\omega)$. [HINT: Use estimates analogous to those used just above to show that $f^{(k)}(\omega)=0$ for $k=1,2,\dots$.]

B.3. The maximum modulus principle

Lemma B.5. *Let Ω be an open connected bounded nonempty subset of $\mathbb{C}$; let f be holomorphic in Ω and continuous in $\overline{\Omega}$, the closure of Ω; let $\partial\Omega$ denote the boundary of Ω; and let $M=\max\{|f(\lambda)|:\lambda\in\overline{\Omega}\}$. Then:*

(1) $\max\{|f(\lambda)|:\lambda\in\partial\Omega\}=M$

(2) $|f(\lambda)|=M$ *for some point $\lambda\in\Omega$ if and only if f is constant in $\overline{\Omega}$.*

Proof. Suppose first that $f(\omega)|=M$ for some point $\omega\in\Omega$ and let $B_r(\omega)\subset\Omega$. Then, in the usual notation, Cauchy's formula implies that

$$|f(\omega)|=\left|\frac{1}{2\pi i}\int_{\Gamma_r}\frac{f(\zeta)}{\zeta-\omega}d\zeta\right|=|\frac{1}{2\pi}\int_0^{2\pi}|f(\omega+re^{i\theta})|d\theta$$

and thus, as $|f(\omega+re^{i\theta})|\leq M$, equality must prevail: i.e., $|f(\omega+\rho e^{i\theta})|=M$, first for $\rho=r$ and then, by the same argument, for $0\leq\rho\leq r$. Therefore the set $A=\{\lambda\in\Omega:|f(\lambda)|=M\}$ is open. Moreover, $B=\{\lambda\in\Omega:|f(\lambda)|<M\}$ is open. Therefore, since $A\cup B=\Omega$, $A\cap B=\emptyset$ and Ω is connected, it follows that either $A=\emptyset$ or $B=\emptyset$.

The proof is now easily completed by invoking Exercise B.1. □

B.4. $\ln(1-\lambda)$ when $|\lambda|<1$

Lemma B.6. *If $|\lambda|<1$, then*

$$1-\lambda=\exp\left\{-\sum_{n=1}^{\infty}\frac{\lambda^n}{n}\right\}.$$

Proof. Let

$$g(\lambda)=\exp\left\{-\sum_{n=1}^{\infty}\frac{\lambda^n}{n}\right\}.$$

Then

$$g'(\lambda)=g(\lambda)\left\{-\sum_{n=1}^{\infty}\lambda^{n-1}\right\}=-\frac{g(\lambda)}{1-\lambda}$$

and

$$g'(0)=-g(0)=-1.$$

Moreover,

$$g''(\lambda) = \frac{(\lambda-1)g'(\lambda) - g(\lambda)}{(\lambda-1)^2} = 0\,.$$

Thus, all the higher order derivatives of $g(\lambda)$ are also equal to zero in $B_1(0)$, and hence the power series expansion

$$g(\lambda) = \sum_{k=0}^{\infty} \frac{g^{(k)}(0)}{k!}\lambda^k$$

reduces to

$$g(\lambda) = g(0) + g'(0)\lambda = 1 - \lambda\,.$$

□

Corollary B.7. If $|\lambda| < 1$, then

$$\ln(1-\lambda) = -\sum_{n=1}^{\infty} \frac{\lambda^n}{n}\,.$$

B.5. Rouché's theorem

Lemma B.8. *Let $f(\lambda)$ be holomorphic in an open set Ω that contains the closed disc $\overline{B_r(\omega)}$ for some $r > 0$ and suppose that $|f(\lambda)| > 0$ if $|\lambda - \omega| = r$ and let $N_f(\omega, r)$ denote the number of zeros of f inside $B_r(\omega)$, counting multiplicities. Then*

$$N_f(\omega, r) = \frac{1}{2\pi i}\int_{\Gamma_r} \frac{f'(\zeta)}{f(\zeta)}d\zeta\,.$$

Proof. The main observation is that if $f(\lambda)$ has a zero of order k at some point $\alpha \in B_r(\omega)$, then $f(\lambda) = (\lambda - \alpha)^k p(\lambda)$, where $p(\lambda)$ is holomorphic in Ω and $p(\alpha) \neq 0$. Thus, if C_ρ denotes a circle of radius ρ that is centered at α and directed counterclockwise, then

$$\frac{1}{2\pi i}\int_{C_\rho} \frac{f'(\zeta)}{f(\zeta)}d\zeta = \frac{1}{2\pi i}\int_{C_\rho} \left\{k + \frac{p'(\zeta)}{p(\zeta)}\right\} d\zeta = k\,,$$

if ρ is taken sufficiently small. In other words, the residue of f'/f at the point α is equal to k. The final formula follows by invoking Theorem 17.10 to add up the contribution from each of the distinct zeros of $f(\lambda)$ inside $B_r(\omega)$; there are only finitely many, thanks to Lemma B.3. □

Theorem B.9. (Rouché) *Let $f(\lambda)$ and $g(\lambda)$ be holomorphic in an open set Ω that contains the closed disc $\overline{B_r(\omega)}$ for some $r > 0$ and suppose that*

$$|f(\lambda) - g(\lambda)| < |f(\lambda)| \quad \textit{if} \quad |\lambda - \omega| = r\,.$$

Then f and g have the same number of zeros inside $B_r(\omega)$, counting multiplicities.

Proof. Under the given assumptions, $|f(\lambda)| > 0$ and $|g(\lambda)| > 0$ for every point λ on the boundary of $B_r(\omega)$. Therefore, the difference between the number of roots of f inside $B_r(\omega)$ and the number of roots of g inside $B_r(\omega)$ is given by the formula

$$\begin{aligned} N_f(\omega, r) - N_g(\omega, r) &= \frac{1}{2\pi i}\int_{\Gamma_r}\left\{\frac{f'(\zeta)}{f(\zeta)} - \frac{g'(\zeta)}{g(\zeta)}\right\}d\zeta \\ &= \frac{1}{2\pi i}\int_{\Gamma_r}\frac{h'(\zeta)}{h(\zeta)}d\zeta\,, \end{aligned}$$

where

$$h(\zeta) = \frac{f(\zeta)}{g(\zeta)} \quad \text{for} \quad \zeta \in \Gamma_r$$

and Γ_r is a circle of radius r directed counterclockwise that is centered at the point ω. In view of the prevailing assumptions on f and g,

$$|1 - h(\zeta)| < 1 \quad \text{for } \zeta \in \Gamma_r\,,$$

and $h(\zeta)$ is holomorphic in the set $\{\lambda \in \mathbb{C} : r - \varepsilon < |\lambda - \omega| < r + \varepsilon\}$ for some $\varepsilon > 0$, thanks to Lemma B.4. Therefore, by Lemma B.6, we can write

$$\begin{aligned} h(\zeta) &= 1 - (1 - h(\zeta)) \\ &= \exp\{\varphi(\zeta)\}\,, \end{aligned}$$

where

$$\varphi(\zeta) = -\sum_{n=1}^{\infty}\frac{(1 - h(\zeta))^n}{n}.$$

Thus,

$$h'(\zeta) = h(\zeta)\varphi'(\zeta) \quad \text{for} \quad \zeta \in \Gamma_r$$

and

$$\begin{aligned} \frac{1}{2\pi i}\int_{\Gamma_r}\frac{h'(\zeta)}{h(\zeta)}d\zeta &= \frac{1}{2\pi i}\int_{\Gamma_r}\varphi'(\zeta)d\zeta \\ &= \frac{1}{2\pi i}\int_0^{2\pi}\frac{d}{d\zeta}\varphi(\zeta(t))\zeta'(t)dt \\ &= \frac{1}{2\pi i}\int_0^{2\pi}\frac{d}{dt}\varphi(\zeta(t))dt \\ &= \frac{1}{2\pi i}\{\varphi(\zeta(2\pi)) - \varphi(\zeta(0)\} = 0\,. \end{aligned}$$

□

B.6. Liouville's theorem

If $f(\lambda)$ is holomorphic in the full complex plane $\mathbb{C}$ and $|f(\lambda)| \leq M < \infty$ for every $\lambda \in \mathbb{C}$, then $f(\lambda)$ is constant.

Proof. Under the given assumptions

$$f(\lambda) = \sum_{n=0}^{\infty} \frac{f^{(n)}(0)}{n!}\lambda^n$$

for every point $\lambda \in \mathbb{C}$. Moreover, the formula

$$\frac{f^{(n)}(0)}{n!} = \frac{1}{2\pi i}\int_{\Gamma_R} \frac{f(\zeta)}{\zeta^{n+1}} d\zeta$$

for the coefficients implies that

$$\begin{aligned} \left|\frac{f^{(n)}(0)}{n!}\right| &= \left|\frac{1}{2\pi}\int_0^{2\pi} \frac{f(Re^{i\theta})}{(Re^{i\theta})^n} d\theta\right| \\ &\leq \frac{M}{R^n} \end{aligned}$$

for every $R \geq 0$. Therefore $f^{(n)}(0) = 0$ for $n \geq 1$. □

Much the same sort of analysis can be used to establish variants of the following sort:

Exercise B.2. Show that if $f(\lambda)$ is holomorphic in the full complex plane $\mathbb{C}$ and

$$|f(\lambda)| = 27 + \sqrt{3}|\lambda|^{3/2}$$

for every point $\lambda \in \mathbb{C}$, then $f(\lambda)$ is a polynomial of degree one.

B.7. Laurent expansions

Let $f(\lambda)$ be holomorphic in the annulus $0 < |\lambda - \omega| < R$, let $0 < r_1 < r_2 < R$ and let Γ_{r_j} be a circle of radius r_j, $j = 1, 2$, centered at ω and directed counterclockwise. Then, for $r_1 < |\lambda - \omega| < r_2$,

$$f(\lambda) = \frac{1}{2\pi i}\int_{\Gamma_{r_2}} \frac{f(\zeta)}{\zeta - \lambda} d\zeta - \frac{1}{2\pi i}\int_{\Gamma_{r_1}} \frac{f(\zeta)}{\zeta - \lambda} d\zeta\,. \tag{B.1}$$

Moreover,

$$\begin{aligned}\frac{1}{2\pi i}\int_{\Gamma_{r_2}} \frac{f(\zeta)}{\zeta-\lambda}d\zeta &= \frac{1}{2\pi i}\int_{\Gamma_{r_2}} \frac{f(\zeta)}{\zeta-\omega-(\lambda-\omega)}d\zeta \\ &= \frac{1}{2\pi i}\int_{\Gamma_{r_2}} \frac{f(\zeta)}{(\zeta-\omega)\left(1-\frac{\lambda-\omega}{\zeta-\omega}\right)}d\zeta \\ &= \sum_{n=0}^{\infty}\left(\frac{1}{2\pi i}\int_{\Gamma_{r_2}} \frac{f(\zeta)}{(\zeta-\omega)^{n+1}}d\zeta\right)(\lambda-\omega)^n\end{aligned}$$

and

$$\begin{aligned}-\frac{1}{2\pi i}\int_{\Gamma_{r_1}} \frac{f(\zeta)}{\zeta-\lambda}d\zeta &= -\frac{1}{2\pi i}\int_{\Gamma_{r_1}} \frac{f(\zeta)}{\left(\frac{\zeta-\omega}{\lambda-\omega}-1\right)(\lambda-\omega)}d\zeta \\ &= \sum_{n=0}^{\infty}\left(\frac{1}{2\pi i}\int_{\Gamma_{r_1}} f(\zeta)(\zeta-\omega)^n d\zeta\right)(\lambda-\omega)^{-(n+1)} \\ &= \sum_{j=-\infty}^{-1}\left(\frac{1}{2\pi i}\int_{\Gamma_{r_1}} \frac{f(\zeta)}{(\zeta-\omega)^{j+1}}d\zeta\right)(\lambda-\omega)^j.\end{aligned}$$

Thus, $f(\lambda)$ can be expressed in the form

$$f(\lambda) = \sum_{j=-\infty}^{\infty} a_j(\lambda-\omega)^j, \tag{B.2}$$

where

$$a_j = \frac{1}{2\pi i}\int_{\Gamma_r} \frac{f(\zeta)}{(\zeta-\omega)^{j+1}}d\zeta \quad \text{for } j = \cdots, -1, 0, 1, \cdots,$$

Γ_r denotes a circle of radius r that is centered at ω and directed counterclockwise and $0 < r < R$. The representation (B.1) is called the **Laurent expansion** of f about the point ω. If $a_{-k} \neq 0$ for some positive integer k and $a_j = 0$ for $j < -k$, then ω is said to be a **pole of order** k of $f(\lambda)$.

B.8. Partial fraction expansions

Theorem B.10. *Let*

$$f(\lambda) = \frac{(\lambda-\alpha_1)^{m_1}\cdots(\lambda-\alpha_k)^{m_k}}{(\lambda-\beta_1)^{n_1}\cdots(\lambda-\beta_\ell)^{n_\ell}},$$

where the $k+\ell$ points $\alpha_1, \cdots, \alpha_k$ and $\beta_1, \cdots, \beta_\ell$ are all distinct and $m_1 + \cdots + m_k \leq n_1 + \cdots + n_\ell$. Then:

(1) *$f(\lambda)$ has a pole of order n_j at the point β_j for $j = 1, \ldots, \ell$.*

(2) *Let*

$$g_j(\lambda) = \sum_{i=-n_j}^{-1} a_{ji}(\lambda - \beta_j)^i, \quad j = 1.\dots, \ell$$

denote the sum of the terms with negative indices in the Laurent expansion of $f(\lambda)$ *at the point* β_j. *Then* $f(\lambda) - g_j(\lambda)$ *is holomorphic in a ball of radius* r_j *centered at* β_j *if* r_j *is chosen small enough.*

(3) $f(\lambda) = g_1(\lambda) + \cdots + g_\ell(\lambda) + c$, *where*

$$c = \lim_{\lambda\to\infty} f(\lambda).$$

Proof. Under the given assumptions

$$f(\lambda) - \{g_1(\lambda) + \cdots + g_\ell(\lambda)\}$$

is holomorphic in all of $\mathbb{C}$ and tends to a finite limit as $\lambda \to \infty$. Therefore

$$|f(\lambda) - \{g_1(\lambda) + \cdots + g_\ell(\lambda)\}| \le M < \infty$$

for all $\lambda \in \mathbb{C}$. Thus, by Liouville's theorem,

$$f(\lambda) - \{g_1(\lambda) + \cdots + g_\ell(\lambda)\} = c$$

for every point $\lambda \in \mathbb{C}$. $\square$

Bibliography

[1] Pedro Alegría. Orthogonal polynomials associated with the Nehari problem. *Portugaliae Mathematica*, 62:337–347, 2005.

[2] Gregory S. Ammar and William B. Gragg. Schur flows for orthogonal Hessenberg matrices. *Fields Inst. Commun.*, 3:27–34, 1994.

[3] Tsuyoshi Ando. Concavity of certain maps on positive definite matrices and applications to Hadamard products. *Linear Algebra Appl.*, 26:203–241, 1979.

[4] Tom M. Apostol. *Mathematical Analysis*. Addison-Wesley, 1957.

[5] Sheldon Axler. Down with determinants. *Amer. Math. Monthly*, 102:139–154, 1995.

[6] Harm Bart, Israel Gohberg, and Marinus A. Kaashoek. *Minimal Factorization of Matrix and Operator Functions*. Birkhäuser, 1979.

[7] Rajendra Bhatia. *Matrix Analysis*. Springer-Verlag, 1997.

[8] Rajendra Bhatia. On the exponential metric increasing property. *Linear Algebra Appl.*, 375:211–220, 2003.

[9] Albrecht Boetcher and Harold Widom. Szegö via Jacobi. *Linear Algebra Appl.*, 419:656–657, 2006.

[10] Jonathan M. Borwein and Adrian S. Lewis. *Convex Analysis and Nonlinear Optimization. Theory and Examples*. Springer, 2006.

[11] Truman Botts. Convex sets. *Amer. Math. Monthly*, 49:527–535, 1942.

[12] Stanley Boylan. Learning with the Rav: Learning from the Rav. *Tradition*, 30:131–144, 1996.

[13] Roger Brockett. Using feedback to improve system identification. *Lecture Notes in Control and Information Sciences*, 329:45–65, 2006.

[14] Juan F. Camino, J. William Helton, Robert E. Skelton, and Ye Jieping. Matrix inequalities: a symbolic procedure to determine convexity automatically. *Integral Equations Operator Theory*, 46:399–454, 2003.

[15] Shalom Carmy. Polyphonic diversity and military music. *Tradition*, 34:6–32, 2000.

[16] Hervé Chapellat and S. P. Bhattacharyya. An alternative proof of Kharitonov's theorem. *IEEE Trans. Automatic Control*, 34:448–450, 1989.

[17] Barry Cipra. Andy Rooney, PhD. *The Mathematical Intelligencer*, 10:10, 1988.

[18] Chandler Davis, W. M. Kahan, and Hans F. Weinberger. Norm preserving dilations and their applications to optimal error bounds. *SIAM J. Numer. Anal.*, 19:445–469, 1982.

[19] Ilan Degani. *RCMS - right correction Magnus schemes for oscillatory ODE's and cubature formulas and oscillatory extensions.* PhD thesis, The Weizmann Institute of Science, 2005.

[20] Chen Dubi and Harry Dym. Riccati inequalities and reproducing kernel Hilbert spaces. *Linear Algebra Appl.*, 420:458–482, 2007.

[21] Harry Dym. *J Contractive Matrix Functions, Reproducing Kernel Hilbert Spaces, and Interpolation.* Amer. Math. Soc, 1989.

[22] Harry Dym. On Riccati equations and reproducing kernel spaces. *Oper. Theory Adv. Appl.*, 124:189–215, 2001.

[23] Harry Dym. Riccati equations and bitangential interpolation problems with singular Pick matrice. *Contemporary Mathematics*, 323:361–391, 2002.

[24] Harry Dym and Israel Gohberg. Extensions of band matrices with band inverses. *Linear Algebra Appl.*, 36:1–24, 1981.

[25] Harry Dym and Israel Gohberg. Extensions of kernels of Fredholm operators. *Journal d'Analyse Mathematique*, 42:51–97, 1982/1983.

[26] Harry Dym and J. William Helton. The matrix multidisk problem. *Integral Equations Operator Theory*, 46:285–339, 2003.

[27] Harry Dym, J. William Helton, and Scott McCullough. The Hessian of a noncommutative polynomial has numerous negative eigenvalue. *Journal d'Analyse Mathematique*, to appear.

[28] Harry Dym and Victor Katsnelson. Contributions of Issai Schur to analysis. *Progr. Math.*, 210:xci–clxxxviii, 2003.

[29] Richard S. Ellis. *Entropy, Large Deviations and Statistical Mechanics.* Springer-Verlag, 1985.

[30] Ludwig D. Faddeev. 30 years in mathematical physics. *Proc. Steklov Institute*, 176:3–28, 1988.

[31] Abraham Feintuch. *Robust Control Theory in Hilbert Space.* Springer, 1998.

[32] J. I. Fujii, M. Fujii, T. Furuta, and R. Nakamoto. Norm inequalities equivalent to Heinz inequality. *Proc. Amer. Math. Soc.*, 118:827–830, 1993.

[33] T. Furuta. Norm inequalities equivalent to Löwner-Heinz theorem. *Rev. Math. Phys.*, 1:135–137, 1989.

[34] Israel Gohberg, Marinus A. Kaashoek, and Hugo J. Woerdeman. The band method for positive and contractive extension problems. *J. Operator Theory*, 22:109–155, 1989.

[35] Israel Gohberg and Mark G. Krein. *Introduction to the Theory of Linear Nonselfadjoint Operators.* American Math. Soc., 1969.

[36] Paul Halmos. *A Hilbert Space Problem Book.* Van Nostrand, 1967.

[37] G. H. Hardy, J. E. Littlewood, and G. Pólya. *Inequalities.* Cambridge University Press, 1959.

[38] Michael Heymann. The pole shifting theorem revisited. *IEEE Trans. Automatic Control*, 24:479–480, 1979.

[39] Alfred Horn. Doubly stochastic matrices and the diagonal of a rotation. *Amer. J. Math.*, 76:620–630, 1953.

[40] Roger Kahn. *Memories of Summer.* University of Nebraska Press, 1997.

[41] Yakar Kannai. An elementary proof of the no-retraction theorem. *American Math. Monthly*, 88:264–268, 1981.

[42] Garrison Kiellor. *Woebegone Boy.* Viking, 1997.

[43] Donald E. Knuth, Tracy Larrabee, and Paul M. Roberts. *Mathematical Writing.* Mathematical Association of America, 1989.

[44] Peter Lancaster and Leiba Rodman. *Algebraic Riccati Equations.* Oxford University Press, 1995.

[45] Peter Lancaster and Miron Tismenetsky. *The Theory of Matrices.* Academic Press, 1985.

[46] Joseph LaSalle and Solomon Lefschetz. *Stability by Liapunov's Direct Method with Applications.* Academic Press, 1961.

[47] Zhiyun Lin, Bruce Francis, and Manfredi Maggiore. Necessary and sufficient graphical conditions for formation control of unicycles. *IEEE Trans. Automatic Control*, 50:121–127, 2005.

[48] David G. Luenberger. *Optimization by Vector Space Methods.* John Wiley & Sons, 1969.

[49] André Malraux. *Antimemoirs.* Holt Rhinehart and Winston, 1968.

[50] Alan McIntosh. The Toeplitz-Hausdorff theorem and ellipticity conditions. *Amer. Math. Monthly*, 85:475–477, 1978.

[51] Henryk Minc. *Permanents.* Addison-Wesley, 1978.

[52] Henryk Minc. *Nonnegative Matrices.* John Wiley, 1988.

[53] Patrick O'Brian. *Master and Commander.* Norton, 1990.

[54] Patrick O'Brian. *The Far Side of the World.* Norton, 1992.

[55] Alex Olshevsky and Vadim Olshevsky. Kharitonov's theorem and Bezoutians. *Linear Algebra Appl.*, 399:285–297, 2005.

[56] Vladimir Peller. *Hankel Operators and Their Applications.* Springer, 1957.

[57] Elijah Polak. *Optimization: Algorithms and Consistent Approximation.* Springer-Verlag, 2003.

[58] Vladimir P. Potapov. The multiplicative structure of J-contractive matrix functions. *Amer. Math. Soc. Transl. (2)*, 15:131–243, 1960.

[59] Aaron Rakeffet-Rothkoff. *The Rav: The World of Rabbi Joseph B. Soloveichik, Volume 2.* Ktav Publishing House, 1999.

[60] Walter Rudin. *Real and Complex Analysis.* McGraw Hill, 1966.

[61] David L. Russell. *Mathematics of Finite-Dimensional Control Systems.* Marcel Dekker, 1979.

[62] Thomas L. Saaty and Joseph Bram. *Nonlinear Mathematics.* Dover, 1981.

[63] Issai Schur. Über eine Klasse von Mittelbildungen mit Anwendungen auf die Determinantenttheorie. *Sitzungsberichte der Berliner Mathematischen Gesellschaft*, 22:9–20, 1923.

[64] Barry Simon. OPUC on one foot. *Bull. Amer. Math. Soc.*, 42:431–460, 2005.

[65] Barry Simon. The sharp form of the strong Szegö theorem. *Contemporary Mathematics*, 387:253–275, 2005.

[66] Teiji Takagi. An algebraic problem related to an analytic theorem of Carathéodory and Fejér on an allied theorem of Landau. *Japanese J. of Mathematics*, 1:83–93, 1924.

[67] E. C. Titchmarsh. *Eigenfunction expansions associated with second-order differential equations. Vol. 2.* Oxford, 1958.

[68] Lloyd N. Trefethen and David Bau, III. *Numerical Linear Algebra.* SIAM, 1997.

[69] Sergei Treil and Alexander Volberg. Wavelets and the angle between past and future. *J. Funct. Anal.*, 143:269–308, 1997.

[70] Robert James Waller. *The Bridges of Madison County.* Warner Books, 1992.

[71] Roger Webster. *Convexity.* Oxford University Press, 1994.

[72] W. Murray Wonham. *Linear Multivariable Control.* Springer-Verlag, 1985.

[73] Kosaku Yosida. *Functional Analysis.* Springer-Verlag, 1965.

[74] Kemin Zhou, John C. Doyle, and Keith Glover. *Robust and Optimal Control.* Prentice Hall, 1996.

Notation Index

Frequently used symbols.

Subject Index

Titles in This Series

TITLES IN THIS SERIES

41 **Georgi V. Smirnov,** Introduction to the theory of differential inclusions, 2002

40 **Robert E. Greene and Steven G. Krantz,** Function theory of one complex variable, third edition, 2006

39 **Larry C. Grove,** Classical groups and geometric algebra, 2002

38 **Elton P. Hsu,** Stochastic analysis on manifolds, 2002

37 **Hershel M. Farkas and Irwin Kra,** Theta constants, Riemann surfaces and the modular group, 2001

36 **Martin Schechter,** Principles of functional analysis, second edition, 2002

35 **James F. Davis and Paul Kirk,** Lecture notes in algebraic topology, 2001

34 **Sigurdur Helgason,** Differential geometry, Lie groups, and symmetric spaces, 2001

33 **Dmitri Burago, Yuri Burago, and Sergei Ivanov,** A course in metric geometry, 2001

32 **Robert G. Bartle,** A modern theory of integration, 2001

31 **Ralf Korn and Elke Korn,** Option pricing and portfolio optimization: Modern methods of financial mathematics, 2001

30 **J. C. McConnell and J. C. Robson,** Noncommutative Noetherian rings, 2001

29 **Javier Duoandikoetxea,** Fourier analysis, 2001

28 **Liviu I. Nicolaescu,** Notes on Seiberg-Witten theory, 2000

27 **Thierry Aubin,** A course in differential geometry, 2001

26 **Rolf Berndt,** An introduction to symplectic geometry, 2001

25 **Thomas Friedrich,** Dirac operators in Riemannian geometry, 2000

24 **Helmut Koch,** Number theory: Algebraic numbers and functions, 2000

23 **Alberto Candel and Lawrence Conlon,** Foliations I, 2000

22 **Günter R. Krause and Thomas H. Lenagan,** Growth of algebras and Gelfand-Kirillov dimension, 2000

21 **John B. Conway,** A course in operator theory, 2000

20 **Robert E. Gompf and András I. Stipsicz,** 4-manifolds and Kirby calculus, 1999

19 **Lawrence C. Evans,** Partial differential equations, 1998

18 **Winfried Just and Martin Weese,** Discovering modern set theory. II: Set-theoretic tools for every mathematician, 1997

17 **Henryk Iwaniec,** Topics in classical automorphic forms, 1997

16 **Richard V. Kadison and John R. Ringrose,** Fundamentals of the theory of operator algebras. Volume II: Advanced theory, 1997

15 **Richard V. Kadison and John R. Ringrose,** Fundamentals of the theory of operator algebras. Volume I: Elementary theory, 1997

14 **Elliott H. Lieb and Michael Loss,** Analysis, 1997

13 **Paul C. Shields,** The ergodic theory of discrete sample paths, 1996

12 **N. V. Krylov,** Lectures on elliptic and parabolic equations in Hölder spaces, 1996

11 **Jacques Dixmier,** Enveloping algebras, 1996 Printing

10 **Barry Simon,** Representations of finite and compact groups, 1996

9 **Dino Lorenzini,** An invitation to arithmetic geometry, 1996

8 **Winfried Just and Martin Weese,** Discovering modern set theory. I: The basics, 1996

7 **Gerald J. Janusz,** Algebraic number fields, second edition, 1996

6 **Jens Carsten Jantzen,** Lectures on quantum groups, 1996

For a complete list of titles in this series, visit the AMS Bookstore at **www.ams.org/bookstore/**.